Lecture Notes in Computer Science   12744

More information about this subseries at http://www.springer.com/series/7407

Maciej Paszynski · Dieter Kranzlmüller ·
Valeria V. Krzhizhanovskaya ·
Jack J. Dongarra · Peter M. A. Sloot (Eds.)

Computational Science – ICCS 2021

21st International Conference
Krakow, Poland, June 16–18, 2021
Proceedings, Part III

 Springer

Editors
Maciej Paszynski (iD)
AGH University of Science and Technology
Krakow, Poland

Valeria V. Krzhizhanovskaya (iD)
University of Amsterdam
Amsterdam, The Netherlands

Peter M. A. Sloot (iD)
University of Amsterdam
Amsterdam, The Netherlands

ITMO University
St. Petersburg, Russia

Nanyang Technological University
Singapore, Singapore

Dieter Kranzlmüller (iD)
Ludwig-Maximilians-Universität München
Munich, Germany

Leibniz Supercomputing Center (LRZ)
Garching bei München, Germany

Jack J. Dongarra (iD)
University of Tennessee at Knoxville
Knoxville, TN, USA

ISSN 0302-9743   ISSN 1611-3349 (electronic)
Lecture Notes in Computer Science
ISBN 978-3-030-77966-5   ISBN 978-3-030-77967-2 (eBook)
https://doi.org/10.1007/978-3-030-77967-2

LNCS Sublibrary: SL1 – Theoretical Computer Science and General Issues

This Springer imprint is published by the registered company Springer Nature Switzerland AG
The registered company address is: Gewerbestrasse 11, 6330 Cham, Switzerland

Preface

Welcome to the proceedings of the 21st annual International Conference on Computational Science (ICCS 2021 - https://www.iccs-meeting.org/iccs2021/).

In preparing this edition, we had high hopes that the ongoing COVID-19 pandemic would fade away and allow us to meet this June in the beautiful city of Kraków, Poland. Unfortunately, this is not yet the case, as the world struggles to adapt to the many profound changes brought about by this crisis. ICCS 2021 has had to adapt too and is thus being held entirely online, for the first time in its history.

These challenges notwithstanding, we have tried our best to keep the ICCS community as dynamic and productive as always. We are proud to present the proceedings you are reading as a result of that.

ICCS 2021 was jointly organized by the AGH University of Science and Technology, the University of Amsterdam, NTU Singapore, and the University of Tennessee.

The International Conference on Computational Science is an annual conference that brings together researchers and scientists from mathematics and computer science as basic computing disciplines, as well as researchers from various application areas who are pioneering computational methods in sciences such as physics, chemistry, life sciences, engineering, arts, and humanitarian fields, to discuss problems and solutions in the area, identify new issues, and shape future directions for research.

Since its inception in 2001, ICCS has attracted an increasing number of attendees and higher quality papers, and this year is not an exception, with over 350 registered participants. The proceedings have become a primary intellectual resource for computational science researchers, defining and advancing the state of the art in this field.

The theme for 2021, "**Computational Science for a Better Future**," highlights the role of computational science in tackling the current challenges of our fast-changing world. This conference was a unique event focusing on recent developments in scalable scientific algorithms, advanced software tools, computational grids, advanced numerical methods, and novel application areas. These innovative models, algorithms, and tools drive new science through efficient application in physical systems, computational and systems biology, environmental systems, finance, and other areas.

ICCS is well known for its excellent lineup of keynote speakers. The keynotes for 2021 were given by

- **Maciej Besta**, ETH Zürich, Switzerland
- **Marian Bubak**, AGH University of Science and Technology, Poland | Sano Centre for Computational Medicine, Poland
- **Anne Gelb**, Dartmouth College, USA
- **Georgiy Stenchikov**, King Abdullah University of Science and Technology, Saudi Arabia
- **Marco Viceconti**, University of Bologna, Italy

- **Krzysztof Walczak**, Poznan University of Economics and Business, Poland
- **Jessica Zhang**, Carnegie Mellon University, USA

This year we had 635 submissions (156 submissions to the main track and 479 to the thematic tracks). In the main track, 48 full papers were accepted (31%); in the thematic tracks, 212 full papers were accepted (44%). A high acceptance rate in the thematic tracks is explained by the nature of these tracks, where organisers personally invite many experts in a particular field to participate in their sessions.

ICCS relies strongly on our thematic track organizers' vital contributions to attract high-quality papers in many subject areas. We would like to thank all committee members from the main and thematic tracks for their contribution to ensure a high standard for the accepted papers. We would also like to thank *Springer, Elsevier,* and *Intellegibilis* for their support. Finally, we appreciate all the local organizing committee members for their hard work to prepare for this conference.

We are proud to note that ICCS is an A-rank conference in the CORE classification.

We wish you good health in these troubled times and look forward to meeting you at the conference.

June 2021

Maciej Paszynski
Dieter Kranzlmüller
Valeria V. Krzhizhanovskaya
Jack J. Dongarra
Peter M. A. Sloot

Organization

Local Organizing Committee at AGH University of Science and Technology

Chairs

Maciej Paszynski
Aleksander Byrski

Members

Marcin Łos
Maciej Woźniak
Leszek Siwik
Magdalena Suchoń

Thematic Tracks and Organizers

Advances in High-Performance Computational Earth Sciences: Applications and Frameworks – IHPCES

Takashi Shimokawabe
Kohei Fujita
Dominik Bartuschat

Applications of Computational Methods in Artificial Intelligence and Machine Learning – ACMAIML

Kourosh Modarresi
Paul Hofmann
Raja Velu
Peter Woehrmann

Artificial Intelligence and High-Performance Computing for Advanced Simulations – AIHPC4AS

Maciej Paszynski
Robert Schaefer
David Pardo
Victor Calo

Biomedical and Bioinformatics Challenges for Computer Science – BBC

Mario Cannataro
Giuseppe Agapito

Mauro Castelli
Riccardo Dondi
Italo Zoppis

Classifier Learning from Difficult Data – CLD^2

Michał Woźniak
Bartosz Krawczyk

Computational Analysis of Complex Social Systems – CSOC

Debraj Roy

Computational Collective Intelligence – CCI

Marcin Maleszka
Ngoc Thanh Nguyen
Marcin Hernes
Sinh Van Nguyen

Computational Health – CompHealth

Sergey Kovalchuk
Georgiy Bobashev
Stefan Thurner

Computational Methods for Emerging Problems in (dis-)Information Analysis – DisA

Michal Choras
Robert Burduk
Konstantinos Demestichas

Computational Methods in Smart Agriculture – CMSA

Andrew Lewis

Computational Optimization, Modelling, and Simulation – COMS

Xin-She Yang
Leifur Leifsson
Slawomir Koziel

Computational Science in IoT and Smart Systems – IoTSS

Vaidy Sunderam
Dariusz Mrozek

Computer Graphics, Image Processing and Artificial Intelligence – CGIPAI

Andres Iglesias
Lihua You
Alexander Malyshev
Hassan Ugail

Data-Driven Computational Sciences – DDCS

Craig Douglas

Machine Learning and Data Assimilation for Dynamical Systems – MLDADS

Rossella Arcucci

**MeshFree Methods and Radial Basis Functions in Computational
Sciences – MESHFREE**

Vaclav Skala
Marco-Evangelos Biancolini
Samsul Ariffin Abdul Karim
Rongjiang Pan
Fernando-César Meira-Menandro

Multiscale Modelling and Simulation – MMS

Derek Groen
Diana Suleimenova
Stefano Casarin
Bartosz Bosak
Wouter Edeling

Quantum Computing Workshop – QCW

Katarzyna Rycerz
Marian Bubak

**Simulations of Flow and Transport: Modeling, Algorithms
and Computation – SOFTMAC**

Shuyu Sun
Jingfa Li
James Liu

**Smart Systems: Bringing Together Computer Vision, Sensor Networks
and Machine Learning – SmartSys**

Pedro Cardoso
Roberto Lam

João Rodrigues
Jânio Monteiro

Software Engineering for Computational Science – SE4Science

Jeffrey Carver
Neil Chue Hong
Anna-Lena Lamprecht

Solving Problems with Uncertainty – SPU

Vassil Alexandrov
Aneta Karaivanova

Teaching Computational Science – WTCS

Angela Shiflet
Nia Alexandrov
Alfredo Tirado-Ramos

Uncertainty Quantification for Computational Models – UNEQUIvOCAL

Wouter Edeling
Anna Nikishova

Reviewers

Ahmad Abdelfattah	Bartosz Balis	Michael Burkhart
Samsul Ariffin Abdul Karim	Krzysztof Banas	Allah Bux
Tesfamariam Mulugeta Abuhay	Dariusz Barbucha	Krisztian Buza
	Valeria Bartsch	Aleksander Byrski
Giuseppe Agapito	Dominik Bartuschat	Cristiano Cabrita
Elisabete Alberdi	Pouria Behnodfaur	Xing Cai
Luis Alexandre	Joern Behrens	Barbara Calabrese
Vassil Alexandrov	Adrian Bekasiewicz	Jose Camata
Nia Alexandrov	Gebrail Bekdas	Almudena Campuzano
Julen Alvarez-Aramberri	Mehmet Belen	Mario Cannataro
Sergey Alyaev	Stefano Beretta	Alberto Cano
Tomasz Andrysiak	Benjamin Berkels	Pedro Cardoso
Samuel Aning	Daniel Berrar	Alberto Carrassi
Michael Antolovich	Sanjukta Bhowmick	Alfonso Carriazo
Hideo Aochi	Georgiy Bobashev	Jeffrey Carver
Hamid Arabnejad	Bartosz Bosak	Manuel Castañón-Puga
Rossella Arcucci	Isabel Sofia Brito	Mauro Castelli
Costin Badica	Marc Brittain	Eduardo Cesar
Marina Balakhontceva	Jérémy Buisson	Nicholas Chancellor
	Robert Burduk	Patrikakis Charalampos

Henri-Pierre Charles
Ehtzaz Chaudhry
Long Chen
Sibo Cheng
Siew Ann Cheong
Lock-Yue Chew
Marta Chinnici
Sung-Bae Cho
Michal Choras
Neil Chue Hong
Svetlana Chuprina
Paola Cinnella
Noélia Correia
Adriano Cortes
Ana Cortes
Enrique
  Costa-Montenegro
David Coster
Carlos Cotta
Helene Coullon
Daan Crommelin
Attila Csikasz-Nagy
Loïc Cudennec
Javier Cuenca
António Cunha
Boguslaw Cyganek
Ireneusz Czarnowski
Pawel Czarnul
Lisandro Dalcin
Bhaskar Dasgupta
Konstantinos Demestichas
Quanling Deng
Tiziana Di Matteo
Eric Dignum
Jamie Diner
Riccardo Dondi
Craig Douglas
Li Douglas
Rafal Drezewski
Vitor Duarte
Thomas Dufaud
Wouter Edeling
Nasir Eisty
Kareem El-Safty
Amgad Elsayed
Nahid Emad

Christian Engelmann
Roberto R. Expósito
Fangxin Fang
Antonino Fiannaca
Christos
  Filelis-Papadopoulos
Martin Frank
Alberto Freitas
Ruy Freitas Reis
Karl Frinkle
Kohei Fujita
Hiroshi Fujiwara
Takeshi Fukaya
Wlodzimierz Funika
Takashi Furumura
Ernst Fusch
David Gal
Teresa Galvão
Akemi Galvez-Tomida
Ford Lumban Gaol
Luis Emilio
  Garcia-Castillo
Frédéric Gava
Piotr Gawron
Alex Gerbessiotis
Agata Gielczyk
Adam Glos
Sergiy Gogolenko
Jorge
  González-Domínguez
Yuriy Gorbachev
Pawel Gorecki
Michael Gowanlock
Ewa Grabska
Manuel Graña
Derek Groen
Joanna Grzyb
Pedro Guerreiro
Tobias Guggemos
Federica Gugole
Bogdan Gulowaty
Shihui Guo
Xiaohu Guo
Manish Gupta
Piotr Gurgul
Filip Guzy

Pietro Hiram Guzzi
Zulfiqar Habib
Panagiotis Hadjidoukas
Susanne Halstead
Feilin Han
Masatoshi Hanai
Habibollah Haron
Ali Hashemian
Carina Haupt
Claire Heaney
Alexander Heinecke
Marcin Hernes
Bogumila Hnatkowska
Maximilian Höb
Jori Hoencamp
Paul Hofmann
Claudio Iacopino
Andres Iglesias
Takeshi Iwashita
Alireza Jahani
Momin Jamil
Peter Janku
Jiri Jaros
Caroline Jay
Fabienne Jezequel
Shalu Jhanwar
Tao Jiang
Chao Jin
Zhong Jin
David Johnson
Guido Juckeland
George Kampis
Aneta Karaivanova
Takahiro Katagiri
Timo Kehrer
Christoph Kessler
Jakub Klikowski
Alexandra Klimova
Harald Koestler
Ivana Kolingerova
Georgy Kopanitsa
Sotiris Kotsiantis
Sergey Kovalchuk
Michal Koziarski
Slawomir Koziel
Rafal Kozik

Bartosz Krawczyk
Dariusz Krol
Valeria Krzhizhanovskaya
Adam Krzyzak
Pawel Ksieniewicz
Marek Kubalcík
Sebastian Kuckuk
Eileen Kuehn
Michael Kuhn
Michal Kulczewski
Julian Martin Kunkel
Krzysztof Kurowski
Marcin Kuta
Bogdan Kwolek
Panagiotis Kyziropoulos
Massimo La Rosa
Roberto Lam
Anna-Lena Lamprecht
Rubin Landau
Johannes Langguth
Shin-Jye Lee
Mike Lees
Leifur Leifsson
Kenneth Leiter
Florin Leon
Vasiliy Leonenko
Roy Lettieri
Jake Lever
Andrew Lewis
Jingfa Li
Hui Liang
James Liu
Yen-Chen Liu
Zhao Liu
Hui Liu
Pengcheng Liu
Hong Liu
Marcelo Lobosco
Robert Lodder
Chu Kiong Loo
Marcin Los
Stephane Louise
Frederic Loulergue
Hatem Ltaief
Paul Lu
Stefan Luding

Laura Lyman
Scott MacLachlan
Lukasz Madej
Lech Madeyski
Luca Magri
Imran Mahmood
Peyman Mahouti
Marcin Maleszka
Alexander Malyshev
Livia Marcellino
Tomas Margalef
Tiziana Margaria
Osni Marques
M. Carmen Márquez
    García
Paula Martins
Jaime Afonso Martins
Pawel Matuszyk
Valerie Maxville
Pedro Medeiros
Fernando-César
    Meira-Menandro
Roderick Melnik
Valentin Melnikov
Ivan Merelli
Marianna Milano
Leandro Minku
Jaroslaw Miszczak
Kourosh Modarresi
Jânio Monteiro
Fernando Monteiro
James Montgomery
Dariusz Mrozek
Peter Mueller
Ignacio Muga
Judit Munoz-Matute
Philip Nadler
Hiromichi Nagao
Jethro Nagawkar
Kengo Nakajima
Grzegorz J. Nalepa
I. Michael Navon
Philipp Neumann
Du Nguyen
Ngoc Thanh Nguyen
Quang-Vu Nguyen

Sinh Van Nguyen
Nancy Nichols
Anna Nikishova
Hitoshi Nishizawa
Algirdas Noreika
Manuel Núñez
Krzysztof Okarma
Pablo Oliveira
Javier Omella
Kenji Ono
Eneko Osaba
Aziz Ouaarab
Raymond Padmos
Marek Palicki
Junjun Pan
Rongjiang Pan
Nikela Papadopoulou
Marcin Paprzycki
David Pardo
Anna Paszynska
Maciej Paszynski
Abani Patra
Dana Petcu
Serge Petiton
Bernhard Pfahringer
Toby Phillips
Frank Phillipson
Juan C. Pichel
Anna
    Pietrenko-Dabrowska
Laércio L. Pilla
Yuri Pirola
Nadia Pisanti
Sabri Pllana
Mihail Popov
Simon Portegies Zwart
Roland Potthast
Malgorzata
    Przybyla-Kasperek
Ela Pustulka-Hunt
Alexander Pyayt
Kun Qian
Yipeng Qin
Rick Quax
Cesar Quilodran Casas
Enrique S. Quintana-Orti

Ewaryst Rafajlowicz
Ajaykumar Rajasekharan
Raul Ramirez
Célia Ramos
Marcus Randall
Lukasz Rauch
Vishal Raul
Robin Richardson
Sophie Robert
João Rodrigues
Daniel Rodriguez
Albert Romkes
Debraj Roy
Jerzy Rozenblit
Konstantin Ryabinin
Katarzyna Rycerz
Khalid Saeed
Ozlem Salehi
Alberto Sanchez
Aysin Sanci
Gabriele Santin
Rodrigo Santos
Robert Schaefer
Karin Schiller
Ulf D. Schiller
Bertil Schmidt
Martin Schreiber
Gabriela Schütz
Christoph Schweimer
Marinella Sciortino
Diego Sevilla
Mostafa Shahriari
Abolfazi
   Shahzadeh-Fazeli
Vivek Sheraton
Angela Shiflet
Takashi Shimokawabe
Alexander Shukhman
Marcin Sieniek
Nazareen
   Sikkandar Basha
Anna Sikora
Diana Sima
Robert Sinkovits
Haozhen Situ
Leszek Siwik

Vaclav Skala
Ewa
   Skubalska-Rafajlowicz
Peter Sloot
Renata Slota
Oskar Slowik
Grazyna Slusarczyk
Sucha Smanchat
Maciej Smolka
Thiago Sobral
Robert Speck
Katarzyna Stapor
Robert Staszewski
Steve Stevenson
Tomasz Stopa
Achim Streit
Barbara Strug
Patricia Suarez Valero
Vishwas Hebbur Venkata
Subba Rao
Bongwon Suh
Diana Suleimenova
Shuyu Sun
Ray Sun
Vaidy Sunderam
Martin Swain
Jerzy Swiatek
Piotr Szczepaniak
Tadeusz Szuba
Ryszard Tadeusiewicz
Daisuke Takahashi
Zaid Tashman
Osamu Tatebe
Carlos Tavares Calafate
Andrei Tchernykh
Kasim Tersic
Jannis Teunissen
Nestor Tiglao
Alfredo Tirado-Ramos
Zainab Titus
Pawel Topa
Mariusz Topolski
Pawel Trajdos
Bogdan Trawinski
Jan Treur
Leonardo Trujillo

Paolo Trunfio
Ka-Wai Tsang
Hassan Ugail
Eirik Valseth
Ben van Werkhoven
Vítor Vasconcelos
Alexandra Vatyan
Raja Velu
Colin Venters
Milana Vuckovic
Jianwu Wang
Meili Wang
Peng Wang
Jaroslaw Watróbski
Holger Wendland
Lars Wienbrandt
Izabela Wierzbowska
Peter Woehrmann
Szymon Wojciechowski
Michal Wozniak
Maciej Wozniak
Dunhui Xiao
Huilin Xing
Wei Xue
Abuzer Yakaryilmaz
Yoshifumi Yamamoto
Xin-She Yang
Dongwei Ye
Hujun Yin
Lihua You
Han Yu
Drago Žagar
Michal Zak
Gabor Závodszky
Yao Zhang
Wenshu Zhang
Wenbin Zhang
Jian-Jun Zhang
Jinghui Zhong
Sotirios Ziavras
Zoltan Zimboras
Italo Zoppis
Chiara Zucco
Pavel Zun
Pawel Zyblewski
Karol Zyczkowski

Contents – Part III

Computational Analysis of Complex Social Systems

Computational Collective Intelligence

Computational Health

Classifier Learning from Difficult Data

Soft Confusion Matrix Classifier for Stream Classification

Pawel Trajdos$^{(\boxtimes)}$ and Marek Kurzynski

Wroclaw University of Science and Technology, Wroclaw, Poland
{pawel.trajdos,marek.kurzynski}@pwr.edu.pl

Abstract. In this paper, the issue of tailoring the soft confusion matrix (SCM) based classifier to deal with stream learning task is addressed. The main goal of the work is to develop a wrapping-classifier that allows incremental learning to classifiers that are unable to learn incrementally. The goal is achieved by making two improvements in the previously developed SCM classifier. The first one is aimed at reducing the computational cost of the SCM classifier. To do so, the definition of the fuzzy neighbourhood of an object is changed. The second one is aimed at effective dealing with the concept drift. This is done by employing the ADWIN-driven concept drift detector that is not only used to detect the drift but also to control the size of the neighbourhood. The obtained experimental results show that the proposed approach significantly outperforms the reference methods.

Keywords: Classification · Probabilistic model · Randomized reference classifier · Soft confusion matrix · Stream classification

1 Introduction

Classification of streaming data is one of the most difficult problems in modern pattern recognition theory and practice. This is due to the fact that a typical data stream is characterized by several features that significantly impede making the correct classification decision. These features include: continuous flow, huge data volume, rapid arrival rate, and susceptibility to change [19]. If a streaming data classifier aspires to practical applications, it must face these requirements and have to satisfy numerous constraints (e.g. bounded memory, single-pass, real-time response, change of data concept) to an acceptable extent. It is not easy, that is why the methodology of recognizing stream data has been developing very intensively for over two decades, proposing new, more and more perfect classification methods [9,24].

Incremental learning is a vital capability for classifiers used in stream data classification [27]. It allows the classifier to utilize new objects generated by the stream to improve the model built so far. It also allows, to some extent, dealing with the concept drift. Some of the well-known classifiers are naturally capable to be trained iteratively. Examples of such classifiers are neural networks, nearest neighbours classifiers, or probabilistic methods such as the naive

© Springer Nature Switzerland AG 2021
M. Paszynski et al. (Eds.): ICCS 2021, LNCS 12744, pp. 3–17, 2021.
https://doi.org/10.1007/978-3-030-77967-2_1

Bayes classifier [11]. Some of the classifiers were tailored to be learned incrementally. An example of such a method is well-known Hoeffding Tree classifier [26]. Those types of classifiers can be easily used in stream classification systems. On the other hand, when a classifier is unable to learn in an incremental way, the options for using for stream classification are very limited [27]. Only one option is to keep a set of objects and rebuild the classifier from scratch whenever it is necessary [11].

To bridge this gap, we propose a wrapping-classifier-based on the soft confusion matrix approach (SCM). The wrapping-classifier may be used to add incremental learning functionality to any batch classifier. The classifier based on the idea of soft confusion matrix has been proposed in [30]. It proved to be an efficient tool for solving such practical problems as hand gesture recognition [22]. An additional advantage in solving the above-mentioned classification problem is the ability to use imprecise feedback information about a class assignment. The SCM-based algorithm was also successfully used in multilabel learning [31].

Dealing with the concept drift using incremental learning only is insufficient. This is because the incremental classifiers deal effectively only with the incremental drift [9]. To handle the sudden concept drift, additional mechanism such as single/multiple window approach [21], forgetting mechanisms [33], drift detectors [2] must be used. In this study, we decided to use ADWIN algorithm [2] to detect the drift and to manage the set of stored objects. We use the ADWIN-based detector because this approach was shown to be an effective method [1,13].

The concept drift may also be dealt with using ensemble classifiers [9]. There are a plethora of ensemble-based approaches [4,12,19] however, in this work we are focused on single-classifier-based systems.

The rest of the paper is organized as follows. Section 2 presents the corrected classifier and gives insight into its two-level structure and the original concepts of RRC and SCM which are the basis of its construction. Section 3 describes the adopted model of concept drifting data stream and provides details of chunk-based learning scheme of base classifiers and online dynamic learning of the correcting procedure and describes the method of combining ensemble members. In Sect. 4 the description of the experimental procedure is given. The results are presented and discussed in Sect. 5. Section 6 concludes the paper.

2   Classifier with Correction

2.1   Preliminaries

Let us consider the pattern recognition problem in which $x \in \mathcal{X}$ denotes a feature vector of an object and $j \in \mathcal{M}$ is its class number ($\mathcal{X} \subseteq \Re^d$ and $\mathcal{M} = \{1, 2, \ldots, M\}$ are feature space and set of class numbers, respectively). Let $\psi(\mathcal{L})$ be a classifier trained on the learning set $\mathcal{L}$, which assigns a class number i to the recognized object. We assume that $\psi(\mathcal{L})$ is described by the canonical model [20], i.e. for given x it first produces values of normalized classification functions (supports) $g_i(x), i \in \mathcal{M}$ ($g_i(x) \in [0,1], \sum g_i(x) = 1$) and then classify object according to the maximum support rule:

$$\psi(\mathcal{L}, x) = i \Leftrightarrow g_i(x) = \max_{k \in \mathcal{M}} g_k(x). \tag{1}$$

To recognize the object x we will apply the original procedure, which using additional information about the local (relative to x) properties of $\psi(\mathcal{L})$ can change its decision to increase the chance of correct classification of x.

The proposed correcting procedure which has the form of classifier $\psi^{(Corr)}(\mathcal{L}, \mathcal{V})$ built over $\psi(\mathcal{L})$ will be called a wrapping-classifier. The wrapping classifier $\psi^{(Corr)}(\mathcal{L}, \mathcal{V})$ acts according to the following Bayes scheme:

$$\psi^{(Corr)}(\mathcal{L}, \mathcal{V}, x) = i \Leftrightarrow P(i|x) = \max_{k \in \mathcal{M}} P(k|x), \tag{2}$$

where *a posteriori* probabilities $P(k|x), k \in \mathcal{M}$ can be expressed in a form depending on the probabilistic properties of classifier $\psi(\mathcal{L})$:

$$P(j|x) = \sum_{i \in \mathcal{M}} P(i, j|x) = \sum_{i \in \mathcal{M}} P(i|x)P(j|i, x). \tag{3}$$

$P(j|i, x)$ denotes the probability that x belongs to the j-th class given that $\psi(\mathcal{L}, x) = i$ and $P(i|x) = P(\psi(\mathcal{L}, x) = i)$ is the probability of assigning x to class i by $\psi(\mathcal{L})$ Since for deterministic classifier $\psi(\mathcal{L})$ both above probabilities are equal to 0 or 1 we will use two concepts for their approximate calculation: randomized reference classifier (RRC) and soft confusion matrix (SCM).

2.2 Randomized Reference Classifier (RRC)

RRC is a randomized model of classifier $\psi(\mathcal{L})$ and with its help the probabilities $P(\psi(\mathcal{L}, x) = i)$ are calculated.

RRC ψ^{RRC} as a probabilistic classifier is defined by a probability distribution over the set of class labels $\mathcal{M}$. Its classifying functions $\{\delta_j(x)\}_{j \in \mathcal{M}}$ are observed values of random variables $\{\Delta_j(x)\}_{j \in \mathcal{M}}$ that meet – in addition to the normalizing conditions – the following condition:

$$\mathbf{E}[\Delta_i(x)] = g_i(x), \ i \in \mathcal{M}, \tag{4}$$

where $\mathbf{E}$ is the expected value operator. Formula (4) denotes that ψ^{RRC} acts – on average – as the modeled classifier $\psi(\mathcal{L})$, hence the following approximation is fully justified:

$$P(i|x) = P(\psi(\mathcal{L}, x) = i) \approx P(\psi^{RRC}(x) = i), \tag{5}$$

where

$$P(\psi^{RRC}(x) = i) = P[\Delta_i(x) > \Delta_k(x), k \in \mathcal{M} \setminus i] \tag{6}$$

can be easily determined if we assume – as in the original work of Woloszynski and Kurzynski [32] – that $\Delta_i(x)$ follows the beta distribution.

2.3 Soft Confusion Matrix (SCM)

SCM will be used to determine the assessment of probability $P(j|i,x)$ which denotes class-dependent probabilities of the correct classification (for $i = j$) and the misclassification (for $i \neq j$) of $\psi(\mathcal{L}, x)$ at the point x. The method defines the neighborhood of the point x containing validation objects in terms of fuzzy sets allowing for flexible selection of membership functions and assigning weights to individual validation objects dependent on distance from x.

The SCM providing an image of the classifier local (relative to x) probabilities $P(j|i,x)$, is in the form of two-dimensional table, in which the rows correspond to the true classes while the columns correspond to the outcomes of the classifier $\psi(\mathcal{L})$, as it is shown in Table 1.

Table 1. The soft confusion matrix of classifier $\psi(\mathcal{L})$

		Classification by ψ			
		1	2	...	M
	1	$\varepsilon_{1,1}(x)$	$\varepsilon_{1,2}(x)$	...	$\varepsilon_{1,M}(x)$
True	2	$\varepsilon_{2,1}(x)$	$\varepsilon_{2,2}(x)$	...	$\varepsilon_{2,M}(x)$
Class	:	:	:	$\ddots$	:
	M	$\varepsilon_{M,1}(x)$	$\varepsilon_{M,2}(x)$	...	$\varepsilon_{M,M}(x)$

The value $\varepsilon_{i,j}(x)$ is determined from validation set $\mathcal{V}$ and is defined as the following ratio:

$$\varepsilon_{i,j}(x) = \frac{|\mathcal{V}_j \cap \mathcal{D}_i \cap \mathcal{N}(x)|}{|\mathcal{V}_j \cap \mathcal{N}(x)|}, \tag{7}$$

where $\mathcal{V}_j, \mathcal{D}_i$ and $\mathcal{N}(x)$ are fuzzy sets specified in the validation set $\mathcal{V}$ and $|\cdot|$ denotes the cardinality of a fuzzy set [7].

The set $\mathcal{V}_j$ denotes the set of validation objects from the j-th class. Formulating this set in terms of fuzzy sets theory it can be assumed that the grade of membership of validation object $x_\mathcal{V}$ to $\mathcal{V}_j$ is the class indicator which leads to the following definition of $\mathcal{V}_j$:

$$\mathcal{V}_j = \{(x_\mathcal{V}, \mu_{\mathcal{V}_j}(x_\mathcal{V}))\}, \tag{8}$$

$$\mu_{\mathcal{V}_j}(x_\mathcal{V}) = \begin{cases} 1 \text{ if } x_\mathcal{V} \in j\text{-th class,} \\ 0 \text{ elsewhere.} \end{cases} \tag{9}$$

The concept of fuzzy set $\mathcal{D}_i$ is defined as follows:

$$\mathcal{D}_i = \{(x_\mathcal{V}, \mu_{\mathcal{D}_i}(x_\mathcal{V})) : x_\mathcal{V} \in \mathcal{V}, \mu_{\mathcal{D}_i}(x_\mathcal{V}) = P(i|x_\mathcal{V})\}, \tag{10}$$

where $P(i|x_\mathcal{V})$ is calculated according to (5) and (6). Formula (10) demonstrates that the membership of validation object $x_\mathcal{V}$ to the set $\mathcal{D}_i$ is not determined by the decision of classifier $\psi(\mathcal{L})$. The grade of membership of object $x_\mathcal{V}$ to $\mathcal{D}_i$ depends on the potential chance of classifying $x_\mathcal{V}$ to the i-th class by the classifier $\psi(\mathcal{L})$. We assume, that this potential chance is equal to the probability $P(i|x_\mathcal{V}) = P(\psi(\mathcal{L}, x_\mathcal{V}) = i)$ calculated approximately using the randomized model RRC of classifier $\psi(\mathcal{L})$.

Set $\mathcal{N}(x)$ plays the crucial role in the proposed concept of SCM, because it decides which validation objects $x_\mathcal{V}$ and with which weights will be involved in determining the local properties of the classifier $\psi(\mathcal{L})$ and – as a consequence – in the procedure of correcting its classifying decision. Formally, $\mathcal{N}(x)$ is also a fuzzy set:

$$\mathcal{N}(x) = \{(x_\mathcal{V}, \mu_{\mathcal{N}(x)}(x_\mathcal{V}))\}, \tag{11}$$

but its membership function is not defined univocally because it depends on many circumstances. By choosing the shape of the membership function $\mu_{\mathcal{N}(x)}$ we can freely model the adopted concept of "locality" (relative to x).

$\mu_{\mathcal{N}(x)}(x_\mathcal{V})$ depends on the distance between validation object $x_\mathcal{V}$ and test object x: its value is equal to 1 for $x_\mathcal{V} = x$ and decreases with increasing the distance between $x_\mathcal{V}$ and x. This leads to the following form of the proposed membership function of the set:

$$\mu_{\mathcal{N}}(x_\mathcal{V}) = \begin{cases} C \exp(-\beta \|x - x_\mathcal{V}\|^2), & \text{if} \quad \|x - x_\mathcal{V}\| < K_d \\ 0 & \text{otherwise} \end{cases} \tag{12}$$

$\|\cdot\|$ denotes Euclidean distance in the feature space $\mathcal{X}$, K_d is the Euclidean distance between x and the K-th nearest neighbor in $\mathcal{V}$, $\beta \in \Re_+$ and C is a normalizing coefficient. The first factor in (12) limits the concept of "locality" (relatively to x) to the set of K nearest neighbors with Gaussian model of membership grade.

Since under the stream classification framework, there should be only one pass over the data [19], K and β parameters cannot be found using the extensive grid search approach just like it was for the originally proposed approach [22,30]. Consequently, in this work, we decided to set β to 1. Additionally, the initial number of nearest neighbours is found using a simple rule of thumb [6]:

$$\hat{K} = \left\lceil \sqrt{|\mathcal{V}|} \right\rceil. \tag{13}$$

To avoid ties, the final number of neighbours K is set as follows:

$$K = \begin{cases} \hat{K} & \text{if} \quad M \mod \hat{K} \neq 0 \\ \hat{K} + 1 & \text{otherwise} \end{cases} \tag{14}$$

Additionally, the computational cost of computing the neighbourhood may be further reduced by using the kd-tree algorithm to find the nearest neighbours [18].

Finally, from (8), (10) and (11) we get the following approximation:

$$P(j|i, x) \approx \frac{\varepsilon_{i,j}(x)}{\sum_{j \in \mathcal{M}} \varepsilon_{i,j}(x)}, \tag{15}$$

which together with (3), (5) and (6) give (2) i.e. the corrected classifier $\psi^{\mathrm{Corr}}(\mathcal{L}, \mathcal{V})$.

2.4   Creating the Validation Set

In this section, the procedure of creating the validation set $\mathcal{V}$ from the training $\mathcal{L}$ is described. In the original work describing SCM [30], the set of labelled data was wplit into the learning set $\mathcal{L}$ and the validation set $\mathcal{V}$. The learning set and the validation set were disjoint $\mathcal{L}' \cap \mathcal{V} = \emptyset$. The cardinality of the validation set was controlled by the γ parameter $|\mathcal{V}| = \gamma |\mathcal{L}|$, $\gamma \in [0, 1]$. The γ coefficient was usually set to 0.6, however to achieve the highest classification quality, it should be determined using the grid-search procedure. As it was said above, in this work we want to avoid using the grid-search procedure. Therefore, we construct the validation set using three-fold cross-validation procedure that allows using of the entire learning set as a validation set. The procedure is described in Algorithm 1.

Algorithm 1: Procedure of training the SCM classifier. Including the procedure of validation set creation.

Data: $\mathcal{L}$ – Initial learning set;
Result: $\mathcal{V}$ – Validation set;
$\mathcal{D}_i$ – Decision sets (see (10));
$\psi(\mathcal{L})$ – Trained classifier.

1 **begin**
2   **for** $k \in \{1, 2, 3\}$ **do**
3   Extract fold specific training and validation set $\mathcal{L}_k$, $\mathcal{V}_k$;
4   Learn the $\psi(\mathcal{L}_k)$ using $\mathcal{L}_k$;
5          $\mathcal{V} := \mathcal{V} \cup \mathcal{V}_k$;
6   Update the class-specific decision sets $\mathcal{D}_i$ using predictions of $\psi(\mathcal{L}_k)$ for instances from $\mathcal{V}_k$ (see (10));
7   **end**
8   Learn the $\psi(\mathcal{L})$ using $\mathcal{L}$;
9 **end**

---

3   Classification of Data Stream

The main goal of the work is to develop a wrapping-classifier that allows incremental learning to classifiers that are unable to learn incrementally. In this section, we describe the incremental learning procedure used by the SCM-based wrapping-classifier.

3.1   Model of Data Stream

We assume that instances from a data stream $\mathcal{S}$ appear as a sequence of labeled examples $\{(x^t, j^t)\}, t = 1, 2, ..., T$, where $x^t \in \mathcal{X} \subseteq \Re^d$ represents a d-dimensional feature vector of an object that arrived at time t and $j^t \in \mathcal{M} = \{1, 2, \ldots, M\}$ is its class number. In this study we consider a completely supervised learning approach which means that the true class number j^t is available after the arrival of the object x^t and before the arrival of the next object x^{t+1} and this information may be used by classifier for classification of x^{t+1}. Such a framework is one of the most often considered in the related literature [3, 25].

In addition, we assume that a data stream can be generated with a time-varying distribution, yielding the phenomenon of concept drift [9]. We do not impose any restrictions on the concept drift. It can be real drift referring to changes of class distribution or virtual drift referring to the distribution of features. We allow sudden, incremental, gradual, and recurrent changes in the distribution of instances creating a data stream. Changes in the distribution can cause an imbalanced class system to appear in a changing configuration.

3.2 Incremental Learning for SCM Classifier

We assumed that the base classifier $\psi(\mathcal{L})$ wrapped by the SCM classifier is unable to learn incrementally. Consequently, an initial training set has to be used to build the classifier. This initial data set is called an initial chunk $\mathcal{B}$. The desired size of the initial bath is denoted by $|\mathcal{B}_{\text{des}}|$. The initial data set is built by storing incoming examples from the data stream. By the time the initial batch is collected, the prediction is impossible. Until then, the prediction is made on the basis of *a priori* probabilities estimated from the incomplete initial batch.

Since $\psi(\mathcal{L})$ is unable to learn incrementally, incremental learning is handled with changing the validation set. Incoming instances are added to the validation set until the ADWIN-based drift detector detects that the concept drift has occurred. The ADWIN-based drift detector analyses the outcomes of the corrected classifier for the instances stored in the validation set [2]. When there is a significant difference between the older and the newer part of the validation set, the detector removes the older part of the validation set. The remaining part of the validation set is then used to correct the outcome of $\psi(\mathcal{L})$. The ADWIN-based drift detector also controls the size of the neighbourhood. Even if there is no concept drift, the detector may detect the deterioration of the classification quality when the neighbourhood becomes too large.

The detailed procedure of the ensemble building is described in Algorithms 2 and 3.

4 Experimental Setup

To validate the classification quality obtained by the proposed approaches, the experimental evaluation, which setup is described below, is performed.

The following base classifiers were employed:

- ψ_{HOE} – Hoeffding tree classifier [26]
- ψ_{NB} – Naive Bayes classifier with kernel density estimation [16].

Algorithm 2: Validation set update controlled by ADWIN detector.

Data: $\mathcal{V}$ – validation set;
x – new instance to add;
Result: Updated validation set

1 **begin**
2 | i= $\psi(\mathcal{L}, \mathcal{V}, x)$;   // Predict object class using corrected classifier
3 | Check the prediction using ADWIN detector;
4 | **if** *ADWIN detector detects drift* **then**
5 | | Ask the detector fot the newer part of the validation set $\mathcal{V}_{\text{new}}$;
6 | | $\mathcal{V} := \mathcal{V}_{\text{new}}$;
7 | $\mathcal{V} := \mathcal{V} \cup x$;
8 **end**

Algorithm 3: Incremental learning procedure of the SCM wrapping-classifier.

Data: x – new instance;
Result: Learned SCM wrapping-classifier

1 **begin**
2 | **if** $|\mathcal{B}| \geq |\mathcal{B}_{\text{des}}|$ **then**
3 | | Train the SCM classifier using the procedure described in Algorithm 1 using $\mathcal{B}$ as a learning set;
4 | | $\mathcal{B} := \emptyset$;
5 | | $\mathcal{V}' := \mathcal{V}$;   // Make a copy of the validation set
6 | | **foreach** *object* $x' \in \mathcal{V}'$ **do**
7 | | | Update the validation set $\mathcal{V}$ using x' and the procedure described in Algorithm 2
8 | | **end**
9 | **else if** *Is SCM classifier trained* **then**
10 | | Update the validation set $\mathcal{V}$ using x and the procedure described in Algorithm 2
11 | **else**
12 | | $\mathcal{B} := \mathcal{B} \cup x$;
13 | **end**
14 **end**

- ψ_{KNN} – KNN classifier [14].
- ψ_{SGD} – SVM classifier built using stochastic gradient descent method [28].

The classifiers implemented in WEKA framework [15] were used. If not stated otherwise, the classifier parameters were set to their defaults. We have chosen the classifiers that offer both batch and incremental learning procedures.

The experimental code was implemented using WEKA [15] framework. The source code of the algorithms is available online[1,2].

During the experimental evaluation, the following classifiers were compared:

1. ψ_{B} – The ADWIN-driven classifier created using the unmodified base classifier (The base classifier is able to update incrementally.) [2].
2. ψ_{nB} – The ADWIN-driven created using the unmodified base classifier with the incremental learning disabled. The base classifier is only retrained whenever ADWIN-based detector detects concept drift.
3. ψ_{S} – The ADWIN-driven approach using SCM correction scheme with online-learning. As described in Sect. 3.

---

[1] https://github.com/ptrajdos/rrcBasedClassifiers/tree/develop.
[2] https://github.com/ptrajdos/StreamLearningPT/tree/develop.

4. ψ_{nS} – The ADWIN-driven approach created using SCM correction scheme but the online-learning is disabled. The SCM-corrected classifier is only retrained whenever ADWIN-based detector detects concept drift.

To evaluate the proposed methods, the following classification-loss criteria are used [29]: Macro-averaged FDR (1- precision), FNR (1-recall), Matthews correlation coefficient (MCC). The Matthews coefficient is rescaled in such a way that 0 is perfect classification and 1 is the worst one. Quality measures from the macro-averaging group are considered because this kind of measures is more sensitive to the performance for minority classes. For many real-world classification problems, the minority class is the class that attracts the most attention [23].

Following the recommendations of [5] and [10], the statistical significance of the obtained results was assessed using the two-step procedure. The first step is to perform the Friedman test [5] for each quality criterion separately. Since multiple criteria were employed, the familywise errors (FWER) should be controlled [17]. To do so, the Holm [17] procedure of controlling FWER of the conducted Friedman tests was employed. When the Friedman test shows that there is a significant difference within the group of classifiers, the pairwise tests using the Wilcoxon signed-rank test [5] were employed. To control FWER of the Wilcoxon-testing procedure, the Holm approach was employed [17]. For all tests, the significance level was set to $\alpha = 0.01$.

The experiments were conducted using 48 synthetic datasets generated using the STREAM-LEARN library[3]. The properties of the datasets were as follows: Datasets size: 30k examples; Number of attributes: 8; Types of drift generated: incremental, sudden; Noise: 0%, 10%, 20%; Imbalance ratio: 0 – 4.

Datasets used in this experiment are available online[4].

To examine the effectiveness of the incremental update algorithms, we applied an experimental procedure based on the methodology which is characteristic of data stream classification, namely, the test-then-update procedure [8]. The chunk size for evaluation purposes was set to 200.

5   Results and Discussion

To compare multiple algorithms on multiple benchmark sets, the average ranks approach is used. In this approach, the winning algorithm achieves a rank equal to '1', the second achieves a rank equal to '2', and so on. In the case of ties, the ranks of algorithms that achieve the same results are averaged.

The numerical results are given in Table 2, 3, 4 and 5. Each table is structured as follows. The first row contains the names of the investigated algorithms. Then, the table is divided into six sections – one section is related to a single evaluation criterion. The first row of each section is the name of the quality

---

[3] https://github.com/w4k2/stream-learn.
[4] https://github.com/ptrajdos/MLResults/blob/master/data/stream_data.tar.xz?
raw=true.

criterion investigated in the section. The second row shows the p-value of the Friedman test. The third one shows the average ranks achieved by algorithms. The following rows show p-values resulting from the pairwise Wilcoxon test. The p-value equal to .000 informs that the p-values are lower than 10^{-3}. P-values lower than α are bolded. Due to the page limit, the raw results are published online[5].

To provide a visualization of the average ranks and the outcome of the statistical tests, the rank plots are used. The rank plots are compatible with the rank plots described in [5]. That is, each classifier is placed along the line representing the values of the achieved average ranks. The classifiers between which there are no significant differences (in terms of the pairwise Wilcoxon test) are connected with a horizontal bar placed below the axis representing the average ranks. The results are visualised on Figs. 1, 2, 3 and 4.

Let us begin with an analysis of the correction ability of the SCM approach when incremental learning is disabled. Although this kind of analysis has been already done [22,30], in this work it should be done again since the definition of the neighbourhood is significantly changed (see Sect. 2.3). To assess the impact of the SCM-based correction, we compare the algorithms ψ_{nB} and ψ_{nS} for different base classifiers. For ψ_{HOE} and ψ_{NB} base classifiers the employment of SCM-based correction allows achieving significant improvement in terms of all quality criteria (see Figs. 1 and 2). For the remaining base classifiers, on the other hand, there are no significant differences between ψ_{nB} and ψ_{nS}. These results confirm observations previously made in [22,30]. That is, the correction ability of the SCM approach is more noticeable for classifiers that are considered to be weaker ones. The previously observed correction ability holds although the extensive grid-search technique is not applied.

In this paper, the SCM-based approach is proposed to be used as a wrapping-classifier that handles the incremental learning for base classifiers that are unable to be updated incrementally. Consequently, now we are going to analyse the SCM approach in that scenario. The results show that ψ_S significantly outperforms ψ_{nB} for all base classifiers and quality criteria. It means that it works great as the incremental-learning-handling wrapping-classifier. What is more, it outperforms ψ_{nS} also for all base classifiers and criteria. It clearly shows that the source of the achieved improvement does not lie in the batch-learning-improvement-ability but the ability to handle incremental learning is also present. Moreover, it handles incremental learning more effective than the base classifiers designed to do so. This observation is confirmed by the fact that ψ_S also outperforms ψ_B for all base classifiers and quality criteria.

---

[5] https://github.com/ptrajdos/MLResults/blob/master/RandomizedClassifiers/ Results_cldd_2021.tar.xz?raw=true.

Table 2. Statistical evaluation for the stream classifiers based on ψ_{HOE} classifier.

	ψ_B	ψ_{nB}	ψ_S	ψ_{nS}	ψ_B	ψ_{nB}	ψ_S	ψ_{nS}	ψ_B	ψ_{nB}	ψ_S	ψ_{nS}
Crit. name	MaFDR				MaFNR				MaMCC			
Friedman p-value	1.213e−28				5.963e−28				5.963e−28			
Average rank	2.000	3.812	1.00	3.188	2.000	3.583	1.00	3.417	2.000	3.667	1.00	3.333
ψ_B		.000	.000	.000		.000	.000	.000		.000	.000	.000
ψ_{nB}			.000	.000			.000	.111			.000	.002
ψ_S				.000				.000				.000

Table 3. Statistical evaluation for the stream classifiers based on ψ_{NB} classifier.

	ψ_B	ψ_{nB}	ψ_S	ψ_{nS}	ψ_B	ψ_{nB}	ψ_S	ψ_{nS}	ψ_B	ψ_{nB}	ψ_S	ψ_{nS}
Crit. name	MaFDR				MaFNR				MaMCC			
Friedman p-value	3.329e−28				3.329e−28				1.739e−28			
Average rank	2.021	3.771	1.00	3.208	2.000	3.708	1.00	3.292	2.000	3.792	1.00	3.208
ψ_B		.000	.000	.000		.000	.000	.000		.000	.000	.000
ψ_{nB}			.000	.000			.000	.001			.000	.000
ψ_S				.000				.000				.000

Table 4. Statistical evaluation for the stream classifiers based on ψ_{KNN} classifier.

	ψ_B	ψ_{nB}	ψ_S	ψ_{nS}	ψ_B	ψ_{nB}	ψ_S	ψ_{nS}	ψ_B	ψ_{nB}	ψ_S	ψ_{nS}
Crit. name	MaFDR				MaFNR				MaMCC			
Friedman p-value	1.883e−27				1.883e−27				1.883e−27			
Average rank	2.000	3.521	1.00	3.479	2.000	3.542	1.00	3.458	2.000	3.500	1.00	3.500
ψ_B		.000	.000	.000		.000	.000	.000		.000	.000	.000
ψ_{nB}			.000	.955			.000	.545			.000	.757
ψ_S				.000				.000				.000

Table 5. Statistical evaluation for the stream classifiers based on ψ_{SGD} classifier.

	ψ_B	ψ_{nB}	ψ_S	ψ_{nS}	ψ_B	ψ_{nB}	ψ_S	ψ_{nS}	ψ_B	ψ_{nB}	ψ_S	ψ_{nS}
Crit. name	MaFDR				MaFNR				MaMCC			
Friedman p-value	3.745e−27				1.563e−27				1.563e−27			
Average rank	2.042	3.500	1.00	3.458	2.021	3.292	1.00	3.688	2.000	3.438	1.00	3.562
ψ_B		.000	.000	.000		.000	.000	.000		.000	.000	.000
ψ_{nB}			.000	.947			.000	.005			.000	.088
ψ_S				.000				.000				.000

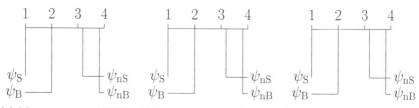

(a) Macro-averaged FDR   (b) Macro-averaged FNR   (c) Macro-averaged MCC

Fig. 1. Ranking plot for the stream classifiers based on ψ_{HOE} classifier.

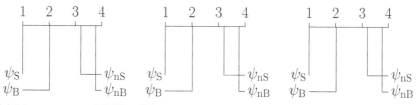

(a) Macro-averaged FDR   (b) Macro-averaged FNR   (c) Macro-averaged MCC

Fig. 2. Ranking plot for the stream classifiers based on ψ_{NB} classifier.

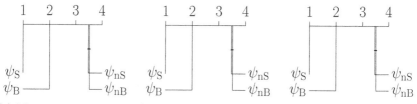

(a) Macro-averaged FDR   (b) Macro-averaged FNR   (c) Macro-averaged MCC

Fig. 3. Ranking plot for the stream classifiers based on ψ_{KNN} classifier.

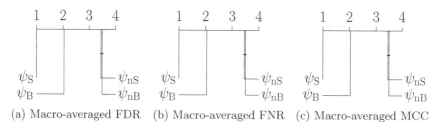

(a) Macro-averaged FDR   (b) Macro-averaged FNR   (c) Macro-averaged MCC

Fig. 4. Ranking plot for the stream classifiers based on ψ_{SGD} classifier.

6   Conclusions

In this paper, we propose a modified SCM classifier to be used as a wrapping-classifier that allows incremental learning of classifiers that are not designed to be incrementally updated. We applied two modifications of the SCM wrapping-classifier originally described in [22,30]. The first one is a modified neighbour-hood definition. The newly proposed neighbourhood does not need an excessive grid-search procedure to be performed to find the best set of parameters. Due to the modified neighbourhood definition, the computational cost of performing the SCM-based correction is significantly smaller. The second modification is to incorporate ADWIN-based approach to create and manage the validation set used by SCM-based algorithm. This modification not only allows the proposed method to effectively deal with the concept drift but also it can shrink the neighbourhood when it becomes too wide.

The experimental results show that the proposed approach outperforms the reference methods for all investigated base classifiers in terms of all considered quality criteria.

The results obtained in this study are very promising. Consequently, we are going to continue our research related to the employment of randomised classifiers in the task of stream learning. Our next step will probably be a proposition of a stream learning ensemble that used the SCM-correction method proposed in this paper.

Acknowledgments. This work was supported by the statutory funds of the Department of Systems and Computer Networks, Wroclaw University of Science and Technology.

References

1. de Barros, R.S.M., de Carvalho Santos, S.G.T.: An overview and comprehensive comparison of ensembles for concept drift. Inf. Fusion **52**, 213–244 (2019). https://doi.org/10.1016/j.inffus.2019.03.006
2. Bifet, A., Gavaldà, R.: Learning from time-changing data with adaptive windowing. In: Proceedings of the 2007 SIAM International Conference on Data Mining. Society for Industrial and Applied Mathematics, April 2007. https://doi.org/10.1137/1.9781611972771.42
3. Brzezinski, D., Stefanowski, J.: Combining block-based and online methods in learning ensembles from concept drifting data streams. Inf. Sci. **265**, 50–67 (2014). https://doi.org/10.1016/j.ins.2013.12.011
4. Brzezinski, D., Stefanowski, J.: Reacting to different types of concept drift: The accuracy updated ensemble algorithm. IEEE Trans. Neural Netw. Learn. Syst. **25**(1), 81–94 (2014). https://doi.org/10.1109/tnnls.2013.2251352
5. Demšar, J.: Statistical comparisons of classifiers over multiple data sets. J. Mach. Learn. Res. **7**, 1–30 (2006)
6. Devroye, L., Györfi, L., Lugosi, G.: A Probabilistic Theory of Pattern Recognition. SMAP, vol. 31. Springer, New York (1996). https://doi.org/10.1007/978-1-4612-0711-5
7. Dhar, M.: On cardinality of fuzzy sets. IJISA **5**(6), 47–52 (2013). https://doi.org/10.5815/ijisa.2013.06.06
8. Gama, J.: Knowledge Discovery from Data Streams, 1st edn. Chapman and Hall/CRC (2010). https://doi.org/10.1201/ebk1439826119

9. Gama, J., Žliobaitė, I., Bifet, A., Pechenizkiy, M., Bouchachia, A.: A survey on concept drift adaptation. CSUR **46**(4), 1–37 (2014). https://doi.org/10.1145/2523813
10. Garcia, S., Herrera, F.: An extension on "statistical comparisons of classifiers over multiple data sets" for all pairwise comparisons. J. Mach. Learn. Res. **9**, 2677–2694 (2008)
11. Giraud-Carrier, C.: A note on the utility of incremental learning. AI Commun. **13**(4), 215–223 (2000)
12. Gomes, H.M., Barddal, J.P., Enembreck, F., Bifet, A.: A survey on ensemble learning for data stream classification. CSUR **50**(2), 1–36 (2017). https://doi.org/10.1145/3054925
13. Gonçalves, P.M., de Carvalho Santos, S.G., Barros, R.S., Vieira, D.C.: A comparative study on concept drift detectors. Expert Syst. Appl. **41**(18), 8144–8156 (2014). https://doi.org/10.1016/j.eswa.2014.07.019
14. Guo, G., Wang, H., Bell, D., Bi, Y., Greer, K.: KNN model-based approach in classification. In: Meersman, R., Tari, Z., Schmidt, D.C. (eds.) OTM 2003. LNCS, vol. 2888, pp. 986–996. Springer, Heidelberg (2003). https://doi.org/10.1007/978-3-540-39964-3_62
15. Hall, M., Frank, E., Holmes, G., Pfahringer, B., Reutemann, P., Witten, I.H.: The WEKA data mining software. SIGKDD Explor. Newsl. **11**(1), 10 (2009). https://doi.org/10.1145/1656274.1656278
16. Hand, D.J., Yu, K.: Idiot's Bayes: not so stupid after all? Int. Stat. Rev./Revue Internationale de Statistique **69**(3), 385 (2001). https://doi.org/10.2307/1403452
17. Holm, S.: A simple sequentially rejective multiple test procedure. Scand. J. Stat. **6**(2), 65–70 (1979). https://doi.org/10.2307/4615733
18. Hou, W., Li, D., Xu, C., Zhang, H., Li, T.: An advanced k nearest neighbor classification algorithm based on KD-tree. In: 2018 IEEE International Conference of Safety Produce Informatization (IICSPI). IEEE, December 2018. https://doi.org/10.1109/iicspi.2018.8690508
19. Krawczyk, B., Minku, L.L., Gama, J., Stefanowski, J., Woźniak, M.: Ensemble learning for data stream analysis: a survey. Inf. Fusion **37**, 132–156 (2017). https://doi.org/10.1016/j.inffus.2017.02.004
20. Kuncheva, L.I.: Combining Pattern Classifiers. Wiley (2014). https://doi.org/10.1002/9781118914564
21. Kuncheva, L.I., Žliobaitė, I.: On the window size for classification in changing environments. IDA **13**(6), 861–872 (2009). https://doi.org/10.3233/ida-2009-0397
22. Kurzynski, M., Krysmann, M., Trajdos, P., Wolczowski, A.: Multiclassifier system with hybrid learning applied to the control of bioprosthetic hand. Comput. Biol. Med. **69**, 286–297 (2016). https://doi.org/10.1016/j.compbiomed.2015.04.023
23. Leevy, J.L., Khoshgoftaar, T.M., Bauder, R.A., Seliya, N.: A survey on addressing high-class imbalance in big data. J. Big Data **5**(1), 1–30 (2018). https://doi.org/10.1186/s40537-018-0151-6
24. Mehta, S., et al.: Concept drift in streaming data classification: algorithms, platforms and issues. Procedia Comput. Sci. **122**, 804–811 (2017). https://doi.org/10.1016/j.procs.2017.11.440
25. Nguyen, H.-L., Woon, Y.-K., Ng, W.-K.: A survey on data stream clustering and classification. Knowl. Inf. Syst. **45**(3), 535–569 (2014). https://doi.org/10.1007/s10115-014-0808-1
26. Pfahringer, B., Holmes, G., Kirkby, R.: New options for Hoeffding trees. In: Orgun, M.A., Thornton, J. (eds.) AI 2007. LNCS (LNAI), vol. 4830, pp. 90–99. Springer, Heidelberg (2007). https://doi.org/10.1007/978-3-540-76928-6_11

27. Read, J., Bifet, A., Pfahringer, B., Holmes, G.: Batch-incremental versus instance-incremental learning in dynamic and evolving data. In: Hollmén, J., Klawonn, F., Tucker, A. (eds.) IDA 2012. LNCS, vol. 7619, pp. 313–323. Springer, Heidelberg (2012). https://doi.org/10.1007/978-3-642-34156-4_29

28. Sakr, C., Patil, A., Zhang, S., Kim, Y., Shanbhag, N.: Minimum precision requirements for the SVM-SGD learning algorithm. In: 2017 IEEE International Conference on Acoustics, Speech and Signal Processing (ICASSP). IEEE, March 2017. https://doi.org/10.1109/icassp.2017.7952334

29. Sokolova, M., Lapalme, G.: A systematic analysis of performance measures for classification tasks. Inf. Process. Manag. **45**(4), 427–437 (2009). https://doi.org/10.1016/j.ipm.2009.03.002

30. Trajdos, P., Kurzynski, M.: A dynamic model of classifier competence based on the local fuzzy confusion matrix and the random reference classifier. Int. J. Appl. Math. Comput. Sci. **26**(1), 175–189 (2016). https://doi.org/10.1515/amcs-2016-0012

31. Trajdos, P., Kurzynski, M.: A correction method of a binary classifier applied to multi-label pairwise models. Int. J. Neur. Syst. **28**(09), 1750062 (2018). https://doi.org/10.1142/s0129065717500629

32. Woloszynski, T., Kurzynski, M.: A probabilistic model of classifier competence for dynamic ensemble selection. Pattern Recognit. **44**(10–11), 2656–2668 (2011). https://doi.org/10.1016/j.patcog.2011.03.020

33. Žliobaitė, I.: Combining similarity in time and space for training set formation under concept drift. IDA **15**(4), 589–611 (2011). https://doi.org/10.3233/ida-2011-0484

Some Proposal of the High Dimensional PU Learning Classification Procedure

Konrad Furmańczyk$^{(\boxtimes)}$ ⓘ, Marcin Dudziński ⓘ,
and Diana Dziewa-Dawidczyk ⓘ

Institute of Information Technology, Warsaw University of Life Sciences,
Warsaw, Poland
{konrad_furmanczyk,marcin_dudzinski,diana_dziewa_dawidczyk}@sggw.edu.pl

Abstract. In our work, we propose a new classification method for positive and unlabeled (PU) data, called the LassoJoint classification procedure, which combines the thresholded Lasso approach in the first two steps with the joint method based on logistic regression, introduced by Teisseyre et al. [12], in the last step. We prove that, under some regularity conditions, our procedure satisfies the screening property. We also conduct some simulation study in order to compare the proposed classification procedure with the oracle method. Prediction accuracy of the proposed method has been verified for some selected real datasets.

Keywords: Positive unlabeled learning · Logistic regression · Empirical risk minimization · Thresholded Lasso

1 Introduction

Learning from positive and unlabeled (PU in short) data is an approach, where training data contains only positive and unlabeled examples, which means that the true labels $Y \in \{0, 1\}$ are not observed directly, since only surrogate variable $S \in \{0, 1\}$ is observable. This surrogate variable equals 1 - if an example is labeled, or 0 - if otherwise. The PU datasets appear in a large number of applications. For example, they often appear while dealing with the so-called under-reporting data from medical surveys, fraud detection and ecological modeling (see, e.g., Hastie and Fithian [6]). Some other interesting examples of the under-reporting survey data may be found in Bekker and Davis [1] and Teisseyre et al. [12].

Suppose that X is a feature vector and, as mentioned earlier, $Y \in \{0, 1\}$ denotes a true class label and $S \in \{0, 1\}$ is a variable indicating, whether an example is labeled or not (then, $S = 1$ or $S = 0$, respectively). We apply a commonly used assumption, called the Selected Completely At Random (SCAR) condition, which states that the labeled examples are randomly selected from a set of positives examples, independently from X, i.e. $P(S = 1|Y = 1, X) = P(S = 1|Y = 1)$. Let $c = P(S = 1|Y = 1)$. The parameter c is called the label frequency and plays a key role in the PU learning problem. The primary

© Springer Nature Switzerland AG 2021
M. Paszynski et al. (Eds.): ICCS 2021, LNCS 12744, pp. 18–25, 2021.
https://doi.org/10.1007/978-3-030-77967-2_2

objective of our note is to introduce a new PU learning classification procedure leading to the estimation of the posterior probability $f(x) = P(Y = 1|X = x)$.

The three basic methods of this estimation have been proposed so far. They consist in minimizing the empirical risk of logistic loss function and are known as: the naive method, the weighted method, and the joint method (the last one has been recently introduced in the paper of Teisseyre et al. [12]). All of these approaches have been thoroughly described in [12]. As the joint method will be applied in our procedure's construction, some details regarding this method will be presented in the next section of our article. We have named our proposed classification method as the LassoJoint procedure, since it is a three-step approach combining the thresholded Lasso procedure with the joint method from Teisseyre et al. [12]. Namely, in its two first steps we perform - for some pre-specified level - the thresholded Lasso procedure, in order to obtain the support for coefficients of a feature vector X, while in the third step we apply - on the previously determined support - the joint method. Apart from the works, where different learning methods applying logistic regression for PU data have been proposed, there are also some other interesting articles, where various machine learning tools in the PU learning problems have been used. In this context, it is worthwhile to mention: the papers of Hou [7] and Guo [5], where the generative adversial networks (GAN) for the PU problem have been employed, the work of Mordelet and Vert [10], where the bagging Support Vector Machine (SVM) approach for the PU data have been applied, and an article of Song and Raskutti [11], where the multidimensional PU problem with regard to the features selection has been investigated, and where the so-called PUlasso design has been established. It turns out that the LassoJoint procedure, which we propose in our work, is computationally simple and efficient in comparison to the other existing methods where the PU problem is considered. The simplicity and efficiency of our approach have been confirmed by the conducted simulation study.

The remainder of the paper is structured as follows. Namely, in Sect. 2 we describe our classification procedure in detail, in particular we also prove that the introduced method is the so-called screening procedure (i.e., it selects with a high probability the most significant predictors and the number of selected features is not greater that the sample size), as the screening property is necessary to apply the joint method in the final step of the procedure. In turn, in Sect. 3 we carry out some numerical study, in order to check the efficiency of the proposed approach, while in Sect. 4 we summarize and conclude our research. The results of numerical experiments on real data are given in Supplement[1].

2   The Proposed LassoJoint Algorithm

In our considerations, we assume that we have a random vector (Y, X), where $Y \in \{0, 1\}$ and $X \in \mathbb{R}^p$ is a feature vector, and that a random sample $(Y_1, X_1), \ldots, (Y_n, X_n)$ is distributed as (Y, X) and independent of it. In addition, we suppose that the coordinates X_{ji} of X_i, $i = 1, \ldots, n$, $j = 1, \ldots, p$, are

---

[1]   https://github.com/kfurmanczyk/ICCS21.

subgaussian with a parameter σ_{jn}^2, i.e. $E \exp(uX_{ji}) \le \exp\left(u^2\sigma_{jn}^2/2\right)$ for all $u \in \mathbb{R}$.

Let: $s_n^2 = \max_{1 \le j \le p} \sigma_{jn}^2$, $\limsup_n s_n^2 < \infty$, and $P(Y = 1 \mid X = x) = q(x^T\beta)$ for some function $0 < q(x) < 1$ and all $x \in \mathbb{R}^p$, where p may depend on n and $p > n$. Put: $I_0 = \{j : \beta_j \ne 0\}$, $I_1 = \{1, \ldots, p\}\backslash I_0$ and $|I_0| = p_0$. We shall assume - as in Kubkowski and Mielniczuk [9] - that the distribution of X satisfies the linear regression condition (LRC), which means that

$$E(X \mid X^T\beta) = u_0 + u_1 X^T\beta \text{ for some } u_0, u_1 \in \mathbb{R}^p.$$

This condition is fulfilled (for all β) by the class of elliptical distributions (such that, e.g., the normal distribution or the multivariate t-Student distribution). Reasoning as in Kubkowski and Mielniczuk [9], we obtain that under (LRC), there exists η satisfying $\beta^* = \eta\beta$, where $\eta \ne 0$ if $\mathrm{cov}(Y, X^T\beta) \ne 0$, and where $\beta^* = \arg\min_\beta R(\beta)$, with R standing for the risk function given by $R(\beta) = -E_{(X,Y)}l(\beta, X, Y)$, where in turn, $l(\beta, X, Y) = Y\log\sigma(X^T\beta) + (1-Y)\log\left(1 - \sigma(X^T\beta)\right)$, with σ denoting logistic function of the form $\sigma(X^T\beta) = \exp(X^T\beta)/\left[1 + \exp(X^T\beta)\right]$. Put: $I_0^* = \{j : \beta_j^* \ne 0\}$, $I_1^* = \{1, \ldots, p\}\backslash I_0^*$. It may be observed that under (LRC), we have $I_0 = I_0^*$ and consequently that $\mathrm{Supp}(I_0) = \mathrm{Supp}(I_0^*)$. In addition, put $H(b) = E(X^TX\sigma'(X^Tb))$ and define a cone $C(d, w) = \{\Delta \in \mathbb{R}^p : \|\Delta_{w^c}\|_1 \le d\|\Delta_w\|_1\}$, where: $w \subseteq \{1, \ldots, p\}$, $w^c = \{1, \ldots, p\}\backslash w$, $\Delta_w = (\Delta_{w_1}, \ldots, \Delta_{w_k})$, for $w = (w_1, \ldots, w_k)$. Furthermore, let κ be a generalized minimal eigenvalue of the matrix $H(\beta^*)$ given by $\kappa = \inf_{\Delta \in C(3, s_0^*)} \frac{\Delta^T H(\beta^*)\Delta}{\Delta^T\Delta}$. Moreover, we also define $\beta_{\min}^*$ and $\beta_{\min}$ as $\beta_{\min}^* := \min_{j \in I_0^*}|\beta_j^*|$ and $\beta_{\min} := \min_{j \in I_0}|\beta_j|$, respectively.

After these preliminaries, we are now in a position to depict the proposed method. Namely, our procedure, called the LassoJoint approach, is a three-step method, which is described as follows:

(1) For available PU dataset (s_i, x_i), $i = 1, \ldots, n$, we perform the ordinary Lasso procedure (see Tibshirani [14]) for some tuning parameter $\lambda > 0$, i.e. we compute the following Lasso estimator of β^*: $\hat{\beta}^{(L)} = \arg\min_{\beta \in R^{p+1}} \hat{R}(\beta) + \lambda\sum_{j=1}^p |\beta_j|$, where $\hat{R}(\beta) = -\frac{1}{n}\sum_{i=1}^n \left[s_i \log\left(\sigma(x_i^T\beta)\right) + (1 - s_i)\log\left(1 - \sigma(x_i^T\beta)\right)\right]$ and subsequently, we obtain the corresponding support $\mathrm{Supp}^{(L)} = \{1 \le j \le p : \hat{\beta}_j^{(L)} \ne 0\}$;

(2) We perform the thresholded Lasso for some prespecified level δ and obtain the support $\mathrm{Supp}^{(TL)} = \{1 \le j \le p : \left|\hat{\beta}_j^{(L)}\right| \ge \delta\}$;

(3) We apply the joint method from Teisseyre et al. [12] for the predictors from $\mathrm{Supp}^{(TL)}$.

Remark. The PU problem is related to incorrect specification of the logistic model. Under the SCAR assumption, we obtain that $P(S = 1|X = x) = cq(x^T\beta)$ and consequently, if $cq() \ne \sigma()$, then in step (1) we are fitting misspecified logistic model to (S, X). Generally speaking, the joint method from [12] consists in fitting the PU data to logistic function and in the minimization, with respect to β and $c = P(S = 1 \mid Y = 1)$, of the following empirical risk $\hat{R}(\beta, c) = -\frac{1}{n} \sum_{i=1}^{n} \left[s_i \log\left(c\sigma(x_i^T \beta)\right) + (1 - s_i) \log\left(1 - c\sigma(x_i^T \beta)\right) \right]$, where σ stands for logistic function and $\{(s_i, x_i)\}$ is the sample of observations from the distribution of a random vector (S, X).

The newly proposed LassoJoint procedure is similar to the LassoSD approach introduced in Furmańczyk and Rejchel [3]. The only difference between these two methods is that, we apply the joint method from [12] in the last step of our procedure - contrary to the procedure from [3], where the authors use multiple hypotheses testing in its final stage. The introduced LassoJoint procedure is determined by the two parameters λ and δ, which may depend on n. The selection of λ and δ is possible, if we impose the following conditions - denoted as the assumptions (A1)–(A4):

(A1) The generalized eigenvalue κ, of the matrix $H(\beta^*)$, is such that $m \leq \kappa \leq M$, for some $0 < m < M$;

(A2) $p_0^2 \log(p) = o(n)$, $\log(p) = o(n\lambda^2)$, $\lambda^2 p_0^2 \log(np) = o(1)$, as $n \to \infty$;

(A3) $p_0 + \frac{c_n^2}{\delta^2} \leq n$, where $c_n = 10 \frac{\sqrt{p_0}}{\kappa} \lambda$;

(A4) $\beta_{\min} \geq \left(\delta + c_n/\sqrt{p_0}\right)/\eta$.

Clearly, in view of (LRC), the condition from (A4) is equivalent to the constraint that $\beta_{\min}^* \geq \delta + c_n/\sqrt{p_0}$. In addition, due to (A2), we get that $c_n \to 0$.

The main strictly theoretical result of our work is the following assertion.

Theorem 1. *(Screening property) Under the conditions (LRC) and (A1)–(A4), we have that with a probability at least $1 - \epsilon_n$, where $\epsilon_n \to 0$:*

(a) $\left| \text{Supp}^{(TL)} \right| \leq p_0 + \frac{c_n^2}{\delta^2} \leq n,$

(b) $I_0 \subset \text{Supp}^{(TL)}.$

The presented theorem states that the proposed LassoJoint procedure is the so-called screening procedure, which means that (in the first two steps) this method selects, with a high probability, the most significant predictors of the model and that the number of selected features is not greater than the sample size n. This screening property guarantees that with a high probability, we may .apply the joint procedure from Teisseyre et al. [12], based on fitting logistic regression to PU data. The proof of Theorem 1 uses the following lemma, which straightforwardly follows from Theorem 4.9 in Kubkowski [8].

Lemma 1. *Under the assumptions (A1)–(A2), we obtain that with a probability at least $1 - \epsilon_n$, with ϵ_n satisfying $\epsilon_n \to 0$, the following property holds:* $\left\| \hat{\beta}^{(L)} - \beta^* \right\|_2 \leq c_n$, *where $c_n \to 0$ and $\|x\|_2 = \sqrt{\sum_{j=1}^{p} x_j^2}$ for $x \in \mathbb{R}^p$.*

Proof. First, we prove the relation stated in (a). By the definition of $\hat{\beta}_j^{(L)}$, we get $\sum_{j \in I_1 \cap \mathrm{Supp}^{(TL)}} \left(\hat{\beta}_j^{(L)} \right)^2 \geq \delta^2 \left| I_1 \cap \mathrm{Supp}^{(TL)} \right|$. Hence,

$$\left| I_1 \cap \mathrm{Supp}^{(TL)} \right| \leq \frac{1}{\delta^2} \sum_{j \in I_1 \cap \mathrm{Supp}^{(TL)}} \left(\hat{\beta}_j^{(L)} \right)^2$$

$$= \frac{1}{\delta^2} \sum_{j \in I_1 \cap \mathrm{Supp}^{(TL)}} \left(\hat{\beta}_j^{(L)} - \beta_j^* \right)^2 \leq \frac{1}{\delta^2} \left\| \hat{\beta}^{(L)} - \beta^* \right\|_2^2,$$

and

$$\left| \mathrm{Supp}^{(TL)} \right| \leq \left| I_0 \cap \mathrm{Supp}^{(TL)} \right| + \left| I_1 \cap \mathrm{Supp}^{(TL)} \right| \leq p_0 + \frac{1}{\delta^2} \left\| \hat{\beta}^{(L)} - \beta^* \right\|_2^2.$$

It follows from the cited lemma that with a probability at least $1 - \epsilon_n$, where $\epsilon_n \to 0$, we have $\left| \mathrm{Supp}^{(TL)} \right| \leq p_0 + \frac{c_n^2}{\delta^2}$. Combining this inequality with (A3), we obtain (a). Thus, we only need to prove the property in (b).

Since $\left\{ \min_{j \in I_0} \left(\hat{\beta}_j^{(L)} \right)^2 \geq \delta^2 \right\} \subseteq \left\{ I_0 \subset \mathrm{Supp}^{(TL)} \right\}$, it is sufficient to show that

$$P \left(\min_{j \in I_0} \left(\hat{\beta}_j^{(L)} \right)^2 \geq \delta^2 \right) \geq 1 - \epsilon_n. \tag{1}$$

Let:
$\hat{\beta}_{I_0}^{(L)} := \left\{ \hat{\beta}_j^{(L)} : j \in I_0 \right\}$, $\beta_{I_0}^* := \left\{ \beta_j^* : j \in I_0 \right\}$. As $\left\| \hat{\beta}^{(L)} - \beta^* \right\|_2^2 \geq \left\| \hat{\beta}_{I_0}^{(L)} - \beta_{I_0}^* \right\|_2^2$, we have from the given lemma that with a probability at least $1 - \epsilon_n$,

$$p_0 \min_{j \in I_0} \left(\hat{\beta}_j^{(L)} - \beta_j^* \right)^2 \leq \left\| \hat{\beta}_{I_0}^{(L)} - \beta_{I_0}^* \right\|_2^2 = \sum_{j \in I_0} \left(\hat{\beta}_j^{(L)} - \beta_j^* \right)^2 \leq c_n^2.$$

Hence, $\min_{j \in I_0} \left| \hat{\beta}_j^{(L)} - \beta_j^* \right| \leq c_n / \sqrt{p_0}$. In addition, by the triangle inequality, we obtain that for $j \in I_0$, $\left| \hat{\beta}_j^{(L)} \right| \geq \left| \beta_j^* \right| - \left| \hat{\beta}_j^{(L)} - \beta_j^* \right|$ and therefore, $\min_{j \in I_0} \left| \hat{\beta}_j^{(L)} \right| \geq \min_{j \in I_0} \left| \beta_j^* \right| - c_n / \sqrt{p_0}$. This and (A4) imply that with a probability at least $1 - \epsilon_n$, $\min_{j \in I_0} \left(\hat{\beta}_j^{(L)} \right) \geq \delta^2$, which yields (1) and consequently (b).

3   Numerical Study

Suppose that: $X_1, \ldots X_p$ are generated independently from $N(0, 1)$, and Y_i, $i = 1, \ldots, n$, are generated from the $\mathrm{binom}(1, p_i)$ distribution, where: $p_i = \sigma(\beta_0 + \beta_1 X_{1i} + \ldots + \beta_p X_{pi}), \beta_0 = 1$. The following high-dimensional models were simulated:

(M1) $p_0 = 5, p = 1.2 \cdot 10^3, n = 10^3, \beta_1 = \ldots = \beta_{p_0} = 1, \beta_{p_0+1} = \ldots = \beta_p = 0$;
(M2) $p_0 = 5, p = 1.2 \cdot 10^3, n = 10^3, \beta_1 = \ldots = \beta_{p_0} = 2, \beta_{p_0+1} = \ldots = \beta_p = 0$;
(M3) $p_0 = 5, p = 10^3, n = 10^3, \beta_1 = \ldots = \beta_{p_0} = 2, \beta_{p_0+1} = \ldots = \beta_p = 0$;

(M4) $p_0 = 20, p = 10^3, n = 10^3, \beta_1 = \ldots = \beta_{p_0} = 2, \beta_{p_0+1} = \ldots = \beta_p = 0;$
(M5) $p_0 = 5, p = 2 \cdot 10^3, n = 2 \cdot 10^3, \beta_1 = \ldots = \beta_{p_0} = 2, \beta_{p_0+1} = \ldots = \beta_p = 0;$
(M6) $p_0 = 5, p = 2 \cdot 10^3, n = 2 \cdot 10^3, \beta_1 = \ldots = \beta_{p_0} = 3, \beta_{p_0+1} = \ldots = \beta_p = 0.$

For all of the specified models, the LassoJoint method was implemented. In its first step, the Lasso method was used with some tuning parameters λ that were chosen either on the basis of 10-fold cross-validation scheme in the first scenario or by putting $\lambda = ((\log p)/n)^{1/3}$ in the second scenario. In the second step, we applied the thresholded Lasso design for $\delta = 0.5 \cdot ((\log p)/n)^{1/3}$.

In the third - and simultaneously - the last step of our procedure, the variables selected by the thresholded Lasso method were employed to the joint method from [12] for the problem of the PU data classification. From the listed models, we randomly selected $c \cdot 100\%$ of the labeled observations of S, for $c = 0.1; 0.3; 0.5; 0.7; 0.9$. Next, we generated a test sample of size 1000 from our models and determined their accuracy percentage based on 100 MC replications of our experiments. The idea of our procedure's accuracy assesment is similar to the idea from Furmańczyk and Rejchel [4]. We applied the 'glmnet' package [2] from the R software and [13] in our computations. The results of our simulation study are collected in Table 1 (the column 'oracle' shows the accuracy of classifier that uses only the significant predictors and the true parameters of logistic models).

Table 1. Results for M1–M6

c	Model	Scen 1	Scen 2	Oracle	Model	Scen 1	Scen 2	Oracle
0.1	M1	0.526	0.597	0.808	M4	0.501	0.531	0.939
0.3	M1	0.492	0.456	0.808	M4	0.530	0.529	0.939
0.5	M1	0.598	0.530	0.808	M4	0.671	0.596	0.939
0.7	M1	0.688	0.591	0.808	M4	0.699	0.656	0.939
0.9	M1	0.743	0.623	0.808	M4	0.792	0.733	0.939
0.1	M2	0.505	0.504	0.887	M5	0.410	0.473	0.885
0.3	M2	0.586	0.514	0.887	M5	0.667	0.514	0.885
0.5	M2	0.705	0.565	0.887	M5	0.770	0.588	0.885
0.7	M2	0.770	0.636	0.887	M5	0.803	0.648	0.885
0.9	M2	0.820	0.698	0.887	M5	0.680	0.694	0.885
0.1	M3	0.516	0.568	0.885	M6	0.537	0.548	0.921
0.3	M3	0.608	0.505	0.885	M6	0.742	0.532	0.921
0.5	M3	0.708	0.567	0.885	M6	0.812	0.594	0.921
0.7	M3	0.778	0.640	0.885	M6	0.853	0.668	0.921
0.9	M3	0.820	0.706	0.885	M6	0.710	0.724	0.921

Real data experiments and all codes in R are presented in Supplement, available on https://github.com/kfurmanczyk/ICCS21.

4   Conclusions

The results of our simulation study show that if c increases, then the percentage of correct classifications increases as well. They also show that the classifications obtained by applying the proposed LassoJoint method display smaller classification errors (and thus - better classification accuracy) for the models with larger signals (i.e., for the M5 and M6 models). Comparing the M3 and M5 models, we can see that with an increase of the number of significant predictors (p_0), the classification accuracy is slightly decreasing. Furthermore, in all cases - except for the situation where $c = 0.1$ - the selection of the tuning parameter λ obtained by using the cross-validation design results in better classification accuracy. In addition, we may observe that in the case when $c = 0.7$ or $c = 0.9$, our Lasso-Joint approach is nearly as good as the 'oracle' method. In turn, for $c = 0.1$ the classification accuracy was low - from 0.41 do 0.60, but in the 'easiest' case, i.e. when $c = 0.9$, the classification accuracy ranged from 0.7 to 0.82. Furthermore, the results of our experiments conducted on real datasets show that if c increases, then the percentage of correct classifications increases as well. In addition, these results show similar classification accuracy among all of the considered classification methods (see Supplement). The proposed new LassoJoint classification method for PU data allows for the relatively low simulation computational costs to analyze data in a high-dimensional case, i.e. when the number of predictors exceeds the size of available sample ($p > n$). We aim to devote our further research to a more detailed analysis of the introduced procedure, in particular to the examination regarding optimal selection of the model parameters.

References

1. Bekker, J., Davis, J.: Learning from positive and unlabeled data: a survey. Mach. Learn. **109**(4), 719–760 (2020). https://doi.org/10.1007/s10994-020-05877-5
2. Friedman, J., Hastie, T., Simon, N., Tibshirani, R.: Glmnet: Lasso and elastic-net regularized generalized linear models. R package version 2.0 (2015)
3. Furmańczyk, K., Rejchel, W.: High-dimensional linear model selection motivated by multiple testing. Statistics **54**, 152–166 (2020)
4. Furmańczyk, K., Rejchel, W.: Prediction and variable selection in high-dimensional misspecified classification. Entropy **22**(5), 543 (2020)
5. Guo, T., et al.: On positive-unlabeled classification in GAN. In: CVPR (2020)
6. Hastie, T., Fithian, W.: Inference from presence-only data; the ongoing controversy. Ecography **36**, 864–867 (2013)
7. Hou, M., Chaib-draa, B., Li, C., Zhao, Q.: Generative adversarial positive-unlabeled learning. In: Proceedings of the Twenty-Seventh International Joint Conference on Artificial Intelligence (IJCAI 2018) (2018)
8. Kubkowski, M.: Misspecification of binary regression model: properties and inferential procedures. Ph.D. thesis, Warsaw University of Technology, Warsaw (2019)
9. Kubkowski, M., Mielniczuk, J.: Active set of predictors for misspecified logistic regression. Statistics **51**, 1023–1045 (2017)
10. Mordelet, F., Vert, J.P.: A bagging SVM to learn from positive and unlabeled examples. Pattern Recogn. Lett. **37**, 201–209 (2013)

11. Song, H., Raskutti, G.: High-dimensional variable selection with presence-only data. arXiv:1711.08129v3 (2018)
12. Teisseyre, P., Mielniczuk, J., Łazęcka, M.: Different strategies of fitting logistic regression for positive and unlabelled data. In: Krzhizhanovskaya, V.V., et al. (eds.) ICCS 2020. LNCS, vol. 12140, pp. 3–17. Springer, Cham (2020). https://doi.org/10.1007/978-3-030-50423-6_1
13. Teisseyre, P.: Repository from https://github.com/teisseyrep/Pulogistic. Accessed 1 Jan 2021
14. Tibshirani, R.: Regression shrinkage and selection via the lasso. J. Roy. Statist. Soc. Ser. B **58**, 267–288 (1996)

Classifying Functional Data from Orthogonal Projections – Model, Properties and Fast Implementation

Ewa Skubalska-Rafajłowicz[1] (ID) and Ewaryst Rafajłowicz[2](✉) (ID)

[1] Department of Engineering Informatics, Wroclaw University of Science and Technology, Wyb. Wyspianskiego 27, 50 370 Wrocław, Poland
ewa.rafajlowicz@pwr.edu.pl
[2] Department of Control Systems and Mechatronics, Wroclaw University of Science and Technology, Wyb. Wyspianskiego 27, 50 370 Wrocław, Poland
ewaryst.rafajlowicz@pwr.edu.pl

Abstract. We consider the problem of functional, random data classification from equidistant samples. Such data are frequently not easy for classification when one has a large number of observations that bear low information for classification. We consider this problem using tools from the functional analysis. Therefore, a mathematical model of such data is proposed and its correctness is verified. Then, it is shown that any finite number of descriptors, obtained by orthogonal projections on any differentiable basis of $L_2(0, T)$, can be consistently estimated within this model.

Computational aspects of estimating descriptors, based on the fast implementation of the discrete cosine transform (DCT), are also investigated in conjunction with learning a classifier and using it on-line. Finally, the algorithm of learning descriptors and classifiers were tested on real-life random signals, namely, on accelerations, coming from large bucket-wheel excavators, that are transmitted to an operator's cabin. The aim of these tests was also to select a classifier that is well suited for working with DCT-based descriptors.

Keywords: Functional data classification · Random element · Bias · Functional data model · Classifying signals · DCT

1 Introduction

Tasks of classifying functional data are difficult for many reasons. The majority of them seems to concern a large number of observations, frequently having an unexpectedly low information content from the point of view of their classification. This kind of difficulty arises in many industrial applications, in which sensors may provide thousands of samples per second (see, e.g., our motivation example at the end of this section).

We focus our attention on classifying data from repetitive processes, i.e., on stochastic processes that have a finite and the same duration $T > 0$ and

© Springer Nature Switzerland AG 2021
M. Paszynski et al. (Eds.): ICCS 2021, LNCS 12744, pp. 26–39, 2021.
https://doi.org/10.1007/978-3-030-77967-2_3

after time T the process, denoted as $\mathbf{X}(t)$, $t \in [0, T]$ is repeated with the same or different probability measures. For simplicity of the exposition, we confine ourselves to two such measures and the problem is to classify samples from $\mathbf{X}(t)$, $[0, T]$ to two classes, having at our disposal a learning sequence $\mathbf{X}_n(t)$, $[0, T]$, $n = 1, 2, \ldots, N$ of correctly classified subsequences. An additional requirement is to classify newly incoming samples almost immediately after the present $[0, T]$ is finished, so as to be able to use the result of classification for making decisions for the next period (also called a pass). This requirement forces us to put emphasis not only on the theoretical but also on the computational aspects of the problem.

An Outline of the Approach and the Paper Organization. It is convenient to consider the whole $\mathbf{X}$ and $\mathbf{X}_n(t)$'s as random elements in a separable Hilbert space. We propose a framework (Sect. 2) that allows us to impose probability distributions on them in a convenient way, namely, by attaching them to a finite number of orthogonal projections, but the residuals of the projections definitely do not act as white noise, since the samples are highly correlated, even when they are far in time within $[0, T]$ interval. After proving the correctness of this approach (Sect. 2, Lemma 1), we propose, in Sect. 3, the method of learning descriptors, which are projections of $\mathbf{X}$ and $\mathbf{X}_n(t)$'s on a countable basis of the Hilbert space. We also sketch proof of the consistency of the learning process in a general case and then, we concentrate on the computational aspect of learning the descriptors (Sect. 4) by the fast discrete cosine transform (DCT) and its joint action together with learning and using a classifier of descriptors. Finally, in Sect. 5, the proposed method was intensively tested on a large number of augmented data, leading to the selection of classifiers that cooperate with the learning descriptors in the most efficient way, from the viewpoint of the classification quality measures.

Motivating Case Study. Large mechanical constructions such as bucket-wheel excavators, used in open pit mines, undergo repetitive excitations that are transmitted to an operator's cabin, invoking unpleasant vibrations, which influence the operator's health in the long term. These excitations can be measured by accelerometers, as samples from functional observations that repeatedly occur after each stroke of the bucket into the ground. Roughly speaking, these functional observations can be classified into two classes, namely, to class I, representing typical, heavy working conditions and to class II, corresponding to less frequent and less heavy working conditions, occurring, e.g., when a sand background material is present (see Fig. 1 for an excerpt of functional data from Class I and II, a benchmark file is publicly available from the Mendeley site [28], see also [29] for its detailed description).

Proper and fast classification can be useful for decision making whether to use more or fewer vibrations damping in the next period between subsequent shocks, invoked by strokes of the bucket into the ground. We refer the reader to [25] to the study on a control system based on magneto-rheological dampers, for which the classifier proposed here can be used as an upper decision level.

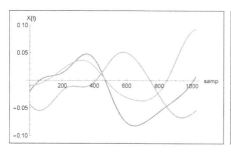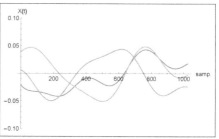

Fig. 1. An excerpt of functional data, representing accelerations vs sample number of an operator's cabin in bucket-wheel excavators. Left panel – curves from Class I (heavy working conditions), right panel – curves from Class II (less onerous working conditions)

Previous Works. Over the last twenty years the problems of classifying functions, curves and signals using methods from functional analysis has attracted considerable attention from researchers. We refer the reader to the fundamental paper [10] on (im-)possibilities of classifying probability density functions with (or without) certain qualitative properties. Function classification, using a functional analogue of the Parzen kernel classifier is developed in [6], while in [5,12] generalizations of the Mahalanobis distance are applied. Mathematical models of functional data are discussed in [18]. The reader is also referred to the next section for citations of related monographs and to [21].

All the above does not mean that problems of classifying functions, mainly sampled signals, were not considered earlier. Conversely, the first attempts at classifying electrocardiogram (ECG) signals can be traced back, at least, to the 1960s, see [1] for the recent review and to [20] for feature selection using the FFT.

The recognition problems for many other kinds of bio-medical signals have been extensively studied. We are not able to review all of them, therefore, we confine ourselves to recent contributions, surveys, and papers more related to the present one.

Electroencephalogram (EEG) signals are rather difficult for an automatic classification, hence the main effort is put on a dedicated feature selection, see [13,14] and survey papers [4,19] the latter being of special interest for human-computer interactions. In a similar vein, in [2] the survey of using electromyography (EMG) signals is provided. For a long time, also studies on applying the EMG signals classification for control of hand prosthesis had been conducted. We refer the reader to recent contributions [17,30] and to [8] for a novel approach to represent a large class of signals arising in a health care system.

Up to now, problems of classifying data from accelerometers, as those arising in our motivating case study, have not received too much attention (see [22], where the recognition of whether a man is going upstairs or downstairs is considered).

Our derivations are based on orthogonal projections. One should notice that classifiers based on orthogonal expansions were studied for a long time, see [15] for one of the pioneering papers on classifiers based on probability density estimation and [9] for the monograph on probabilistic approaches to pattern recognition. Observe, however, that in our problem we learn the expansion coefficients in a way that closer to nonparametric estimation of a regression function with non-random (fixed design) cases (see, e.g., [23]). Furthermore, in our case observation errors are correlated, since they arise from the truncation of the orthogonal series with random coefficients.

2   Model of Random Functional Data and Problem Statement

Constructing a simple mathematical description of random functional data, also called random elements, is a difficult task, since in infinite dimensional Hilbert spaces an analogue of the uniform distribution does not exists (see monographs: [3,11,16,27] for basic facts concerning probability in spaces of functions). Thus, it is not possible to define probability density function (p.d.f.) with respect to this distribution. As a way to get around this obstacle, we propose a simple model of random elements in the Hilbert space $L_2(0, T)$ of all squared integrable functions, where $T > 0$ is the horizon of observations.

V1) Let us assume that $\mathbf{v}_k(t)$, $t \in [0, T$, $k = 1, 2, \ldots$ is a selected orthogonal and complete, infinite sequence of functions in $L_2(0, T)$, which are additionally normalized, i.e., $||\mathbf{v}_k|| = 1$, $k = 1, 2, \ldots$, where for $g \in L_2(0, T)$ its squared norm $||g||^2$ is defined as $<g, g>$, while $<g, h> = \int_0^T g(t)\, h(t)\, dt$ is the standard inner product in $L_2(0, T)$.

Within this framework, any $g \in L_2(0, T)$ can be expressed as

$$g = \sum_{k=1}^{\infty} < g, \mathbf{v}_k > \mathbf{v}_k, \tag{1}$$

where the convergence is understood in the L_2 norm. For our purposes we consider a class of random elements, denoted further as $\mathbf{X}$, $\mathbf{Y}$ etc. that can be expressed as follows

$$\mathbf{X} = \sum_{k=1}^{K} \theta_k \mathbf{v}_k + \sum_{k=K+1}^{\infty} \alpha_k \mathbf{v}_k, \tag{2}$$

where

- $1 \leq K < \infty$ is a preselected positive integer that splits[1] the series expansion of $\mathbf{X}$ into two parts, namely, the first one that we later call an informative part and the second one, which is either much less informative or noninformative at all from the point of view of classifying $\mathbf{X}$,

---

[1] For theoretical purposes K is assumed to be fixed and a priori known. Later, we comment on the selection of K in practice.

- coefficients θ_k, $k = 1, 2 \ldots, K$ are real-valued random variables that are drawn according to exactly one of cumulative, multivariate distribution functions $F_I(\bar{\theta})$ or $F_{II}(\bar{\theta})$, $\bar{\theta} \overset{def}{=} [\theta_1, \theta_2, \ldots, \theta_K]$,
- α_k, $k = (K+1), (K+2), \ldots$ are also random variables (r.v.'s), having properties that are specified below.

We shall write

$$\mathbf{X}(t) = \sum_{k=1}^{K} \theta_k \, \mathbf{v}_k(t) + \sum_{k=K+1}^{\infty} \alpha_k \, \mathbf{v}_k(t), \quad t \in [0, T], \tag{3}$$

when the dependence of $\mathbf{X}$ on t has to be displayed.

Distribution functions $F_I(\bar{\theta})$ and $F_{II}(\bar{\theta})$, as well as those according to α_k, $k = (K+1), (K+2), \ldots$ are drawn, are not known, but we require that the following assumptions hold.

R1) The second moments of θ_k, $k = 1, 2 \ldots, K$ exist and they are finite. The variances of θ_k's are denoted as σ_k^2.

R2) The expectations $\mathbb{E}(\alpha_k) = 0$, $k = (K+1), (K+2), \ldots$, where $\mathbb{E}$ denotes the expectations with respect to all θ_k's and α_k's. Furthermore, there exists a finite constant $0 < C_0 < \infty$, say, such that

$$\mathbb{E}(\alpha_k^2) \leq \frac{C_0}{k^2}, \quad k = (K+1), (K+2), \ldots. \tag{4}$$

R3) Collections θ_k, $k = 1, 2 \ldots, K$ and α_k, $k = (K+1), (K+2), \ldots$ are mutually uncorrelated in the sense that $\mathbb{E}(\theta_k \, \alpha_l) = 0$ for all $k = 1, 2 \ldots, K$ and $l = (K+1), (K+2), \ldots$. Furthermore, $\mathbb{E}(\alpha_j \, \alpha_l) = 0$ for $j \neq l$, $j, l = (K+1), (K+2), \ldots$.

To motivate assumption R2), inequality (4), notice that expansion coefficients of smooth, e.g., continuously differentiable, functions into the trigonometric series decay as $O(k^{-1})$, while the second order differentiability yields $O(k^{-2})$ rate of decay.

To illustrate the simplicity of this model, consider $\bar{\theta}$ that drawn at random from the K-variate normal distribution with the expectation vector $\bar{\mu}_c$ and the covariance matrix Σ_c^{-1}, where c stands for class label I or II. Consider also sequence α_k, $k = (K+1), (K+2), \ldots$ of the Gaussian random variables, having the zero expectations, that are mutually uncorrelated and uncorrelated also with $\bar{\theta}$. Selecting the dispersions of α_k's of the form: σ_0/k, $0 < \sigma_0 < \infty$, we can draw at random $\bar{\theta}$ and α_k's for which R1)–R3) hold. Thus, it suffices to insert them into (3). We underline, however, that in the rest of the paper, the gaussianity of $\bar{\theta}$ and α_k's are not postulated.

Lemma 1 (Model correctness). *Under V1), R1) and R2) model (2) is correct in the sense that $\mathbb{E}||\mathbf{X}||^2$ is finite.*

Indeed, applying V1), and subsequently R1) and R2), we obtain

$$\mathbb{E}||\mathbf{X}||^2 = \sum_{k=1}^{K} \mathbb{E}(\theta_k^2) + \sum_{k=(K+1)}^{\infty} \mathbb{E}(\alpha_k^2) \leq \sum_{k=1}^{K} \mathbb{E}(\theta_k^2) + C_0\,\gamma_K, \qquad (5)$$

where $\gamma_K \overset{def}{=} \sum_{k=(K+1)}^{\infty} k^{-2} < \infty$, since this series is convergent.   •

Lemma 2 (Correlated observations). *Under V1), R1) and R2) observations $\mathbf{X}(t')$ and $\mathbf{X}(t'')$ are correlated for every t', $t'' \in [0, T]$ and for their covariance we have:*

$$\mathbb{C}ov(\mathbf{X}(t'), \mathbf{X}(t'')) = \sum_{k=(K+1)}^{\infty} \mathbb{E}(\alpha_k^2)\,\mathbf{v}_k(t')\,\mathbf{v}_k(t''), \qquad (6)$$

and, for commonly bounded basis functions, its upper bound is given by

$$|\mathbb{C}ov(\mathbf{X}(t'), \mathbf{X}(t''))| \leq c_0^2\,\gamma_K, \quad c_0 \overset{def}{=} \sup_{k}\ \sup_{t\in[0,T]} |\mathbf{v}_k(t)|. \qquad (7)$$

Problem Statement.   Define a residual random element $\mathbf{r}_K$ as follows: $\mathbf{r}_K = \sum_{k=(K+1)}^{\infty} \alpha_k\,\mathbf{v}_k$ and observe that (by R2)) $\mathbb{E}(\mathbf{r}_K) = 0$, $\mathbb{E}||(\mathbf{r}_K)||^2 \leq C_0\,\gamma_K < \infty$. Define also an informative part of $\mathbf{X}$ as $\mathbf{f}_{\bar{\theta}} = \sum_{k=1}^{K} \theta_k\,\mathbf{v}_k$ and assume that we have observations (samples) of $\mathbf{X}$ at equidistant points $t_i \in [0, T]$, $i = 1, 2, \ldots, m$ which are of the form

$$x_i = \mathbf{X}(t_i) = \mathbf{f}_{\bar{\theta}}(t_i) + \mathbf{r}_K(t_i), \quad i = 1, 2, \ldots, m. \qquad (8)$$

Having these observations, collected as $\bar{x}$, at our disposal, the problem is to classify $\mathbf{X}$ to class I or II. These classes correspond to unknown information on whether $\bar{\theta}$ in (8) was drawn according to F_I or F_{II} distributions, which are also unknown.

The only additional information is that contained in samples from learning sequence $\{(\mathbf{X}^{(1)}, j_1), (\mathbf{X}^{(2)}, j_2), \ldots, (\mathbf{X}^{(N)},, j_N)\}$. The samples from each $\mathbf{X}^{(n)}$ have exactly the same structure as (8) and they are further denoted as $\bar{x}^{(n)} = [x_1^{(n)}, x_2^{(n)}, \ldots x_m^{(n)}]^{tr}$, while $j_n \in \{I, II\}$, $n = 1, 2, \ldots, N$ are class labels attached by an expert.

Thus, the learning sequence is represented by collection $\mathcal{X}_N \overset{def}{=} [\bar{x}^{(n)}, n = 1, 2, \ldots, N]$, which is an $m \times N$ matrix and the sequence of labels $\mathcal{J} \overset{def}{=} \{j_n \in \{I, II\}, n = 1, 2, \ldots, N\}$ Summarizing, our aim is to propose a nonparametric classifier that classifies random function $\mathbf{X}$, represented by $\bar{x}$, to class I or II and a learning procedure based on $\mathcal{X}_N$ and the corresponding j_n's.

3   Learning Descriptors from Samples and Their Properties

The number of samples in $\bar{x}$ and $\bar{x}_n$'s is frequently very large (when generated by electronic sensors, it can be thousands of samples per second). Therefore, it is impractical to build a classifier directly from samples. Observe that the orthogonal projection of $\mathbf{X}$ on the subspace spanned by $\mathbf{v}_1$, $\mathbf{v}_2$, ... $\mathbf{v}_K$ is exactly $\mathbf{f}_{\bar{\theta}}$. Thus, the natural choice of descriptors of $\mathbf{X}$ would be $\bar{\theta}$, but it is not directly accessible. We do not have also a direct access to $\bar{\theta}^{(n)}$'s constituting $\mathbf{X}^{(n)}$'s. Hence, we firstly propose a nonparametric algorithm of learning $\bar{\theta}$ and $\bar{\theta}^{(n)}$'s from samples. We emphasize that this algorithm formally looks like as well known algorithms of estimating regression functions (see, e.g., [23, 26]), but its statistical properties require re-investigation, since noninformative residuals $\mathbf{r}_N$ have a different correlation structure than that which appears in classic nonparametric regression estimation problems.

Denote by $\hat{\theta}_k$ the following expression

$$\hat{\theta}_k = \frac{T}{m} \sum_{i=1}^{m} x_i \, \mathbf{v}_k(t_i), \quad k = 1, 2, \ldots, K \tag{9}$$

further taken as the learning algorithm for $\theta_k = <\mathbf{X}, \mathbf{v}_k>$.

Asymptotic Unbiasedness. It can be proved that for continuously differentiable $\mathbf{X}(t)$, $t \in [0, T]$ and $\mathbf{v}_k$'s we have

$$|\mathbb{E}_{\bar{\theta}}(\hat{\theta}_k) - \theta_k| \leq \frac{T L_1}{m} \tag{10}$$

where $\mathbb{E}_{\bar{\theta}}$ is the expectation with respect to α_k's, conditioned on $\bar{\theta}$ and $L_1 > 0$ is the maximum of $|\mathbf{X}'(t)|$ and $|\mathbf{v}_k'(t)|$, $t \in [0, T]$.

One can largely reduce errors introduced by approximate integration by selecting $\mathbf{v}_k$'s that are orthogonal in the summation sense on sample points, i.e.,

$$\frac{T}{m} \sum_{i=1}^{m} \mathbf{v}_l(t_i) \, \mathbf{v}_k(t_i) = 0 \text{ for } k \neq l, \quad k, l = 1, 2, \ldots . \tag{11}$$

The well known example of such basis is provided by the cosine series

$$\mathbf{v}_1(t) = 1, \; \mathbf{v}_2(t) = \sqrt{2} \, \cos(\pi \, t/T), \; \mathbf{v}_3(t) = \sqrt{2} \, \cos(2 \, \pi \, t/T), \ldots \tag{12}$$

computed at equidistant t_i's.

Lemma 3 (Bias). *For all $k = 1, 2, \ldots, K$ we have: 1) if $\mathbf{X}(t)$ and $\mathbf{v}_k(t)$'s are continuously differentiable $t \in [0, T]$, then $\hat{\theta}_k$ is asymptotically unbiased, i.e., $\mathbb{E}_{\bar{\theta}}(\hat{\theta}_k) \to \theta_k$ as $m \to \infty$, 2) if for $\mathbf{v}_k$, $k = 1, 2, \ldots, K$ and m conditions (11) hold, then $\hat{\theta}_k$ is unbiased for m finite, i.e., $\mathbb{E}_{\bar{\theta}}(\hat{\theta}_k) = \theta_k$.*

Variance and Mean Square Error (MSE). Analogously, assuming that $\mathbf{v}_k$'s and $\mathbf{X}(t)|$ are twice continuously differentiable, we obtain:

$$\mathbb{V}\text{ar}_{\bar{\theta}}(\hat{\theta}_k) \leq \frac{T L_2}{m^2} \, \gamma_K, \tag{13}$$

where L_2 is the maximum of $\mathbf{X}''(t)|$ and $|\mathbf{v}_k''(t)|$, $t \in [0, T]$. Thus, the conditional mean squared error of learning $\hat{\theta}_k$ is not larger than $\frac{T L_1}{m} + \frac{T L_2}{m^2} \gamma_K$ and it can be reduced by enlarging m.

Lemma 4 (Consistency). *For all $k = 1, 2, \ldots, K$ we have:*

$$\mathbb{E}_{\bar{\theta}} \left(\hat{\theta}_k - \theta_k \right)^2 \rightarrow 0, \ as \ m \rightarrow \infty, \tag{14}$$

i.e., $\hat{\theta}_k$ is consistent in the MSE sense, hence also in the probability.

Notice also that this is the worst case analysis in the class of all twice differentiable functions $\mathbf{X}(t)|$ and $|\mathbf{v}_k(t)|$, which means that L_1 and L_2 depend on k.

Observe that replacing x_i's in (9) by $x_i^{(n)}$'s we obtain estimators $\hat{\theta}_k^{(n)}$ of the descriptors $\theta_k^{(n)}$ in the learning sequence. Obviously, the same upper bounds (10) and (13) hold also for them.

4   A Fast Algorithm for Learning Descriptors and Classification

The above considerations are, to a certain extent, fairly general. By selecting (12) as the basis, one can compute all $\hat{\theta}_k$'s in (9) simultaneously by the fast algorithm, being the fast version of the discrete cosine transform (see, e.g., [7] and [24]). The action of this algorithm on $\bar{x}$ (or on $\bar{x}^{(n)}$'s) is further denoted as $\mathcal{FDCT}(\bar{x})$. Notice, however, that for vector $\bar{x}$, containing m samples, also the output of the $\mathcal{FDCT}(\bar{x})$ contains m elements, while we need only $K < m$ of them, further denoted as $\hat{\bar{\theta}} = [\hat{\theta}_k, \ k = 1, 2, \ldots, K]^{tr}$. Thus, if $\text{Trunc}_K[.]$ denotes the truncation of a vector to its K first elements, then

$$\hat{\bar{\theta}} = \text{Trunc}_K[\mathcal{FDCT}(\bar{x})], \tag{15}$$

is the required version of the learning of all the descriptors at one run, at the expense of $O(m \log(m))$ arithmetic operations.

Remark 1. If K is not known in advance, it is a good point to select it by applying $\text{Trunc}_K[\mathcal{FDCT}(.)]$ to $\bar{x}^{(n)}$'s together with the minimization of one of the well known criterions such as the AIC, BIC etc. Notice also that K plays the role of a smoothing parameter, i.e., smaller K provides a less wiggly estimate of $\mathbf{X}(t)$, $t \in [0, T]$.

A Projection-Based Classifier for Functional Data. The algorithm: $\text{Trunc}_K[\mathcal{FDCT}(.)]$ is crucial for building a fast classifier from projections, since it will be used many times both in the learning phase as well as for fast recognition of forthcoming observations of $\mathbf{X}$'s. The second ingredient that we need is a properly chosen classifier for K dimensional vectors $\bar{\theta}$. Formally, any reliable and fast classifier can be selected, possibly excluding the nearest neighbors classifiers, since they require to keep and look up the whole learning sequence, unless its special edition is not done. For the purposes of this paper we select the support vector machine (SVM) classifier and the one that is based on the logistic regression (LReg) classifier. We shall denote by $\texttt{Class}[\bar{\theta}, \{\bar{\theta}^{(n)}, \mathcal{J}\}]$ the selected classifier that – after learning it from the collection of descriptors $\{\bar{\theta}^{(n)}\}$, $n = 1, 2, \ldots, N$ and correct labels $\mathcal{J}$ – classifies descriptor $\bar{\theta}$ of new $\mathbf{X}$ to I or II class.

A Projection-Based Classification Algorithm (PBCA)
Learning

1. Convert available samples of random functions into descriptors:

$$\bar{\theta}^{(n)} = \texttt{Trunc}_K[\mathcal{FDCT}(\bar{x}^{(n)})], \quad n = 1, 2, \ldots, N$$

   and attach class labels j_n to them in order to obtain $(\bar{\theta}^{(n)}, j_n)$, $n = 1, 2, \ldots, N$.
2. Split this sequence into the learning sequence of the length $1 < N_l < N$ with indexes selected uniformly at random (without replacements) from $n = 1, 2, \ldots, N$. Denote the set of this indexes by $\mathcal{J}_l$ and its complement by $\mathcal{J}_v$.
3. Use $\bar{\theta}^{(n)}$, $n \in \mathcal{J}_l$ to learn classifier $\texttt{Class}[., \{\bar{\theta}^{(n)}, \mathcal{J}_l\}]$, where dot stands for a dummy variables, representing a descriptor to be classified.
4. Verify the quality of this classifier by testing it on all descriptors with indexes from $\mathcal{J}_v$, i.e., compute

$$\hat{j}_{n'} = \texttt{Class}[\bar{\theta}^{(n')}, \{\bar{\theta}^{(n)}, \mathcal{J}_l\}], \quad n' \in \mathcal{J}_v. \tag{16}$$

5. Compare the obtained class labels $\hat{j}_{n'}$ with proper ones $j_{n'}$, $n' \in \mathcal{J}_v$ and count the number of true positive (TP), true negative (TN), false positive (FP) and false negative (FN) cases. Use them to compute the classifier quality indicators such as *accuracy, precision, F1 score, ...* and store them.
6. Repeat steps 2–5 a hundred times, say, and assess the quality of the classifier, using the collected indicators. If its quality is satisfactory, go to the on-line classification phase. Otherwise, repeat steps 2–5 for different K.

On-Line Classification

Acquisition: collect samples $x_i = \mathbf{X}(t_i)$, $i = 1, 2, \ldots, m$ of the next random function and form vector $\bar{x}$ from them.
Compute descriptors: $\bar{\theta} = \texttt{Trunc}_K[\mathcal{FDCT}(\bar{x})]$.
Classification: Compute predicted class label $\hat{j}$ for descriptors $\bar{\theta}$ as $\hat{j} = \texttt{Class}[\bar{\theta}, \{\bar{\theta}^{(n)}, \mathcal{J}_l\}]$.
Decision: if appropriate, make a decision corresponding to class $\hat{j}$ and go to the Acquisition step.

Even for a large number of samples from repetitive functional random data the PBCA is relatively fast for the following reasons.

- The most time-consuming Step 1 is performed only once for each (possibly long) vector of samples from the learning sequence. Furthermore, the fast $\mathcal{FDCT}$ algorithm provides the whole vector of m potential descriptors in one pass. Its truncation to K first descriptors is immediate and it can be done many times, without running the $\mathcal{FDCT}$ algorithm. This advantage can be used for even more advanced task of looking for a sparse set of descriptors, but this topic is outside the scope of this paper.
- Steps 2–5 of the learning phase are repeated many times for the validation and testing reasons, but this is done off-line and for descriptor vectors of the length $K << m$. The total execution time of the validation and testing phase depends on the time of learning $\texttt{Class}[., \{\bar{\theta}^{(n)}, \mathcal{J}_l\}]$ that depends on a particular choice of the classifier $\texttt{Class}$. For the SVM and LogReg classifiers and for K about dozens, it takes a few seconds on a standard PC with 3 GHz CPU clock.
- The execution time of the on-line usage phase is fast, since it uses the fast version of DCT only once for the incoming vector of samples $\bar{x}$ at the expense of $O(m \log(m))$ operations, while the already trained recognizer has to classify $\bar{\theta}$ of the length K only.

5 Testing on Accelerations of the Operator's Cabin

The PBCA was tested on samples of a function (signal), representing the accelerations of an operator's cabin (see Fig. 1), mounted on a bucket-wheel excavator. The aim of testing was not only to check the correctness of the algorithm, but also to select a suitable classifier.

We had 44 000 samples, acquired with the rate 512 Hz and grouped into portions of $T = 2$ s. duration each. The resulting $\bar{x}^{(n)}$'s of the length $m = 1024$ samples, representing the learning sequence $\mathbf{X}_n$, $n = 1, 2, \ldots, N = 43$, were extended by adding labels of their proper classifications. A low-pass filter with the cutoff[2] frequency 5 Hz was applied before using $\mathcal{FDCT}$. The number of $K = 16$ of estimated descriptors $\hat{\theta}_k^{(n)}$, $k = 1, 2, \ldots, K$ was selected as the first K elements of $\mathcal{FDCT}$ sequences.

As one can notice, 44 000 samples occurred to be low informative for functional data classification. Therefore, for the aim of our tests, we had to use augmented data. In the augmentation process we used a silent, nice feature of the projection-based descriptors and the linearity of (9) with respect to samples. Namely, instead of augmenting raw samples, we augmented $\hat{\theta}_k^{(n)}$, $k = 1, 2, \ldots, K$ by adding to each of them pseudo-random errors that had Gaussian distribution with zero mean and dispersion $\sigma_a = 0.018$. Taking into account that most of $\hat{\theta}_k^{(n)}$'s was of the order ± 0.5, the interval $\pm 3\sigma_a$ has the length of 10.8% of their amplitudes. In this way the augmented testing sequence, containing $N' = 43\,000$ examples, having $K = 16$ descriptors, was generated.

The following classifiers were tested as part of the PBCA:

---

[2] From earlier experiments [25], it was known that frequencies of importance are less than 2.5 Hz.

LogR – the logistic regression classifier,
SVM – the support vector machine,
DecT – the decision tree classifier,
gbTr – the gradient boosted trees,
RFor – the random forests classifier,
5NN – the 5 nearest neighbors[3] classifier.

Table 1. Left table – a summary of learning and testing the PBCA on the augmented acceleration data for different classifiers (abbreviations explained in the text). Right table – an example of the confusion matrix when the LogR classifier was used.

Classifier	$LogR$	SVM	$DecT$	$gbTr$	$RFor$	$5NN$				
Accuracy	0.91	**0.94**	0.84	0.92	0.91	0.90				
Cohen κ	0.76	**0.82**	0.60	0.77	0.73	0.70			*Pred. class*	
MCC [6]	0.76	**0.82**	0.61	0.77	0.73	0.71			I	II
Precision	**0.96**	0.94	0.93	0.94	0.92	0.90		I	30303	2697
Recall	0.92	**0.98**	0.86	0.96	0.96	**0.98**		II	1220	8780
Specificity	**0.88**	0.80	0.60	0.79	0.72	0.65				
FScore	0.94	**0.96**	0.89	0.95	0.94	0.94				

MCC is the abbreviation for the Matthews Correlation Coefficient.

The results of learning and testing are summarized in Table 1. Its right panel contains just one example of the confusion matrix – for illustration only. The left panel summarizes all the extensive simulations. It contains the values of indicators that are the most frequently used for assessing the quality of classifiers.

The analysis of these quality indicators allows recommending the SVM and the LogR classifiers as the decision unit, applied after learning descriptors. Also the CPU time of $7.5\,10^{-6}$ s, used for the SVM and LogR classifier to recognize a new example, was slightly better than for the rest of classifiers displayed in Table 1, which needed about $10\text{--}15\,10^{-6}$ s, as the average of 30000 simulation experiments.

6   Concluding Remarks

The mathematical model of random infinite-dimensional data is proposed that allows us to impose arbitrary probability distribution on a finite dimensional space of descriptors. Its correctness is proved and the learning algorithm for these descriptors is proposed and investigated. In particular, it was shown that the learning algorithm is consistent in the MSE sense for any finite number of the descriptors.

The fast version of the learning algorithm is tested from the view point of its cooperation with a finite dimensional classifier. The winners are the SVM and

---

[3] The 5 NN classifier was tested for comparisons only. We do not recommend its usage with the PCBA, since it requires storing all the learning sequence, unless its editing (condensation) is not done.

logistic regression classifiers, as tested on augmented real data. By passing, a new approach to data augmentation is proposed. Namely, instead of augmenting raw observations, we use random perturbation of estimated descriptors, which leads to essential computational savings. On the other hand, the descriptors estimated from the raw learning sequence are sufficient for learning the classifiers, which means a kind of raw data compression when they are disregarded.

Further research in this direction is desirable. One can consider extending them by including ensembles of classifiers and neural network-based recognizers.

From the practical point of view, it would be also of interest to consider the classification of signals from accelerometers to more than two classes, taking into account the kind of background that is met by a bucket-wheel excavator. This is, however, outside the scope of this paper, since it requires cumbersome data labeling by experts.

Further directions of research may include also other applications, e.g., a human motion classification, based on a motion capture cameras, a computer-aided laparoscopy training and theoretical aspects such as classifying random elements by learning their derivatives.

Acknowledgements. The authors express their thanks to Professor P. Moczko and Dr. J. Więckowski from the Faculty of Mechanical Engineering, Wroclaw University of Science and Technology for permission to use data from the bucket-wheel excavator.

References

1. Abdulla, L., Al-Ani, M.: A review study for electrocardiogram signal classification. UHD J. Sci. Technol. **4**(1), 103–117 (2020). https://doi.org/10.21928/uhdjst. v4n1y2020.pp103-117
2. Ahsan, M.R., Ibrahimy, M.I., Khalifa, O.O., et al.: EMG signal classification for human computer interaction: a review. Eur. J. Sci. Res. **33**(3), 480–501 (2009)
3. Aneiros, G., Bongiorno, E.G., Cao, R., Vieu, P., et al.: Functional Statistics and Related Fields. CONTRIB.STAT.CONTRIB.STAT., Springer, Cham (2017). https://doi.org/10.1007/978-3-319-55846-2
4. Azlan, W.A., Low, Y.F.: Feature extraction of electroencephalogram (EEG) signal - a review. In: 2014 IEEE Conference on Biomedical Engineering and Sciences (IECBES), pp. 801–806 (2014). https://doi.org/10.1109/IECBES.2014.7047620
5. Berrendero, J.R., Bueno-Larraz, B., Cuevas, A.: On Mahalanobis distance in functional settings. J. Mach. Learn. Res. **21**(9), 1–33 (2020)
6. Biau, G., Bunea, F., Wegkamp, M.H.: Functional classification in Hilbert spaces. IEEE Trans. Inf. Theory **51**(6), 2163–2172 (2005). https://doi.org/10.1109/TIT. 2005.847705
7. Britanak, V., Yip, P.C., Rao, K.R.: Discrete Cosine and Sine Transforms: General Properties, Fast algorithms and Integer Approximations. Elsevier, Amsterdam (2010)
8. Cyganek, B., Woźniak, M.: Tensor based representation and analysis of the electronic healthcare record data. In: 2015 IEEE International Conference on Bioinformatics and Biomedicine (BIBM), pp. 1383–1390 (2015). https://doi.org/10.1109/ BIBM.2015.7359880

9. Devroye, L., Györfi, L., Lugosi, G.: A Probabilistic Theory of Pattern Recognition. SMAP, vol. 31. Springer, New York (2013). https://doi.org/10.1007/978-1-4612-0711-5
10. Devroye, L., Lugosi, G.: Almost sure classification of densities. J. Nonparametric Stat. **14**(6), 675–698 (2002). https://doi.org/10.1080/10485250215323
11. Ferraty, F., Vieu, P.: Nonparametric Functional Data Analysis: Theory and Practice. SSS, Springer, New York (2006). https://doi.org/10.1007/0-387-36620-2
12. Galeano, P., Joseph, E., Lillo, R.E.: The Mahalanobis distance for functional data with applications to classification. Technometrics **57**(2), 281–291 (2015)
13. Gandhi, T., Panigrahi, B.K., Anand, S.: A comparative study of wavelet families for EEG signal classification. Neurocomputing **74**(17), 3051–3057 (2011)
14. Garrett, D., Peterson, D.A., Anderson, C.W., Thaut, M.H.: Comparison of linear, nonlinear, and feature selection methods for EEG signal classification. IEEE Trans. Neural Syst. Rehabil. Eng. **11**(2), 141–144 (2003). https://doi.org/10.1109/TNSRE.2003.814441
15. Greblicki, W., Pawlak, M.: Classification using the Fourier series estimate of multivariate density functions. IEEE Trans. Syst. Man Cybern. **11**, 726–730 (1981)
16. Horváth, L., Kokoszka, P.: Inference for Functional Data with Applications. SSS, vol. 200. Springer, New York (2012). https://doi.org/10.1007/978-1-4614-3655-3
17. Kurzynski, M., Wolczowski, A.: EMG and MMG signal recognition using ensemble of one-feature classifiers with pruning via clustering method. In: 2019 International Conference on Advanced Technologies for Communications (ATC), pp. 38–43. IEEE (2019)
18. Ling, N., Vieu, P.: Nonparametric modelling for functional data: selected survey and tracks for future. Statistics **52**(4), 934–949 (2018). https://doi.org/10.1080/02331888.2018.1487120
19. Lotte, F., Congedo, M., Lécuyer, A., Lamarche, F., Arnaldi, B.: A review of classification algorithms for EEG-based brain–computer interfaces. J. Neural Eng. **4**(2), R1–R13 (2007). https://doi.org/10.1088/1741-2560/4/2/r01
20. Mironovova, M., Bíla, J.: Fast Fourier transform for feature extraction and neural network for classification of electrocardiogram signals. In: 2015 Fourth International Conference on Future Generation Communication Technology (FGCT), pp. 1–6 (2015). https://doi.org/10.1109/FGCT.2015.7300244
21. Mueller, H.G., et al.: Peter Hall, functional data analysis and random objects. Ann. Stat. **44**(5), 1867–1887 (2016)
22. Preece, S.J., Goulermas, J.Y., Kenney, L.P.J.: A comparison of feature extraction methods for the classification of dynamic activities from accelerometer data. IEEE Trans. Biomed. Eng. **56**(3), 871–879 (2009). https://doi.org/10.1109/TBME.2008.2006190
23. Rafajłowicz, E.: Nonparametric orthogonal series estimators of regression: a class attaining the optimal convergence rate in L2. Stat. Probab. Lett. **5**, 219–224 (1987)
24. Rafajłowicz, E., Skubalska-Rafajłowicz, E.: FFT in calculating nonparametric regression estimate based on trigonometric series. J. Appl. Math. Comput. Sci. **3**(4), 713–720 (1993)
25. Rafajłowicz, W., Więckowski, J., Moczko, P., Rafajłowicz, E.: Iterative learning from suppressing vibrations in construction machinery using magnetorheological dampers. Autom. Constr. **119**, 103326 (2020)
26. Rutkowski, L., Rafajłowicz, E.: On optimal global rate of convergence of some nonparametric identification procedures. IEEE Trans. Autom. Control AC **34**, 1089–1091 (1989)

27. Srivastava, A., Klassen, E.P.: Functional and Shape Data Analysis. SSS, vol. 1. Springer, New York (2016). https://doi.org/10.1007/978-1-4939-4020-2
28. Więckowski, J.: Data from vibration in SchRs1200, Mendeley Data, V1. http://dx.doi.org/10.17632/htddgv2p3b.1. Accessed Jan 2021
29. Więckowski, J., Rafajlowicz, W., Moczko, P., Rafajlowicz, E.: Data from vibration measurement in a bucket wheel excavator operator's cabin with the aim of vibrations damping. Data in Brief **106836**, (2021)
30. Wozniak, M., Połap, D., Nowicki, R.K., Napoli, C., Pappalardo, G., Tramontana, E.: Novel approach toward medical signals classifier. In: 2015 International Joint Conference on Neural Networks (IJCNN). pp. 1–7 (2015). https://doi.org/10.1109/IJCNN.2015.7280556

Clustering and Weighted Scoring Algorithm Based on Estimating the Number of Clusters

Jakub Klikowski[ID] and Robert Burduk[(✉)][ID]

Department of Systems and Computer Networks, Wroclaw University of Science and Technology, Wroclaw, Poland
{jakub.klikowski,robert.burduk}@pwr.edu.pl

Abstract. Imbalanced datasets are still a big method challenge in data mining and machine learning. Various machine learning methods and their combinations are considered to improve the quality of the classification of imbalanced datasets. This paper presents the approach with the clustering and weighted scoring function based on geometric space are used. In particular, we proposed a significant modification to our earlier algorithm. The proposed change concerns the use of automatic estimating the number of clusters and determining the minimum number of objects in a particular cluster. The proposed algorithm was compared with our earlier proposal and state-of-the-art algorithms using highly imbalanced datasets. The performed experiments show that the proposed modification is statistically better for a larger number of reference classifiers than the original algorithm.

Keywords: Imbalanced data · Ensemble of classifiers · Class imbalance · Decision boundary · Scoring function

1 Introduction

Machine learning methods can be divided into several groups, which include, among others, supervised learning, unsupervised learning, association rules, or time series analysis. It is vital for supervised learning to have data for which the class labels are known. In real problems, the number of objects in each class rarely is the same. If the imbalanced ratio expressed as the majority class's quotient to the minority class is much greater than 1, we deal with imbalanced data. Such data is also called skew data. Many practical problems concern imbalanced data because they arise directly from the problem's characteristics and the available training data [10]. Examples of practical applications where there are skews include: network intrusion detection [2,14], source code fault detection [6], or in general fraud detection [1].

In the supervised classification of skew data, methods belonging to two main trends are used. These are data-level [7,13,25] and algorithm-level [28] methods. The data-level methods use a resampling process that can be performed by

© Springer Nature Switzerland AG 2021
M. Paszynski et al. (Eds.): ICCS 2021, LNCS 12744, pp. 40–49, 2021.
https://doi.org/10.1007/978-3-030-77967-2_4

oversampling, undersampling, and hybrid in nature. The algorithm-level methods concern the modification of known machine learning algorithms to increase minority class classification performance [9,12,16].

In this article, we consider the algorithm-level approach. In particular, we present a significant modification of our previous algorithm presented in [17]. The proposed modification uses the Silhouette Value [23] to estimate the number of clusters automatically. Additionally, we took into account the number of necessary objects to designate one cluster. Taking the above into account, the main objectives of this work are summarized as follows:

– A proposal of a new clustering and weighted scoring algorithm based on estimating the number of clusters.
– The proposed algorithm has considered the minimum number of objects in each cluster.
– A new experimental setup on highly imbalanced datasets compares the proposed algorithm with the previous one and other state-of-the-art algorithms for supervised classification.

The paper is structured as follows: the Sect. 2 introduces the base concept of ensemble of classifiers and presents the proposed algorithm. In the Sect. 3 the experiments that were carried out are presented, while results and the discussion appear in the Sect. 4. Finally, we conclude the paper in the Sect. 5.

2   Proposed Method

2.1   Ensemble of Classifiers

The ensemble of classifiers (EoC) is widely discussed in the literature to solve the problem of skew data [9,15,20]. The use of the EoC belongs to the algorithm-level approach to solving the imbalanced data problem.

In general, the idea of EoC determination is to build a predictive model by integrating multiple base classification models $\Psi_1, \Psi_2, \ldots, \Psi_L$, where L is the number of classifiers in the EoC. The procedure for creating an EoC can be divided into three major steps: generation – a phase where individual classifiers are trained, selection – a phase where only a few (or even one) individual models from the previous step are selected for inclusion in the EoC and combining the base classifier outputs.

The idea presented in the following article is the construction of the EoC, diversified by the disjoint division of problem classes into clusters, introducing the integration rule based on the geometric characteristics of its components.

The first of the two steps necessary to build an effective EoC is the selection of models for its pool [19], required to make as independent decisions as possible. The strategy adopted in the discussed method is the use of the homogeneous [27] ensemble, built on the basis of linear classifiers, where each model is learned from a combination of [8] class clusters determined using the *K-Means* [5] algorithm.

2.2 Clustering and Weighted Scoring Algorithm Based on Estimating the Number of Clusters

This paper presents a certain extension of the CWS method developed by Ksieniewicz and Burduk [17]. The implemented changes focus on extending this approach's main idea and expanding the research to a larger pool of datasets. To better justify the new proposed algorithm, the whole procedure will be described step by step.

The Clustering and Weighted Scoring with Estimating the Number of Clusters *CWS-ENC* is based on an approach that uses the original procedure to determine objects' score function. The value obtained depends heavily on the position of object x in the geometric space. The scoring function [17] is expressed by the Eq. 1:

$$wsf_l(x) = 1 - \frac{sf_l(x)}{\sum_{l=1}^{L} sf_l(x)},\tag{1}$$

Where one of the main components is $sf_l(x)$. This is a function used to determine the distance from the decision boundaries of models (Ψ_l) forming *EoC* to clusters' centroids. Clusters are created with the K-means algorithm. The procedure to calculate the distance is described by the Eq. 2, where $\|\Psi_l(x)\|$ is result of base classifier (Ψ_l) decision function on object x, C is the number of clusters, and d_c means the distance from the object x to clusters' centroids expressed by any distance metric. Preliminary experiments have shown that the method gets the best results using the "Manhattan" distance metric.

$$sf_l(x) = \|\Psi_l(x)\| + \sum_{c=1}^{C} d_c,\tag{2}$$

An essential new change of this algorithm is forming clusters (Algorithm 1). The original idea [17] assumed that the number of clusters is chosen arbitrarily or experimentally. However, this requires additional preparation before starting a new experiment or assumes that it is optimal for all datasets. The proposed solution implies that the number of clusters should be selected dynamically based on the clusters' evaluation consistency using the Silhouette Coefficient [11] metric. This is the maximum value of the mean Silhouette Value $sv(x)$ [23] over the entire dataset and is described by the Eq. 3:

$$sv(x) = \frac{b(x) - a(x)}{max\{a(x), b(x)\}},\tag{3}$$

Where x is an object from cluster C, $b(x)$ is a mean dissimilarity of x to cluster C and $a(x)$ is a mean dissimilarity of x to other clusters. In practice, this means that the different number of clusters in the range from 2 to K_{max} are created and evaluated using the metric Silhouette Coefficient. Then the most optimal option is selected. In the performed preexperiments, the value of K_{max} was set as 5.

Moreover, a particular heuristic was introduced for minority class clusters. This rule blocks the possibility of creating new clusters when the currently divided set of samples does not reach the threshold for the minimum number of objects. This avoids situations where the algorithm tries to divide several objects from the minority class into multiple clusters in the strongly imbalanced dataset. Such an enforced segmentation only increases the imbalance ratio, which results in the deterioration of the classifier's predictive ability. This threshold's value is a parameter of the method and was set at 25 samples in the preexperiments.

The procedure of the ENC-CWS method is described in more detail by the Algorithm 1. In the first step, the whole dataset is divided into subsets D_i composed by objects from one class. Then for each subset D_i the quantity of the objects is checked, when this number is less than the predetermined threshold S_{min} then the algorithm creates only one cluster for this data. When the number of objects is more, then follows the procedure to determine the correct number of clusters. Using the K-means method, the data is clustered where the cluster number M_i changes from 2 to the predetermined K_{max}. For each set of clusters, consistency is measured using the Silhouette Coefficient metric. Then, the setup with K^{M_i} clusters that obtains the best score is selected and these clusters are stored. Next, the centroids of the created clusters are determined. After this the learning of new models with base classifier is performed. It is done using combinations of clusters in one-to-one manner, but from different classes. Then a weighted scoring function is computed for each sample.

3   Experimental Evaluation

The experimental analysis aims to verify the predictive performance of the *CWS-ENC* method for imbalance problems. In the following, the research questions are formulated:

RQ1: How does a method employing the idea of weighted score function based on objects geometric position handle datasets with varying imbalance levels in a comparison to the selected approaches?

RQ2: Do the modifications made bring improvements over the original version of the method?

3.1   Setup

The experimental evaluation was implemented in the Python programming language. Some elements from the *scikit-learn* [21] and *stream-learn* [18] libraries were used to perform the experiments. The project implementation with the results is available on the GitHub code repository[1]. The conducted experimental analysis aims to verify whether the introduced modifications will improve classification quality compared to the previous method variant. In the following is the list of approaches that were compared with the proposed method:

---

[1] Repository link: https://github.com/w4k2/cws-enc.

Algorithm 1: CWS-ENC – for binary problem

Input: D – Learning set
　　　　x – object
　　　　K_{max} – maximum number of clusters
　　　　S_{min} – minimum number of samples

Output: The ensemble classifier decision

1　Divide D into D_i subsets of data, where $i \in 1, 2$ is the number of class labels.
2　If size of D_i is greater than S_{min} determine the best number of clusters M_i from 2 to K_{max} using the Silhouette Coefficient metric for each class. Otherwise M_i is equal 1.
3　Divide D_i into K^{M_i} clusters using the K-means clustering algorithm separately for each $i - th$ class.
4　Find the cluster centroids $C_1^i, \ldots, C_{M_i}^i$ as the means of the points in the respective clusters.
5　Train base classifier $\Psi_1, \ldots, \Psi_L$ using each combination of clusters from different class labels, i.e. one cluster from each class label, $L = M_1 * M_2$.
6　Calculate weighted scoring functions for the object x:

$$ws f_l(x) = 1 - \frac{s f_l(x)}{\sum_{l=1}^{L} s f_l(x)},$$

　　where

$$s f_l(x) = \|\Psi_l(x)\| + \sum_{c=1}^{2} d_c.$$

7　The ensemble classifier decision:

$$\widehat{\Psi}(x) = sign\left(\sum_{l=1}^{L} ws f_l(x)\Psi_l(x)\right),$$

where $\Psi(x)$ is the prediction returned by base classifier $\Psi(x) \in \{-1, 1\}$.

- **CWS-ENC** (*Clustering and Weighted Scoring with Estimating the Number of Clusters*)—*EoC* proposed in this work and explained in Sect. 2.
- **CWS** (*Clustering and Weighted Scoring*)—*EoC* with the pool diversified by pairs of clusters and integrated geometrically by the rules proposed by Ksieniewicz and Burduk [17].
- **SVC** (*Support Vector Machine*)—the base model with the scaled gamma and linear kernel [22].
- **CMV** (*Clustering and Majority Vote*)—*EoC* identical with *CWS* but integrated using the majority vote [24].
- **CSA** (*Clustering and Support Accumulation*)—*EoC* identical with *CWS* and *CMV* but integrated using the support accumulation rule [26].

The testing procedure consist in evaluating datasets with the *Stratified K-fold Crossvalidation*, where the K is equal 5. The classification quality was expressed in the form of six metrics - *balanced accuracy score* (BAC), *F1-score* (F-1), *G-mean* (GMN), *precision* (PRE), *recall* (REC) and *specificity* (SPE). Next, for the obtained results, statistical analysis was performed using the *Wilcoxon rank test* with the significance level *alpha* = 0.05 [4]. 58 imbalanced binary datasets were used to conduct the study, which are described in Table 1.

Table 1. Overview of real datasets used in experimental evaluation (KEEL [3])

Dataset name	IMB. RATIO	SAMPLES	FEATURES	Dataset name	IMB. RATIO	SAMPLES	FEATURES
abalone-21_vs_8	40	581	8	glass4	15	214	9
abalone-3_vs_11	32	502	8	glass5	23	214	9
abalone9-18	16	731	8	glass6	6.4	214	9
cleveland-0_vs_4	13	177	13	led7digit-0-2-4-5-6-7-8-9_vs_1	11	443	7
dermatology-6	17	358	34	lymphography-normal-fibrosis	24	148	18
ecoli-0-1-3-7_vs_2-6	39	281	7	new-thyroid1	5.1	215	5
ecoli-0-1-4-6_vs_5	13	280	6	newthyroid2	5.1	215	5
ecoli-0-1-4-7_vs_2-3-5-6	11	336	7	poker-9_vs_7	30	244	10
ecoli-0-1-4-7_vs_5-6	12	332	6	shuttle-6_vs_2-3	22	230	9
ecoli-0-1_vs_2-3-5	9.2	244	7	shuttle-c2-vs-c4	20	129	9
ecoli-0-1_vs_5	11	240	6	vowel0	10	988	13
ecoli-0-2-3-4_vs_5	9.1	202	7	winequality-red-3_vs_5	68	691	11
ecoli-0-2-6-7_vs_3-5	9.2	224	7	winequality-red-8_vs_6	35	656	11
ecoli-0-3-4-6_vs_5	9.2	205	7	winequality-red-8_vs_6-7	46	855	11
ecoli-0-3-4-7_vs_5-6	9.3	257	7	winequality-white-3_vs_7	44	900	11
ecoli-0-3-4_vs_5	9	200	7	winequality-white-9_vs_4	33	168	11
ecoli-0-4-6_vs_5	9.2	203	6	yeast-0-2-5-6_vs_3-7-8-9	9.1	1004	8
ecoli-0-6-7_vs_3-5	9.1	222	7	yeast-0-2-5-7-9_vs_3-6-8	9.1	1004	8
ecoli-0-6-7_vs_5	10	220	6	yeast-0-3-5-9_vs_7-8	9.1	506	8
ecoli2	5.5	336	7	yeast-0-5-6-7-9_vs_4	9.4	528	8
ecoli3	8.6	336	7	yeast-1-2-8-9_vs_7	31	947	8
ecoli4	16	336	7	yeast-1-4-5-8_vs_7	22	693	8
glass-0-1-4-6_vs_2	11	205	9	yeast-1_vs_7	14	459	7
glass-0-1-5_vs_2	9.1	172	9	yeast-2_vs_4	9.1	514	8
glass-0-1-6_vs_2	10	192	9	yeast-2_vs_8	23	482	8
glass-0-1-6_vs_5	19	184	9	yeast3	8.1	1484	8
glass-0-4_vs_5	9.2	92	9	yeast4	28	1484	8
glass-0-6_vs_5	11	108	9	yeast5	33	1484	8
glass2	12	214	9	yeast6	41	1484	8

4   Results

The obtained results are presented in a Table 2, on which the exact values of the mean ranks and advantages with statistical significance are printed. Some numbers indicate that the method performance is statistically better under the rank values than the other methods. It can be easily seen that the proposed approach obtains statistical superiority over the methods *SVC*, *CMV* and *CSA* for most of the metrics. The exceptions are precision and specificity.

Much better readability of the average results is presented by the radar plot showing the graphical form results. The advantage in the scores obtained for the

Table 2. Results for mean ranks and statistical significance

	CWS-ENC (1)	CWS (2)	SVC (3)	CMV (4)	CSA (5)
BAC	3.724	3.379	2.974	2.241	2.681
	3,4,5	4,5	4	–	–
F-1	3.603	3.319	3.034	2.336	2.707
	3,4,5	4,5	4	–	–
GMN	3.655	3.336	2.810	2.397	2.802
	3,4,5	3,4	–	–	–
REC	3.586	3.457	2.586	2.595	2.776
	3,4,5	3,4,5	–	–	–
PRE	3.267	3.138	3.198	2.578	2.819
	4	4	4	–	–
SPE	2.716	2.681	3.776	2.672	3.155
	–	–	all	–	4

CWS-ENC method is easily seen here. All rankings for this method except the specificity metric are more or less better. There is also a noticeable improvement in quality over the method without the proposed modifications. Unfortunately, this advantage does not achieve statistical significance.

The proposed method performs poorly for the specificity metric compared to the others. The *SVC* dominates in this metric and has the best result with a statistically significant advantage over the rest of the methods. However, it is essential to note that the strong ability to classify majority class data is associated with a high decrease in the recall metric and slightly for BAC, F-1, and GMN metrics. This is a typical performance of a method that predicts too much bias toward the majority class when dealing with imbalanced data. Overall, the *CMV* approach received the weakest result (Fig. 1).

4.1   Lessons Learned

In summary, the research questions stated above will be answered:

RQ1: How does a method employing the idea of weighted score function based on objects geometric position handle datasets with varying imbalance levels?
Tests performed on 58 datasets whose imbalance level varies between 5 and 68 allows for a substantial study of binary imbalanced problems. The obtained results and their statistical analysis show that the weighted score function based on objects' geometric position is the right solution for imbalanced data classification. The presented method improves the classification quality expressed in different metrics compared to the *CMV* or *CSA* methods. For minority class and aggregate metrics, this advantage is statistically significant.

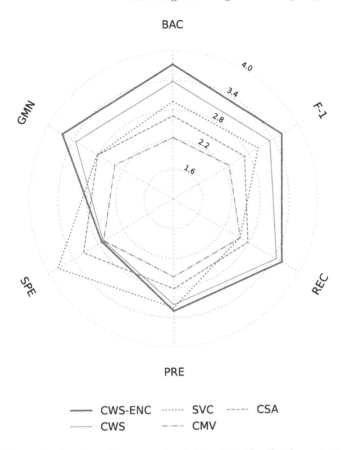

Fig. 1. Radar plot of mean ranks obtained by the Friedman test.

RQ2: Do the modifications made bring improvements over the original version of the method?
The implemented changes bring a visible improvement in the obtained results. The analysis of ranking tests shows that the proposed modification to dynamically select the number of clusters and threshold for the minimum number of samples achieves predictive performance better than the original approach. Unfortunately, the advantage does not have statistical significance, which may be because, despite some adjustments, the methods have many common traits.

5   Conclusions

In this article was proposed a new method, based on an existing approach [17]. Introduced changes brought a noticeable performance improvement. Extended testing on a larger collection of imbalanced datasets and statistical analysis showed the presented method's good classification quality. The proposed algorithm modification significantly statistically improves minority class classification performance. In the datasets used, the majority class was marked in the confusion matrix as a negative class and the minority class as a positive class. The algorithm proposed in the article increases the value of the classification quality measure, which is REC, and reduces the value of SPE. Changes in these two measures' values, expressed as the statistical test's mean ranks indicate that the proposed algorithm identifies objects from the minority class more accurately. The results obtained has significant potential for further development and broader research on the imbalanced dataset.

Future work in the following directions is worth considering:

– Perform experiments for multi-class problems.
– Use more and different linear base classifiers for testing.

Acknowledgements. This work was supported by the Polish National Science Centre under the grant No. 2017/25/B/ST6/01750 as well as by the statutory funds of the Department of Systems and Computer Networks, Faculty of Electronics, Wroclaw University of Science and Technology.

References

1. Abdallah, A., Maarof, M.A., Zainal, A.: Fraud detection system: a survey. J. Netw. Comput. Appl. **68**, 90–113 (2016)
2. Abdulhammed, R., Faezipour, M., Abuzneid, A., AbuMallouh, A.: Deep and machine learning approaches for anomaly-based intrusion detection of imbalanced network traffic. IEEE Sens. Lett. **3**(1), 1–4 (2018)
3. Alcalá-Fdez, J., et al.: Keel data-mining software tool: data set repository, integration of algorithms and experimental analysis framework. J. Multiple-Valued Logic Soft Comput. **17**, 255–287 (2011)
4. Alpaydin, E.: Introduction to Machine Learning. MIT Press, Cambridge (2014)
5. Basu, S., Banerjee, A., Mooney, R.: Semi-supervised clustering by seeding. In: Proceedings of 19th International Conference on Machine Learning (ICML 2002). Citeseer (2002)
6. Choraś, M., Pawlicki, M., Kozik, R.: Recognizing faults in software related difficult data. In: Rodrigues, J.M.F., et al. (eds.) ICCS 2019. LNCS, vol. 11538, pp. 263–272. Springer, Cham (2019). https://doi.org/10.1007/978-3-030-22744-9_20
7. Fotouhi, S., Asadi, S., Kattan, M.W.: A comprehensive data level analysis for cancer diagnosis on imbalanced data. J. Biomed. Inform. **90**, 103089 (2019)
8. Fred, A., Lourenço, A.: Cluster ensemble methods: from single clusterings to combined solutions. In: Okun, O., Valentini, G. (eds.) Supervised and Unsupervised Ensemble Methods and Their Applications. SCI, vol. 126, pp. 3–30. Springer, Heidelberg (2008). https://doi.org/10.1007/978-3-540-78981-9_1
9. Galar, M., Fernandez, A., Barrenechea, E., Bustince, H., Herrera, F.: A review on ensembles for the class imbalance problem: bagging-, boosting-, and hybrid-based approaches. IEEE Trans. Syst. Man Cybern. Part C (Appl. Rev.) **42**(4), 463–484 (2011)
10. Haixiang, G., Yijing, L., Shang, J., Mingyun, G., Yuanyue, H., Bing, G.: Learning from class-imbalanced data: review of methods and applications. Expert Syst. Appl. **73**, 220–239 (2017)

11. Kaufmann, L., Rousseeuw, P.J.: Finding Groups in Data: An Introduction to Cluster Analysis. Wiley, New York (1990)
12. Klikowski, J., Ksieniewicz, P., Woźniak, M.: A genetic-based ensemble learning applied to imbalanced data classification. In: Yin, H., Camacho, D., Tino, P., Tallón-Ballesteros, A.J., Menezes, R., Allmendinger, R. (eds.) IDEAL 2019. LNCS, vol. 11872, pp. 340–352. Springer, Cham (2019). https://doi.org/10.1007/978-3-030-33617-2_35
13. Koziarski, M., Woźniak, M., Krawczyk, B.: Combined cleaning and resampling algorithm for multi-class imbalanced data with label noise. arXiv preprint arXiv:2004.03406 (2020)
14. Kozik, R., Choras, M., Keller, J.: Balanced efficient lifelong learning (B-ELLA) for cyber attack detection. J. UCS **25**(1), 2–15 (2019)
15. Krawczyk, B., Woźniak, M.: Leveraging ensemble pruning for imbalanced data classification. In: 2018 IEEE International Conference on Systems, Man, and Cybernetics (SMC), pp. 439–444. IEEE (2018)
16. Krawczyk, B., Woźniak, M., Schaefer, G.: Cost-sensitive decision tree ensembles for effective imbalanced classification. Appl. Soft Comput. **14**, 554–562 (2014)
17. Ksieniewicz, P., Burduk, R.: Clustering and weighted scoring in geometric space support vector machine ensemble for highly imbalanced data classification. In: Krzhizhanovskaya, V.V., et al. (eds.) ICCS 2020. LNCS, vol. 12140, pp. 128–140. Springer, Cham (2020). https://doi.org/10.1007/978-3-030-50423-6_10
18. Ksieniewicz, P., Zyblewski, P.: stream-learn-open-source python library for difficult data stream batch analysis. arXiv preprint arXiv:2001.11077 (2020)
19. Kuncheva, L.I.: Combining Pattern Classifiers: Methods and Algorithms. Wiley, Hoboken (2004)
20. Lopez-Garcia, P., Masegosa, A.D., Osaba, E., Onieva, E., Perallos, A.: Ensemble classification for imbalanced data based on feature space partitioning and hybrid metaheuristics. Appl. Intell. **49**(8), 2807–2822 (2019)
21. Pedregosa, F., et al.: Scikit-learn: machine learning in Python. J. Mach. Learn. Res. **12**, 2825–2830 (2011)
22. Platt, J.C.: Probabilistic outputs for support vector machines and comparisons to regularized likelihood methods. In: Advances in Large Margin Classifiers, pp. 61–74. MIT Press (1999)
23. Rousseeuw, P.J.: Silhouettes: a graphical aid to the interpretation and validation of cluster analysis. J. Comput. Appl. Math. **20**, 53–65 (1987)
24. Ruta, D., Gabrys, B.: Classifier selection for majority voting. Inf. Fusion **6**(1), 63–81 (2005)
25. Szeszko, P., Topczewska, M.: Empirical assessment of performance measures for preprocessing moments in imbalanced data classification problem. In: Saeed, K., Homenda, W. (eds.) CISIM 2016. LNCS, vol. 9842, pp. 183–194. Springer, Cham (2016). https://doi.org/10.1007/978-3-319-45378-1_17
26. Woźniak, M.: Hybrid Classifiers: Methods of Data, Knowledge, and Classifier Combination. SCI, vol. 519. Springer, Heidelberg (2013). https://doi.org/10.1007/978-3-642-40997-4
27. Woźniak, M., Graña, M., Corchado, E.: A survey of multiple classifier systems as hybrid systems. Inf. Fusion **16**, 3–17 (2014)
28. Zhang, C., et al.: Multi-imbalance: an open-source software for multi-class imbalance learning. Knowl.-Based Syst. **174**, 137–143 (2019)

Exact Searching for the Smallest Deterministic Automaton

Wojciech Wieczorek[1] , Łukasz Strąk[2(✉)] , Arkadiusz Nowakowski[2] ,
and Olgierd Unold[3]

[1] University of Bielsko-Biala, Willowa 2, 43-309 Bielsko-Biala, Poland
wwieczorek@ath.bielsko.pl
[2] University of Silesia in Katowice, Bankowa 14, 40-007 Katowice, Poland
{lukasz.strak,arkadiusz.nowakowski}@us.edu.pl
[3] Wroclaw University of Science and Technology, Wyb. Wyspianskiego 27,
50-370 Wroclaw, Poland
olgierd.unold@pwr.edu.pl

Abstract. We propose an approach to minimum-state deterministic
finite automaton (DFA) inductive synthesis that is based on using satisfiability modulo theories (SMT) solvers. To that end, we explain how
DFAs and their response to input samples can be encoded as logic formulas with integer variables, equations, and uninterpreted functions. An
SMT solver is then tasked with finding an assignment for such a formula,
from which we can extract the automaton of a required size. We provide
an implementation of this approach, which we use to conduct experiments on a series of benchmarks. The results showed that our method
outperforms in terms of CPU time other SAT and SMT approaches and
other exact algorithms on prepared benchmarks.

Keywords: Grammatical inference · Automata identification ·
Satisfiability modulo theories · Exact search

1 Introduction

In his notable paper [7] Gold proved that the following problem is NP-complete.

INSTANCE: Finite alphabet Σ, two finite subsets $S_+, S_- \subseteq \Sigma^*$, integer
$K > 0$.

QUESTION: Is there a K-state deterministic finite automaton (DFA) A that
recognizes a language $L \subseteq \Sigma^*$ such that $S_+ \subseteq L$ and $S_- \subseteq \Sigma^* - L$?

This problem is important both from theoretical and practical points of view.
In the theory of grammatical inference [11], we can pose interesting questions.
For instance: What if instead of a finite state automaton, a regular expression
[1] or a context-free grammar is required [13, 19]? What if we are given all words
up to a certain length [23]? What if we are given infinitely many words [8]?
The problem of finding the smallest deterministic automaton (the optimization
version of the above given instance) is also crucial in practice, since searching

© Springer Nature Switzerland AG 2021
M. Paszynski et al. (Eds.): ICCS 2021, LNCS 12744, pp. 50–63, 2021.
https://doi.org/10.1007/978-3-030-77967-2_5

for a small acceptor compatible with examples and counter-examples is generally a good idea in grammatical inference applications [25].

The purpose of the present proposal is threefold. The first objective is to devise an algorithm for the smallest deterministic automaton problem. It entails preparing satisfiability modulo theories (SMT) logical formula before starting the searching process. The second objective is to implement this encoding by means of an available SMT solver to get our approach working. The third objective is to investigate to what extent the power of SMT solvers makes it possible to tackle the regular inference problem for large-size instances and to compare our approach with existing ones. Particularly, we will refer to the following classical and new exact DFA identification methods: Bica [17], Exbar [15], Zakirzyanov et al.'s SAT encoding [26], and Smetsers et al.'s SMT encoding [21]. In view of the possibility of future comparisons with other methods, the Python implementation of our method is given via GitHub.[1]

This paper is organized into five sections. In Sect. 2, we present necessary definitions and facts originated from automata, formal languages, and constraint programming. Section 3 describes our inference algorithm based on solving an SMT formula. Section 4 shows experimental results of our approach. Concluding comments are made in Sect. 5.

2   Preliminaries

We assume the reader to be familiar with basic regular language and automata theory, e.g., from Hopcroft et al. textbook [12], so that we introduce only some notations and notions used later in the paper.

2.1   Words and Languages

An *alphabet* is a finite, non-empty set of symbols. We use the symbol Σ for an alphabet. A *word* is a finite sequence of symbols chosen from an alphabet. For a word w, we denote by $|w|$ the length of w. The *empty word* ε is the word with zero occurrences of symbols. Let x and y be words. Then xy denotes the *catenation* of x and y, that is, the word formed by making a copy of x and following it by a copy of y. As usual, Σ^* denotes the set of words over Σ. A word w is called a *prefix* of a word u if there is a word x such that $u = wx$. It is a *proper* prefix if $x \neq \varepsilon$. A set of words all of which are chosen from some Σ^*, where Σ is a particular alphabet, is called a *language*.

2.2   Deterministic Finite Automata

A *deterministic finite automaton* (DFA) is a five-tuple $A = (\Sigma, Q, s, F, \delta)$ where Σ is an alphabet, Q is a finite set of states, $s \in Q$ is the initial state, $F \subseteq Q$ is a set of final states, and δ is a relation from $Q \times \Sigma$ to Q such that $((q, a), r_1) \in \delta$ and $((q, a), r_2) \in \delta$ implies $r_1 = r_2$ for every pair $(q, a) \in Q \times \Sigma$.

---

[1] https://github.com/wieczorekw/wieczorekw.github.io/tree/master/SMT4DFA.

Members of δ are called *transitions*. A transition $((q,a),r) \in \delta$ with $q, r \in Q$ and $a \in \Sigma$, is usually written as $\delta(q,a) = r$. Relation δ specifies the moves: the meaning of $\delta(q,a) = r$ is that automaton A in the current state q reads a and moves to next state r. If for given q and a there is no such r that $((q,a),r) \in \delta$, simply saying $\delta(q,a)$ is undefined, the automaton stops and we can assume it enters the rejecting state. Moving into a state that is not final is also regarded as rejecting but it may be only an intermediate state.

It is convenient to define $\bar{\delta}$ as a relation from $Q \times \Sigma^*$ to Q by the following recursion: $((q, ya), r) \in \bar{\delta}$ if $((q, y), p) \in \bar{\delta}$ and $((p, a), r) \in \delta$, where $a \in \Sigma$, $y \in \Sigma^*$, and requiring $((t, \varepsilon), t) \in \bar{\delta}$ for every state $t \in Q$. The *language accepted* by automaton A is then

$$L(A) = \{x \in \Sigma^* \mid \text{there is } q \in F \text{ such that } ((s, x), q) \in \bar{\delta}\}.$$

Two automata are *equivalent* if they accept the same language.

A *sample* S will be an ordered pair $S = (S_+, S_-)$ where S_+, S_- are finite languages with an empty intersection (have no common word). S_+ will be called the *positive part of S* (*examples*), and S_- the *negative part of S* (*counterexamples*).

Let S be a sample over an alphabet Σ, P be the set of all prefixes of $S_+ \cup S_-$, and let $A = (\Sigma, Q, s, F, \delta)$ be a DFA. Let $f \colon P \to Q$ be bijective. For simplicity of notation, we will write q_p instead of $f(p) = q$ for $q \in Q$ and $p \in P$. An automaton A is an *augmented prefix tree acceptor* (APTA) if:

- $s = q_\varepsilon$,
- $F = \{q_p \mid p \in S_+\}$,
- $\delta = \{((q_p, a), q_r) \mid p, r \in P \text{ such that } r = pa\}$,
- $\{F, R, N\}$ is a partition of Q, where F is the set of final states, $R = \{q_p \mid p \in S_-\}$ is the set of rejecting states, and $N = Q - (F \cup R)$ is the set of neutral states.

Clearly, $L(A) = S_+$ and transitions (as arcs) along with states (as vertices) form a tree with the q_ε root, whose edges are labelled by symbols taken from an alphabet Σ (see Fig. 1 as an example).

From now on, if the states of a DFA $A = (\Sigma, Q, s, F, \delta)$ have been partitioned into final, F, rejecting, R, and neutral, N, states, then we will say that $x \in \Sigma^*$ is: (a) *recognized by accepting* (or simply *accepted*) if there is $q \in F$ such that $((s, x), q) \in \bar{\delta}$, (b) *recognized by rejecting* if there is $q \in R$ such that $((s, x), q) \in \bar{\delta}$, and (c) *rejected* if it is not accepted. So, when we pass an automaton according with consecutive symbols (the letters of a word) and stop after reading the last symbol at a neutral state, then the word is rejected but is not recognized by rejecting.

Many exact and inexact DFA learning algorithms work starting from building an APTA and then folding it up into a smaller hypothesis by merging various pairs of compatible nodes. The merging operation takes two states, q_1 and q_2, and replaces them with a single state, q_3, in such a way that every state incoming to q_1 or q_2 now incomes to q_3 and every state outcoming from q_1 or q_2 now outcomes

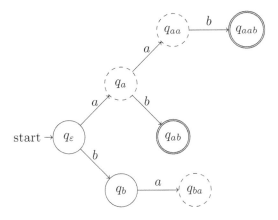

Fig. 1. An APTA accepting ab, aab and 'rejecting' a, aa, ba. The set of its states, Q, is partitioned into $F = \{q_{ab}, q_{aab}\}$, $R = \{q_a, q_{aa}, q_{ba}\}$, $N = \{q_\varepsilon, q_b\}$.

from q_3. It should be noted that the effect of the merge is that a DFA will possibly lose the determinism property through this. Therefore, not all merges will be admitted.

2.3   Satisfiability Modulo Theories

In computer science and mathematical logic, the satisfiability modulo theories (SMT) problem is a decision problem for logical formulas with respect to combinations of background theories—possibly other than Boolean algebra—expressed in classical first-order logic with equality [14]. Examples of theories typically used in computer science are the theory of real numbers, the theory of integers, the theory of uninterpreted functions, and the theories of various data structures such as lists, arrays, bit vectors and so on. Therefore, SMT can be thought of as a form of the constraint satisfaction problem.

Translating a problem into an SMT formula is often the first choice when we are dealing with applications in fields that require determining the satisfiability of formulas in more expressive logics than SAT. The concrete syntax of the SMT standard is very complex. The description of SMT-LIB language [2] is a good reference as a basic background. There are a lot of research publications on the construction of SMT solvers. The reader is referred to an introductory textbook on the topic [14]. Decision procedures for logical formulas with respect to equations and uninterpreted functions use the results of Tarjan [22] and Dawney et al. [5]. Linear arithmetic, on the other hand, may be solved with recipes given by Dutertre and Moura [6]. In our implementation, Z3 library [16] was used, which is released under MIT License and available through a web page.[2]

---

[2] https://github.com/Z3Prover/z3.

3   Proposed Encoding for the Induction of Deterministic Automata

Our translation reduces DFA identification into an SMT instance. Suppose we are given a sample S with nonempty S_+ and nonempty S_- over an alphabet Σ, and a nonnegative integer K. We want to find a $(K + 1)$-state DFA $A = (\Sigma, \{q_0, q_1, \ldots, q_K\}, s, F, \delta)$ such that every $w \in S_+$ is recognized by accepting and every $w \in S_-$ is recognized by rejecting.

3.1   Direct Encoding

Let P be the set of all prefixes of $S_+ \cup S_-$. The integer variables will be $x_p \in \mathbb{Z}$. Assume further that $\{f_a\}_{a \in \Sigma}$ is a family of uninterpreted functions $\mathbb{Z} \times \mathbb{Z} \to \mathbb{B}$ indexed by Σ and $g \colon \mathbb{Z} \to \mathbb{B}$ is also an uninterpreted function. The interpretation of these variables and functions is as follows. The value of x_p is the index of a state which is reached after passing from the initial state, s, through transitions determined by consecutive symbols of a prefix p in a resultant automaton A. The value of $f_a(i, j)$ is $\top$ if $((q_i, a), q_j) \in \delta$, $f_a(i, j) = \bot$ otherwise. Finally, we let $g(i) = \top$ if $q_i \in F$ and $\bot$ if not.

Let us now see how to describe the constraints of the relationship between an automaton A and a sample S in terms of SMT.

1. Naturally, the number of states is already fixed up

$$0 \leq x_p \leq K, \quad \text{for } p \in P.$$

2. Every example has to be recognized by accepting

$$g(x_p) = \top, \quad \text{for } p \in S_+.$$

3. Every counter-example has to be recognized by rejecting

$$g(x_p) = \bot, \quad \text{for } p \in S_-.$$

4. Finally, for every pair of prefixes (p, pa) ($a \in \Sigma$, $p, pa \in P$), there has to be exactly one transition from q_{x_p} to $q_{x_{pa}}$ on a symbol a. We can guarantee this by requiring

$$f_a(x_p, x_{pa}) = \top,$$

$$f_a(x_p, i) \implies x_{pa} = i, \quad \text{for } 0 \leq i \leq K.$$

The conjunction of above-mentioned clauses makes the desired SMT formula.

Theorem 1. *The above encoding of (1)–(4) is proper, i.e., if there exists a DFA with at most $K + 1$ states that matches a given sample S, then a DFA A determined by variables x and functions f, g accepts all examples and rejects all counter-examples.*

Outline of the Proof. First note that if there exists a DFA with at most $K+1$ states that matches a given sample S, then the SMT formula has to be satisfiable. Let A be a DFA determined by x, f, and g. Take any $u \in S_+$, $u = a_0 a_1 \cdots a_{m-1}$, $m \geq 0$. Because of (4) there is exactly one path q_ε, q_{a_0}, $q_{a_0 a_1}$, $\ldots$, q_u in A that is traversed on reading the sequence of symbols $a_0, a_1, \cdots, a_{m-1}$. On account of (2) the last state $q_u \in F$. So u is recognized by accepting. Now, take any $w \in S_-$, $w = a_0 a_1 \cdots a_{m-1}$, $m \geq 0$. Similarly, based on (4) and (3), we can conclude that w is recognized by rejecting. Naturally, $s = q_\varepsilon$. Because every x_p is not bigger than K, the automaton A has at most $K + 1$ states.

3.2 Symmetry Breaking

So as to improve the speed of search process we follow the symmetry-breaking advice [4]. In case there is no k-state DFA for a given sample S, the corresponding (unsatisfiable) SMT instance will solve the problem many times: once for each permutation of the state indices. Therefore, we build an APTA A for S, and then construct the graph G whose vertices are the states of A and there are edges between vertices that can be merged (i.e., x_p and x_r can get the same state index). There are two types of constraints which must be respected during graph building:

- Consistency constraints (on pairs of states): $q_p \in F$ cannot be merged with $q_r \in R$.
- Determinization constraints (on couple of pairs of states): if $\delta(q_{p_1}, a) = q_{r_1}$ and $\delta(q_{p_2}, a) = q_{r_2}$, then merging q_{p_1} with q_{p_2} implies that q_{r_1} and q_{r_2} must also be merged in order to produce a deterministic automaton.

Notice that in any valid solution to an SMT instance, all vertices in an independent set[3] in G must get a different index. So, we fix vertices in a large independent set I to numbers (states indices) in a preprocessing step. For finding an independent set a greedy algorithm analyzed in [9] is used. Obviously, if there is an independent set of size n in G ($|I| = n$), then there is no k-state DFA for S with $k < n$.

3.3 Iterative SMT Solving

The translation of DFA identification into SMT (direct encoding with symmetry breaking predicates) uses a fixed set of states. To prove that the minimal size of a DFA equals K, we have to show that the translation with K states is satisfiable and that the translation with $K - 1$ states is unsatisfiable. Algorithm 1 is used to determine the minimal size.

---

[3] The independent set problem and the well-known clique problem are complementary: a clique in G is an independent set in the complement graph of G and vice versa.

Algorithm 1. Determine minimum-state DFA for a given sample

---
function MINDFA(S_+, S_-, Σ)
    $K \leftarrow |I|$ as shown in Section 3.2
    loop
        construct an SMT formula as shown in Sections 3.1 and 3.2
        solve the formula (we use Z3)
        if the formula is satisfiable **then**
            decode a DFA using variables x and function g
            return the DFA
        else
            $K \leftarrow K + 1$
        end if
    end loop
end function

---

4 Experimental Results

In this section, we describe some experiments comparing the performance of our approach implemented[4] in Python (SMT) with Pena and Oliveira's C implementation[5] of backtrack search (BICA), our effective implementation[6] of Lang's algorithm in C++ (EXBAR), Zakirzyanov et al.'s translation-to-SAT approach implemented[7] in Java (SAT), and Smetsers et al.'s translation-to-SMT approach implemented[8] in Python (Z3GI), when positive and negative words are given. For these experiments, we used a set of 70 samples based on randomly generated regular expressions.

4.1 Brief Description of Other Approaches

The Bica [17] algorithm incrementally builds a hypothesis DFA by examining the nodes in the APTA in breadth-first order. For each of these checks, the algorithm decides whether it is possible to identify that node with one node in the hypothesis, or whether the hypothesis DFA needs to be changed. In order to do this efficiently, it applies advanced search techniques which general idea is based on conflict diagnosis. First, the full set of restrictions is created and entered into a constraint database. This database of constraints is then used to keep a set of constraints that define the solution. When a conflict is reached, the algorithm diagnoses which earlier decision is the cause of the conflict, and backtracks all the way to the point where that bad decision was taken.

    The Exbar [15] algorithm also starts with building the APTA. In the beginning, all nodes in the APTA are blue (candidates for merging), except for the

---

[4] https://github.com/wieczorekw/wieczorekw.github.io/tree/master/SMT4DFA.
[5] We have obtained Linux executable file from the authors.
[6] https://github.com/lazarow/exbar.
[7] https://github.com/ctlab/DFA-Inductor.
[8] https://gitlab.science.ru.nl/rick/z3gi/tree/lata.

red (nodes already in a hypothesis) root. Then, the algorithm considers the blue nodes that are adjacent to the red nodes and for a selected candidate decides whether to accept it (changing its color into red) or merge with another previously accepted node. The order in which the blue nodes are picked and merged matters, hence Exbar first chooses nodes that can be disposed of in as few ways as possible. The best kind of blue node is the node that cannot be merged with any red node. The next best kind is a node that has only one possible merge and so forth. Additionally, the algorithm tries all possible merges in order to avoid latent conflicts. Exbar searches a deterministic automaton that has at most N states (the red nodes limit) and is consistent with all positive and negative examples. If DFA is not found, then the maximum number of the red nodes is increased.

The headmost idea of SAT encoding comes from [10], where the authors based on transformation from DFA identification into graph coloring proposed in [3]. In another work [26] BFS-based symmetry breaking predicates were proposed, instead of original max-clique predicates, which improved the translation-to-SAT technique what was demonstrated with the experiments on randomly generated input data. The core idea is as follows. Consider a graph H, the complement of a graph G described in Sect. 3.2. Finding minimum-size DFA is equivalent to a graph coloring (i.e., such an assignment of labels traditionally called 'colors' to the vertices of a graph H that no two adjacent vertices share the same color) with a minimum number of colors. The graph coloring constraints, in turn, can be efficiently encoded into SAT [24].

Suppose that $A = (\Sigma, Q = \{0, 1, \ldots, K{-}1\}, s = 0, F, \delta)$ is a target automaton and P is the set of all prefixes of $S_+ \cup S_-$. An SMT encoding proposed in [21] uses four functions: $\delta \colon Q \times \Sigma \to Q$, $m \colon P \to Q$, $\lambda^A \colon Q \to \mathbb{B}$, $\lambda^T \colon S_+ \cup S_- \to \mathbb{B}$, and the following five constraints:

$$m(\varepsilon) = 0,$$

$$x \in S_+ \iff \lambda^T(x) = \top,$$

$$\forall xa \in P \colon x \in \Sigma^*, \, a \in \Sigma \quad \delta(m(x), a) = m(xa),$$

$$\forall x \in S_+ \cup S_- \quad \lambda^A(m(x)) = \lambda^T(x),$$

$$\forall q \in Q \quad \forall a \in \Sigma \quad \bigvee_{r \in Q} \delta(q, a) = r.$$

The encoding has been also implemented using Z3Py, the Python front-end of an efficient SMT solver Z3. The main difference between this and our proposals lies in the way they determine state indices and ensure determinism. Smetsers et al. [21] used for these purposes functions m and δ, we used integer variables x and the family of functions f along with the collection of implications. Besides, our δ (decoded from family f) is a relation, while their δ is a surjection, which always defines a completely specified automaton.

4.2 Benchmarks

As far as we know all common benchmarks are too hard to be solved by exact algorithms without some heuristic non-exact steps. Thus, our own algorithm was used for generating problem instances. This algorithm builds a set of words with the following parameters: size N of a regular expression to be generated, alphabet size A, the number $|S|$ of words actually generated and their minimum, d_{min}, and maximum, d_{max}, lengths. The algorithm is arranged as follows. First, using Algorithm 2 construct a random regular expression E.

Algorithm 2. Generate random expression

---

    function GEN(N, star_parent ← False by default)
        if $N \leq 1$ **then**
            return randomly chosen symbol from Σ
        else
            if star_parent **then**
                operator ← choose randomly: concatenation or alternation
            else
                operator ← choose randomly: concat., alt. or repetition
            end if
            if operator is repetition **then**
                return Gen($N - 1$, True)*
            else if operator is alternation **then**
                r ← choose a random integer from range $[1, \max(1, N - 2)]$
                return (Gen(r) | Gen($N - r - 1$))
            else
                r ← choose a random integer from range $[1, \max(1, N - 2)]$
                return (Gen(r) Gen($N - r - 1$))
            end if
        end if
    end function

Next, obtain corresponding minimum-state DFA M. Then, as long as a sample S is not symmetrically structurally complete[9] with respect to M repeat the following steps: (a) using the Xeger library for generating random strings from a regular expression, get two words u and w; (b) truncate as few symbols from the end of w as possible in order to achieve a counter-example $\bar{w}$, if it has succeeded, add u to S_+ and $\bar{w}$ to S_-. Finally, accept $S = (S_+, S_-)$ as a valid sample if it is not too small, too large or highly imbalanced.

In this manner we produced 70 sets with: $N \in [30, 80]$, $A \in \{2, 5, 6, 7, 8\}$, $|S| \in [60, 3000]$, $d_{min} = 0$, and $d_{max} = 850$. For the purpose of diversification samples' structure, 10 of them (those with $A = 6$, and $N = 26, 27, \ldots, 35$) were generated with Σ in Algorithm 2 extended by one or two randomly generated

---

[9] Refer to Chap. 6 of [11] for the formal definition of this concept.

words of length between 10 and 20. The file names with samples[10] have the form
'aAwords$N$.txt'.

4.3   Performance Comparison

In all experiments, we used Intel Xeon W-2135 CPU, 3.7 GHz processor, under
Ubuntu 20.04 LTS operating system with 32 GB RAM. The time limit (TL) was
set to 3600 s. The results are listed in Table 1.

Table 1. Execution times of exact solving DFA identification in seconds. The sign '–'
means that the execution was impossible.

Problem	SAT	EXBAR	BICA	Z3GI	SMT
a2words30	7.01	0.01	0.41	1.71	2.69
a2words36	19.91	2.98	2858.36	201.53	29.68
a2words40	22.99	3.70	–	TL	74.07
a2words42	0.37	0.01	0.10	0.30	0.44
a2words44	0.48	0.02	0.19	0.83	1.08
a2words45	5.07	0.61	1.15	TL	32.56
a2words47	0.50	0.05	0.20	2017.24	1.19
a2words48	3.84	0.96	3.28	232.11	8.27
a2words49	72.39	1.84	–	61	60.76
a2words50	0.82	0.06	0.19	21.91	1.58
a6words26	29.47	TL	92.08	33.28	12.15
a6words27	33.88	TL	TL	9.63	9.67
a6words28	2.58	12.31	13.58	2.36	2.46
a6words29	104.38	589.84	–	9.65	28.85
a6words30	3.20	0.19	0.40	0.70	1.54
a6words31	2.33	0.15	8.56	1.59	3.34
a6words32	10.22	0.4	0.46	2.57	5.62
a6words33	5.30	TL	2.69	5.36	7.37
a6words34	87.92	2114.58	213.85	18.17	16.54
a6words35	82.18	0.04	0.60	1.89	8.21
a8words60	118.90	TL	23.08	7.73	15.69
a8words61	1588.92	11.67	–	83.31	134.43
a8words62	42.34	1.92	TL	25.11	77.03
a8words63	86.44	TL	–	16.50	41.92
a8words64	35.68	1.09	0.35	0.91	7.85
a8words65	TL	TL	–	TL	TL
a8words66	121.37	1.35	2.96	3.03	12.70
a8words67	43.19	34.72	TL	9.75	34.38
a8words68	TL	105.04	–	172.39	243.05
a8words69	427.59	23.43	36.30	0.78	34.61
a5words40	8.32	0.93	4.37	1.24	3.78
a5words41	509.23	4.46	–	TL	111.65
a5words42	4.71	0.08	1.06	0.38	2.73
a5words43	TL	133.03	–	427.64	320.10

---

[10] https://github.com/lazarow/exbar/tree/master/samples.

Table 2. Execution times of exact solving DFA identification in seconds.

Problem	SAT	EXBAR	BICA	Z3GI	SMT
a5words44	**TL**	0.15	–	2.49	227.65
a5words45	95.95	0.76	1.31	4.16	32.04
a5words46	568.48	0.03	2.32	8.20	27.43
a5words47	**TL**	0.05	–	1.82	130.99
a5words48	7.03	0.20	–	1.16	4.38
a5words49	67.32	**TL**	**TL**	25.16	491.73
a6words50	10.38	0.09	0.77	3.25	3.77
a6words51	328.00	**TL**	–	37.97	76.26
a6words52	1.10	0.02	0.17	0.32	0.82
a6words53	**TL**	5.02	–	8.68	143.01
a6words54	66.08	0.36	9.56	6.47	18.09
a6words55	848.91	**TL**	–	11.41	43.55
a6words56	**TL**	0.93	–	36.25	147.75
a6words57	135.24	**TL**	–	36.60	179.41
a6words58	1374.14	184.24	–	25.16	83.36
a6words59	16.90	0.54	0.91	1.87	5.17
a7words60	135.74	0.04	0.58	1.63	10.71
a7words61	1.48	0.00	0.15	0.7	1.28
a7words62	149.90	13.63	**TL**	26.85	35.88
a7words63	9.00	0.51	1.03	0.89	5.36
a7words64	0.62	0.00	0.10	0.17	0.57
a7words65	7.47	0.05	0.52	0.70	2.75
a7words66	412.56	0.12	3.8	4.17	62.81
a7words67	3147.21	2829.06	–	33.46	488.95
a7words68	158.95	71.04	–	96.68	90.91
a7words69	114.08	**TL**	**TL**	44.94	674.51
a8words70	349.94	193.38	**TL**	57.11	272.24
a8words71	**TL**	**TL**	–	5.67	444.03
a8words72	1238.59	3.69	–	17.43	56.11
a8words73	**TL**	16.52	–	1.96	370.62
a8words74	46.02	26.78	0.37	29.18	7.67
a8words75	200.96	**TL**	87.07	17.35	34.16
a8words76	1206.09	164.72	–	15.36	114.76
a8words77	61.51	**TL**	1649.78	936.65	20.76
a8words78	**TL**	150.43	–	**TL**	260.69
a8words79	2368.12	**TL**	–	**TL**	TL
Mean	751.56	867.25	686.88	378.18	187.40

First, note that two of the analyzed solutions are written in Python (SMT and Z3GI) and use Z3 library that is implemented in C++, the remaining algorithms are implemented in C (BICA), C++ (EXBAR) and Java (SAT). While the comparison of the execution times of algorithms in Python is not objectionable, doubts may arise when comparing Python implementations with other programming languages. First of all, note that Python is a scripting language, while C, C++, and Java are non-scripting languages. As shown by empirical research [18], in the initialization phase of a C and C++ program, programs show up to three and four times advantage over Java and five to ten times faster than script languages. In the internal data structure search phase, the advantage of C and C++ over Java is about two-fold, and the scripting languages are comparable or may even be faster than Java.

While computing the mean values, all TL cells were substituted by 3600. The dash sign (only in BICA column) means that the program we obtained form the authors was not able to execute on a certain file due to the "Too many collisions. Specify a large hash table." error. Therefore, for the comparison between SMT and BICA, we took only those rows which do not contain the dash sign.

In order to determine whether the observed mean difference between SMT and remaining methods is a real CPU time decrease we used a paired samples t test [20, pp. 1560–1565] for SMT vs SAT, SMT vs EXBAR, SMT vs BICA, and SMT vs Z3GI. As we can see from Table 2, p value is low in all cases, so we can conclude that our results did not occur by chance and that using our SMT encoding is likely to improve CPU time performance for prepared benchmarks. It must be mentioned, however, that the difference between the two SMT-based approaches (SMT and Z3GI) is not of strong significance. The p-value is 0.07. Usually p-value should be below 0.05 to tell that the rejection of the hypothesis H_0 (the two means are equal) is strong, or the result is highly statistically significant. On the other hand, the execution times would seem to suggest that for large or complex data sets, SMT performs better. In Table 3 there is no entry in which SMT exceeds the time limit and Z3GI does not, but there are four entries (a2words40 with 382 words for which a minimum-size DFA has 13 states, a2words45 with 485 words for which a minimum-size DFA has 14 states, a5words41 with 1624 words for which a minimum-size DFA has 11 states, and a8words78 with 1729 words for which a minimum-size DFA has 4 states) in which Z3GI exeeds the time limit and SMT does not.

Table 3. Obtained p values from the paired samples t test.

SMT vs SAT	SMT vs EXBAR	SMT vs BICA	SMT vs Z3GI
1.51e−04	1.11e−04	2.13e−03	7.10e−02

5   Conclusions

We presented an efficient translation from DFA identification into an SMT instance. By performing this new encoding, we can use the advanced SMT solvers more efficiently than using the earlier approach shown in the literature. In experimental results, we show that our approach outperforms the current state-of-the-art satisfiability-based methods and the well-known backtracking algorithms, which was confirmed by an appropriate statistical test.

It is possible to verify whether the DFAs found are equivalent to the given regexes used in benchmark generation, but it seems useless because the samples are too small to infer a target automaton. On the other hand, the size of a proper sample is beyond the scope of exact algorithms, such as our SMT-based or SAT-based ones. For such big data, we use—in the GI field—heuristic search. In some sense, then, the minimal-size DFA is an overgeneralization, but one should remember that thanks to the parameter K, we can also obtain less general automata. This parameter can be regarded as the degree of data generalization. The smallest K for which our SMT formula is satisfiable, will give the most general automaton. As K increases, we obtain a set of less general automata. What is more, usually the running time for larger K is growing short.

We see here a new area of research. One may ask, for example, whether for any K an obtained DFA is equivalent to an original regex. Perhaps it is a matter of a number of factors including the size of data, the density of data, etc.

References

1. Angluin, D.: An application of the theory of computational complexity to the study of inductive inference. Ph.D. thesis, University of California (1976)
2. Barrett, C., Fontaine, P., Tinelli, C.: The SMT-LIB standard: version 2.6. Technical report, Department of Computer Science, The University of Iowa (2017)
3. Coste, F., Nicolas, J.: Regular inference as a graph coloring problem. In: Workshop on Grammar Inference, Automata Induction, and Language Acquisition, ICML 1997 (1997)
4. Crawford, J.M., Ginsberg, M.L., Luks, E.M., Roy, A.: Symmetry-breaking predicates for search problems. In: Proceedings of the Fifth International Conference on Principles of Knowledge Representation and Reasoning, pp. 148–159. Morgan Kaufmann Publishers Inc. (1996)
5. Downey, P.J., Sethi, R., Tarjan, R.E.: Variations on the common subexpression problem. J. ACM **27**(4), 758–771 (1980)
6. Dutertre, B., de Moura, L.: A fast linear-arithmetic solver for DPLL(T). In: Ball, T., Jones, R.B. (eds.) CAV 2006. LNCS, vol. 4144, pp. 81–94. Springer, Heidelberg (2006). https://doi.org/10.1007/11817963_11
7. Gold, E.M.: Complexity of automaton identification from given data. Inf. Control **37**, 302–320 (1978)
8. Gold, E.M.: Language identification in the limit. Inf. Control **10**, 447–474 (1967)
9. Halldórsson, M.M., Radhakrishnan, J.: Greed is good: approximating independent sets in sparse and bounded-degree graphs. Algorithmica **18**(1), 145–163 (1997)

10. Heule, M.J.H., Verwer, S.: Exact DFA identification using SAT solvers. In: Sempere, J.M., García, P. (eds.) ICGI 2010. LNCS (LNAI), vol. 6339, pp. 66–79. Springer, Heidelberg (2010). https://doi.org/10.1007/978-3-642-15488-1_7
11. de la Higuera, C.: Grammatical Inference: Learning Automata and Grammars. Cambridge University Press, New York (2010)
12. Hopcroft, J.E., Motwani, R., Ullman, J.D.: Introduction to Automata Theory, Languages, and Computation, 2nd edn. Addison-Wesley, Boston (2001)
13. Imada, K., Nakamura, K.: Learning context free grammars by using SAT solvers. In: Proceedings of the ICMLA 2009, pp. 267–272. IEEE Computer Society (2009)
14. Kroening, D., Strichman, O.: Decision Procedures - An Algorithmic Point of View. TTCSTTCS, 2nd edn. Springer, Heidelberg (2016). https://doi.org/10.1007/978-3-662-50497-0
15. Lang, K.J.: Faster algorithms for finding minimal consistent DFAs. Technical report, NEC Research Institute (1999)
16. de Moura, L., Bjørner, N.: Z3: an efficient SMT solver. In: Ramakrishnan, C.R., Rehof, J. (eds.) TACAS 2008. LNCS, vol. 4963, pp. 337–340. Springer, Heidelberg (2008). https://doi.org/10.1007/978-3-540-78800-3_24
17. Pena, J.M., Oliveira, A.L.: A new algorithm for exact reduction of incompletely specified finite state machines. IEEE Trans. CAD Integr. Circuits Syst. **18**(11), 1619–1632 (1999). https://doi.org/10.1109/43.806807
18. Prechelt, L.: An empirical comparison of C, C++, Java, Perl, Python, Rexx and Tcl. IEEE Comput **33**(10), 23–29 (2000)
19. Sakakibara, Y.: Learning context-free grammars using tabular representations. Pattern Recogn. **38**(9), 1372–1383 (2005)
20. Salkind, N.J.: Encyclopedia of Research Design. SAGE Publications Inc, Thousand Oaks (2010)
21. Smetsers, R., Fiterău-Broștean, P., Vaandrager, F.: Model learning as a satisfiability modulo theories problem. In: Klein, S.T., Martín-Vide, C., Shapira, D. (eds.) LATA 2018. LNCS, vol. 10792, pp. 182–194. Springer, Cham (2018). https://doi.org/10.1007/978-3-319-77313-1_14
22. Tarjan, R.E.: Efficiency of a good but not linear set union algorithm. J. ACM **22**(2), 215–225 (1975)
23. Trakhtenbrot, B.A., Barzdin, Y.M.: Finite Automata: Behavior and Synthesis. North-Holland Publishing Company, Amsterdam (1973)
24. Walsh, T.: SAT v CSP. In: Dechter, R. (ed.) CP 2000. LNCS, vol. 1894, pp. 441–456. Springer, Heidelberg (2000). https://doi.org/10.1007/3-540-45349-0_32
25. Wieczorek, W.: Grammatical Inference. SCI, vol. 673. Springer, Cham (2017). https://doi.org/10.1007/978-3-319-46801-3
26. Zakirzyanov, I., Shalyto, A., Ulyantsev, V.: Finding all minimum-size DFA consistent with given examples: SAT-based approach. In: Cerone, A., Roveri, M. (eds.) SEFM 2017. LNCS, vol. 10729, pp. 117–131. Springer, Cham (2018). https://doi.org/10.1007/978-3-319-74781-1_9

Learning Invariance in Deep Neural Networks

Han Zhang[1,2] and Tomasz Arodz[1(✉)] (iD)

[1] Department of Computer Science, Virginia Commonwealth University, Richmond, VA 23284, USA
tarodz@vcu.edu
[2] Department of Computer Science and Technology, School of Data Science and Artificial Intelligence, Dongbei University of Finance and Economics, Dalian, China
hanzhang@dufe.edu.cn

Abstract. One of the long-standing difficulties in machine learning involves distortions present in data – different input feature vectors may represent the same entity. This observation has led to the introduction of invariant machine learning methods, for example techniques that ignore shifts, rotations, or light and pose changes in images. These approaches typically utilize pre-defined invariant features or invariant kernels, and require the designer to analyze what type of distortions are to be expected. While specifying possible sources of variance is straightforward for images, it is more difficult in other domains. Here, we focus on learning an invariant representation from data, without any information of what the distortions present in the data, only based on information whether any two samples are distorted variants of the same entity, or not. In principle, standard neural network architectures should be able to learn the invariance from data, given sufficient numbers of examples of it. We report that, somewhat surprisingly, learning to approximate even a simple types of invariant representation is difficult. We then propose a new type of layer, with a richer output representation, one that is better suited for learning invariances from data.

Keywords: Invariant learning · Deep learning · Autoencoder

1 Introduction

Machine learning deals with many application scenarios which pose difficulties for the learning method. These include supervised learning problems with highly skewed distribution of cardinalities across classes [9], and data with concept drift [5]. But even in the absence of the above problems, learning may be difficult – one example involves scenarios where data becomes distorted, in some unknown way, prior to being captured; that is, the same object can be represented by multiple different feature vectors [19]. Methods that provide invariance to specific types

Supported by NSF grant IIS-1453658.

M. Paszynski et al. (Eds.): ICCS 2021, LNCS 12744, pp. 64–74, 2021.
https://doi.org/10.1007/978-3-030-77967-2_6

of distortions, such as image translation or rotation, have a long history in pattern recognition. Approaches based on Fourier transform [10], Zernike moments [8], and Radon transform [1], have been used to generate invariant features that can be used in downstream learning methods. Invariant kernels have also been proposed [6]. In constructing deep learning models, convolutional and pooling layers [12] offer limited amount of invariance to translation [13], and dedicated methods for achieving more robust translational invariance [11,18], or rotation invariance [3] have been proposed. What unites these approaches is that the type of invariance the model can handle is inherent in the construction, and is fixed. Learning the invariance from training data is an alternative approach. It is common in the language understanding domain, where word embeddings are learned, using large text corpus, to provide very similar vector representation to words that have the same meaning, but are different [14–16]. Outside of language modeling, the first method to learn invariances from training data, Augerino [2], has been proposed recently; it operates by specifying a parametric distribution over various augmentations of the input, and learning which augmentations, to which the model should be invariant, are useful. One should note that learning representations that are invariant over individual samples, as above, is different from learning representations that are invariant, with respect to some distribution characteristics, over several populations, often from different domains, as is common in domain adaptation approaches [20]. It is also distinct from the problem of learning equivariant representations, that is, representations in which variation on input leads to an equivalent variation in the representation [17].

Here, we propose a different method for learning invariant neural networks from data. Instead of defining a priori which transformations of input the network should ignore, or learning it through analyzing various input augmentations, we consider a scenario in which the training set consists not only of input samples, but also includes information which samples are the same, up to some transformation. For example, the training set may consist of several horizontal and vertical mirror version of an image, and based on the information that these images are essentially the same, the network should learn to become invariant to horizontal and vertical axial symmetry. More formally, for samples $x \in \mathcal{X}$, given an unknown family of transformations $\{v_\theta : \mathcal{X} \to \mathcal{X}\}$ parameterized by some vector θ, we aim to train a model $F_\beta(x)$, where β are trainable model weights, such that $F_\beta(x) = F_\beta(x')$ if and only if a θ exists such that $x' = v_\theta(x)$. That is, the representation resulting from model F should be the same for the same entity, irrespective of the distortion. On the other hand, two samples that are not distorted variants of the same underlying entity should have different representation. In order to construct network F that produces the invariant representation, we consider an auto-encoder architecture, consisting of an encoder and a decoder network. In our design, the intermediate layer resulting from an encoder network is partitioned into two parts: the invariant part representing the desired F_β, and the variance part that is needed during training for the decoder to reconstruct the auto-encoder's input and can be discarded after training. To facilitate learning the invariance, we assume the training set consists of triples

$(x, x', d) \in \mathcal{X} \times \mathcal{X} \times \mathbb{R}_+$, where d captures whether samples x and x' are the same, up to a distortion, or not; we use this information to equip the auto-encoder with an invariance-promoting loss.

The rest of the paper is organized as follows. In Sect. 2, we describe the autoencoder architecture in more detail. In Sect. 3, we provide experimental evidence that autoencoders constructed using standard neural network building blocks neural network have difficulties in learning invariance from data. In Sect. 4, we propose a new, richer layer that leads to much more effective learning of invariance from data. In Sect. 5, we show that an autoencoder built using the new layers can learn nontrivial types of invariance.

2   Architecture for Learning Invariant Representations

An auto-encoder is a neural network that attempts to copy its inputs x to its outputs y, that is $y \approx x$. Internally, it has a latent intermediate layer z that describes a code used to represent some aspect of the input. Since the network's output y is supposed to be similar to input x, we typically are most interested in z, the intermediate layer, not in y. An auto-encoder consists of two parts: an encoder f that transforms the inputs x to intermediate code z ($z = f(x)$) and a decoder g that produces a reconstruction of inputs from the intermediate code $y = g(z) = g(f(x))$. Function f and g are represented by multi-layer network. We want the outputs y to be as close to the inputs x as possible; to achieve that, we use mean-squared error of the reconstruction as the loss

$$L_{auto)}(x, y) = L(x, g(f(x))) = \|x - g(f(x))\|_2^2. \tag{1}$$

The idea of auto-encoders can be applied to dimensionality reduction, feature learning, and pre-training of a deep neural network – in all these cases, the useful part is the encoder, which is for example trained to produce low-dimensional, feature-rich representation of the input. The decoder is added to ascertain that all the relevant information from the input is represented in the result of the encoder, z.

We extend the auto-encoder to include an additional loss ℓ_{inv} operating on the intermediate code $z = f(x)$, to train the network to learn representations with desired properties. Here, we want some part of the code z to be invariant. At the same time, the remaining part of z will capture the information about the distortion, since auto-encoder needs full information about the sample in order to reconstruct it faithfully. In this way, the invariant representation will not be trivial, for example, all null.

Fully-connected feed-forward neural networks with at least one hidden layer and with a nonlinear activation function are universal approximators [4,7], and are thus capable of modeling arbitrary well-behaved functions. In principle, we should be able to train and encoder-decoder pair that provides invariance in the intermediate code $z = f(x)$, although the network may need to be wide.

3   Learning Invariant Representations Is Hard

Before attempting to training networks to learn unknown invariance based on triples $(x, x', d) \in \mathcal{X} \times \mathcal{X} \times \mathbb{R}_+$, we test whether training the network to produce an invariant representation can be achieved using standard building blocks in the simplified scenario when the desired invariance is known, and the invariant representation can be pre-defined by the network designer. That is, given any input x, the desired code $z = f(x)$ is known in advance. In this simplified scenario, the network does not have to come up with the invariant representation on its own. Instead of triples $(x, x', d) \in \mathcal{X} \times \mathcal{X} \times \mathbb{R}$, the network is given pairs (x, z^*), where z^* is the what we desire the code z to be, and can use them to learn the invariant mapping in a supervised way.

To test the ability of standard neural architectures to learn pre-defined invariant representations, we focused on invariance to circular translation. The design of an encoder-decoder with the latent code that is invariant to this transformation is straightforward: the discrete Fourier transform can act as an encoder that transforms input vector x into the frequency domain vector z, where the modulus is invariant to circular shift in the input, and the phase is not invariant. Then, the inverse Fourier transform can be used as a decoder that can put back the modulus and phase to reconstruct the original input. That is, the additional loss operating on the output of the encoder is just

$$\ell_{inv}(z, x) = \|z - \text{DFT}(x)\|_2^2,$$

where $z^* = \text{DFT}(x)$ can be pre-calculated numerically and provided to the network as a supervised training signal. The complete training loss for the network $g \circ f$ is then composed of the auto-encoder and the invariance terms

$$L(x) = \|x - g(f(x))\|_2^2 + \lambda \|f(x) - \text{DFT}(x)\|_2^2,$$

where λ are user-defined coefficients, we used $\lambda = 8$ in the experiments, indicating that we focus more on the intermediate layer than on the auto-encoder reconstruction error.

We constructed two datasets, one with 10 features, and one with 20 features. All feature values for all samples are sampled independently from a uniform univariate distribution on $[-1, 1]$. The target supervised signal for the intermediate code z, denoted by z^*, is calculated by performing discrete Fourier transform on each sample, $z^* = \text{DFT}(x)$, which results in a vector of complex numbers of the same dimensionality as the input x. Then, we define four quantities, each being a real-valued vector of the same dimensionality as the input vector x

$$z^* = \text{DFT}(x),$$
$$modulus = |z^*|$$
$$phase = \text{Phase}(z^*)$$
$$cosine = \cos(phase)$$
$$sine = \sin(phase).$$

Of these four vectors, the modulus is known to be invariant to input translation. We conducted two experiments, representing the desired input representation $z^* = \mathrm{DFT}(x)$ as $[modulus, phase]$ or as $[modulus, sine, cosine]$. In each experiment, the network's output z is compared to the desired output z^*, and the discrepancy in the form of the mean-square error is used as the loss that should be minimized during training. In addition to the training set consisting of 60,000 samples, we also created a separate test set of 10,000 samples following the same protocol. We use stochastic gradient descent (SGD) with batch size 128 to minimize the loss $L(x)$.

We tested four accessible nonlinear activation functions: unipolar sigmoid, bipolar sigmoid, ReLU, and SELU. We tested networks with depth varying between 2 and 20 layers, and we made the networks wide by using $8d$ neurons in each layer, where d is the dimensionality of input vectors x. The results Fig. 1 show that SELU activation function is better than the other three, but none of the four activation functions leads to low MSE.

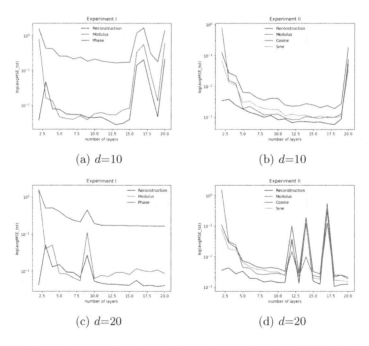

Fig. 1. MSE for each part of code z with activation function SELU. Experiment I (left) denotes learning modulus-phase, while Experiment II (right) denotes learning modulus-sine-cosine. We tested inputs with $d = 10$ and $d = 20$ input features.

To exclude the scenario where the joint task of learning the reconstruction and the invariance makes the problem challenging, we compared the MSE for the full auto-encoder with the result for only training the encoder f to approximate $z^* = \mathrm{DFT}(x)$ using $\|f(x) - \mathrm{DFT}(x)\|_2^2$ as the only term in the loss. We also trained only the decoder g, that is, we minimized $\|x - g(\text{DFT}(x))\|_2^2$. Results in Fig. 2 show that the difficulty comes mostly from training the encoder.

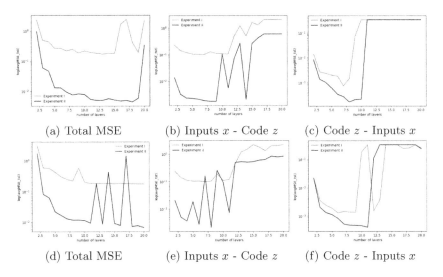

(a) Total MSE   (b) Inputs x - Code z (c) Code z - Inputs x

(d) Total MSE   (e) Inputs x - Code z (f) Code z - Inputs x

Fig. 2. MSE for training the full auto-encoder (left), just the encoder (center), and just the decoder (right). Plots on top show the results of the dataset of 10 features and 20 features on the bottom.

The results of experiments in this section show that deep and relatively wide networks, with the most popular activation functions, with up to 20 layers and with width exceeding the input dimensionality by a factor of 8, are not suited well to learn invariant representations, even in a simplified scenario where the exact form of the representation is known a priori and can be used as a target training signal in a supervised way.

4   Proposed New Layer for Learning Invariances

We hypothesize that using layers with a linear transformation followed by a single, fixed nonlinearity applied element-wise is to restrictive to learn complicated invariant representation. To alleviate that problem, we define an extended layer based on a richer set of transformations

$$layer(O_1) = \begin{cases} O_1 & = \textit{Output from previous layer,} \\ O_2 & = \sqrt{X_{Even}^2 + X_{Odd}^2}, \\ O_3 & = T_{Even}\big/O_2, \\ O_4 & = T_{Odd}\big/O_2, \\ O_5 & = O_3 * O_4, \\ Outputs & = [O_1, O_2, O_3, O_4, O_5], \end{cases}$$

where

$$X = O_1 W + b$$
$$X_{Even} = \textit{Choose the even columns of } X,$$
$$X_{Odd} = \textit{Choose the odd columns of } X.$$

The layer consists of a skip connection similar to those used in residual networks (O_1), a 2-norm (O_2), a normalized linear transformation (O_3 and O_4), and a normalized quadratic transformation (O_5), concatenated together. Weights W and biases b are the trainable parameters of the layer. The number of columns of O_2, O_3, O_4, and O_5 is all half the number of O_1, thus the dimensionality of the output of the layer is three times the dimensionality of input.

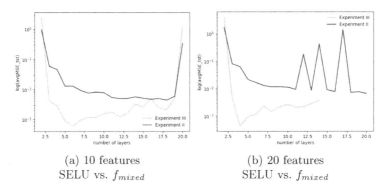

(a) 10 features
SELU vs. f_{mixed}

(b) 20 features
SELU vs. f_{mixed}

Fig. 3. Comparison of an auto-encoder with standard layers that use SELU activation function (blue) and auto-encoder using the new proposed layers (yellow) on the task of supervised learning of translation invariance, for $d = 10$ (left) and $d = 20$ (right) input features. (Color figure online)

The results in Fig. 3 show that the new layer is much more capable of approximating the invariance. In the simple case when the invariant representation, z^*, is known a priori, the mean-squared error of approximating it is lower by more than an order of magnitude for the new layer (yellow in Fig. 3) compared to standard layer with SELU activation function (blue in Fig. 3), which performed best compared to RELU and sigmoids. Notably, the best results with the new layer are achieved for relatively shallow networks, while for SELU-based layers a deep network is needed.

5   Learning Invariant Representations from Data

The experiments above assumed that we know what type of transformation – e.g., translation – is present in the input data, and thus we have a way of calculating the desired invariant representation and training the network in a supervised way. While this approach is useful in evaluating inherent ability of different architectures to capture invariance, it is far from our goal of learning invariance from data – if the desired transformation, for example DFT, is known a priori, there is no need for the network to learn to approximate it, it can be used directly as a pre-processing step.

Our goal is to show the network examples of input samples that are the same but have been transformed, and samples that are not the same. We want to train the network to discover what the invariant transformation is based on the above information alone, without defining the specific type of invariance upfront. To this end, the network is presented on input with triples $(x, x', d) \in \mathcal{X} \times \mathcal{X} \times \mathbb{R}_+$, where d is null if one sample is a transformed version of the other, and not null otherwise. We then train the auto-encoder $g(f(x))$, and we focus on part of the intermediate code $z = f(x)$, denoted $z_{inv}(x)$, to capture the invariant representation. We expect that given a triple (x, x', d), the intermediate codes $z_x = z_{inv}(x)$ and $z_{x'} = z_{inv}(x')$ for the two samples will have $\hat{d}(x, x') = \|z_x - z_{x'}\|$ similar to d. We are most interested in preserving small distances; thus, we use the inverse of squared Euclidean distance as the invariance-promoting term in the loss

$$\ell_{inv}(d, \hat{d}(x, x')) = \left(\frac{1}{\delta + \alpha d^2} - \frac{1}{\delta + \alpha \hat{d}(x, x')^2} \right)^2, \qquad (2)$$

where δ and α are hyperparameters.

5.1   Experimental Validation on Translation Invariance

We conducted a series of experiments to validate the ability of the new layer to learn invariance from data.

Our first experiment involves learning translation invariance. We create a dataset in which samples come in pairs, the first sample is random as described before, and the second sample is a shifted version of the first sample, with the amount of shift selected randomly. For example, if we have a sample $1, 2, 3, 4, 5$, and we want to shift this sample by 2 positions, this sample is then changed to $4, 5, 1, 2, 3$. As the true distance, d^* we use the Euclidean distance between modulus of Fourier transforms of the samples; thus, we have null distance if two samples are shifted versions of each other, and positive distance otherwise.

The results presented in Fig. 4, left panel, show that in the absence of the true desired values of the intermediate code, and with access to pairwise distance data instead, the auto-encoder is still able to learn invariant representation equivalent to the Fourier transform of the input.

We also created a dataset in which each sample is composed of two parts, left and right, and a circular shift occurs independently within each part. The parts are of equal size, that is, if the sample has 10 features, each part consists of 5 features. If both parts of the sample are shifted version of another sample, the true distance d^* is null. We also create samples which are the same, concerning invariance, only in one part – those samples have $d^* > 0$ and allow us to detect

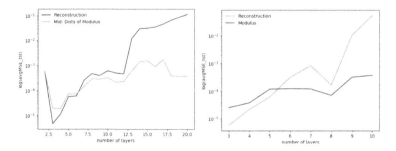

Fig. 4. MSE for auto-encoder reconstruction (red) and for learning invariant intermediate code (blue) for simple shift invariance (left) and for two-part shift invariance
(right). (Color figure online)

if invariance for both parts is appropriately learned. The results in Fig. 4, right
panel, show that our architecture can successfully learn this type of invariance
– the MSE is below 10^{-4}.

5.2   Experimental Validation Beyond Translation Invariance

To move beyond simple shift-invariance, we tested invariance to an unknown
set of permutations of dimensions. Specifically, prior to experiments with data
of dimensionality d, we created a random cycle over a graph with d nodes, one
per input dimension, and performed a cyclic shift of dimensions by a random
number of steps along that cycle – the network should learn to be invariant to
this set of permutations of data dimensions. As another example, we combined
cyclic translation with multiplication of the input by a scalar. The results in
Fig. 5 show that both types of invariances are learned successfully.

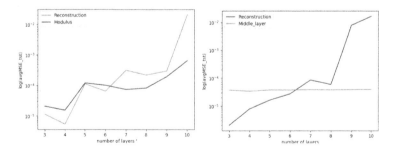

Fig. 5. MSE for auto-encoder reconstruction (red) and for learning invariant intermediate code (blue) for a set of fixed, unknown permutations of dimensions (left), and for
translation and scaling (right). (Color figure online)

The above experiments show that the proposed approach can be used to
learn representations invariant to various transformations: shifting, scaling, and

shuffling dimensions, and it can still achieve excellent performance even if the desired output of the invariant transformation is not known a priori, but has to be learned from examples.

6   Conclusions and Future Work

We focused on constructing neural networks that are invariant to transformations in the input samples. Instead of known, pre-defined type of invariance, we consider a more flexible scenario where invariance is learned from data. First, we showed that standard neural networks are poorly suited to capture invariance, leading to the need for approaches such as dataset augmentation with rotated, translated, or scaled versions of input images [2]. We then propose a new, richer layer, and show that it is more capable of learning invariance. We then show that the proposed new approach is effective in learning invariance form data, by utilizing information about which samples represent the same input subjected to some unknown transformation. These results open the avenue to creating neural networks that can be robust to various changes in the input – our future work will focus on exploring practical applications of this new type of networks. One possible application is in analyzing molecular profiles, such as gene expression, where several similar but different expression patterns can be functionally similar, for example if they represent utilization of two alternative biological pathways.

Acknowledgements. T.A. is supported by NSF grant IIS-1453658.

References

1. Arodź, T.: Invariant object recognition using radon-based transform. Comput. Inform. **24**(2), 183–199 (2012)
2. Benton, G., Finzi, M., Izmailov, P., Wilson, A.G.: Learning invariances in neural networks. In: Advances in Neural Information Processing Systems, vol. 33, pp. 17605–17616 (2020)
3. Cheng, G., Han, J., Zhou, P., Xu, D.: Learning rotation-invariant and fisher discriminative convolutional neural networks for object detection. IEEE Trans. Image Process. **28**(1), 265–278 (2018)
4. Cybenko, G.: Approximation by superpositions of a sigmoidal function. Math. Control Signals Syst. **2**(4), 303–314 (1989). https://doi.org/10.1007/BF02551274
5. Gama, J., Žliobaitė, I., Bifet, A., Pechenizkiy, M., Bouchachia, A.: A survey on concept drift adaptation. ACM Comput. Surv. **46**(4), 1–37 (2014)
6. Haasdonk, B., Burkhardt, H.: Invariant kernel functions for pattern analysis and machine learning. Mach. Learn. **68**(1), 35–61 (2007)
7. Hornik, K.: Approximation capabilities of multilayer feedforward networks. Neural Netw. **4**(2), 251–257 (1991)
8. Khotanzad, A., Hong, Y.H.: Invariant image recognition by Zernike moments. IEEE Trans. Pattern Anal. Mach. Intell. **12**(5), 489–497 (1990)

9. Krawczyk, B.: Learning from imbalanced data: open challenges and future directions. Prog. Artif. Intell. **5**(4), 221–232 (2016). https://doi.org/10.1007/s13748-016-0094-0
10. Lai, J.H., Yuen, P.C., Feng, G.C.: Face recognition using holistic Fourier invariant features. Pattern Recogn. **34**(1), 95–109 (2001)
11. Laptev, D., Savinov, N., Buhmann, J.M., Pollefeys, M.: Ti-pooling: transformation-invariant pooling for feature learning in convolutional neural networks. In: Proceedings of the IEEE Conference on Computer Vision and Pattern Recognition, pp. 289–297 (2016)
12. LeCun, Y., Bottou, L., Bengio, Y., Haffner, P.: Gradient-based learning applied to document recognition. Proc. IEEE **86**(11), 2278–2324 (1998)
13. Lenc, K., Vedaldi, A.: Understanding image representations by measuring their equivariance and equivalence. In: Proceedings of the IEEE Conference on Computer Vision and Pattern Recognition, pp. 991–999 (2015)
14. Mikolov, T., Sutskever, I., Chen, K., Corrado, G.S., Dean, J.: Distributed representations of words and phrases and their compositionality. In: Advances in Neural Information Processing Systems, pp. 3111–3119 (2013)
15. Panahi, A., Saeedi, S., Arodz, T.: word2ket: space-efficient word embeddings inspired by quantum entanglement. In: International Conference on Learning Representations (2019)
16. Pennington, J., Socher, R., Manning, C.: GloVe: global vectors for word representation. In: Proceedings of the 2014 Conference on Empirical Methods in Natural Language Processing, pp. 1532–1543 (2014)
17. Qi, G.J., Zhang, L., Lin, F., Wang, X.: Learning generalized transformation equivariant representations via autoencoding transformations. IEEE Trans. Pattern Anal. Mach. Intell. (2020, in press). https://www.computer.org/csdl/journal/tp/5555/01/09219238/1nMMelzChbO
18. Shen, X., Tian, X., He, A., Sun, S., Tao, D.: Transform-invariant convolutional neural networks for image classification and search. In: Proceedings of the 24th ACM International Conference on Multimedia, pp. 1345–1354 (2016)
19. Wood, J.: Invariant pattern recognition: a review. Pattern Recogn. **29**(1), 1–17 (1996)
20. Zhao, H., Des Combes, R.T., Zhang, K., Gordon, G.: On learning invariant representations for domain adaptation. In: International Conference on Machine Learning, pp. 7523–7532. PMLR (2019)

Mimicking Learning for 1-NN Classifiers

Przemysław Śliwiński[1]([✉]) [ID], Paweł Wachel[1] [ID], and Jerzy W. Rozenblit[2] [ID]

[1] Wrocław University of Science and Technology, 50-370 Wrocław, Poland
{przemyslaw.sliwinski,pawel.wachel}@pwr.edu.pl
[2] University of Arizona, Tucson, AZ 85721, USA
jerzyr@arizona.edu

Abstract. We consider the problem of mimicking the behavior of the nearest neighbor algorithm with an unknown distance measure. Our goal is, in particular, to design and update a learning set so that two NN algorithms with various distance functions ρ_p and ρ_q, $0 < p, q < \infty$, classify in the same way, and to approximate the behavior of one classifier by the other. The autism disorder-related motivation of the problem is presented.

1 Introduction and Problem Statement

We investigate a problem of mimicking the behavior of a nearest neighbor binary classifier with a distance measure function $\rho_p(\mathbf{x}, \mathbf{X}) = \|\mathbf{x} - \mathbf{X}\|_p$, $p \in (0, \infty)$. The training set consists of pairs $\mathcal{S} = \{(\mathbf{X}_n, Y_n)\}$, where $\mathbf{X}_n \in R^2$ is a two-dimensional vector (a pattern), and $Y_n \in \{0, 1\}$ denotes a class the vector belongs to.

Unlike most of the classification problems, where the goal is to construct an effective classifier (and investigate its properties), we assume that there exists a fixed classifier, $g_q(\cdot)$, implementing the nearest neighbor rule based on a distance measure ρ_q, with unknown q. A fundamental assumption here is that neither its classification rule nor the distance function can be replaced.

Within such a framework, the following problems are examined:

1. Given a training set $\mathcal{S} = \{(\mathbf{X}_n, Y_n)\}$, $n = 1, \ldots, N$, design a new (augmented) training set $\mathcal{A} \supset \mathcal{S}$ that allows the unknown classifier $g_q(\cdot)$ to approximate (mimic) the behavior of a known classifier $g_p(\cdot)$.
2. For the known classifier $q_p(\cdot)$ and a new data pair $\{(\mathbf{X}_{N+1}, Y_{N+1})\}$, modify the training set $\mathcal{A}$ to make the unknown one, $g_q(\cdot)$, work in the same way.

Our problem formulation is based on the autistic perception model (presented in more detail in Sect. 4).

2 1-NN Classifiers and On-Grid Learning Sets

Here we shortly recall a rule implemented in the nearest neighbor classifier and its basic asymptotic properties; see [2, Ch. 5.1] and the works cited therein.

© Springer Nature Switzerland AG 2021
M. Paszynski et al. (Eds.): ICCS 2021, LNCS 12744, pp. 75–80, 2021.
https://doi.org/10.1007/978-3-030-77967-2_7

2.1   Nearest Neighbor Classifier

Let $\mathbf{x}$ be a new pattern and let

$$S_{\rho,p}(\mathbf{x}) = \left\{ \left(\mathbf{X}_{(1)}(\mathbf{x}), Y_{(1)} \right), \ldots, \left(\mathbf{X}_{(N)}(\mathbf{x}), Y_{(N)} \right) \right\}$$

be a sequence in which the training pairs from S are sorted w.r.t. increasing distances $\rho_p(\mathbf{x}, \mathbf{X}_n)$. The NN rule assigns $\mathbf{x}$ to the class indicated by the first pattern in the ordered sequence:

$$g_S(\mathbf{x}) = Y_{(1)}.$$

In spite of its simplicity, the algorithm has relatively good asymptotic properties. In particular, the following upper bound holds for the expected error probability

$$L_{NN} = \lim_{N \to \infty} P\left\{ g_S(\mathbf{X}) \neq Y \right\} \leq 2L^*,$$

where $L^* = E\{2\eta(\mathbf{X})(1 - \eta(\mathbf{X}))\}$ and $\eta(\mathbf{x}) = P(Y = 1|\mathbf{X} = \mathbf{x})$ are *the Bayes error* and *a posteriori probability* of error for an arbitrary distribution of $\mathbf{X}$, respectively. In other words, the error of the NN classifier is asymptotically at most twice as large as of the optimal classifier.[1] The NN algorithm is universal in the sense that its asymptotic performance does not depend on the choice of the distance measure ρ if it is an arbitrary norm in R^d, $d < \infty$, [2, Pr. 5.1].

The formal results presented above are thus encouraging, however, we are interested in a non-asymptotic behavior of the NN classifier and in distance measures that are not derived from norms, that is, we admit ρ_p, ρ_q with $p, q < 1$.

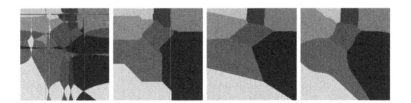

Fig. 1. Voronoi diagrams for various distance measures $\rho_p, p = 1/4, 1, 2$ and 4, from left to right. $N = 8$ (for improved visibility the Voronoi cells have different colors, in spite of the fact that we examine a dichotomy problem, *i.e.* the binary classifiers only) (Color figure online)

2.2   Voronoi Cells

For two-dimensional patterns, a working principle of the nearest neighbor classifier can conveniently be illustrated in the form of the Voronoi diagrams where each training pattern $\{\mathbf{X}_n\}$ determines a set (called a Voronoi cell) of its closest neighbors (with respect to the selected distance measure).[2]

---

[1] Hence, if L^* is small, the performance of the NN algorithm can be acceptable, [2, Ch. 2.1 and 5.2].

[2] See https://github.com/Bahrd/Voronoi for the Python scripts.

2.3   On-Grid Training Set

The various shapes of the Voronoi cells – as clearly seen in Fig. 1 – imply that for the same training set S, the classifiers with various distance functions L_q will make different decisions. In particular, if the parameter q in ρ_q is unknown, these decisions can be difficult to predict, especially for $q < 1$.

Observe, however, that if the training patterns $\{\mathbf{X}_n\}$, $n = 1, \dots, N \times N$, form a grid,[3] that is, they are all located at the crossing of lines parallel to OX and OY axes, and determined by their coordinates, then, for any $p \in (0, \infty)$, the shapes of the resulting Voronoi diagrams will be the same.

Fig. 2. The original Voronoi diagram for $\rho_p, p = 1/4$, $N = 8$ (left) and its (identical) distance measure-independent approximations, generated for $q = 1, 2$, and 4

We will only present a sketch of the proof of this property here. Consider a case $p = 2$ and take a pair of points. Then, using the *ruler and compass*, construct the line which is equidistant to them. Observe that:

– The line splits the plane and creates two Voronoi cells,
– The construction is not based on the radius of the drawn circles but on their mirror symmetry with respect to the constructed line. Since all circles in ρ_p possess this symmetry property, the splitting line will have the same location for any p.

Repeating the construction for all pairs of adjacent points (corresponding with patterns $\{\mathbf{X}_n\}$) will yield the Voronoi diagram the same for all ρ_p, $p \in (0, \infty)$; see Figs. 2 and 3.

Remark 1. Observe that, by virtue of the construction, the same behavior of the NN classifiers for various values p holds only if points are located on the grid. It seems to be a serious restriction, however in what follows we assume that forming a grid is a necessary condition for two classifiers with different p and q to behave in the same way.

---

[3] Note that, due to randomness of the patterns $\{X_n\}$, the grid points are not equidistant.

3   Proposed Solutions

The solution to the first and second problems can now be immediately derived as follows:

1. Given a set S of N learning pairs, add $N(N-1)$ new pairs that create an $N \times N$ on-grid set $\mathcal{A}$ with new points classified according to the known classifier $g_p(\cdot)$. If the approximation appears too crude (*i.e.* the shapes of the new Voronoi cells are not sufficiently similar to the original ones), then one can add new L points and create a denser grid $(L+N) \times (L+N)$, *cf.* Figs. 3. The obvious expense is the quadratic growth of the number of training points.
2. In order to add a new pattern $(\mathbf{X}_{N \times N+1}, Y_{N \times N+1})$ to the existing on-grid set $\mathcal{A}$, the set of accompanying patterns has to be added as well in order to maintain a grid structure of the training set. That is, the new set of patterns $\{(\mathbf{X}_{N \times N+n}, Y_{N \times N+n})\}$, $n = 1, \ldots, N+1$, need to be added with each new pattern classified by the known classifier $g_p(\cdot)$.

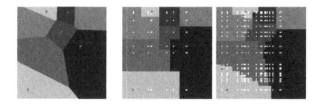

Fig. 3. The diagram of the $q_2(\cdot)$ classifier with $N = 8$ training pairs and its distance measure-independent approximations: 'crude' for $N = 8 \times 8$ and 'fine' for $N = 16 \times 16$ grid points (red dots – the initial set **S**, white – augmented on-grid sets **A**) (Color figure online)

4   Autistic Learning Context and Final Remarks

The specific assumptions made in this note are derived from observations of autistic persons and from hypotheses based on the phenomena published in the literature; see *e.g.*: [1], [4, 7–9]. For instance:

- The attention to details and perception of minute changes, that is characteristic for autistic persons, can correspond to the distance measures ρ_q with $q < 1$. Note that q can be unknown, different for each person, and (because, for instance, of varying fatigue level) can also vary in time, *cf.* [8].
- An increased attention can result in sensory overload and chronic fatigue-like state, decreasing learning abilities and increasing impatience can be represent by the *greedy* 1-NN algorithm (rather than by the k-NN one[4]).

---

[4] The k-NN algorithm can however be used to model a known classifier (a 'teacher').

– The binary classification can be applied to generic, yet still useful classes, like known vs. unknown/unpleasant scenes/situations/objects.
– On-grid patterns points can be seen, in general, as a serious restriction, however, in a controlled environment, the new samples can be, in principle, provided and augmented by a therapist.

We believe that our approach, starting from observations, to hypotheses, to models and algorithms (together with their formal and simulation-based verification), resulting eventually in treatment and therapy proposals, is compatible with the *computational psychiatry* methodology;[5] *cf.* [3,5,6]. Early examination of formal properties of the proposed models can also limit a number of actual experiments the autistic persons are involved in. The primary reasons behind this motivation is that trials and tests are:

– Expensive, as gathering numerous and representative autistic cohort without disturbing their routine (and subsequently, interfere with the outcome of the experiment) is laborious.
– Time-consuming, because getting the autistic person used to the new environment requires sometimes a months-long preparation period.
– Difficult to assess, since autistic persons may not be able to communicate the results, and therefore some indirect and noninvasive measurement methods have to be applied.

Acknowledgments. Authors want to thank the Reviewers for their comments and suggestions.

References

1. Association, A.P., et al.: Diagnostic and statistical manual of mental disorders (DSM-5®). American Psychiatric Pub (2013)
2. Devroye, L., Györfi, L., Lugosi, G.: A Probabilistic Theory of Pattern Recognition. SMAPSMAP, vol. 31. Springer, New York (1996). https://doi.org/10.1007/978-1-4612-0711-5
3. Galitsky, B.: Computational Autism. HCIS, 1st edn. Springer, Cham (2016). https://doi.org/10.1007/978-3-319-39972-0
4. Maurer, D., Werker, J.F.: Perceptual narrowing during infancy: a comparison of language and faces. Dev. Psychobiol. **56**(2), 154–178 (2014)
5. Palmer, C.J., Lawson, R.P., Hohwy, J.: Bayesian approaches to autism: towards volatility, action, and behavior. Psychol. Bull. **143**(5), 521 (2017)
6. Śliwiński, P.: On autistic behavior model. In: 2019 Spring Simulation Conference (SpringSim), pp. 1–8. IEEE (2019)

---

[5] This relatively new scientific discipline aims at developing and examining theoretical and computational models that could serve as a basis for new therapies and/or medicines.

7. Wiggins, L.D., Robins, D.L., Bakeman, R., Adamson, L.B.: Breif report: sensory abnormalities as distinguishing symptoms of autism spectrum disorders in young children. J. Autism Dev. Disord. **39**(7), 1087–1091 (2009)
8. Wilson, R.C., Shenhav, A., Straccia, M., Cohen, J.D.: The eighty five percent rule for optimal learning. Nat. Commun. **10**(1), 1–9 (2019)
9. Yang, J., Kanazawa, S., Yamaguchi, M.K., Motoyoshi, I.: Pre-constancy vision in infants. Curr. Biol. **25**(24), 3209–3212 (2015)

Application of Multi-objective Optimization to Feature Selection for a Difficult Data Classification Task

Joanna Grzyb$^{(\boxtimes)}$ ⬥, Mariusz Topolski ⬥, and Michał Woźniak ⬥

Faculty of Electronics, Department of Systems and Computer Networks,
Wrocław University of Science and Technology, Wybrzeże Wyspiańskiego 27, 50-370
Wrocław, Poland
{joanna.grzyb,mariusz.topolski,michal.wozniak}@pwr.edu.pl

Abstract. Many different decision problems require taking a compromise between the various goals we want to achieve into account. A specific group of features often decides the state of a given object. An example of such a task is the feature selection that allows increasing the decision's quality while minimizing the cost of features or the total budget. The work's main purpose is to compare feature selection methods such as the classical approach, the one-objective optimization, and the multi-objective optimization. The article proposes a feature selection algorithm using the Genetic Algorithm with various criteria, i.e., the cost and accuracy. In this way, the optimal Pareto points for the nonlinear problem of multi-criteria optimization were obtained. These points constitute a compromise between two conflicting objectives. By carrying out various experiments on various base classifiers, it has been shown that the proposed approach can be used in the task of optimizing difficult data.

Keywords: Multi-objective optimization · Feature selection · Cost-sensitive · Classification

1 Introduction

The development of modern information technologies leading to increasingly faster digitization of every aspect of human life makes information systems necessary to process more data. With technological progress, the problem of acquiring and storing large amounts of training data for machine learning models disappeared. As a consequence, the volume of features describing a given object was increased. This, in turn, caused a deterioration in the quality of the classification.

The reason why more information does not mean better classification is the so-called *Curse of Dimensionality* also known as *small n, large p* [1] and it was first described by Richard Bellman [2]. As dimensions are added to the feature set, the distances between specific points are constantly increasing. Additionally, the number of objects needed for correct generalization is increasing. Non-parametric classifiers, such as Neural Networks or those that use radial basis

© Springer Nature Switzerland AG 2021
M. Paszynski et al. (Eds.): ICCS 2021, LNCS 12744, pp. 81–94, 2021.
https://doi.org/10.1007/978-3-030-77967-2_8

functions, where the number of objects required for valid generalization grows exponentially, are worse at this problem [3].

The *Hughes phenomenon* arises from the the *Curse of Dimensionality* [4]. For a fixed number of samples, the recognition accuracy may increase with increasing features but decreases if the number of attributes exceeds a certain optimal value. This is due to the distance between the samples and the noise in the data or irrelevant features.

The above-mentioned *Curse of Dimensionality* makes it challenging to classify objects in a task when a specific decision is made based on the analysis of multiple criteria. In some complex problems, we use not one but many criteria to make the final decision. There is no general concept of optimality in such tasks because there are various objective criteria that are often contradictory: the optimal solution for one criterion may differ from the optimal one for another criterion. Therefore, there is no single optimality criterion for measuring the solution quality for many such real problems. One popular approach to deal with such a task is the Pareto optimization [5]. The Pareto optimality concept, named in honor of the Italian scientist Vilfredo Pareto, is a compromise of many goals of solving some complex and challenging problem [6].

Finding or approximating a set of non-dominated solutions and choosing among them is the main topic of multi-criteria optimization and multi-criteria decision making. There are many methods used in the multi-criteria optimization task, including Simplex models, methods based on graphs, trees, two-phase Simplex, etc. Decision trees are compared in [7]. The lower cost limit obtains similar or better quality for some data sets because it uses a greedy approach that does not guarantee globally optimal solutions.

Modern research in multi-criteria optimization uses methods that, apart from improving classification accuracy, also affect supervised classifiers' generalization ability. One such approach uses an unsupervised grouping procedure based on ascending Growing Hierarchical Self-Organising Maps (GHSOM) to select features [8]. In [9], Jiang et al. proposed a wrapper framework for Test-Cost-Sensitive Feature Selection (TCSFS) where the difference between the accuracy and the total cost create the evaluation function. The two-objective optimization problem was transformed into one-objective. However, the method reaches good accuracy and a low total test cost. The same objectives but in the form of the two-objective optimization case were considered in [10] as the Two-Archive Multi-Objective Artificial Bee Colony Algorithm (TMABC-FS). For solving the Cost-Sensitive Feature Selection (CSFS) problem, this method is a good alternative.

For comparison with above-mentioned techniques, other methods of feature selection are used in tasks with one criterion, such as PCA (*Principal Component Analysis*), ICA (*Independent Component Analysis*) [11], LDA (*Linear Discriminant Analysis*) applied to the linear combination [12] and PCA modification in the form of CCPCA (*Centroid Class Principal Component Analysis*) [13] or GPCA (*Gradient Component Analysis*) [14]. Also some statistical methods are applied to select features: *ANOVA* [15] i.e. the analysis of the variance and *Pear-*

son's correlation coefficient [16,17] which are better at a single criterion task. Here it is also worth mentioning *wrapper* methods which make selection based on the analysis of the results of a specific classifier [15,18].

Multi-criteria optimization is widely used in many fields and is gaining increasing interest. It is a promising field of research. Important applications of the multi-criteria optimization include the minimization of various types of error rates in the machine learning (false positive, false negative) [19,20], the optimization of delivery costs and inventory costs in logistics [21], the optimization of building design in terms of health, energy efficiency and cost criteria [22]. Other applications can be found in medicine. For example, when looking for new therapeutic drugs, we maximize a drug's potency while minimizing the cost of synthesis and undesirable side effects [23].

One should notice that considering the cost of feature acquisition, we encounter additional feature selection limitations. In some cases, if an additional maximum budget is given, some of the features may not be taken into account. Therefore, features' cost should be considered as an additional factor introducing difficulties in the data for the feature selection task. Considering the importance of cost-sensitive data classification, we decided to tackle this problem in our paper. It is possible to obtain a lower total test cost and the accuracy comparable (not worse) to other methods. Hence our contributions are:

- The method proposal of using the two-objective optimization *NSGAII* algorithm and ranking its solutions according to the accuracy, the total cost, or *PROMETHEE*.
- The experimental evaluation of the proposed method with the classical approach and the one-objective optimization *Genetic Algorithm*.

2   Methods

Ikram and Cherukuri mentioned that *Chi-square* is the best method for multi-class problems [24] and a few of our chosen data sets are multi-class, we used this method as the feature selection technique. The *Chi-square* test statistic applied to the *Select K-best* function choose K-features from the data sets, which are the most relevant to the classification process. This is the classical approach, and we refer to it as *FS* (Feature Selection).

Many papers address the problem of feature selection using one-objective optimization such as *Genetic Algorithms* (*GA*) [25]. *GA* searches a population, and through the iteration, it evaluates and uses genetic operators (selection, mutation, crossover) to finds the best solutions [26]. In our experiment, we used *GA* with two different objective functions. Firstly, the maximum accuracy score has been applied and we indicate it as *GA-a*.

$$\text{maximize } g_1 = accuracy \qquad (1)$$

Secondly, we aggregated the accuracy score and selected features' total cost to obtain a cost-sensitive classifier marked as *GA-ac*.

$$\text{maximize } g_2 = \frac{accuracy}{cost} \qquad (2)$$

However, using only one-objective can be insufficient, and even the aggregating process is not enough. Hence, the better approach is to used the multi-objective optimization where each criterion is considered separately [25]. From several algorithms we chose *NSGAII* (Non-dominated Sorting Genetic Algorithm II) [27] – the updated multi-objective version of *GA*. It was used in the feature selection problem [28]. In our experiment, *NSGAII* has been applied with two fitness functions, and each of them is treated independently. The accuracy has to be maximized and the total cost - minimized.

$$\begin{cases} \text{maximize } f_1 = accuracy \\ \text{minimize } f_2 = cost \end{cases} \tag{3}$$

The diagram of the general genetic algorithm is in Fig. 1. The binary representation is an example of 6 features of the *liver* data set. The bit string is a vector of features called an individual or a solution, where 1 means that the feature was selected by the algorithm and 0 - not selected. In the beginning, random sampling is performed, so the initial set of solutions is created. Then, the binary tournament random mating selection is used. N-individuals are selected in each tournament, where $n = 2$ in our case. Individuals are compared with each other, and the winner is taken to the next generation population. It is a simple and efficient solution ensuring diversity [29]. Next, two genetic operators are applied to produce new offsprings: the binary point crossover and the bit-flip mutation. The selection is used to choose significant solutions to create the population and genetic operators explore the search space. As shown in Fig. 1, the crossover swaps the part of the bit string, and the mutation replaces the bit with the opposite value. The search is over when the algorithm reaches the population size.

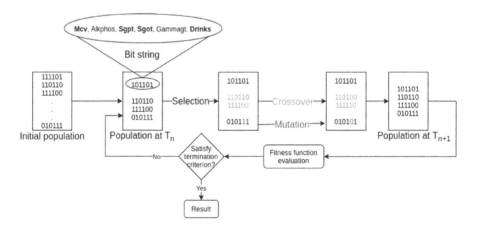

Fig. 1. Diagram of the genetic algorithm and operators

The *NSGAII* algorithm returns the non-dominated set of solutions (called the Pareto front) from which one solution must be chosen to contain a subset of the best features. These features are used to learn the classifier and obtain the best performance and the lowest total cost. However, the best features are not the same when there are different expectations. Sometimes the total test cost or the accuracy is more important, but sometimes there is a need to have good two criteria. Therefore, we applied three approaches to ranking solutions. Firstly, the criterion based on the maximum accuracy was chosen as the *NSGA-a* method. Secondly, the solution with the minimum cost was selected as *NSGA-c*. Lastly, the *PROMETHEE II* [30] approach was implemented as *NSGA-p*. This approach is a pair-wise comparison that returns the ranking of solutions from the best to the worst. It requires criteria weights and the preference function. Then, based on outranking flows (positive and negative), the complete ranking is obtained [31].

We compare all methods described above and believe the multi-objective optimization approach can achieve a more inexpensive total test cost without an accuracy drop compared to other methods. To prove our hypothesis, we conduct the experimental evaluation in the following section.

3   Experimental Evaluation

Our research tries to find answers to two research questions:

RQ1: How do feature selection methods work for each classifier?
RQ2: Can feature selection method based on multi-objective optimization outperform classical methods based on one-objective optimization?

3.1   Setup

Data sets with the corresponding features' cost are obtained from the UCI Machine Learning Repository [32]. All of them are medical data sets where the total cost of tests is important, so classification is not a trivial task. The information about the number of examples, attributes, and classes is in Table 1. The aim of the first data set *heart-disease* is to predict if a patient has heart disease. The *hepatitis* data set contains information about patients with Hepatitis disease and decision 1 or 2 (die or survive). The *liver-disorders* data set has the smallest number of features. Based on them, the decision of a person who suffers from alcoholism is made. The *pima-indians-diabetes* contains only female medical data from the Pima Indians group (Native Americans), in which class says if a person has diabetes or not. The last data set *thyroid-disease* is the biggest one and it has many features. It contains three classes that decide if an individual is normal or suffers from hyperthyroidism or hypothyroidism (1, 2, or 3). We consider only these data sets in our experiment since we do not know where the feature cost is.

The project is implemented in the Python programming language and it is available in the GitHub repository[1] along with results from the experiment.

---

[1] https://github.com/joannagrzyb/moofs.

Table 1. Data sets

Data sets	Number of examples	Number of attributes	Number of classes
heart-disease	303	13	4
hepatitis	155	19	2
liver-disorders	345	6	2
pima-indians-diabetes	768	8	2
thyroid-disease	7200	21	3

A few libraries were used: Pymoo [33], Matplotlib [34], Pandas [35], Numpy [36] and scikit-learn [37]. From the last one we used following classifiers with the default parameters:

- Decision Tree Classifier - *CART*
- Support Vector Machines - *SVM*
- Gaussian Naive Bayes - *GNB*
- K-Nearest Neighbors Classifier - *kNN*

Before experiments, all data sets must be preprocessed. First, missing values were replaced with the most frequent ones. Then, data and features' cost were normalized. The number of examples is relatively small, so Repeated Stratified K-Fold (5 splits × 2 repeats) cross-validation was used to avoid overfitting. Lastly, mechanisms of feature selection described in Sect. 2 were used. The population size of all genetic algorithms was set to 100.

After the preprocessing experiments were run, and based on the accuracy score, which measures the performance, the comparison of methods is made.

3.2   Experiment

While multi-objective optimization is used, the algorithm finds many solutions, and it returns only the Pareto set. Figure 2 shows three non-dominated solutions in the form of black dots, each of which has corresponding values of the accuracy score and the total cost of the selected features. The greater the accuracy, the higher the cost, so choosing which solution to choose is crucial. Hence, we applied three different criteria described in Sect. 2 to select the best solution. In the *PROMETHEE* approach, the weight 0.4 was used for the accuracy and the weight 0.6 for the cost.

The experiment compares six feature selection methods along with four classifiers tested on five data sets. The number of features changes from 1 to the maximum features in the data set. 24 micro charts are showed for each data set, in which an orange line represents the accuracy and a blue line - the total cost. The cost of each feature has been normalized to a value from 0 to 1, so the total cost is the sum of these values, not the original ones from the UCI and it is on the second y-axis on the right. The y-axis on the left contains the accuracy

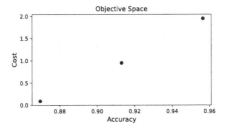

Fig. 2. Pareto front, data set: *hepatitis*

score, and the x-axis - number of features. All approaches to the feature selection problem with the abbreviation coming up in figures are presented in Table 2.

Table 2. Methods' abbreviation

Methods	Objectives	Criteria	Abbr
Select K-best (Chi-square) feature selection	-	-	FS
Genetic algorithm	Max. accuracy	-	GA-a
	Max. (accuracy/cost)	-	Ga-ac
Non-dominated sorting genetic algorithm II	Max. accuracy min. cost	Max. accuracy	NSGA-a
		Min. cost	NSGA-c
		PROMETHEE	NSGA-p

Figure 3a shows results for the *heart* data set. For optimization methods, the total cost has a shape similar to the exponential function, unlike the classical approach *FS*, in which the total cost grows very fast. For *SVM* and *kNN*, the accuracy is stable for all methods, so it is cost-effective to choose the smaller number of features because the accuracy is almost the same, but the total cost is much smaller for *GA-ac, NSGA-c, NSGA-p*. The optimal number of features, in this case, is 4 or 5. For remaining classifiers, the tendency is not the same and too many features lead to a deterioration of the classification's quality.

Figure 3b shows results for the *hepatitis* data set. It can be seen that using all features to learn the classifier is not always the best idea. This data set contains 19 features, and for *GNB* many features disturb in good classification. We can obtain the same accuracy level for other classifiers but a much smaller total cost using only 5 features and optimization methods with the cost criterion. Figure 4 shows values of the accuracy and the total cost for all tested approaches using *SVM* classifier in the *hepatitis*. The accuracy is very stable among all methods and through a different number of selected features. Furthermore, as we observed earlier, the total cost is much smaller for three methods *GA-ac, NSGA-c* and *NSGA-p*.

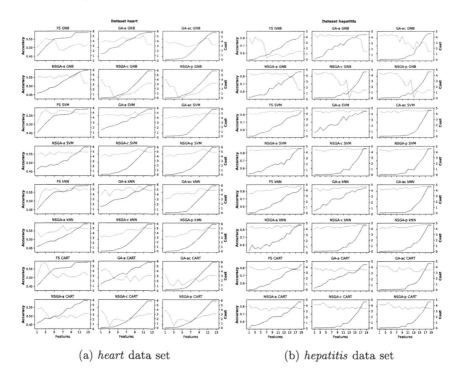

(a) *heart* data set   (b) *hepatitis* data set

Fig. 3. Accuracy and cost

The *liver* data set (Fig. 5a) has the smallest number of features, so the difference between classical approaches and optimization ones is not very large, especially for the cost, which is almost linear. However, the accuracy for non-classical methods has a bigger value than for *FS*. The optimal feature number is 3 for *SVM* and *kNN* where methods prefer the accuracy (*GA-a*, *GA-ac*, *NSGA-a*) and also they keep the low total cost. Even if you do not need the cost-sensitive classifier, it is better not to use *FS* in that case because it has a smaller performance.

In the *pima* data set in Fig. 5b the cost is the smallest for methods with the cost criterion (*GA-ac*, *NSGA-c*, *NSGA-p*), but in the same time they obtain the smallest accuracy. As in the previous case, the optimal number of features is 3. In that point, the *GA-a* and *NSGA-a* for *GNB* achieve the highest accuracy over 75% and the lowest cost under 1.

As in the previous data set, the *thyroid* data set (Fig. 6) has the similar cost shape of *GA-ac*, *NSGA-c* and *NSGA-p* methods and they achieve very small total cost. For them, along with SVM and kNN, the optimal number of selected features is 9 with the cost close to 0 and the accuracy around 95%. Overall, for this data set, the classification quality is much bigger than in other data sets because thyroid has a few thousands times more instances.

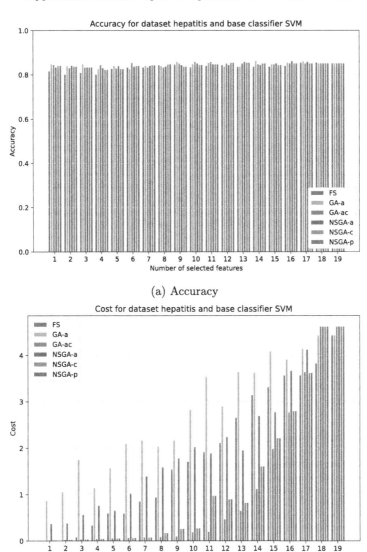

(a) Accuracy

(b) Cost

Fig. 4. Bar charts for data set *hepatitis* and *SVM* classifier

3.3   Lessons Learned

After conducted experiments, we can answer the research question posed at the beginning of this section:

RQ1: How do feature selection methods work for each classifier?

GNB is not recommended as the accuracy is usually smaller than for

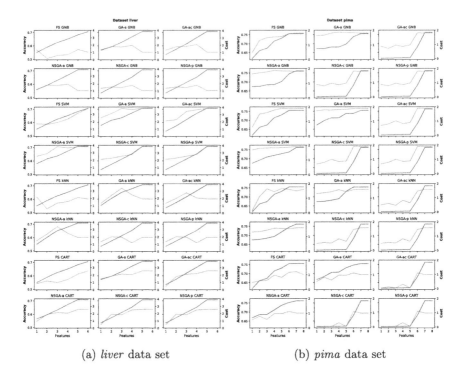

(a) *liver* data set   (b) *pima* data set

Fig. 5. Accuracy and cost

other classifiers and it has greater discrepancy among different number of features. Otherwise *SVM* and *kNN* gives the stable accuracy score among all feature selection techniques in *heart-disease*, *hepatitis* and *thyroid* data sets. Unlike these data, the accuracy varies in each technique through different number of selected features from 5% to 10% in *liver-disorders* and *pima-indians-diabetes* data sets. *CART* is similar to other methods.

RQ2: Can feature selection method based on multi-objective optimization outperform classical methods based on one-objective optimization ?

Suppose there is a need to have a good performance classification and a low total cost. In that case, it is worth considering the multi-objective optimization algorithms to select the best features. The quality of the classification depends on the data. However, it can be useful in the medical environment where some tests can be costly. Thanks to this approach, a person can select only tests that give good results during the classification of the disease, and at the same time, the cost will be low.

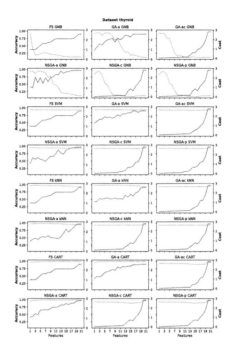

Fig. 6. Accuracy and cost for data set thyroid

4 Conclusions

The work's purpose was to propose a new method and compare a few different feature selection approaches in the classification data with feature cost. The selection of K-best features with the Chi-square statistic, *Genetic Algorithm* in the form of the one-objective and multi-objective optimization were applied to classifiers, such as Gaussian Naive Bayes, Support Vector Machines, K-Nearest Neighbors and Decision Tree Classifier. In our method, we expected a much lower cost and accuracy close to other methods.

Based on the conducted experiments, we gather two criteria to consider if the aim is to achieve good classification and the low total cost of features. Depending on the data set, the one-objective method or the two-objective method should be used to obtain better results than the classical approach, so our proposition is quite effective and feasible.

Further research directions include increasing the number of data sets and taking into account various optimization criteria. It will be interesting to have data with more features and patterns used to learn and test the developed methods. Although *Chi-square* is one of the best feature selection methods, other methods can be used, which may be more effective with different optimization criteria.

Acknowledgements. This work was supported by the Polish National Science Centre under the grant No. 2019/35/B/ST6/04442.

References

1. Fort, G., Lambert-Lacroix, S.: Classification using partial least squares with penalized logistic regression. Bioinformatics **21**(7), 1104–1111 (2005)
2. Bellman, R.E.: Adaptive Control Processes: A Guided Tour, vol. 2045. Princeton University Press (2015)
3. Jimenez, L.O., Landgrebe, D.A.: Hyperspectral data analysis and supervised feature reduction via projection pursuit. IEEE Trans. Geosci. Remote Sens. **37**(6), 2653–2667 (1999)
4. Hughes, G.: On the mean accuracy of statistical pattern recognizers. IEEE Trans. Inf. Theor. **14**(1), 55–63 (1968)
5. Klinger, A.: Letter to the editor–improper solutions of the vector maximum problem. Oper. Res. **15**(3), 570–572 (1967)
6. Vakhania, N., Werner, F.: A brief look at multi-criteria problems: multi-threshold optimization versus pareto-optimization. In: Multi-criteria Optimization-Pareto-optimal and Related Principles. IntechOpen (2020)
7. Penar, W., Wozniak, M.: Cost-sensitive methods of constructing hierarchical classifiers. Exp. Syst. **27**(3), 146–155 (2010)
8. De la Hoz, E., De La Hoz, E., Ortiz, A., Ortega, J., Martínez-Álvarez, A.: Feature selection by multi-objective optimisation: application to network anomaly detection by hierarchical self-organising maps. Knowl. Based Syst. **71**, 322–338 (2014)
9. Jiang, L., Kong, G., Li, C.: Wrapper framework for test-cost-sensitive feature selection. IEEE Trans. Syst. Man Cybern. Syst. **51**, 1747–1756 (2021)
10. Zhang, Y., Cheng, S., Shi, Y., Gong, D., Zhao, X.: Cost-sensitive feature selection using two-archive multi-objective artificial bee colony algorithm. Exp. Syst. Appl. **137**, 46–58 (2019)
11. Karande, K.J., Badage, R.N.: Facial feature extraction using independent component analysis. In: Annual International Conference on Intelligent Computing, Computer Science and Information Systems, ICCSIS 2016, pp. 28–29 (2016)
12. Vyas, R.A., Shah, S.M.: Comparision of PCA and LDA techniques for face recognition feature based extraction with accuracy enhancement. Int. Res. J. Eng. Technol. (IRJET) **4**(6), 3332–3336 (2017)
13. Topolski, M.: The modified principal component analysis feature extraction method for the task of diagnosing chronic lymphocytic leukemia type B-CLL. J. Univ. Comput. Sci. **26**(6), 734–746 (2020)
14. Topolski, M.: Application of the stochastic gradient method in the construction of the main components of PCA in the task diagnosis of multiple sclerosis in children. In: Krzhizhanovskaya, V.V., et al. (eds.) ICCS 2020. LNCS, vol. 12140, pp. 35–44. Springer, Cham (2020). https://doi.org/10.1007/978-3-030-50423-6_3
15. Bommert, A., Sun, X., Bischl, B., Rahnenführer, J., Lang, M.: Benchmark for filter methods for feature selection in high-dimensional classification data. Comput. Stat. Data Anal. **143**, 106839 (2020)
16. Cai, J., Luo, J., Wang, S., Yang, S.: Feature selection in machine learning: a new perspective. Neurocomputing **300**, 70–79 (2018)

17. Risqiwati, D., Wibawa, A.D., Pane, E.S., Islamiyah, W.R., Tyas, A.E., Purnomo, M.H.: Feature selection for EEG-based fatigue analysis using Pearson correlation. In: 2020 International Seminar on Intelligent Technology and Its Applications (ISITIA), pp. 164–169. IEEE (2020)
18. Remeseiro, B., Bolon-Canedo, V.: A review of feature selection methods in medical applications. Comput. Biol. Med. **112**, 103375 (2019)
19. Yevseyeva, I., Basto-Fernandes, V., Ruano-OrdáS, D., MéNdez, J.R.: Optimising anti-spam filters with evolutionary algorithms. Exp. Syst. Appl. **40**(10), 4010–4021 (2013)
20. Wang, P., Emmerich, M., Li, R., Tang, K., Bäck, T., Yao, X.: Convex hull-based multiobjective genetic programming for maximizing receiver operating characteristic performance. IEEE Trans. Evol. Comput. **19**(2), 188–200 (2014)
21. Geiger, M.J., Sevaux, M.: The biobjective inventory routing problem – problem solution and decision support. In: Pahl, J., Reiners, T., Voß, S. (eds.) INOC 2011. LNCS, vol. 6701, pp. 365–378. Springer, Heidelberg (2011). https://doi.org/10.1007/978-3-642-21527-8_41
22. Hopfe, C.J., Emmerich, M.T.M., Marijt, R., Hensen, J.: Robust multi-criteria design optimisation in building design. In: Proceedings of Building Simulation and Optimization, Loughborough, UK, pp. 118–125 (2012)
23. Rosenthal, S., Borschbach, M.: Design perspectives of an evolutionary process for multi-objective molecular optimization. In: Trautmann, H., et al. (eds.) EMO 2017. LNCS, vol. 10173, pp. 529–544. Springer, Cham (2017). https://doi.org/10.1007/978-3-319-54157-0_36
24. Thaseen, I.S., Kumar, C.A.: Intrusion detection model using fusion of chi-square feature selection and multi class SVM. J. King Saud Univ. Comput. Inf. Sci. **29**(4), 462–472 (2017)
25. Enguerran, G., Abadi, M., Alata, O.: An hybrid method for feature selection based on multiobjective optimization and mutual information. J. Inf. Math. Sci. **7**(1), 21–48 (2015)
26. dos S Santana, L.E.A., de Paula Canuto, A.M.: Filter-based optimization techniques for selection of feature subsets in ensemble systems. Exp. Syst. Appl. **41**(4), 1622–1631 (2014)
27. Deb, K., Pratap, A., Agarwal, S., Meyarivan, T.: A fast and elitist multiobjective genetic algorithm: NSGA-II. IEEE Trans. Evol. Comput. **6**(2), 182–197 (2002)
28. Singh, U., Singh, S.N.: Optimal feature selection via NSGA-II for power quality disturbances classification. IEEE Trans. Ind. Inf. **14**(7), 2994–3002 (2017)
29. Razali, N.M., Geraghty, J., et al.: Genetic algorithm performance with different selection strategies in solving TSP. In: Proceedings of the World Congress on Engineering, vol. 2, pp. 1–6. International Association of Engineers Hong Kong (2011)
30. Kou, G., Yang, P., Peng, Y., Xiao, F., Chen, Y., Alsaadi, F.E.: Evaluation of feature selection methods for text classification with small datasets using multiple criteria decision-making methods. Appl. Soft Comput. **86**, 105836 (2020)
31. Behzadian, M., Kazemzadeh, R.B., Albadvi, A., Aghdasi, M.: PROMETHEE: a comprehensive literature review on methodologies and applications. Eur. J. Oper. Res. **200**(1), 198–215 (2010)
32. Lichman, M., et al.: UCI Machine Learning Repository (2013)
33. Blank, J., Deb, K.: Pymoo: multi-objective optimization in Python. IEEE Access **8**, 89497–89509 (2020)
34. Hunter, J.D.: Matplotlib: a 2D graphics environment. Comput. Sci. Eng. **9**(3), 90–95 (2007)

35. McKinney, W.: Data structures for statistical computing in Python. In: van der Walt, S., Millman, J. (eds.) Proceedings of the 9th Python in Science Conference, pp. 56–61 (2010)
36. Oliphant, T.E.: A Guide to NumPy, vol. 1. Trelgol Publishing USA (2006)
37. Pedregosa, F., et al.: Scikit-learn: machine learning in Python. J. Mach. Learn. Res. **12**, 2825–2830 (2011)

Deep Embedding Features for Action Recognition on Raw Depth Maps

Jacek Trelinski and Bogdan Kwolek$^{(\boxtimes)}$

AGH University of Science and Technology,
30 Mickiewicza, 30 -059 Krakow, Poland
{tjacek,bkw}@agh.edu.pl

Abstract. In this paper we present an approach for embedding features for action recognition on raw depth maps. Our approach demonstrates high potential when amount of training data is small. A convolutional autoencoder is trained to learn embedded features, encapsulating the content of single depth maps. Afterwards, multichannel 1D CNN features are extracted on multivariate time-series of such embedded features to represent actions on depth map sequences. In the second stream the dynamic time warping is used to extract action features on multivariate streams of statistical features from single depth maps. The output of the third stream are class-specific action features extracted by TimeDistributed and LSTM layers. The action recognition is achieved by voting in an ensemble of one-vs-all weak classifiers. We demonstrate experimentally that the proposed algorithm achieves competitive results on UTD-MHAD dataset and outperforms by a large margin the best algorithms on 3D Human-Object Interaction Set (SYSU 3DHOI).

Keywords: Data scarcity · Convolutional neural networks · Feature embedding

1 Introduction

People have an innate tendency to recognize and even predict other people's intentions based on their actions [1] and understanding actions and intentions of other people is one of most vital social skills we have [2]. In recent years, deep learning-based algorithms have shown high potential in modeling high-level abstractions from intricate data in many areas such as natural language processing, speech processing and computer vision [3]. After seminal works [4,5] that showed potential and effectiveness of deep learning in human activity recognition, many related studies have been published in this area [6,7]. Most of the present state-of-the-art methods for action recognition either aims at improving the recognition performance through modifications of the backbone CNN network, or they investigate different trade-offs between computational efficiency and performance, c.f. work done in Amazon [8]. However, while deep learning-based algorithms have achieved remarkable results, putting this technology into

© Springer Nature Switzerland AG 2021
M. Paszynski et al. (Eds.): ICCS 2021, LNCS 12744, pp. 95–108, 2021.
https://doi.org/10.1007/978-3-030-77967-2_9

practice can be difficult in many applications for human activity analysis because training deep models requires large datasets and specialized and energy-intensive equipment. In order to cope with such challenges, massively parallel processing capabilities offered by photonic architectures were investigated recently to achieve energy-efficient solutions for real-time action recognition [9].

Several recent methods treat the problem of human action recognition as a generic classification task and try to transfer best practice from ImageNet classification with difference that the input are frame sequences instead of single frames. However, human activities are complex, ambiguous, have different levels of granularity and differ in realization by individuals, including action dynamics. Difficulties in recognition involve many factors such as non-rigid shape of humans, temporal structure, body movement, and human-object interaction, etc. Due to such factors, environmental complexities and plenty another challenges, current algorithms have poor performance in comparison to human ability to recognize human motions and actions [10,11]. As shown in a recent study [12], humans predict actions using grammar-like structures, and this may be one of the reasons of not sufficient recognition performance of current end-to-end approaches that neglect such factors. Moreover, as showed in the discussed study, losing time-information is a feature that can help grouping actions together in the right way. One of the important conclusions of this work is that time may rather confuse than help in recognition and prediction.

3D-based approaches to human action recognition provide higher accuracy than 2D-based ones. Most of the present approaches to action recognition on depth maps are based on the skeleton [13,14]. The number of approaches based on depth maps only, particularly deep learning-based is very limited [11]. One reason of lower interest on such research direction is that depth data is difficult as well as the presence of noise in raw depth map sequences. Despite that skeleton-based methods usually achieve better results than algorithms using only depth maps, they can fail in many scenarios due to skeleton extraction failure. Moreover, in scenarios involving interaction with objects, where detection of objects shapes, 6D poses, etc., is essential, skeleton only-based methods can be less useful. Depth maps acquired from wall-mounted or ceiling-mounted sensors permit accurate detection of patient mobility activities and their duration in intensive care units [15] as well as events like human falls [16].

Traditional approaches to activity recognition on depth maps rely on the handcrafted feature-based representations [17,18]. In contrast to handcrafted representation-based approaches, in which actions are represented by engineered features, learning-based algorithms are capable of discovering the most informative features automatically from raw data. Such deep learning-based methods permit processing images/videos in their raw forms and thus they are capable of automating the process of feature extraction and classification. These methods employ trainable feature extractors and computational models with multiple layers for action representation and recognition.

In this work we propose an approach that, despite limited amount of data, permits achieving high classification scores in action recognition on the basis of raw depth data. To cope with limited and difficult data for learning the action classifier we utilize multi-stream features, which are extracted using DTW, TimeDistributed and LSTM layers (TD-LSTM), and convolutional autoencoder followed by a multi-channel, temporal CNN (1D-CNN). In order to improve model uncertainty the final decision is taken on the basis of several models that are simpler but more robust to the specifics of noisy data sequences.

2   The Algorithm

A characteristic feature of the proposed approach is that it does not require skeleton. Thanks to using depth maps only, our algorithm can be employed on depth data provided by stereo cameras, which can deliver the depth data for persons being at larger distances to the sensors. It is well known that the Kinect sensor fails to estimate the skeleton in several scenarios. In the next Section, we demonstrate experimentally that despite no use of the skeleton, our algorithm achieves better accuracies than several skeleton-based algorithms. In the proposed approach, various features are learned in different domains, like single depth map, time-series of embedded features, time-series warped by DTW (dynamic time warping), and final decision is taken on the basis of voting of one-vs-all weak classifiers. In the proposed approach multi-stream features are processed to extract action features in sequences of depth maps. Action features are extracted using DTW, TimeDistributed and LSTM layers (TD-LSTM), and convolutional autoencoder followed by a multi-channel, temporal CNN (1D-CNN). In consequence, to cope with variability in the observations as well as limited training data, particularly in order to improve model uncertainty the final decision is taken on the basis of several models that are simpler but more robust to the specifics of the noisy data sequences.

The algorithm was evaluated on UTD-MHAD and SYSU 3DHOI datasets. Since in SYSU 3DHOI dataset the performers are not extracted from depth maps, we extracted the subjects. For each depth map we determined a window surrounding the person, which has then been scaled to the required input shape.

In Subsect. 2.1 we present features describing the person's shape in single depth maps. Afterwards, in Subsect. 2.2 we outline features representing multi-variate time-series. Then, in Subsect. 2.3 we detail embedding actions using neural network with TimeDistributed and LSTM layers. In Subsect. 2.4 we discuss multi-class classifiers to construct ensemble. Finally, in Subsect. 2.5 we describe the ensemble as well as our algorithm that for each classified action determines classifiers for voting.

2.1   Embedding Action Features Using CAE and Multi-channel, Temporal CNN

Embedding Frame-Features. Since current datasets for depth-based action recognition have insufficient number of sequences to learn deep models with adequate generalization capabilities, we utilize a convolutional autoencoder (CAE) operating on single depth maps to extract informative frame-features. Time-series of such features representing actions in frame sequences are then fed to multi-channel, temporal CNN that is responsible for extraction embedded features. Because the number of frames in the current benchmark datasets for RGB-D-based action recognition is pretty large, deep feature representations can be learned. Given an input depth map sequence $x = \{x_1, x_2, \ldots, x_T\}$, we encode each depth map x_i using a CNN backbone f into a feature $f(x_i)$, which results in a sequence of embedded feature vectors $f(x) = \{f(x_1), f(x_2), \ldots, f(x_T)\}$. The dimension of such embedding for a depth map sequence is $T \times D_f$, where D_f is size of the embedded vector.

An autoencoder is a type of neural network that projects a high-dimensional input into a latent low-dimensional code (encoder), and then carries out a reconstruction of the input using such a latent code (the decoder) [19]. To achieve this the autoencoder learns a hidden representation for a set of input data, by learning how to ignore less informative information. This means that the autoencoder tries to generate from such a reduced encoding an output representation that is close as possible to its input. When the hidden representation uses fewer dimensions than the input, the encoder carries out dimensionality reduction. An autoencoder consists of an internal (hidden) layer that stores a compressed representation of the input, as well as an encoder that maps the input into the code, and a decoder that maps the code to a reconstruction of the original input. The encoder compresses the input and produces the code, whereas the decoder reconstructs the input using only this code. Learning to replicate its input at its output is achieved by learning a reduction side and a reconstructing side. Autoencoders are considered as unsupervised learning technique since no explicit labels are needed to train them. Once such a representation with reduced dimensionality is learned, it can then be taken as input to a supervised algorithm that can then be trained on the basis of a smaller labeled data subset.

We extracted frame-features using encoder/decoder paradigm proposed in [20]. We implemented a convolutional autoencoder in which the input depth map is first transformed into a lower dimensional representation through successive convolution operations and rectified linear unit (ReLU) activations and afterwards expanded back to its original size using deconvolution operations. The mean squared error, which measures how close the reconstructed input is to the original input has been used as the loss function in the unsupervised learning. The network has been trained using Adam optimizer with learning rate set to 0.001. After training, the decoding layers of the network were excluded from the convolutional autoencoder. The network trained in such a way has been used to extract low dimensional frame-features. The depth maps acquired by the sensor were projected two 2D orthogonal Cartesian planes to represent top and side view of the maps. On training subsets we trained a single CAE for all classes. The convolutional autoencoder has been trained on depth maps of size $3 \times 64 \times 64$. The CAE network architecture is shown in Fig. 1. The network consists of two encoding layers and two associated decoding layers. The size of depth map embedding is equal to 100.

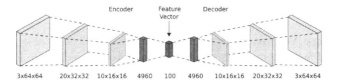

Fig. 1. Architecture of convolutional autoencoder.

Features of Time-Series. Embedding Action Features Using Multi-channel, Temporal CNN. On the basis of depth map sequences representing human actions the CAE that was discussed above produces multivariate time-series. Having on regard that depth map sequences differ in length, such variable length time-series were interpolated to a common length. In multi-channel, temporal CNNs (MC CNNs) the 1D convolutions are applied in the temporal domain. In this work, the time-series (TS) of frame-features that were extracted by the CAE have been used to train a multi-channel 1D CNN. The number of channels is equal to 100, see Fig. 1. The multivariate time-series were interpolated to the length equal to 64. Cubic-spline algorithm has been utilized to interpolate the TS to such a common length.

The first layer of the MC CNN is a filter (feature detector) operating in time domain. Having on regard that the amount of the training data in current datasets for depth-based action recognition is quite small, the neural network consists of two convolutional layers, each with 8×1 filter, 4×1 and 2×1 max pools, and strides set to 1 with no padding, respectively, see Fig. 2. The number of neurons in the dense layer is equal to 100. The number of output neurons is equal to number of the classes. Nesterov Accelerated Gradient (Nesterov Momentum) has been used to train the network, in 1000 iterations, with momentum set to 0.9, dropout equal to 0.5, learning rate equal to 0.001, and L1 parameter set to 0.001. After the training, the output of the dense layer has been used to embed the features, which are referred to as 1D-CNN features.

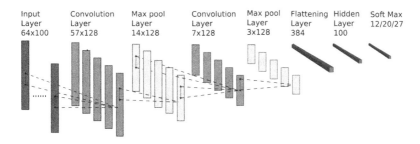

Fig. 2. Flowchart of the multi-channel CNN for multivariate time-series modeling.

2.2   DTW-based Action Features

Frame-Feature Vector. For each depth frame we calculate also handcrafted features describing the person's shape. Similarly to learned frame-features that have been described in Subsect. 2.1, we project the acquired depth maps onto three orthogonal Cartesian views to capture the 3D shape and motion information of human actions. Only pixels representing the extracted person in depth maps are utilized for calculating the features. The following vectors of frame-features were calculated on such depth maps:

1. correlation (xy, xz and zy axes),
2. $x-$coordinate for which the corresponding depth value represents the closest pixel to the camera, $y-$coordinate for which the corresponding depth value represents the closest pixel to the camera.

This means that the person shape in each depth map is described by 3 and 2 features, respectively, depending on the chosen feature set. A human action represented by a number of depth maps is described by a multivariate time-series of length equal to number of frames and dimension 2 or 3 in dependence on the chosen feature set.

DTW-based Features. Dynamic time warping (DTW) is an effective algorithm for measuring similarity between two temporal sequences, which may vary in speed and length. It calculates an optimal match between two given sequences, e.g. time series [21]. In time-series classification one of the most effective algorithms is 1-NN-DTW, which is a special k-nearest neighbor classifier with $k = 1$ and a dynamic time warping for distance measurement. In DTW the sequences are warped non-linearly in time dimension to determine the best match between two samples such that when the same pattern exists in both sequences, the distance is smaller. Let us denote $D(i, j)$ as the DTW distance between subsequences $x[1 : j]$ and $y[1 : j]$. Then the DTW distance between x and y can be determined by the dynamic programming algorithm according to the following iterative equation:

$$D(i, j) = \min\{D(i - 1, j - 1), D(i - 1, j), D(i, j - 1)\} + |x_i, y_j| \qquad (1)$$

The time complexity of calculation of DTW distance is $O(nm)$, where n and m are the length of x and y, respectively.

   We calculate the DTW distance between all depth maps sequences in the training subset. For each depth map sequence the DTW distances between multivariate time-series were calculated for the feature sets 1 and 2. The DTW distances between a given sequence and all remaining sequences from the training set were then used as features. This means that the resulting feature vector has size $n_t \times 2$, where n_t denotes the number of training depth map sequences. The DTW distances have been calculated using library [22].

2.3 Embedding Actions Using Neural Network Consisting of TimeDistributed and LSTM Layers

The neural network operates on depth map sequences, where each sample has a 64×64 data format, across 30 time-steps. The frame batches of size 30 were constructed by sampling with replacement. In first three layers we employ TimeDistributed wrapper to apply the same Conv2D layer to each of the 30 time-steps, independently. The first TimeDistributed layer wraps 32 convolutional filters of size 5×5, with padding set to 'same'. The second TimeDistributed layer wraps 32 convolutional filters of size 5×5. The third TimeDistributed layer wraps the max pooling layer in window of size 4×4. Afterwards, TimeDistributed layer wraps the flattening layer. Next, two TimeDistributed layers wrap dense layers with 256 and 128 neurons, respectively. At this stage the output shape is equal to (None, 30, 128). Finally, we utilize 64 LSTMs and then 64 global average pooling filters, see Fig. 3. The resulting features are called TD-LSTM. The neural networks have been trained using adadelta with learning rate set to 0.001. The loss function was categorical crossentropy and the models were trained as one-vs-all. The motivation of choosing such approach is due to redundant depth maps, i.e. the same human poses in different actions.

2.4 Multi-class Classifiers to Construct Ensemble

The features described in Subsects. 2.1 – 2.3 were used to train multi-class classifiers with softmax encoding, see Fig. 3. Having on regard that for each class an action-specific classifier to extract depth map features has been trained, the number of such classifiers is equal to the number of actions to be recognized. The convolutional autoencoder operating on sequences of depth maps delivers time-series of CAE-based frame-features, on which we determine 1D-CNN features (Subsect. 2.1). Similarly to features mentioned above, the DTW-based features (Subsect. 2.2) are also common features for all classes. The base networks of TimeDistributed-LSTM network (Subsect. 2.3) operating on sequences of depth maps deliver class-specific action features. The discussed TD-LSTM features are of size 64, see Fig. 3, and they are then concatenated with action features mentioned above. The multi-class classifiers delivering at the outputs the softmax-encoded class probability distributions are finally used in an ensemble responsible for classification of actions.

2.5 Ensemble of Classifiers

Figure 3 depicts the ensemble for action classification. The final decision is calculated on the basis of voting of the classifiers. In essence, the final decision is taken using an ensemble of individual models. One advantage of this approach is its interpretability. Because each class is expressed by one classifier only, it is possible to gain knowledge about the discriminative power of individual classifiers. As we can see, for each class the action features that are common for all actions are concatenated with class-specific features, and then used to train multi-class classifiers.

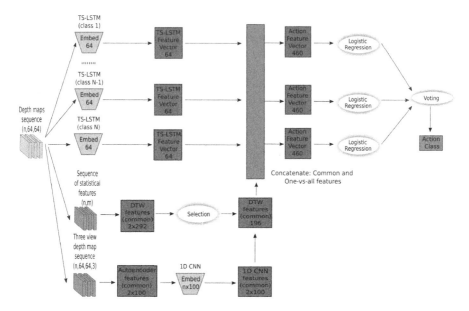

Fig. 3. Ensemble operating on features extracted by DTW, features embedded by CAE and 1D-CNN, which are then concatenated with class-specific features that are embedded by TimeDistributed and LSTM neural networks.

Having on regard that not all classifiers do not contribute equally in decision making we selected the classifiers individually for each testing example. During the classification of each action we initially perform a classification using all classifiers trained in advance and additionally we employ a k-NN classifier. The k-NN classifiers operate only on TD-LSTM features, whereas the logistic regression (LR) classifiers operate on the concatenated features. We consider each class-specific LR classifier with corresponding k-NN with k set to three and inspect their decisions. If decisions of both classifiers are the same then the LR classifier will take in the final voting about the action category. In a case when less than three LR classifiers were selected for the final voting then all LR classifiers attend in the final voting. The discussed algorithm has been compared with an algorithm based on differential evolution (DE), which is responsible for determining the weights for the soft voting.

3   Experimental Results

The proposed algorithm has been evaluated on two publicly available benchmark datasets: UTD-MHAD dataset [23] and SYSU 3D Human-Object Interaction Set (SYSU 3DHOI) [24]. The datasets were selected having on regard their frequent use by action recognition community in the evaluations and algorithm comparisons.

The UTD-MHAD dataset contains 27 different actions performed by eight subjects (four females and four males). All actions were performed in an indoor environment with a fixed background. Each performer repeated each action four times. The dataset consists of 861 data sequences and it was acquired using the Kinect sensor and a wearable inertial sensor.

The SYSU 3D Human-Object Interaction (3DHOI) dataset was recorded by the Kinect sensor and comprises 480 RGB-D sequences from 12 action classes, including calling with cell phone, playing with a cell phone, pouring, drinking, wearing backpack, packing a backpack, sitting on a chair, moving a chair, taking something from a wallet, taking out a wallet, mopping and sweeping. Actions were performed by 40 subjects. Each action involves a kind of human-object interactions. Some motion actions are quite similar at the beginning since the subjects operate or interact with the same objects, or actions start with the same sub-action, such as standing still. The above mentioned issues make this dataset challenging following the evaluation setting in [25], in which depth map sequences with the first 20 subjects were used for training and the rest for testing.

Table 1 presents experimental results that were achieved on the UTD-MHAD dataset. As we can observe, the ensemble consisting of weak classifiers operating on only one-vs-all features, which were embedded using the LSTMs achieves relatively low accuracy in comparison to remaining results, i.e. the recognition performances in row #3 are lower than remaining performances. The DTW features if used alone or when combined with the features embedded by the LSTMs permit achieving better results in comparison to results presented in row #3. The features embedded by CAE and 1D-CNN, see results in first row, permit to achieve better results in comparison to results, which we discussed above. Concatenating the above mentioned features with the features embedded by LSTMs leads to slightly better results, cf. results in the first and fourth row. The best results were achieved by the ensemble consisting of weak classifiers operating on one-vs-all features (LSTM-based), concatenated with features embedded by CAE and 1D-CNN, and concatenated with DTW features. Although the features embedded by LSTMs achieve relatively poor results, when combined with the other features they improve the recognition accuracy significantly. The discussed results were achieved by the logistic regression classifiers. They were obtained

Table 1. Recognition performance on UTD-MHAD dataset.

Common	One-vs-all	Accuracy	Precision	Recall	F1-score
1D-CNN	–	0.8558	0.8593	0.8558	0.8474
DTW	–	0.7930	0.8096	0.7930	0.7919
–	TD-LSTM	0.6419	0.6833	0.6419	0.6322
1D-CNN	TD-LSTM	0.8581	0.8649	0.8581	0.8504
DTW	TD-LSTM	0.8256	0.8455	0.8256	0.8242
DTW 1D-CNN	TD-LSTM	**0.8814**	**0.8844**	**0.8814**	**0.8747**

on the basis of soft voting in the ensemble, which gave slightly better results in comparison to hard voting. Logistic regression returns well calibrated predictions by default as it directly optimizes the Log loss and therefore it has been chosen to built the ensemble. Figure 4 depicts the confusion matrix for the best results achieved on the discussed dataset.

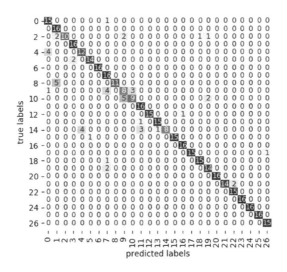

Fig. 4. Confusion matrix on UTD-MHAD dataset.

Table 2 presents experimental results that were achieved using feature selection. As we can notice, our feature selection algorithm permits achieving better results in comparison to results shown in Table 1. The Differential Evolution allows achieving the best classification performance.

Table 2. Recognition performance on UTD-MHAD dataset with selected classifiers for voting.

Common	Voting using selected classifiers				Differential evolution(DE)			
	Accuracy	Precision	Recall	F1-score	Accuracy	Precision	Recall	F1-score
–	0.6535	0.6992	0.6535	0.6426	0.6721	0.7112	0.6721	0.6601
TD-LSTM	0.8628	0.8688	0.8628	0.8548	0.8581	0.8649	0.8581	0.8504
DTW	0.8326	0.8557	0.8326	0.8304	0.8302	0.8497	0.8302	0.8279
DTW TD-LSTM	0.8860	0.8883	0.8860	0.8784	**0.8907**	**0.8919**	**0.8907**	**0.8833**

Table 3 presents the recognition performance of the proposed method compared with previous methods. Most of current methods for action recognition on UTD-MHAD dataset are based on skeleton data. Methods based on skeleton modality usually achieve better results in comparison to methods relying on depth data only. Despite the fact that our method is based on depth modality, we evoked the recent skeleton-based methods to show that it outperforms many of them.

Table 3. Comparative recognition performance of the proposed method with recent algorithms on MHAD dataset.

Method	Modality	Accuracy [%]
JTM [26]	Skeleton	85.81
SOS [27]	Skeleton	86.97
Kinect & inertial [23]	Skeleton	79.10
Struct. SzDDI [28]	Skeleton	89.04
WHDMMs+ConvNets [29][28]	Depth	73.95
Proposed method	Depth	**89.07**

Table 4 illustrates results that were achieved on the 3DHOI dataset. As we can observe, the ensemble consisting of weak classifiers operating on only one-vs-all features, which were embedded using the LSTMs achieves comparable results with results that were obtained using DTW features, and whose performances are lower in comparison to remaining results. Combining DTW features with features embedded by LSTMs leads to better results in comparison to results achieved using only features embedded by LSTMs, compare results in row #5 with results in row #3. The features embedded by CAE and 1D-CNN, see results in first row, permit to achieve better results in comparison to results, which we discussed above. Combining features embedded by CAE and 1D-CNN with features embedded by LSTMs leads to further improvement of the recognition performance, see results in row #4. The best results were achieved by the ensemble consisting of weak classifiers operating on one-vs-all features (LSTM-based) concatenated with features embedded by CAE and 1D-CNN, and concatenated with DTW features. The discussed results were achieved by the logistic regression classifiers. They were obtained on the basis of soft voting in the ensemble, which gave slightly better results in comparison to hard voting. Figure 5 illustrates the confusion matrix.

Table 5 presents results that were obtained using feature selection. As we can observe, both our algorithm and DE improve results presented in Table 4. Results achieved by our algorithm are superior in comparison to results achieved by differential evolution.

Table 6 presents results achieved by recent algorithms on 3DHOI dataset in comparison to results achieved by our algorithm. As we can observe, our algorithm achieves the best results on this challenging dataset. It is worth noting that method [30] relies on depth and skeleton modalities, whereas [25] additionally utilizes RGB images jointly with the skeleton data.

Table 4. Recognition performance on SYSU 3DHOI dataset.

Common	One-vs-all	Accuracy	Precision	Recall	F1-score
1D-CNN	–	0.8114	0.8197	0.8114	0.8104
DTW	–	0.4781	0.4889	0.4781	0.4546
–	TD-LSTM	0.4781	0.4800	0.4781	0.4627
1D-CNN	TD-LSTM	0.8553	0.8591	0.8553	0.8550
DTW	TD-LSTM	0.5044	0.5318	0.5044	0.4872
DTW 1D-CNN	TD-LSTM	**0.8947**	**0.8953**	**0.8947**	**0.8941**

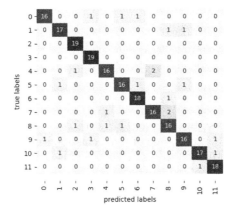

Fig. 5. Confusion matrix on 3DHOI dataset.

Table 5. Recognition performance on 3D HOI dataset with selected classifiers for voting.

Common	Voting using selected classifiers				Differential evolution(DE)			
	Accuracy	Precision	Recall	F1-score	Accuracy	Precision	Recall	F1-score
–	0.5482	0.5521	0.5482	0.5318	0.5263	0.5325	0.5263	0.5140
1D-CNN	0.8640	0.8709	0.8640	0.8638	0.8684	0.8743	0.8684	0.8688
DTW	0.5351	0.5619	0.5351	0.5151	0.5351	0.5602	0.5351	0.5172
DTW 1D-CNN	**0.9035**	**0.9033**	**0.9035**	**0.9024**	0.8904	0.8909	0.8904	0.8895

Table 6. Comparative recognition performance of the proposed method with recent algorithms on 3DHOI dataset.

Method	Modality	Acc. [%]
MSRNN [25]	Depth+RGB+skel	79.58
PTS [30]	Depth+skeleton	87.92
Proposed method	Depth	**90.35**

4 Conclusions

In this paper we presented an approach to encapsulate the content of raw depth maps sequences with human actions. The algorithm has been designed to recognize actions in scenarios when amount of training data is small. It achieves considerable gain in action recognition accuracy on challenging 3D Human-Object Interaction Set (SYSU 3DHOI). On UTD-MHAD dataset it outperforms recent methods on raw depth maps and outperforms most recent methods on skeleton data. The novelty of the proposed method lies in multi-stream features, which are extracted using dynamic time warping, TimeDistributed and LSTM layers, and convolutional autoencoder followed by a multi-channel, temporal CNN. The main methodological results show that despite data scarcity the proposed approach builds classifiers that are able to cope with difficult data and outperforms all the other methods in terms of accuracy.

Acknowledgment. This work was supported by Polish National Science Center (NCN) under a research grant 2017/27/B/ST6/01743.

References

1. Blakemore, S.J., Decety, J.: From the perception of action to the understanding of intention. Nature Rev. Neurosci. **2**(8), 561–567 (2001)
2. Blake, R., Shiffrar, M.: Perception of human motion. Annu. Rev. Psychol. **58**(1), 47–73 (2007)
3. Pouyanfar, S., et al.: A survey on deep learning: Algorithms, techniques, and applications. ACM Comput. Surv. **51**(5), 1–36 (2018)
4. Yang, J.B., Nguyen, M.N., San, P., Li, X.L., Krishnaswamy, S.: Deep convolutional neural networks on multichannel time series for human activity recognition. In: Proceedings of the 24th International Conference on Artificial Intelligence, AAAI Press, pp. 3995–4001 (2015)
5. Lane, N.D., Georgiev, P.: Can deep learning revolutionize mobile sensing? In: 16th International Workshop on Mobile Computing Systems and Applications ACM, pp. 117–122 (2015)
6. Beddiar, D.R., Nini, B., Sabokrou, M., Hadid, A.: Vision-based human activity recognition: a survey. Multimedia Tools Appl. **79**(41), 30509–30555 (2020)
7. Majumder, S., Kehtarnavaz, N.: Vision and inertial sensing fusion for human action recognition: a review. IEEE Sens. J. **21**(3), 2454–2467 (2021)
8. Martinez, B., Modolo, D., Xiong, Y., Tighe, J.: Action recognition with spatial-temporal discriminative filter banks. In: IEEE/CVF International Conference on Computer Vision (ICCV), IEEE Computer Society, pp. 5481–5490 (2019)
9. Antonik, P., Marsal, N., Brunner, D., Rontani, D.: Human action recognition with a large-scale brain-inspired photonic computer. Nature Mach. Intell. **1**(11), 530–537 (2019)
10. Liang, B., Zheng, L.: A survey on human action recognition using depth sensors. In: International Conference on Digital Image Computing: Techniques and Applications, pp. 1–8 (2015)
11. Wang, L., Huynh, D.Q., Koniusz, P.: A comparative review of recent Kinect-based action recognition algorithms. IEEE Trans. Image Process. **29**, 15–28 (2020)

12. Wörgötter, F., Ziaeetabar, F., Pfeiffer, S., Kaya, O., Kulvicius, T., Tamosiunaite, M.: Humans predict action using grammar-like structures. Sci. Rep. **10**(1), 3999 (2020)
13. Ali, H.H., Moftah, H.M., Youssif, A.A.: Depth-based human activity recognition: a comparative perspective study on feature extraction. Future Comput. Inf. J. **3**(1), 51–67 (2018)
14. Ren, B., Liu, M., Ding, R., Liu, H.: A survey on 3D skeleton-based action recognition using learning method. arXiv, 2002.05907 (2020)
15. Yeung, S., et al.: A computer vision system for deep learning-based detection of patient mobilization activities in the ICU. NPJ Digit. Med. **2**(1), 1–5 (2019)
16. Haque, A., Milstein, A., Fei-Fei, L.: Illuminating the dark spaces of healthcare with ambient intelligence. Nature **585**(7824), 193–202 (2020)
17. Yang, X., Zhang, C., Tian, Y.L.: Recognizing actions using depth motion maps-based histograms of oriented gradients. In: Proceedings of the 20th ACM International Conference on Multimedia, pp. 1057–1060. ACM (2012)
18. Xia, L., Aggarwal, J.: Spatio-temporal depth cuboid similarity feature for activity recognition using depth camera. In: CVPR, pp. 2834–2841 (2013)
19. Hinton, G., Salakhutdinov, R.: Reducing the dimensionality of data with neural networks. Science **313**(5786), 504–507 (2006)
20. Masci, J., Meier, U., Cireşan, D., Schmidhuber, J.: Stacked convolutional auto-encoders for hierarchical feature extraction. ICANN **I**, 52–59 (2011)
21. Paliwal, K., Agarwal, A., Sinha, S.: A modification over Sakoe and Chiba's dynamic time warping algorithm for isolated word recognition. Signal Proc. **4**(4), 329–333 (1982)
22. Meert, W., Hendrickx, K., Craenendonck, T.V.: DTAIdistance, ver. 2.0. https://zenodo.org/record/3981067 (2021)
23. Chen, C., Jafari, R., Kehtarnavaz, N.: UTD-MHAD: A multimodal dataset for human action recognition utilizing a depth camera and a wearable inertial sensor. In: IEEE ICIP, pp. 168–172 (2015)
24. Hu, J., Zheng, W., Lai, J., Zhang, J.: Jointly learning heterogeneous features for RGB-D activity recognition. In: CVPR, pp. 5344–5352 (2015)
25. Hu, J., Zheng, W., Ma, L., Wang, G., Lai, J., Zhang, J.: Early action prediction by soft regression. IEEE Trans. PAMI **41**(11), 2568–2583 (2019)
26. Wang, P., Li, W., Li, C., Hou, Y.: Action recognition based on joint trajectory maps with convolutional neural networks. Knowledge-Based Syst. **158**, 43–53 (2018)
27. Hou, Y., Li, Z., Wang, P., Li, W.: Skeleton optical spectra-based action recognition using convolutional neural networks. IEEE Trans. CSVT **28**(3), 807–811 (2018)
28. Wang, P., Wang, S., Gao, Z., Hou, Y., Li, W.: Structured images for RGB-D action recognition. In: ICCV Workshops, pp. 1005–1014 (2017)
29. Wang, P., Li, W., Gao, Z., Zhang, J., Tang, C., Ogunbona, P.: Action recognition from depth maps using deep convolutional neural networks. IEEE Trans. Hum. Mach. Syst. **46**(4), 498–509 (2016)
30. Wang, X., Hu, J.F., Lai, J.H., Zhang, J., Zheng, W.S.: Progressive teacher-student learning for early action prediction. In: CVPR, pp. 3551–3560 (2019)

Analysis of Variance Application in the Construction of Classifier Ensemble Based on Optimal Feature Subset for the Task of Supporting Glaucoma Diagnosis

Dominika Sułot[1]([✉])(iD), Paweł Zyblewski[2](iD), and Paweł Ksieniewicz[2](iD)

[1] Department of Biomedical Engineering, Wroclaw University of Science
and Technology, Wrocław, Poland
`dominika.sulot@pwr.edu.pl`
[2] Department of Systems and Computer Networks,
Wroclaw University of Science and Technology, Wrocław, Poland
`{pawel.zyblewski,pawel.ksieniewicz}@pwr.edu.pl`

Abstract. The following work aims to propose a new method of constructing an ensemble of classifiers diversified by the appropriate selection of the problem subspace. The experiments were performed on a numerical dataset in which three groups are present: healthy controls, glaucoma suspects, and glaucoma patients. Overall, it consists of medical records from 211 cases described by 48 features, being the values of biomarkers, collected at the time of glaucoma diagnosis. To avoid the risk of losing information hidden in the features, the proposed method – for each base classifier – draws a separate subset of the features from the available pool, according to the probability determined by the ANOVA test. The method was validated with four base classifiers and various subspace sizes, and compared with existing feature selection methods. For all of the presented base classifiers, the method achieved superior results in comparison with the others. A high generalization power is maintained for different subspace sizes which also reduces the need to optimize method hyperparameters. Experiments confirmed the effectiveness of the proposed method to create an ensemble of classifiers for small, high-dimensional datasets.

Keywords: Analysis of variance · Glaucoma classification · Subspace selection · Non-uniform random subspace · Classifier ensembles

1 Introduction

Sight is the main and most important human sense. Using its capabilities, we are receiving most of the information coming to us from the surrounding world. Therefore, the narrowing of the field of view – often irrevocably leading to total blindness – make it much more difficult to perform everyday activities. One of the main factors leading to such disorders is glaucoma. It is estimated that over

M. Paszynski et al. (Eds.): ICCS 2021, LNCS 12744, pp. 109–117, 2021.
https://doi.org/10.1007/978-3-030-77967-2_10

70 million people are suffering from this disease, but due to its asymptomatic course, even over 50% of those affected may not be aware of it [15]. The most common type of glaucoma is *primary open-angle glaucoma*, which does not give any symptoms until the later stages, and its effects are irreversible. This form of the disease will be taken into consideration in this article and will be later abbreviated as *glaucoma*.

The current research is focusing on the early detection of glaucoma, at the stage preceding irreversible changes – especially – visual field narrowing [4,9,10]. Biomarkers are the most commonly used for this purpose. They include, among others, *intraocular pressure* (IOP), *retinal nerve fiber layer* (RNFL) *thickness*, parameters concerning the position and shape of the *lamina cribrosa*, size and shape of the optic nerve disc and many others. Their observation and analysis can have an impact on the early detection of developing glaucoma and can help in trying to control the disease [2].

The dynamic development of machine learning methods allows to support medical diagnostics [3,7]. However, due to a large number of available biomarkers in the case of glaucoma, automatic classification of medical cases into disease groups, using pattern recognition methods, states a non-trivial problem. It results from limited datasets and a large number of features describing each analyzed sample. This problem is widely known as a *curse of dimensionality* [1]. There are different methods to cope with this problem [11], mainly based on the feature selection, i.e. selecting the best subset of the available feature space, which will be used for further analysis. However, this solution is not always effective, and by rejecting a certain number of features, the information contained in them is lost. Additionally, sometimes long-term optimization of parameters is needed to obtain a satisfactory result. An example of such a technique may be an application of statistical methods to rank the features [5] or to use *Principal Component Analysis* to reduce the size of the dataset [16].

The development of these methods and, at the same time, the solution to the problem of rejecting input features is the use of *ensemble learning* to train classifiers based on different subspaces. The most common type of such processing is using the *Random Subspace* method, also known as feature bagging [6]. It consists in drawing and returning features for a separate subspace for a single classifier.

The following paper proposes a novel method that includes the basics of the two above-mentioned methods. It allows to automatically build an ensemble of classifiers on a non-randomly drawn subspace of features, where the probability of drawing depends on the ranking obtained with the use of *Analysis of Variance* (ANOVA) [14]. Thus, each classifier is learned from a smaller amount of data, avoiding *the curse of dimensionality*, and at the same time, no features are rejected, minimizing the risk of rejecting valuable information contained in them. Such an approach leads to an increase in the overall diversity of the trained pool and allows to achieve high classification quality. The proposed method was named ANOVA *Subspace Ensemble* (ANOVA SSE). Finally, the proposed method was tested on a numerical medical dataset, and a built ensemble was able to classify glaucoma progression groups.

The main contributions of this study are:

- a novel solution to diversify a homogeneous pool of classifiers based on the analysis of variance,
- experimental evaluation of the method in the context of standard solutions to the problem,
- statistical analysis of the obtained results.

2 Methods

The paper focuses on a method, that aims to extract a set of feature subsets from a dataset and at the same time, creating a classifier ensemble, in which each classifier will be trained on a different subset. The method is based on the ANOVA test, used to calculate the probability with which a given feature will be drawn from the entire set of features. From the results of ANOVA, F-value for each feature is taken, and transformed in the way, that the sum of an array created from this set of F-values will be equal to one. This operation is performed so that the F-value vector may be interpreted as a discrete probability distribution.

The created array is passed to a function that generates the random sample from a features vector, as a probability for each entry. Finally, the function, according to the given probability, will draw from the set of features, increasing the probability of drawing the features that obtained the greater F-value. Then, as many subsets as there are classifiers in the ensemble are drawn, and each classifier is trained on a separate subset.

Further, in the training process, weight is calculated for each classifier in the ensemble based on its *balanced accuracy* on the training set. The aforementioned procedure, which is the classifier fit function, is described in the form of pseudocode in Algorithm 1. At the final prediction, the supports of each are multiplied by the weights and then the standard accumulation of the supports is done. The method uses the information contained in all the features, not rejecting any of them, while favoring features that have higher F-value, thus maximizing the quality of the created ensemble.

3 Dataset

The dataset used in this study is a retrospective data collection, described in more detail in [8]. It consists of a set of biomarker values for each of the patients. These values are typically acquired during glaucoma diagnosis and are commonly used. The set includes intraocular pressure (IOP), retinal nerve fiber layer thickness (RNFL), optic disk morphology parameters, and many others. An experienced ophthalmologist assigned each patient individually to one of three groups: healthy controls, glaucoma suspects, and glaucoma patients, based on the collected data and images acquired with optical coherence tomography. The entire

---

Algorithm 1: The fit function for the ANOVA SSE method

Input: X as an array of training examples with y containing corresponding
       labels and n as an subspace size and k as a size of ensemble

Output: A list of trained base classifier (*ensemble*) and corresponding *weights*

n_features := number of features avaiable in X;

p := the list of F-value for each feature in X obtained from ANOVA;

$p := p/\text{sum}(p)$;

$f := [0, 1, ..., n\_features]$;

classfiers := a list of k base classifiers;

for *classifier in classifiers* **do**

    | *ss_indexes* := a list of n drawn numbers from f with the probability of p;

    | train *classifier* on all samples from X but only on features with *ss_indexes*;

    | append trained *classifier* to a list of *ensemble*;

    | append weight, calculated as a balance accuracy score for *classifier*

    | determined on those training samples, to a *weights* list ;

end

---

collection contains data from 211 patients (69 controls, 72 glaucoma suspects, 70 glaucoma patients), and for each of them, there are 48 features available. Each patient from whom the data was derived gave their written consent and the studies were approved by the *Bioethical Committee of the Wroclaw Medical University (KB–332/2015)*.

4 Experiment Design

The whole experiment was conducted using *Python* language and *scikit-learn 0.23.2* [12] package. The implementation of both the proposed method and experimental code, to preserve the possibility of replication of performed experiments, is publicly available in Github repository[1].

The performance evaluation, as well as the comparison of the various methods, was based on the *balanced accuracy* score, which is calculated as the arithmetic mean of specificity and sensitivity. The *t-test with non-parametric correction* was used to check whether the results obtained with the different methods are statistically dependent [13]. The 5×5 repeated cross-validation protocol was used to obtain reliable results, both for the proposed and reference methods.

Four base classifiers were used and validated in the experiments: *Multilayer Perceptron* (MLP), *k-Nearest Neighbors* (kNN), *Classification and Regression Trees* CART and *Support Vector Machines* classifier SVC. The parameters of the individual classifiers used for the experiments are presented in Table 1.

For comparison, models based on the two most common feature selection methods were calculated, i.e. *Random Subspace* and the method of selecting only the k most differentiable features based on the ANOVA test (*k-best*).

---

[1] https://github.com/w4k2/anova_sse.

Table 1. The parameters of the classifiers that were set during the computational experiments.

Classifier	Parameters
MLP	Hidden layers = 100; activation = the rectified linear activation function; solver = Adam; alpha = 0.0001; constant learning rate = 0.001; maximum number of iterations = 20; beta 1 = 0.9; beta 2 = 0.999; epsilon = 1e−8
KNN	Number of neighbours = 5; uniform weights; leaf size = 30; Euclidean metric
DTC	Gini impurity criterion; without maximum depth; the minimum number of samples required to split an internal node = 2; the minimum number of samples required to be at a leaf node = 1
SVC	Linear kernel; with the enabled probability estimates; number of iteration = 1; one vs rest decision function shape

In addition, the results for simple, single models, that were built on a full available feature space, are also presented.

The aim of the experiments was to verify the effectiveness of the proposed method both as a method for selecting a subspace on which classifier ensemble was to be trained, as well as a method operating on a small part of the feature subspace from the initial data set (e.g. due to the acceleration of learning). Therefore, for solutions based on feature selection, experiments were performed for several sizes of subspace ranging from 1 to 48.

5 Experimental Evaluation

The obtained results are shown in Table 2 and Fig. 1. They are the mean values and standard deviations calculated across folds. Additionally, the approaches for which the given method is statistically better are marked under the results in the table. What may be observed, with a size of feature subspace greater than 18, none of the considered methods is statistically better than the proposed one. Comparing them using the same base classifier leads to the observation that the proposed method receives statistically better or no worse results than the other two feature-selection methods.

Furthermore, as can be seen in Fig. 1, it always get better results than the base solution trained on the all features, regardless of the number of attributes used. By analyzing plots, it can also be concluded that the obtained balanced accuracy maintains its stability even for a large number of features. Which means that a long-term optimization of the subspace size is not needed to get a good result, unlike the method based on the selection of k-best features.

Additionally, in cases where random subspace achieves high balanced accuracy, the proposed method is also able to achieve a similar level of accuracy, but for a much smaller number of features. The graphs also show a black, dashed

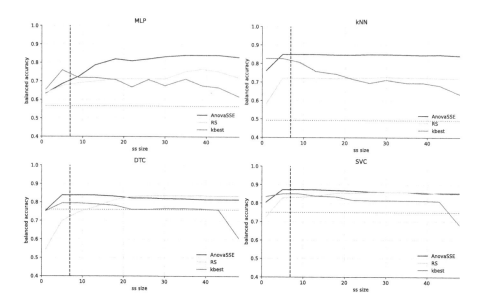

Fig. 1. The dependence of the balanced accuracy on the size of the feature subspace for various methods based on the selection of features and for the classifiers learned on the whole set of features (dotted red). Additionally, a black vertical line marks the number of features corresponding to the square root of all available in the set. (Color figure online)

vertical line, which is the place where the number of features is equal to the root of all available features. Here our method in each case achieves better results than random subspace.

Summarizing the results presented in Table 2, the proposed method shows the potential for being used to create an ensemble of classifiers based on different subspaces of features, while the size of these subspaces is not critical and does not significantly affect the results. Ultimately, the maximum score obtained with this method is .872 using SVC as a base classifier and feature subspace size of 9. This model is statistically superior to almost any model based on the same subspace size.

The results averaged over all four base classifiers are presented on the left side of Table 3. They show that the proposed method always obtained statistically better results than the other two, with the size of the feature subspace greater than 9. In the case of size 9, there is no statistical difference between the proposed one and the method consisting in using 9 best features based on the ANOVA test. From the obtained results it can be concluded that for the considered problem of glaucoma progression group classification the best choice of a base classifier is SVC, which is always statistically better or not worse than the other achieving the highest result of averaged balanced accuracy up to .843.

Table 2. The mean value and the standard deviation for all of the considered methods based of feature selection for four different base classifiers. The results are presented for different sizes of the used feature subspace. The numbers of the methods for which the method obtained statistically better results are also shown under the results. *ss size* is an abbreviation for the utilized size of subspace.

ss size	K-BEST				RANDOM SUBSPACE				ANOVA SSE			
	MLP (1)	kNN (2)	DTC (3)	SVC (4)	MLP (5)	kNN (6)	DTC (7)	SVC (8)	MLP (9)	kNN (10)	DTC (11)	SVC (12)
9	.781 ±.062 5,9	.807 ±.056 5, 6, 9	.794 ±.052 5, 9	.850 ±.057 3, 5–7, 9	.675 ±.048 —	.721 ±.063 —	.746 ±.065 —	.831 ±.059 5–7, 9	.686 ±.029 —	.849 ±.064 5–7, 9	.836 ±.054 5–7, 9	.872 ±.055 1–3, 5–9
18	.668 ±.061 —	.742 ±.065 —	.782 ±.059 1	.833 ±.055 1, 2, 5, 6	.736 ±.048 1	.723 ±.063 —	.814 ±.062 1, 5, 6	.853 ±.060 1–3, 5, 6, 9	.786 ±.046 1, 2	.847 ±.055 1, 2, 5, 6, 9	.832 ±.058 1–3, 5, 6	.868 ±.055 1–3, 5, 6, 9
30	.717 ±.068 —	.694 ±.066 —	.766 ±.058 —	.813 ±.049 1, 2, 5, 6	.712 ±.070 —	.723 ±.073 —	.835 ±.055 1, 2, 5, 6	.859 ±.056 1–3, 5, 6	.832 ±.052 1, 2, 5, 6	.849 ±.050 1–3, 5, 6	.819 ±.062 2, 5, 6	.861 ±.060 1–3, 5, 6
39	.712 ±.063 —	.694 ±.056 —	.764 ±.061 2	.810 ±.059 1, 2, 6	.752 ±.059 —	.723 ±.071 —	.830 ±.060 1, 2, 5, 6	.862 ±.058 1–3, 5, 6	.840 ±.061 1, 2, 5, 6	.847 ±.056 1, 2, 5, 6	.814 ±.060 1, 2	.853 ±.068 1–3, 5, 6
48	.616 ±.043 —	.634 ±.061 —	.605 ±.055 —	.681 ±.066 1, 3	.719 ±.042 1–3	.721 ±.066 1, 3	.833 ±.054 1–6	.849 ±.054 1–6	.846 ±.052 1–6	.852 ±.052 1–6	.814 ±.059 1–6	.861 ±.069 1–6

Table 3. The mean value and the standard deviation for all of the considered methods averaged over base classifiers (left side) and of the considered base classifiers averaged over methods (right side). The results are presented for different sizes of the used feature subspace. The numbers/letters of the methods for which the method obtained statistically better results are also shown under the results. *ss size* is an abbreviation for the utilized size of subspace.

ss size	K-BEST (1)	RANDOM SUBSPACE (2)	ANOVA SSE (3)	MLP (a)	kNN (b)	DTC (c)	SVC (d)
9	.808 ±.044 2	.743 ±.048 —	.811 ±.042 2	.714 ±.033 —	.793 ±.051 a	.793 ±.044 a	.851 ±.051 a, b, c
18	.756 ±.044 —	.781 ±.043 —	.833 ±.043 1, 2	.730 ±.035 —	.771 ±.048 —	.809 ±.050 a, b	.852 ±.050 a, b
30	.748 ±.034 —	.782 ±.046 —	.840 ±.046 1, 2	.754 .044 —	.755 ±.050 —	.803 ±.049 —	.841 ±.053 a, b
39	.745 ±.038 —	.792 ±.050 1	.839 ±.049 1, 2	.768 ±.051 —	.755 ±.050 —	.803 ±.049 —	.841 ±.053 a, b
48	.634 ±.048 —	.780 ±.041 1	.843 ±.047 1, 2	.727 ±.034 —	.735 ±.044 —	.751 ±.042 —	.797 ±.048 a, b, c

6   Conclusions

This paper takes up the topic of generating an ensemble of classifiers on the basis of a high-dimensional dataset. The proposed method tries to solve this problem with the use of all available features, so as not to reject important information that is hidden in these features that less differentiating classes. The method is presented on a demanding dataset, which is above all small but also contains many features that define each object. This set includes three classes: healthy controls, glaucoma suspects, and glaucoma patients. The proposed method shows that, in comparison with other methods based also on feature selection, it can achieve the highest results, which was confirmed by statistical tests that further support the benefits of using non-uniform feature selection. An additional advantage is that the method is characterized by only small fluctuations in balanced accuracy when changing the size of the feature subspace. This reduces the need for a time-consuming process to search parameters to find the optimal size. Additionally, in solutions where the random subspace method turns out to be effective, the proposed method allows achieving similar results of accuracy with a much smaller size of the feature subspace, speeding up the learning process.

Acknowledgments. The work of D. Sułot was supported by *InterDok – Interdisciplinary Doctoral Studies Projects* at *Wroclaw University of Science and Technology*, a project co-financed by the European Union under the European Social Fund, while P. Ksieniewicz and P. Zyblewski were supported by the *Polish National Science Centre* under the grant No. 2017/27/B/ST6/01325.

References

1. Bellman, R.: Curse of dimensionality. Adaptive control processes: a guided tour. Princeton, NJ **3**, 2 (1961)
2. Beykin, G., Norcia, A.M., Srinivasan, V.J., Dubra, A., Goldberg, J.L.: Discovery and clinical translation of novel glaucoma biomarkers. Prog. Retinal Eye Res. **80**, 100875 (2020)
3. Goecks, J., Jalili, V., Heiser, L.M., Gray, J.W.: How machine learning will transform biomedicine. Cell **181**(1), 92–101 (2020)
4. Gupta, K., Thakur, A., Goldbaum, M., Yousefi, S.: Glaucoma precognition: Recognizing preclinical visual functional signs of glaucoma. In: Proceedings of the IEEE/CVF Conference on Computer Vision and Pattern Recognition (CVPR) Workshops (June 2020)
5. Guyon, I., Elisseeff, A.: An introduction to variable and feature selection. J. Mach. Learn. Res. **3**, 1157–1182 (2003)
6. Ho, T.K.: The random subspace method for constructing decision forests. IEEE Trans. Pattern Anal. Mach. Intell. **20**(8), 832–844 (1998)
7. Jackowski, K., Jankowski, D., Ksieniewicz, P., Simić, D., Simić, S., Woźniak, M.: Ensemble classifier systems for headache diagnosis. In: Piętka, E., Kawa, J., Wieclawek, W. (eds.) Information Technologies in Biomedicine, vol. 4, pp. 273–284. Springer, Cham (2014) https://doi.org/10.1007/978-3-319-06596-0_25

8. Krzyżanowska-Berkowska, P., Czajör, K., Robert, I.D.: Associating the biomarkers of ocular blood flow with lamina cribrosa parameters in normotensive glaucoma suspects. comparison to glaucoma patients and healthy controls. PLoS One **16**(3), e0248851 (2021)
9. Krzyżanowska-Berkowska, P., Czajor, K., Syga, P., Iskander, D.R.: Lamina cribrosa depth and shape in glaucoma suspects. comparison to glaucoma patients and healthy controls. Curr. Eye Res. **44**(9), 1026–1033 (2019)
10. Kurysheva, N.I., Parshunina, O.A., Shatalova, E.O., Kiseleva, T.N., Lagutin, M.B., Fomin, A.V.: Value of structural and hemodynamic parameters for the early detection of primary open-angle glaucoma. Curr. Eye Res. **42**(3), 411–417 (2017)
11. Mwangi, B., Tian, T.S., Soares, J.C.: A review of feature reduction techniques in neuroimaging. Neuroinformatics **12**(2), 229–244 (2014)
12. Pedregosa, F., et al.: Scikit-learn: machine learning in python. J. Mach. Learn. Res. **12**, 2825–2830 (2011)
13. Santafe, G., Inza, I., Lozano, J.A.: Dealing with the evaluation of supervised classification algorithms. Artif. Intell. Rev. **44**(4), 467–508 (2015)
14. Tabachnick, B.G., Fidell, L.S.: Experimental designs using ANOVA. Thomson/Brooks/Cole Belmont, CA (2007)
15. Weinreb, R.N., Aung, T., Medeiros, F.A.: The pathophysiology and treatment of glaucoma: a review. Jama **311**(18), 1901–1911 (2014)
16. Wold, S., Esbensen, K., Geladi, P.: Principal component analysis. Chemom. Intell. Lab. Syst. **2**(1–3), 37–52 (1987)

Multi-objective Evolutionary Undersampling Algorithm for Imbalanced Data Classification

Szymon Wojciechowski$^{(\boxtimes)}$ ⓘ

Department of Systems and Computer Networks, Faculty of Electronics,
Wrocław University of Science and Technology,
Wybrzeże Wyspiańskiego 27, 50-370 Wrocław, Poland
szymon.wojciechowski@pwr.edu.pl

Abstract. The classification of imbalanced data is an important topic of research conducted in recent years. One of the commonly used techniques for dealing with this problem is undersampling, aiming to balance the training set by selecting the most important samples of the original set. The selection procedure proposed in this paper is using the multi-objective genetic algorithm NSGA-2, for a search of the optimal subset of the learning set. The paper presents a detailed description of the considered method. Moreover, the proposed algorithm has been compared with a selection of reference algorithms showing promising results.

Keywords: Imbalanced data · Multi-objective optimization · Evolutionary algorithms

1 Introduction

Machine learning applications has an impact on our lives at almost every level, from influencing sources of information, through robotics applications, to medical diagnostics. It is quite easy to observe newcoming challenges such as fake news detection [9]. Regardless of the application, many real-life classification problems are charged with the difficulty of imbalanced class distribution. This topic has been extensively studied, which led to many techniques for dealing with imbalanced data.

Those methods are used to reduce the bias towards the majority class resulting from the *prior* probability of a problem. Most commonly used classification methods are susceptible to any disproportions in the class distribution among training sets. Two main groups of solutions aiming to reduce negative influence coming from class imbalance are (*a*) data preprocessing methods – being the main topic of this paper, (*b*) algorithm-level solutions, and (*c*) hybrid methods, employing mostly ensemble approaches. Solutions described as *preprocessing methods* may be further divided into sample generation methods (*oversampling*), sample selection methods (*undersampling*) or methods combining these

© Springer Nature Switzerland AG 2021
M. Paszynski et al. (Eds.): ICCS 2021, LNCS 12744, pp. 118–127, 2021.
https://doi.org/10.1007/978-3-030-77967-2_11

two approaches. The purpose of all techniques is to reduce the imbalance ratio between the classes. Generation of new samples may rely on random duplication of existing patterns or the creation of new synthetic representations for the minority class, such as Synthetic Minority Oversampling Technique (SMOTE) and its modifications [7]. This group of neighborhood-based methods does not include information about the mutual distribution of classes. More sophisticated algorithms such as RBO [8] create a potential function based on the radial function density distribution of classes. As for undersampling techniques, various strategies can be distinguished: Random Undersampling (RUS), NearMiss, or Evolutionary Undersampling (EUS) [6].

The last of the algorithms mentioned above is particularly interesting because of the *genetic algorithm* (GA) optimization on which it is built on. When dealing with data imbalanced data, the evaluation of created models requires various metrics, especially those that are compromising trade-off between *Sensitivity* and *Specificity* or *Sensitivity* and *Precision*. The multi-objective optimization allows combining those in a way, to create the set of feasible solutions, which are placed on non-dominated *Pareto-optimal front*. To understand this idea, solutions must be considered as points in the solution space (M-dimensional, for M objective functions). At each step algorithm evaluates new solutions, leaving only those which are maximizing all cost functions simultaneously. One way to achieve this is by sorting solutions based on crowding distance [5]. As the final step, provided solutions are placed on the a convex curve, beyond which procedure was not able to find more acceptable solutions.

In most cases, this leads to a situation where user input on the importance of the metrics is required. This can be found as an advantage in real-life problems. However, on most of the benchmark datasets, it is impossible to obtain this information, which forces the selection of one of the solutions from *Pareto-optimal front*. Fortunately, *Multi-criteria Decision Making* (MCDM) techniques can be used for this task, providing a procedural way of selecting the best solution.

The main contribution of this work is the proposal of Multi-objective Evolutionary Undersampling algorithms (MEUS). Experimental studies presented in this paper are comparing proposed algorithm with other undersampling methods.

The rest of the paper is organized as follows. Section 2 provides a detailed description of the proposed algorithm, which consists of a formulation of a multi-objective optimization problem for sample selection and pseudocode presentation of the complete procedure. Section 3 presents the results comparing the proposed algorithm with reference methods. The article concludes in Sect. 4 where the observations and further directions for the research of this topic are discussed.

2   Algorithms

The proposed algorithm's idea is based on evolutionary undersampling guided for classification measures (EUSCM-MS) [6]. The optimization procedure will target the optimal subset of majority class samples from training set $\mathcal{TS}$ that will maximize the classification measure calculated for the validation set $\mathcal{VS}$. Both training and validation set are separated from learning set $\mathcal{LS}$. The learning set splitting method SPLIT can be treated as a parameter to the algorithm. In this study the stratified split with train-test ratio of 4:1 was used.

The optimization algorithm used by the method proposed in this article is multi-objective version of genetic algorithm - NSGA-2 [4]. The optimization problem that will be considered as a part of proposed method is presented in generalized formulation:

$$\begin{aligned} \underset{s}{\text{minimize}} \quad & f_i(s) \quad i = 1, 2, \ldots M \\ \text{subject to} \quad & g(s) = 0 \end{aligned} \tag{1}$$

It is assumed that the solution s is a binary word, for which each bit represents the inclusion of sample from $\mathcal{TS}$ into new training set $\mathcal{TS}'$. A feasible solution is balancing complete learning set by removing $N_{maj} - N_{min}$ majority samples out from it, which imposes a constraint

$$g(s) = N_{min} - \sum_{i=0}^{i=|s|} s_i. \tag{2}$$

To provide a complete formulation of the algorithm's optimization process it is required to define objective function. As it was already mentioned, the idea behind optimized under-sampling is to maximize the metric for the classifier model ψ trained on $\mathcal{TS}'$. Evaluation is based on the classifier predictions of $\mathcal{VS}$, which can be counted into four groups: TP and TN for correct predictions of minority and majority class respectively, FN and FP for error type I and II, respectively.

Considering the preference of each class, some basic metrics describing the ratio of the properly predicted samples to the number of class samples in the validation set can be defined:

True Positive Rate (TPR or $Sensitivity$)

$$TPR = \frac{TP}{TP + FN}, \tag{3}$$

True Negative Rate (TNR or $Specificity$)

$$TNR = \frac{TN}{TN + FP} \tag{4}$$

and Positive Predictive Value (PPV or $Precision$)

$$PPV = \frac{TP}{TP + FP}. \tag{5}$$

However, to give a fair trade-off between those values, geometric mean ($Gmean$) is often used as a aggregated metric for comparing classifier performance on imbalanced data. There are two known definitions for this metric, one including *Precision*:

$$Gmean = \sqrt{TPR \times PPV} \tag{6}$$

and second considering *Sensitivity*:

$$Gmean_s = \sqrt{TPR \times TNR}. \tag{7}$$

Let EVAULUATE($\mathcal{TS}, \mathcal{VS}$) denote a function which trains classifier using $\mathcal{TS}$ then tests the model using $\mathcal{VS}$ creating a vector of one or more metric values calculated according to aforementioned equations. Results obtained for training set subset $\mathcal{TS}'$ which is based on s will be used to define objective functions $f_i(s)$ of Eq. 1.

The output of proposed multi-objective optimization approach, is the set of solution placed on non-dominated *Pareto-optimal front* for defined objective functions. An example of solution space for *Specificity* and *Sensitivity* is presented in Fig. 1. The blue dashed curve marks the *Pareto-optimal front*.

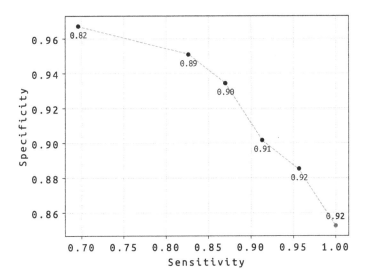

Fig. 1. Example of Pareto-optimal solutions for *Specificity* and *Sensitivity* provided from optimization of training set selection. (Color figure online)

It can be observed that the solutions are creating a convex curve, which tends to maximize both metrics. A $Gmean_s$ calculated for the solution is presented near each point. Two far right solutions are maximizing this measure, however only one of them can be selected for creating a final model. For that reason, as a final step of the procedure, *Simple Additive Weighting* (SAW) [12] algorithm is used to select a single solution. As for the discussed example, the choosen solution is marked with red. The complete procedure of MEUS is presented in Algorithm 1.

Algorithm 1. Pseudocode of the proposed MEUS method.

Input:
  $\mathcal{LS}$ – learning set
Symbols:
  $\mathcal{TS}$ – training set
  $\mathcal{TS}_{min}$ – $\mathcal{TS}$ subset including minority samples
  $\mathcal{TS}_{maj}$ – $\mathcal{TS}$ subset including majority samples
  $\mathcal{VS}$ – validation set
  S – set of solutions found by multi-objective optimization
  O – set of objective function values for each solution
  W – scores calculated by SAW
  s – selected final solution
Output:
  $\mathcal{TS}'$ – balanced subset of training set

1: $\mathcal{TS}, \mathcal{VS} \leftarrow \text{SPLIT}(\mathcal{LS})$
2: $S, O \leftarrow \text{NSGA-2}(\mathcal{TS}, \mathcal{VS})$
3: $W \leftarrow \text{SAW}(O)$
4: $s = S_i \Leftrightarrow i = argmax(H)$
5: $\mathcal{TS}' \leftarrow \mathcal{TS}_{min} \cup \{\mathcal{TS}_{majj} \Leftrightarrow s_j = 1\}$

---

3 Experimental Evaluation

This section describes the experimentation setup: implementation of classifiers and selection of datasets used for evaluation. The main motivation for this experimental study is to determine answers for the following questions:

– *RQ1:* Can multi-objective approach improve the method based on single-objective genetic algorithm?
– *RQ2:* Are the algorithms based on evolutionary optimization better than the reference methods?

All the algorithms used in the experiments were evaluated in an experimental environment implemented in Python. Base classifiers are implemented in *sklearn* package [11], undersampling methods in *imblearn* package [10] and optimization algorithms in *pymoo* package [3]. The implementation of the MEUS algorithm proposed in this paper and additional results of experiments are available as additional materials in on-line repository[1]. Datasets used for evaluation are collected from the imbalanced category of keel repository [1]. The number of samples per majority and minority class and Imbalance Ratio (IR) are presented in Table 1.

To provide reliable results the experiment protocol used for classifiers validation was 5×2 CV [2] and all the results were tested for statistical significance employing Wilcoxon singed-rank test with the significance level of $\rho < 0.05$. In a series of experiments, MEUS will be used with three combinations of objective functions:

---
[1] https://github.com/w4k2/moo-undersampling.

Table 1. Datasets summary

Dataset	N_{maj}	N_{min}	IR
glass1	138	76	1.82
wisconsin	444	239	1.86
pima	500	268	1.87
iris0	100	50	2.00
glass0	144	70	2.06
yeast1	1055	429	2.46
haberman	225	81	2.78
vehicle2	628	218	2.88
vehicle1	629	217	2.90
vehicle3	634	212	2.99
glass-0-1-2-3_vs_4-5-6	163	51	3.20
vehicle0	647	199	3.25
ecoli1	259	77	3.36
newthyroid2	180	35	5.14
new-thyroid1	180	35	5.14
ecoli2	284	52	5.46
segment0	1979	329	6.02
glass6	185	29	6.38
yeast3	1321	163	8.10
ecoli3	301	35	8.60
page-blocks0	4913	559	8.79

- MEUS-SS - *Sensitivity* and *Specificity*,
- MEUS-SP - *Sensitivity* and *Precision*,
- MEUS-SSP - *Sensitivity*, *Specificity* and *Precision*.

The proposed method will be compared with other reference approaches: *Random Undersampling* (RUS), *NearMiss* (NM), evolutionary undersampling guided for *Gmean* (EUS-GM), evolutionary undersampling guided for $Gmean_s$ (EUS-GMS) and finally with the original unbalanced learning set (None).

To provide a wide range of experiments, all undersampling methods were combined with three different base classifiers: *Classification and Regression Trees* (CART), *Naive Bayes Classifier* (GNB), *k-Nearest Neighbors Classifier* (KNN).

Both MEUS and EUS approaches are based on GA approach for optimization. In case of each preprocessing method, optimization algorithm is sharing the same parameters characteristic for evolutionary approach:

- **Population**
  The population is constructed from 50 individuals (solutions) at each epoch. The initial population is guaranteed to meet the balancing constraint.

– **Mutation**

The mutation is implemented as a bit-flip operation on two opposite bits in solution. This approach guarantees that the new individual will represent a feasible solution. All experiments were carried out with a mutation probability of 25%.

– **Crossing**

For the crossing procedure, two parents are paired randomly. First, a new individual is created from bit-wise `AND` operation between both parents. The created solution is not guaranteed to be feasible, but the sum of bits will always be lower or equal the designated value. To repair this individual, random '0' bits are flipped until a solution is feasible. All experiments were carried with the crossing probability of 25%.

– **Termination**

The algorithm is guaranteed to stop after 10000 iterations. However, it can finish faster if after 20 epochs best solution was not improved by the tolerance limit. The tolerance limit is higher for multi-objective optimization (0.0025) than for single-objective (0.0001) which is justified by algorithm-specific behavior.

4   Results

The comparison of preprocessing methods combined with CART base classifier is presented in Table 2. Results are provided as an average of $Gmean_s$ scores obtained in the experiment. Also, the presented table shows the result for the non-parametrical Wilcoxon paired test (fold-wise) listed below each result, denoting reference method to which the score was better and the null hypothesis was rejected.

It can be observed that in most cases, all methods achieve a better result than NM and the classifier trained on the dataset without preprocessing, with a exception of *glass-0-1-2-3_vs_4-5-6* where basic CART was better than other models, and significantly better then proposed MEUS-SS method. It should be also noticed, that in almost every case, optimization based methods were better than RUS except for *ecoli3* dataset. The most interesting aspect observed in the presented results is the relationship between single-objective optimization algorithms and multi-objective optimization algorithms. The only case in which it is possible to indicate a statistically significant advantage of multi-objective optimization is the set *glass1* for which MEUS-SP was better than EUS-GMS. In conclusion to *RQ1*, based on the considered example it cannot be unequivocally stated that multi-objective optimization gives better results than single-objective optimization.

It is particularly interesting that, contrary to what could have been expected, the precision-based optimization (EUS-GM, MEUS-SP, MEUS-SSP) in some cases gives better results for the $Gmean_s$ which does not include this metric. One possible explanation for this phenomenon is that the *Precision* grows with the decreasing number of predictions made for the minority class, thus favoring the

Table 2. $Gmean_s$ results for CART base classifier

Preprocessing dataset	NONE[1]	EUS-GMS[2]	EUS-GM[3]	MEUS-SS[4]	MEUS-SP[5]	MEUS-SSP[6]	NM[7]	RUS[8]
ecoli1	0.813	0.865 (1, 7)	0.852 (1, 7)	0.853 (1, 7)	0.868 (1, 7)	0.864 (1, 7)	0.798	0.867 (1, 7)
ecoli2	0.833 (7)	0.849 (7)	0.860 (7)	0.850 (7)	0.842 (7)	0.838 (7)	0.738	0.833 (7)
ecoli3	0.725 (7)	0.797 (1, 7)	0.800 (1, 7)	0.780 (7)	0.781 (1, 7)	0.815 (1, 7)	0.490	0.832 (1, 7)
glass-0-1-2-3_vs_4-5-6	0.901 (4)	0.867	0.880	0.874	0.876	0.870	0.899	0.899
glass0	0.776 (7)	0.781 (7)	0.784 (7)	0.792 (7)	0.790 (7)	0.804 (7)	0.707	0.760
glass1	0.712 (7)	0.692 (7)	0.720 (7)	0.742 (2, 7)	0.727 (7)	0.722 (7)	0.627	0.725 (7)
glass6	0.864 (8)	0.888 (8)	0.882	0.887 (8)	0.877	0.891 (8)	0.870	0.856
haberman	0.531	0.577 (1)	0.595 (1)	0.580 (1)	0.604 (1)	0.591 (1)	0.577 (1)	0.580
iris0	1.000	0.998	0.998	1.000	1.000	1.000	1.000	1.000
new-thyroid1	0.927	0.925	0.925	0.936	0.931	0.936	0.934	0.921
newthyroid2	0.896	0.931	0.933	0.956 (1, 7)	0.952 (1, 7)	0.956 (1, 7)	0.886	0.923
page-blocks0	0.896 (7)	0.936 (1, 7)	0.938 (1, 7, 8)	0.932 (1, 7)	0.935 (1, 7)	0.938 (1, 7, 8)	0.854	0.926 (1, 7)
pima	0.651	0.674	0.676	0.679 (1)	0.680 (1)	0.679 (1)	0.666	0.664
segment0	0.980 (7)	0.984 (7)	0.982 (7)	0.984 (7)	0.986 (7)	0.983 (7)	0.885	0.981 (7)
vehicle0	0.900	0.916	0.920	0.917	0.915	0.921	0.840	0.906
vehicle1	0.641	0.717 (1, 7)	0.717 (1, 7)	0.705 (1, 7)	0.713 (1, 7)	0.706 (1, 7)	0.634	0.703 (1, 7)
vehicle2	0.926	0.928	0.935 (8)	0.930	0.941 (1, 7, 8)	0.937	0.921	0.928
vehicle3	0.639	0.709 (1, 7)	0.715 (1, 7, 8)	0.708 (1, 7)	0.710 (1, 7)	0.702 (1, 7)	0.602	0.696 (1, 7)
wisconsin	0.938	0.949 (1, 7, 8)	0.947 (7, 8)	0.948 (7, 8)	0.947 (7)	0.947 (7, 8)	0.934	0.937
yeast1	0.606 (7)	0.661 (1, 7)	0.657 (1, 7)	0.651 (1, 7)	0.658 (1, 7)	0.667 (1, 4, 7)	0.567	0.643 (1, 7)

majority class's predictions. In that case, it should also increase the *Sensitivity*. It is worth noticing that only the training set is balanced. Therefore the imbalance still remains in the validation set. As a result, *Precision* may turn out to be a better optimization criterion than *Sensitivity*.

Table 3 shows the results for the mean rankings for all tested methods to provide a more general comparison of all methods. The Wilcoxon test results are provided below each of the presented values as in the case of the previously discussed results. It is not possible to clearly state that evolutionary undersampling can benefit from multi-objective optimization. However, it can be observed that in the case of CART and GNB, MEUS-SSP achieves the best results. An interesting relationship occurs in the case of KNN for $Gmean_s$, where EUS-GMS turned out to be the best preprocessing method. However, there is also a statistical relationship between a given method and MEUS-SSP. Also, the only metric for which the evolutionary method managed to achieve a better result than the RUS was *Precision*, which had no significant impact on *Gmean*.

Table 3. Ranking results

Preprocessing	None[1]	EUS-GMS[2]	EUS-GM[3]	MEUS-SS[4]	MEUS-SP[5]	MEUS-SSP[6]	NM[7]	RUS[8]
CART								
Sensitivity	1.275	5.125 1	5.450 1, 7	5.025 1	5.100 1	4.600 1	3.850	5.575 1, 7
Specificity	7.825 2, 3, 4, 5 5, 6, 7, 8	3.725 7, 8	4.650 2, 7, 8	4.500 7, 8	4.850 2, 7, 8	6.225 2, 3, 4, 5 7, 8	1.950	2.275 7
Precision	7.725 2, 3, 4, 5 6, 7, 8	3.600 7, 8	4.700 2, 7, 8	4.525 7, 8	4.975 2, 7, 8	6.325 2, 3, 4, 5 7, 8	1.925	2.225 7
G-mean_s	2.675 7	4.800 1, 7	5.400 1, 7	5.575 1, 7, 8	5.925 1, 7, 8	6.075 1, 2, 7, 8	2.025	3.525 1, 7
G-mean	4.275 7	4.250 7, 8	5.200 2, 7, 8	4.775 7, 8	5.575 2, 7, 8	6.375 1, 2, 4, 7 8	2.275	3.275 7
GNB								
Sensitivity	4.200 7	5.125 6, 7	5.200 6, 7	5.075 6, 7	5.575 6, 7	3.625 7	2.375	4.825 1, 7
Specificity	4.625	4.275 7, 8	4.325 8	4.800 7, 8	4.450 8	6.575 2, 3, 4, 5 7, 8	3.875	3.075
Precision	4.475 7, 8	4.575 7, 8	4.675 7, 8	4.950 7, 8	4.850 7, 8	6.675 1, 2, 3, 4 5, 7, 8	2.725	3.075 7
G-mean_s	3.275 7	5.375 1, 7, 8	4.825 1, 7, 8	6.100 1, 3, 7, 8	5.600 1, 7, 8	5.575 1, 7, 8	2.275	2.975 7
G-mean	4.025 7	4.925 1, 7, 8	5.375 1, 7, 8	4.950 1, 7, 8	5.400 1, 7, 8	5.325 1, 7, 8	2.575	3.425 7
KNN								
Sensitivity	1.025	6.375 1, 4, 5, 6 7	5.700 1, 4, 5, 7	4.375 1, 7	4.525 1, 7	4.700 1, 7	2.775 1	6.525 1, 3, 4, 5 6, 7
Specificity	7.725 2, 3, 4, 5 6, 7, 8	3.725 8	3.225 8	4.525 2, 3, 8	4.775 2, 3, 8	4.925 2, 3, 8	4.425	2.675
Precision	7.700 2, 3, 4, 5 6, 7, 8	4.075 8	3.525 8	4.500 3, 8	4.750 3, 8	4.975 3, 8	3.600	2.875
G-mean_s	2.725	5.825 1, 3, 4, 5 7	4.925 1, 7	4.900 1, 7	5.050 1, 7	4.975 1, 7	2.375	5.225 1, 7
G-mean	4.675 7	5.175 7	4.175 7	4.550 7	4.900 7	5.175 7	2.675	4.675 7

In the results for GNB as base classier a significant advantage of the proposed methods over the reference ones can be observed. It is also worth paying attention to the high MEUS-SSP result for *Specificity* and *Precision*, but very low *Sensitivity*, also noticeable for CART.

Based on the conducted experiments *RQ2* can be answered. The proposed undersampling methods based on evolutionary optimization are in many cases outperforming reference algorithms. Same as it was observed before, it cannot be unequivocally stated that multi-objective optimization gives better results than single-objective optimization because of statistical analysis, however it can be also observed that the proposed multi-objective approach is improving the ranking results.

5   Conclusions

A series of experiments showed that the proposed method achieves very good results for the metrics used to evaluate the classifiers. There is also a noticeable statistical relationship between the single-objective and multi-objective approaches.

One of the advantages of the proposed methods, which was not discussed in results evaluation, is that it creates many acceptable Pareto-optimal solutions. Under benchmark circumstances, it is necessary to choose one solution using the SAW method. However, it could be possible to obtain information from the expert in which optimized metric she/he is interested in, particularly in a real-life application. It should also be noticed that the metrics used in this study were limited to the basic ones. The expert can define his/her own set of metrics for which the solutions will be optimized or provide weights from MCDM.

In the perspective of further research, the parameters of GA should be optimized for best performance. Moreover, the definition of fitness function could be extended to static analysis of the set, e.g., some overlapping regions or statistical analysis metrics.

Acknowledgements. This work was supported by the Polish National Science Centre under the grant No. 2019/35/B/ST6/04442.

References

1. Alcala-Fdez, J., et al.: Keel data-mining software tool: data set repository, integration of algorithms and experimental analysis framework. J. Multiple Valued Logic Soft Comput. **17**, 255–287 (2010)
2. Alpaydin, E.: Combined 5×2 cv F test for comparing supervised classification learning algorithms. Neural Comput. **11**(8), 1885–1892 (1999)
3. Blank, J., Deb, K.: Pymoo: multi-objective optimization in Python. IEEE Access **8**, 89497–89509 (2020)
4. Deb, K., Pratap, A., Agarwal, S., Meyarivan, T.: A fast and elitist multiobjective genetic algorithm: NSGA-ii. IEEE Trans. Evol. Comput. **6**(2), 182–197 (2002)
5. Deb, K.: Multi-objective Optimisation Using Evolutionary Algorithms an Introduction, pp. 3–34. Springer, London (2011). https://doi.org/10.1007/978-0-85729-652-8_1
6. García, S., Herrera, F.: Evolutionary undersampling for classification with imbalanced datasets: proposals and taxonomy. Evol. Comput. **17**(3), 275–306 (2009)
7. Kovács, G.: An empirical comparison and evaluation of minority oversampling techniques on a large number of imbalanced datasets. Appl. Soft Comput. **83**, 105662 (2019)
8. Koziarski, M., Krawczyk, B., Woźniak, M.: Radial-based oversampling for noisy imbalanced data classification. Neurocomputing **343**, 19–33 (2019)
9. Ksieniewicz, P., Zyblewski, P., Choraś, M., Kozik, R., Giełczyk, A., Woźniak, M.: Fake news detection from data streams. In: 2020 International Joint Conference on Neural Networks (IJCNN), pp. 1–8 (2020)
10. Lemaître, G., Nogueira, F., Aridas, C.K.: Imbalanced-learn: a Python toolbox to tackle the curse of imbalanced datasets in machine learning. J. Mach. Learn. Res. **18**(17), 1–5 (2017)
11. Pedregosa, F., et al.: Scikit-learn: machine learning in Python. J. Mach. Learn. Res. **12**, 2825–2830 (2011)
12. Zanakis, S.H., Solomon, A., Wishart, N., Dublish, S.: Multi-attribute decision making: a simulation comparison of select methods. Eur. J. Oper. Res. **107**(3), 507–529 (1998)

Missing Value Imputation Method Using Separate Features Nearest Neighbors Algorithm

Tomasz Orczyk$^{(\boxtimes)}$ [ID], Rafał Doroz [ID], and Piotr Porwik [ID]

Faculty of Science and Technology, University of Silesia in Katowice, Bedzinska 39, 41-200 Sosnowiec, PL, Poland
tomasz.orczyk@us.edu.pl

Abstract. Missing value imputation is a problem often meet when working with medical and biometric data sets. Prior to working on these datasets, missing values have to be eliminated. It could be done by imputing estimated values. However, imputation should not bias data, nor alter the class balance. This paper presents an innovative approach to the problem of imputation of missing values in the training data for the classification. Method uses the *k-NN* classifier on a separate features to impute missing values. The unique approach used in this method allows using data from incomplete vectors to impute another incomplete vectors, unlike in conventional methods, where only complete vectors could be used in the imputation process. The paper also describes a test protocol, where the Cross Validation with a Set Substitution method is used as an evaluation tool for scoring missing value imputation methods.

Keywords: Missing data · Data imputation · *k-NN*

1 Introduction

Data classification is a crucial part of numerous systems, including a decision support systems, and an identity verification systems. Numerous classification algorithms require a complete data vectors on the input, while a real life collected data sets (especially in the fields of medicine and biometry [20]), often contain some amount of missing or undetermined values [11]. In order to work with such data we need to either, propose an algorithm that is able to utilize incomplete data vectors, or we need to impute missing values [24]. If we decide to impute, we must be able to determine the quality of such imputed data set. Evaluating the data imputation algorithm is not a trivial task. It may be done statistically or experimentally using a classifier and a benchmark data set in a cross validation run [22]. However if a data set will be imputed prior to the cross validation, there is a high risk of overfitting the imputed data, leading to a good, but false score. We can impute a training set, an input data, or both. Yet different approaches have to be used for imputing a training data set, and for imputing

© Springer Nature Switzerland AG 2021
M. Paszynski et al. (Eds.): ICCS 2021, LNCS 12744, pp. 128–141, 2021.
https://doi.org/10.1007/978-3-030-77967-2_12

an input data. Missing value imputation methods may be divided into a single and a multiple imputation methods [13]. By a single imputation method, we understand an algorithm that for one incomplete vector creates one complete vector. It is in contrast with a multiple imputation methods which, for each missing value, create a set of complete vectors with imputed values. They reflect values distribution in the whole set. Both imputation methods can be either a value duplication based (so-called hot deck), or a value generation based (e.g. mean substitution, regression). Value duplication based methods use existing feature values for imputation, which makes them particularly useful for text and enumerable type features. Value generation based methods, calculate the missing value from other existing values in the set, thus are usable for numeric and enumerable types of values, and especially for real values. The simplest possible method of a single imputation of missing values is using a class-clustered or a global mean (or median) value.

It is also a known method to use a k-NN classifier to impute missing values [23]. This can be done either as choosing an existing value from the remaining samples, or as an average of most similar (nearest neighboring) samples. The k-NN classifier can't operate on the incomplete vectors and thus, in conventional implementation, can only use a complete vectors as a reference ones. In this paper we propose data imputation method, which will be a single value imputation method, and will be using a modified k-NN classifier operating on separate features from the original feature space for approximating missing values in the training (labeled) data set. The single-dimensional subspaces allow to use as much data as it is possible, even values from other incomplete vectors. Its efficacy will be proven in statistical, and experimental manner.

2   Background Works

One of the most significant works on the topic of missing value imputation have been published in 1987 by Little and Rubin [18]. This paper defines basic types of missing values: Missing Completely At Random (MCAR), Missing At Random (MAR), and Not Missing At Random (NMAR), sometimes called Non-Ignorable (NI). These terms mainly apply to the missing data in surveys, but could be used for other data sets. The crucial point of this categorization is the reason for which the value is missing. Missingness of a MCAR data does not depend on its value, nor on any other observed values, missingness of a MAR data does not depend on the missing value itself, but is relied to other observed values, and missingess of NMAR/NI data depends on the missing value itself (thus it is significant, i.e. denial of answer is also an answer). Little and Rubin, in their original paper, also proposed some basic methods for dealing with missing values, like mean/mode substitution, and regression. In recent years, more imputations methods have been introduced, often using machine learning and genetic based algorithms. Most authors concentrate on Missing Completely At Random data, where 30% or less values are missing. A comprehensive review on recent advances in the field of missing value imputation can be found in [7, 8, 14, 17]. From these papers it can be seen, that there is no single, defined test protocol for imputation algorithms. Different authors, use different methods to compare proposed algorithm with other, so it is impossible to directly compare with published results. One common thing is, they may be divided into two groups: direct (statistical) evaluation of the imputed data sets, and a wrapped evaluation of a classification results on the imputed data. According to the [17] most authors use direct evaluation methods, and not use cross validation for missing values simulation.

3   Method Description

The proposed method is a single imputation method for MCAR values, based on a regression variant of the k-NN classifier, applied independently to each feature, and it will be called k-NNI for short. It consists of the following stages:

1. First we need to define a data structures used by the algorithm. Let a single data vector be defined as $\mathbf{f} = [f_1, \ldots, f_F, c]$. Data vector consists of some features f_n, $n = 1, \ldots, F$, and the class label c. In practice, we have a collection of many vectors $\mathbf{f}$. Let there are P such vectors, then set of vectors $\mathbf{f}_p$, $p = 1, \ldots, P$ can be presented in the matrix form. This matrix contains all input data vectors and class labels (Fig. 1 STAGE 1):

$$\mathbf{O} = \begin{bmatrix} \mathbf{f}_1 \\ \vdots \\ \mathbf{f}_P \end{bmatrix} = \begin{bmatrix} f_{1,1} & \cdots & f_{F,1} & c_1 \\ \vdots & \cdots & \vdots & \vdots \\ f_{1,P} & \cdots & f_{F,P} & c_P \end{bmatrix}, \tag{1}$$

where $f_{n,p}$ denotes the n-th feature of the p-th vector.

2. Taking into account the class labels $c_1, \ldots, c_P$, matrix $\mathbf{O}$ can be divided into several matrices $\mathbf{O}^c$. Each matrix $\mathbf{O}^c$ contains vectors that belong only to one class (Fig. 1 STAGE 2).

$$\mathbf{O}^c = \{\mathbf{f}_n \in \mathbf{O} : c_n = c\}. \tag{2}$$

3. Each matrix $\mathbf{O}^c$ is scanned for a missing values, row by row, and column by column. For each missing value, a column within which a missing value is found, becomes a target variable column T (Fig. 1 STAGE 3).
4. Row with a missing value becomes a query vector $\mathbf{V}$ (Fig. 1 STAGE 4):

$$\mathbf{V} = \begin{bmatrix} f_1 & \cdots & f_F \end{bmatrix}. \tag{3}$$

Remaining rows form an auxiliary matrix $\mathbf{X}^{c,n}$. This matrix contains R data vectors, each described by the $F - 1$ features:

$$\mathbf{X}^{c,n} = \begin{bmatrix} f_{1,1} \cdots f_{i-1,1} & f_{i+1,1} \cdots f_{F,1} & T_1 \\ \vdots & \vdots & \vdots \\ f_{1,R} \cdots f_{i-1,R} & f_{i+1,R} \cdots f_{F,R} & T_R \end{bmatrix}, \tag{4}$$

5. Each pair of query vector $\mathbf{V}$, and matrix $\mathbf{X}^{c,n}$ is used as an input data for k-NN. Instead of using complete vectors as a classifier input. A regression variant of the k-NN is applied to each feature - target value pair ($\{f_n, T\}$). This produces the predicted target values according to each feature separately. The $F - 1$ values obtained from each feature are then averaged, and resulting value Ψ is a prediction of the missing value. This process will be explained in details later in this chapter.

6. The predicted value is imputed to a copy of the original data set $\mathbf{Z}$, which will be called the imputed data set. This ensures that the values imputed in one iteration will not be used as a reference values in consequent iterations. The illustration of the missing value calculation using the k-NNI regression process is shown on Fig. 2.

The k-NN classifiers used within the k-NNI are using an Euclidean distance metric. On a 1D data, this metric effectively becomes an absolute difference of values.

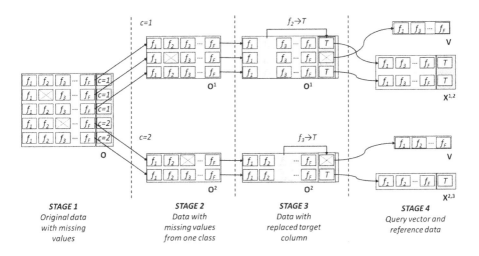

Fig. 1. Feature projection for k-NNI imputation (row indices omitted).

The partial decision of the classifier (ψ_n) is computed as follows:

$$\psi_n(\mathbf{V}, n) = \frac{\sum_{i=1}^{k}(\hat{T}_i)}{k},\tag{5}$$

where:
$\hat{T}_i$ – i-th target value from the reference set sorted in ascending order according to the distance of n-th element from the classified vector ($\mathbf{V}$) to the feature from n-th column of the matrix ($\mathbf{X}^{c,n}$);
k – a number of the analyzed nearest neighbors.

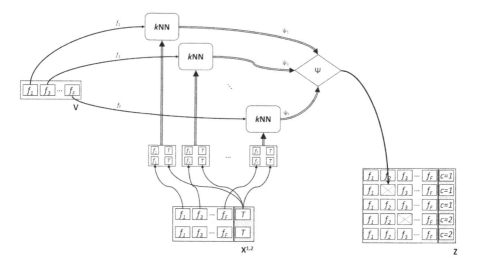

Fig. 2. Missing value calculation in the *k-NNI* (single missing value, row indices omitted).

The final decision of the classifier is calculated as an arithmetic mean of a partial decisions over all features:

$$\Psi(\mathbf{V}) = \frac{\sum_{n=1}^{F}(\psi_n(\mathbf{V}))}{F}. \tag{6}$$

The returned target value is a mean of the mean of all nearest neighbors target values from all features existing in the classified vector.

This algorithm has no learning phase, so adding new reference samples is made at almost no cost. It classifies each feature separately, so it does not require a data normalization, and it does not require a removal nor any special handling of the incomplete data vectors. Its classification speed may be improved by presorting each feature's values in ascending (or descending) order, to find the nearest neighbors even faster.

Source code in *C#* for algorithm described above is freely available at GitHub: https://github.com/torczyk/SFkNNI.

4   Experiment Description

The purpose of this experiment is to evaluate the quality of the imputed data. By the imputation quality we will understand how well the imputed data resembles the original data. For this purpose a set of six benchmark databases, containing only complete vectors of real data, have been degraded by removing values of a random features in a random data vectors. Multiple sets of such *degraded data sets* have been prepared, with a different placement and different amount of missing values. These *degraded data sets* have been imputed using the proposed method, and compared with an arbitrary chosen, state of the art, single imputation method. In a real life, missing values can occur in both, the labeled reference (training) data, as well as in the classified data, but in this paper we will limit imputation to the reference data set.

4.1   Data Characteristics

For the purpose of the experiment, a six benchmark databases form the UCI Repository [3] has been used: Wine Data Set (WINE) [1], Breast Cancer Wisconsin (WDBC) [19], Cardiotocography Data Set (CTG) [6], Diabetic Retinopathy Debrecen/Messidor (MESS) [2], Mesothelioma's disease (MESO) [12], and Cryotherapy (CRYO) [15]. Basic characteristics of the data sets, together with a reference overall classification accuracy (i.e. accuracy obtained on a leave-one-out cross validation using Naive Bayes classifier), has been presented in the Table 1. All these data sets contain no missing values, and will be called the *original data sets*.

Table 1. The reference data sets characteristics.

DB name	Num. of vectors	Num. of features	Num. of classes	Ref. accuracy
WINE	178	13	3	0.972
WDBC	569	32	2	0.935
CTG	90	7	3	0.820
MESS	1151	20	2	0.566
MESO	324	34	2	0.978
CRYO	90	7	2	0.856

From each of the *original data sets*, *degraded data sets* have been created, by removing values at random positions (the class labels have been left unaltered). These sets were generated for 7 thresholds of the randomly missing values: 5%, 10%, 15%, 20%, 25%, 30%, and 35%. For each threshold level 10 *degraded data sets* have been generated, with a different missing values placement, giving a total of $6 \times 70 = 420$ *degraded data sets*. These *degraded data sets* have been imputed by means of five algorithms: *k-NNI* with a parameter $k = 3$, *Predictive Mean Matching (PMM)* as implemented in the *MICE package* in *R environment*, with a parameter $k = 1$ (for single imputation), Mean imputation, Median imputation, and a *k-NN* imputation as implemented in the *VIM package* in *R environment*, with default parameters, creating *imputed data sets* for further experiments.

4.2   Test Protocol

Proposed imputation algorithm will be evaluated in two ways. In the first step, the quality of the imputed data will be evaluated statistically. The quality score for this stage will be the Normalized Root Mean Square Error [21] for the *imputed data sets* against the *original data set*, defined as:

$$NRMS = \frac{\|\mathbf{X}_{imp} - \mathbf{X}_{ori}\|_F}{\|\mathbf{X}_{ori}\|_F}, \qquad (7)$$

where:
$\mathbf{X}_{imp}$ – imputed data set,
$\mathbf{X}_{ori}$ – original data set,
$\|\ \|_F$ – Frobenius norm.
This measure shows how much the imputed data set differs from the original data set. It can have values from 0 to 1, and lower is better.

In the next step the *imputed data set* will be used to perform an actual classification, using an arbitrary chosen classifier – Naive Bayes [16]. To avoid using imputed data for both training and testing the classifier [9], a modification of the regular leave-one-out cross validation [10] method has been used. Vectors from the imputed data sets will be used solely for training the classifier, while for testing classifier, the corresponding vectors from the original data set will be applied. As for each database, all the *degraded* and the *imputed data sets* are derived from the same *original data set*, order of data vectors is maintained within these data sets. Thus it is possible to draw only a vector identifiers instead of the actual data vectors in the Leave One Out Cross Validation. Using these identifiers, a corresponding data vectors are taken from two data sets - the *complete* for testing vectors, and the *imputed* for training vectors.

Classification accuracy will be tested by means of the Overall Classification Accuracy, defined as follows:

$$OCA = \frac{TP}{card(\mathbf{X})}, \qquad (8)$$

where:
TP – number of correctly classified samples,
$card(\mathbf{X})$ – cardinality of data set $\mathbf{X}$.
This is the measure that shows what percentage of classified samples was classified correctly. It takes values from 0 to 1, and higher value is better.

4.3   Test Results

Normalized Root Mean Square Deviation. The NRMS of imputed values in 6 different databases, by means of the proposed method and 4 other imputation methods has been presented in Tables 2–7.

Comparison of average NRMS of the imputed values by means of all tested imputation methods has been shown in Table 8. As can be seen in Table 8, the NRMS results look promising, but what we really care about is usefulness of these data sets in the classification process.

Table 2. NRMS of the imputed *WINE* data set.

Imput. method	Missing values in dataset						
	5%	10%	15%	20%	25%	30%	35%
k-NNI	**0.040**	**0.067**	**0.077**	**0.088**	**0.103**	**0.116**	**0.116**
PMM	0.059	0.085	0.116	0.128	0.144	0.171	0.181
Mean	0.081	0.119	0.139	0.169	0.203	0.207	0.219
Median	0.084	0.120	0.145	0.174	0.211	0.211	0.223
k-NN	0.044	0.068	0.084	0.094	0.109	0.124	0.131

Table 3. NRMS of the imputed *WDBC* data set.

Imput. method	Missing values in dataset						
	5%	10%	15%	20%	25%	30%	35%
k-NNI	0.058	0.083	0.104	0.126	0.130	0.153	0.176
PMM	**0.013**	**0.025**	**0.031**	**0.042**	**0.046**	**0.060**	**0.079**
Mean	0.104	0.160	0.190	0.234	0.250	0.287	0.323
Median	0.109	0.168	0.200	0.247	0.263	0.302	0.340
k-NN	0.039	0.066	0.085	0.118	0.133	0.167	0.197

Table 4. NRMS of the imputed *CTG* data set.

DB name	Missing values in dataset						
	5%	10%	15%	20%	25%	30%	35%
k-NNI	0.034	0.048	0.059	0.067	0.075	0.083	0.090
PMM	0.025	0.039	**0.048**	**0.058**	**0.067**	**0.076**	**0.085**
Mean	0.046	0.065	0.080	0.091	0.102	0.112	0.121
Median	0.047	0.067	0.082	0.093	0.104	0.115	0.124
k-NN	**0.021**	**0.036**	0.050	0.061	0.073	0.085	0.097

Table 5. NRMS of the imputed *MESS* data set.

Imput. method	Missing values in dataset						
	5%	10%	15%	20%	25%	30%	35%
k-NNI	0.115	0.154	0.190	0.219	0.246	0.268	0.295
PMM	**0.079**	**0.114**	**0.141**	**0.163**	**0.186**	**0.209**	**0.226**
Mean	0.144	0.196	0.241	0.277	0.310	0.336	0.369
Median	0.149	0.204	0.249	0.284	0.321	0.347	0.382
k-NN	0.089	0.139	0.186	0.228	0.269	0.300	0.343

Table 6. NRMS of the imputed *MESO* data set.

Imput. method	Missing values in dataset						
	5%	10%	15%	20%	25%	30%	35%
k-NNI	**0.078**	0.111	0.136	**0.161**	0.175	0.190	0.209
PMM	0.105	0.166	0.192	0.227	0.248	0.271	0.296
Mean	0.077	**0.110**	**0.135**	**0.161**	**0.174**	**0.189**	**0.209**
Median	0.077	0.111	0.137	0.166	0.177	0.193	0.216
k-NN	0.087	0.129	0.154	0.181	0.207	0.220	0.246

Table 7. NRMS of the imputed *CRYO* data set.

Imput. method	Missing values in dataset						
	5%	10%	15%	20%	25%	30%	35%
k-NNI	**0.122**	0.196	0.193	**0.244**	**0.371**	**0.401**	0.431
PMM	0.207	0.334	0.388	0.658	0.616	0.647	0.661
Mean	0.136	0.210	0.198	0.280	0.392	0.438	0.422
Median	0.137	0.205	0.188	0.271	0.382	0.435	**0.410**
k-NN	0.127	**0.193**	**0.190**	0.272	0.387	0.426	0.521

Table 8. Average NRMS of the imputed data sets (lower is better).

Imput. method	Dataset name					
	WINE	WDBC	CTG	MESS	MESO	CRYO
k-NNI	**0.087**	0.119	0.065	0.212	**0.151**	**0.280**
PMM	0.126	**0.042**	**0.057**	**0.160**	0.215	0.502
Mean	0.162	0.221	0.088	0.268	**0.151**	0.297
Median	0.167	0.233	0.090	0.277	0.154	0.290
k-NN	0.094	0.115	0.060	0.222	0.175	0.302

Overall Classification Accuracy. In the next experiment, the quality of the imputed data sets have been evaluated by means of a *leave-one-out cross validation* using a *Naive Bayes* classifier in the *KNIME environment*. Tables 9–14 show the *Overall Classification Accuracy* on the *imputed data sets* used as the training data sets. Test vectors were taken from the *original data set*.

For an easier comparison, an average overall classification accuracy has been presented in Table 15. Results confirm good quality of imputed data, and according to OCA values, but require further analysis.

Table 9. Overall classification accuracy (±std. deviation) on the *WINE* database.

Imput. method	Missing values in dataset						
	5%	10%	15%	20%	25%	30%	35%
k-NNI	**0.976**	**0.975**	**0.978**	**0.976**	**0.976**	**0.976**	**0.974**
	(±0.004)	(±0.007)	(±0.006)	(±0.006)	(±0.004)	(±0.006)	(±0.005)
PMM	0.974	0.97	0.971	0.97	0.967	0.967	0.971
	(±0.005)	(±0.007)	(±0.006)	(±0.009)	(±0.007)	(±0.012)	(±0.013)
Mean	0.972	0.972	0.97	0.969	0.966	0.962	0.955
	(±0.003)	(±0.003)	(±0.005)	(±0.012)	(±0.008)	(±0.007)	(±0.013)
Median	0.971	0.971	0.97	0.964	0.956	0.95	0.944
	(±0.002)	(±0.003)	(±0.005)	(±0.008)	(±0.006)	(±0.008)	(±0.014)
k-NN	0.974	0.972	0.974	0.971	0.972	0.969	0.966
	(±0.005)	(±0.006)	(±0.007)	(±0.009)	(±0.005)	(±0.004)	(±0.01)

Table 10. Overall classification accuracy (±std. deviation) on the *WDBC* database.

Imput. method	Missing values in dataset						
	5%	10%	15%	20%	25%	30%	35%
k-NNI	0.93	**0.933**	0.93	**0.931**	**0.932**	**0.931**	0.932
	(±0.001)	(±0.002)	(±0.002)	(±0.002)	(±0.002)	(±0.002)	(±0.003)
PMM	**0.932**	**0.933**	**0.931**	0.93	0.931	**0.931**	0.929
	(±0.001)	(±0.003)	(±0.002)	(±0.003)	(±0.002)	(±0.003)	(±0.002)
Mean	0.929	0.93	0.93	0.93	0.93	0.928	0.926
	(±0.001)	(±0.001)	(±0.001)	(±0.002)	(±0.002)	(±0.001)	(±0.002)
Median	0.927	0.928	0.928	0.925	0.924	0.917	0.917
	(±0.001)	(±0.001)	(±0.002)	(±0.002)	(±0.003)	(±0.003)	(±0.005)
k-NN	0.93	0.931	0.93	0.93	**0.932**	**0.931**	**0.933**
	(±0.001)	(±0.002)	(±0.002)	(±0.001)	(±0.002)	(±0.002)	(±0.002)

Table 11. Overall classification accuracy (±std. deviation) on the *CTG* database.

Imput. method	Missing values in dataset						
	5%	10%	15%	20%	25%	30%	35%
k-NNI	0.812	0.807	0.811	**0.814**	**0.812**	**0.816**	**0.815**
	(±0.002)	(±0.002)	(±0.008)	(±0.01)	(±0.01)	(±0.009)	(±0.011)
PMM	**0.817**	**0.816**	**0.816**	0.81	0.805	0.808	0.804
	(±0.002)	(±0.004)	(±0.003)	(±0.004)	(±0.005)	(±0.006)	(±0.005)
Mean	0.813	0.808	0.804	0.79	0.782	0.774	0.766
	(±0.001)	(±0.003)	(±0.003)	(±0.005)	(±0.006)	(±0.006)	(±0.006)
Median	0.814	0.81	0.807	0.798	0.791	0.787	0.778
	(±0.002)	(±0.002)	(±0.002)	(±0.004)	(±0.004)	(±0.007)	(±0.004)
k-NN	0.819	**0.816**	0.814	0.807	0.801	0.793	0.779
	(±0.001)	(±0.002)	(±0.003)	(±0.005)	(±0.006)	(±0.009)	(±0.005)

Table 12. Overall classification accuracy (±std. deviation) on the *MESS* database.

Imput. method	Missing values in dataset						
	5%	10%	15%	20%	25%	30%	35%
k-NNI	**0.566**	**0.57**	**0.571**	**0.577**	**0.583**	**0.586**	**0.591**
	(±0.001)	(±0.003)	(±0.002)	(±0.004)	(±0.004)	(±0.005)	(±0.004)
PMM	**0.566**	0.566	0.566	0.565	0.564	0.565	0.564
	(±0.001)	(±0.002)	(±0.001)	(±0.003)	(±0.004)	(±0.004)	(±0.003)
Mean	0.565	0.566	0.565	0.569	0.571	0.575	0.579
	(±0.001)	(±0.001)	(±0.001)	(±0.002)	(±0.004)	(±0.005)	(±0.004)
Median	**0.566**	0.569	0.569	0.575	0.578	0.581	0.584
	(±0.001)	(±0.003)	(±0.002)	(±0.002)	(±0.005)	(±0.004)	(±0.003)
k-NN	**0.566**	0.568	0.568	0.572	0.575	0.575	0.577
	(±0.001)	(±0.002)	(±0.002)	(±0.002)	(±0.003)	(±0.003)	(±0.002)

Table 13. Overall classification accuracy (±std. deviation) on the *MESO* database.

Imput. method	Missing values in dataset						
	5%	10%	15%	20%	25%	30%	35%
k-NNI	0.981	0.978	0.975	0.973	0.968	0.968	0.961
	(±0.001)	(±0.004)	(±0.005)	(±0.006)	(±0.005)	(±0.008)	(±0.007)
PMM	0.98	**0.98**	**0.979**	**0.98**	**0.977**	**0.977**	**0.977**
	(±0.002)	(±0.004)	(±0.007)	(±0.006)	(±0.009)	(±0.008)	(±0.006)
Mean	**0.994**	0.979	0.964	0.953	0.94	0.931	0.913
	(±0.001)	(±0.009)	(±0.01)	(±0.007)	(±0.01)	(±0.012)	(±0.018)
Median	0.974	0.965	0.955	0.939	0.904	0.876	0.852
	(±0.005)	(±0.006)	(±0.01)	(±0.024)	(±0.032)	(±0.054)	(±0.056)
k-NN	0.979	0.972	0.969	0.96	0.944	0.926	0.898
	(±0.002)	(±0.004)	(±0.006)	(±0.012)	(±0.021)	(±0.029)	(±0.039)

Table 14. Overall classification accuracy on the *CRYO* database.

Imput. method	Missing values in dataset (std. deviation)						
	5%	10%	15%	20%	25%	30%	35%
k-NNI	0.847	0.846	0.842	0.838	0.837	0.84	**0.836**
	(±0.007)	(±0.008)	(±0.016)	(±0.017)	(±0.02)	(±0.014)	(±0.028)
PMM	0.853	0.852	**0.852**	0.85	0.841	0.846	0.83
	(±0.007)	(±0.009)	(±0.009)	(±0.013)	(±0.012)	(±0.018)	(±0.034)
Mean	0.838	0.831	0.831	0.827	0.821	0.823	0.814
	(±0.006)	(±0.016)	(±0.016)	(±0.019)	(±0.018)	(±0.019)	(±0.023)
Median	0.853	0.852	0.848	0.853	0.84	0.84	0.818
	(±0.005)	(±0.012)	(±0.009)	(±0.009)	(±0.014)	(±0.016)	(±0.032)
k-NN	**0.858**	**0.856**	0.851	**0.852**	**0.854**	**0.848**	0.831
	(±0.005)	(±0.007)	(±0.008)	(±0.013)	(±0.017)	(±0.015)	(±0.024)

Table 15. Average classification accuracy of the imputed values (higher is better).

Imput. method	Dataset name					
	WINE	WDBC	CTG	MESS	MESO	CRYO
k-NNI	**0.976**	**0.931**	**0.812**	**0.578**	0.972	0.841
PMM	0.970	**0.931**	0.811	0.565	**0.979**	0.846
Mean	0.967	0.929	0.791	0.570	0.953	0.827
Median	0.961	0.924	0.798	0.575	0.923	0.843
k-NN	0.971	**0.931**	0.804	0.572	0.950	**0.850**

4.4 Summary

Both NRMS and OCA are normalized values, thus they may be compared over different datasets, using the rank method.

From observation of raw NRMS and accuracy results it may not be clear, which method outperforms rest. To make results more comparable, ranks have been calculated for all tests. As can be seen, for both measures, the proposed *k-NNI* method has the lowest rank (Table 16), and thus can be seen as superior to other methods in the comparison.

Table 16. Average rank of imputation method according to NRMS and OCA (lower is better).

Method	NRMS	OCA
k-NNI	**2.1**	**2.0**
PMM	2.7	2.5
Mean	3.5	4.1
Median	4.2	3.9
k-NN	2.6	2.5

Results of the Overall Classification Accuracy can be also compared in pairs using a Bayesian Signed-Rank Test [4] in the form of ROPE [5] diagrams (Fig. 3).

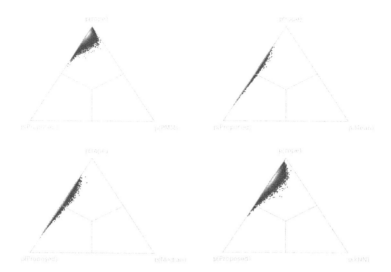

Fig. 3. ROPE diagrams for proposed method vs. PMM, Mean, Median, and kNN imputation.

5   Conclusions

We may conclude that in the aspect of both the NRMS and the overall classification accuracy, the proposed *k-NNI* is superior to all compared methods, including the state of the art *PMM*. This was confirmed using ROPE diagram, yet they show that the advantage of the proposed method over the PMM method is not statistically meaningful. The detailed analysis shows, that the *k-NNI* gains more advantage over other methods on data sets containing a significant number of the missing values ($\geq 20\%$). In conjunction with the speed, and ease of implementation of the proposed method, this should be considered as a satisfactory result. Additionally, the separate processing of each feature allows parallel processing of multiple features at the same time. It also eliminates the need for a normalization or any other form of a scaling of a data. The empirical results prove, that the imputed data does not overfit the classifier, i.e. the classification accuracy of the degraded, and then imputed data set, is not better than of the original data set. It is also stable in the function of a number of missing values.

References

1. Aeberhard, S., Coomans, D., De Vel, O.: Comparison of classifiers in high dimensional settings. Department of Mathematics and Statistics, James Cook University of North Queensland, Australia, Technical report 92-02 (1992)
2. Antal, B., Hajdu, A.: An ensemble-based system for automatic screening of diabetic retinopathy. Knowl. Based Sys. **60**, 20–27 (2014)
3. Asuncion, A., Newman, D.: UCI machine learning repository (2007)
4. Benavoli, A., Corani, G., Demsar, J., Zaffalon, M.: Time for a change: a tutorial for comparing multiple classifiers through Bayesian analysis. arXiv e-prints (June 2016). https://arxiv.org/abs/1606.04316

5. Benavoli, A., Mangili, F., Corani, G., Zaffalon, M., Ruggeri, F.: A Bayesian Wilcoxon signed-rank test based on the Dirichlet process. In: Proceedings of the 30th International Conference on Machine Learning, ICML 2014, pp. 1–9 (2014). http://www.idsia.ch/~alessio/benavoli2014a.pdf
6. Ayres-de Campos, D., Bernardes, J., Garrido, A., Marques-de Sa, J., Pereira-Leite, L.: SisPorto 2.0: a program for automated analysis of cardiotocograms. J. Matern. Fetal Med. 9(5), 311–318 (2000)
7. Chhabra, G., Vashisht, V., Ranjan, J.: A review on missing data value estimation using imputation algorithm. J. Adv. Res. Dyn. Control Sys. 11, 312–318 (2019)
8. Donders, A.R.T., Van Der Heijden, G.J., Stijnen, T., Moons, K.G.: A gentle introduction to imputation of missing values. J. Clin. Epidemiol. 59(10), 1087–1091 (2006)
9. Dong, Y., Peng, C.Y.J.: Principled missing data methods for researchers. Springerplus 2(1), 222 (2013)
10. Efron, B.: The Jackknife, the Bootstrap, and Other Resampling Plans, vol. 38 (1982)
11. Enders, C.K.: Applied Missing Data Analysis. Guilford Press (2010)
12. Er, O., Tanrikulu, A.C., Abakay, A., Temurtas, F.: An approach based on probabilistic neural network for diagnosis of Mesothelioma's disease. Comput. Electr. Eng. 38(1), 75–81 (2012)
13. Gelman, A., Hill, J.: Data Analysis Using Regression and Multilevel/Hierarchical Models. Cambridge University Press, Cambridge (2006)
14. Hox, J.J.: A review of current software for handling missing data. Kwantitatieve methoden 20, 123–138 (1999)
15. Khozeimeh, F., Alizadehsani, R., Roshanzamir, M., Khosravi, A., Layegh, P., Nahavandi, S.: An expert system for selecting wart treatment method. Comput. Biol. Med. 81, 167–175 (2017)
16. Lewis, D.D.: Naive (Bayes) at forty: the independence assumption in information retrieval. In: Nédellec, C., Rouveirol, C. (eds.) ECML 1998. LNCS, vol. 1398, pp. 4–15. Springer, Heidelberg (1998). https://doi.org/10.1007/BFb0026666
17. Lin, W.-C., Tsai, C.-F.: Missing value imputation: a review and analysis of the literature (2006–2017). Artif. Intell. Rev. 53(2), 1487–1509 (2019). https://doi.org/10.1007/s10462-019-09709-4
18. Little, R., Rubin, D.: Statistical Analysis with Missing Data. Wiley Series in Probability and Statistics. Wiley (1987). https://books.google.pl/books?id=w40QAQAAIAAJ
19. Mangasarian, O.L., Street, W.N., Wolberg, W.H.: Breast cancer diagnosis and prognosis via linear programming. Oper. Res. 43(4), 570–577 (1995)
20. Porwik, P., Doroz, R., Wrobel, K.: A new signature similarity measure. In: 2009 World Congress on Nature and Biologically Inspired Computing (NaBIC), pp. 1022–1027. IEEE (2009)
21. Razavi-Far, R., Cheng, B., Saif, M., Ahmadi, M.: Similarity-learning information-fusion schemes for missing data imputation. Knowl. Based Sys. 187, 104805 (2020)
22. Schmitt, P., Mandel, J., Guedj, M.: A comparison of six methods for missing data imputation. J. Biom. Biostat. 6(1), 1 (2015)
23. Troyanskaya, O., et al.: Missing value estimation methods for DNA microarrays. Bioinformatics 17(6), 520–525 (2001)
24. Zhang, Z.: Missing data imputation: focusing on single imputation. Ann. Transl. Med. 4(1), 9 (2016)

On Validity of Extreme Value Theory-Based Parametric Models for Out-of-Distribution Detection

Tomasz Walkowiak⬤, Kamil Szyc⬤, and Henryk Maciejewski[✉]⬤

Wroclaw University of Science and Technology, Wrocław, Poland
{tomasz.walkowiak,kamil.szyc,henryk.maciejewski}@pwr.edu.pl

Abstract. Open-set classifiers need to be able to recognize inputs that are unlike the training or known data. As this problem, known as out-of-distribution (OoD) detection, is non-trivial, a number of methods to do this have been proposed. These methods are mostly heuristic, with no clear consensus in the literature as to which should be used in specific OoD detection tasks. In this work, we focus on a recently proposed, yet popular, Extreme Value Machine (EVM) algorithm. The method is unique as it uses parametric models of class inclusion, justified by the Extreme Value Theory, and as such is deemed superior to heuristic methods. However, we demonstrated a number of open-set text and image recognition tasks, in which the EVM was outperformed by simple heuristics. We explain this by showing that the parametric (Weibull) model in EVM is not appropriate in many real datasets, which is due to unsatisfied assumptions of the Extreme Value Theorem. Hence we argue that the EVM should be considered another heuristic method.

Keywords: Open-set classification · Out-of-distribution detection · Extreme Value Machine · Extreme Value Theory

1 Introduction

Machine learning systems deployed for real-world recognition tasks often have to deal with data that come from categories unseen during training. This occurs especially in image or text recognition, where it is usually infeasible to collect training examples that correspond to all categories which can be encountered at prediction time. Hence it is important that classifiers can detect such examples as unrecognized and not silently assign them to one of the known classes. However, most state-of-the-art models for image recognition operate as closed-set classifiers, i.e., they tend to assign any example to some of the known classes. An illustration of such behavior by the well-known ResNet model is shown in Fig. 1. Such misclassification errors limit adoption of closed-set models in problems where new categories emerge over time (incremental learning problems) or can lead to accidents in safety-critical computer vision applications, which is a crucial concern in AI Safety [1].

© Springer Nature Switzerland AG 2021
M. Paszynski et al. (Eds.): ICCS 2021, LNCS 12744, pp. 142–155, 2021.
https://doi.org/10.1007/978-3-030-77967-2_13

Ligature   Jellyfish (99%)   Highway   Dam (99%)

Fig. 1. Images of unknown class (Ligature, Highway, not available in training data) recognized by the ResNet-50 model as a known class (Ligature recognized as Jellyfish, Highway recognized as Dam). Examples from [12].

To deal with this problem, several methods have been proposed to recognize when inputs to classifiers are unlike the training examples. In different studies, such inputs are referred to as anomalous, outliers, or out-of-distribution (OoD) examples with regard to the training data. Classifiers that incorporate such detection methods are known as open-set classifiers. A recent comprehensive survey of open-set recognition methods is given in [6]. Closed-set classifiers fail to reject OoD examples, as they approximate posterior probabilities $P(c_i|x)$ for an input sample x, where $c_i \in \{c_1, c_2, \ldots, c_M\}$ are the categories known in training data and assign any sample to the class maximizing $P(c_i|x)$. Open-set classifiers attempt to reject unrecognized inputs that are reasonably far from known data. This is, broadly, done by constructing decision boundaries based on distributions of training data or by building abating probability models, where the probability of class membership decreases as observations move from training/known data.

An example of the former approach is the '1-vs-set' model proposed by [23], and examples of the latter are W-SVM [22], or PI-SVM (probability-of-inclusion SVM) by [13] (all these methods are open-set versions of the SVM model). Junior et al. [14] proposed an open-set version of the Nearest-Neighbours classifier, with a threshold on class similarity scores used to realize the rejection option. Bendale and Bould [2] proposed an open-set version the nearest-class-mean model [17], with rejection based on the thresholded Mahalanobis distance, see also [15]. Specific models for open-set recognition with deep CNNs include the Openmax [3] and OoD methods with outlier-exposure [9,10], which rely on the observation that OoD differs in terms of the distribution of softmax probabilities as compared with known (in-distribution) examples.

In contrast to all these methods, which can be seen as heuristic procedures, with no theoretical justification, Rudd et al. [20] proposed a theoretically sound classifier - the Extreme Value Machine (EVM). Its parametric model of the probability of inclusion uses the Weibull distribution, which is justified by the Extreme Value Theory. The authors claim that this leads to the superior performance of EVM on some open-set benchmark studies reported as compared to heuristic methods.

The motivation of this research comes from the observation we made that for a number of datasets in text or image recognition the EVM is surpassed by simpler, heuristic models. The main contributions of this work are the following. We analyzed Extreme Value Theory assumptions, which justify the adoption of the Weibull distribution by the EVM method. We showed that these assumptions often do not hold in real recognition problems and illustrated this in a number of text and image classification studies. We empirically compared the EVM with simple OoD detection methods based on the LOF (Local Outlier Factor) and explained what properties of the training data lead to low performance of the EVM. We conclude that the theoretical soundness of EVM in many real-life studies can be questioned, and hence the method should be considered another heuristic procedure.

The paper is organized as follows. In Sect. 2, we explain how open-set classifiers perform out of distribution detection using a probability of inclusion-based and density-based methods. Then we provide details on the EVM (probability of inclusion-based) and the LOF (density-based), which we later use in the comparative study. We also provide OoD evaluation metrics. In Sect. 3 we report results of the numerical study comparing EVM with LOF on both text and image data and provide results of goodness-of-fit tests, which show that the EVM Weibull model is not appropriate. We discuss this concerning the EVT assumptions. Finally, we discuss the type of inter-class separation which most likely leads to the low performance of the EVM.

2   Methods

2.1   Out of Distribution Detection

In order to realize open-set recognition, classifiers must be able to reject as unrecognized the samples which are out-of distribution with regard to the training data of known classes. This allows reducing the open-space risk [23], i.e. misclassification of these OoD samples by assigning them to one of the known classes.

The key difference between open-set classifiers is how the rejection option is implemented.

A commonly used approach to reduce the open-space risk is to implement the *probability of inclusion* model. An input sample x is then classified as $c_i = \arg\max_{c \in C} P(c|x)$ providing $P(c_i|x) > \delta$, and labelled as unrecognized otherwise. The models of the probability of class inclusion attempt to model $P(c_i|x)$ as a decreasing function of the distance between x and the training data X_i pertaining to class c_i. Such models are referred to as compact abating probability (CAP) models [22]. The Extreme Value Machine which is the focus of this work is based on this idea; in Sect. 2.3 we explain how EVM constructs the CAP model for $P(c_i|x)$.

Another approach is realized by distance-based methods, where rejection is done by directly using distance to known data. An input sample x is classified as $c_i = \arg\max_{c \in C} P(c|x)$ providing $d(x, X_i) < \delta$, where $d(x, X_i)$ is some measure of distance between x and the known training data X_i pertaining to class c_i. For $d(x, X_i) \geq \delta$, x is unrecognized. This idea is implemented e.g. by the open-set version of the nearest class mean classifier [2,17], where $d(x, X_i)$ is calculated as Mahalanobis distance.

Density-based methods can be seen as conceptually related to the distance-based methods, however the measure $d(x, X_i)$ used to realize rejection of OoD samples is calculated as some measure of outlierness of x with regard to the known data X_i. This can be based on the density or the outlierness factor such as the Local Outlier Factor (LOF) [4]. The latter is used in the empirical study as an alternative method compared to the Extreme Value Machine.

2.2 Local Outlier Factor

The Local Outlier Factor [4] is based on an analysis of the local density of points. It works by calculating the so-called *local reachability distance*, defined as an average distance between a given point, its neighbors, and their neighbors. The relative density of a point against its neighbors is used to indicate the degree of the object being an OoD. The local outlier factor is formally defined as the average of the ratio of the local reachability of an object to its k-nearest neighbors. If the LOF value for a given point is larger than some threshold, the point is assumed to be OoD. In the case of the open set classification problem, the LOF threshold could be calculated based on the assumption that the training data include a given portion of outliers (called contamination in code[1]).

2.3 Extreme Value Machine

Extreme Value Machine constructs a compact abating probability model of $P(c_i|x)$ that x belongs to c_i. This popular model is justified by the Extreme Value Theory, and as such, deemed superior by the authors as compared with heuristic models.

Technically, to construct the CAP model for a class $c_i \in C = \{c_1, c_2, \ldots, c_M\}$, we create the radial inclusion function for each point $x_i \in X_i$, where X_i represents the training data for class c_i. Given a fixed point $x_i \in X_i$, τ closest training examples from classes other than c_i are selected, denoted here as $\{t_1, \ldots, t_\tau\}$, and the margin distances from x_i to these examples are calculated as

$$m_{ij} = \frac{\|x_i - t_j\|}{2}, \quad j = 1, \ldots, \tau \tag{1}$$

---

[1] http://scikit-learn.org/stable/modules/generated/sklearn.neighbors.
LocalOutlierFactor.html.

Then the parametric model of the margin distance from x_i is estimated by fitting the Weibull distribution to the data $\{m_{i1}, m_{i2}, \ldots, m_{i\tau}\}$. This step is justified by the authors by the Extreme Value Theory, and is later analyzed in terms of validity of the underlying assumptions in Sect. 3.3. The fitted Weibull model is described by the scale λ_i and shape κ_i parameters, and hence the Weibull survival function $(1 - CDF)$ is postulated as the radial class inclusion function:

$$\Psi(x_i, x) = e^{-\left(\frac{\|x_i - x\|}{\lambda_i}\right)^{\kappa_i}} \tag{2}$$

This can be interpreted as the CAP model of the decreasing probability of inclusion of the sample x in the class represented by training example x_i.

Given this model, the open-set classification of an input x is done as follows. The probability that x is associated with the class c_i is estimated as $\hat{P}(c_i|x) = \Psi(x_j, x))$, where $x_j = \arg\max_{x_k \in X_i} \Psi(x_k, x)$ (i.e. x_j is the training example in X_i closest to x). Finally, the open-set classification of x is done as $c_i = \arg\max_{c \in C} \hat{P}(c|x)$ if $\hat{P}(c_i|x) > \delta$, and x is considered unknown otherwise.

Remarks on the Extreme Value Machine Implementation.

It should be noticed that the 'official' implementation of the EVM[2] uses the libMR[3] library for the Weibull model fitting (libMR is provided by the authors of [24]). Given the sample $\{m_{i1}, m_{i2}, \ldots, m_{i\tau}\}$, libMR first performs linear transformation: $\eta_{ij} = -m_{ij} - \max\{m_{i1}, m_{i2}, \ldots, m_{i\tau}\} + 1$, $j = 1, \ldots, \tau$, and then returns the parameters (λ_i, κ_i) of the Weibull model fitted to $\{\eta_{i1}, \eta_{i2}, \ldots, \eta_{i\tau}\}$. The parameters (λ_i, κ_i) are used in Eq. 2.

In the empirical study in Sect. 3, we verify the goodness of this fit and show that in all the datasets considered the Weibull model is *not* appropriate for $\{m_{i1}, m_{i2}, \ldots, m_{i\tau}\}$ (the original margin distances) and for $\{\eta_{i1}, \eta_{i2}, \ldots, \eta_{i\tau}\}$ (the transformed margin distances).

2.4 OoD Evaluation Metric

In the next section, we want to empirically compare the performance of the EVM and LOF methods in the task of OoD detection. In the evaluation of OoD detection algorithms, we follow the approach used in [11]. OoD detection is treated as binary classification, with the OoD examples defined as the positive class and the in-distribution examples as the negative class. As the OoD detection quality metric we used the area under the receiver operating characteristic curve (AUROC). It could be used since EVM and LOF (and other OoD methods, Sect. 2.1) use a rejection threshold value which affects the false positive and true negative rates. Technically, the ROC curve shows the False Positive Rate (FPR) on the x-axis and the True Positive Rate (TPR) on the y-axis across multiple

---

[2] https://github.com/EMRResearch/ExtremeValueMachine.
[3] https://github.com/Vastlab/libMR.

thresholds. In the OoD problem, the FPR measures the fraction of in-distribution examples that are misclassified as outliers. The TPR measures the fraction of OoD examples that are correctly labeled as outliers.

The performed experiments data were divided into three data sets: training, testing, and outlier one. The first two are classical data sets used in closed-set classification and represent in-distribution data. The training data set was used to built OoD models, where test and outlier ones (as a negative and positive class) are used to evaluate OoD algorithms. In performed experiments, the number of outliers was set to be equal to the size of the test data. It could be noticed that in the presented approach, OoD detection algorithms have no knowledge about OoD world. They built their models based on in-distribution data only.

3 Computational Experiments

In this section, we empirically compare the EVM and LOF methods in the task of OoD in image classification and text documents classification. Since results of this study (Sect. 3.2) show that the theoretically-justified EMV can be out-performed by a heuristic procedure, we verify using goodness-of-fit tests if the EVM margin distances (Eq. (1)) in these datasets follow the Weibull distribution (Sect. 3.3). Next in Sect. 3.4, we show that the EVM model with the Weibull model replaced by some other distributions (e.g. normal) realizes similar performance. Finally, we visually illustrate the way how the EVM and LOF form the in-distribution and out-of-distribution areas, using the CIFAR-10 dataset projected onto the 2D space of the first two PCA components. This allows us to partly explain the difference in the performance of OoD by the EVM and LOF in our experiments.

3.1 Data Sets

To evaluate the OoD detection algorithms, we used two different sources of data: text documents and images.

For the text documents case, we used the corpus of articles extracted from the Polish language Wikipedia (Wiki). It consists of 9, 837 documents assigned to 34 subject categories (classes). The corpus is divided into training [19] and testing [18] set. As the OoD example, we randomly selected articles from the Polish press news [25] dataset (Press).

Several approaches to represent documents by feature vectors were developed during the past years. For our study, we have the most classical one - TF-IDF [21] and one of the most recent approaches - BERT [5]. The TF-IDF uses a bag of word model [7] where a feature vector consists of a set of frequencies of words (terms). To limit the size of feature vectors, we focused only on the most frequent terms. The term frequency (TF) representation is modified by the Inverted Document Frequency (IDF) [21], giving the $TF - IDF$ one. In performed experiments, we used single words as well as 2-, and 3-grams. The vector space was limited to 1000 terms. Moreover, the final TF-IDF vectors were

L2 normalized. The most frequent terms and corresponding IDFs were set up on the training set and used for $TF - IDF$ feature calculation for all data (i.e., training, testing, and outliers).

The second method, $BERT$ [5], uses state-of-the-art deep-learning algorithms (i.e. Transformers), resulting in a context-aware language modeling approach. In this study, we used the Polbert[4] [16], a pre-trained BERT model for Polish. The Polbert network with additional classification layers was tuned on the Wiki data set. Only the embedding layer of the BERT was frozen. Since the Polbert is capable of analyzing up to 512 subwords, longer texts were cut-off. The closed set accuracy was 94.21%. As a feature vector (768-dimensional), we used the first (with index zero) token from the last Transformer layer (i.e., the one before the classification layers).

For the case of images, we used the CIFAR-10 database[5]. It contains $60,000$ 32×32 color images divided in 10 classes, i.e. airplane, automobile, bird, cat, deer, dog, frog, horse, ship, and truck. There are $5,000$ images per class in the training set and $1,000$ in the test set. The ResNet-101 [8] CNN model was trained from scratch for the classification task, and it achieved 95.15% final accuracy. The 2048-dimensional feature vectors were extracted from this model. As features, we used the output of the average global pooling layer (called "avgpool").

As out-of-distribution data, the MNIST[6] and the CIFAR-100[7] benchmark data sets were chosen. For each test in this paper, the number of OoD examples was equal to the number of images used in the CIFAR-10 test set. The MNIST dataset contains $70,000$ 28×28 grayscale images of handwritten digits. We transformed them into three RGB channels and added extra padding (to keep 32×32 size) to make them fit the trained CNN model. The CIFAR-100 set has 100 classes with 600 images per class. None of the CIFAR-100 classes appear in CIFAR-10.

3.2 Comparison of OoD Detection Methods

We compared the EVM with the LOF algorithm in the context of image and text data. Since we wanted to observe the effect of input space modifications on the quality of OoD detection, we also used the standardized versions of each data set (we used the popular z-score normalization with the mean and standard deviation for each variable estimated on the train data set).

In Table 1 we compare the AUCROC measure for the EVM and LOF method over different data sets.

Despite its theoretical justification, the EVM is clearly outperformed by the heuristic LOF algorithm in most test cases, except for the text data with BERT feature vectors.

To explain this, we verified if the theoretically-grounded Weibull distribution used in EVM is appropriate for data encountered in real OoD studies.

---

[4] https://huggingface.co/dkleczek/bert-base-polish-cased-v1.

[5] https://www.cs.toronto.edu/~kriz/cifar.html.

[6] http://yann.lecun.com/exdb/mnist/.

[7] https://www.cs.toronto.edu/~kriz/cifar.html.

Table 1. AUCROC for EVM and LOF over different data sets. Standardised data sets are denoted by '+stand'

Data set	EVM	LOF
Wiki.vs.Press.TF-IDF	0.792937	**0.827835**
Wiki.vs.Press.TF-IDF+stand	0.521511	**0.793559**
Wiki.vs.Press.BERT	**0.943888**	0.904297
Wiki.vs.Press.BERT+stand	**0.942756**	0.904234
CIFAR-10.vs.CIFAR-100	0.796409	**0.888728**
CIFAR10.vs.CIFAR100+stand	0.879586	**0.893454**
CIFAR10.vs.MNIST	0.897874	**0.984625**
CIFAR10.vs.MNIST+stand	0.972384	**0.982649**

3.3 Weibull Distribution Testing

The main theoretical assumption of the EVM is that margin distances (see Sect. 2.3) follow the Weibull distribution. We empirically verified this by using the Kolmogorov-Smirnov goodness of fit test, with the null hypothesis that the margin distances have the Weibull distribution estimated by the EVM implementation. It is important to state that the margin distances were scaled by the implementation as mentioned in Sect. 2.3. In Table 2 we present mean p-values of Kolmogorov-Smirnov tests for all training examples in each data set. Assuming the test significance level of 5%, we conclude that the Weibull distribution is not appropriate (p-value < 5%) or marginally accepted (p-value = 0.087, Wiki.BERT data) in four out of six test cases. Clearly, the datasets with the highest p-values (i.e. Wiki.BERT or CIFAR-10+stand, p-value > 5%, Weibull distribution appropriate) correspond to the OoD test cases in which EVM showed the best performance, as shown in Table 1.

The detailed analysis of p-values for CIFAR-10 data set is shown on histograms in Fig. 2a and 2b. We can notice that normalization of CIFAR-10 data changes the distribution of margin distances: for a majority of training examples in the raw dataset (Fig. 2a), the Weibull model does not fit the data (most of p-values < 5%), whereas for standardized data (Fig. 2b) the Weibull model is appropriate (most of p-values > 5%). This clearly leads to improved performance of the EVM in OoD detection as shown in Table 1 (AUCROC increased from 0.897 to 0.972).

This analysis proves that margin distances do not follow the Weibull distribution in many real datasets, contrary to the theoretical justification given in [20] (Theorem 2). The justification given in [20] is grounded on the Fisher-Tippett-Gnedenko (or Extreme Value) Theorem, which states that for a series of n i.i.d. random variables, their maximum M_n is asymptotically Weibull-distributed, i.e. for some constants a_n, b_n $Pr(\frac{M_n - b_n}{a_n} < z) \to G(z)$ as $n \to \infty$, where $G(z)$ is under some assumptions the Weibull distribution. Hence the underlying assumption needed for the margin distances (Eq. 1) to follow the Weibull distribution is

Table 2. Mean p-values from Weibull goodness-of-fit tests for different datasets

Data set	Mean p-value
Wiki.TF-IDF	0.003366
Wiki.TF-IDF+stand	0.010634
Wiki.BERT	0.086883
Wiki.BERT+stand	0.111862
CIFAR-10	0.043452
CIFAR-10+stand	0.215405

that they can be treated as the maximum from a series of i.d.d. random variables, which was not shown, but only postulated in [20].

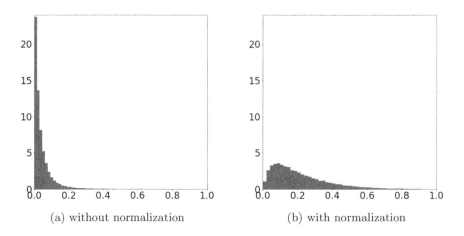

(a) without normalization   (b) with normalization

Fig. 2. Histograms of Weibull goodness-of fit test p-values for raw and standardized CIFAR-10 features.

3.4   EVM as a Heuristic OoD Procedure

The analysis reported in the previous section leads to the conclusion that the margin distances do not follow the Weibull distribution. Therefore, we believe that the very Weibull distribution is not the key to the EVM performance. To confirm this, we substituted the Weibull distribution by some other distributions and repeated the previous OoD detection experiments using this modified EVM. More specifically, we followed the EVM algorithm as described in the original paper [20], but fitted the parametric model directly to the margin distances (m_{ij}, Eq. (1)), and not to the transformed $\{\eta_{i1}, \eta_{i2}, \ldots, \eta_{i\tau}\}$. We tried four alternative CDFs: the Weibull Minimum Extreme Value (Weib min), Normal, Gamma, and empirical CDF (ECDF). The achieved AUROC values are compared with the original EVM in Table 3. These results suggest that the parametric distribution type, as well as transformation applied by the libMR[8] have a minor influence on the EVM performance. None of the analyzed distributions clearly outperformed other models. This confirms that the assumption of the Weibull distribution is not essential for the performance of the EVM.

Table 3. AUCROC for the original EVM and its modifications based on other parametric models

Data set	EVM	Weib min	Normal	Gamma	ECDF
Wiki.vs.Press.TF-IDF	**0.792937**	0.559025	0.588428	0.761878	0.523205
Wiki.vs.Press.TF-IDF+stand	0.521511	**0.717006**	0.551327	0.673260	0.518462
Wiki.vs.Press.BERT	0.943888	0.935413	**0.949327**	0.940496	0.922417
Wiki.vs.Press.BERT+stand	0.942756	0.930245	**0.947738**	0.941205	0.920710
CIFAR-10.vs.CIFAR-100	0.796409	**0.887941**	0.828649	0.772505	0.751740
CIFAR-10.vs.CIFAR-100+stand	**0.879586**	0.832547	0.878440	0.870419	0.817091
CIFAR-10.vs.MNIST	0.897874	**0.957250**	0.926712	0.872924	0.859684
CIFAR-10.vs.MNIST+stand	0.972384	**0.978867**	0.978688	0.972445	0.971501

3.5   Low Dimensional Example

To illustrate the behavior of OoD methods, we performed a set of numerical experiments on CIFAR-10 images projected by PCA onto two-dimensional space. Projected images were used to built EVM and LOF models. Next, the 2D space, in the area of CIFAR-10 data values, was equally sub-sampled, forming a X-Y grid. Each of the grid points was assigned to the OoD or in-distribution class by the EVM and LOF algorithm using different values of rejection threshold, as presented in Fig. 3 and Fig. 4. The training data points are marked in colors corresponding to the original class. Black dots represent grid data marked by the corresponding algorithm as in-distributions, whereas white dots represent OoD points. Notice, that background is also white, so the area without any black dots represents OoD space.

In Fig. 3a, we can notice that the in-distribution area (black points) covers not only training data but also the area around them. So, all test examples are likely to be correctly recognized as in-distribution, but OoD examples laying between classes will be incorrectly recognized as in-distribution. Hence, the in-distribution areas (shown by black points in Fig. 3a) are apparently too wide. We can narrow them by increasing the threshold. However, as we can notice in Fig. 3a, in the case of some classes (like the plane, ship, bird, and frog), this only slightly squeezes the in-distribution areas, while in other cases (like deer and truck), training data gets marked as OoD (white dots inside gray and blue area)

---

[8] https://github.com/Vastlab/libMR.

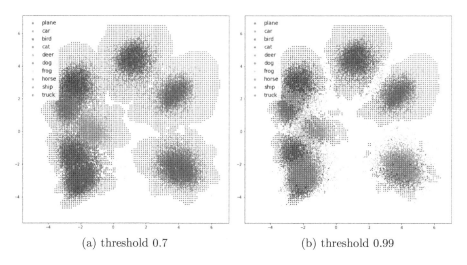

(a) threshold 0.7   (b) threshold 0.99

Fig. 3. OoD detection by the EVM for PCA projected CIFAR-10 data. The color points represent training data (CIFAR-10 images projected on 2D). Black points (forming an X-Y grid) are in-distribution data detected by EVM. White points (visible on colored areas, especially in picture (b)) correspond to data detected by EVM as OoD. Notice, that background is also white, so the area without any black dots is the OoD space.

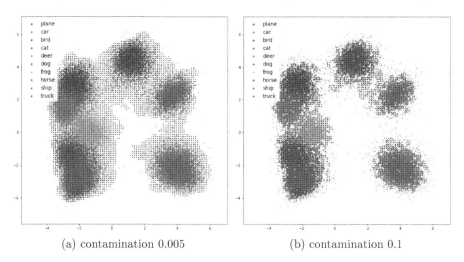

(a) contamination 0.005   (b) contamination 0.1

Fig. 4. OoD detection by LOF for PCA projected CIFAR-10 data. The meaning of white and black dots the same as in the previous figure.

as OoD objects. It is an undesirable behavior of the EVM, resulting in a large number of wrong decisions. It could be noticed that such problems occur when one class is close to another, or when classes partly overlap (as cars and trucks in our example).

A similar analysis done for the LOF method (Fig. 4) does not reveal such undesirable behavior. Figure 4a is similar to Fig. 3a. Moreover, after enlarging the contamination parameter (this results in decreasing the rejection threshold), the in-distribution areas (dark points) fit now closely to training data (Fig. 4b). However, a close look at Fig. 4a shows that LOF in-distribution areas do not extend beyond training points in directions opposite to other classes (see, for example, top of the plane class - marked by blue), contrary to directions to other classes (observe the bottom area of the 'plane' class and compare it with the area above). Such behavior is caused by the fact that LOF has no knowledge about individual classes and sees the whole training data as one 'in-distribution' set.

4   Conclusion

In this paper, we showed that the theoretical assumptions underlying the popular Extreme Value Machine are not fulfilled in the context of many real datasets. Inter-class distances (margin distances) in practice often do not follow the Weibull distribution, as assumed by the EVM. We compared the EVM with another popular OoD detection method - LOF and showed that EVM should not be generally considered superior to this heuristic method. Both these methods attempt to model the local similarities around the training examples as the 'in-distribution' space. However, the EVM takes into account the distances to other nearest classes, while LOF is focused only on local similarities.

Since the theoretical soundness of EVM in many real-life studies can be questioned, we argue that the method should be considered another heuristic OoD procedure.

Several data-related factors affect EVM performance. First, for high- dimensional data (the curse of dimensionality effect), the inter-class borders are sampled very roughly. (In our experiments, data dimensionality was between 768 and 2048). Secondly, the EVM builds a border between OoD space and 'in-distribution' space using the distances to the nearest points from other classes. When classes are well separated (large inter-class gap), this leads to a high probability of inclusion of out-of-distribution examples lying far from the known data. Hence, models of in-distribution areas tend to be over-extended, as compared e.g. with the LOF model.

A modification of the EVM is worth investigating, in which the models of the probability of inclusion are built using not only distances to other classes but also in-class distances. We believe this would address some problems observed in the EVM.

Acknowledgements. The research reported in this paper was partly sponsored by National Science Centre, Poland (grant 2016/21/B/ST6/02159).

References

1. Amodei, D., Olah, C., Steinhardt, J., Christiano, P., Schulman, J., Mané, D.: Concrete problems in AI safety. arXiv preprint arXiv:1606.06565 (2016)
2. Bendale, A., Boult, T.: Towards open world recognition. In: Proceedings of the IEEE Conference on Computer Vision and Pattern Recognition, pp. 1893–1902 (2015)
3. Bendale, A., Boult, T.E.: Towards open set deep networks. In: Proceedings of the IEEE Conference on Computer Vision and Pattern Recognition, pp. 1563–1572 (2016)
4. Breunig, M.M., Kriegel, H.P., Ng, R.T., Sander, J.: LOF: identifying density-based local outliers. ACM SIGMOD Rec. **29**, 93–104 (2000)
5. Devlin, J., Chang, M.W., Lee, K., Toutanova, K.: BERT: pre-training of deep bidirectional transformers for language understanding. arXiv preprint arXiv:1810.04805 (2018)
6. Geng, C., Huang, S., Chen, S.: Recent advances in open set recognition: a survey. IEEE Trans. Pattern Anal. Mach. Intell. (2020)
7. Harris, Z.S.: Distributional structure. Word **10**(2–3), 146–162 (1954)
8. He, K., Zhang, X., Ren, S., Sun, J.: Deep residual learning for image recognition. In: Proceedings of the IEEE Conference on Computer Vision and Pattern Recognition, pp. 770–778 (2016)
9. Hendrycks, D., Gimpel, K.: A baseline for detecting misclassified and out-of-distribution examples in neural networks. arXiv preprint arXiv:1610.02136 (2016)
10. Hendrycks, D., Mazeika, M., Dietterich, T.: Deep anomaly detection with outlier exposure. arXiv preprint arXiv:1812.04606 (2018)
11. Hendrycks, D., Mazeika, M., Dietterich, T.: Deep anomaly detection with outlier exposure. In: Proceedings of the International Conference on Learning Representations (2019)
12. Hendrycks, D., Zhao, K., Basart, S., Steinhardt, J., Song, D.: Natural adversarial examples. arXiv preprint arXiv:1907.07174 (2019)
13. Jain, L.P., Scheirer, W.J., Boult, T.E.: Multi-class open set recognition using probability of inclusion. In: Fleet, D., Pajdla, T., Schiele, B., Tuytelaars, T. (eds.) ECCV 2014. LNCS, vol. 8691, pp. 393–409. Springer, Cham (2014). https://doi.org/10.1007/978-3-319-10578-9_26
14. Mendes Jr., P., et al.: Nearest neighbors distance ratio open-set classifier. Mach. Learn. **106**(3), 359–386 (2016). https://doi.org/10.1007/s10994-016-5610-8
15. Kamoi, R., Kobayashi, K.: Why is the Mahalanobis distance effective for anomaly detection? arXiv preprint arXiv:2003.00402 (2020)
16. Kłeczek, D.: Polbert: attacking Polish NLP tasks with transformers. In: Ogrodniczuk, M., Kobyliński, Ł. (eds.) Proceedings of the PolEval 2020 Workshop, pp. 79–88. Institute of Computer Science, Polish Academy of Sciences (2020)
17. Mensink, T., Verbeek, J., Perronnin, F., Csurka, G.: Distance-based image classification: generalizing to new classes at near-zero cost. IEEE Trans. Pattern Anal. Mach. Intell. **35**(11), 2624–2637 (2013)
18. Młynarczyk, K., Piasecki, M.: Wiki test - 34 categories. CLARIN-PL digital repository (2015). http://hdl.handle.net/11321/217
19. Młynarczyk, K., Piasecki, M.: Wiki train - 34 categories. CLARIN-PL digital repository (2015). http://hdl.handle.net/11321/222
20. Rudd, E.M., Jain, L.P., Scheirer, W.J., Boult, T.E.: The extreme value machine. IEEE Trans. Pattern Anal. Mach. Intell. **40**(3), 762–768 (2017)

21. Salton, G., Buckley, C.: Term-weighting approaches in automatic text retrieval. Inf. Process. Manag. **24**(5), 513–523 (1988). https://doi.org/10.1016/0306-4573(88)90021-0
22. Scheirer, W.J., Jain, L.P., Boult, T.E.: Probability models for open set recognition. IEEE Trans. Pattern Anal. Mach. Intell. **36**(11), 2317–2324 (2014)
23. Scheirer, W.J., de Rezende Rocha, A., Sapkota, A., Boult, T.E.: Toward open set recognition. IEEE Trans. Pattern Anal. Mach. Intell. **35**(7), 1757–1772 (2012)
24. Scheirer, W.J., Rocha, A., Micheals, R.J., Boult, T.E.: Meta-recognition: The theory and practice of recognition score analysis. IEEE Trans. Pattern Anal. Mach. Intell. **33**(8), 1689–1695 (2011)
25. Walkowiak, T., Malak, P.: Polish texts topic classification evaluation. In: Proceedings of the 10th International Conference on Agents and Artificial Intelligence - Volume 2: ICAART, pp. 515–522. INSTICC, SciTePress (2018). https://doi.org/10.5220/0006601605150522

Clustering-Based Ensemble Pruning in the Imbalanced Data Classification

Paweł Zyblewski$^{(\boxtimes)}$ (iD)

Department of Systems and Computer Networks, Faculty of Electronics,
Wrocław University of Science and Technology,
Wybrzeże Wyspiańskiego 27, 50-370 Wrocław, Poland
`pawel.zyblewski@pwr.edu.pl`

Abstract. Ensemble methods in combination with data preprocessing techniques are one of the most used approaches to dealing with the problem of imbalanced data classification. At the same time, the literature indicates the potential capability of classifier selection/ensemble pruning methods to deal with imbalance without the use of preprocessing, due to the ability to use expert knowledge of the base models in specific regions of the feature space. The aim of this work is to check whether the use of ensemble pruning algorithms may allow for increasing the ensemble's ability to detect minority class instances at the level comparable to the methods employing oversampling techniques. Two approaches based on the clustering of base models in the diversity space, proposed by the author in previous articles, were evaluated based on the computer experiments conducted on 41 benchmark datasets with a high *Imbalance Ratio*. The obtained results and the performed statistical analysis confirm the potential of employing classifier selection methods for the classification of data with the skewed class distribution.

Keywords: Imbalanced data · Classifier ensemble · Ensemble pruning · Multistage organization

1 Introduction

When dealing with real-life binary classification problems we can often encounter cases, in which the number of samples belonging to one class (also known as majority class) significantly exceeds the number of samples in the other class (called minority class). However, classical pattern recognition algorithms usually assume a balanced distribution of problem instances. Therefore, in the case of skewed class distribution, they tend to display a significant bias towards the majority class. In the literature, three main types of approaches have been distinguished in order to deal with the imbalanced data classification problems [9]:

- *Data-level methods* based on the modification of the training set in such a way as to reduce the bias towards the majority class.

© Springer Nature Switzerland AG 2021
M. Paszynski et al. (Eds.): ICCS 2021, LNCS 12744, pp. 156–171, 2021.
https://doi.org/10.1007/978-3-030-77967-2_14

- *Algorithm-level methods* modifying classical pattern recognition algorithms in order to adapt them to deal with imbalance.
- *Hybrid methods* combining the two above-mentioned approaches.

One of the frequently used approaches to imbalanced data classification is the classifier ensemble [11]. Here, the methods based on the Static and Dynamic Classifier Selection (DCS) are particularly noteworthy [3], as they take into account the base model expertise in specific regions of the feature space. Ksieniewicz in [10] proposed the *Undersampled Majority Class Ensemble* (UMCE), which generates the classifier pool by dividing an unbalanced problem into a series of balanced ones. The *Dynamic Ensemble Selection Decision-making* (DESD) algorithm was presented by Chen et al. [2] in order to employ the weighting mechanism to select base classifiers that are experts in minority class recognition. Wojciechowski and Woźniak [18] employed Decision Templates in order to integrate the classifier pool decisions in case of imbalanced data classification. Klikowski and Woźniak [7] proposed the *Genetic Ensemble Selection* (GES) for imbalanced data, which generated diverse classifier pool based on feature selection using a genetic algorithm.

This article deals with the concept closely related to the classifier selection, knows as ensemble pruning. Zhou in [20] proposed the following taxonomy of such methods:

- *Ranking-based pruning* selecting a certain number of top ranked classifiers, according to a chosen metric [14].
- *Optimization-based pruning* treating the classifier selection problem as an optimization task [17,21].
- *Clustering-based pruning* clustering-based pruning which groups base models making similar decisions, and then selecting prototype classifiers from each cluster to constitute the pruned ensemble.

In this work, the application of clustering-based ensemble pruning algorithms for the imbalanced classification task will be considered. Such methods consist of two steps. First, using the selected clustering algorithm, the base classifiers are grouped in such a way that each cluster contains models that have a similar impact on the ensemble performance. For this purpose, clustering methods such as e.g. *k-means* clustering [4], hierarchical agglomerative clustering [5] and spectral clustering [19] were used. The most important element of clustering-based ensemble pruning methods is defining the space in which clustering takes place. The Euclidean distance was used by Lazarevic and Obradovic [13], while employing the pairwise diversity matrix was proposed by Kuncheva [11].

Then, from each of the clusters, a single model (also knows as the prototype classifier) is selected to be included in the pruned ensemble. For this purpose, e.g., the classifier with the highest accuracy score in [4] or the model farthest from the other clusters [5] can be selected. This step also includes the problem of selecting the number of clusters. It can be determined by evaluating the method on the validation set [4] or, in the case of fuzzy clustering methods, automatically selected using membership values of statistical indexes [8].

The main goal of this work is to examine whether the use of expert classifier knowledge in a given feature space region will allow establishing an ensemble capable of dealing with the imbalanced data classification without the need of using preprocessing techniques.

The main contributions of the following work are as follows:

– Employing proposed Clustering-based Ensemble Pruning methods for the imbalanced data classification problem.
– Experimental evaluation of the proposed algorithms on benchmark datasets and comparison with methods using data preprocessing.

2   Clustering-Based Pruning and Multistage Voting Organization

This section presents ensemble pruning algorithms based on clustering in the one-dimensional diversity space, which were proposed by Zyblewski and Woźniak in [22,23].

Clustering-Based Pruning (CPR)
Clustering is performed in the one-dimensional M measure space, which is calculated based on classifier diversity measures. In this work, 5 different diversity metrics were used, namely *the entropy measure E, measurement of interrater agreement k, averaged disagreement measure (Dis_{av}), Kohavi-Wolpert variance (KW)*, and the averaged Q statistic (Q_{av}) [12]. The M measure for a given classifier Ψ_i is defined as a difference between the diversity of the whole classifier pool Π and pool without said classifier

$$M(\Psi_i) = Div(\Pi) - Div(\Pi - \Psi_i) \tag{1}$$

An examples of the resulting clustering spaces for each of the diversity measures is shown in Fig. 1.

Then the *k-means* clustering algorithm is employed in order to group base classifier with similar effect on the ensemble performance. Finally, from each cluster, a prototype model with the highest *balanced accuracy score* is selected to be a part of the pruned ensemble.

Clustering-Based Multistage Voting Organization
Additionally, the *Random Sampling Multistage Organization* (RSMO) algorithm was proposed, that is a modification of the multistage organization, which was first described in [6]. This proposal is based on the *Multistage Organization with Majority Voting* (MOMV) as detailed by Ruta and Gabrys in [16]. This approach can be compared to static classifier selection and ensemble pruning as it selects models for the first voting layer based on sampling with replacement. Sampling usage is based on the assumption that the classifiers within a given cluster have a similar effect on the ensemble performance, therefore they do not have to be all used in the classification process. Then, the final decision is produced by two layers of majority voting. An example of such an organization is shown in Fig. 2.

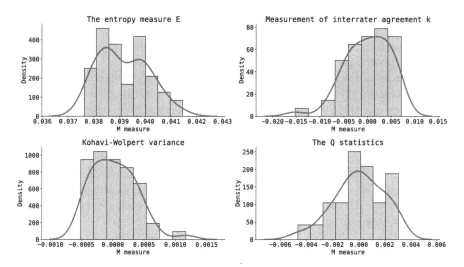

Fig. 1. Histograms and density estimation plots for M measure based on each ensemble diversity metric calculated on the glass2 dataset. *Disagreement measure* was omitted due to the results identical to the *Kohavi-Wolper variance*.

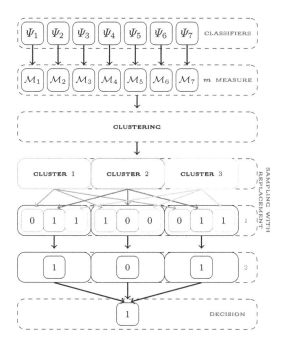

Fig. 2. Example of clustering-based multistage voting organization with 9 classifiers and 3 clusters. The number of groups in the first layer corresponds to the number of clusters. Then, using sampling with replacement, a single classifier from each cluster is selected to be a part of each group.

Computational and Memory Complexity Analysis
The proposed method includes the stage of determining the M measure value of each base classifier, the clustering of models in the diversity space and the selection of prototype classifiers.

In order to obtain the M measure value for each base classifier, first, the ensemble diversity must be calculated. The complexity of this process is $O(n)$ or $O(n^2)$, where n is the number of base classifiers, depending on whether the non-pairwise or pairwise measure is used. Then, the M measure calculation process has the complexity of $O(n)$.

The *k-means* algorithm was used for clustering in diversity space. Therefore, the complexity of clustering is $O(ncde)$, where c is the number of clusters, d is the number of data dimensions, and e describes the number of iterations/epochs of the algorithm [1]. As the clustering space is one-dimensional, complexity is reduced to $O(nce)$.

3   Experimental Evaluation

The research was carried out on 41 imbalanced datasets presented in Table 1, where #I is the number of instances, #F is the number of features and IR denotes the *Imbalance Ratio*. However, it should be noted that the experiments could only be carried out on those datasets for which the *k-means* clustering algorithm was able to find the desired number of clusters (from 2 to 7) for a given classifier and diversity measure.

Table 1. Imbalanced datasets characteristics.

Dataset	#I	#F	IR	Dataset	#I	#F	IR
ecoli-0-1_vs_2-3-5	244	7	9	glass2	214	9	12
ecoli-0-1_vs_5	240	6	11	glass4	214	9	15
ecoli-0-1-3-7_vs_2-6	281	7	39	glass5	214	9	23
ecoli-0-1-4-6_vs_5	280	6	13	led7digit-0-2-4-5-6-7-8-9_vs_1	443	7	11
ecoli-0-1-4-7_vs_2-3-5-6	336	7	11	page-blocks-1-3_vs_4	472	10	16
ecoli-0-1-4-7_vs_5-6	332	6	12	shuttle-c0-vs-c4	1829	9	14
ecoli-0-2-3-4_vs_5	202	7	9	shuttle-c2-vs-c4	129	9	20
ecoli-0-2-6-7_vs_3-5	224	7	9	vowel0	988	13	10
ecoli-0-3-4_vs_5	200	7	9	yeast-0-2-5-6_vs_3-7-8-9	1004	8	9
ecoli-0-3-4-6_vs_5	205	7	9	yeast-0-2-5-7-9_vs_3-6-8	1004	8	9
ecoli-0-3-4-7_vs_5-6	257	7	9	yeast-0-3-5-9_vs_7-8	506	8	9
ecoli-0-4-6_vs_5	203	6	9	yeast-0-5-6-7-9_vs_4	528	8	9
ecoli-0-6-7_vs_3-5	222	7	9	yeast-1_vs_7	459	7	14
ecoli-0-6-7_vs_5	220	6	10	yeast-1-2-8-9_vs_7	947	8	31
ecoli4	336	7	16	yeast-1-4-5-8_vs_7	693	8	22
glass-0-1-4-6_vs_2	205	9	11	yeast-2_vs_4	514	8	9
glass-0-1-5_vs_2	172	9	9	yeast-2_vs_8	482	8	23
glass-0-1-6_vs_2	192	9	10	yeast4	1484	8	28
glass-0-1-6_vs_5	184	9	19	yeast5	1484	8	33
glass-0-4_vs_5	92	9	9	yeast6	1484	8	41
glass-0-6_vs_5	108	9	11				

The evaluation of the proposed methods is based on six metrics used in the case of imbalanced classification problems, i.e. *balanced accuracy score*, *G-mean*, *F_1 score*, *precision*, *recall*, and *specificity*. As base classifiers, *Gaussian Naïve Bayes* Classifier (GNB) and *Classification and Regression Tree* (CART), based on the *scikit-learn* implementation [15], were used. The fixed size of the classifier pool was set to 50 base models, generated using a stratified version of bagging. This bagging generates each bootstrap sampling with replacement majority and minority classes separately while maintaining the original *Imbalance Ratio*. The size of each bootstrap is set to half the size of the original training set. The proposed approaches were evaluated on the basis of 5 times repeated 2-fold cross-validation. The ensemble's decision is based on support accumulation. Statistical analysis of the obtained results was performed using the Wilcoxon rank-sum test ($p = 0.05$). All experiments have been implemented in *Python* programming language and can be repeated using the code on *Github*[1].

Research Questions
The conducted research aims to answer two main questions:

1. Is the static classifier selection able to improve the results obtained by combining the entire classifier pool for the task of imbalanced data classification?
2. Can the use of static classifier selection in the problem of imbalanced data classification result in performance comparable with the use of preprocessing techniques?

Goals of the Experiments
Experiment 1 – Comparison with Standard Combination
The aim of the first experiment is to compare the proposed methods with a combination of the entire classifier pool. Support accumulation (SACC) and majority voting (MV) of all 50 base models were used as reference methods. The best of the proposed methods is then used in Experiment 2.

Based on the preliminary study, the following pairs of the *diversity measure:number of clusters* were selected for this experiment:

- CPR GNB – E: 2, k: 2, KW: 2, Dis_{av}: 2, Q_{av}: 3,
- CPR CART – E: 5, k: 3, KW: 3, Dis_{av}: 3, Q_{av}: 5,
- RSMO GNB – E: 6, k: 6, KW: 6, Dis_{av}: 4, Q_{av}: 5,
- RSMO CART – E: 7, k: 7, KW: 7, Dis_{av}: 7, Q_{av}: 3.

Experiment 2 – Comparison with Preprocessing Techniques
In the second experiment, the methods selected in Experiment 1 are compared with the combination of the whole classifier pool generated using preprocessing methods. Preprocessing is performed separately for each of the bootstraps generated by stratified bagging. *Random Oversampling* (ROS), SMOTE, SVM-SMOTE (SVM) and *Boderline*-SMOTE (B2) were selected as the preprocessing techniques.

---

[1] https://github.com/w4k2/iccs21-ensemble-pruning.

3.1   Experiment 1 – Comparison with Standard Combination

Clustering-Based Pruning

Figure 3 shows radar plots with the average ranks achieved by each method on all evaluation metrics. For the *gaussian naïve bayes* classifier, the advantage of the proposed methods over the combination of the entire available classifier pool can be observed. The only exception is *recall*, where GNB CPR-E2 is comparable to the reference methods, while the other proposed approaches display a slightly lower average rank value.

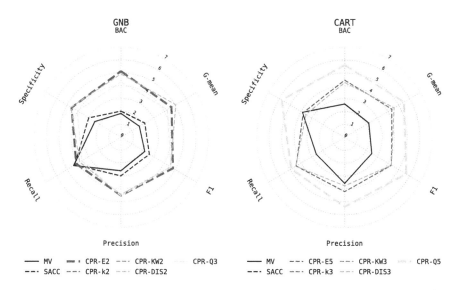

Fig. 3. Visualization of the mean ranks achieved by each method.

These observations are confirmed by Table 2. The numbers under the average rank of each ensemble method indicate which algorithms were statistically significantly worse than the one in question. It presents the results of the performed statistical analysis, on the basis of which it can be concluded that the proposed methods achieve statistically significantly better average ranks than the combination of the entire classifier pool for each of the metrics, except *recall*, where no statistically significant differences were reported. Worth noting is also the identical performance of methods based on measures k, KW, and Dis_{av}.

Particularly promising results can be observed when using CART as the base classifier. In this case, the measure of diversity Q_{av} performs best. Based on the statistical analysis, it achieves statistically significantly better results than the combination of the entire classifier pool, as well as the pruning algorithms using other measures of diversity for the clustering space construction. This is true for every metric except *recall*.

Table 2. Results of Wilcoxon statistical test on global ranks for proposed methods in comparison to the combination of the whole classifier pool.

GNB

	MV (1)	SACC (2)	CPR-E2 (3)	CPR-K2 (4)	CPR-KW2 (5)	CPR-DIS2 (6)	CPR-Q3 (7)
BAC	1.839	2.018	5.125	4.911	4.911	4.911	4.286
	—	—	1, 2	1, 2	1, 2	1, 2	1, 2
$G-mean$	1.696	2.196	4.661	5.054	5.054	5.054	4.286
	—	1	1, 2	1, 2	1, 2	1, 2	1, 2
$F_1 score$	2.196	2.625	4.804	4.589	4.589	4.589	4.607
	—	—	1, 2	1, 2	1, 2	1, 2	1, 2
$Precision$	2.607	3.000	4.518	4.446	4.446	4.446	4.536
	—	—	1, 2	1, 2	1, 2	1, 2	1, 2
$Recall$	4.393	4.304	4.143	3.839	3.839	3.839	3.643
	—	—	—	—	—	—	—
$Specificity$	2.429	3.000	4.589	4.643	4.643	4.643	4.054
	—	1	1, 2	1, 2	1, 2	1, 2	1, 2

CART

	MV (1)	SACC (2)	CPR-E5 (3)	CPR-K3 (4)	CPR-KW3 (5)	CPR-DIS3 (6)	CPR-Q5 (7)
BAC	2.586	2.586	4.448	4.259	4.259	4.259	5.603
	—	—	1, 2	1, 2	1, 2	1, 2	all
$G-mean$	2.224	2.224	4.362	4.569	4.569	4.569	5.483
	—	—	1, 2	1, 2	1, 2	1, 2	all
$F_1 score$	2.500	2.500	4.328	4.328	4.328	4.328	5.690
	—	—	1, 2	1, 2	1, 2	1, 2	all
$Precision$	3.569	3.569	4.207	3.759	3.759	3.759	5.379
	—	—	—	—	—	—	all
$Recall$	2.603	2.603	4.448	4.483	4.483	4.483	4.897
	—	—	1, 2	1, 2	1, 2	1, 2	1, 2
$Specificity$	3.879	3.879	3.810	3.552	3.552	3.552	5.776
	—	—	—	—	—	—	all

Based on the results of the statistical analysis, the GNB CPR-E2 and CART CPR-Q5 methods were selected for the next experiment. These approaches displayed the highest average ranks as well as a good ability to recognize the minority class.

Random Sampling Two Step Voting Organization

Figure 4 and Table 3 show the comparison of two-step majority voting compared to the reference methods. In the case of GNB RSMO, the most notable is the approach using the Q_{av} diversity measure, which is the most balanced in terms of all evaluation metric. Is also statistically comparable with the reference methods in terms of the ability to recognize the minority class.

As in the case of GNB, when we use the CART decision tree as the base classifier, the most interesting relationships are represented by the method based on the Q diversity measure. We can see that the CART RSMO-Q3 algorithm achieves highest average ranks in terms of all evaluation metrics. Additionally is is statistically significantly better than reference methods and most of the RSMO approaches using different diversity measures to establish the clustering space.

On the basis of the obtained results, the GNB RSMO-Q5 and CART RSMO-Q3 methods were selected for Experiment 2.

3.2   Experiment 2 – Comparison with Preprocessing Techniques

Clustering-Based Pruning
Figure 5 shows the results of comparing the methods selected in Experiment 2 with the approaches employing preprocessing techniques.

When the base classifiers is GNB, it can be noticed that, despite achieving average rank values for each of the metrics, the proposed methods are never statistically significantly worse than the reference approaches using preprocessing (Table 4). Additionally, GNB CPR-E2 shows statistically higher *precision* than that achieved by using Random Oversampling and SMV-SMOTE.

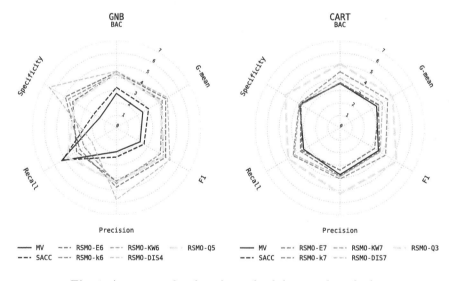

Fig. 4. Average rank values for each of the tested methods.

Table 3. Results of Wilcoxon statistical test on global ranks for proposed methods in comparison to the combination of the whole classifier pool.

GNB

	MV (1)	SACC (2)	RSMO-E6 (3)	RSMO-K6 (4)	RSMO-KW6 (5)	RSMO-DIS4 (6)	RSMO-Q5 (7)
BAC	2.750	3.232	4.339	4.446	4.304	4.518	4.411
	—	—	1, 2	1, 2	1, 2	1	1, 2
$G-mean$	2.518	3.071	4.250	4.714	4.393	4.875	4.179
	—	—	1, 2	1, 2	1, 2	1, 2	1, 2
$F_1 score$	2.268	2.643	4.500	4.714	4.196	5.125	4.554
	—	—	1, 2	1, 2	1, 2	1, 2, 5	1, 2
$Precision$	1.929	2.357	4.286	4.768	4.482	5.786	4.393
	—	—	1, 2	1, 2	1, 2	all	1, 2
$Recall$	5.286	5.179	3.839	3.518	3.107	2.339	4.732
	3, 4, 5, 6	3, 4, 5, 6	5, 6	6	6	—	3, 4, 5, 6
$Specificity$	1.607	2.179	4.196	4.929	4.607	6.500	3.982
	—	1	1, 2	1, 2, 3, 7	1, 2	all	1, 2

CART

	MV (1)	SACC (2)	RSMO-E7 (3)	RSMO-K7 (4)	RSMO-KW7 (5)	RSMO-DIS7 (6)	RSMO-Q3 (7)
BAC	3.569	3.569	3.638	3.603	4.500	3.983	5.138
	—	—	—	—	1, 2	—	1, 2, 3, 4, 6
$G-mean$	3.483	3.483	3.707	3.672	4.569	4.052	5.034
	—	—	—	—	1, 2	—	1, 2, 3, 4, 6
$F_1 score$	3.655	3.655	3.293	3.776	4.362	3.845	5.414
	—	—	—	—	3	—	all
$Precision$	3.741	3.741	3.379	3.862	4.086	3.914	5.276
	—	—	—	—	—	—	all
$Recall$	3.448	3.448	3.621	3.759	4.466	4.328	4.931
	—	—	—	—	1, 2	—	1, 2, 3, 4
$Specificity$	3.828	3.828	3.707	3.724	3.879	3.828	5.207
	—	—	—	—	—	—	all

The ensemble pruning methods seem to perform better when using the CART decision tree as the base classifier. Again, none of the reference methods achieved statistically significantly better average ranks than the proposed approach. At the same time, however, CART CPR-Q5 achieves a statistically significantly better rank value than ROS for BAC, *G-mean* and F_1 *score*. This method is also statistically significantly better than *Borderline*-SMOTE in terms of F_1 *score* and *specificity*.

Random Sampling Two Step Voting Organization
The results of the statistical analysis for the comparison of the proposed RSMO methods with the preprocessing-based approaches shown in Table 5. The average rank values for each of the metrics for are shown in Fig. 6.

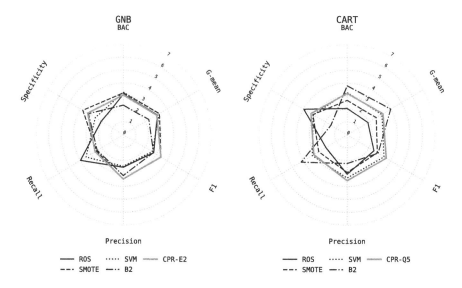

Fig. 5. Visualization of the mean ranks achieved by each method.

Table 4. Results of Wilcoxon statistical test on global ranks for the selected methods in comparison to the preprocessing techniques.

GNB					
	ROS (1)	SMOTE (2)	SVM (3)	B2 (4)	CPR-E2 (5)
BAC	3.125	3.232	3.286	2.286	3.071
	4	4	4	—	—
$G-mean$	3.089	3.286	3.268	2.321	3.036
	4	4	4	—	—
$F_1 score$	2.768	3.429	2.625	2.750	3.429
	—	3	—	—	—
$Precision$	2.518	3.446	2.446	3.161	3.429
	—	1, 3	—	—	1, 3
$Recall$	3.982	2.500	3.464	2.429	2.625
	2, 4, 5	—	2, 4	—	—
$Specificity$	2.054	3.768	2.607	3.250	3.321
	—	1, 3	—	1	1
CART					
	ROS (1)	SMOTE (2)	SVM (3)	B2 (4)	CPR-Q5 (5)
BAC	2.052	2.672	3.276	3.793	3.207
	—	—	1, 2	1, 2	1
$G-mean$	1.897	2.655	3.172	4.000	3.276
	—	1	1	1, 2, 3	1
$F_1 score$	2.448	2.759	3.379	2.828	3.586
	—	—	1, 2	—	1, 4
$Precision$	3.034	2.897	3.328	2.207	3.534
	4	4	4	—	4
$Recall$	1.948	2.603	3.190	4.207	3.052
	—	1	1, 2	all	—
$Specificity$	3.966	3.138	3.103	1.483	3.310
	2, 3, 4	4	4	—	4

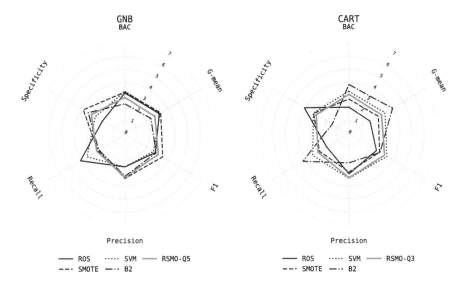

Fig. 6. Average rank values for each of the tested methods

When the base classifier is GNB, the two-step majority voting methods achieve results comparable to *Borderline*-SMOTE, however, they are statistically significantly worse in terms of *recall* than Random Oversampling and SVM-SMOTE. When two-step voting is used in conjunction with the CART decision tree, the proposed method achieves statistically significantly better *precision* than the reference methods. However, it is statistically significantly inferior to *Borderline*-SMOTE in terms of *G-mean* and *recall*.

3.3 Lessons Learned

Based on the preliminary experiment determining the appropriate number of clusters for a given diversity measure, it can be concluded that the classifier pool generation using stratified bagging probably does not allow for achieving a high ensemble diversity in the case of GNB. This is indicated by the fact that the methods using this classifier perform best when the clustering space is divided into just two groups. Decision trees, which show a greater tendency to obtain diverse base models, do much better in this respect. It is also worth noting that in the case of CART, due to no tree depth limitation, the results of the majority vote were in line with the accumulation of support.

Table 5. Results of Wilcoxon statistical test on global ranks for the selected methods in comparison to the preprocessing techniques.

GNB

	ROS (1)	SMOTE (2)	SVM (3)	B2 (4)	RSMO-Q5 (5)
BAC	3.196	3.268	3.321	2.357	2.857
	4	4	4	—	—
$G-mean$	3.196	3.321	3.268	2.429	2.786
	4	4	4	—	—
F_1score	2.839	3.500	2.661	2.893	3.107
	—	3	—	—	—
$Precision$	2.518	3.411	2.518	3.232	3.321
	—	1, 3	—	—	—
$Recall$	4.071	2.482	3.464	2.393	2.589
	all	—	2, 4, 5	—	—
$Specificity$	2.054	3.804	2.714	3.357	3.071
	—	1, 3	—	1	—

CART

	ROS (1)	SMOTE (2)	SVM (3)	B2 (4)	RSMO-Q3 (5)
BAC	2.086	2.707	3.310	3.828	3.069
	—	—	1, 2	1, 2	—
$G-mean$	1.966	2.724	3.310	4.000	3.000
	—	1	1	1, 2, 3	—
F_1score	2.552	2.828	3.483	2.862	3.276
	—	—	1, 2	—	—
$Precision$	3.069	2.931	3.414	2.224	3.362
	4	4	4	—	4
$Recall$	2.017	2.672	3.293	4.207	2.810
	—	1	1, 2	all	—
$Specificity$	4.069	3.207	3.241	1.517	2.966
	all	4	4	—	4

Regardless of the base classifier used, the results obtained with the use of the measures of diversity k, KW, and Dis_{av} were exactly the same. On this basis, it can be concluded that the diversity spaces generated on their basis coincide. An example of this can be seen in the example shown in Fig. 1, where all three spaces have the same distribution density (where the space based on k is a mirror image of the spaces based on KW and Dis_{av}).

Experiment 1 proved that by a skillful selection of a small group of classifiers, in the imbalanced data classification problem, it is possible to achieve a better performance than that achieved by combining the decisions of the entire classifier pool.

Experiment 2 was able to confirm that thanks to employing the classifier selection methods to the problem of imbalanced data classification, it is possible to obtain results statistically not worse (and sometimes statistically significantly better) than those achieved by the ensembles using preprocessing techniques.

Additionally, from the obtained results, it can be concluded that the use of the two-stage majority voting structure may allow, in the case of imbalanced data classification task, to improve the ensemble performance when compared to the traditional combination of the classifier pool. This is due to the division of classifiers into clusters containing models that make similar errors on problem instances. Thanks to this, after the first voting level, we obtain predictions reflecting the expert knowledge of the base models in each of the recognized feature space regions.

The results of the experiments seem to indicate the averaged Q statistic as the best measure of ensemble diversity for the generation of one-dimensional clustering space. However, according to the research carried out by Kuncheva and Whitaker [12], one cannot indicate the superiority of Q statistics over the other diversity measures.

4   Conclusions

The main purpose of this work was to examine whether the use of static classifier selection/ensemble pruning methods in the imbalanced data classification problems allows for increasing the ensemble's ability to detect minority class instances. The research was conducted using two proposed ensemble pruning methods, based on the base models clustering in one-dimensional diversity space. The obtained results and the performed statistical analysis confirmed that the careful selection of base models in the case of imbalanced data may increase the ability of the pruned ensemble to recognize the minority class. In some cases, such ensembles may even outperform larger classifier pools generated using data oversampling techniques such as *Random oversampling* and *Borderline*-SMOTE.

Future research may include attempts to modify existing classifier selection methods (both static and dynamic) for the purpose of classifying imbalanced data.

Acknowledgment. This work was supported by the *Polish National Science Centre* under the grant No. 2017/27/B/ST6/01325.

References

1. Bora, D.J., Gupta, D., Kumar, A.: A comparative study between fuzzy clustering algorithm and hard clustering algorithm. arXiv preprint arXiv:1404.6059 (2014)
2. Chen, D., Wang, X.-J., Wang, B.: A dynamic decision-making method based on ensemble methods for complex unbalanced data. In: Cheng, R., Mamoulis, N., Sun, Y., Huang, X. (eds.) WISE 2020. LNCS, vol. 11881, pp. 359–372. Springer, Cham (2019). https://doi.org/10.1007/978-3-030-34223-4_23

3. Cruz, R.M., Sabourin, R., Cavalcanti, G.D.: Dynamic classifier selection: recent advances and perspectives. Inf. Fusion **41**(C), 195–216 (2018)
4. Qiang, F., Shang-xu, H., Sheng-ying, Z.: Clustering-based selective neural network ensemble. J. Zhejiang Univ. Sci. A **6**(5), 387–392 (2005). https://doi.org/10.1631/jzus.2005.A0387
5. Giacinto, G., Roli, F., Fumera, G.: Design of effective multiple classifier systems by clustering of classifiers. In: 15th International Conference on Pattern Recognition, ICPR 2000 (2000)
6. Ho, T.K., Hull, J.J., Srihari, S.N.: Decision combination in multiple classifier systems. IEEE Trans. Pattern Anal. Mach. Intell. **16**(1), 66–75 (1994)
7. Klikowski, J., Ksieniewicz, P., Woźniak, M.: A genetic-based ensemble learning applied to imbalanced data classification. In: Yin, H., Camacho, D., Tino, P., Tallón-Ballesteros, A.J., Menezes, R., Allmendinger, R. (eds.) IDEAL 2019. LNCS, vol. 11872, pp. 340–352. Springer, Cham (2019). https://doi.org/10.1007/978-3-030-33617-2_35
8. Krawczyk, B., Cyganek, B.: Selecting locally specialised classifiers for one-class classification ensembles. Pattern Anal. Appl. **20**(2), 427–439 (2015). https://doi.org/10.1007/s10044-015-0505-z
9. Krawczyk, B.: Learning from imbalanced data: open challenges and future directions. Prog. Artif. Intell. **5**(4), 221–232 (2016). https://doi.org/10.1007/s13748-016-0094-0
10. Ksieniewicz, P.: Undersampled majority class ensemble for highly imbalanced binary classification. In: Proceedings of the 2nd International Workshop on Learning with Imbalanced Domains: Theory and Applications. Proceedings of Machine Learning Research, PMLR, ECML-PKDD, Dublin, Ireland, vol. 94, pp. 82–94, 10 September 2018
11. Kuncheva, L.I.: Combining Pattern Classifiers: Methods and Algorithms. Wiley, Hoboken (2004)
12. Kuncheva, L.I., Whitaker, C.J.: Measures of diversity in classifier ensembles and their relationship with the ensemble accuracy. Mach. Learn. **51**(2), 181–207 (2003)
13. Lazarevic, A., Obradovic, Z.: The effective pruning of neural network classifiers. 2001 IEEE/INNS International Conference on Neural Networks, IJCNN 2001 (2001)
14. Margineantu, D.D., Dietterich, T.G.: Pruning adaptive boosting. In: Proceedings of the 14th International Conference on Machine Learning, ICML 1997, San Francisco, CA, USA, pp. 211–218. Morgan Kaufmann Publishers Inc. (1997)
15. Pedregosa, F., et al.: Scikit-learn: machine learning in Python. J. Mach. Learn. Res. **12**, 2825–2830 (2011)
16. Ruta, D., Gabrys, B.: A theoretical analysis of the limits of majority voting errors for multiple classifier systems. Pattern Anal. Appl. **2**(4), 333–350 (2002)
17. Ruta, D., Gabrys, B.: Classifier selection for majority voting. Inf. Fusion **6**(1), 63–81 (2005)
18. Wojciechowski, S., Woźniak, M.: Employing decision templates to imbalanced data classification. In: de la Cal, E.A., Villar Flecha, J.R., Quintián, H., Corchado, E. (eds.) HAIS 2020. LNCS (LNAI), vol. 12344, pp. 120–131. Springer, Cham (2020). https://doi.org/10.1007/978-3-030-61705-9_11
19. Zhang, H., Cao, L.: A spectral clustering based ensemble pruning approach. Neurocomputing **139**, 289–297 (2014)
20. Zhou, Z.H.: Ensemble Methods: Foundations and Algorithms. Chapman & Hall CRC, Boca Raton (2012)

21. Zhou, Z.H., Wu, J., Tang, W.: Ensembling neural networks: many could be better than all. Artif. Intell. **137**(1–2), 239–263 (2002)
22. Zyblewski, P., Woźniak, M.: Clustering-based ensemble pruning and multistage organization using diversity. In: Pérez García, H., Sánchez González, L., Castejón Limas, M., Quintián Pardo, H., Corchado Rodríguez, E. (eds.) HAIS 2019. LNCS (LNAI), vol. 11734, pp. 287–298. Springer, Cham (2019). https://doi.org/10.1007/978-3-030-29859-3_25
23. Zyblewski, P., Woźniak, M.: Novel clustering-based pruning algorithms. Pattern Anal. Appl. **23**(3), 1049–1058 (2020). https://doi.org/10.1007/s10044-020-00867-8

Improvement of Random Undersampling to Avoid Excessive Removal of Points from a Given Area of the Majority Class

Małgorzata Bach[(✉)] and Aleksandra Werner

Department of Applied Informatics, Faculty of Automatic Control, Electronics and Computer Science, Silesian University of Technology, Gliwice, Poland
{malgorzata.bach,aleksandra.werner}@polsl.pl

Abstract. In this paper we focus on class imbalance issue which often leads to sub-optimal performance of classifiers. Despite many attempts to solve this problem, there is still a need to look for better ones, which can overcome the limitations of known methods. For this reason we developed a new algorithm that in contrast to traditional random undersampling removes maximum k nearest neighbors of the samples which belong to the majority class. In such a way, there has been achieved not only the effect of reduction in size of the majority set but also the excessive removal of too many points from the given area has been successfully prevented. The conducted experiments are provided for eighteen imbalanced datasets, and confirm the usefulness of the proposed method to improve the results of the classification task, as compared to other undersampling methods. Non-parametric statistical tests show that these differences are usually statistically significant.

Keywords: Classification · Imbalanced dataset · Sampling methods · Undersampling · K-Nearest Neighbors methods.

1 Introduction

Uneven class distribution can be observed in datasets concerning many areas of human life – medicine [21,29], engineering [12,24], banking, telecommunications [15,35], scientific tasks such as pattern recognition [30], etc. Many other examples along with the exhaustive state-of-the-art which refers to development of research in learning of imbalanced problem is included in [16,17]. Unfortunately, a lot of learning systems are not adapted to operate on imbalanced data, and although many techniques have already been proposed in literature it is still an unresolved issue and requires further studies. The charts presented in [16] confirm these facts and show that the number of publications on the problem of class imbalance has increased in recent years.

This work was supported by Statutory Research funds of Department of Applied Informatics, Silesian University of Technology, Gliwice, Poland.

© Springer Nature Switzerland AG 2021
M. Paszynski et al. (Eds.): ICCS 2021, LNCS 12744, pp. 172–186, 2021.
https://doi.org/10.1007/978-3-030-77967-2_15

We have dealt with this issue for several years, which resulted in the publication of the scientific papers on that subject [2,4]. The problem has not only been analyzed in the context of specific real data using well-known balancing algorithms, but we have also tried to develop our own data sampling methods that could help reduce problems arising from the skewed data distribution. For example, our method presented in paper [3] is oriented toward finding and thinning clusters of examples from the majority class. However, while this method in many cases outperformed the other compared ones, the results were not fully satisfactory. Therefore, we decided to do further research and analyses, the effect of which is KNN_RU algorithm presented in the article. The combination of random undersampling and the idea of the nearest neighbors allows to remove maximum k nearest neighbors of the samples which belong to the majority class, and thus prevent an excessive removal of too many points from the given area. To investigate the impact of such a method on the result of the binary classification task, we conducted experiments, where 6 classifiers were applied for eighteen datasets of various imbalanced ratio. We also confronted our approach with four other methods belonging to the data-level balancing category and test whether the differences between them are statistically significant. To assess classifiers' accuracy, we applied a number of metrics advisable for classification of imbalanced data. Decision of analyzing scores of multiple estimation methods instead of one was motivated by the fact that no single metric is able to comprise all the interesting aspects of the analyzed model.

The structure of this paper is as follows. Section 2 overviews the ideas which address the imbalanced data challenge. In Sect. 3 the proposed algorithm of undersampling is outlined. The experiment details are described in Sect. 4. There are also a short characteristic of analyzed data together with information about applied classifiers and performance metrics used for evaluation. The results of the performed tests and the discussion of the outcomes are given in this part of the paper too, while the conclusions are given in Sect. 5.

2   Learning from Imbalanced Data

Related literature, e.g. [3,14,22,31], provides information on solutions that counteract the effects of data imbalance. They can be categorized into three major groups: data-level, algorithmic level and cost-sensitive methods.

Data-level approaches tackle class imbalance by adding (oversampling) or removing (undersampling) instances to achieve a more balanced class distribution. Solutions of this category can be used at a preprocessing stage before applying various learning algorithms. They are independent of the selected classifier. Compared to the methods of the other two groups, data level solutions usually require significantly less computing power. They are also characterized by simplicity and speed of operation, which is especially important in the case of large datasets. Importantly, resampling methods can also extend standard ensemble classifiers to prepare data before learning component classifiers. The results presented in [14] show that such extensions applied e.g. to the bagging method significantly improve the outcomes.

The simplest method from the group of undersampling techniques is random undersampling (RU) which tries to balance class distribution by random elimination of majority class examples. However, one of the drawbacks is, it can discard data that is potentially important for learning. To overcome this limitation heuristic approaches are used to identify and remove less significant training examples. These may be borderline examples or examples which are suspicious of being noisy, and their removal can make the decision surface smoother.

One of the most commonly used classes of heuristic undersampling methods is based on *k-Nearest Neighbors* algorithm (KNN). In Wilson's *Edited Nearest Neighbor* (ENN) method undersampling of the majority class is done by removing samples whose class label differs from the class of the majority of their k nearest neighbors [5,36].

Neighborhood Cleaning Rule (NCL) algorithm for a two-class problem can be described as follows: for each example in the training set its three nearest neighbors are found. If tested example x_i belongs to the dominant class and the classification given by its three nearest neighbors contradicts the original class of x_i, then x_i is removed. Otherwise, if x_i belongs to the minority class and its three nearest neighbors misclassify x_i as a dominant, then the nearest neighbors that belong to the majority class are removed [27].

Tomek link (T-link) algorithm can also be used to reduce majority class [34]. Tomek link can be defined as a pair of minimally distant nearest neighbors of the opposite classes. Formally, a pair of examples x_i and x_j is called a Tomek link if they belong to different classes and are each other's nearest neighbors. Tomek link can be used both as a method of undersampling and data cleaning, in the first case only the majority class examples being a part of Tomek link are eliminated, while in the second case the examples of both classes are removed.

Many other informed undersampling methods can be found in the literature, but because our new solution is a kind of the hybrid of random undersampling and the k-Nearest Neighbor algorithm, we decided to present only the solutions based on mentioned concepts.

3   KNN_RU Algorithm Outline

One of the problems with random undersampling is that there is no possibility to control what objects are removed and thus there is a danger of losing valuable information about the majority class. Accordingly, the method works well only when the removal does not change the distribution of the majority class objects. In other case heuristic methods should be used, which try to reject the least-significant examples of the majority class. Unfortunately, these methods also have some drawbacks, namely they usually do not allow to influence the number of removed elements because it only comes from the nature of the dataset. Therefore, sometimes only a small number of observations meets the criteria taken into account in the individual algorithm and is removed from the set.

In order to solve the described problems an attempt was made to create the method which would reduce undesirable effects occurring during random

---

Algorithm. KNN_RU method for undersampling

---

function KNN_RU (S_{maj}, P, k)

    $l = |S_{maj}|$; // l is the number of examples from the majority class

    ToRemove $\leftarrow$ matrix (nrow=l, ncol=k);

    for i = 1 **to** l

        Calculate the distance between i^{th} element of S_{maj} and other samples;

        Sort the distance and determine nearest neighbors based on

        the k minimum distance;

        Save indexes of the found neighbors in the ith row of ToRemove matrix;

        $Z = \lfloor P * l \rfloor$; //$Z$ is the number of examples to be removed from S_{maj}

        if (length(unique(ToRemove)) >= Z) **then**

            $R \leftarrow$ sample(unique(ToRemove), Z, replace = FALSE)

        else $R \leftarrow$ unique(ToRemove);

  return $S_{maj} - R$ // The subset of the majority class

---

elimination and at the same time allow to determine the number of observations which should be removed from the majority class.

The proposed solution KNN Random Undersampling (KNN_RU) is similar to the traditional random undersampling. The difference is that removing instances is not based on the full set of majority objects, but on k nearest neighbors of each of the samples belonging to the majority class. The ability to control the number of analyzed neighbors and the percentage of undersampling let you fine-tune the algorithm to find such a set of majority objects which allows to achieve the satisfactory accuracy of classification.

The following notations are established to make presentation of the algorithm more clear. S is the training dataset with m examples (i.e., $|S| = m$) defined as: $S = \{(x_i, y_i)\}$, $i = 1, ..., m$, where $x_i \subset X$ is an instance in the n-dimensional feature space, and $y_i \subset Y = \{1, ..., C\}$ is a class identity label associated with instance x_i[1]. $S_{min} \subset S$ and $S_{maj} \subset S$ are the set of minority and majority class examples in S, respectively.

The arguments of the function are: the set of elements belonging to the dominant class – S_{maj}, the number of nearest neighbors analyzed for each majority object – k, and the percentage of undersampling to carry out – P. The result of the function is a subset of the majority class.

Algorithm works as follows. For each element of S_{maj} subset its k nearest neighbors are found and their indexes are stored in the auxiliary matrix. In the next step the duplicated indexes are identified. If the total number of unique indexes is greater than Z, where Z is the number of examples which should be removed, then Z random objects from the found set are selected for discarding. However, if due to the nature of the dataset too many objects' indexes are repeated and consequently the number of unique indexes is less than or equal to Z, then all found nearest neighbors are removed. In this case the assumed percentage of undersampling – P may not be achieved. Such a situation can

---

[1] For the two-class classification problem $C = 2$.

occur primarily when the value of parameter Z is very large while parameter k relatively small.

Due to the fact that in the proposed method for each sample at most k of its neighbors are removed, it reduces the risk of removing too many points from a certain area in comparison with the standard random undersampling method. Consequently, it also decreases danger of losing important information.

4   Experiments

The KNN_RU method was compared with several generally known balancing techniques in order to verify whether the proposed algorithm can effectively solve the problem of class imbalance in practice. To make the comparisons the original random undersampling and three heuristic methods: *ENN*, *NCL*, and *Tomek links* were used. All these methods are briefly described in the previous section.

The outline of the performed experiments was as follows:

- Each analyzed dataset was undersampled with five methods. The obtained subsets were treated by six classifiers, such as Naive Bayes, Rule Induction, k-Nearest Neighbor, Random Forests, Support Vector Machines and Neural Networks, and the precision of classification was measured by 6 metrics.
- The tested undersampling methods used parameters k (number of nearest neighbors) and/or P (percentage of undersampling). The sampling of the datasets was performed for the odd values of $k = 1$, 3, 5, 7 and $P = 10\%$, 20%, 30% ... until full balance was achieved. It allowed to find the balancing level that gave the best precision of classification.
- To make the analyses more complete, the results were also compared with those based on the original set of data (i.e. without balancing).

Experimental Environment

The presented research was performed using the RapidMiner ver. 9.8 and R software environments.

A lot of conventional classification algorithms are often biased towards the majority class and consequently cause higher misclassification rate for the minority examples. All objects are often assigned to the dominant, i.e. negative, class regardless of the values of the feature vector. In [33] authors included a brief introduction to some well-developed classifier learning methods and indicated the deficiency of each of them with regard to the problem of class imbalance. Unfortunately, there are no clear guidelines which classifiers should be used in relation to imbalanced data, therefore descriptions of tests carried out using various classifiers can be found in the literature. Thus, six classifiers based on different paradigms and varying in their complexity were used in presented study: Naive Bayes [20], Rule Induction [25], k-Nearest Neighbor [1], Random Forests [6], Support Vector Machines [9], and Neural Networks [8].

With regard to the classifiers used, the following parameter values were set: (a) Laplace correction was used for NB classifier; (b) The entropy was taken into account as the criterion for selecting attributes and numerical splits for RI classifier; (c) One neighbor was selected for determining the output class in the case of KNN. Additionally Mixed Measures were used to enable the calculation of distances for both nominal and numerical attributes. For numerical values the Euclidean distance was calculated. For nominal values a distance of 0 was taken if both values were the same, and a distance of 1 was taken otherwise; (d) 'Number of trees' parameter which specifies the number of random trees to generate in RF was set to 100; (e) There was used a feed-forward neural network trained by a back propagation algorithm (multi-layer perceptron). 'Training cycles' parameter which specifies the number of cycles used for the neural network training was set to 500. The 'hidden layer size' parameter was set to -1^2. The default settings were applied for the remaining parameters.

Five independent 5-fold cross-validation experiments were conducted and the final gained results were the average values of these tests[3]. It was stratified validation, which means that each fold contained roughly the same proportions of examples from each class. To optimize parameters of used undersampling algorithms double cross-validations were carried out – the inner one to guide the search for optimal parameters while an outer one to validate those parameters on an independent validation set.

To evaluate tested methods 18 datasets which considered clinical cases, the biology of the shellfish population, proteins in yeast's cell and criminological investigations, was taken from KEEL[4] and UCI[5] repositories. When a dataset was not two-class, each class was successively considered as the positive, while the remaining were merged, thus forming one negative majority class. For that reason some file names have consecutive numbers in their suffix (Table 1). The ratio between the number of negative and positive instances, IR, ranged from 1.82 to 41.4 depending on the dataset.

Regarding performance measures, we chose ones, which are recommended as the most valuable for evaluating imbalanced data classifications [23]: sensitivity, specificity, Balanced Accuracy (BAcc), Geometric Mean (GMean), F-Measure, and Cohen's Kappa statistic.

Results and Discussion

As mentioned in the paragraph outlining the experiments, the classification tasks were performed for the original datasets as well as for the ones which were undersampled with the use of various methods.

The results in terms of F-Measure showed that KNN_RU got the best result in 73 out of 108 tested combinations (18 datasets * 6 classifiers). In the next 5 cases KNN_RU gave the best result ex aequo with the other tested methods. Considering the Kappa metrics the proposed method outperformed the remaining ones in 70 cases and in 11 other cases more than one method achieved the

---

[2] I.e. the layer size was calculated as: 1+(Number of attributes+Number of classes)/2.

[3] We used 5-fold cross-validation instead of 10-fold cross-validation because one of the tested datasets (Glass5) had fewer than 10 examples of the minority class.

[4] http://www.keel.es/datasets.php.

[5] http://archive.ics.uci.edu/ml/index.html.

Table 1. Datasets summary descriptions

Name	Instances	Features	IR	Name	Instances	Features	IR
Abalone	731	8	16.4	Glass4	214	9	15.46
Breast	483	9	18.32	Glass5	214	9	22.78
Ecoli1	336	7	3.36	Glass6	214	9	6.38
Ecoli2	336	7	5.46	Vowel0	988	13	9.98
Ecoli3	336	7	8.6	Yeast1	1484	8	2.46
Ecoli4	336	7	15.8	Yeast3	1484	8	8.1
Glass0	214	9	2.06	Yeast4	1484	8	28.1
Glass1	214	9	1.82	Yeast5	1484	8	32.73
Glass2	214	9	11.59	Yeast6	1484	8	41.4

same best result as KNN_RU. Table 2 summarizes the number of the best results of the balancing methods in terms of the F-Measure, Kappa, BAcc, and GMean metrics. More detailed results for the BAcc and Kappa metrics can be found on the Gitlab[6].

Table 2. Summary of the classification best results in terms of the analyzed metrics between KNN_RU and the other balancing methods

Metrics	KNN_RU	Equal results	Other methods
F-Measure	73	5	30
Kappa	70	11	27
BAcc	61	7	40
GMean	56	8	44

The proposed KNN_RU solution in most cases gives better results than the other tested methods. However, the obtained results are not always equally good. There may be several possible reasons for this state of affairs. One of them is the fact of different data characteristics. In [19,28] it is concluded that class imbalance itself does not seem to be a big problem, but when it is associated with highly overlapped classes it can significantly reduce the number of correctly classified examples of minority class.

To analyze the problem more thoroughly the scatterplots were generated for all tested datasets. Two exemplary plots for Glass2 and Glass4 are presented in Fig. 1(A)–(B)[7]. Both sets have the same collection of attributes and are charac-

---

[6] https://gitlab.aei.polsl.pl/awerner/knn_ru.
[7] To facilitate visualization and enable the presentation of an exemplary dataset in a two-dimensional space there was performed dimensionality reduction via principal component analysis.

terized by a similar degree of imbalancing (IR = 11.59 and IR = 15.46, respectively), but they vary in level of class overlapping (higher for Glass2). The results obtained for the Glass2 dataset are much worse than those for Glass4, although the imbalance ratio for the first set is slightly lower. The maximum value of BAcc that was achieved for the tested combinations of classifiers and undersampling methods does not exceed 0.8, while for Glass4 the values are much higher (they reach even 0.9516).

Considering the Kappa metrics, for the Glass2 set KNN_RU algorithm gives the best results for 5 out of 6 tested classifiers. However, the differences are small. None of the tested undersampling methods gives satisfactory results for the Glass2 set. Particularly poor results are obtained for Naive Bayes and SVM classifiers. In these cases, Kappa does not exceed 0.1 for any of the tested undersampling methods.

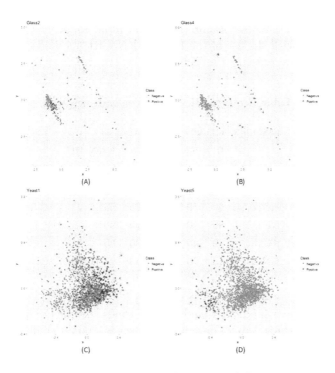

Fig. 1. Scatterplot for Glass2 (A), Glass4 (B), Yeast1 (C) and Yeast5 (D) datasets.

Figure 1(C)–(D) presents the scatterplots for Yeast1 and Yeast5 datasets. In this case, the sets differ substantially in degree of imbalancing (IR = 2.46 and IR = 32.73 respectively). Although the size of the minority class in the Yeast1 set is over 13 times larger than in the Yeast5 one, the results obtained for the first of mentioned dataset are much worse in many of the analyzed cases. This confirms that the difficulty in separating the small class from the dominant one is a very important issue and that the classification performance cannot be stated explicitly taking into consideration only degree of imbalancing, since other factors such as sample size and separability are equally valid. In regard to the Yeast1 set, the maximum achieved value of Kappa is a little over 0.4 (0.416 for NN classifier) while for Yeast5 more then 0.6 (0.625 for RF classifier). It is also noteworthy that for the Yeast5 set the proposed KNN_RU algorithm gives the best results for 5 out of 6 tested classifiers. On the remaining one KNN_RU gives the best result ex aequo with RU methods.

It should be emphasized that in about 80% of the analyzed cases the results which were found to be optimal for the KNN_RU method were achieved for a lower level of undersampling in comparison with the RU method. It is important because any intrusion of the source dataset by its under- and/or over-sampling can cause undesirable data distortion. The major drawback of undersampling is that it can discard potentially useful data that could be important for the learning process. Therefore, it is significant to obtain satisfactory classification accuracy with the least possible interference in input data. The average levels of undersampling with the use of the RU and KNN_RU methods for the Glass2 and Abalone sets are presented in Table 3.

Table 3. Average level of undersampling (%)

Classifier	Glass2		Abalone	
	RU	KNN_RU	RU	KNN_RU
NB	75	60	10	10
RI	70	55	80	60
KNN	60	30	45	55
RF	80	40	90	90
SVM	90	40	80	80
NN	60	55	45	20

In the case of the T-link, ENN, and NCL methods the number of removed elements comes from the nature of data and the user has no possibility to influence the level of undersampling. For the analyzed datasets these methods tended to remove less number of examples than the RU and KNN_RU methods. However, it can be concluded that in most cases the number of samples removed using the analyzed heuristic methods was not optimal. It means that the tested classifiers often achieved weaker performance than in the case of undersampling using the RU and/or KNN_RU methods. Taking into consideration BAcc measure only in 6 of the analyzed cases the tested heuristic methods of undersampling gave the best results. There were: (a) the combination of the Glass2 dataset, the KNN classifier and the T-link undersampling method; (b) the Ecoli1 dataset, the RI classifier and T-link; (c) the Ecoli2 dataset with the NN classifier sampled using NCL; (d) the Yeast4, the RI classifier and ENN; (e) the Yeast5 dataset, the NB classifier and NCL ex aequo with ENN undersampling method; (f) the Abalone dataset, the KNN classifier and NCL.

It is well known that the choice of the evaluation metrics can affect the assessment of which tested methods are considered to be the best. The results presented in Table 2 confirm this fact. The various combinations of the classifiers and undersampling methods are often ranked differently by various evaluation measures. For example, according to BAcc for the Abalone dataset and the Random Forest classifier the traditional method of random undersampling (RU) proved to be the best. However, according to the Kappa metrics the proposed KNN_RU algorithm gave better results than RU. Figure 2 presents the results of all analyzed performance measures for the mentioned Random Forest classifier and the Abalone dataset.

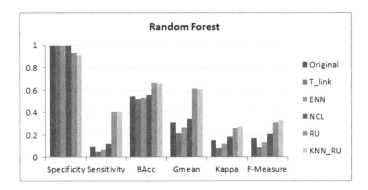

Fig. 2. Performance measures for Random Forest classifier and Abalone dataset.

To conduct a more complete comparison of the obtained results, the statistical tests [10] were applied. The undersampling methods for each dataset and classifier were ranked separately. The best method received rank 1, the second best – rank 2, and so on. In case of ties, average ranks were assigned. For instance, if two methods reached the same best result they both got rank 1.5. Then, average ranks across all datasets were calculated.

To compare the obtained average ranks, the omnibus Friedman test [13,32] with Iman-Davenport extension [18] was applied. It allowed to check whether there were any significant differences, with 95% confidence level, between the tested methods. There was used the Conover-Iman post-hoc test[8] to find the particular pairwise comparisons which caused these differences.

Table 4 shows the results of the comparison of the tested undersampling methods calculated for the Kappa metrics. It presents the average ranks for each method and the Conover-Iman p-value between KNN_RU and the method from a given row. In all cases the average ranks for KNN_RU are lower than those obtained for the other methods, and that allows to treat the proposed

---

[8] This test is considered to be more powerful than the Bonferroni-Dune one.

Table 4. Average ranks for the Kappa metrics

Classifier	Under-sampling method	Avarage rank	p-Conover-Iman	Classifier	Under-sampling method	Avarage rank	p-Conover-Iman
NB Iman-Davenport = 32.435773 p<0.000001	KNN_RU	1.583		RF Iman-Davenport = 42.389222; p<0.000001	KNN_RU	1.361	
	Original	3.917	<0.000001		Original	4.833	<0.000001
	T-link	4.944	<0.000001		T-link	4.722	<0.000001
	ENN	3.889	<0.000001		ENN	4.389	<0.000001
	NCL	4.667	<0.000001		NCL	3.694	<0.000001
	RU	2	0.23156		RU	2	**0.04982**
RI Iman-Davenport= 9.96058; p<0.000001	KNN_RU	1.861		SVM Iman-Davenport = 38.42044; p<0.000001	KNN_RU	1.306	
	Original	4.44	<0.000001		Original	5	<0.000001
	T-link	4.194	**0.000003**		T-link	4.139	**0.000001**
	ENN	3.806	**0.000068**		ENN	4.472	<0.000001
	NCL	4.111	**0.000006**		NCL	4.028	<0.000001
	RU	2.583	0.123512		RU	2.056	**0.027813**
KNN Iman-Davenport = 8.040611; p = 0.000003	KNN_RU	1.806		NN Iman-Davenport = 13.31763; p<0.000001	KNN_RU	1.361	
	Original	4.361	**0.000001**		Original	4.694	<0.000001
	T-link	4.167	**0.000001**		T-link	4.444	<0.000001
	ENN	3.583	**0.000459**		ENN	3.722	**0.000003**
	NCL	4.139	**0.000007**		NCL	3.889	**0.000001**
	RU	2.944	**0.021884**		RU	2.889	**0.001783**

solution as a control method. The Iman-Davenport test for each used classifier gives p-value less than 0.05. It means that it rejected the hypothesis that the compared undersampling methods were equivalent. The results in bold indicate the methods which are significantly worse than the KNN_RU one. It can be seen that the proposed undersampling method is significantly better than the heuristic ones. However, the comparison with random undersampling (RU) shows that although KNN_RU is better, i.e. has lower average rank in 2 cases: for the NB and RI classifiers, the statistical significance of the observed differences cannot be confirmed.

Better results obtained with the KNN_RU method compared to heuristic ones can be explained by the fact that it allows to influence the number of objects that should be removed. On the other hand, compared to the classic random method of undersampling, the proposed solution reduces the risk of removing too many points from a certain area.

It should be noted that although the method used in KNN_RU to remove objects is slightly more sophisticated than in the case of RU, it does not guarantee cleaning the decision surface, reducing class overlapping or removing noisy examples. Therefore, we decided to create a hybrid solution, which would first use a method that better detects and removes borderline or noisy cases and would apply the KNN_RU solution in the next step.

At the beginning, tests were carried out using the Glass2 dataset, which appeared to be one of the most problematic. Combinations T_link + KNN_RU and NCL + KNN_RU were tested. The obtained results confirmed validity of the conception of hybridization the methods. Table 5 presents the results with the use of basic version of KNN_RU and its combinations. One can see that KNN_RU preceded by T_link or NCL improves values of the classification metrics. Better results are highlighted in bold.

Table 5. Classification results for the Glass2 dataset

Classifier	KNN_RU				T_link + KNN_RU				NCL + KNN_RU			
	F-measure	BAcc	Kappa	Gmean	F-measure	BAcc	Kappa	Gmean	F-measure	BAcc	Kappa	Gmean
NB	0.196	0.634	0.065	0.583	**0.197**	**0.637**	**0.067**	**0.587**	**0.198**	**0.637**	**0.067**	**0.587**
RI	0.263	0.606	0.114	0.52	0.219	0.604	**0.136**	**0.573**	0.252	**0.620**	**0.162**	**0.584**
KNN	0.432	0.705	0.327	0.665	0.426	**0.799**	**0.35**	**0.798**	0.428	**0.746**	**0.365**	**0.729**
RF	0.182	0.559	0.112	0.456	**0.305**	**0.664**	**0.224**	**0.635**	**0.283**	**0.609**	**0.217**	**0.521**
SVM	0.105	0.527	0.073	0.242	**0.187**	**0.597**	0.054	**0.552**	0.173	**0.578**	0.034	**0.492**
NN	0.457	0.765	0.393	0.756	**0.542**	**0.827**	**0.486**	**0.824**	0.474	0.712	**0.432**	0.670

The averages for each measure (average across a column) were calculated. Analyzing the values obtained in this way the improvement was observed for both tested combinations. It was approximately from 3% for NCL + KNN_RU and BAcc up to 23% for T_link + KNN_RU and GMean. The use of the hybrid version of KNN_RU in most cases improves the results also in the remaining datasets.

5   Conclusions

Many researchers suggest that random undersampling is one of the more effective resampling methods [11, 26]. However, this method has a drawback, namely it does not allow to control which samples from the majority class are thrown away. To reduce – at least partially – this disadvantage we propose the KNN_RU algorithm which combines random approach with k-nearest neighbors analysis.

Six metrics for classification performance evaluation were examined and as it was shown in the experimental section the choice of quality metrics had an impact on the way the various undersampling methods were ranked. Nevertheless, the outcomes of classification experiments conducted with KNN_RU on eighteen datasets in most cases outperformed the results obtained for four compared undersampling methods.

To make a comparison of the KNN_RU method more comprehensive we also contrasted it with the previously developed one [3], which was mentioned in the

Introduction section. For the majority of tested data sets, the advantage of the current solution over the previous one has also been confirmed[9]. In our experiments all resampling methods were used only to perform the undersampling task, which means that chosen samples were removed exclusively from the majority class. However, some methods based on the idea of k-nearest neighbors could be used to remove examples from both classes. It means that each example misclassified by its nearest neighbors can be removed from the training set, regardless of the class it belongs to. It should insure more detailed data cleaning, and in consequence improve the accuracy of the classifications. Therefore, we plan to analyze the next combinations of the proposed method with other resampling methods which provide undersampling and have data cleaning capabilities.

References

1. Aha, D., Kibler, D.: Instance-based learning algorithms. Mach. Learn. **6**, 37–66 (1991)
2. Bach, M., Werner, A.: Cost-sensitive feature selection for class imbalance problem. In: Advances in Intelligent Systems and Computing . ISAT 2017. AISC, vol. 655, pp. 182–194. Springer, Cham (2018). https://doi.org/10.1007/978-3-319-67220-5_17
3. Bach, M., Werner, A., Palt, M.: the proposal of undersampling method for learning from imbalanced datasets. Procedia Comput. Sci. **159**(2019), 125–134 (2019). https://doi.org/10.1016/j.procs.2019.09.167
4. Bach, M., Werner, A., Żywiec, J., Pluskiewicz, W.: The study of under- and over-sampling methods' utility in analysis of highly imbalanced data on osteoporosis. Inf. Sci. Life Sci. Data Analysis **381**, 174–190 (2016). https://doi.org/10.1016/j.ins.2016.09.038, ISSN: 0020-0255, Elseviere
5. Beckmann, M., et al.: A KNN undersampling approach for data balancing. J. Intell. Learn. Syst. Appl. **7**, 104–116 (2015). https://doi.org/10.4236/jilsa.2015.74010
6. Breiman, L.: Random forest. In: Machine Learning. Springer, vol. 45(1), pp. 5–32 (2001). https://doi.org/10.1007/978-1-4419-9326-7_5
7. Chawla, N.: Data mining for imbalanced datasets: an overview, The Data Mining and Knowledge Discovery Handbook, pp. 853–867. Springer (2005). https://doi.org/10.1007/978-0-387-09823-4_45
8. Cheng, B., Titterington, D.M.: Neural networks: a review from a statistical perspective. Stat. Sci. **9**, 2–54 (1994)
9. Cortes, C., Vapnik, V.: Support-vector network. Mach. Learn. **20**, 273–297 (1995)
10. Derrac, J., et al.: A practical tutorial on the use of nonparametric statistical tests as a methodology for comparing evolutionary and swarm intelligence algorithms. Swarm Evol. Comput. **1**, 3–18 (2011)
11. Dittman, D., et al.: Comparison of data sampling approaches for imbalanced bioinformatics data. In: Proceedings of the 27 International Florida Artificial Intelligence Research Society Conference (2014)
12. Duan, L., et al.: A new support vector data description method for machinery fault diagnosis with unbalanced datasets. Expert Syst. Appl. **64**, 239–246 (2016)

---

[9] The complete comparisons are on Gitlab: https://gitlab.aei.polsl.pl/awerner/knn_ru.

13. Friedman, M.: The use of ranks to avoid the assumption of normality implicit in the analysis of variance. J. Am. Stat. Assoc. **32**(200), 675–701 (1937)
14. Galar, M., et al.: A review on ensembles for the class imbalance problem: bagging-, boosting-, and hybrid-based approaches. IEEE Trans. Syst. Man, Cybern., Part C: Appl. Rev. **42**(4), 463–484 (2012)
15. Chun, G.: Analysis of imbalanced data set problem: the case of churn prediction for telecommunication. Artif. Intell. Res. **6**(2), 93 (2017). https://doi.org/10.5430/air.v6n2p93
16. Haixiang, G., et al.: Learning from class imbalanced data: review of methods and applications. Expert Syst. Appl. **73**, 220–239 (2017). https://doi.org/10.1016/j.eswa.2016.12.035
17. Kaur, H., et al.: A systematic review on imbalanced data challenges in machine learning: applications and solutions. ACM Comput. Surv. (2019). https://dl.acm.org/doi/abs/10.1145/3343440
18. Iman, R., Davenport, J.: Approximations of the critical region of the fbietkan statistic. Commun. Stat.-Theor. Meth. **9**(6), 571–595 (1980)
19. Japkowicz, N.: Class imbalances: are we focusing on the right issue? ICML-KDD'2003 Workshop: Learning from Imbalanced Data Sets (2003)
20. John, G., Langley, P.: Estimating continuous distributions in Bayesian classifiers. In: 11th Conference on Uncertainty in Artificial Intelligence, San Mateo, pp. 338–345 (1995)
21. Krawczyk, B., et al.: Evolutionary undersampling boosting for imbalanced classification of breast cancer malignancy. Appl. Soft Comput. **38**, 714–726 (2016)
22. Lopez, V., et al.: An insight into classification with imbalanced data: empirical results and current trends on using data intrinsic characteristics. Inf. Sci. **250**, 113–141 (2013). https://doi.org/10.1016/j.ins.2013.07.007
23. Luque, A., et al.: The impact of class imbalance in classification performance metrics based on the binary confusion matrix. Patt. Recogn. **91**, 216–231 (2019)
24. Mao, W., et al.: Online sequential prediction of bearings imbalanced fault diagnosis by extreme learning machine. Mech. Syst. Signal Process. **83**, 450–473 (2017)
25. Michalak, M., Sikora, M., Wróbel, Ł.: Rule quality measures settings in a sequential covering rule induction algorithm - an empirical approach. In: Proceedings of the Federated Conference on Computer Science and Information Systems, pp. 109–118 (2015). https://doi.org/10.15439/2015F388
26. Mishra, S.: Handling imbalanced data: SMOTE vs. Random undersampling. IRJET **4**(08)((2017). ISSN: 2395 0072
27. Prati, R.C., Batista, G.E., Monard, M.C.: Data mining with imbalanced class distributions: concepts and methods. In: 4th Indian International Conference on AI (2009). ISBN 9780972741279
28. Prati, R.C., Batista, G.E.A.P.A., Monard, M.C.: Class imbalances *versus* class overlapping: an analysis of a learning system behavior. In: Monroy, R., Arroyo-Figueroa, G., Sucar, L.E., Sossa, H. (eds.) MICAI 2004. LNCS (LNAI), vol. 2972, pp. 312–321. Springer, Heidelberg (2004). https://doi.org/10.1007/978-3-540-24694-7_32
29. Richardson, A., Lidbury, B.: Enhancement of hepatitis virus immunoassay outcome predictions in imbalanced routine pathology data by data balancing and feature selection before the application of support vector machines. BMC Med. Info. Decis. Mak. **17**(1), 121 (2017)
30. Sandhan, T., Choi, J,Y.: Handling imbalanced datasets by partially guided hybrid sampling for pattern recognition. In: 22nd International Conference on Pattern Recognition, pp. 1449–1453 (2014). https://doi.org/10.1109/ICPR.2014.258

31. SCI2S Research Material on Classification with Imbalanced Datasets, A University of Granada Research Group, October 2020. http://sci2s.ugr.es/imbalanced
32. SCI2S Research Material on the Use of Non-Parametric Tests for Data Mining and Computational Intelligence, October 2020. A University of Granada Research Group. http://sci2s.ugr.es/sicidm
33. Sun, et al.: Classification of imbalanced data: a review. Int. J. Pattern Recogn. Artif. Intell. **23**(4), 687–719, World Scientific (2009)
34. Tomek, I.: Two modifications of CNN. IEEE Trans. Syst. Man Commun. **SMC-6**, 769–772 (1976)
35. Hou, W.-H., et al.: A novel dynamic ensemble selection classifier for an imbalanced data set: an application for credit risk assessment Knowledge-Based Systems (2020). https://doi.org/10.1016/j.knosys.2020.106462
36. Wilson, D.L.: Asymptotic properties of nearest neighbor rules using edited data. IEEE Trans. Syst. Man Cybern. **2**(3), 408–420 (1972)

Predictability Classes for Forecasting Clients Behavior by Transactional Data

Elizaveta Stavinova[✉], Klavdiya Bochenina, and Petr Chunaev

National Center for Cognitive Technologies, ITMO University,
Saint Petersburg 199034, Russia
estavinova@icloud.com, chunaev@itmo.ru

Abstract. Nowadays, the task of forecasting the client's behavior using his/her digital footprints is highly demanded. There are many approaches to predict the client's next purchase or the next location visited that focus on achieving the best possible prediction quality in terms of different quality metrics. Within such approaches, the quality is however usually evaluated on the entire set of clients, without dividing them into classes with a different predictability rate of client's behavior. In contrast to the approaches of this type, we propose a method for the identification of the client's behaviour predictability class by means of a foreign trip in the next month by using only client's historical transactional data. In a sense, this allows us to estimate the quality of forecasting the client's foreign trip before the actual prediction procedure. Our experiments show that the approach is rather efficient and that the predictability classes obtained quite agree with the prediction quality classes found within the actual forecasting.

Keywords: Event forecasting · Predictability · Transactional data

1 Introduction

Forecasting the client's behavior is an extremely popular topic nowadays with different applications—from the prediction of the next purchase [19] to that of the future location [12]. There are many approaches that focus on achieving the best possible prediction quality in terms of different quality metrics. It is common that the predictions are made and evaluated due to training the forecasting models on a part of the data with further testing them on some test data. Within such approaches, the quality is however usually evaluated on the entire set of clients, without dividing them into classes with a different predictability rate of client's behavior. In contrast to the approaches of this type, we are interested in the identification of the client's behaviour predictability class by means of

This research was financially supported by the Russian Science Foundation, Agreement 17-71-30029 with co-financing of Bank Saint Petersburg.

The original version of this chapter was revised: the missing funding information was added. The correction to this chapter is available at
https://doi.org/10.1007/978-3-030-77967-2_62

a foreign trip in the next month (a particular case of an *event*) by using only client's historical transactional data.

To be more precise, we propose a classification method that exploits the idea that all clients can be divided into classes based on the predictability rate of a foreign trip in the next month according to their transactional history. Once the clients are divided into the classes, a bidirectional Long Short-Term Memory (LSTM) network [16] is trained on some data to identify the predictability class before the actual forecasting. This trained network can be further used to identify the predictability class of a new client or to rearrange the client's class in the case of changing his/her travel behavior. The fact that we know that a client belongs to one of the predictability classes gives us the opportunity to estimate the trip prediction quality for a client before the actual forecasting.

Earlier works in the field of predictability were dedicated to the measurement of the time series predictability [9] and the predictability of the features [8]. It should be mentioned that the goal of the time series predictability analysis is to estimate how possible is it to capture the time series patterns while the feature predictability analysis is aimed to the selection of such features which are useful for the predictive model. The novelty of our research consists in the estimation of the event predictability and the usage of it for the assessment of the model performance before the prediction. By event predictability we mean the predictability of time series consisting of categorical values that are binary labels indicating the event.

Let us mention that we use the LSTM network for making the predictions because of its advantage in remembering time dependencies which are important for this kind of task [5]. This network was already used for the prediction of the client's location [17] and proved its applicability to this problem.

It is important to note that our classification method seems to be useful for financial organizations to develop beneficial rates or personalized advertising campaigns for those clients whose travel behavior is more predictable.

The paper is organized as follows. Section 2 contains the related work on the topic of predictability and different measures which were developed to estimate it. Section 3 describes the methodology of the research connected with the forecasting the foreign trips and defines the predictability phenomena. Finally, Sect. 4 presents the data and gives a look at how the raw transactional data are processed to be used further. Section 4 presents the results and their analysis.

2   Related Work

The first definition of an unpredictable random variable was proposed in [2]. A random variable x_t is called unpredictable with respect to an information set Ω_{t-1} if the conditional distribution $F_{x_t}(x_t|\Omega_{t-1})$ and the unconditional distribution $F_{x_t}(x_t)$ of x_t coincide, i.e.

$$F_{x_t}(x_t|\Omega_{t-1}) = F_{x_t}(x_t). \tag{1}$$

In particular, if Ω_{t-1} consists of the past realizations of x_t, then (1) indicates that the knowledge about the past realizations does not improve the prediction quality of x_t. Note that the unpredictability of x_t in this sense is an inherent property of x_t that is independent of a prediction algorithm.

It is important to understand the differences between the concepts of the predictability and the prediction quality. These concepts are closely related but differ in the time moment of their determination as the predictability is determined before the prediction while the prediction quality is done after the model responses have been received. Thus, the predictability rate is an individual measure that is inherent to each object (time series, feature, event, etc.) and the prediction quality is an aggregated one, that is based on the model predictions. Further in this section a literature review on measures that are used to estimate the predictability of various objects is presented.

2.1 Predictability Measures for Time Series

A significant part of all works on predictability measures is represented by studies dedicated to measuring the predictability of a time series before the actual forecasting. In most cases, the predictability measure should answer the question: is it worth to make predictions for a given time series, or in other words, does this time series contain some patterns that could be inferred and learned by a forecasting model? A first measure that tries to answer the above-mentioned question is now called the Kaboudan coefficient and is proposed in [9]. In addition, the author conduct experiments on the usage of this measure on time series from the financial sphere. To measure the predictability, the coefficient exploits the idea of comparing the original time series with a time series obtained by random shuffling the values of the original one.

The coefficient was further used as a base for other predictability measures. For example, a modified coefficient is proposed in [3] to overcome the following two problems of the initial one: (a) the dependence between the coefficient value and the time series length and (b) the narrow range of the coefficient distribution in the case of a long-term series. A few years later, the authors of [14] further modify the Kaboudan coefficient and apply it for measuring the predictability rate of financial time series.

In addition to the papers applying predictability measures to financial time series, there is one on the analysis of time series describing the streamflow observed in river basins [20]. Its authors propose to estimate the predictability rate of a univariate time series via the so-called coefficient of efficiency:

$$CE(n, Q) = 1 - \frac{\sum_{i=1}^{n}(Q_i - \hat{Q}_i)^2}{\sum_{i=1}^{n}(Q_i - \bar{Q})^2},$$

where n is the test period size, Q_i is the actual series value, $\hat{Q}_i$ is the predicted value, $\bar{Q}$ is the average value of the observation period series.

Moreover the concept of predictability is used to analyze the efficiency of the genetic programming models applied to time series prediction [1,21]. The authors of [18] also work with it to determine the most appropriate predictors in the problem of genomic sequence identification. As for the multivariate time series, there is a measure that allows to estimate the expected decrease in the prediction error of a multivariate model in relation to a univariate one [13].

2.2 Predictability Measures for Features

The purpose of the features predictability estimation consists in the selection of such a feature set that describes the object behavior in the best way and allows the model to make predictions of a desired quality. In fact, if a feature does not contain information about the future, its usage may add randomness to model responses and prevent it from making an accurate prediction. In the case of large input data dimensions, it becomes necessary to match the initial features with a certain feature set of a smaller dimension which can help the model to make more stable and efficient forecasts. However, standard dimensionality reduction algorithms are focused on the preserving the data properties that are not related to the predictability, and therefore there is a possibility of missing important information contained in the data.

One of the works devoted to the determination of the features predictability is [8], which identifies the most useful features for predicting the remaining time of the system performance. The authors define the predictability rate as a function that depends on the prediction horizon, the model class, the model parameters and the required accuracy threshold. The proposed predictability measure pulls the threshold and the accuracy achieved by the model into a single value between 0 and 1. Next, the set of features and the model providing the best predictability are chosen by brute force.

There are a few works in the field of the predictive features extraction developed for multivariate time series. The recently proposed method called Forecastable Component Analysis [6] is in fact one of the methods for reducing the dimension of time-dependent signals. Predictive Feature Analysis [15] is an unsupervised learning algorithm that aims to select only those input signals that behave as predictable as possible. Another approach to the dimension reduction of the input data is called Slow Feature Analysis [22]. It explicitly uses time dependencies in the data and distinguishes slowly changing features which can be regarded as predictable ones.

3   Method

3.1 Predictability Measure

In this work we face the task of event prediction and therefore adapt (1) for providing the following definition of event predictability. If the distribution function of the label at the moment t does not depend on its previous values, i.e.

$$F(label_t | \Omega_{t-1}) = F(label_t), \tag{2}$$

where F is the distribution function, Ω_{t-1} is previous values of the label, then the label is called *unpredictable*. Here a label means a binary indicator of the event at time step t (1 is when the event takes place while 0 when it does not).

As for the predictability rate, it can be estimated by the prediction error-based approach. Often the predictability rate is called the sample predictability rate because it is computed using the forecast errors. Here, we use the following simple sample predictability rate:

$$C(n, Q, m) = 1 - \frac{1}{n} \sum_{i=1}^{n} |Q_i - \hat{Q}_i| \in [0, 1],$$

where n is the test period size, Q_i is the actual event indicator, $\hat{Q}_i$ is the predicted probability of the event, m is a forecasting model.

The proposed coefficient is used in our method to divide all clients into classes according to the predictability rate of a foreign trip in the next month. This division is done after the model predictions are obtained for the train data. Basing on the values of C, we will determine two predictability classes, separated by the value of median($\mathbf{C}$), where $\mathbf{C}$ is the set of C-values for the entire dataset of clients. The motivation for this division in two classes is to distinguish objects with high and low predictability.

3.2   Predictability Class Identification

We identify the client's behavior predictability class by computing the value of C for a chosen dataset and a model. Thus, for the identification we should obtain model predictions and compare them with the real data to know the predictability rate. After that, we can estimate how predictable is the client's behavior. But in practice, it will be very useful to skip the step of using a model and having only the event indicators for the past few months claim if this client has predictable trip behavior or not. That is why we developed a method for the predictability class identification by the sequence of the event indicators.

In fact, this method is the way of solving the classification task. The goal is to identify which predictability class a client belongs to having only feature vectors of the train period. This task is as close to practice as possible, since at a certain point in time we have access only to the observation period data without possibility to access the data from the future (i.e. from the test period).

To solve this problem of the sequence classification we use a Bidirectional LSTM network [16]. The *input* of this network is a set of categorical sequences consisting of the event indicators (or the number of the events) with the step of one month. The length of the sequence is chosen to be six. The network *outputs* the predictability class for each sequence from the input. To train this model, we extracted the sequences describing the client's trip behavior throughout the last six months of the train period and estimated these clients predictability class using the prediction model, its answers and the coefficient C. After training the classification model on the data for the last six months, the quality is measured using the next six months of the transaction history. The main idea here is obtaining the trained model which can be further used in the case of new data arriving when we should recalculate the client predictability class or identify the predictability class for a new client.

4   Results

4.1   Data Description and Processing

For our experiments we use three transactional datasets from Russian banks.

Firstly, we describe two datasets that contain transactions made by clients of a large regional Russian bank and that cannot be made publicly available due to the bank's policy. The first dataset ($D1$) contains the transaction history of almost 3,000 clients over two years (from January 2016 to January 2018). The second dataset ($D2$) consists of transactions made by 6,000 clients in 2018 (from January to December). Although $D2$ has the shorter observation period, it contains three times more transactions than in the first dataset. Thus, $D1$ is sparse and may have the incomplete transaction history for some clients. Also, one of the possible explanations of this fact is the difference in the clients samples: the first sample of clients can be more passive in their spending behavior.

Now we describe the third dataset used that is publicly available and serves for the purposes of reproducibility of our results. Namely, this dataset (called Raiffeisen below) is from a Kaggle competition initiated by the Raiffeisen bank. This dataset as well as the implementation of the approaches described in this paper can be accessed through Github[1]. The dataset contains the transaction history of 10,000 clients and describes their spendings during 2017 (from January to December).

The structure of the datasets is very similar so we can process them in a rather unified way. The first stage of the data processing is the categorization of transactions based on the Merchant Category Code (MCC). All MCCs are divided into 87 categories. Then all transactions are divided into the two groups according to the location: made in Russia and abroad. This is done using the ISO country code that is specified in the transaction details. We use feature vectors that characterize the client spending in each month to formalize his/her behavior. These vectors contain the information about the amount of money spent in roubles by the client in a particular month in each category. After that, a Yeo-Johnson transformation [23] is applied to these vectors to create a monotonic transformation of the data. The vectors obtained have 88 dimensions, where the first 87 coordinates correspond to 87 categories of MCC codes and the 88th coordinate is the label of the client's location in this month. The goal is to predict the client's location label in the next month. In this problem statement the month has the abroad location label if a client has at least one abroad transaction during this month. It should be mentioned that transactions made in Internet are marked as home transactions.

As for the Raiffeisen dataset, the prediction of a foreign trip is difficult there since only approximately 0.08% of clients took at least one foreign trip during 2017. It is likely that training the model on such a small amount of data may result in a poor forecasting quality. That is why it was decided to predict not the foreign trip but the fact of the transaction in a particular category.

---

[1] https://github.com/stavinova/predictability-classes.git.

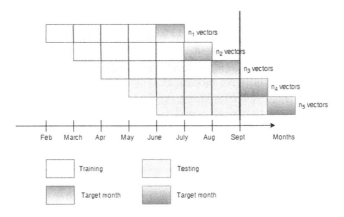

Fig. 1. The time-based cross validation procedure (the months correspond to $D1$). (Color figure online)

The restaurants category is chosen to be predicted so that the task of forecasting it is rather challenging and, at the same time, there is enough information to successfully complete the training process (the frequency of restaurants-related transactions is 0.41). This goal reformulation let us use the data related to all clients and to train the model of a sufficient quality. Moreover, this task provides some changes in the feature vectors. They contain the information about the number of transactions made by the client in a particular month in each category. No transformation is performed with these vectors and, of course, there is no need in the client's location label, thus, the vectors have 87 dimensions.

Note that in what follows, an event means a foreign trip for the datasets $D1$ and $D2$ and a visit to a restaurant for the Raiffeisen dataset.

After the stage of data preprocessing and feature vectors extraction, all vectors are divided into subsequences of vectors as follows. Each subsequence contains seven vectors (i.e., months) for $D1$ and five vectors in case of $D2$ and the Raiffeisen dataset, all elements of the subsequence describe the spending of one client and this subsequence is continuous during these months. Next, the train period ($D1$: from November 2016 to August 2017, $D2$: from January 2018 to June 2018, Raiffeisen: from January 2017 to June 2017) and the test period ($D1$: from September 2017 to January 2018, $D2$: from July 2018 to December 2018, Raiffeisen: from July 2017 to December 2017) are defined. All subsequences which end in months of the test period are set aside for testing the quality of the model that is trained using the feature vectors from the training period. Model training process is based on the time-based cross validation procedure which is shown in Fig. 1.

The LSTM network is used to predict an event for a client. The number of hidden layers and the number of neurons are determined via several experiments with different settings. Six experiments are conducted with different network configurations: 1, 2, 3 hidden layers with 32, 64 neurons in each hidden layer. A combination of two hidden layers with 64 neurons is selected basing on the performance in the experiments. The learning rate value is set 0.001 basing on another series of experiments.

In the current problem statement, the considered event is a foreign trip (or, in the case of the Raifeissen dataset, a visit to a restaurant) which will probably be made by a client in the next month. It is necessary to answer the question: how predictable is this event? Or more precisely, how predictable is the fact of a foreign trip (or a visit to a restaurant) in the next month for a client in terms of the described LSTM model with feature vectors as inputs?

4.2 Predictability Measurement

Earlier in this paper, we introduced a measure for estimating the predictability rate which we call the coefficient C. This coefficient is based on the values which are predicted by the model for the event forecasting. Our goal is to calculate the values of this coefficient for each client in all datasets. The coefficient values are computed by the following procedure. Firstly, the model based on the LSTM network is trained on feature vectors from the training period. After receiving the model responses, the values of the coefficient C for each client are calculated using the actual event indicators from the test period, predicted probabilities of the event and the length of the test period for a client. Figure 2 shows the distribution of coefficient values for the three datasets.

The next step is the creation of predictability classes based on the values of the coefficient C. We decide to divide the clients into classes with the value of C separated by median(C), where C is the set of C-values for the entire dataset of clients. This division provides the classes of almost equal size, moreover, this parameter should not be readjusted for different samples. Note that median(C) = 0.87 for $D1$, median(C) = 0.68 for $D2$ and median(C) = 0.59 the Reiffeisen dataset. Two classes, of low and high predictability, are further formed according to the value of the coefficient C. The distribution of the clients by predictability classes is shown in Table 1. The clients from the high predictability class have more predictable behavior in terms of LSTM model since the model answers are closer to the actual situation comparing to those for the low predictability class.

After defining the predictability classes, the study of the forecast quality for each class is conducted. The precision and recall metrics are used for the

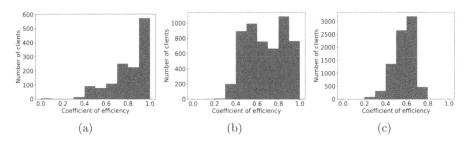

| | (a) | (b) | (c) |

Fig. 2. Distribution of the coefficient C values: (a) $D1$, (b) $D2$, (c) Raiffeisen. (Color figure online)

Table 1. The predictability classes obtained

Class name	Coefficient C values	$D1$	$D2$	Raiffeisen
High predictability	$[\mathrm{median}(\mathbf{C}), 1]$	674	2687	4030
Low predictability	$[0, \mathrm{median}(\mathbf{C}))$	675	2687	4031

quality assessment. Figure 3 shows the precision-recall curves for the predictability classes described in Table 1. Moreover, the frequency of events in the test set is shown to compare the classifiers quality with the random guessing. For all datasets the proposed classifiers are better than the random guessing. Figure 3 shows that the clients from $D1$ have far less foreign trips comparing to $D2$ where the average trip frequency is 0.47.

The high predictability class in the dataset $D1$ has perfect prediction quality but the foreign trip frequency for them are almost zero (0.001). That means that almost the half of the $D1$ clients did not take a foreign trip in the test period. Thus, the high predictability class in $D1$ consists of the clients who stayed at home for the test period and are correctly classified by the model, and that is why the curve corresponding to the low predictability class almost coincides with the curve for all the clients.

As for the dataset $D2$, the division into the predictability classes gives us a class with almost perfect precision-recall curve and a class with the curve located a little higher than the random guessing. It should be mentioned that the trip frequencies for these two classes are close to each other (0.43 and 0.51, respectively). This means that the proposed division of $D2$ into these predictability classes is useful. Moreover, it shows that in the case of $D2$ the model is able to predict trips not only by their frequency.

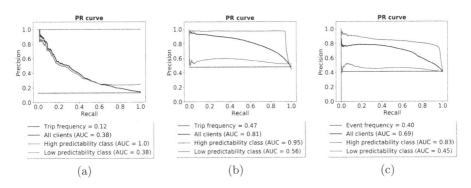

(a)   (b)   (c)

Fig. 3. Precision-recall curves for different predictability classes obtained after the model prediction and the forecast for six months ahead: (a) $D1$, (b) $D2$, (c) Raiffeisen. (Color figure online)

Table 2. Confusion matrices for predictability class identification on: (a) $D1$, (b) $D2$, (c) Raiffeisen

(a)		Forecast	
		High	Low
Actual	High	548	1
	Low	336	267

(b)		Forecast	
		High	Low
Actual	High	2050	544
	Low	170	2458

(c)		Forecast	
		High	Low
Actual	High	1048	305
	Low	366	877

The prediction quality for the Raiffeisen dataset classes is quite similar to the quality of $D2$ classes. The trip frequencies for both predictability classes are close to each other, too.

4.3 Predictability Class Identification

Earlier in this paper, we introduced the approach of division the clients into two predictability classes using the coefficient C. That approach assumes that the model predictions are already obtained and the coefficient values are computed after the forecast moment. But now we would like to focus on the task of the client predictability class identification before the forecast moment. For that purpose, we solve the problem of the client's predictability class identification using the data related exclusively to the train period. This problem is solved using the method proposed in the previous section. The confusion matrices that allow us to estimate the Accuracy of the predictability class identification for three different datasets are presented in Table 2. The Accuracy metric of the predictability class identification achieved on the dataset $D1$ is 70.74%, on the second dataset $D2$ is 86.32%, while that on the Raiffeisen dataset is 74.15%.

It is possible to see that the results for $D1$ are too optimistic as the classifier marks 336 clients as high predictable while their actual behavior is poorly predictable. As for $D2$, the results are more balanced and accurate. In the case of the Raiffeisen dataset, the results are quite balanced but the Accuracy is not as high as on $D2$.

4.4 Analysis of the Results

To analyze the effectiveness of the division into predictability classes before the prediction moment, we evaluate the prediction quality for classes obtained by the identification algorithm. Figure 4 shows the prediction quality for the estimated predictability classes compared with the overall prediction quality. To understand the predictability class identification quality, we can compare Fig. 3 and Fig. 4 because they should be as close to each other as possible.

In case of $D1$ the estimated high predictability class became the class with low predictability and drawn in Fig. 4 by the red color. This situation appeared because the high predictability class in the original division consists of the clients who stayed at home during the test period and a few clients who took a trip which is guessed by the model. But now according to Table 2(a), the estimated high

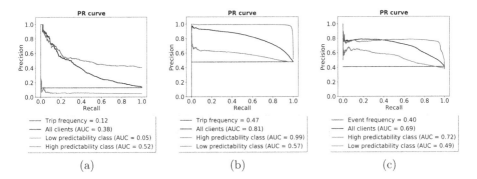

Fig. 4. Precision-recall curve for the estimated classes: (a) $D1$, (b) $D2$, (c) Raiffeisen. (Color figure online)

predictability class contains not only high predictable clients but also 336 low predictable clients. In this situation, the trip frequency for this class is increased (0.04) and the model does not guess the correct answer for the new clients. That is why the precision-recall curve is located on the value of 0.04 in this case. The precision-recall curve for another predictability class is situated higher than on Fig. 3 because it becomes smaller than the original one according to Table 2. As for the dataset $D2$, the curves represented in Fig. 3 and Fig. 4 are very close to each other, thus, the class identification algorithm works at the high quality level. Identification results for the Raiffeisen dataset are worse than on $D2$ but still precise enough to distinguish two classes of high and low predictability which can be useful in practise.

It is worth saying that the task of the event prediction using only the transactional data is highly depends on the quality of the data. This fact is demonstrated by the three different datasets. In the case of $D2$ and Raiffeisen, our approach of estimating the predictability class before the prediction moment can be used to infer a high predictability class with the prediction quality higher than the overall one. In the case of $D1$, we showed that the high predictability class consists of the clients almost without trips whose behavior is guessed by the model.

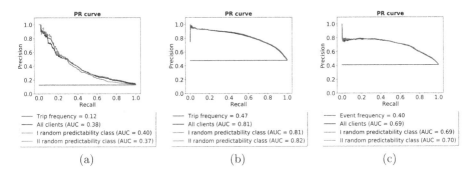

Fig. 5. Precision-recall curves for random classes: (a) $D1$, (b) $D2$, (c) Raiffeisen. (Color figure online)

To demonstrate the effectiveness of the proposed division into predictability classes, we performed the random division of all clients into two classes and estimated the prediction quality for them. The results are shown in Fig. 5. It can be seen that the precision-recall curves in this case are very close to the original one which is obtained for the data without division. It emphasizes the usability of the proposed approach as it suggests a reasonable division of clients while the random division fails.

5 Conclusion

We have proposed an approach for measurement the predictability of an event, for example, a foreign trip or a visit to a restaurant in the next month by a client after obtaining the model predictions. For a predictability estimation we proposed a special measure comparing the actual event indicator and the probability of this event estimated by the model. Using the coefficient values we divided the clients into two groups according to the event predictability for them after the prediction moment. After that, we proposed a classification method for the identification of the predictability class for a client before the prediction by the usage of the historical data. The method helps to infer the class of clients whose behavior (events) can be predicted with high quality (on the condition that these clients have a complete transaction history). If we know that a client belongs to one of the predictability classes, we can further estimate the prediction quality for this or another client, similar in behavior to the former, before the actual forecasting.

Our future work will be related to testing our approach in the situations where data concept drift or another non-stationarity is present and where specific forecasting methods should be applied, see e.g. [4,7,10,11]. It seems that in such cases one should be careful about a possible predictability rate drift of the client's behavior, too.

References

1. Chen, S.H., Navet, N.: Failure of genetic-programming induced trading strategies: Distinguishing between efficient markets and inefficient algorithms. In: Chen, S.H., Wang, P.P., Kuo, T.W. (eds.) Computational Intelligence in Economics and Finance, pp. 169–182. Springer, Heidelberg (2007). https://doi.org/10.1007/978-3-540-72821-4_11
2. Clements, M., Hendry, D.: Forecasting Economic Time Series. Cambridge University Press (1998)
3. Duan, M.: Time series predictability. Ph.D. thesis, Marquette University (2002)
4. Gama, J.a., Žliobaitundefined, I., Bifet, A., Pechenizkiy, M., Bouchachia, A.: A survey on concept drift adaptation. ACM Comput. Surv. 46(4) (2014). https://doi.org/10.1145/2523813
5. Gers, F.A., Schraudolph, N.N., Schmidhuber, J.: Learning precise timing with LSTM recurrent networks. J. Mach. Learn. Res. 3, 115–143 (2002)

6. Goerg, G.: Forecastable component analysis. In: International Conference on Machine Learning, pp. 64–72 (2013)
7. Janardan, Mehta, S.: Concept drift in streaming data classification: algorithms, platforms and issues. Procedia Comput. Sci. **122**, 804–811 (2017). 5th International Conference on Information Technology and Quantitative Management, ITQM 2017
8. Javed, K., Gouriveau, R., Zemouri, R., Zerhouni, N.: Features selection procedure for prognostics: an approach based on predictability. IFAC Proc. Vol. **45**(20), 25–30 (2012)
9. Kaboudan, M.: A measure of time series' predictability using genetic programming applied to stock returns. J. Forecast. **18**(5), 345–357 (1999)
10. Krawczyk, B., Minku, L.L., Gama, J., Stefanowski, J., Woźniak, M.: Ensemble learning for data stream analysis: a survey. Inf. Fusion **37**, 132–156 (2017)
11. Mehmood, H., Kostakos, P., Cortes, M., Anagnostopoulos, T., Pirttikangas, S., Gilman, E.: Concept drift adaptation techniques in distributed environment for real-world data streams. Smart Cities **4**(1), 349–371 (2021)
12. Moon, G., Hamm, J.: A large-scale study in predictability of daily activities and places. In: MobiCASE, pp. 86–97 (2016)
13. Peña, D., Sánchez, I.: Measuring the advantages of multivariate vs. univariate forecasts. J. Time Ser. Anal. **28**(6), 886–909 (2007)
14. Prelipcean, G., Popoviciu, N., Boscoianu, M.: The role of predictability of financial series in emerging market applications. In: Proceedings of the 9th WSEAS International Conference on Mathematics & Computers in Business and Economics, MCBE 2008, pp. 203–208 (2008)
15. Richthofer, S., Wiskott, L.: Predictable feature analysis. In: 2015 IEEE 14th International Conference on Machine Learning and Applications (ICMLA), pp. 190–196. IEEE (2015)
16. Schuster, M., Paliwal, K.K.: Bidirectional recurrent neural networks. IEEE Trans. Signal Process. **45**(11), 2673–2681 (1997)
17. Stavinova, E., Bochenina, K.: Forecasting of foreign trips by transactional data: a comparative study. Procedia Comput, Sci. **156**, 225–234 (2019)
18. Teodorescu, H.N., Fira, L.I.: Analysis of the predictability of time series obtained from genomic sequences by using several predictors. J. Intelli. Fuzzy Syst. **19**(1), 51–63 (2008)
19. Vaganov, D., Funkner, A., Kovalchuk, S., Guleva, V., Bochenina, K.: Forecasting purchase categories with transition graphs using financial and social data. In: Staab, S., Koltsova, O., Ignatov, D.I. (eds.) SocInfo 2018. LNCS, vol. 11185, pp. 439–454. Springer, Cham (2018). https://doi.org/10.1007/978-3-030-01129-1_27
20. Wang, W., Van Gelder, P.H., Vrijling, J.: Measuring predictability of daily streamflow processes based on univariate time series model. In: Proceedings of the iEMSs 4th Biennial Meeting - International Congress on Environmental Modelling and Software: Integrating Sciences and Information Technology for Environmental Assessment and Decision Making, iEMSs 2008, pp. 1378–1385 (2008)
21. Wilson, G., Banzhaf, W.: Fast and effective predictability filters for stock price series using linear genetic programming. In: IEEE Congress on Evolutionary Computation. pp, 1–8. IEEE (2010)
22. Wiskott, L., Sejnowski, T.J.: Slow feature analysis: Unsupervised learning of invariances. Neural Comput. **14**(4), 715–770 (2002)
23. Yeo, I.K., Johnson, R.A.: A new family of power transformations to improve normality or symmetry. Biometrika **87**(4), 954–959 (2000)

A Non-intrusive Machine Learning Solution for Malware Detection and Data Theft Classification in Smartphones

Sai Vishwanath Venkatesh[1]([✉])(iD), D. Prasannakumaran[2], Joish J. Bosco[1],
R. Pravin Kumaar[1], and Vineeth Vijayaraghavan[1]

[1] Solarillion Foundation, Chennai, India
{saivishwanathv,pravin.kumaar99,vineethv}@ieee.org
[2] SSN College of Engineering, Chennai, India
prasannakumaran18110@cse.ssn.edu.in

Abstract. As smartphones strive to provide more versatility and functionality to satiate their growing demand, more user data becomes vulnerable and exposed to attackers. Successful mobile malware attacks could steal a user's location, photos, or even banking information. Due to the lack of post-attack strategies, firms also risk going out of business due to data theft. Thus, there is a need to not only detect malware intrusion in smartphones but to also identify the data that has been stolen in order to assess, aid in recovery and prevent future attacks. In this paper, we propose such a machine learning solution (https://github.com/PrasannaKumaran/AndroidDataTheft) which is accessible, non-intrusive and can perform intrusion detection and stolen data classification for any app under supervision. We do this with Android usage data obtained from publicly available data collection framework–*SherLock*. We test the performance of our architecture for multiple users on real-world data collected using the same framework. Our architecture exhibits less than 9% inaccuracy in detecting malware and can classify the type of data that is being stolen with 83% certainty.

Keywords: Data classification · Malware detection · Cybersecurity · Smartphone

1 Introduction

Currently, Android has more than 1.6 billion active users, which accounts for more than 70% of the global market share of mobile operating systems. As a result, the application market for Android is flooded with apps. We define *malicious app* or *malware* as Android applications that present themselves to the user as benign, but secretly steal user information in the background. Although the Android application store (Google Play Store) verifies apps for malicious intent upon release, it does not aggressively track updates from these verified apps and cannot account for third-party apps downloaded independently by

© Springer Nature Switzerland AG 2021
M. Paszynski et al. (Eds.): ICCS 2021, LNCS 12744, pp. 200–213, 2021.
https://doi.org/10.1007/978-3-030-77967-2_17

the user. A report released in 2020 by McAfee Advanced Threat Research and Mobile Malware Research [18] suggests that malware developers roll out malware through verified apps in Google Play as updates to shield themselves from preliminary verification. Undetected malware attacks can steal sensitive information from users such as photos, documents and browsing data. Data breaches are extremely disastrous for small and midsize firms and businesses. A report by the U.S. Securities and Exchange Commission [21] states that 60% of small firms can not recuperate from data breaches and go out of business within 6 months. The IBM "Cost of a Data Breach Report 2020" [12] suggests that companies establish an incident response (IR) plan to determine the damage done by the breach and contain it as soon as possible. It goes on to state that companies with an IR plan save an average of $2 million in the event of a data breach. Furthermore, the report projects an increase in the costs of data breaches due to the COVID-19 pandemic and the increase in digital reliability. This calls for a need to not only detect malicious attacks but also identify the stolen data to assess the damage, strategically recover and prevent future attacks. Performing this can help in understanding malware trends and aid in malware prevention research.

We propose a novel two-stage machine learning approach to detect malicious attacks for any app under supervision and identify the data stolen by the attack to aid in assessment and recovery.

The course of this paper is as follows: Sect. 2 discusses relevant research in the field of malware detection. In Sect. 3, we describe the dataset used in our study extensively. We elucidate the steps taken to make the data computationally feasible in Sects. 4 and 5. Later, in Sect. 6 we outline our model architecture and describe the parameters of its evaluation. In Sect. 7 we report and discuss our findings. Finally we conclude our work and discuss future scope to this research in Sect. 8.

2 Related Work

Mobile malware detection has been an active and broad area of research for the past several years. Static analysis was one of the first major mobile malware detection approaches proposed [10,19]. Here, the source code of the target malware is analyzed to identify semantic signatures. Although static analysis can detect malware even before running the app, static analysis systems fail when the malware uses obfuscation techniques such as code encryption and repackaging. Dynamic analysis techniques [6,9] address code obfuscation and encryption in malware detection by executing the source code of the application in an isolated environment to analyze runtime characteristics based on frequency. However, this proves to be a bottleneck in systems that use dynamic analysis as clean and noiseless data is hard to achieve and implement in real-world scenarios. Static and dynamic methods additionally require super-user (root) access since they require source code to be executed. Furthermore, Moser et al. [16] suggest that the rate of developing rule-based solutions can not match the fast rate of new malware released to the world. Thus, these solutions will fail to perform for new malware since they are rule-based and discrete solutions.

Machine learning approaches were introduced to swiftly aid in detecting new malware as they are released. Notable works using these approaches include [4,5,8,22] that outperform static and dynamic methods by modelling network usage for detection. Bläsing et al. [6] used various anomaly detection methods to detect malware using system and network data collected. Ronen et al. [17] goes on to detect and classify the family of detected malware by analysing Dalvik bytecode from Android devices. However, these works fail to address the security risk for any end user trying to obtain bytecode. This exposes the phone to further vulnerabilities due to the need for root access. There is a need for non-intrusive malware detection systems based on low privilege information such as usage statistics. This would allow easier user applicability and ensure better security over super-user vulnerabilities.

We propose modeling malware on usage statistics data and we consider one of the largest and most granular datasets for mobile sensor and software sampling - Sherlock dataset. As a result of the dataset's versatility, it is suitable for a multitude of use-cases. Since it does not require root access to probe its data, it is safe and reproducible for malware detection. Zheng et al. [24] explored usage patterns, relationship between mobile usage and the state (benign/malicious) of the application for this data. Wassermann et al. [23] used low-level system features from this dataset coupled with sampling techniques to deal with the inherent class imbalance and detect malicious actions performed on a smartphone.

Although current research tackles malware detection extensively it fails to address data theft classification to aid damage assessment and recovery from data breaches.

We use the SherLock dataset to develop a machine-learning based malware detection pipeline that is capable of identifying the type of data stolen.

3   Dataset

The SherLock Dataset [15], spanning over 10 billion records and involving over 50 volunteers is the result of a real-world data collection experiment to obtain low-level Android usage data alongside emulated malware. Such statistics do not require root access, therefore making any solution developed on the dataset more secure under real-world circumstances since rooting exposes a mobile phone to further vulnerabilities.

The experiment introduces two data collection agents to the mobile phones provided to the volunteers – *Sherlock* and *Moriarty*. *Moriarty* emulates malicious actions on the volunteer's mobile phones randomly through the course of the experiment, generating distinct labels between malicious and benign actions. Meanwhile, *Sherlock* logs usage attributes and statistics in the background.

Table 1. Categories of data theft

Malware service type	Target information
Contacts	Phonebook data
GPS	User coordinates (latitude and longitude)
URL	Web address of every page visited by the user recently
Audio records	Audio records collected during the session
Contacts	Names and phone numbers
BrowserInfo	Account details, bookmarks and browser history
Photos	Images from gallery

3.1 Sherlock Data Collection Agent

One of the ways Sherlock logs phone attributes is through *Pull Probes* which extract data periodically at a constant sampling rate. For our experiments we consider the most frequently sampled pull probe in Sherlock named *T4*, which has a sampling rate of 5 s. T4 probes Global System Features as well as Local Application Features.

Global System Features (GSF): These features pertain to attributes with a global scope in the Android system such as network traffic, CPU and memory utilization, I/O interrupts and Wi-Fi related data. There are a total of 128 Global System Features.

Local Application Features (LAF): Alongside Global System Features, Linux-level data [1] for every running application is sampled. This includes process-specific features such as the scheduling priority, number of bytes transferred, number of threads and kernel-level features used by an application at the time instant. There are a total of 56 Local Application Features.

Local Application Features used in context with Global System Features together provide a rich feature set to determine if a given app exhibits malicious behaviour.

3.2 Moriarty Malicious Agent

Moriarty presents itself to the user as a benign application, such as a game or a browser depending on the version of the app but covertly performs malicious actions. The malware emulated by each version is dissimilar to its precursor and targets different vulnerabilities in each version as illustrated in Table 1. The malware used by Moriarty are behavioural copies of malware found in the real-world.

The app contains labels indicating whether an action executed is benign or malicious. Furthermore, the details of malicious actions such as the type of data stolen, number of bytes transmitted and time taken to transfer the stolen information are logged along with the labels. To collect sufficient information for the experiment, the volunteers were reminded to use the Moriarty app at least once every couple of days.

For our experiments we have considered a computationally feasible subset of the SherLock dataset. It consists of data collected during the first quarter of 2016, with over 300 million records, spanning across 5 users.

4   Data Pre-processing

We aim to enable efficient data merging between Local Application Features (LAF) and Global System Features (GSF). Let g and n denote the number of GSF and LAF. Assuming there are m apps running at the same time, each Global System Feature would correspond to multiple LAF at that instant of time. The vector space of application data (LAF at time t), denoted by Ω_t for any time instant t is represented in Eq. (1).

$$\Omega_t = \begin{Bmatrix} \omega_{11} & \omega_{12} & \cdots & \omega_{1n} \\ \omega_{21} & \omega_{22} & \cdots & \omega_{2n} \\ \vdots & \vdots & \vdots & \vdots \\ \omega_{m1} & \omega_{m2} & \cdots & \omega_{mn} \end{Bmatrix} \tag{1}$$

Consequently, if a relational join operation between GSF and LAF was performed it would lead to the generation of GSF duplicates for every running application with a shape of $(m, g + n)$. The size of this data denoted by S_{np} is $m * (g + n)$ memory units. With the dataset spanning over 300 million records, it becomes essential to reduce memory consumption to expedite the data handling and modeling process. Therefore to overcome duplicates, Ω_t is transformed into a row vector of shape $(1, m * n)$ by performing $PIVOT$ operation represented in Eq. (2), thus obtaining a functional dependency with time.

$$PIVOT(\Omega_t) := \{\omega_{ij} \mid i \in M \ and \ j \in N\} \tag{2}$$

M = Set of all applications on the device
N = Set of local application features

As a result of using $PIVOT(\Omega_t)$ to merge with GSF as opposed to using Ω_t, we obtain a shape of $(1, g + m * n)$ and size of this data denoted by S_p is $1 * (g + m * n)$. The size comparison of the data obtained from merging GSF with and without pivot operation is illustrated in Eq. (3). Therefore, with an increase in the number of applications the overall throughput decreases.

$$g + m * n << g * m + m * n$$
$$\implies \ S_p << S_{np} \tag{3}$$

For the first quarter of 2016 in SherLock, $g = 128$ features and $n = 56$ features with an average of $m = 55$ apps running at any given time.

We observed that S_{np} / S_p was 3.2 indicating that the pivot operation was effective in reducing the size of the merged data. We obtain a dataset with 14,234 features and 5.81 million records on merging this data with Moriarty labels.

5   Feature Selection

We strive to reduce the dataset to its most informative features for smooth and utilitarian processing. On closer inspection of the 14,234 features, we discovered that 12,726 features had more than 70% null values in them, and we obtain 1508 features as a result of their removal. However, this remains significantly large for us to process, considering that we have 5.8 million records.

To further reduce the feature set, we pursue a feature selection method that ensures relevance towards our objective – malware detection and target classification. We considered LightGBM [13] as it has proven to be fast and scalable especially when implemented on high dimensional datasets [7]. Using this technique we reduce our feature space to 150 and 100 important features for malicious detection and target classification respectively. With the features reduced to less than 15% of 1508 features, we now implement stepwise forward selection [11] – an iterative method to determine the least number of features required to obtain any given model's best performance. Using stepwise forward selection we reduce the features required to detect malware to 10 features and the features required to determine the data targeted by malware to 16 features.

As a result of our feature selection approach, the feature set is reduced to approximately 0.1% of the original feature set. Table 2 lists the most important features that were considered for modeling.

Table 2. Features selected for proposed architecture

Model stage	GSF	LAF
Malware detector	totalmemory_used_size, totalmemory_freesize, traffic_totalrxpackets	dalvikprivatedirty_Moriarty, dalvikprivatedirty_WhatsApp, dalvikpss_Samsung Push Service, otherpss_SherLock, rss_SherLock, uidrxbytes_Moriarty, num_threads_SherLock
Target classifier	–	utime_SherLock, rss_SherLock, utime_Moriarty, stime_Moriarty, importance_SherLock, lru_SherLock, dalvikprivatedirty_SherLock, vsize_Hangouts, num_threads_Moriarty, rss_Hangouts, otherpss_Hangouts, dalvikpss_Hangouts, num_threads_SherLock, utime_Unified Daemon, otherprivatedirty_Context Service, vsize_Chrome

6   Experimental Framework

Knowing the kind of data the malware steals could be of more use during data breach assessment compared to just detecting the presence of a malicious action. We propose a two-stage architecture illustrated in Fig. 1 to classify data targeted by a positively detected malware. Our approach detects if a malicious action occurs in the first stage and if positively detected, classifies the data targeted during the malicious action in the second stage.

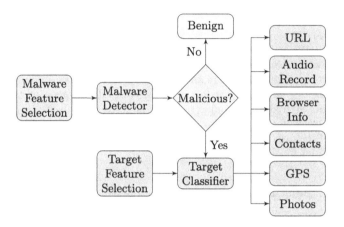

Fig. 1. Two-stage architecture

Malware Detection: We primarily consider supervised tree-based models (Extra Trees, Random Forest, Decision Tree and XGBoost) [2,3] for malicious detection since they have proven to be effective for the data in use [14,23,24]. Anomaly detection methods are suggested by Mirsky et al. [15] due to the sparse frequency of malicious records observed in the data as compared to benign (1:90). We aim to identify if anomaly detection methods are effective as per prior assumption, therefore we consider a tree-based anomaly and outlier detection method–Isolation Forest.

Target Classification: We pass the values detected as malicious in the first stage to further classify the data targeted in this stage. This is a multi–class classification problem to determine the type of data targeted by the malware as seen in Table 1. We consider Extra Trees, XGBoost and K-Nearest Neighbours for this task.

6.1   Evaluation Metrics

Malware Detection: We propose using *False Omission Rate* (FOR) and *False Positive Rate* (FPR) to evaluate the performance of a malware detector. Accuracy and True Positive Rates as considered by [4,20,22,24] are not ideal metrics as they evaluate the model's performance using the true positive values of the majority class. These values are generally high for highly imbalanced data such as *SherLock* and therefore compensate for the impreciseness in classifying the minority class.

We aim to reduce the number of instances where a malware is misclassifed as benign. Therefore we consider *False Omission Rate* (FOR) and *False Positive Rate* (FPR) to evaluate the performance of the malware detector.

For. Illustrated in Eq. (4) this metric indicates the fraction of benign actions that are misclassified

$$FOR = \frac{Number\ of\ benign\ records\ predicted\ as\ malicious}{Total\ number\ of\ malicious\ predictions}$$

$$= \frac{False\ Negatives}{False\ Negatives + True\ Negatives} \tag{4}$$

FPR. Illustrated in Eq. (5) this metric indicates the fraction of malicious records that go undetected by the malware classifier

$$FPR = \frac{Number\ of\ malicious\ records\ falsely\ predicted\ as\ benign}{Total\ number\ of\ malicious\ records}$$

$$= \frac{False\ Positives}{False\ Positives + True\ Negatives} \tag{5}$$

Although each metric can be used individually, we propose using both FOR and FPR in conjunction to discover a detector with an overall good-fit for detecting presence of malware. A lower FOR signifies the success of the first stage of our architecture (malware detection). Meanwhile, a lower FPR signifies a smaller error that will cascade to the next stage. Ideally, both FOR and FPR need to be minimised to improve performance in data classification stage of our proposed two-stage architecture.

Target Classification: Target Classification is a multi-class classification task that involves predicting what kind of data has been stolen by the malware. The different types of malware stolen were given equal importance and hence equal weights were considered for all the classes. Therefore, the average F1-score is the metric of choice used to evaluate the model in this stage.

7   Results and Discussions

To evaluate the performance of our proposed architecture, we consider training and testing on all the users combined. Each user has been proportionally sampled (stratified) while splitting the data into 75% for training and 25% for testing. Since the proposed architecture consists of two stages, it cascades performance at each level. We report the results at each stage for a deeper understanding of our model's performance.

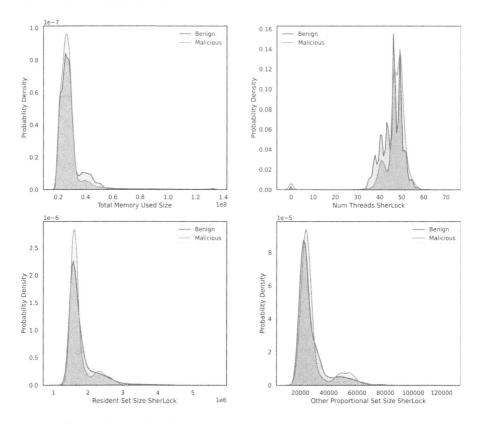

Fig. 2. Density distributions of some of the most important features

7.1   Malware Detection

Tree-based classifiers display superior performance for this task as illustrated in Table 3. This is due to their ability to capture discrete and categorical information more accurately.

However, contrary to our prior assumption, the tree-based outlier detection method – Isolation Forest fails to detect malware with an FOR of 0.79. On observing the density distributions of some of the most important features (Fig. 2) we discover an overlap between malicious and benign distributions. Anomaly detection methods are effective to identify outliers from distributions [20]. Since unsupervised and anomaly detection methods rely on the malware to exist outside benign distribution, these methods may fail to detect malicious activity for this data.

Table 3. Malware detection results

Classifier	False omission rate	False positive rate
Decision tree	0.063	0.222
Extra trees	**0.087**	**0.019**
Random forest	0.088	0.058
XGBoost	0.646	0.296
Isolation forest	0.793	0.976

With the least FOR of all the models considered (illustrated in Table 3), Decision Tree and Extra Trees are the best malware detectors with 6.3% and 8.7% FOR respectively. However, on closer inspection of the Decision Tree detector we observe that it can only achieve this accuracy at the cost of 22.2% FPR. Since this is not desirable for a performance cascading architecture as discussed in Sect. 6.1, we use Extra Trees to determine if an action is malicious before we classify its target in the next stage of our two-stage model.

7.2   Progressive Learning

To achieve good detection, it is necessary for any user to be trained using the SherLock framework before the user can successfully monitor a newly installed app from the market. The time taken by each user to train the detector with SherLock would desirably need to be reduced, which can be done by minimising the required train data for the detection task. Our detector tackles this problem by combining all the users we have and performs stratified training and testing. Our detector exhibits the same accuracy with a decrease in train size as the number of users it has learned from increases. This is visualized in Fig. 3 where we consider a threshold of 0.15 FOR to analyse the change in required train data for an Extra Trees detector trained on 1–5 users. To achieve the threshold FOR when our detector had trained only on a single user, the detector required atleast 76% train data. However, our detector reduces the percentage of train data required from each user as it learns from more users. When the model was trained on 5 users, it required only 52.5% of the train data to achieve the threshold FOR.

Table 4. Target classification results

Models	Average F1 score	Class-wise F1 score (Support)					
		Audio record (5)	BrowserInfo (13)	Contacts (2,343)	GPS (522)	Photos (662)	URL (91)
Extra trees	0.82	0.44	0.58	0.99	0.99	0.99	0.90
XGBoost	**0.83**	**0.44**	**0.64**	**0.99**	**0.98**	**0.99**	**0.89**
K-Nearest neighbors	0.79	0.40	0.50	0.99	0.98	0.99	0.86

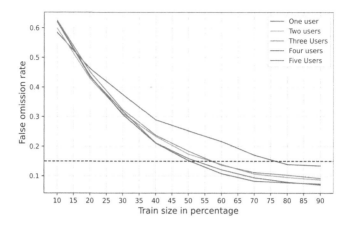

Fig. 3. Progressive learning with increase in users

7.3   Target Classification

Table 4 illustrates the results for classifying the target of the malicious actions predicted by the first stage. Due to the non-linearity posed by the data stream, we considered tree-based algorithms such as Extra Trees and XGBoost. Although XGBoost and Extra Trees display comparable performances, we prefer XGBoost to be integrated with our final pipeline since it has proven to be more scalable than the latter and displays the highest average performance of the models considered for the second stage.

With less than 9% inaccuracy in detecting malware from the first stage, we can predict with 83% certainty on what kind of data is being stolen when we use an Extra Trees detector (Table 3) coupled with an XGBoost classifier (Table 4).

Furthermore, by using our feature selection approach we maintain the aforementioned model performance with the feature set reduced to approximately 0.1% of the original set.

Stepwise forward selection for malware detection (illustrated in Fig. 4) reveals that we only require 10 features to determine if an action is malicious to achieve a minimum FOR and FPR of 0.087 and 0.019 respectively. Figure 5 illustrates stepwise forward selection for target classification and suggests that we require only 16 features to categorize the type of data stolen. As a result of using such a small feature set, we minimize our throughput and processing time drastically.

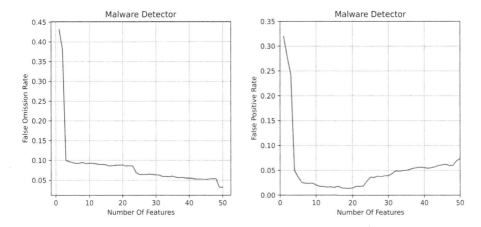

Fig. 4. Stepwise forward selection convergence – malware detection

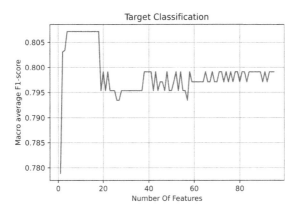

Fig. 5. Stepwise forward selection convergence – target classification

8   Conclusion

In this paper, we propose and successfully test a two-stage machine learning model on the SherLock dataset to detect malicious actions in a smartphone and identify the type of data it steals. We successfully reduce one of the largest datasets for malware classification (SherLock) to 0.1% salient features of its initial feature set using our data preprocessing techniques. Furthermore, we go on to propose using False Omission Rate and False Positive Rate in conjunction to evaluate malware detectors. With just 8.7% inaccuracy in detecting malware from the first stage, our model can predict the kind of data stolen with 83% certainty when we use an Extra Trees detector coupled with an XGBoost classifier. We exhibit our detector's robustness with the gradual decrease in the required train data from one user to achieve the aforementioned performance by training on more users and data. Anomaly detection techniques for malware fail, since malicious actions do not lie outside benign distributions as conventionally expected.

Although the proposed model reduces the percentage of train data required by a user to the minimum, malware detection is still dependent on user behaviour to work. There exists the need for a truly user-independent machine learning solution for malware detection to enhance user experience and ergonomics.

References

1. Linux manual. https://man7.org/linux/man-pages/man5/proc.5.html
2. scikit-learn. https://scikit-learn.org/stable
3. Xgboost. https://xgboost.readthedocs.io/en/latest/
4. Arora, A., Garg, S., Peddoju, S.K.: Malware detection using network traffic analysis in android based mobile devices. In: 2014 Eighth International Conference on Next Generation Mobile Apps, Services and Technologies, pp. 66–71 (2014). https://doi.org/10.1109/NGMAST.2014.57
5. Bekerman, D., Shapira, B., Rokach, L., Bar, A.: Unknown malware detection using network traffic classification. In: 2015 IEEE Conference on Communications and Network Security (CNS), pp. 134–142 (2015). https://doi.org/10.1109/CNS.2015.7346821
6. Bläsing, T., Batyuk, L., Schmidt, A., Camtepe, S.A., Albayrak, S.: An android application sandbox system for suspicious software detection. In: 2010 5th International Conference on Malicious and Unwanted Software, pp. 55–62 (2010). https://doi.org/10.1109/MALWARE.2010.5665792
7. Chen, C., Zhang, Q., Ma, Q., Yu, B.: LightGBM-PPI: predicting protein-protein interactions through LightGBM with multi-information fusion. Chemometr. Intell. Lab. Syst. **191**, 54–64 (2019). https://doi.org/10.1016/j.chemolab.2019.06.003
8. Chen, Z., et al.: Machine learning based mobile malware detection using highly imbalanced network traffic. Inf. Sci. 433–434 (2017). https://doi.org/10.1016/j.ins.2017.04.044
9. Enck, W., et al.: Taintdroid: an information-flow tracking system for realtime privacy monitoring on smartphones. ACM Trans. Comput. Syst. **32**(2) (2014). https://doi.org/10.1145/2619091
10. Felt, A.P., Chin, E., Hanna, S., Song, D., Wagner, D.: Android permissions demystified. In: Proceedings of the 18th ACM Conference on Computer and Communications Security, pp. 627–638, CCS 2011. Association for Computing Machinery, New York (2011). https://doi.org/10.1145/2046707.2046779
11. Han, J., Kamber, M., Pei, J.: Data Mining: Concepts and Techniques (Jan 2012). https://doi.org/10.1016/C2009-0-61819-5
12. IBM: Cost of a Data Breach Report 2020 (2020). https://www.ibm.com/security/digital-assets/cost-data-breach-report/
13. Ke, G., et al.: LightGBM: a highly efficient gradient boosting decision tree. In: Proceedings of the 31st International Conference on Neural Information Processing Systems, pp. 3149–3157, NIPS2017. Curran Associates Inc., Red Hook (2017)
14. Memon, L., Bawany, N., Shamsi, J.: A comparison of machine learning techniques for android malware detection using apache spark. J. Eng. Sci. Technol. **14**, 1572–1586 (2019)
15. Mirsky, Y., Shabtai, A., Rokach, L., Shapira, B., Elovici, Y.: Sherlock vs moriarty: a smartphone dataset for cybersecurity research, pp. 1–12 (Oct 2016). https://doi.org/10.1145/2996758.2996764
16. Moser, A., Kruegel, C., Kirda, E.: Limits of static analysis for malware detection, pp. 421–430 (Jan 2008). https://doi.org/10.1109/ACSAC.2007.21

17. Ronen, R., Radu, M., Feuerstein, C., Yom-Tov, E., Ahmadi, M.: Microsoft malware classification challenge. CoRR abs/1802.10135 (2018). http://arxiv.org/abs/1802.10135

18. Samani, R.: McAfee mobile threat report q1 (2020). https://www.mcafee.com/content/dam/consumer/en-us/docs/2020-Mobile-Threat-Report.pdf

19. Schmidt, A., et al.: Static analysis of executables for collaborative malware detection on android. In: 2009 IEEE International Conference on Communications, pp. 1–5 (2009). https://doi.org/10.1109/ICC.2009.5199486

20. Shabtai, A., Kanonov, U., Elovici, Y., Glezer, C., Weiss, Y.: "andromaly": a behavioral malware detection framework for android devices. J. Intell. Inf. Syst. **38**(1), 161–190 (2012). https://doi.org/10.1007/s10844-010-0148-x

21. U.S. Securities and Exchange Commission: The need for greater focus on the cybersecurity challenges facing small and midsize businesses (2015). https://www.sec.gov/news/statement/cybersecurity-challenges-for-small-midsize-businesses.html

22. Wang, S., et al.: Trafficav: an effective and explainable detection of mobile malware behavior using network traffic, pp. 1–6 (Jun 2016). https://doi.org/10.1109/IWQoS.2016.7590446

23. Wassermann, S., Casas, P.: BIGMOMAL: big data analytics for mobile malware detection. In: Proceedings of the 2018 Workshop on Traffic Measurements for Cybersecurity, pp. 33–39, WTMC 2018. Association for Computing Machinery, New York (2018). https://doi.org/10.1145/3229598.3229600

24. Zheng, Y., Srinivasan, S.: Mobile app and malware classifications by mobile usage with time dynamics. In: Barolli, L., Takizawa, M., Xhafa, F., Enokido, T. (eds.) AINA 2019. AISC, vol. 926, pp. 595–606. Springer, Cham (2020). https://doi.org/10.1007/978-3-030-15032-7_50

Analysis of Semestral Progress in Higher Technical Education with HMM Models

Ewa Lach[✉][ID], Damian Grzechca[ID], Andrzej Polański[ID], Jerzy Rutkowski, and Michał Staniszewski[ID]

Faculty of Automatic Control, Electronics and Computer Science, Silesian University of Technology, Akademicka 16, 44-100 Gliwice, Poland
{Ewa.Lach,Damian.Grzechca,Andrzej.Polanski,Jerzy.Rutkowski, Michal.Staniszewski}@polsl.pl

Abstract. Supporting educational processes with Hidden Markov Models (Hmms) has great potential. In this paper, we explore the possibility of identifying students' learning progress with HMMs. Students' grades are used to train the HMMs to find out if the analysis of obtained models lets us detect patterns emerging from student's results. We also try to predict the final students' results on the basis of their partial grades. A new, classification approach for this problem, using properties of HMMs is proposed: High and Low State Model (HLSM).

Keywords: Education · Hidden Markov Model · Students classification

1 Introduction

Mathematical modeling of educational processes is a wide area, which can concern different aspects of learning/cognitive processes, such as mastering of skills, strategies, behaviors, studies on influence of many factors on effectiveness of learning/teaching, analyses and design of plans for educational scenarios. It can also concern different types of educational processes and different levels of education. Mathematical or statistical modeling of learning/cognitive processes has been a part of researches in psychometrics, psychology and cognitive psychology, over decades (e.g. [1–4]). There is also a lot of on-going research activity in modeling educational processes. New studies in the area of mathematical and statistical modeling of educational processes head towards including many explanatory variables of educational processes in one model, e.g., expertise, motivation, organization, discovering distinct phases of the learning progress, as well as distinguishing levels of students competence (e.g. [5–9]). Recently published approaches also include a wide range of different mathematical and statistical tools, among which applications of artificial intelligence and machine learning algorithms for developing educational programs and scenarios, are especially interesting [10]. Studies in the area of applying formal models for educational

M. Paszynski et al. (Eds.): ICCS 2021, LNCS 12744, pp. 214–228, 2021.
https://doi.org/10.1007/978-3-030-77967-2_18

processes highly benefit from wide availability of electronic records now routinely kept in educational institutions. Supporting educational processes with mathematical modeling obviously has a great potential. Research towards aiding educational practice with mathematical modeling is considerably improving human education. Mathematical modeling allows for better understanding the learning process, factors behind it and contributes to better defining and measuring learning achievements and indexes of student progress. Formulating mathematical models can also help developing educational plans towards personalizing, profiling with respect to background, motivation level [11]. There are, nevertheless, many challenges in developing and applying models of educational processes. There are numerous factors that can influence learning processes. It might be disputable whether the factors are indeed important or which are stronger than others. Both factors influencing learning and indexes of the quality of the learning process, mastery, practice, experience, are not directly observed [12]. Researchers must design systems and models for their estimations. Students groups in studies are always heterogeneous, which poses difficulties for formulating conclusions of experiments and for repeatability of research results. Acquisition of skills is a dynamic process, whose dynamics is not trivial and again can be very heterogeneous. Learning processes are different for different levels, types and targets of education. One of the very promising mathematical approaches for modeling learning processes is by using hidden Markov models (HMM), which allow for capturing the key elements of the learning or skills acquisition processes. States of HMM can represent unobservable states of students learning progress while emission matrices can represent different testing procedures, which generate observations available to the tutor/researcher. HMM can be used for predicting results of education/learning and for estimating dynamics of the process. A special version of the HMM model of learning progress is the Bayesian Knowledge Tracing model (BKT) [13], with two unobservable states. There are numerous studies devoted to different aspects of applications of HMMs for modeling educational processes [5,7,14,15].

On the basis of the collected data (grades of students from technical university for selected courses) we have pursued a research involving using HMM for modeling educational progress of students. We have fitted HMM models for the data on semestral courses for students groups. We performed statistical analyses of the obtained HMM models and we compared them with simpler descriptive models of the students learning progress. We used elaborated models for predicting exam results. In particular, we proposed new classifier (based on HMMs): High and Low State Model (HLSM). Comparisons between models and statistical analyses allowed us to study some questions regarding modeling of educational processes: "Can one observe (in semestral/yearly) horizon the dynamics of the learning process by using HMM?", "How does the scenario of the tutoring process influence the process of skills acquisition?".

2 Educational Data

In our research, we study achievements of the students from technical university: The Silesian University of Technology, the Faculty of Automatic Control, Electronics and Computer Science, from different disciplines of studies: Informatics, Biotechnology and interdisciplinary studies: Control, Electronics, and Information Engineering (CEIE). The results were gathered during selected technical curses. A characteristic feature of courses at technical university is very strong feedback given to students throughout their duration. The feedback is given by partial grades, i.e., scores of assignments, short tests and scores of reports from laboratory exercises. We use two types of scores: partial grades, collected weekly or every two, or three weeks during laboratories or exercise classes and semestral/final scores collected as either result of examination sessions or as results of one or two semestral credit tests. Both types of scores are routinely kept as an element of administrative documentation of the education activity of the Faculty of Automatic Control, Electronics and Computer Science. Much of the available data, especially older records, is still stored on paper. However, there is a clear trend of switching to electronic recording, which will allow for the expansion of data related to education and increase the possibilities of their analysis.

2.1 Description of Courses

In our work, we used data from the following five courses:

Introduction to System Dynamics (SD) is a one-semester course for third semester of CEIE, at the engineering level of studies. The course includes 15 lectures (30 h) and 7 laboratories (15 h). The course is taught in English. The final score is the result of the written exam. The aim of the course is making students familiar with problems and methods related to modeling physical dynamical systems and dynamical systems as engineering constructions. During the classes part students are obligated to solve three different tasks which are marked by a tutor: modeling physical and engineering systems by using the method of balances (2 classes), Lagrange equations I and II (2 classes) and electromechanical analogies I and II (1 or 2 classes depending on schedule).

Fundamentals of Computer Programming (FoCP) is a one–semester course taught in Polish at the engineering level, in the first year of Informatics studies. The course includes 15 lectures (30 h) and 15 laboratories (30 h). The final score is the result of the written exam. FoCP course provides knowledge required to understand, design and write computer programs in the C++ language.

Statistical Inference (SI). It is a two-semester course given to the students of the discipline Bio-engineering, specializing in bioinformatics, at the MSc. level of studies. The course includes 15 lectures (30 h) and 15 laboratories (30 h). The course is taught in Polish. Topics covered by the course include statistical learning theory, construction and validation of classifiers, as well as numerous examples of using statistical learning and classification in biotechnology and bioinformatics.

Introduction to Electric and Electronic Circuits (CT), formerly Circuit Theory Course. It is a two-semester course provided for students of CEiE, at the engineering level of studies. The course is taught in English. The aim of the course is to provide students a trade-off between theory (mainly lectures), practical problems (computational problems solved on the classroom tutorials), and practice (physical observations and measurements in the laboratory). The laboratory exercises give an understanding of electric and electronic circuit connections – students translate circuit diagrams into real circuit connections. First semester of CT covers topics related to DC (Direct Current Circuit analysis), the second semester is mainly focused on AC (Alternating Current Analysis).

Biostatistics and Biometry (Bib). It is a one-semester course for students of the discipline Bio-engineering, at the engineering level of studies. The course includes 15 lectures (30 h), 7 laboratories (15 h) and 7 classroom exercises (15 h). The final score is the result (average) of two semestral credit tests. The course is taught in Polish. Topics covered by the course include distributions of random variables, data pre-processing, parameter estimation, parametric and non-parametric statistical tests.

2.2   Description of the Collected Data

SD. Data contain students' classroom exercises test marks recorded bi-weekly as well as final/semestral scores. Grades come from 206 different students attending the course over the years 2015–2018 with pass rate 78%. Students received partial grades from the eight-element set.

FoCP. Data contain results of students laboratory tests (10 short tests) recorded weekly as well as final/semestral scores. Grades come from 872 different students attending the course over the years 2016–2019 with pass rate 36%. Students received partial grades from the three-element set.

SI. Data contain results of students six laboratory tests as well as final/semestral scores. Grades come from 74 different students attending the course over one year (2015) with pass rate 49%. Students received partial grades from the five-element set.

CT. Data contain students' classroom exercises test marks recorded weekly (15 tests) as well as final/semestral scores. Grades come from 62 different students attending the course over one year (2016/17) with pass rate 35%. Students received partial grades from the twelve-element set.

Bib. Data contain students' classroom exercises tests marks recorded weekly as well as final/semestral scores. Grades come from 96 different students attending the course over one year (2015) with pass rate 45%. Students received partial grades from the eleven-element set.

3   Methodology of the Analysis of the Educational Progress of Students with the Use of HMMs

HMMs are particularly suitable for analyzing the process of learning. They can model the hypothesis that students' levels of understanding of the material can be represented by a number of hidden states. Finding the unobservable states of HMMs in which students can be during the learning process and estimating possible transitions between states, can shed light to the behavioral patterns emerging in education. The choice of the HMMs for modeling learning process is also concordant to the assumption that during the educational process, not only what grades are given to students is important, but also in which order they are given. Accordingly, the fact that a student gets grades: $\{3, 4, 5\}$, and not $\{5, 4, 3\}$ provides important (additional) information about the student's learning process and is often taken into account by the teacher.

The language of HMMs is as follows. N is a number of states, M–number of observations, $S = \{S_1, S_2, \ldots, S_N\}$–set of states, $V = \{1, 2, \ldots, M - 1\}$–set of observations, T–sequence length, $Q = \{q_1; q_2; \ldots; q_T\}$–sequence of states, $O = \{o_1; o_2; \ldots; o_T\}$–sequence of observations. Hidden Markov models (HMMs) are stochastic models identifying processes unfolding over time: movement through a sequence of states Q, that are not directly observed. Each state is described by probabilities of observed variables, that for all states generate Emission probabilities matrix B:

$$b_{j(k)} = P(o_t = k | q_t = S_j),\tag{1}$$

where $1 \leq t \leq T$, $1 \leq j \leq N$, $1 \leq k \leq M$, $\sum_{k=1}^{M} b_{j(k)} = 1$. State Transition Probability matrix A describes the likelihood of moving from one state to each other state in the next time period:

$$a_{ij} = P(q_t = S_j | q_{t-1} = S_i),\tag{2}$$

where $1 \leq i, j \leq N$, $\sum_{j=1}^{N} a_{ij} = 1$. Π is a vector of initial distribution transitions probabilities. The Fig. 2 shows the matrices A and B of an illustrative six-states model. HMM can be estimated with the Baum-Welch algorithm [16] using the sequences of observable variables O. The Viterbi algorithm [17] can be used to estimate Q for selected HMM and O.

In our research, we use partial grades of students as available observations to estimate/train the HMMs to further find out if the analysis of obtained models lets us detect patterns emerging from student's results. As neither the true number of unobservable states nor values of model parameters (the matrices A and B) are known they are estimated on the basis of available observations using the Baum-Welch algorithm, which estimation quality depends on the initial values of parameters. For the first try, we use the same values for each parameter:

$$a_{ij} = 1/N, \quad b_{j(k)} = 1/M\tag{3}$$

If a Baum-Welch algorithm cannot differentiate between states, we set initial differentiating values. First, we change transition probabilities, and if it does not help we change emission probabilities. The second initial values for transition probabilities use the assumption that it is easier to stay in the same state than to move to another. We use:

$$a_{ij} = \begin{cases} 0.8 & \text{for } i = j \\ \frac{0.2}{N-1} & \text{for } i \neq j \end{cases} \tag{4}$$

To define the second proposition of initial values for emission probabilities, we assume, since observables are grades, that the probability that similar grades will occur together is greater than that there will be more distant grades in one state. Additionally, to solve the problem with the division of scores into states, if the obtained value are not an integer, greater variety among higher grades is assumed. We use:

$$b_{j(k)} = \begin{cases} \frac{2}{M+fMdN} & \text{for } \begin{aligned} & k > (fMdN * (j-1) + (M - (fMdN * N))) \text{ and} \\ & k \leq (fMdN * j + (M - (fMdN * N))) \end{aligned} \\ \frac{1}{M+fMdN} & \text{for } \begin{aligned} & k \leq (fMdN * (j-1) + (M - (fMdN * N))) \text{ and} \\ & k > (fMdN * j + (M - (fMdN * N))) \end{aligned} \end{cases} \tag{5}$$

where $fMdN = floor(M/N)$ and function $floor(x)$ rounds the real value x to the next lower integer, in the $-\infty$ direction.

As part of the analysis of the educational process, we look at the state transition probability matrix A and emission probabilities matrix B of obtained HMM models, looking for patterns and rules. We compare HMMs for all students (HMMall), students who passed (HMMpass), and failed (HMMfail) courses.

We also compare the goodness-of-fit, which describes how well a model fits a set of observations, as assessed by the log-likelihood function (l). The function (l) is a logarithmic transformation of the likelihood function, that measure the probability that the model describe this particular data (sequences of observations). When fitting HMM models, it is possible to increase the likelihood by adding states, but doing so may result in overfitting. We compare log-likelihood for training and testing dataset.

The number of hidden states is estimated by multiple runs of the Baum-Welch algorithm with different values of N ($2 \leq N \leq M$) and using Akaike information criterion (AIC) [18] and Bayesian information criterion (BIC) [19]. BIC and AIC are used to compare models for a given set of data, trading-off between the goodness-of-fit of the model and the simplicity of the model. by introducing a penalty term for the number of parameters in the model. The penalty term is larger in BIC than in AIC.

$$AIC = (-2)L + 2k \quad BIC = (-2)L + ln(n)k, \tag{6}$$

where: L–maximum log-likelihood, k–number of independently adjusted parameters within the model, n–the sample size (T*(number of sequences)). The model with the lowest BIC or AIC criterion is preferred.

To test if adding states makes the HMM models overfitted in practice we use the cross-validation technique to randomly divide the data in half 200 times, calculating the mean L for the training and testing dataset for HMMs with different values of N ($2 \leq N \leq M$).

3.1 Results of Fitting HMMs to Students Partial Scores Data

In this part of the work, we have fitted HMMs to all data on students' partial scores to find patterns and conclusions regarding students passing the courses. First, initial values of transition probabilities needed to be set according to the second type of initial transition probabilities (Eq. 4), because the training algorithm wasn't able to vary states with all probabilities set equally. For models with three and more states for courses SD and CT initial values for emission probabilities needed to be varied also (Eq. 5). Without that emission probabilities for the second and subsequent states were identical.

Table 1 presents AIC and BIC calculated for HMMs of five analyzed courses. It shows that for our data adding states to models increases the value of AIC and BIC. Considering this selection criterion, we should use two-state models to analyze our data.

Table 1. BIC and AIC for HMM models

	FoCP		Bib		CT		SD		SI	
N	AIC	BIC	AIC	BIC	AIC	BIC	AIC	BIC	AIC	BIC
2	34.46	98.26	104.97	215.97	104.33	218.06	53.89	148.17	44.61	89.70
3	52.02	166.87	137.08	316.39	133.90	316.25	79.83	235.39	64.23	141.51
4	70.44	249.09	172.46	428.60	167.82	426.63	109.78	336.06	88.01	203.95
5			211.99	553.52	204.86	547.97	143.74	450.16	115.85	276.86
6			255.48	690.93	249.08	684.32	181.68	577.67	147.62	360.16
7			303.24	841.15	295.89	831.10	223.64	718.63		
8			354.93	1003.83	347.31	990.33	269.57	872.98		
9			410.79	1179.23	402.79	1161.46				
10			470.34	1366.85	462.41	1344.57				
11			534.10	1567.22	526.41	1539.89				
12			601.54	1779.81	593.33	1745.97				

The same result was obtained for experiments comparing the mean L for the training and testing set for successive N values. We can see in Fig. 1(b–e) an increasing trend for L, calculated for the training set, and a decreasing trend for the testing set for all courses. The same behavior can be observed for the HMMs trained with data from earlier years and tested with data from subsequent years (Fig. 1a). So we can observe that with more states we get models that are better fitted for the training datasets, but less universal.

Finally, we looked at the matrices for HMMs with successive numbers of states. For two states, there is a clear division of emission probabilities for individual states. As the number of states increases, this division becomes blurred. With two states, one state has a high probability of having higher grades, the other state has a high probability of having lower grades. Adding additional

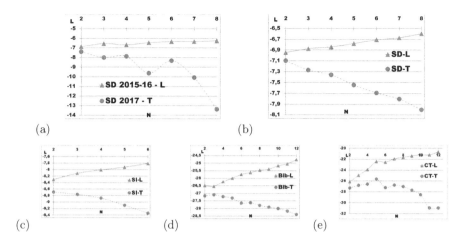

(a)   (b)

(c)   (d)   (e)

Fig. 1. Maximum log-likelihood values calculated for the training (suffix 'L') and testing (suffix 'T') datasets for a subsequent number of states of HMM models.

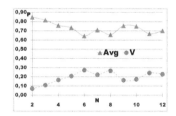

Fig. 2. State Transition Probability matrix A and Emission probabilities matrix B for six-sates HMM model for course SD.

Fig. 3. Average probability for all courses (Avg) of staying in the current state and its variance (V) for a subsequent number of HMM states.

states does not result in the emergence of intermediate grade distributions. Instead, there are states for which we cannot make conclusive deductions about students' abilities based on their grading probability distribution. Within one state there is, for example, the probability at the level of 0.1–0.3 for $1/3$–$2/3$ grades (e.g. Fig. 2). There may also be states with one or two grades with high probability, but the likelihood of leaving this state is greater than for the other states (e.g. Fig. 2). We observe, also, a decreasing trend in the average probability of staying in the current state (a_{ii}) with a simultaneous increase in the variance of a_{ii} (Fig. 3). This means frequent movement between states with a large variety of probabilities describing the transition between states. There are states in which the probability of staying is close to 0 and states from which it is impossible to get out $(a_{ii} = 1)$. This means that, from an analytical point of

Table 2. Values for courses' HMMs.

Course	FoCP			SD			Bib			SI			CT			FoCP				SD			
	All	Pass	Fail	All	Pass	Fail	All	Pass	Fail	All	Pass	Fail	All	Pass	Fail	2016	2017	2018	2019	2015	2016	2017	2018
GHPr	0.70	0.80	0.59	0.73	0.72	0.52	0.96	0.98	0.63	0.80	0.95	0.57	0.86	0.96	0.71	0.72	0.72	0.66	0.73	0.93	0.81	0.77	0.74
GLPr	0.92	0.75	0.97	0.90	0.93	0.87	0.61	0.58	0.94	0.95	0.84	0.92	0.94	0.82	0.98	0.89	0.95	0.92	0.93	0.91	0.88	0.79	0.94
HS->HS	0.92	0.94	0.91	1.00	1.00	0.70	0.64	0.67	0.95	0.84	0.80	0.95	0.86	0.90	0.78	0.94	0.90	0.94	0.89	1.00	1.00	1.00	1.00
LS->LS	0.95	0.92	0.96	0.83	0.77	0.85	0.83	0.81	0.76	0.76	0.78	1.00	0.98	0.93	0.96	0.95	0.94	1.00	0.94	0.60	0.57	0.66	0.84
L	−7.89	−6.68	−8.31	−6.97	−6.83	−7.10	−26.65	−25.97	−26.82	−8.39	−7.85	−8.48	−24.74	−26.72	−22.66	−7.79	−8.05	−7.80	−7.82	−8.08	−9.55	−6.78	−8.26
L-var	7.39	7.58	5.57	1.61	1.50	1.57	9.56	5.37	9.10	2.82	5.45	1.64	69.07	58.98	82.31	7.23	7.80	7.30	7.38	3.33	1.86	0.79	3.27

view, it is difficult to establish any patterns in student behavior for HMMs with more states.

The above results prompted us to focus on the analysis of two-state HMMs. For each course, the two-state model divides the student skills into the high state HS (with a high probability of obtaining higher grades) and the low state LS (with a high probability of obtaining poor grades). Regardless of the course type, number, and type of intermediate grades, we can observe the presence of the high state and low one.

In order to compare two-state HMMs, additional values were introduced: $GHPr$ (the probability of good grades in HS) and $GLPr$ (the probability of poor grades in LS). These values are calculated by dividing the emission matrix B into two parts and summing the probabilities of the columns from the different parts for each state: for HS for good grades, for LS for poor grades.

We can observe similarities and differences in student behavior for different courses (Table 2). For example, the $GHPr$ and $GLPr$ values are similar for 4 out of 5 courses. Except for Bib, the $GLPr$ value is greater than the $GHPr$ and is around 0.93, which tells us that in LS, a student has a high chance of getting a poor grade (around 0.93). This chance is greater than the chance of getting a good grade in HS (for two courses it is around 0.71, for the other two it is around 0.83). As a participant of the Bib course, student has a good chance of a good grade in HS (0.96) and as much as 0.39 in LS. This shows that students get many more good grades during Bib course and average Bib students have a greater chance of being assigned to LS, than for the rest courses. The likelihood of a transition between these states varies with course. For FoCP, students have a high probability of remaining in the current state (about 0.93); for SD there is a 0.17 probability of a transition to HS from LS and no possibility to leave HS; for CT, there is a 0.14 probability of going to LS from HS and only 0.02 of leaving the LS. BiB and SI students have more options for transitions between states. This may be influenced by the existence of dependencies between subsequent graded tasks (without understanding the previous graded task, it is impossible to understand the next topic) or the lack of it (each mark grades a separate topic) and the ease of catching up. Looking at the variance of L for all courses, we can assume that the behavior of the students of CT is more varied then for rest courses. The comparison of the values of $GHPr$, $GLPr$, state transition probabilities and L for HMMs from individual years for two courses: FoCP and SD shows that students behave similarly within the same courses (Table 2).

Interesting conclusions can be drawn from the comparison of HMMs for all students, students who passed, and failed courses (Table 2). It can be expected that: the $GHPr$ will be higher for HMMpass and lower for HMMfail, and the opposite for the $GLPr$, the probability of staying in HS will be higher for HMMpass and lower for HMMfail, and the opposite for the probability of staying in LS. As for L, greater fitness for HMMpass and HMMfail is expected. The results, however, are not so obvious. The above-described situation occurs only for the FCoP, for the others, there are greater or lesser deviations from the expected results. This would indicate that, within a given course, passing it or not more or less can be explained by partial grades. On this basis, we can assume that other factors determine to a greater extent the passing or failing of some courses. As for the HMMs log-likelihood - L, some models achieved a slight improvement, but also some models received a lower L. On this basis, we can conclude that there is a variation in behavior among students who pass or fail a course (this is especially visible for students who fail: 4 out of 5 courses recorded lower L for HMMfail).

4   Predicting Students Final Results

Analyzing student partial grades the problem of predicting with them student's exam results was explored. In order to prepare a comprehensive comparison of classification results, five different classical approaches plus a new High and Low State Model (HLSM), based on results from Sect. 3, were applied.

The obtained course data is randomly split in the proportion 67:33 200 times [24]. 67% of the data is used as a training dataset, the rest as a testing dataset. In addition, we check whether models trained on data from previous years for a specific course can be used to predict exam results in subsequent years for that course.

4.1   Models

In order to prepare comprehensive comparison of classification results, five different classical approaches were applied.

1. Support vector machine (SVM) is mainly used for two-class classification. SVM finds classes by applying the best hyperplane to distinguish one class from another (by means of largest margin). In this context, margin refers to the maximum width of a plate parallel to the hyperplane without internal data points. SVM was applied with the Gaussian kernel function [20].
2. K nearest neighbour (KNN) can give the k closest points in the feature points. Classification results are based on a neighbour's plural voting, where the object is assigned to the class most frequently found among its k closest neighbors. The number of neighbours in the algorithm was set to one with a single standardization [21].
3. Linear regression (LR) models can be treated as a linear relationship of response with respect to one or more predictive factors. LR was applied along with binomial distribution [22].

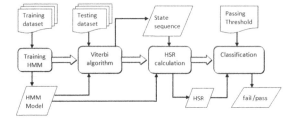

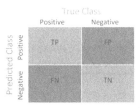

Fig. 4. HLSM classification process.   **Fig. 5.** The confusion matrix.

4. Naive Bayes classifier (NB) in statistics is known as a probabilistic classifier which is based on Bayes theorem assuming strong (or rather naive) independence between given features [23].

5. HMM classic classification method (HMMC) consists of training separate HMMs for each class. The tested object is assigned to the class (HMM model) for which the log-likelihood is the highest [7].

6. High and Low State Model (HLSM) is a new classification model proposed by the authors for the selected problem. It is based on the analysis of two-state HMM models trained on the partial grades received by students during specified technical courses. For each of the analyzed courses, the two-state HMM models were characterized by one state with an assigned high probability of receiving higher partial grades and a low probability of receiving lower partial grades (HS), and a second state with opposite probabilities (LS). The proposed HLSM model assumes that by finding out the degree of membership to a HS of a student's behavior, we can determine the chance of a student to pass an exam. The classification process starts with the training of the two-state HMM model. Then, as a result of comparing the probabilities in the emission matrix for the extreme one or two grades, the high and low states are identified. Next, we use the Viterbi algorithm to calculate for a given sequence of students' grades, the most likely path (sequence of states) through the trained HMM model. For a generated sequence of states, we calculate how many times the student is in a high state in relation to all states in which he is staying during the entire course: High State Rate (HSR). Finally, we use HSR to predict whether the student will pass or not the final exam comparing this value to the passing threshold (PT). The diagram in the Fig. 4 shows the whole classification process.

For all classifying models, we can define four parameters that assess their performance: True Positive (TP), False Positive (FP), True Negative (TN), False Negative (FN) as presented in the confusion matrix (Fig. 5). Unfortunately, none of these values can be used alone to judge the quality of a classifier. This results in the definition of various measures that are using TP, FP, TN, and FN values in different configurations. In the paper, we will use the area under the ROC curve (AUC) that can give an idea about the usefulness of the classifier in general. Greater AUC means a better model for a specific classification. The ROC

(Receiver Operating Characteristic) curve (Fig. 6(b)) shows in a graphical way the trade-off between sensitivity and specificity for every possible cut-off, where:

$$Sensitivity = TP/(TP + FN), \quad Specificity = TN/(TN + FP) \quad (7)$$

In addition to comparing models, the ROC curve was used, also, to determine the cut-off point for HLSM: passing threshold (PT). We use Youden's index J.

$$J = Sensitivity + Specificity - 1 \quad (8)$$

4.2 Results

During the conducted experiments the highest AUC values were obtained by the LR and HLSM models (Fig. 6). For three courses (SD, FoCP and SI) LR model gets the highest results, for two (CT and Bib) - HLSM. For the course SD, the lead of a model LR is very small (0.003) and when we look at the classification for the following semesters of SD, we can see that HLSM has the best results for the three years, and only in one semester LR is doing better. If we average the results for subsequent years of SD, HLSM has a lead of 0.049. For FoCP and SI, LR has a lead (about 0.023), higher than for SD. The analysis of AUC for the following semesters of FoCP shows LR leading for 3 out of 4 semesters (0.028 on average). Only in 2019, HLSM has an AUC higher by 0.002. However, for CT and Bib, HLSM has a greater difference in AUC value to LR (Bib-0.178, CT-0.099). Because of that, when averaging the results for all courses, HLSM has the highest value. The situation is similar when predicting behaviors of students based on models generated for previous years. The LR and HLSM models achieve the best results alternately. When averaging the results LR is better for the course SD, HLSM for FoCP.

When comparing the predictions based on the models for previous years for course SD, the HMMC model presented interesting behaviors (out of 9 classifications, it achieves the best results twice, four times the lowest). Ultimately, if we compare the averaged values of the classification for models for previous years for the SD course with the result of the classification for SD, HMMC recorded the largest decrease in AUC value. In the remaining experiments, the HMMC obtained positive results, but never exceeded the HLSM.

The courses are also characterized by different values of AUC. For SD, Bib, and SI, the AUC is quite low (0.61-0.64) for FoCP it is higher (0.747) and CT has the best chance of a good classification (0.831). This is in line with the correlation calculated between the partial grades and the passing of the course (accordingly: Bib-0.19, SD-0.27, SI-0.28, FoCP-0.42, CT-0.62).

To determine if our data wasn't skewed by chance, we calculated statistical significance for the AUC values (t-test and Wilcoxon rank-sum test) and obtained results indicating that our experiments are statistically significant at the 0.05 significance level.

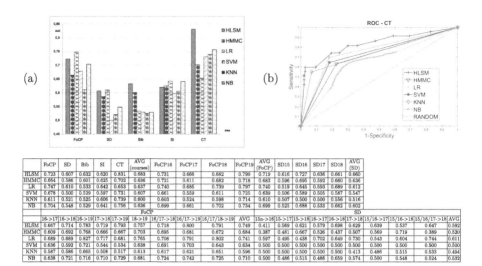

	FoCP	SD	Bib	SI	CT	AVG (courses)	FoCP16	FoCP17	FoCP18	FoCP19	AVG (FoCP)	SD15	SD16	SD17	SD18	AVG (SD)
HLSM	0.723	0.607	0.632	0.620	0.831	0.683	0.731	0.666	0.682	0.799	0.719	0.616	0.727	0.636	0.661	0.660
HMMC	0.664	0.586	0.601	0.625	0.702	0.636	0.721	0.611	0.682	0.718	0.683	0.596	0.695	0.592	0.660	0.636
LR	0.747	0.610	0.533	0.642	0.653	0.637	0.740	0.685	0.739	0.797	0.740	0.519	0.645	0.593	0.689	0.612
SVM	0.678	0.500	0.529	0.597	0.731	0.607	0.661	0.559	0.611	0.725	0.639	0.506	0.589	0.505	0.587	0.547
KNN	0.611	0.521	0.525	0.606	0.739	0.600	0.603	0.524	0.598	0.714	0.610	0.507	0.500	0.500	0.556	0.516
NB	0.704	0.548	0.529	0.641	0.756	0.636	0.699	0.661	0.702	0.734	0.699	0.525	0.688	0.533	0.662	0.602

	FoCP										SD									
	16->17	16->18	16->19	17->18	17->19	18->19	16/17->18	16/17->19	16/17/18->19	AVG	15n->16	15->17	15->18	16->17	16->18	17->18	15/16->17	15/16->18	15/16/17->18	AVG
HLSM	0.667	0.714	0.783	0.719	0.793	0.757	0.718	0.800	0.791	0.749	0.411	0.569	0.621	0.579	0.698	0.629	0.639	0.537	0.647	0.592
HMMC	0.609	0.692	0.768	0.666	0.667	0.703	0.695	0.681	0.672	0.684	0.387	0.481	0.667	0.526	0.437	0.507	0.569	0.719	0.389	0.520
LR	0.689	0.827	0.717	0.681	0.765	0.708	0.791	0.802		0.741	0.597	0.495	0.438	0.702	0.649	0.730	0.543	0.604	0.744	0.611
SVM	0.636	0.592	0.721	0.544	0.534	0.638	0.691	0.703	0.643	0.634	0.500	0.500	0.500	0.500	0.500	0.500	0.500	0.500	0.500	0.500
KNN	0.587	0.586	0.665	0.508	0.517	0.613	0.617	0.621	0.651	0.596	0.500	0.500	0.500	0.500	0.500	0.413	0.486	0.515	0.533	0.494
NB	0.638	0.721	0.716	0.710	0.729	0.681	0.724	0.742	0.725	0.710	0.500	0.486	0.515	0.486	0.659	0.574	0.500	0.548	0.524	0.532

Fig. 6. AUC values calculated for different classifiers for five courses and for following semesters of SD and FoCP and an illustrative ROC curve for course CT (b).

5   Conclusion

When looking at the trained HMM models for different courses, it can be noticed that, while one can always distinguish high and low states for each course, the models differ significantly. As shown in Sect. 3, it is possible to distinguish patterns for different courses based on generated HMMs. These observations can be interpreted that by using HMM models, we can draw conclusions about the educational process in individual courses.

The performed experiments also showed that it is possible, to some extent, to predict the students' final results based on their partial grades. In particular, the proposed new classifier, based on HMMs, HLSM achieved good results.

Acknowledgment. This publication was supported by the Department of Graphics, Computer Vision and Digital Systems, under statue research project (Rau6, 2021), Silesian University of Technology (Gliwice, Poland). The authors would like to thank for involvement in data collection especially to Prof. J. Polańska, Prof. S. Deorowicz, Prof. K. Simiński, Dr. F. Binczyk, Dr. A. Papież and Dr. J. Żyła.

References

1. Brainerd, C.J.: Markovian interpretations of conservation learning. Psychol. Rev. **86**(3), 181–213 (1979). https://doi.org/10.1037/0033-295X.86.3.181
2. Brainerd, C.J.: Developmental invariance in a mathematical model of associative learning. Child Dev. **51**(2), 349–363 (1980). https://doi.org/10.2307/1129267
3. Brainerd, C.J.: Three-state models of memory development: a review of advances in statistical methodology. J. Exp. Child Psychol. **40**(3), 375–394 (1985). https://doi.org/10.1016/0022-0965(85)90072-4

4. Cleeremans, A., McClelland, J.L.: Learning the structure of event sequences. J. Exp. Psychol. Gener. **120**(3), 235–253 (1991). https://doi.org/10.1037/0096-3445. 120.3.235

5. Geigle, C., Zhai, C.: Modeling student behavior with two-layer hidden Markov models. JEDM **9**(1), 1–24 (2017). https://doi.org/10.5281/zenodo.3554623

6. Kassarnig, V., et al.: Academic performance and behavioral patterns. EPJ Data Sci. **7**, 10 (2018). https://doi.org/10.1140/epjds/s13688-018-0138-8

7. Witteveen, D., Attewell, P.: The college completion puzzle: a hidden Markov model approach. Res. High. Educ. **58**, 449–467 (2017). https://doi.org/10.1007/s11162-016-9430-2

8. Tenison, C., Anderson, J.R.: Modeling the distinct phases of skill acquisition. J. Exp. Psychol. Learn. Mem. Cogn. **42**(5), 749–767 (2016). https://doi.org/10.1037/xlm0000204

9. Logue, A.W., Douglas, D., Watanabe-Rose, M.: Corequisite mathematics remediation: results over time and in different contexts. Educ. Eval. Policy Anal. **41**(3), 294–315 (2019). https://doi.org/10.3102/0162373719848777

10. Chi, M.T.H., et al.: Translating the ICAP theory of cognitive engagement into practice. Cogn. Sci. **42**, 1777–1832 (2018). https://doi.org/10.1111/cogs.12626

11. D'Mello, S., Olney, A., Williams, C., Hays, P.: Gaze tutor: a gaze-reactive intelligent tutoring system. Int. J. Hum. Comput. Stud. **70**(5), 377–398 (2012). https://doi.org/10.1016/j.ijhcs.2012.01.004

12. Beck, J.E., Chang, K.: Identifiability: a fundamental problem of student modeling. In: Conati, C., McCoy, K., Paliouras, G. (eds.) UM 2007. LNCS (LNAI), vol. 4511, pp. 137–146. Springer, Heidelberg (2007). https://doi.org/10.1007/978-3-540-73078-1_17

13. Corbett, A.T., Anderson, J.R.: Knowledge tracing: modeling the acquisition of procedural knowledge. User Model. User-Adap Inter. **4**, 253–278 (1994). https://doi.org/10.1007/BF01099821

14. Wang, G., Tang, Y., Li, J., Hu, X.: Modeling student learning behaviors in ALEKS: a two-layer hidden Markov modeling approach. In: Penstein Rosé, C., et al. (eds.) AIED 2018, Part II. LNCS (LNAI), vol. 10948, pp. 374–378. Springer, Cham (2018). https://doi.org/10.1007/978-3-319-93846-2_70

15. Hu, Q., Rangwala, H.: Course-specific Markovian models for grade prediction. In: Phung, D., Tseng, V.S., Webb, G.I., Ho, B., Ganji, M., Rashidi, L. (eds.) PAKDD 2018, Part II. LNCS (LNAI), vol. 10938, pp. 29–41. Springer, Cham (2018). https://doi.org/10.1007/978-3-319-93037-4_3

16. Baum, L.E., Petrie, T.: Statistical inference for probabilistic functions of finite state Markov chains. Ann. Math. Stat. **37**(6), 1554–1563 (1966). https://doi.org/10.1214/aoms/1177699147

17. Viterbi, A.: Error bounds for convolutional codes and an asymptotically optimum decoding algorithm. IEEE Trans. Inf. Theor. **13**(2), 260–269 (1967). https://doi.org/10.1109/TIT.1967.1054010

18. Akaike, H.: A new look at the statistical model identification. IEEE Trans. Autom. Control **19**(6), 716–723 (1974). https://doi.org/10.1109/TAC.1974.1100705

19. Schwarz, G.: Estimating the dimension of a model. Ann. Stat. **6**(2), 461–464 (1978). https://doi.org/10.1214/aos/1176344136

20. Christianini, N., Shawe-Taylor, J.: An Introduction to Support Vector Machines and Other Kernel-Based Learning Methods. Cambridge University Press, Cambridge (2000). https://doi.org/10.1017/CBO9780511801389

21. Altman, N.S.: An introduction to kernel and nearest-neighbor nonparametric regression. Am. Stat. **46**(3), 175–185 (1992). https://doi.org/10.1080/00031305.1992.10475879
22. Collett, D.: Modeling Binary Data. Chapman and Hall, New York (2002)
23. Hastie, T., Tibshirani, R., Friedman, J.: The Elements of Statistical Learning. SSS, Springer, New York (2009). https://doi.org/10.1007/978-0-387-84858-7
24. Xu, Y., Goodacre, R.: Splitting training and validation set: a comparative study of cross-validation, bootstrap and systematic sampling for estimating the generalization performance of supervised learning. J. Anal. Test. **2**, 249–262 (2018). https://doi.org/10.1007/s41664-018-0068-2

Vicinity-Based Abstraction: VA-DGCNN Architecture for Noisy 3D Indoor Object Classification

Jakub Walczak[1]([⊠]) [ID], Adam Wojciechowski[1] [ID], Patryk Najgebauer[2] [ID], and Rafał Scherer[2] [ID]

[1] Institute of Information Technology, Lodz University of Technology, Łódź, Poland
{jakub.walczak,adam.wojciechowski}@p.lodz.pl
[2] Department of Intelligent Computer Systems, Częstochowa University of Technology, Al. Armii Krajowej 36, 42-200 Częstochowa, Poland
{patryk.najgebauer,rafal.scherer}@pcz.pl

Abstract. One of the outstanding benchmark architectures for point cloud processing with graph-based structures is Dynamic Graph Convolutional Neural Network (DGCNN). Though it works well for classification of nearly perfectly described digital models, it leaves much to be desired for real-life cases burdened with noise and 3D scanning shadows. Therefore we propose a novel, feature-preserving vicinity abstraction (VA) layer for the EdgeConv module. This allowed for enriching the global feature vector with the local context provided by the k-NN graph. Rather than processing a point together with its neighbours at once, local information is aggregated before further processing, unlike in the original DGCNN. Such an approach enabled a model to learn accumulated information instead of max-pooling features from local context at the end of each EdgeConv module. Thanks to this strategy mean- and overall classification accuracy increased by 9.4pp and 4.4pp, respectively. Furthermore, thanks to processing aggregated information rather than the entire vicinity, the new VA-DGCNN model converges significantly faster than the original DGCNN.

Keywords: 3D object classification · Point clouds · DGCNN

1 Introduction

Classification of real three-dimensional objects, registered as a point cloud, is currently a hot research topic, mainly due to the popularity of ubiquitous depth sensors and emerging applications of 3D data automatic processing. Whereas artificial, synthetic point sets [14] are handled quite well by the current benchmark methods, processing real-world data, collected in non-laboratory poorly constrained environments, still leaves much to be desired. Such data imposes many challenges [12], usually not related to synthetic data such as uneven object sampling, noise presence, missing points resulting from scanning shadows, or remarkable classes imbalance, are just cases in point.

© Springer Nature Switzerland AG 2021
M. Paszynski et al. (Eds.): ICCS 2021, LNCS 12744, pp. 229–241, 2021.
https://doi.org/10.1007/978-3-030-77967-2_19

A challenge with approximating a function of point clouds is twofold: processing imperfect real-life data and achieving immutable results regardless of any permutation of point ordering. In other words, 3D object spatial point distribution can be depicted by any permutation of data. Unlike in the case of images, sound waves, or other well-ordered structures, point clouds imply permutation equivariance property of a model. As pointed out by the authors of [11], representation of a function on sets is burdened with some limitations out of which the essential one is the minimum size of latent space required to make a model able to represent and generalize a set. The authors concluded that the latent space dimension should be at least as large as the input size of a set [11].

Due to substantial cardinality of point cloud data sets, their efficient classification is recently performed with graph-based approaches. Such a hierarchical point feature regression allows for retrieving core features constellation and performing efficient classification afterwards. Nevertheless, incomplete and noised real-world data sets impose severe challenges on a core features aggregation process. In the paper, we aim at improving the performance of real-life 3D data classification. To this end, we developed a novel method for processing real-life three-dimensional objects described by point clouds. The proposed method introduces multi-layer perceptron-based (MLP-based) features integration rather than a coarse statistical features aggregation. Such a subtle modification exceeds the accuracy of the state-of-the-art methods on an established real-world data set (S3DIS [1]).

Compared to the literature, our method is faster as it uses vicinity-aware aggregation of the local context. At the same time, it demonstrates better accuracy because it learns more descriptive features. These studies provide new insights, showing that the noisy, incomplete point clouds require aggregation of the local context before hierarchical features learning. We also investigate the associated problem of a local context range determination and show the tradeoff between the size of the neighbourhood and the processing speed.

The rest of the paper is organized as follows. Section 2 presents related works and the DGCNN architecture. A novel VA-DGCNN method is proposed in Sect. 3. Research methodology, in turn, is outlined in Sect. 4. Section 5 contains results of experiments. The article is concluded in Sect. 6.

2   Related Works

Point clouds are one of the simplest representation of 3D shapes coming, for example, from LIDARs or time-of-flight cameras. Current methods allow processing point clouds directly without intermediate mesh or denoising. Neural networks innate input are regular structures (sequences, vectors, images or volumes), whereas point clouds are unstructured and of various size. The first network able to process such data was PoinNet [7]. PointNet takes a point from a point cloud and uses max pooling symmetric function. Qi et al. [8] in their further study elaborated an extension of the basic PointNet network, introducing hierarchical neural network for processing local features with a new approach called PointNet++.

Since then, among point cloud classification-driven deep neural network models, graph-based solutions became one of the most efficient ones. As they surpassed the previous approaches, we compare our method in Sect. 4 only with these state-of-the-art graph-based networks. Graph-based networks usually consider points as vertices of a graph and construct then directed graph edges relying on features of a point vicinity. More than often, such methods perform convolution and pooling operation in spatial domain. Convolution is normally implemented within MLP-inspired module over spatial neighbourhood, whereas pooling aggregates information from a vicinity by forming a new coarsened graph. A graph spread over a point cloud reflects well its sparsity and characteristics. Moreover, it is inherently permutation-invariant, assuring nontrivial property crucial for unorganised point cloud classification [9]. In consequence, dominant graph-based approaches [4] like DGCNN [13] or closely related LDGCNN [15] exceed 92% for overall accuracy and 90% for mean accuracy on the well-known benchmark data set ModelNet40 [14]. Other competitive graph-based models like HGCNN [16], Dynamic Points Agglomeration Module (DPAM) [5], Kernel Correlation Network (KCNet) [10], or ClusterNet [2] – utilizing rigorously rotation-invariant module to extract point rotation-invariant features, can also be analysed; however they reveal inferior classification accuracy in comparison with DGCNN [4].

The classification accuracy deteriorates when performing on non-synthetic, real world data sets like S3DIS [1]. Such a database of real scanned objects, affected by uneven object sampling, noise presence, missing points – resulting from scanning shadows, imposes severe disturbances reducing quality of semantic segmentation $mIoU$ to about 56% or 64% for DGCNN and DPAM respectively [4].

Intrigued by real data (S3DIS) experiments of semantic segmentation, we conducted a study on the classification accuracy of such data. It revealed that for S3DIS database, DGCNN substantially deteriorates classification accuracy, achieving barely 60.2% and 80.3% for mean accuracy and overall accuracy respectively. It confirmed our assumptions about the vulnerability of DGCNN to noised and incomplete data. We claim that the drastic deterioration of classification accuracy results from coarse, averaged aggregation of neighbouring points features. For supporting the above hypothesis, we selected the DGCNN network as it is a well-established baseline solution with still superior accuracy.

DGCNN is an architecture constituted by stacked layers of, so-called, Edge-Conv modules. They are, in turn, followed by MLP and MaxPool layers. A single EdgeConv module selects at first k nearest neighbours for each analysed point. Then, vectors between point's neighbours and a point itself, as well as points' centroid coordinates itself, are processed by MLP (see Fig. 1).

A k-NN graph for the input point cloud $\mathcal{D}$ of N points and F features/coordinates, is defined as a matrix $\mathbf{V}$ of size $N \times k \times F$. For the first iteration, when $F = 3$ (features are input coordinates) an entry for $j-$th neighbour of $i-$th point is denoted as: $\mathbf{V}_{ij} = \{\mathbf{x}_j^{(x)}, \mathbf{x}_j^{(y)}, \mathbf{x}_j^{(z)}\}$ such that $||\mathbf{x}_i - \mathbf{x}_j||_2 \leq ||\mathbf{x}_i - \mathbf{V}_{i(j+1)}||_2$. MLP is defined as in Eq. 1.

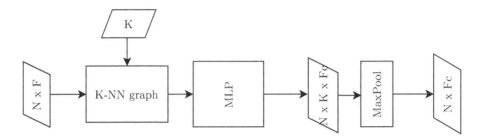

Fig. 1. EdgeConv layer of DGCNN architecture. The input is a matrix of N points of F features (coordinates), k denotes the number of nearest neighbours used to construct the graph, and F_c is the number of resulting features on the layer output.

$$mlp(\cdot) = ReLU(bn(\cdot * G)) \tag{1}$$

where G is a matrix of convolution kernels, $bn(\ \cdot\ )$ is a batch normalization, namely, scaling throughout batches, and $ReLU$ is defined as $ReLU(\cdot) = \max\{0, \cdot\}$

Later on, the maximum feature out of all neighbours is extracted for each point. Out of resulting values, a vector of length F_c of maxima throughout neighbours is drawn for each analysed point. Such an approach lets a model learn local context well. However, it takes local information into consideration in the last but one layer, MaxPool, where the vicinity context is aggregated into a single representative per point. We argue that processing a point and the associated vicinity information at once, may deteriorate classification results due to the loss accumulated information provided by the local context. Introducing an additional layer of abstraction may enrich global feature vector with local information at a scale [3] – here defined by the vicinity of size k. Additionally, such vicinity-aware feature aggregation may speed up the model training.

3   Proposed Method

The classical DGCNN is constructed by stacked layers of edge-convolution modules (EdgeConv, see Fig. 1), followed by a multilayer perceptron, where the maximum value of features throughout the entire vicinity k is extracted for the every point $\{p_1, p_2, p_3, ..., p_N\}$. In such an architecture, features are processed without any regard to local context provided by the vicinity until the last but one layer, i.e. MaxPool, which, in the simplest manner, aggregates local context to a single value per point.

In order to take into consideration the vicinity context from the very beginning, we decided to introduce a simple single-layer module of shared vicinity abstraction (VA) (Eq. 2) just between the k-NN graph and MLP (see Fig. 2). VA acts as if it combines neighbours' features so that the information is aggregated in place and more general traits might be retrieved for further processing. As such, it may be thought of as generating a kind of overlapping super-points for each patch defined by a point and its vicinity

$$\mathcal{D}_{VA} = VA(\mathbf{V}) = \text{ReLU}(\mathbf{W}^T\mathbf{V} + \mathbf{h}) , \qquad (2)$$

where $\mathcal{D}_{VA}$ is a point cloud of points after VA layer, $\mathbf{W}$ are weights associated with the layer, $\mathbf{V}$ is a matrix of neighbours of dimensions $(N \times k \times F)$, and $\mathbf{h}$ is the layer bias.

$$\mathcal{D}' = \text{EdgeConv}(\mathcal{D}, k) , \qquad (3)$$

where k is an assumed number of nearest neighbours. After VA layer, each point of the input point cloud $\mathcal{D}$ is transformed into multidimensional feature space: $\mathcal{D}_{VA} = \{\mathbf{x}_1^{VA}, \mathbf{x}_2^{VA}, \mathbf{x}_3^{VA}, ..., \mathbf{x}_N^{VA}\}$ where $\mathbf{x}_i^{VA} \in \mathbb{R}^F$ is a resulting vector of features (Eq. 2, Fig. 3) which is passed to convolution layers. This describes a single EdgeConv module where an input point cloud matrix $\mathcal{D}$ of size $N \times F$ (N points of F initial features) is transformed to vicinity-aware point cloud matrix $\mathcal{D}'$ of size $N \times F_c$ (Eq. 3). Besides classification quality boosting, such a strategy will allow a model to converge faster as instead of processing the entire vicinity, we process only combined features extracted from that vicinity. Such features carry more semantic information with lower memory requirements.

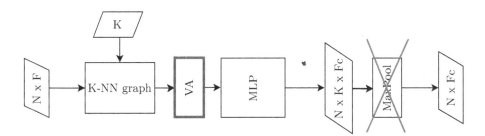

Fig. 2. Proposed modification of EdgeConv layer taking into account the vicinity context before features processing.

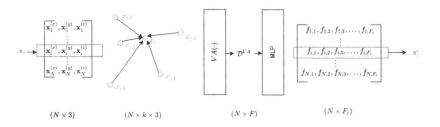

Fig. 3. Shapes of matrices (bottom row) while processing through $EdgeConv$ with VA

It should be noted that VA is a highly permutation-changeable module. There is, however, stipulated an intrinsic order relation among points by means of a distance metric used to collect nearest neighbours (Euclidean in this context). Therefore, VA does not need to show permutation-equivariant properties as neighbours of a point do not permute for that point.

4   Research Methodology

In this section we compare our approach with DGCNN on the real-life S3DIS dataset [1]. The rationale behind comparing only with DGCNN is that it is superior to the PointNet NN family and every other previous method on every 3D data set.

4.1   Used Data

Classification models often are evaluated on either ModelNet10 or ModelNet40 [14] data sets. In these popular benchmark data sets, individual objects instances (point clouds), were constructed by means of sampling digital, CAD models. These data sets seem to be not useful for evaluation of real-life application models due to the fact they consist of synthetic, noiseless, complete object structures rather than depth data acquisition samples, affected by numerous disturbances.

Therefore, we decided to use S3DIS [1] database collected with the Matterport Camera on six different areas (locations) having, in total, 273 single room scans labeled with respect to instances of 14 classes (see Fig. 4). This set represents real data raising real problems, like noise presence or scanning shadows, just to name a few. Moreover, S3DIS is a database of extremely uneven distribution of labels (see Fig. 5). More than 39% of data is tagged as *clutter*, i.e. an indefinite object type not belonging to any of 13 remaining classes. On the other hand, class *stairs* has barely 0.2% of representatives. Some classes have distinguishable geometrical context. On the other hand, three classes: *door*, *window*, *wall* are significantly less geometrically distinguishable than others, thus more challenging for classification.

For regularization purposes dataset was augmented by rotating points along vertical axis and by shifting.

4.2   Evaluation Metrics

In order to indicate competitive advantages reached by the proposed method, standard evaluation metrics for point cloud classification were used [4]. Two measures, namely mean accuracy (mAcc) and overall accuracy (oAcc) can be easily derived from class-wise confusion matrix $\mathcal{C}$ of size $A \times A$ (see Eq. 4 and 5).

$$mAcc = \sum_{i \leq A} \frac{\mathcal{C}_{i,i}}{\sum_{j \leq A} \mathcal{C}_{i,j}} \,, \qquad (4)$$

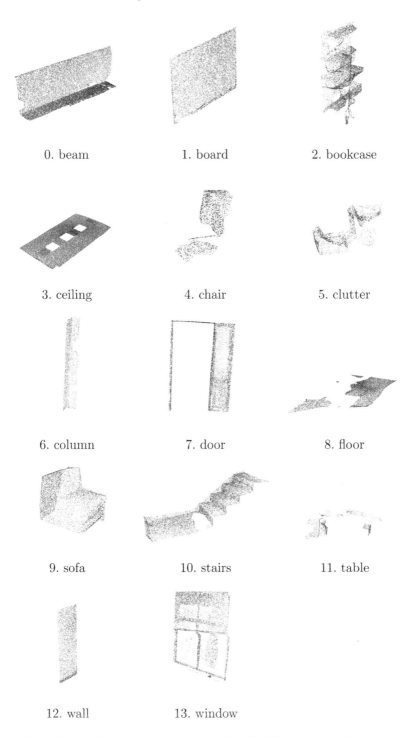

0. beam   1. board   2. bookcase

3. ceiling   4. chair   5. clutter

6. column   7. door   8. floor

9. sofa   10. stairs   11. table

12. wall   13. window

Fig. 4. Examples of class instances present in the S3DIS data set with their reference numbers (0–13).

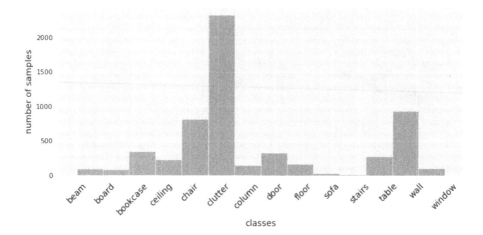

Fig. 5. Histogram of classes across the S3DIS database showing its unbalance.

where i is an iterator over rows (the actual labels) of a confusion matrix $\mathcal{C}$ and j is an iterator over columns (the predicted labels).

$$oAcc = \frac{\sum \mathrm{diag}(\mathcal{C})}{\sum_{i,j} \mathcal{C}_{i,j}} \ , \tag{5}$$

where $0 \leq i,j \leq A$. It should be noted that for data sets of extremely non-uniform classes distribution, $mAcc$ is more informative in drawing general conclusions. Such a situation occurs in the S3DIS database, where classes are relatively unbalanced. One of the classes (*clutter*) is on average about ten times over-represented than others. For multi-class classification mean- and overall accuracies may be thought to represent the same aspects of classification as macro- and micro-averaged F1-score respectively, which means that false positives and false negatives are taken into account as well.

For training deep learning models not only accuracy itself is crucial for evaluation but also the time required by a model to converge. This is the reason why in Table 2 the convergence time (T_c) is also juxtaposed throughout the methods being compared.

4.3   Experiments

Experiments were conducted using 5-fold cross validation. Prior to model training, the entire S3DIS database was split into three subsets: a training one, a validation one, and a test one, respectively of 60%, 20%, and 20%. Splits were generated randomly for each fold, yet keeping the original distribution of classes (see Fig. 5) in each split. Splits were created at the beginning and each object was sampled to contain 1,024 points. Sampling was done by random drawing with replacement. The same data set was used for the state-of-the-art DGCNN and the proposed novel VA-DGCNN model. During the training phase, each object was randomly permuted every time prior to passing it to the network. All hyperparameters and the general complexity were the very same for both DGCNN and VA-DGCNN (Table 1). The solution was implemented in PyTorch [6]. Sufficiently small learning rate (see Table 1) was used to avoid gradient explosion. First EdgeConv layer uses convolutions of filters 3×1 for the first convolution and 64×1 for the rest. For the second EdgeConv filters of size 64×1 and 128×1 were applied. Classifier is built as four-layer MLP with 128, 1024, 512, 256 hidden neurons respectively. Learning was conducted for relatively long time to be sure it converged. Moreover, quality measures for the validation set were tracked to be sure that the model does not overfit.

Table 1. Hyperparameters used for both DGCNN and VA-DGCNN. We perfomed experiments for various k and chose $k = 10$ as the best accuracy-speed tradeoff

Hyperparameter	Value
Learning rate	0.0005
Batch size	32
k	10
1. EdgeConv features	[3, 64, 64, 64]
2. EdgeConv features	[64, 128]
Classification module neurons	[128, 1024, 512, 256]

Besides that studies of the impact of the vicinity size on the classification results were performed. Several possible vicinities k = 2, 5, 10, 15, 20, 25 were tested as to determine the optimal value with the most reasonable time to quality metrics tradeoff.

Table 2. Comparison between the benchmark DGCNN method and the novel VA-DGCNN with the proposed vicinity abstraction component

Measure		DGCNN	VA-DGCNN
$mAcc$	[%]	72.8 ± 7.7	82.2 ± 2.6
$oAcc$	[%]	85.7 ± 3.1	90.1 ± 0.9
T_c	$[\cdot 10^3 \text{ sec}]$	9.7 ± 0.2	4.6 ± 0.2

5   Results

As it might be noticed in Fig. 6, vicinity increase causes nearly logarithmic increase of quality metrics at the cost of linear time growth. An inflection point

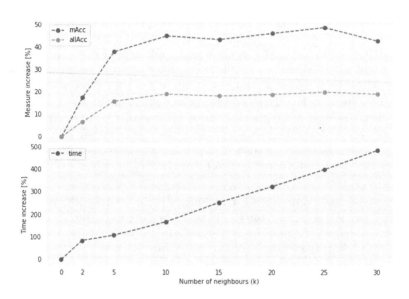

Fig. 6. $oAcc$ and $mAcc$ values for a test split vs. vicinity size (upper chart). Time vs. vicinity size (lower chart)

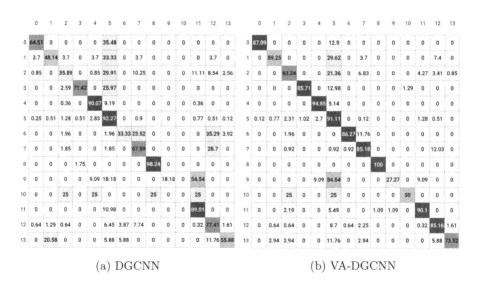

(a) DGCNN

	0	1	2	3	4	5	6	7	8	9	10	11	12	13
0	64.51	0	0	0	0	35.48	0	0	0	0	0	0	0	0
1	3.7	48.14	3.7	0	3.7	33.33	0	3.7	0	0	0	0	3.7	0
2	0.85	0	35.89	0	0.85	29.91	0	10.25	0	0	0	11.11	8.54	2.56
3	0	0	2.59	71.42	0	25.97	0	0	0	0	0	0	0	0
4	0	0	0.36	0	90.07	9.19	0	0	0	0	0	0.36	0	0
5	0.25	0.51	1.28	0.51	2.83	92.27	0	0.9	0	0	0	0.77	0.51	0.12
6	0	0	1.96	0	0	1.96	33.33	23.52	0	0	0	0	35.29	3.92
7	0	0	1.85	0	0	1.85	0	67.59	0	0	0	0	28.7	0
8	0	0	0	1.75	0	0	0	0	98.24	0	0	0	0	0
9	0	0	0	0	9.09	18.18	0	0	0	18.18	0	54.54	0	0
10	0	0	25	0	25	0	0	0	25	0	0	25	0	0
11	0	0	0	0	0	10.98	0	0	0	0	0	89.01	0	0
12	0.64	1.29	0.64	0	0	6.45	3.87	7.74	0	0	0	0.32	77.41	1.61
13	0	20.58	0	0	0	5.88	5.88	0	0	0	0	0	11.76	55.88

(b) VA-DGCNN

	0	1	2	3	4	5	6	7	8	9	10	11	12	13
0	87.09	0	0	0	0	12.9	0	0	0	0	0	0	0	0
1	0	59.25	0	0	0	29.62	0	3.7	0	0	0	0	7.4	0
2	0	0	63.24	0	0	21.36	0	6.83	0	0	0	4.27	3.41	0.85
3	0	0	0	85.71	0	12.98	0	0	0	0	0	1.29	0	0
4	0	0	0	0	94.85	5.14	0	0	0	0	0	0	0	0
5	0.12	0.77	2.31	1.02	2.7	91.11	0	0.12	0	0	0	1.28	0.51	0
6	0	0	1.96	0	0	0	86.27	11.76	0	0	0	0	0	0
7	0	0	0.92	0	0	0.92	0.92	85.18	0	0	0	0	12.03	0
8	0	0	0	0	0	0	0	0	100	0	0	0	0	0
9	0	0	0	0	9.09	54.54	0	0	0	27.27	0	9.09	0	0
10	0	0	25	0	0	25	0	0	0	0	50	0	0	0
11	0	0	2.19	0	0	5.49	0	0	1.09	1.09	0	90.1	0	0
12	0	0.64	0.64	0	0	8.7	0.64	2.25	0	0	0	0.32	85.16	1.61
13	0	2.94	2.94	0	0	11.76	0	2.94	0	0	0	0	5.88	73.52

Fig. 7. Confusion matrices for the original and the proposed model. Rows represent the actual labels and columns predicted labels, respectively. The values of the confusion matrix are expressed as percent by row.

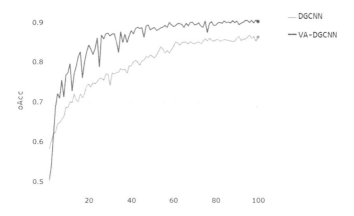

Fig. 8. *oAcc* values for validation split for a single fold

is located for $k = 25$. For wider neighbourhood quality measures tend to lower. In fact, the neighbourhood of size $k = 10$ may be said to be the optimal one as it ensures high $mAcc$ and $allAcc$ keeping processing time in acceptable bounds. In Table 2 a juxtaposition of evaluation measures for DGCNN and VA-DGCNN was presented. As it may be seen therein, the VA component enhanced remarkably the results of classification on the S3DIS data set. Mean accuracy $mAcc$, which is a more informative indicator for non-balanced data sets, yields the value of 82.2% which was an improvement by 9.4pp with respect to the original DGCNN. Also, overall accuracy ($oAcc$) improved by 4.4pp. This confirmed better classification results also for classes having much more representatives than the others (see confusion matrix in Fig. 7). Looking at the confusions matrices (Fig. 7), one may clearly see that confusing objects like *door*, *wall*, or *window* were better distinguishable having applied the proposed VA module. Also classes of a few samples, like *stairs* or *sofa* were better classified in the VA-DGCNN architecture. There was still a confusion between *board* and *bookcase*, yet diminished for the proposed VA-based solution. Having a look at the last row of Table 2 one may notice that the VA-DGCNN converged faster by virtually 53% (see also Fig. 8).

6   Conclusions

The results of the conducted experiments clearly confirmed what stated at the beginning of the paper, namely, aggregation of the local context prior to the actual features distribution learning improves results of object classification on noisy, real-life database like S3DIS. Thanks to the learning of aggregated local vicinity features, training process converges faster and more descriptive features are learnt. By applying the proposed, simple vicinity-abstraction layer, many benefits were reached. At first, thanks to the applied VA module, it was possible to boost $mAcc$ by 9.4pp and $oAcc$ by 4.4pp. Secondly, the convergence time decreased by more than a half, from around 9,700 s to 4,600 s It is also clear that provided evidences relates only to indoor scans and outdoor point clouds would need further investigation due to dramatically different characteristic. As a recap, it was confirmed that enriching the global matrix with the aggregated features of the local context, enhanced the results of classification. An interesting aspect that will be considered in further studies relates to semantic segmentation of noisy data making use of the proposed strategy.

Acknowledgement. The research was co-funded by the Polish National Center for Research and Development under the LIDER XI program.

References

1. Armeni, I., Sax, A., Zamir, A.R., Savarese, S.: Joint 2D–3D-Semantic Data for Indoor Scene Understanding. arXiv e-prints (Feb 2017)
2. Chen, C., Li, G., Xu, R., Chen, T., Wang, M., Lin, L.: Clusternet: deep hierarchical cluster network with rigorously rotation-invariant representation for point cloud analysis. In: Proceedings of the IEEE Conference on Computer Vision and Pattern Recognition, pp. 4994–5002 (2019)
3. Fang, H., Lafarge, F.: Pyramid scene parsing network in 3D: improving semantic segmentation of point clouds with multi-scale contextual information. ISPRS J. Photogram. Remote Sens. **154**, 246–258 (2019)
4. Guo, Y., Wang, H., Hu, Q., Liu, H., Liu, L., Bennamoun, M.: Deep learning for 3D point clouds: A survey. arXiv preprint arXiv:1912.12033 (2019)
5. Liu, J., Ni, B., Li, C., Yang, J., Tian, Q.: Dynamic points agglomeration for hierarchical point sets learning. In: Proceedings of the IEEE International Conference on Computer Vision, pp. 7546–7555 (2019)
6. Paszke, A., Get al.: Automatic differentiation in PyTorch (2017)
7. Qi, C.R., Su, H., Mo, K., Guibas, L.J.: PointNet: deep learning on point sets for 3D classification and segmentation. In: Proceedings of the IEEE Conference on Computer Vision and Pattern Recognition, pp. 652–660 (2017)
8. Qi, C.R., Yi, L., Su, H., Guibas, L.J.: PointNet++: deep hierarchical feature learning on point sets in a metric space. In: Advances in neural information processing systems, pp. 5099–5108 (2017)
9. Segol, N., Lipman, Y.: On universal equivariant set networks. arXiv preprint arXiv:1910.02421 (2019)
10. Shen, Y., Feng, C., Yang, Y., Tian, D.: Mining point cloud local structures by kernel correlation and graph pooling. In: Proceedings of the IEEE Conference on Computer Vision and pattern Recognition, 4548–4557 (2018)
11. Wagstaff, E., Fuchs, F.B., Engelcke, M., Posner, I., Osborne, M.: On the limitations of representing functions on sets. arXiv preprint arXiv:1901.09006 (2019)
12. Walczak, J., Poreda, T., Wojciechowski, A.: Effective Kd-like partition and shifted mahalanobis distance based regression. Remote Sens. **11**, 2465 (2019)
13. Wang, Y., Sun, Y., Liu, Z., Sarma, S.E., Bronstein, M.M., Solomon, J.M.: Dynamic graph CNN for learning on point clouds. ACM Trans. Graph. (TOG) **38**(5), 1–12 (2019)
14. Wu, Z., et al.: 3D ShapeNets: a deep representation for volumetric shapes. In: The IEEE Conference on Computer Vision and Pattern Recognition (CVPR) (Jun 2015)

15. Zhang, K., Hao, M., Wang, J., de Silva, C.W., Fu, C.: Linked dynamic graph CNN: Learning on point cloud via linking hierarchical features. arXiv preprint arXiv:1904.10014 (2019)
16. Zhang, Y., Rabbat, M.: A graph-CNN for 3D point cloud classification. In: 2018 IEEE International Conference on Acoustics, Speech and Signal Processing (ICASSP), pp. 6279–6283. IEEE (2018)

Grid-Based Concise Hash for Solar Images

Rafał Grycuk$^{(\boxtimes)}$ and Rafał Scherer

Czestochowa University of Technology, Al. Armii Krajowej 36, Czestochowa, Poland
{rafal.grycuk,rafal.scherer}@pcz.pl

Abstract. Continuous full-disk observations of the solar chromosphere and corona are provided nowadays by the Solar Dynamics Observatory. Such data are crucial for analysing the Sun-Earth system and life on our planet. Part of the data is an enormous number of high-resolution images. We create a compact grid-based solar image hash to classify or retrieve similar solar images. To compute the hash, we design intermediate hand-crafted features. Then, we use a convolutional autoencoder to encode the descriptors to the form of a concise hash.

Keywords: Fast image hash · Solar activity analysis · Solar image description · CBIR of solar images

1 Introduction

The NASA Solar Dynamics Observatory spacecraft has been providing solar data since 2010. Its part, the Atmospheric Imaging Assembly (AIA) delivers continuous full-disk observations of the solar chromosphere and corona in seven extreme ultraviolet (EUV) channels with the 12-second cadence in the form of high-resolution 4096×4096 pixel images. The images are relatively similar to each other, and general-purpose visual features are not suitable for their description. Moreover, the images are denoted only by their timestamp.

Semantic hashing [18] aims at generating compact vectors which values reflect semantic content of the objects. Thus, to retrieve similar objects we can search for similar hashes which is much faster and takes much less memory than operating directly on the objects. In [18] a multilayer neural network was used to generate hashes. Learned semantic hashes [20] are gaining in popularity in image retrieval. Our initial attempts showed that computing hashes from full-disk solar images would not be viable taking into account the size of the solar image collections (in terms of resolution and the number of images). Therefore, we developed the aforementioned hand-crafted intermediate descriptors.

A full-disk content-based image retrieval system is described in [1]. The authors checked eighteen image similarity measures with various image features resulting in one hundred and eighty combinations. The experiments shed light on what metrics are suitable for comparing solar images to retrieve or classify various phenomena.

© Springer Nature Switzerland AG 2021
M. Paszynski et al. (Eds.): ICCS 2021, LNCS 12744, pp. 242–254, 2021.
https://doi.org/10.1007/978-3-030-77967-2_20

A general-purpose retrieval engine Lucene is used to retrieve solar images in [2]. Each image is a document consisting of 64 elements (rows of each image), and every image-document is unique. The solar images are then queried by setting some wild-card characters in the query strings that allows to search for similar solar events. The Lucene engine is compared in [3] with distance-based image retrieval methods, however, without a clear winner. It turned out that every tested method has its pros and cons in terms of accuracy, speed and applicability. The trade-off between accuracy and speed is significant, and for accurate results, the retrieval time was several minutes.

A sparse model representation of solar images was developed in [8]. The method used the sparse representation from [13] and outperformed previous solar image retrievals in accuracy and speed. In [10], some solar image parameters are chosen to track multiple solar events across images with 6-minute cadence. Sparse codes for AIA images are used also in [9], where ten texture-based image parameters are used to create the code. The parameters are computed for regions determined by a 64×64 grid for nine wavelengths. For each wavelength, a dictionary of k elements is learned, and then a sparse representation is computed. To overcome the curse of dimensionality affecting the solar data, they use the Minkowski norm and choose the right value of p parameter. Finally, the authors used a 256-dimensional descriptor what is an efficient and accurate outcome comparing to the previous approaches. In [11], a method for image retrieval with fuzzy sets and boosting is developed.

To automate solar image retrieval and enable their fast classification, we propose a fast and concise solar image hash generated from one-dimensional hand-crafted features by a fully convolutional autoencoder. The hash has only eleven real-valued elements, and the experiments showed that such compactness is sufficient to describe the images. In the dataset, the images are annotated only by their timestamp. It is very hard to make any meaning to data or explain the trained system [14]. We treat the timestamp as a measure of similarity. After training, our algorithm allows retrieving images by their visual similarity, regardless of the timestamp proximity. The paper is organized as follows. Section 2 introduces the method for generating learned solar hashes. Experiments on the SDO solar image collection are described in Sect. 3. Section 4 concludes the paper.

2   Grid-Based Image Hash for Solar Image Retrieval

Solar images are relatively similar to each other, and general-purpose descriptors are usually not applicable in their retrieval. Therefore, we present a novel grid-based algorithm for the solar image hashing. The proposed hash can be used for image retrieval of solar images in large solar image datasets. The solar images were taken from the Solar Dynamics Observatory (SDO), where they are post-processed and published in the form of Web API by [12]. There are many resolutions available, and we use the high definition 2048×2048 images; thus creating image descriptors requires a significant amount of memory. In our experimental environment, we used GeForce RTX 2080 Ti 11GB GDDR6 graphics card, which allowed us to use 11 GB of memory. Initially, we tried to design directly a full-disc autoencoder. Setting a higher mini-batch value caused an out-of-memory exception. Moreover, the learning time in this simulation took several days versus several minutes as in the presented approach. Therefore, we decided to apply some preprocessing stage (calculation of grid-based descriptor) and then use the autoencoder (see Sect. 2.3) to reduce the hand-crafted vectors to 11-element real-valued hashes without losing significant information about the active regions. The presented algorithm is composed of four main stages: active region detection, calculating solar image hand-crafted descriptors, encoding to hash, and retrieval.

2.1   Active Region Detection

In the first step, we need to obtain the Active Regions (AR). They are brighter regions of the solar images, and they are essential in detecting solar flares. ARs have various shapes, and they change due to the Sun's rotation movement. The presented method determines the positions and shapes of Active Regions. During this process, at first, we change image colour space from RGB to grayscale. Therefore we reduce the number of colour channels from three to one. As a result, every pixel of our grayscale solar image will have intensities values in the range [0..255]. In the next stage, we use the Gaussian blur in order to remove insignificant, small regions. Then, the pixel intensities are filtered by using the threshold th, provided as the algorithm parameter. Thus, the obtained image is properly preprocessed for the thresholding stage. Subsequently, every pixel intensity is compared with the provided threshold parameter th value. If the value is greater or equal, we determine that pixel is a part of the active region. The value th parameter was determined empirically to 180, and it was obtained for the given solar image dataset [12]. In the next step, we apply the morphological operations, namely erosion and dilation to the thresholded image, obtained in the previous step. The morphological operation erosion eradicates separated small objects (pixels). These objects can be referred as "islands". After this operation, only substantive (important) objects remain. The dilation operation, on the other hand, makes objects more visible; thus, it fills in small holes in objects. These two types of operations can enhance the important areas of the active regions. More informations about morphological operations can be found in [5,19]. The process of active region detection is described in the form of pseudo-code in Algorithm 1. Figure 1 presents an input image (left) and active regions detected in the image (right). The applied operations allow detecting active regions of the solar image. The accurate detection of these regions is vital to subsequent stages of the algorithm. The location and the shape of active regions are significant in detecting the Coronal Mass Ejection (CME) and thus, in the solar flare prediction.

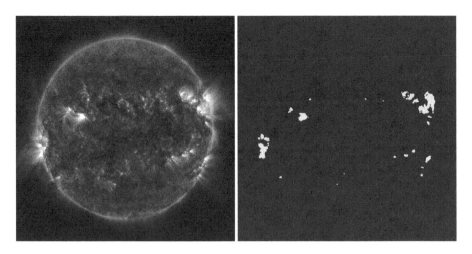

Fig. 1. Active region detection process, left image is the input, right image is the output.

INPUT: $SolarImage$
OUTPUT: $ActiveRegionDetectedImg$
$GrayScaleImg := ConvertGrayScale(SolarImage)$
$BlurredImg := Blur(GrayScaleImg)$
$ThreshImg := Threshold(BlurredImg)$
$ErodedImg := Erode(ThreshImg)$
$ActiveRegionDetectedImg = Dilate(ErodedImg)$
Algorithm 1: Active region detection steps.

2.2   Calculating Grid-Based Descriptor

In this section, we describe the process of calculating the grid-based descriptor. We take an image with active region detected (AR image) as input, and we obtain a grid-based descriptor of 100 length. The descriptor length was determined empirically, but this can be changed by adjusting the parameters $gridSizeN$ and $gridSizeM$. Values of these variables are equal 10 for both of them, which gives us a 100-element descriptor vector. In the first step, we take the AR image and divide it into cells (sub-images) using the grid. During this process, we slice image at x-axis and afterwards for each slice (cells) we perform slicing at y-axis. Therefore, we obtain an image grid, where grid cells contain sub-images. By using the parameters ($gridSizeN$ and $gridSizeM$) we can define a number of grid cells both for x and y axis. In the next step, we calculate a sum of active region pixels for each grid cell. As a result of this process, we obtain $n \times m$ **DM** matrix, where each element contains a grid cell sum. In the last step of this stage, we perform a matrix normalization and vectorization and provide a **DV** vector of size $n * m$, e.g. if matrix **DM** is 10×10 then **DV** size is 100. It should be noted that **DV** size depends on values of $gridSizeN$ and $gridSizeM$ parameters. It also should be noted that both grid size parameters was obtained empirically, and there

INPUT: ARI - active region detected image
$gridSizeN$ - grid size in x-axis
$gridSizeM$ - grid size in y-axis
OUTPUT: $GridBasedDescriptorVector$
Local Variables: $ImageCells$ - list for containing grid image cells
$SumMatrix$ - matrix sums of pixels in the cells
$HSlices := DivideIntoHorizontalSlices(ARI, gridSizeN)$
foreach $HSlice \in HSlices$ **do**
$\quad CellsForHSlice := DivideSlicesVerticallyIntoCells(HSlice, gridSizeM)$
$\quad$ **foreach** $CellForHSlice \in CellsForHSlice$ **do**
$\quad\quad |\quad ImageCells.Add(CellForHSlice)$
$\quad$ **end**
end
foreach $ImageCell \in ImageCells$ **do**
$\quad CellSum := CalculateSumOfActiveRegionPixelsInCell(ImageCell)$
$\quad SumMatrix.SetCellSum(CellSum)$
end
$GridBasedDescriptorVector = VectorizeMatrix(SumMatrix)$
Algorithm 2: Algorithm for calculating grid-based descriptor.

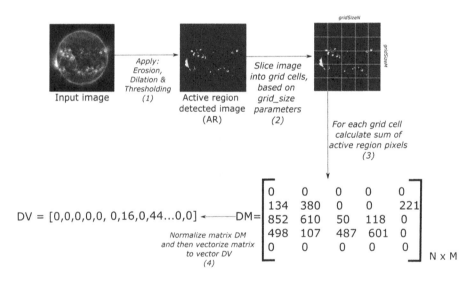

Fig. 2. Steps for calculation of the grid-based descriptor.

values have significant impact on the results. Scaling the values for both axis allows to determine the most suitable values for given solar image resolution. The entire process of calculating the grid-based descriptor is presented in the form of pseudo-code in Algorithm 2. Let us analyze the consecutive steps of this stage in the visual form; see Fig. 2. In the first step, the input image is subjected to morphological operations of erosion and dilation, and then the thresholding process is applied. As a result, we obtain the active region detected image. This process is defined by Eq. 1. In the next step, we slice the image into the grid cells. Based on the previously obtained grid, we calculate **DM** matrix. This stage can be performed by using Eq. 2. Afterwards, we normalize this matrix and then vectorize it in order to obtain the **DV** vector

$$
t(ARI, i, j, th) = \begin{cases} 1, & ARI_{i,j} \geq th \\ 0, & \text{otherwise} \end{cases}, \tag{1}
$$

where th is threshold value and ARI is active region image.

$$
\mathbf{DM}(ARI, th)_{k,l} = \sum_{i=k*csx}^{(k+1)*csx-1} \sum_{j=l*csy}^{(l+1)*csy-1} t(ARI, i, j, th), \tag{2}
$$

where csx is the cell size in x-axis and csy is the cell size in y-axis. The process of calculating grid-based descriptor allows reducing the data volume during the encoding stage significantly. The aim of this process is to obtain a hand-crafted, intermediate mathematical representation of AR images used in the next step.

2.3   Hash Generation

In this section, the hash generation (autoencoding) process is described. We use previously obtained grid-based image descriptors to reduce the descriptor length and in order to obtain the latent space a one-dimensional hash. To this task, we used a convolutional autoencoder to encode our hand-crafted image descriptors in its latent space. Autoencoders are used for network user identification [6,16] or for image reconstruction and improvement [15,17]. We used the unsupervised convolutional neural network as it does not require labelled data for training (it was not provided in the Web API or the dataset). The autoencoder architecture is presented in Table 1. As can be seen in Table 1, a convolutional autoencoder was used for hash generation process, and the table should be analyzed top to bottom, where the top layer is input. Afterwards, we have a set of encoding layers; it is composed of 3-layer groups, where every group contains three layers ($Conv1D$, $ReLU$, $MaxPool1D$). The $kernel$ parameter used in $Conv1D$ and $MaxPool1D$ layers is equal 2. After three convolutional layers with pooling, we have the latent space, bottleneck layer, which is encoded layer for the hash generation. Then, the autoencoder has decoding groups which are composed of $MaxUnPool1D$ (upsampling) layers, convolutional layers and $ReLU$. There is also a padding layer, which allows obtaining the same shape of decoded data as the input one. For the hash generation, we only use the encoding layers. The reason why we used a one-dimensional autoencoder is that our grid-based image descriptors are one-dimensional vectors for decreasing computational complexity. This process allows reducing the hash length without significant loss of important information about the active regions of the solar image. For the loss function, we used the mean squared error function. We empirically proved that 50 epochs are sufficient to obtain the required level of generalization and to prevent the network over-fitting. After the training process is finished, every image descriptor is provided to the

Table 1. Tabular representation of the convolutional autoencoder model.

Layer (type)	Output shape	Filters (in, out)	Kernel size	Params no.
$Input1d(InputLayer)$	[1, 1, 100]			
$Conv1d\_1(Conv1D)$	[1, 64, 100]	1, 64	2	192
$ReLU\_1$	[1, 64, 100]			
$Max\_pooling1d\_1(MaxPool1D)$	[2, 64, 48]		2	
$Conv1d\_2(Conv1D)$	[2, 32, 48]	64, 32	2	4128
$ReLU\_2$	[2, 32, 48]			
$Max\_pooling1d\_2(MaxPool1D)$	[2, 32, 24]		2	
$Conv1d\_3(Conv1D)$	[1, 1, 23]	32, 1	2	65
$ReLU\_3$	[1, 1, 11]			
$Encoded(MaxPool1D)$	[1, 1, 11]		2	
$up\_sampling1d\_1(MaxUnPool1D)$	[1, 1, 22]		2	
$ConvTranspose1d\_1(ConvTranspose1D)$	[1, 1, 23]	1, 32	2	96
$ReLU\_4$	[1, 1, 23]			
$ConstPadding_1(ConstPad1D)$	[1, 1, 24]		2	
$up\_sampling1d\_2(MaxUnPool1D)$	[2, 32, 48]		2	
$ConvTranspose1d\_2(ConvTranspose1D)$	[2, 64, 49]	32, 64	2	4160
$ReLU\_5$	[2, 64, 49]			
$up\_sampling1d\_3(MaxUnPool1D)$	[1, 64, 98]		2	
$ConvTranspose1d\_3(ConvTranspose1D)$	[1, 1, 99]	61, 1	2	129
$ReLU\_6$	[1, 1, 99]			
$ConstPadding_1(ConstPad1D)$	[1, 1, 100]		2	
$Decoded(Tanh)$	[1, 1, 100]			

latent space (encoded) layers of the autoencoder. As a result of this process, we obtained encoded a fast image hash as a 11-tuple of real-value elements. As can be seen in Table 2, the presented method provides the image hashes, where hashes of consecutive images are similar, which is highly desirable, because consecutive solar corona images have similar active regions. The obtained hash can be used for content-based solar image retrieval applications. It also should be noted that presented autoencoder architecture was selected in order to obtain the most suited generalization level.

2.4   Retrieval

In the last stage of the presented method, we use previously obtained hashes for image retrieval. After previous steps, we can assume that every solar image has a hash assigned in our image database. The retrieval step allows executing the

Table 2. Two examples of similar image pairs with their corresponding 11-element hashes. They show how similar are the vectors for semantically similar images.

Hash values			
Pair 1		Pair 2	
2015-01-01 00:00:00	2015-01-01 00:06:00	2015-03-02 04:06:00	2015-03-02 04:12:00
0.18948224	0.18948224	0.09669617	0.09669617
0.38965224	0.34482240	0.09669617	0.09669617
0.18948224	0.18948224	0.09669617	0.09669617
0.18948224	0.18948781	0.10213041	0.10213816
0.18943328	0.18936896	0.10843775	0.10845160
0.18830273	0.18814683	0.11238195	0.11103466
0.18947415	0.18947072	0.10204165	0.10199506
0.19400954	0.19399905	0.11846152	0.11823325
0.18971351	0.18972327	0.10581696	0.10593924
0.18960209	0.18960896	0.09670250	0.09670459
0.18948224	0.18948224	0.09669617	0.09669617

image query by comparing distances between the query image hash and hashes created for all images stored in the dataset. The retrieval step requires to have a solar image database with a hash generated for every image. In the next step, we calculate the distances between the query image hash and every hash in the database. The distance d is calculated by the cosine distance measure (for more see [7])

$$\cos(QH_j, IH_j) = \sum_{j=0}^{n} \frac{(QH_j \bullet IH_j)}{\|QH_j\| \, \|IH_j\|}, \tag{3}$$

where $\bullet$ is dot product, QH_j is the query image hash, and IH_j a consecutive image hash. After calculating the cosine distance, the images stored in the database are sorted in ascending order by distance to the query (query hash). The last step of the presented method allows to take n images closest to the query and return them to the user as the retrieved images. During query execution, the n parameter is required. The entire process is presented as pseudo-code in Algorithm 3. Alternatively, we can also retrieve images based on a threshold. In such a case, we must provide a threshold parameter instead of n and then retrieve images only if their cosine distance to the query is below the threshold. The proposed method also allows applying such an approach. Nevertheless, we prefer the first method because it is more suited for the system user.

INPUT: $ImageHashes$, $QueryImage$, n
OUTPUT: $RetrievedImages$
foreach $ImageHash \in ImageHashes$ **do**
$\quad | \quad QueryImageHash = CalculateHash(QueryImage)$
$\quad | \quad D[i] = Cos(QueryImageHash, ImageHash)$
end
$SortedDistances = SortAscending(D)$
$RetrievedImages = TakeFirst(n)$

Algorithm 3: Image retrieval steps.

3 Experimental Results

This section describes simulation results along with a solution for the evaluation of unlabelled images. Due to lack of labelled data, unsupervised learning was used for descriptors encoding. Therefore, the evaluation of the proposed method with state of the art approaches is difficult. In order to resolve this problem, we use the Sun's rotation movement to determine a set of similar images (SI). We assumed that consecutive images within a small time window should have similar active regions. Those regions are slightly shifted between consecutive images. The Web API provides solar images with 6 min cadence window. Due to the nature of the Sun movement, we can assume the similarity of consecutive images. The only condition is adjusting the difference time window. Based on experiments, we determined that images within a 48 h window can be treated as similar. Let us take under consideration an image taken at 2012-02-15, 00:00:00. Based on the above assumptions, we can assume that every image in 24 h before and in 24 h after is similar. Only for evaluation purposes, images are identified by the timestamps. The process of determining similar images is presented in Table 3. By using the proposed method for determining image similarity we performed series of experiments and we obtained the similar images (SI). The single experiment can be described by the following steps:

1. Execute image query and obtain the retrieved images.
2. For every retrieved image, compare its timestamp with the query image timestamp.
3. If the timestamp is the 48 h window, the image is similar to the query.

After defining similar images (SI), we can define performance measures *precision* and *recall* [4,21] based on following sets:

– SI - set of similar images,
– RI - set of retrieved images for query,
– $PRI(TP)$ - set of positive retrieved images (true positive),
– $FPRI(FP)$ - false positive retrieved images (false positive),
– $PNRI(FN)$ - positive, not retrieved images,
– $FNRI(TN)$ - false, not retrieved images (TN).

Table 3. Defining image similarity. Based on experiments, we determined that images within a 48-h window can be treated as similar. This allows to evaluate the method.

Timestamp	SI (similar image)/ NSI (not similar image)
2012-02-13, 23:54:00	NSI
2012-02-14, 00:00:00	SI
2012-02-14, 00:06:00	SI
2012-02-14, 00:12:00	SI
2012-02-14, 00:18:00	SI
2012-02-14, 00:24:00	SI
2012-02-14, 00:30:00	SI
....	SI
2012-02-15, 00:00:00	QI (query image)
....	SI
2012-02-15, 23:24:00	SI
2012-02-15, 23:30:00	SI
2012-02-15, 23:36:00	SI
2012-02-15, 23:42:00	SI
2012-02-15, 23:48:00	SI
2012-02-15, 23:54:00	SI
2012-02-16, 00:00:00	NSI

Afterwards, we can define *precision* and *recall* for CBIR systems

$$precision = \frac{|PRI|}{|PRI + FPRI|}, \tag{4}$$

$$recall = \frac{|PRI|}{|PRI + PNRI|}. \tag{5}$$

$$F_1 = 2 * \frac{precision * recall}{precision + recall}. \tag{6}$$

We present the experiment results in Table 4. The presented results proved the effectiveness of the method. Our approach obtains a high value of the *precision* measure. Most of the images close to the query are correctly retrieved. The farther from the query then more positive, not retrieved images (PNRI) are retrieved. This phenomenon is caused by the Sun's rotation, and thus more missing active regions are detected between images. In the 48 h cadence, the significant active region can change its position; this may have a significant impact on the hash. Therefore, the distance to the query will be increased. The simulation environment was created in Python using Pytorch on the following hardware: Intel Core I9-9900k 3.6 GHz, 32 GB RAM, GeForce RTX 2080 Ti 11 GB, Windows Server 2016. The presented solution is available on the BitBucket repository

Table 4. Experiment results for the proposed algorithm, performed on AIA images obtained from [12]. Due to lack of space, we present only a part of all queries.

Timestamp	RI	SI	PRI (TP)	FPRI (FP)	PNRI (FN)	Precision	Recall	F_1
2015-01-01 00:00:00	164	241	159	5	82	0.97	0.66	0,78
2015-01-03 01:00:00	372	481	330	42	151	0.89	0.69	0,77
2015-01-09 16:00:00	372	481	336	36	145	0.90	0.7	0,79
...	...	...	...	...	...	...	...	...
2015-05-12 00:36:00	362	481	330	32	151	0.91	0.69	0,78
2015-05-18 07:36:00	337	481	331	6	150	0.98	0.69	0,81
2015-05-25 18:36:00	349	481	317	32	164	0.91	0.66	0,76
...	...	...	...	...	...	...	...	...
2015-08-09 13:18:00	344	481	305	39	176	0.89	0.63	0,74
2015-08-11 05:24:00	327	481	315	12	166	0.96	0.65	0,77

under the following link: https://bitbucket.org/rafal-grycuk/novel_grid-based_image_hash_for_content/src/src/master. The hash creation time took approximately 17.6 min, for 83,819 images. The encoding stage took approximately 1.5 h. The average retrieval time is approximately 300 ms.

4   Conclusions

In this paper, we proposed a novel grid-based image hash for fast content-based solar image retrieval and classification. Initially, we tried to make hashes directly from full-disc images. It turned out to be infeasible having general-purpose GPUs at our disposal. For this reason, we decided to design intermediate hand-crafted features. To this end, we apply morphological operations for preprocessing and active regions detection and then the grid for descriptor calculation. Only after this step, we use an unsupervised convolutional autoencoder to encode the descriptors to the concise hash form. The process of the second encoding allows reducing the description length significantly; in our experiments, over ten times compared to the hand-crafted descriptor obtained in the first stage. Reducing the hash length is, of course, significant for the speed of calculating the distances between hashes, that is, the similarity of solar images. As solar AIA images are unlabelled, we treat images generated in a short time to each other (up to several hours) as similar. In fact, at other time, the Sun configuration could be similar. Therefore, our precision and recall measures which rely on the image content solely will have even higher values in practice. The presented approach has various potential applications. It can be used for searching, classifying and retrieving solar flares, which has crucial importance for many aspects of life on Earth.

References

1. Banda, J., Angryk, R., Martens, P.: Steps toward a large-scale solar image data analysis to differentiate solar phenomena. Sol. Phys. **288**(1), 435–462 (2013)

2. Banda, J.M., Angryk, R.A.: Scalable solar image retrieval with lucene. In: 2014 IEEE International Conference on Big Data (Big Data), pp. 11–17. IEEE (2014)
3. Banda, J.M., Angryk, R.A.: Regional content-based image retrieval for solar images: traditional versus modern methods. Astron. Comput. **13**, 108–116 (2015)
4. Buckland, M., Gey, F.: The relationship between recall and precision. J. Am. Soc. Inf. Sci. **45**(1), 12 (1994)
5. Dougherty, E.R.: An Introduction to Morphological Image Processing. SPIE, Bellingham (1992)
6. Gabryel, M., Grzanek, K., Hayashi, Y.: Browser fingerprint coding methods increasing the effectiveness of user identification in the web traffic. J. Artif. Intell. Soft Comput. Res. **10**(4), 243–253 (2020). https://doi.org/10.2478/jaiscr-2020-0016
7. Kavitha, K., Rao, B.T.: Evaluation of distance measures for feature based image registration using alexnet. arXiv preprint arXiv:1907.12921 (2019)
8. Kempoton, D., Schuh, M., Angryk, R.: Towards using sparse coding in appearance models for solar event tracking. In: 2016 19th International Conference on Information Fusion (FUSION), pp. 1252–1259 (2016)
9. Kempton, D.J., Schuh, M.A., Angryk, R.A.: Describing solar images with sparse coding for similarity search. In: 2016 IEEE International Conference on Big Data (Big Data), pp. 3168–3176. IEEE (2016)
10. Kempton, D.J., Schuh, M.A., Angryk, R.A.: Tracking solar phenomena from the SDO. Astrophys. J. **869**(1), 54 (2018)
11. Korytkowski, M., Senkerik, R., Scherer, M.M., Angryk, R.A., Kordos, M., Siwocha, A.: Efficient image retrieval by fuzzy rules from boosting and metaheuristic. J. Artif. Intell. Soft Comput. Res. **10**(1), 57–69 (2020). https://doi.org/10.2478/jaiscr-2020-0005
12. Kucuk, A., Banda, J.M., Angryk, R.A.: A large-scale solar dynamics observatory image dataset for computer vision applications. Sci. Data **4**, 1–9 (2017)
13. Mairal, J., Bach, F., Ponce, J., Sapiro, G.: Online learning for matrix factorization and sparse coding. J. Mach. Learn. Res. **11**(1), 19–60 (2010)
14. Mikołajczyk, A., Grochowski, M., Kwasigroch, A.: Towards explainable classifiers using the counterfactual approach - global explanations for discovering bias in data. J. Artif. Intell. Soft Comput. Res. **11**(1), 51–67 (2021). https://doi.org/10.2478/jaiscr-2021-0004
15. Najgebauer, P., Scherer, R., Rutkowski, L.: Fully convolutional network for removing DCT artefacts from images. In: 2020 International Joint Conference on Neural Networks (IJCNN), pp. 1–8 (2020). https://doi.org/10.1109/IJCNN48605.2020.9207249
16. Nowak, J., Holotyak, T., Korytkowski, M., Scherer, R., Voloshynovskiy, S.: Fingerprinting of URL logs: continuous user authentication from behavioural patterns. In: Krzhizhanovskaya, V.V., et al. (eds.) ICCS 2020, Part IV. LNCS, vol. 12140, pp. 184–195. Springer, Cham (2020). https://doi.org/10.1007/978-3-030-50423-6_14
17. Pawlak, M., Panesar, G.S., Korytkowski, M.: A novel method for invariant image reconstruction. J. Artif. Intell. Soft Comput. Res. **11**(1), 69–80 (2021). https://doi.org/10.2478/jaiscr-2021-0005
18. Salakhutdinov, R., Hinton, G.: Semantic hashing. Int. J. Approximate Reasoning **50**(7), 969–978 (2009). https://doi.org/10.1016/j.ijar.2008.11.006. Special Section on Graphical Models and Information Retrieval
19. Serra, J.: Image Analysis and Mathematical Morphology. Academic Press Inc., Cambridge (1983)

20. de Souza, G.B., da Silva Santos, D.F., Pires, R.G., Marananil, A.N., Papa, J.P.: Deep features extraction for robust fingerprint spoofing attack detection. J. Artif. Intell. Soft Comput. Res. **9**(1), 41–49 (2019). https://doi.org/10.2478/jaiscr-2018-0023
21. Ting, K.M.: Precision and recall. In: Sammut, C., Webb, G.I. (eds.) Encyclopedia of Machine Learning, p. 781. Springer, Boston (2011). https://doi.org/10.1007/978-0-387-30164-8_652

Machine Learning Algorithms for Conversion of CVSS Base Score from 2.0 to 3.x

Maciej Nowak$^{(\boxtimes)}$, Michał Walkowski , and Sławomir Sujecki

Department of Telecommunications and Teleinformatics, Wroclaw University
of Science and Technology, 50-370 Wroclaw, Poland
{maciej.nowak,michal.walkowski,slawomir.sujecki}@pwr.edu.pl

Abstract. The Common Vulnerability Scoring System (CVSS) is the industry standard for describing the characteristics of software vulnerabilities and measuring their severity. However, not all publicly known vulnerabilities have criticality rating in CVSS 3.x, which is the latest and most advanced version of the standard. This is due to the large time gap between the publication of the CVSS 2.0 and CVSS 3.x standards, the large number of the detected and published vulnerabilities at the time, and significant differences in the method of determining vulnerability criticality and assigning vector properties to evaluation components. Consequently, organizations using CVSS to prioritize vulnerabilities use both CVSS versions and abandoned the full transition to CVSS 3.x standard. In this paper authors introduce machine learning algorithms for performing conversions from CVSS 2.0 to CVSS 3.x, scores, which should significantly facilitate the upgrade to CVSS 3.x standard for all stakeholders. The considered case corresponds to a real world application with a large potential impact of the research.

Keywords: Security · CVSS score · Well-known vulnerabilities · Machine learning

1 Introduction

The Common Vulnerability Scoring System (CVSS) is the industry standard for describing the characteristics of a software vulnerability and measuring its severity [6]. CVSS was first introduced as a research project by the National Infrastructure Advisory Council (NIAC) in 2005 [18], and afterwards accepted by other organizations [26]. It consists of three categories: Base (B_S), Temporal (T_S) and Environmental (E_S). The B_S category is marked by vulnerability properties that are time invariable. The B_S category properties include access complexity, access vector and the assessment of the extent to which the vulnerability may threaten confidentiality, integrity and system availability. The T_S

Supported by organization Wroclaw University of Science and Technology.

M. Paszynski et al. (Eds.): ICCS 2021, LNCS 12744, pp. 255–269, 2021.
https://doi.org/10.1007/978-3-030-77967-2_21

category describes properties that may change over time. In particular, the T_S category contains information on the existence of a public exploit and patches availability, whilst the E_S category includes information on the systems environment affected by the detected vulnerability.

The standard update to version 3.0 was published in 2015 [10] and significantly differs from the previous version in the method of determining the vulnerability criticality and assigning vector properties to the evaluation component. The following changes were introduced in the B_S category in the latest CVSS version (Table 1):

- the measurement parameters of confidentiality, availability and integrity, have been changed to: none, low or high,
- a physical value (P) has been added to the attack vector, which indicates the security vulnerability in which the attacker must have a physical access to the system in order to be able to exploit it,
- the required permissions (PR) parameter has been added to indicate whether administrative permissions or other need to be achieved on the target system, in order to successfully exploit the vulnerability.

Table 1. Difference between CVSS 2.0 and CVSS 3.1. properties of B_S category

CVSS 2.0	CVSS 3.1
Access vector (AV)	Attack vector (AV)
Access complexity (AC)	Attac complexity (AC)
Authentication (Au)	Privileges required (PR)
	User interaction (UI)
Confidentiality impact (C)	Confidentiality impact (C)
Integrity impact (I)	Integrity impact (I)
Availability impact (A)	Availability impact (A)
	Scope (S)

In the T_S category, the name of the vector *Exploitability* was changed to *Exploit Code Maturity*, whereas the rest of the vectors remained unchanged. In the E_S category, the implemented change involves the replacement of two indicators with so-called modified base scores. Essentially, each of the base metrics can be modified by the organization in order to reflect differences in the tested environment (Table 2).

After the publication of the CVSS 3.0 standard, a large number of companies criticized it for inaccurate parameter descriptions and thereby allowing for their different interpretation [14]. In 2019, the CVSS 3.1 update was released, which specified the description of the metrics, making them more understandable for the recipient. However, the method of calculating the vulnerability assessment, as well as the parameters form and vector representation, have not changed [11].

Table 2. Difference between environmental category for CVSS 2.0 and CVSS 3.1.

CVSS 2.0	CVSS 3.1
Collateral damage potential (CDP)	
Target distribution (TD)	
	Attack vector (MAV)
	Attack complexity (MAC)
	Privileges required (MPR)
	User interaction (MUI)
	Scope (MS)
	Confidentiality impact (MC)
	Integrity impact (MI)
	Availability impact (MA)
	Confidentiality impact (MC)
	Integrity impact (MI)
	Availability impact (MA)
Confidentiality requirement (CR)	Confidentiality requirement (CR)
Integrity requirement (IR)	Integrity requirement (IR)
Availability requirement (AR)	Availability requirement (AR)

Since there is a 10-year difference between the publications of the CVSS 2.0 and CVSS 3.0 standard, it was not possible to convert all scores from CVSS 2.0 to the newer version automatically. Consequently, only new vulnerabilities contain both criticality assessments (currently 73179), the remaining 73856 vulnerabilities, which accounts for 51% of the known vulnerabilities, so far have not been assigned the CVSS score according to the latest version of the standard [20]. Consequently, organizations using CVSS in order to prioritize vulnerability fixes, use both versions [8]. The described situation also causes delays in publishing the vulnerability with its assessment [6], which additionally increases the time gap between the vulnerability publication and its repair [26]. Not every factor responsible for the criticality of the given vulnerability will not be considered, if the CVSS 2.0 is used. Moreover, in case the person responsible for maintaining the IT service obtains the same vulnerability with two different assessments (CVSS 2.0 and CVSS 3.x), the use of these both standards causes a dissonance in vulnerabilities prioritization.

The main objective of this work is to calculate all components of the CVSS vector for B_S category for all the remaining 73856 vulnerabilities, which have not so far been assigned the base score. The novel contribution of this work consists in performing automatic conversions from CVSS 2.0 to CVSS 3.x base score, i.e. B_S category, using machine learning algorithms. The value of this contribution is further enhanced by the fact that the machine learning algorithms are applied to difficult data derived from CVSS database and that the considered case corresponds to a real world application with a large potential impact of the research. This paper is divided into the following sections:

- Related Work - section presents other work related to the present topic. It is devoted to a brief description of work related to machine learning algorithms and CVSS predictions.
- Methods – section presents the description of methods used for the preparation of the database and the machine learning algorithms.
- Results - section contains the discussion of the results describing the advantages of the proposed machine algorithms.
- Conclusions - section gives the summary of the presented results and introduces fields for further research.

2 Related Work

A large number of authors criticized the currently used CVSS standard explicitly and attempted to improve the criticality assessment of the detected vulnerabilities, [7,15,21,22]. Despite the aforementioned criticism, the CVSS standard is widely used by corporations to prioritize vulnerability fixes, supporting the vulnerability management process [8]. However, as noticed in [26] each organization approaches the problem differently. Therefore, in order to help organizations to streamline the process by more effective adaptation of the vulnerability assessment to the vulnerability assessment more efficiently to the consequences of its use, new methods were proposed, for instance, by introduction of solutions enabling the vulnerability predictions [12,27] or facilitating vulnerability assessment using patterns and machine learning [24]. Additionally, the developed algorithms for automatic estimation of the CVSS B_S score, which is obtained from the vulnerability description in order to accelerate the publication process in public National Vulnerability Databases (NVD) [6,13] and to include other categories such as T_S [23] and E_S [17,26]. Despite the indicated works, corporations are still forced to use both standards in order to prioritize the vulnerability correctly [8].

However, to the best knowledge of the authors nobody tried so far to use machine learning to convert scores from CVSS 2.0 to CVSS 3.x standard by estimating all the component values of the result vector included in the final assessment, in order to facilitate the stakeholder vulnerability management process only by the latest version of the standard.

3 Methods

In this section first the description of methods used for the preparation of the database is given and then is followed by the description of the machine learning algorithms.

Open-source software suite: Vulnerability Management Center (VMC) [5,25, 26], was used to prepare a database for known vulnerabilities and weaknesses considering information collected from the public sources. The VMC software also provides an Application Programming Interface (API) for Python that was used to write a script for selection of training data of the vulnerability critical-ity assessment for both CVSS 2.0 and CVSS 3.x standards. This generated a database with 73179 items. For each item, the script transformed an alphanu-meric value of the vector to a numerical form using the CVSS 2.0 specification, cf. an example for Access Vector field given in Table 3 [10]. Similarly, the data has been converted from an alphanumeric form of the CVSS 3.x vector to the numbers of the corresponding classes, cf. an example given in Table 4 for Attack Vector field. Thus one obtains a 7 element vector for B_S CVSS 2.0 standard and an 8 element vector for CVSS 3.x standard. The allowed values of the vectors' elements are defined by the respective CVSS 2.0 and 3.x standards. The B_S scores that are known for both CVSS 2.0 and 3.x standards establish a mapping between both sets of vectors. An initial analysis of this mapping revealed how-ever, that there are many instances of CVSS 2.0 vectors that map spuriously on different CVSS 3.x vectors. As a consequence the machine learning algorithms applied to this mapping performed moderately. Therefore, in order to reduce the number of vectors that map spuriously and to improve the performance of the machine learning algorithms the CVSS 2.0 vector was augmented by 50 elements. The additional elements of the CVSS 2.0 vector were derived from the description fields of each vulnerability, which are available from the CVSS 2.0 vulnerability database. Performing the statistical analysis of the available descriptions, the 50 most frequently occurring keywords were selected, ordered and converted into numbers. The details of this process can be found at [16]. The number corresponding to a keyword was obtained by counting the number of occurrences and dividing by 100. Thus obtained vectors were concatenated with the existing 7 element CVSS 2.0 vector forming a 57 element extended CVSS 2.0 vector. The necessary text processing was carried out using library Natural Language Toolkit (NLTK) [2].

After extending the CVSS 2.0 vector a new mapping was established between the B_S scores that are known for both CVSS 2.0 and 3.x standards, which is the subject of the subsequent machine learning analysis. Table 5 gives an insight into the nature of this mapping. As can be seen there is a large imbalance in the learning data. The greatest imbalance occurs for the AV parameter, where 841 CVSS 2.0 vectors were obtained for the class with index 0, while class 3 contains 53732 vectors. The difficulty of the problem is further increased by a large number of the considered classes.

Table 3. Enumerated members for CVSS 2.0 field Access Vector (AV) and their asso-ciated numerical values as described by the CVSS 2.0 specification.

Metric value	Numerical value
Local (AV:L)	0.395
Adjacent network (AV:A)	0.646
Network (AV:N)	1.0

Table 4. Enumerated members for CVSS 3.x field Attack Vector (AV) and their proposed associated numerical values.

Metric value	Numerical value
Network (N)	3
Adjacent (A)	2
Local (L)	1
Physical (P)	0

Table 5. The number of CVSS 2.0 vectors that map onto a specific class of the CVSS 3.x vector.

CVSS 3.x	Class index	No of CVSS 2.0 vectors	CVSS 3.x	Class index	No of CVSS 2.0 vectors
AV	1	841	UI	1	26449
	2	1573		2	46730
	3	17033	C	1	14074
	4	53732		2	15932
AC	1	6021		3	43173
	2	67158	I	1	12818
PR	1	4452		2	22831
	2	19380		3	37530
	3	49347	A	1	1883
S	1	12156		2	28440
	2	61023		3	42856

For the defined mapping the training set formation procedure was performed in two stages. In stage one undersampling [9] was performed. This stage involves the following 2 steps:

- for each CVSS 3.x vector class the median of CVSS 2.0 extended vectors is obtained by selecting the most frequently occurring CVSS 2.0 extended vector.
- Once the median is known the 80 most correlated vectors with the median were selected,

In stage two all the vectors selected for in stage one were included in a common training set. Thus $22 \cdot 80 = 1760$ vectors were obtained, where 23 turned out to be common and hence were removed. After completing stage two, even though the obtained data is imbalanced it is related to a constant selection rule.

Vectors that are not used for training constitute a testing set. The testing set must contain vectors from each class of all the CVSS 3.x vector components. For the initial optimization the testing sets containing 100 vectors were used.

The used testing sets were selected randomly and the selection procedure was repeated 5 times. The statistics obtained from the results were used to select the most appropriate classification models for conversion from B_S CVSS 2.0 to B_S CVSS 3.x scores.

In the final stage the estimation of the classification effectiveness for the selected algorithms was conducted carrying out 30 trials with testing sets, containing randomly selected 1000 vectors.

Preprocessing was carried out using the Principal Component Analysis (PCA). Then, the obtained data was processed using six machine learning algorithms:

- Naive Bayes classifier (NB),
- k-nearest neighbors algorithm (kNN) with euclidean metric (E) and cosine metric (C), 3 closest neighbors were attained
- Kernel Support Vector Machine (KSVM) in three configurations - with a linear kernel function and the two others, using Gaussian Radial Basis Functions (GRBF). KSVM (GRBF) has been split into a non-trained kernel (UGRBF) version and a trained kernel (TGRBF) version, which adopts interrelated 2-fold Cross-validation (2-fold CV), Misclassifcation Ratio (MCR) and NM algorithm with 30 starting points set.

The detailed description of these algorithms can be found in [1,3,4]. More detailed information regarding TGRBF is provided in [19]. The results obtained using the described algorithms are presented in the next section.

4 Results

In this section the accuracy of CVSS 3.x score prediction using algorithms which were described in Sect. 3 when applied to the CVSS classification problem (Sect. 3) is studied. The original data has been preprocessed by PCA. The algorithm's usefulness for the classification of the CVSS 3.x vector was analyzed using algorithms described in Sect. 3. All classification models were implemented in MATLAB 2015b and were run on a computing server with two CPUs [Intel Xeon®X5650, 2.66 GHz].

Figure 1 shows the average accuracy for the selected machine learning algorithms calculated by performing five test trials on 100 randomly selected vectors. The accuracy was calculated by counting the number of correct predictions and dividing by the number of all testing vectors, i.e. 100 in this instance.

Results from Fig. 1 show that for the AV parameter the highest average classification efficiency was obtained for KSVM (TGRBF) and is equal to 93.2%. Thus KSVM (TGRBF) can be qualified for the next stage of the research.

Considering the AC parameter high average efficiency - 95.8% was recorded for 3 algorithms: kNN (C), kNN (E) and KSVM (lin). The linear class separability algorithms results indicate significant advantage over more complex algorithms - KSVM (URGBF) and (TGRBF), whose efficiency is equal to 94.8%. NB obtains the lowest efficiency level. However, it still exceeds 94%. Consequently,

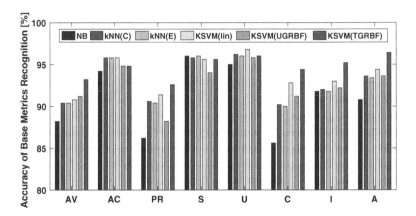

Fig. 1. The comparison of the average accuracy of the CVSS 3.x vector parameters recognition with extended CVSS 2.0 vector, using six algorithms: NS, kNN (C), kNN (E), KSVM (lin), KSVM (UGRBF) and KSVM (TGRBF)

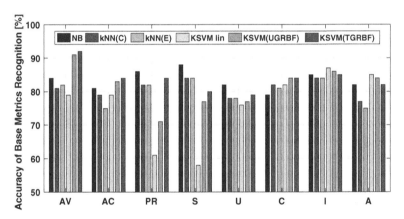

Fig. 2. The comparison of the average accuracy of the CVSS 3.x vector parameters recognition with original CVSS 2.0 vector, using six algorithms: NS, kNN (C), kNN (E), KSVM (lin), KSVM (URGBF) and KSVM (TGRBF)

also kNN (C), kNN (E) and KSVM (lin) can be selected for the next stage of the research.

Considering the PR parameter a significant decrease in the average efficiency with respect to the classification of other parameters of the CVSS 3.x vector, is visible. Thus, the classification constitutes a concern for NB and KSVM (UGRFB). Due to kernel's parameters tuning, KSVM (TRGBF) obtains a 4.4% higher efficiency than the version with no adaptation - 92.6%. Consequently, KSVM (UGRFB) algorithm proved to be best suited for this task.

With respect to S parameter all five algorithms obtained an average efficiency of over 95.5%. The KSVM (URGBF) attained the worst performance. Due to the kernel's adaptation, the SVM (TGRBF) obtained the same efficiency as the KSVM with the linear function - 95.6%. The highest efficiency corresponds to simple algorithms - NB and kNN (E) 96% each. Thus, they can be selected for further analysis.

All the algorithms classify the U parameter with high accuracy, obtaining an efficiency result in the range of 95%. For kNN (E), the highest average accuracy, of all the algorithms marked, was 96.8%. The above-mentioned average constitutes the highest average efficiency classification attained among all parameters. Therefore, kNN (E) proves to be the only suitable candidate for further analysis.

Similarly to PR, C seems to be difficult to classify, therefore two algorithms were selected for further tests - KSVM (TGRBF) with an average efficiency equal to 94.4% and KSVM (lin) - 92.8%. The remaining algorithms, especially the kNN and NB families, are marked by a noticeably lower efficiency – in the range of (85.6–90.2)%.

For I parameter the classification with KSVM (TGRBF) provides an average efficiency in the range of 95.2% and is 2.2% higher than the second KSVM (lin) in order. The remaining algorithms constitute a similar efficiency level – in the range of (91.8–92.2)%. Thus the KSVM (TGRBF) becomes the only candidate.

In the case of the A parameter the obtained results of the average classification efficiency indicate a 2% advantage of KSVM (TGRBF) over KSVM (lin) with an exceptional result of 94.4%. Therefore, both algorithms can be accepted for further analysis. The remaining algorithms also provide a satisfying average classification efficiency, with the worst result indicating 90.8% for NB.

For the sake of comparison the calculations presented in Fig. 1 were repeated for the short (7 element) CVSS 2.0 vector. The results of these calculations are summarised in Fig. 2. Comparing the results from Fig. 2 and Fig. 1 clearly shows the advantages of using the extended CVSS 2.0 vector for the considered machine learning problem.

The selection of the best value for the number of PCs is performed in order to maximize the classification accuracy and to obtain repeatability of the results. If several algorithms were initially selected whilst considering the results from Fig. 1, then after PCs analysis only the best performing algorithms were selected. These algorithms are listed in Table 6 together with the corresponding number of PCs and the average classification accuracy with standard deviation.

Quality of classification was estimated by calculating MCR, Precision, Recall and F1-score. The classification results obtained with algorithms tested are compared using 10 repetitions of 5-fold CV. The results obtained are presented in Table 7.

Next, the classification statistics for the selected algorithms is calculated. Figure 3 shows the calculated statistics of classifying the CVSS 3.x vector parameters for the selected in Table 6 algorithms. For this calculations the testing sets consisted of 1000 elements while the tests were repeated 30 times. The greatest dispersion of the classification accuracy occurs for the parameters PR and C. For the KSVM(lin) algorithm, used for PR classification, the efficiency range equals (79.9–93.5%, considering the outlier. The median is equal to 88.35%. The median

Table 6. The comparison of the best candidates (algorithms) for classification, including the number of PCs and the indicated average accuracy obtained from the 5-time testing trial of the algorithms with a randomly selected 100-element testing set.

Base metrics	Algorithm	Number of PCs/ average accuracy	Alternative algorithm	Number of PCs average accuracy
AV	KSVM(TGRBF)	31/93.2 ± 2.64%	–	–
AC	kNN(E)	26/95.8 ± 1.87%	–	–
PR	KSVM(TGRBF)	11/92.6 ± 2.83%	KSVM(lin)	52/91.4 ± 4.44%
S	NB	33/96.0 ± 1.60%	–	–
U	KSVM(lin)	46/96.8 ± 1.76%	–	–
C	KSVM(TGRBF)	8/94.4 ± 1.43%	KSVM(lin)	51/92.8 ± 1.72%
I	KSVM(TGRBF)	36/95.2 ± 0.74%	–	–
A	KSVM(TGRBF)	22/96.4 ± 1.72%	–	–

Table 7. Precision, Recall, F1-score and MCR calculated for algorithms listed in Table 6

CVSS 3.x	Class index	Precision	Recall	F1 score	MCR [%]
AV	1	0.1707	0.4487	0.2443	14.16 ± 0.78
	2	0.7926	0.6921	0.7387	
	3	0.7144	0.9497	0.8145	
	4	0.9550	0.9340	0.9444	
AC	1	0.9529	0.9433	0.9480	9.34 ± 0.20
	2	0.5809	0.6279	0.6034	
PR	1	0.9361	0.9588	0.9472	18.15 ± 1.12
	2	0.7143	0.6791	0.6931	
	3	0.3930	0.3979	0.3847	
S	1	0.8698	0.9271	0.8976	15.35 ± 0.83
	2	0.7670	0.6340	0.6941	
U	1	0.9281	0.9592	0.9434	7.09 ± 0.33
	2	0.9307	0.8809	0.9051	
C	1	0.9863	0.9885	0.9874	10.13 ± 0.50
	2	0.7434	0.8475	0.7904	
	3	0.9316	0.8784	0.9039	
I	1	0.9931	0.9641	0.9784	7.40 ± 0.29
	2	0.8056	0.9183	0.8579	
	3	0.9345	0.8953	0.9144	
A	1	0.9867	0.9585	0.9724	7.76 ± 0.21
	2	0.2724	0.8021	0.4049	
	3	0.9629	0.9031	0.9320	

value is higher - 88.6%, when using KSVM (TGRBF). However, the values range
dispersion is larger and corresponds to (72.9–92.1)%. Despite the higher median
value when compared with KSVM(lin), KSVM (TGRBF) performed worse. In
the case of parameter C, the situation is similar. The alternative KSVM (lin)
algorithm, except for the lower adjacents, indicates a higher rate for minimum,
maximum and median values - 87.1%, 94.1% and 90.95%, respectively, in comparison to 82.3%, 91.8% and 88.8% obtained for KSVM(TGRBF). It is noted
that 5 outliers were obtained for KSVM (TGRBF). The classification accuracy
calculated using 1000 vectors for the remaining CVSS 3.x parameters is also
lower than the one shown in Table 6. The median values for these parameters
are as follows: AV - 89.75%, AC - 93.3%, S - 95.0%, U - 94.25%, I - 92.6%, A
- 94.80%. The dispersion of values for these parameters is significantly smaller
than for PR and C parameters.

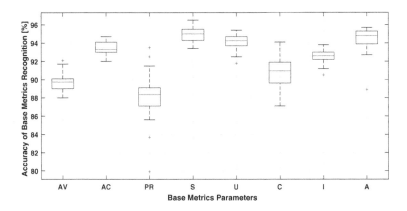

Fig. 3. The classification statistics of CVSS 3.x vector parameters using the program
developed considering Table 6

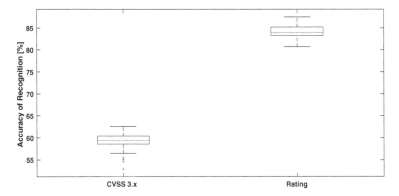

Fig. 4. The efficiency statistics of the B_S CVSS 2.0 to B_S CVSS 3.x conversion and
Qualitative Severity Rating Scale levels.

Full statistical analysis for algorithms used is shown in Table 8.

Table 8. Precision, Recall, F1-score and MCR calculated for algorithms listed in Table 6 obtained from the 30-time testing trial of the algorithms with a randomly selected 1000-element testing set.

CVSS 3.x	Class index	Precision	Recall	F1 score	Accuracy
AV	1	0.2167	0.2273	0.2198	89.75 ± 1.0
	2	0.7296	0.8274	0.7674	
	3	0.8201	0.9429	0.8672	
	4	0.9626	0.9334	0.9474	
AC	1	0.9592	0.9693	0.9642	93.4 ± 0.73
	2	0.6541	0.5854	0.6163	
PR	1	0.9790	0.9582	0.9683	88.12 ± 2.49
	2	0.7565	0.8086	0.7744	
	3	0.3219	0.3100	0.3137	
S	1	0.9851	0.9558	0.9702	94.94 ± 0.71
	2	0.7689	0.9097	0.8330	
U	1	0.9454	0.9629	0.9539	94.15 ± 0.81
	2	0.9345	0.9064	0.9197	
C	1	0.9813	0.9834	0.9823	90.72 ± 1.75
	2	0.7198	0.8059	0.7537	
	3	0.9407	0.9149	0.9269	
I	1	0.9828	0.9508	0.9665	92.44 ± 0.77
	2	0.8489	0.8377	0.8420	
	3	0.9141	0.9387	0.9260	
A	1	0.9601	0.9589	0.9595	94.47 ± 1.30
	2	0.1961	0.4974	0.2686	
	3	0.9646	0.9481	0.9562	

Comparing results from Tables 7 and 8 one can observe that AV, PR and S from validated using real tests perform better than what cross-validation would suggest (100% - MCR[%] vs. accuracy and F1-score). Additionally, with 10-fold CV when compared with 5-fold CV MCR is reduced by 0.6%, 0.35% and 2.1%, respectively for AV, PR and S, whilst other parameters stay unchanged. This implies that AV, PR and S are very sensitive to the number of samples of the learning set. These observations particularly apply to the first class in the case of AV and third class for PR.

Finally, the statistics of B_S CVSS 2.0 to CVSS 3.x conversion accuracy for the and the Qualitative Severity Rating Scale levels and the final CVSS 3.x score as described in the CVSS 3.x documentation [10] was conducted taking into consideration 30 repetitions of a randomly selected set of 1000 testing vectors, are shown in Fig. 4. Due to the conversion of B_S CVSS 2.0 to 3.x considering 8 parameters related to the formula described in the CVSS 3.1 documentation, the expected accuracy value can be approximated by the product of all accuracy values obtained for each component of CVSS 3.x vector. Thus obtained value using the numbers shown in Fig. 3 agrees with the results shown in Fig. 4 to within + 5%. Further, it is noted that the median accuracy of converting B_S CVSS 2.0 to 3.x is equal to 59.35%.

For conversion accuracy to Qualitative Severity Rating Scale levels the median accuracy equals 83.95% and all values are included in the range (80.7–87.5)%.

The results described above were confirmed by performing an extreme test - testing with a set consisting of 50,000 vectors. This trial yielded a B_S 2.0 to 3.x conversion efficiency, equal to 59.82% and conversion to QSRS levels indicating 84.12%, which do not differ significantly from the ones shown in Fig. 4.

5   Conclusions

The paper presents machine learning algorithms for converting B_S CVSS 2.0 scores to B_S CVSS 3.x ones. In order to improve the learning accuracy the CVSS 2.0 vector was augmented by 50 keywords selected from the description field of each vulnerability. The results obtained confirmed the improved performance of the machine learning algorithms when applied to the extended CVSS 2.0 vectors. Further it was found that different machine learning algorithms are best suited for different CVSS 3.x vector components. For the CVSS 3.x components AV, C, I, A the best performing algorithm is KSVM(TGRBF) while for parameters PR, U and C best results are obtained with KSVM(lin). Finally, for AC parameter kNN algorithm performed best while NB algorithm did best for S parameter. The future research will focus on improvement of classification for parameters AV, PR and C.

Acknowledgments. The authors wish to thank Wroclaw University of Science and Technology (statutory activity) for financial support and Agata Szewczyk for proofreading and translation. This publication was created as a part of the Regional Security Operations Center (RegSOC) project (Regionalne Centrum Bezpieczeństwa Cybernetycznego), cofinanced by the National Centre for Research and Development as part of the CyberSecIdent-Cybersecurity and e-Identity program.

References

1. Barber, D.: Bayesian Reasoning and Machine Learning. Cambridge University Press, Cambridge (2012)
2. Bird, S., Klein, E., Loper, E.: Natural Language Processing with Python: Analyzing Text with the Natural Language Toolkit. O'Reilly Media Inc., Newton (2009)
3. Bishop, C.M.: Pattern Recognition and Machine Learning. Springer, Berlin (2006)

4. Bonaccorso, G.: Machine Learning Algorithms. Packt Publishing Ltd., Birmingham (2017)
5. DSecure.me: VMC: Vulnerability Management Center (2021). Accessed 2 Jan 2021. https://github.com/DSecureMe/vmc
6. Elbaz, C., Rilling, L., Morin, C.: Fighting n-day vulnerabilities with automated CVSS vector prediction at disclosure. In: Proceedings of the 15th International Conference on Availability, Reliability and Security, pp. 1–10 (2020)
7. F-Secure: Vulnerability Management Tool (2021). Accessed 2 Jan 2021. https://www.f-secure.com/us-en/business/solutions/vulnerability-management/radar
8. Fall, D., Kadobayashi, Y.: The common vulnerability scoring system vs. rock star vulnerabilities: why the discrepancy? In: ICISSP, pp. 405–411 (2019)
9. Fernández, A., García, S., Galar, M., Prati, R.C., Krawczyk, B., Herrera, F.: Learning from Imbalanced Data Sets. Springer, Cham (2018). https://doi.org/10.1007/978-3-319-98074-4
10. FIRST: Common Vulnerability Scoring System v3.0: Specification Document (2017). Accessed 2 Jan 2021. https://www.first.org/cvss/v3.0/specification-document
11. FIRST: Common Vulnerability Scoring System v3.1: Specification Document (2019). Accessed 2 Jan 2021. https://www.first.org/cvss/v3.1/specification-document
12. Hovsepyan, A., Scandariato, R., Joosen, W., Walden, J.: Software vulnerability prediction using text analysis techniques. In: Proceedings of the 4th International Workshop on Security Measurements and Metrics, pp. 7–10 (2012)
13. Jacobs, J., Romanosky, S., Adjerid, I., Baker, W.: Improving vulnerability remediation through better exploit prediction. J. Cybersecur. 6(1), tyaa015 (2020)
14. Klinedinst, D.J.: CVSS and the Internet of Things (2015). Accessed 2 Jan 2021. https://insights.sei.cmu.edu/cert/2015/09/cvss-and-the-internet-of-things.html
15. Luers, A.L., Lobell, D.B., Sklar, L.S., Addams, C.L., Matson, P.A.: A method for quantifying vulnerability, applied to the agricultural system of the Yaqui Valley, Mexico. Glob. Environ. Change 13(4), 255–267 (2003)
16. Maciej, N., Walkowski, M., Sujecki, S.: CVSS 2.0 extended vector database (2021). Accessed 21 Jan 2021. https://github.com/mwalkowski/cvss-2-extended-vector-database
17. Mell, P., Hu, V., Lippmann, R., Haines, J., Zissman, M.: An overview of issues in testing intrusion detection systems (2003)
18. Mell, P., Scarfone, K., Romanosky, S.: Common vulnerability scoring system. IEEE Secur. Priv. 4(6), 85–89 (2006)
19. Nowak, M.R., et al.: Recognition of pharmacological bi-heterocyclic compounds by using terahertz time domain spectroscopy and chemometrics. Sensors 19(15), 3349 (2019)
20. NVD: National Vulnerability Database (2021). Accessed 2 Jan 2021. https://nvd.nist.gov/
21. Qualys: Vulnerability Management Tool (2021). Accessed 2 Jan 2021. https://www.qualys.com/apps/vulnerability-management/
22. Rapid7: Vulnerability Management Tool (2021). Accessed 2 Jan 2021. https://www.rapid7.com/products/nexpose/
23. Ruohonen, J.: A look at the time delays in CVSS vulnerability scoring. Appl. Comput. Inf. 15(2), 129–135 (2019)
24. Tavabi, N., Goyal, P., Almukaynizi, M., Shakarian, P., Lerman, K.: Darkembed: exploit prediction with neural language models. In: Proceedings of the AAAI Conference on Artificial Intelligence, vol. 32 (2018)

25. Walkowski, M., Krakowiak, M., Oko, J., Sujecki, S.: Distributed analysis tool for vulnerability prioritization in corporate networks. In: 2020 International Conference on Software, Telecommunications and Computer Networks (SoftCOM), pp. 1–6. IEEE (2020)

26. Walkowski, M., Krakowiak, M., Oko, J., Sujecki, S.: Efficient algorithm for providing live vulnerability assessment in corporate network environment. Appl. Sci. **10**(21), 7926 (2020)

27. Younis, A.A., Malaiya, Y.K.: Using software structure to predict vulnerability exploitation potential. In: 2014 IEEE Eighth International Conference on Software Security and Reliability-Companion, pp. 13–18. IEEE (2014)

Applicability of Machine Learning to Short-Term Prediction of Changes in the Low Voltage Electricity Distribution Network

Piotr Cofta[1]⬮, Tomasz Marciniak[1(✉)]⬮, and Krzysztof Pałczyński[2]⬮

[1] UTP University of Science and Technology, Al. Prof. S. Kaliskiego 7,
85-796 Bydgoszcz, Poland
{piotr.cofta,tomasz.marciniak}@utp.edu.pl
[2] UTP University of Science and Technology, Doctorate School,
Al. Prof. S. Kaliskiego 7, 85-796 Bydgoszcz, Poland

Abstract. Low voltage electricity distribution network actively maintains the stability of its key parameters, primarily against the predictable regularity of seasonal changes. This makes long-term coarse prediction practical, but it hampers the accuracy of a short-term fine-grained one. Such predictability can further improve the stability of the network. This paper presents the outcome of research to determine whether Machine Learning (ML) algorithms can improve the accuracy of the prediction of next-second values of three network parameters: voltage, frequency and harmonic distortions. Four ML models were tested: XGBoost Regressor, Dense neural networks (both one and two layer) and LSTM networks, against static predictors. Real data collected from the actual network were used for both training and testing. The challenging nature of this data is due to the network executing corrective measures, thus making parameter values return to their means. This results in non-normal distribution with strong long-term memory impact, but with no viable correlation to use for short-term prediction. Still, results indicate improvements of up to 20%, even for non-optimized ML algorithms, with some scope for further improvements.

Keywords: Machine learning · Power distribution network · Prediction

1 Introduction

The distribution network is a complex system that actively maintains the values of its key parameters (voltage, frequency, distortions etc.) within ranges required by regulators. In this process, it uses both proactive and reactive measures, the latter based on the predictions of changes. The more accurate the prediction, the more effective and efficient the network is in maintaining required values of its parameters.

© Springer Nature Switzerland AG 2021
M. Paszynski et al. (Eds.): ICCS 2021, LNCS 12744, pp. 270–277, 2021.
https://doi.org/10.1007/978-3-030-77967-2_22

Currently, the network uses coarse predictions that benefit from the cyclical and seasonal nature of changes in demand. However, the emergence of smart networks creates opportunities for additional, fine-grained near-instantaneous proactive measures, where the prediction of the next few seconds will be required.

The objective of this research is to conduct the preliminary screening of the class of ML models to determine whether their use stands a chance of increasing the predictability of changes for the time horizon of a second, or few seconds, as compared with simple static predictors. This paper addresses this question through a preliminary study in the applicability of ML algorithms for forecasting the next-second value of three time series composed of voltage, frequency and harmonic distortions. Four ML regressors were tested: XGBoost Regressor, Dense neural networks (both one and two layer) and LSTM networks, using actual time series.

The analysis demonstrated the challenging nature of those time series when it comes to forecasting. Those time series have no normal distribution, and there is a long-term memory dependence with tendencies to switch monotonicity in order to return to their means. Both correlation and autocorrelation analysis found out that there is little to none linear dependencies between the current and past value resulting in signal being very difficult to forecast.

The research demonstrated that despite those challenges, the use of ML algorithms improves the accuracy of prediction by up to 20%, as compared to static predictors, with XGBoost and dense neural networks slightly outperforming the LSTM neural networks. The novelty of this research lies in the following:

– Addressing short-term predictability, as contrasted with more popular long-term predictability;
– Establishing the reference performance with the use of static predictors;
– Analysis of the use of selected ML algorithms for predictability that demonstrated their relative advantage over the reference performance;
– Indicating the most promising ML algorithms.

2 Literature Review

Liu et al. in [5] focused on improving the time series prediction accuracy using wavelet filtering and neural network. Their method eliminates effect of measurement noise, thus improving prediction precision, specifically when wavelet filtering is used.

Górriz et al. in [3] suggested a new method of time series forecasting with the use of Independent Component Analysis (ICA) algorithms and Savitzky-Colay filtering as pre-processing to introduce data to an Artificial Neural Network (ANN) based on radial basis functions (RBF). The presence of pre-processing improved the prediction.

Tao et al. in [8] proposed a chaotic time series prediction method, based on the RBF network. This CIFCA-ROLSA method includes Iterative Fuzzy Clustering Algorithm (CIFCA) and Regularized Orthogonal Least Squares Algorithm (ROLSA). The proposed method was verified on the basis of the known Rollser chaotic system. The results shown in the paper were worse for a ROLSA-only network than for a CIFCA-ROLSA network.

Zhao et al. in [13] compares the results achieved by back propagation (BP) neural network and the RBF neural network in chaotic time series prediction. RBF and BP neural networks are compared based on the Logistic equation which is a known chaotic equation. The prediction performance of the RBF is better than the BP. The use of RBF to predict the Shanghai Composite Index shows that neural networks can be effectively used to short-term stock price forecasts.

Park et al. in [6] proposed a method based on BiLinear Recurrent Neural Network (BLRNN) for time series prediction, enhanced by applying the multi-resolution learning algorithm to BLRNN training to make it more reliable for predicting time series data. The normalized mean square error (NMSE) was used to assess the efficiency of long-term prediction. Both on the Mackey-Glass Series data and on the Sunspot Series data, the predictor proposed by the author achieves better results than the Multiple-Layer Perceptron Neural Network (MLPNN).

Yang et al. [11] suggested genetic algorithms (GAs) for time series prediction. The results showed that there is a correlation between the observed and predicted deformations, and the resulting models have interpretative forms that can be easily used for further analysis and inform decisions. The proposed method has been tested with promising results.

Yu et al. in [12] proposed using hybrid prediction algorithm using wavelet analysis of time series to obtain trend prediction for a spacecraft, where forecasting results can be obtained by adding the predicted value from each layer. In order to verify the results obtained in the model, they were compared with mean absolute error (MAE) and root mean square error (RMSE). The results indicate a high accuracy.

Wu et al. in [10] proposed a method for energy consumption prediction using BP neural networks in combination with exogenous series. For comparison of effectiveness, the tested model (called NARX) was compared with the normal TDNN model, showing better predictions by the NARX model.

Grant et al. in [4] demonstrated benefits of the Cellular Simultaneous Recurrent Neural Network (CSRN) to identify and predict the dynamics of a 12-bus voltage power system, comparing with a standard single SNR. Two types of disturbances were assessed, including perturbations in power system generators and the least stable loads. The method was also assessed in the event of a transmission line failure.

Qiu et al. in [7] proposed a model based on support vector machine (SVM) to predict the voltage breakdown of rod-plane air gaps, using the binary classifier. The predicted results correlate with the 29 experimental values. The influence of atmospheric parameters on the breakdown voltage obtained on the basis of the analysis of the predicted results are almost the same as those obtained experimentally.

Adebayo et al. [2] proposed a model capable of predicting a loss of a bus voltage. The critical bus voltage stability index (CBVSI), is based on the system load bus voltage deviation, the maximum load capacity, and the total number of steps needed to achieve the minimum allowable load on each receive bus. Simulation results against the conventional fast voltage stability index (FVSI) show that the CBVSI-based method can serve as an alternative tool for power system engineers in the operation and planning of the system.

3 Analysis

The time series used for the analysis were selected from the database created from the measurement of parameters of the electricity at a fixed point at the University campus in Bydgoszcz, Poland. The ND20 meter [1] was used to measure three parameters: one phase voltage (V1, V), frequency (F, Hz) and total harmonic distortion of the voltage (THDV1, %). The rated basic error for ND20 is V1: $+-0.2\%$, F: $+-0.2\%$, THDV1: $+-5\%$. Data from two months were used: July 2019 and December 2019, being representatives of two typical seasonal patterns. For each time series, the 'delta' series was created that consisted of second-to-second changes in the specific parameter. Minimal data cleansing was performed prior to the analysis.

3.1 Statistical Properties of Time Series

The analysis of statistical properties shows that time series satisfy regulatory requirements. Shapiro-Wilk test indicate that they are not sampled from normal distribution. Augmented Dickey-Fuller (ADF) test indicated that they are stationary.

Hurst exponent proved to have significant impact on the understanding of the behavior of time series [9]. Its value, which was below 0.5 for all time series, suggests significant long-term memory dependence and tendencies to switch monotonicity in order to preserve oscillations centered in its mean.

Strong long-term memory dependence, suggested by the value of Hurst exponent, turned out to be difficult to extract. The values of extracted seasonality are around four orders of magnitude smaller than the residual (error).

Autocorrelation and partial autocorrelation of each time series indicated some temporal, short-term memory dependence. Values of correlation with lag of 1 h and 24 h are significantly above the error threshold, although it did not later translate into any improvements in the forecast.

Pearson's and Spearman's coefficients indicated that acquired data contain certain correlations between themselves, albeit they is not significant. Nonetheless, machine learning algorithms deployed in this research managed to use these connections.

3.2 Normalization and Reference Metric

Time series were normalized using the Z-score normalization. Normalized data has been divided into non-overlapping feature vectors containing 10 samples of each signal or the differential signal based on the type of experiment conducted. In total, 4,853,209 vectors were created, of which 3,397,245 formed a training set.

The reference metric consisted of the outcome of applying two static predictors (persistence and moving average), each in two variants, to the time series. The persistent predictor always returns the last observed value. The moving average always returns the average over the set number of samples. Errors were calculated using mean absolute error metric (MAE).

Machine learning algorithms' performance was calculated as a difference between the MAE of the most accurate static predictor in regards to the forecasted series (V1, F, THDV1 or their derivatives) and the MAE of a given algorithm, relative to the MAE of the static predictor. Higher values of performance are better.

3.3 Selection and Parametrization of Methods

This research focused on exploring simple versions of the selected few algorithms, rather than elaborating on a single one. Three algorithms were selected, each one theoretically satisfying the characteristics of time series:

- XGBoost Regressor,
- Dense Neural Networks with one and two linear layers (DNN and DNN2, respectively),
- LSTM (Long Short-Term Memory) Neural Networks.

Each model was used to predict the values of only one time series (V1, F, THDV1 or their derivatives) despite using input made out of every signals' samples. It is because prediction of three time series' elements at the same time requires using a combination of loss functions for every metric in order to train the models, leading to the loss in precision.

Both XGBoost and Artificial Neural Networks are well-regarded for their robustness and ability to learn and recognize non-linear patterns. XGBoost Regressor was evaluated with 100 estimators, each having maximum depth equal to 6. Regularization L1 was not performed. Weight of L2 regularization was equal to 1. Minimum sum of instance weight needed in child was set to 1.

Artificial Neural Networks were trained using learning rate equal to 1e−5. The loss function used during the training calculated a mean squared error. First network (DNN) was made of one linear layer with 30 inputs (one for every value of three signals) and one output. Second network (DNN2) contained two linear layers. The first layer had 30 inputs and 60 outputs and the second one had 60 inputs and one output. Layers were separated with a ReLU activation function.

LSTM Neural Network required splitting input data in three channels for every signal (V1, F, THDV1). Every channel was trained with separate LSTM layer with 10 gates and 10 outputs with ReLU activation function. Results of LSTM layers processing were concatenated into one-dimensional, 30-element array which is an input to the linear layer with one output.

Additional analysis was conducted to see whether selected ML algorithms would allow for the prediction of subsequent values, i.e., 11th, 12th etc.

3.4   Results

The results are presented in Table 1, showing relative gain for each combination of a time series and ML algorithms. Higher values indicate greater improvement over the best static predictor for the same timeseries.

The results indicated that the use of ML for predictions outperformed static predictors for all time series. The maximum improvement was 20%, achieved by both XGBoost Regressor and DNN for differential frequency. It indicates that there is indeed an added benefit of using ML for prediction.

Table 1. Results of forecasting using machine learning algorithms.

	XGBoost	DNN	DNN2	LSTM
$V1$	7.7%	7.9%	1.6%	6.1%
$\Delta V1$	13.2%	14.0%	14.0%	12.4%
F	17.8%	17.8%	17.8%	17.8%
ΔF	20.0%	20.0%	17.8%	17.8%
$THDV1$	7.9%	7.4%	7.1%	7.4%
$\Delta THDV1$	8.0%	7.5%	8.8%	8.7%

Predictions were more accurate for differential time series, which confirms early indications regarding the nature of the time series. Considering differential time series, the dense neural network with one layer (DNN) was the most accurate predictor for the differential voltage and frequency time series while the DNN2 with two layers was the most accurate for differential harmonic distortions.

For longer-term prediction, there is a decrease in performance for all models, for both absolute and differential time series, most visible for the frequency time series. It indicates that while there may be a value in using ML for short-term prediction, this value diminishes for the longer-term ones.

4   Conclusions

The second-by second characteristics of the low-voltage distribution network makes short-term fine-grained prediction difficult. The network is not only governed by the law of physics, but it also actively responds to changes in its key parameters, to maintain its stability. Statistical analysis of data sampled throughout summer and winter of 2019 indicated that voltage (V1), frequency (F) and distortions (THDV1) show signs of such self-correction, with tendencies to switch monotonicity in order to preserve oscillations centered in their respective means.

The research demonstrated that the use of ML algorithms for short-term prediction can improve the accuracy of the next second prediction by up to 20%, comparing to static predictors. However, this advantage diminishes for predictions with a longer time horizon. Non-optimized DNN and DNN2 performed slightly better comparing to non-optimized LSTM and XBoost Regressor, specifically for differential time series. Considering that dense neural networks are the simplest models in comparison with XGBoost and LSTM neural networks, and that no optimization has been performed, results are encouraging. It can be expected that certain optimization, as well as certain tuning of the model, may improve the accuracy of prediction even further.

References

1. Lumel homepage. https://www.lumelcom.pl (December 2020). Accessed 7 Dec 2020
2. Adebayo, I.G., Sun, Y.: Performance evaluation of voltage stability indices for a static voltage collapse prediction. In: 2020 IEEE PES/IAS PowerAfrica, pp. 1–5. IEEE (2020)
3. Górriz, J.M., Puntonet, C.G., Salmerón, M., Lang, E.: Time series prediction using ica algorithms. In: 2003 Proceedings of the Second IEEE International Workshop on Intelligent Data Acquisition and Advanced Computing Systems: Technology and Applications, pp. 226–230. IEEE (2003)
4. Grant, L.L., Venayagamoorthy, G.K.: Voltage prediction using a cellular network. In: IEEE PES General Meeting, pp. 1–7. IEEE (2010)
5. Liu, Z., Pei, X., Huang, G.: Real time prediction method of sensor output time series. In: 2009 9th International Conference on Electronic Measurement & Instruments, pp. 2–969. IEEE (2009)
6. Park, D.C.: A time series data prediction scheme using bilinear recurrent neural network. In: 2010 International Conference on Information Science and Applications, pp. 1–7. IEEE (2010)
7. Qiu, Z., Ruan, J., Xu, W., Huang, C.: Breakdown voltage prediction of rod-plane gap in rain condition based on support vector machine. In: 2016 IEEE International Conference on High Voltage Engineering and Application (ICHVE), pp. 1–4. IEEE (2016)
8. Tao, D., Hongfei, X.: Chaotic time series prediction based on radial basis function network. In: Eighth ACIS International Conference on Software Engineering, Artificial Intelligence, Networking, and Parallel/Distributed Computing (SNPD 2007), vol. 1, pp. 595–599. IEEE (2007)
9. Wang, N., Li, Y., Zhang, H.: Hurst exponent estimation based on moving average method. In: Advances in Wireless Networks and Information Systems, pp. 137–142. Springer (2010). https://doi.org/10.1007/978-3-642-14350-2_17
10. Wu, B., Cui, Y., Xiao, D., Zhang, C.: Prediction of energy consumption time series using neural networks combined with exogenous series. In: 2015 11th International Conference on Natural Computation (ICNC), pp. 37–41. IEEE (2015)
11. Yang, C.X., Zhu, Y.F.: Using genetic algorithms for time series prediction. In: 2010 Sixth International Conference on Natural Computation, vol. 8, pp. 4405–4409. IEEE (2010)

12. Yu, H., Liu, J., Wang, M., Hu, S.L., Guo, R.: The trend prediction for spacecraft state based on wavelet analysis and time series method. In: 2014 11th International Computer Conference on Wavelet Actiev Media Technology and Information Processing (ICCWAMTIP), pp. 88–91. IEEE (2014)
13. Zhao, H.: A chaotic time series prediction based on neural network: evidence from the shanghai composite index in China. In: 2009 International Conference on Test and Measurement, vol. 2, pp. 382–385. IEEE (2009)

Computational Analysis of Complex
Social Systems

A Model for Urban Social Networks

Stefano Guarino$^{(\boxtimes)}$, Enrico Mastrostefano, Alessandro Celestini,
Massimo Bernaschi, Marco Cianfriglia, Davide Torre,
and Lena Rebecca Zastrow

Istituto per le Applicazioni Del Calcolo "Mauro Picone",
Consiglio Nazionale Delle Ricerche, Rome, Italy
{s.guarino,e.mastrostefano,a.celestini,m.bernaschi,m.cianfriglia,
d.torre,l.zastrow}@iac.cnr.it

Abstract. Defining accurate and flexible models for real-world networks
of human beings is instrumental to understand the observed properties
of phenomena taking place across those networks and to support computer simulations of dynamic processes of interest for several areas of
research – including computational epidemiology, which is recently high
on the agenda. In this paper we present a flexible model to generate
age-stratified and geo-referenced synthetic social networks on the basis
of widely available aggregated demographic data and, possibly, of estimated age-based social mixing patterns. Using the Italian city of Florence as a case study, we characterize our network model under selected
configurations and we show its potential as a building block for the simulation of infections' propagation. A fully operational and parametric
implementation of our model is released as open-source.

Keywords: Urban social network · Graph model · Simulator ·
Epidemic

1 Introduction and Background

The definition of networks that encode in a suitable way patterns of connection
and interaction among individuals of a population is a widely studied problem.
Among the many reasons, finding accurate models for real-world social networks
is instrumental to study the dynamics of disease spreading [9] or propaganda [14].
Many simulation-based social studies complain about the lack of reliable data
and therefore model social networks by using well-known random graph models [1]. At the other hand of the spectrum, with a focus on physical interactions,
a growing body of research makes use of extensive and often purposely collected
data – e.g., surveys and questionnaires, activity location, traffic and mobility
data – either to extract setting-specific contact matrices [3,24] or for tuning
agent-based simulators [3,10]. While simple random models cannot capture all
the subtleties of real networks [3], a recent call to action raised the attention
towards the need for accurate yet flexible and replicable approaches [32].

In this paper, we present a novel framework for the definition of a data-driven
urban social network, where each edge of the graph represents a "strong tie" [18]

© Springer Nature Switzerland AG 2021
M. Paszynski et al. (Eds.): ICCS 2021, LNCS 12744, pp. 281–294, 2021.
https://doi.org/10.1007/978-3-030-77967-2_23

between two geo-referenced and age-stratified individuals. We tell apart intra-household (*e.g.*, kinship) edges from friendship edges. The former are defined quite naturally by drawing a *clique* (*i.e.*, a complete subgraph) for each household, where the breakdown of the population into households is entirely inferred from the available data. Friendship edges are instead drawn based on three guiding elements: (i) the available contact data (*e.g.*, extracted thanks to [35]), used to trigger an age-based social mixing structure in the network; (ii) the existence of an inverse power-law dependence of friendship upon physical distance [7,15]; (iii) a vertex-intrinsic social fitness [8] that models the individual propensity to have friends. Our network model may be of help in any application setting that requires to gather and elaborate information on the urban social fabric. It may be used as a standalone tool, to characterize urban social relation patterns in connection with the geography and the demography of a given territory – like we show in Sect. 3. Moreover, it is instrumental in increasing the plausibility of simulations of dynamic processes that may be influenced by agents' preferences and personal relations – like we show in Sect. 4 for an epidemic use case.

To guarantee usability and reproducibility, the source code of the software used to simulate instances of our urban social network is publicly available[1]. The model depends on a combination of data-driven and configuration parameters that make it adaptable to different use cases. In [13] we provide a detailed analysis that may guide potential users and that, overall, speaks in favor of certain configurations, on which we will focus in this paper. Of special interest are the combinations of parameters that allow reproducing a few empirical and sociological findings of urban social networks. First of all, the distance-based penalization shall have exponent in the range [0.5, 2] [19, 26]. Further, the graph shall have a heavy- but not fat-tailed degree distribution [15, 16, 19, 20] and be (mostly) connected, as typically observed in urban areas [15, 20, 27, 31]. Since social ties comparable to kinship are rare [18], these properties will be enforced while keeping "small" the average number of friends. As a consequence, in our graph acquaintances correspond to short, but >1, paths, and the network is quite sparse, a necessary feature in most practical applications.

A review of related work follows. We then describe in details our network model (Sect. 2), characterize the network obtained for the city of Florence, Italy, under two selected configurations (Sect. 3), and present an epidemic use case (Sect. 4). Finally, we discuss strengths and current limitations of our model, and we identify suitable directions for future work (Sect. 5).

1.1   Related Work

We construct our synthetic population following an intermediate approach between *Synthetic Reconstruction* (SR) [5] and *Combinatorial Optimization* (CO) [33]. In SR the attributes of each agent are drawn from joint-distributions deduced from aggregate and survey data. In CO a sample of real individuals is available for different sub-areas of the territory of interest, and the whole

---

[1] The source code is released under the GPL v3 at gitlab.com/cranic-group/usn.

population is obtained through replication/resampling methods. Extensions and modifications to these methods are well surveyed in [30].

One element of novelty of our network model is the usage of age-based social mixing data to infer *friendship* links. Computational social scientists often rely on rather simple graph models [1] or, possibly, on exponential random graphs [29]. Sample data (*e.g.*, surveys, questionnaires, diaries), possibly integrated with mobile/traffic/wearable sensor data [11,17,24], are extensively used to model physical contacts. To this end, some authors extract *contact matrices* for specific settings, such as households, schools and workplaces [3,24], others use agent-based simulators to reproduce synthetic interactions [3,10]. We relied on the recently released SOCRATES [35] Data Tool[2] to extract data for Italy from Polymod [24]. The tool allows easily specifying parameters such as age breaks, gender, day of the week, duration or location of the contacts, and it produces a social contact matrix drawing from the best public survey datasets for the selected country.

The introduction of a penalization for "long" edges is not peculiar to our model. While there is wide evidence that geographical factors alone cannot explain the structure of real-world spatial social networks [15,20,31], the dependence of friendship on distance is widely assumed to follow an inverse power-law with exponent $\beta \in [0.5, 2]$ [7,15,16,19,20,26,31,34] – and this surprisingly holds even for online relationships [12]. In particular, $\beta < 1$ seems to work better for short range contacts (<20 km) [16] and for urban networks [34]. The impact of this penalization upon communities, path lengths, degree distribution and other topological properties of the network has already been the object of study [4,36], but previous modeling efforts assumed some simple (*e.g.*, uniform) spatial distribution, instead of using data-driven vertex locations.

Previous empirical findings did play a role, more generally, in guiding our modeling choices. Real-world spatial social networks are usually "small-worlds" [15,31], with a single giant connected component [15,20,27], average degree in the range 5 to 20, and high clustering coefficient [15,16,20,20,31]. In line with sociological studies [18], but contrary to other real-world networks [6,25], such networks do not present very large hubs [15,19,20,27]. Their degree distribution is right skewed and relatively long-tailed [15,20], and it has been, at times, approximated by a power-law with a large (5 to 8) exponent [19,27] or by a Log-normal distribution [16]. Within cities, population density impacts on the frequency of close-range contacts, but usually not on the overall size of each person's network [7]. While geographical proximity and community structure appear to be related [7,15,34], some authors argue that only small clusters (<30 members) are geographically bounded [26] whereas the large ones may span across very large areas of a city [15].

---

[2] https://lwillem.shinyapps.io/socrates_rshiny/.

2   Graph Model

Our urban social network is represented by an unweighted undirected graph $G = (V, E)$, where V is the vertex set of size $N = |V|$ and E is the edge set. In particular, we have $E = E_H \sqcup E_F$, where E_H is the set of *household edges*, E_F is the set of *friendship edges* and $\sqcup$ denotes the disjoint union. In the following, we explain how V, E_H and E_F are defined in our model. We will often use the expression *household graph* to denote the subgraph $G_H = (V, E_H)$ and the expression *friendship graph* to denote the subgraph $G_F = (V, E_F)$.

2.1   Vertex Set

Each vertex $u \in V$ is characterized by three attributes: a fitness score $f_u \geq 0$, an age label $g_u \in \{0, \ldots, n-1\}$, and a tile label $t_u \in \{0, \ldots, T-1\}$.

Fitness. Inspired by previous work, that modelled degree heterogeneity by means of a vertex-intrinsic fitness [8], we make use of a *sociability fitness* attribute f_u. Our model does not put restrictions upon the choice of f_u, but the probability of a friendship edge between u and v is set proportional to f_u and f_v (see Sect. 2.3). The distribution of f_u shall thus be chosen considering its impact on the degree distribution of the friendship graph. For the scope of this paper, we consider $f_u \sim 1 + \mathcal{LN}\left(\ln(2), \frac{1}{4}\right)$, where $\mathcal{LN}$ denotes a Lognormal distribution[3]. This distribution has been chosen empirically in an attempt to mimic two main aspects of real-world spatial social networks: only a few people have very few social links and the hubs are limited in both number and size. In general, Lognormally distributed data occur across different domains [23] and recent work suggests that the sociability of real-world social networks makes no exception [16,20]. Other choices may be preferred, some of which (*e.g.*, a Pareto, a uniform and a constant distribution) are already supported by our simulator.

Age. The age labels define a stratification of the population into age-groups, *i.e.*, a partition of the vertex set V into n disjoint subsets $V_0, \ldots, V_{n-1}$. For the scope of this paper, we consider four age-groups: *children* (0 to 17), *young* people (18 to 34), *adults* (35 to 64) and *elderly* people (65+). The proportion of each group is determined according to census data at the provincial level made available by the Italian Institute of Statistics (ISTAT)[4] and for each vertex u the age label g_u is independently drawn. Any desired age-stratification can be easily specified in the simulator's configuration file – statistics for many other countries are provided, for instance, by the United Nations Statistics Division (UNSD)[5].

Tile. We decompose the territory of interest into a regular lattice of T square tiles of side l and we set the tile label t_u equal to the unique index of the

---

[3] Throughout this paper, we use the parameterization $\mathcal{LN}\left(\lambda, \sigma^2\right)$ where λ and σ^2 are the mean and variance of the associated Normal distribution.

[4] ISTAT data used in this paper are available at https://www.demo.istat.it/pop2020.

[5] https://unstats.un.org/unsd/demographic-social/census/censusdates/.

tile where u resides. The side l is a configuration parameter, set as $l = 1$ Km for the scope of this paper. Approximating the position of each vertex with its tile is instrumental in simplifying the computation of pairwise distances and of the household structure, as better explained in the following. A module of the simulator is responsible for extracting the shape file of the city of interest. We resort to the `overpass` API of the well known OpenStreetMap database[6] to find the minimal grid that contains the city's boundary; we then select only the tiles of the grid whose center lies inside it. We get population density data for the whole city from the WorldPop Project[7], which provides data of the world population for 100 m $\times$ 100 m square cells, and we map those data to our tiles.

2.2 Household Edges

To group individuals into households we follow a heuristic approach, imposing that: (i) all members of a household live in the same tile; (ii) children are younger than their parents; (iii) partners have, on average, a similar age. The algorithm is based on the concept of *household role*, represented as a pair of the form (household-type, role) taking values in {(singles, single), (single-parent, parent), (single-parent, child), (couples, peer), (two-parents, parent), (two-parents, child), (various, various)}[8]. For instance, $r_u = $ (single-parent, parent) means that u is a parent in a household of type single-parent, where $r_u[0] = $ single-parent is the household-type and $r_u[1] = $ parent is the role. We make use of two conditional distributions: $\Pr[r \mid g]$ is the probability that an individual has role r given that she/he belongs to age-group g; $\Pr[k \mid h]$ is the probability that a household of type h has k members. These can be obtained for Italy based on ISTAT aggregate national data, and, *e.g.*, from the UNSD for other countries. At a high level, the heuristics works as follows:

- Extract a role r for each vertex u, based on $\Pr[r \mid g_u]$.
- For all u such that $r_u[0] \in $ {single-parent, two-parents}:
  - if $r_u[0] = $ two-parents, select a random partner v for u such that $t_v = t_u$, $g_v \in [g_u - 1, g_u + 1]$ and $r_v[0] = r_u[0]$;
  - extract the total number of members k_u for the household of u, based on $\Pr[k \mid r_u[0]]$, and compute their total number of children c_u.
- For $i = 1, \ldots, \max_u c_u$:
  - for all u such that $c_u \geq i$, select a random w such that $t_w = t_u$, $g_w < g_u$, $r_w[0] = r_u[0]$ and $r_w[1] = $ child, and assign w to the household of u.
- For all u such that $r_u[0] = $ couples, select a random partner v for u such that $t_v = t_u$, $g_v \in [g_u - 1, g_u + 1]$ and $r_v[0] = r_u[0]$.
- Randomly compose the households of type various, based on $\Pr[k \mid \text{various}]$.

In our simulations, the number of individuals left out of any household by the heuristics is negligible, and the empirical distributions of household types and

---

[6] https://www.openstreetmap.org/.

[7] https://www.worldpop.org/.

[8] The pair (various, various) covers all cases other than the previous ones.

members per type almost perfectly match the expected ones (see [13] for details). The household edges E_H are finally obtained as the union of all the *cliques* that connect all members of the same household.

2.3   Friendship Edges

All friendship edges are drawn independently at random. For each pair $(u, v) \in V \times V$, the probability $\Pr[u, v] = \Pr[(u, v) \in E_F]$ is defined as:

$$\Pr[u, v] = \frac{\mu \cdot N}{2} \cdot \frac{m_{g_u, g_v} \cdot s_{g_u, g_v}}{\sum_{i \leq j} (m_{i,j} \cdot s_{i,j})} \cdot \frac{d(u, v)^{-\beta} \cdot f_u \cdot f_v}{\sum_{u' \in V_{g_u}, v' \in V_{g_v}} (d(u', v')^{-\beta} \cdot f_{u'} \cdot f_{v'})} \quad (2.1)$$

where:

- μ is the average number of friends – a configuration parameter;
- $m_{i,j} = |V_i| \cdot |V_j|$ if $i \neq j$ and $m_{i,i} = \frac{|V_i| \cdot (|V_i| - 1)}{2}$ – deduced from the data-driven age-stratification;
- $s_{i,j}$ is the age-based social mixing for groups i and j – a data-driven coefficient, computed from aggregated social contact data as explained in Sect. 1.1;
- $d(u, v) = \max\left\{\frac{l}{2}, d^*(t_u, t_v)\right\}$ is the approximated distance between u and v, where $d^*(t_u, t_v)$ is the distance between the centers of the tiles t_u and t_v – a data-driven value, except for l which is a configuration parameter;
- β is the exponent that determines the level of penalty imposed to long edges – a configuration parameter.

A thorough description of (2.1) is presented in [13]. Here, we just highlight that $\Pr[u, v]$ is normalized in such a way to guarantee that the data-driven age-based social mixing induced by the coefficients $s_{i,j}$ is respected, up to a scaling factor. Indeed, the expected number of friendship edges between groups i and j is

$$\mathbf{E}[|E_F(i, j)|] = \frac{\mu \cdot N}{2} \cdot \frac{m_{i,j} \cdot s_{i,j}}{\sum_{i \leq j} (m_{i,j} \cdot s_{i,j})}$$

It follows quite easily that $\mathbf{E}[|E_F|] = \frac{\mu \cdot N}{2}$, hence the average degree of the friendship graph is exactly μ, regardless of all other parameters. The expected degree of a specific vertex u is proportional, besides to μ, to f_u and to the average of f_v for all other $v \in V$, weighted by $d(u, v)^{-\beta}$.

3   Network Analysis

Potential sources of information for real friendship patterns, *e.g.*, telephone data [11] or online social networks [21], are usually hard to acquire, private and/or not entirely representative/dependable. Instead of a direct validation of our model against real data, we therefore present a characterization of the urban social network obtained for the city of Florence under selected configurations. We refer the interested reader to [13] for an extensive experimental analysis.

In the following, we use age-stratification and household composition data from ISTAT, spatial population density from WorldPop, and age-based social mixing coefficients from [24], collected through the SOCRATES Data Tool. We additionally take $f_u \sim 1 + \mathcal{LN}(\ln(2), 0.25)$, $\mu = 10$ and we consider both $\beta = 0.5$ and $\beta = 2$. It may be useful to know that, based on our data, Florence counts 363060 residents – roughly, 15% children, 17% young people, 43% adults, 25% elderly people – and is contained in a 15 Km $\times$ 12 Km grid.

3.1 Topology of the Graph

In Table 1, we overview the global topological properties of the graph. We recognize the typical positive assortativity of social networks and a global clustering several orders of magnitude greater than in the equivalent Erdos-Renyi graph. However, a closer inspection highlights that the large number of small cliques introduced in the household graph plays a paramount role in the formation of triangles, whereas the friendship graph, despite the geographical and age-based homophily, shows limited transitivity. Regardless of β, the average shortest path length has a value of the order $\frac{\ln(N)}{\ln(\langle\mathrm{deg}\rangle)}$, typical of small world networks.

Table 1. Social graph for Florence with $\mu = 10$ and $f_u \sim 1 + \mathcal{LN}(\ln(2), 0.25)$: average metrics over 10 independent runs (the negligible variance is omitted).

	$\langle\mathrm{deg}\rangle$	$\langle\mathrm{dist}\rangle$	C	C_{loc}	ρ	# comp	Giant %
$\beta = 0.5$	11.812	5.2633	0.0156	0.0325	0.2106	924.9	99.74%
$\beta = 2$	11.815	5.3199	0.0148	0.0438	0.2605	2333.1	99.28%

$\langle\mathrm{deg}\rangle$: average degree; $\langle\mathrm{dist}\rangle$: average path length; C: global clustering coefficient; C_{loc}: average local clustering coefficient; ρ: degree assortativity; # comp.: number of connected components; giant %: percentage of nodes in the giant component.

From Fig. 1a we see that, as expected, the right tail of the degree distribution is heavy but not fat (*i.e.*, subexponential but not power-law) – as a matter of fact, the frequency of degrees $\geq \mu$ is well-fitted by a Lognormal distribution. Comparing the two regimes for β, we see that $\beta = 2$ yields a larger portion of loosely connected vertices compensated by the presence of greater hubs. The rationale is that only when the dependence on the distance is weak the individuals living in central and denser areas connect to peripheral vertices, that remain otherwise isolated. $\beta = 2$ thus favors the assortativity and the average local clustering, but causes a greater number of connected components. In any case, the giant component consistently covers more than 99% of the graph.

For what concerns the organization of our network in communities, we consider modularity-based clusters obtained with the Louvain algorithm [25]. From Fig. 1b, we see that when $\beta = 0.5$ the network *de facto* consists of $\approx$20 clusters of comparable size. The relatively low modularity of the obtained partition

(≈ 0.27) indicates that these clusters are significantly intertwined. Conversely, when $\beta = 2$ most nodes of the network lie in few well-defined giant communities, surrounded by a multitude of communities of variable size.

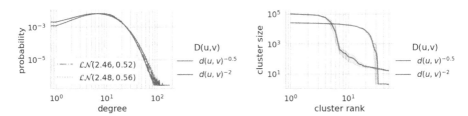

(a) Degree distribution with $\mathcal{LN}$ fit.   (b) Size of the largest 50 communities.

Fig. 1. Social graph for Florence with $\mu = 10$ and $f_u \sim 1 + \mathcal{LN}(\ln(2), 0.25)$: average features with a 95% confidence interval over 10 independent runs.

3.2   Socio-Geography of the Graph

Since our model incorporates a penalization for long edges, we expect some indication of correlation between topological properties and population density. The first, almost obvious, finding is that setting $\beta = 0.5$ significantly favors the creation of long edges at the expenses of very short ones, as shown in Fig. 2a. Notably, the distribution of the edges' physical length does not depend on the chosen μ and f_u, but it is entirely controlled by β.

In Fig. 2b we show the mean and max intra-cluster distance for the first 50 clusters of the graph. In line with empirical findings [26], only very small communities are geographically bounded – this is especially visible for $\beta = 0.5$ due to the sudden drop in community size emerged in Fig. 1b. Remarkably, when $\beta = 2$ the mean intra-cluster distance is often comparable to the tile side l (set to 1 Km as per Sect. 2.1), meaning that, even in large clusters, most adjacent vertices are at one tile of distance or less. When $\beta = 0.5$, instead, the mean distance consistently lies between $2l$ and $3l$.

It is reasonable to expect that vertices that are closer, on average, to other vertices will generally have a greater degree. This is confirmed by Figs. 3a and 3b, two heatmaps where the color gradient indicates the average degree of each tile. In particular, with $\beta = 2$, most tiles are far below average whereas the tiles surrounded by a densely populated area have a high average degree. The introduction of a social fitness attribute makes it possible to achieve the heterogeneity of sociable individuals within each tile. Yet, on average, the vertices having a favorable position in the territory will have a greater degree and the main hubs will be individuals with large f_u living in densely populated areas.

Finally, in Fig. 3c we plot the graph's adjacency matrix for $\beta = 0.5$ (the case $\beta = 2$ being completely alike), where nodes are ordered by their age-group. The

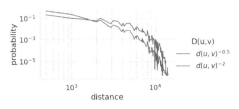
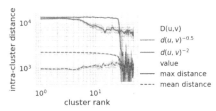

(a) Physical distance of adjacent vertices.   (b) Mean/max intra-cluster distance.

Fig. 2. Social graph for Florence with $\mu = 10$ and $f_u \sim 1 + \mathcal{LN}(\ln(2), 0.25)$: average features with a 95% confidence interval over 10 independent runs.

observable assortativity by age, inherited by the data-driven coefficients $s_{i,j}$, is clear and in qualitative agreement with previous work on social mixing patterns [10,17,22,28]. In analogous contact matrices, it is often possible to identify sub-diagonals which account for parent-children contacts [10,22]. Such sub-diagonals are, in our case, non-detectable having just four age-groups.

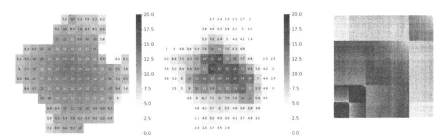

(a) Average degree of each tile for $\beta = 0.5$.   (b) Average degree of each tile for $\beta = 2$.   (c) Adjacency matrix for $\beta = 0.5$, nodes sorted by age.

Fig. 3. Social graph for Florence with $\mu = 10$ and $f_u \sim 1 + \mathcal{LN}(\ln(2), 0.25)$.

4   Epidemic Use Case

To further assess the practical relevance of our model, we simulated a SIR epidemic upon the giant connected component of the synthetic graphs obtained with $\mu = 10$, $f_u \sim 1 + \mathcal{LN}(\ln(2), 0.25)$ and, unless otherwise specified, $\beta = 0.5$. We consider a discrete-time *synchronous cellular automaton* in which the dynamic follows a *reactive process* [2]: at each time step, each infected individual spreads the disease to each of its neighbors with probability λ and recovers with probability δ. For the scope of this use case, we arbitrarily set $\delta = 0.1$ and $\lambda = 0.03$. If I_t denotes the set of infected individuals at time t, we assume that $|I_0| = 100$, *i.e.*, <0.03% of the population is infected at time 0. Albeit typical epidemic simulations consider possibly dynamic and denser networks of contacts, our network of strong ties may be interpreted as a coarse-grained model for highly-infectious, frequent and close contacts. In the following, we aim at showing that the parametric and data-driven nature of our model allows drawing high-level indications about the impact of several socio-demographic and geographic features on the epidemic.

In Fig. 4a we show the evolution of the fraction of infected and recovered individuals for different combinations of β and λ. When $\beta = 2$, the higher frequency of edges in densely populated areas favors a quicker spread of the infection, but the existence of loosely connected areas makes the total number of infected nodes slightly lower with respect to the case $\beta = 0.5$. We also see that, if household edges are three times more likely to transmit the disease than friendship edges (*i.e.*, $\lambda_H = 3\lambda_F$), but the overall average infection probability is still $\langle \lambda \rangle = 0.03$, the epidemic is a bit slower but, eventually, equally pervasive. As shown in Fig. 4b for $\lambda_F = \lambda_H = 0.03$, people living in households of size ≥ 3 have a significantly greater chance of catching the infection, probably due to the combined effect of having, on average, a greater degree and of the presence of children and young people in the household.

Since our model incorporates a data-driven age-based social mixing, it naturally lends itself to an analysis of the evolution of the epidemic inside single age-groups. From Figs. 4c and 4d, we see that children and young people experience a higher and earlier peak, and they are the only age-groups that reach a 90% prevalence of infected individuals. The younger groups are the drivers of the epidemic and the most infected, despite not being the largest groups, but probably due to their strong internal cohesion. In contrast with Fig. 4c, where the individuals in I_0 are chosen uniformly at random, in Fig. 4d all 100 individuals of I_0 are elderly people. A bit surprisingly, in this scenario we only notice a time shift in the epidemic, suggesting that the qualitative behavior of the epidemic depends on when the contagion reaches the younger individuals.

In Fig. 5 we consider the average time, over 10 independent runs, of the first infection occurring in each tile. The whole city center is reached in just a few days both if the infection starts from a central and densely populated tile (Fig. 5a) and if it starts from a peripheral and sparsely populated tile (Fig. 5b). Yet, some areas may be preserved if isolated within one or even two weeks. The starting position of the infection does play a role in our model, with an approximate 50% delay in the time of the first infection for most tiles if the infection starts in the periphery. In that case, the epidemic does not propagate locally but, apparently, it reaches the center before moving back outskirt.

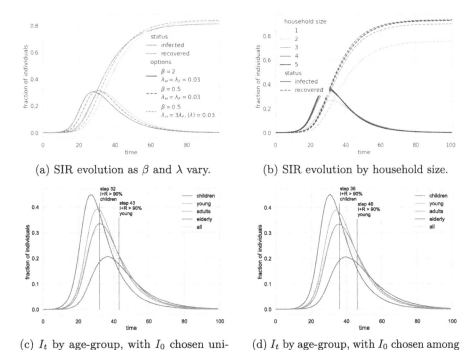

(a) SIR evolution as β and λ vary.   (b) SIR evolution by household size.

(c) I_t by age-group, with I_0 chosen uniformly at random.   (d) I_t by age-group, with I_0 chosen among elderly people only.

Fig. 4. Evolution of the fraction of infected and recovered individuals for different system parameters, within different households and within different age-groups.

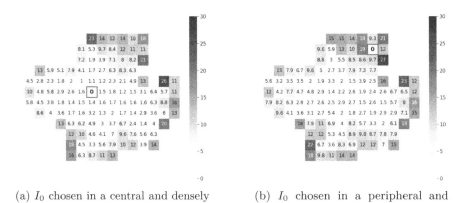

(a) I_0 chosen in a central and densely populated tile.   (b) I_0 chosen in a peripheral and sparsely populated tile.

Fig. 5. Time step of the first infected individual per tile.

5 Discussion and Conclusions

We have implemented a probabilistic model that mimics the strong social ties among a set of nodes representing the population of a given territory, organized into households. Our model is based on just a few clear assumptions: (*i*) not all individuals are equally sociable; (*ii*) the geographical distance and the age difference play a role in the probability that two individuals become friends; (*iii*) it is often fundamental to rely only on data that are already widely available. The simulator provides a way to recreate synthetic social networks within an arbitrary territory for which the social mixing patterns can be inherited from any already existing dataset, thus addressing the common circumstance where aggregated demographic data and some estimate of the age mixing patterns are the only available information. Since the evolution of a social network is significantly slower than most processes of interest occurring on it, our model is, by construction, static. Exploring possible approaches to generate a dynamic interaction network on top of our static social network is among the first directions for future work.

We evaluated our urban social network for the city of Florence, focusing on two configurations selected in light of previous empirical findings – for a more detailed analysis of the model we refer the reader to [13]. With only 10 friends on average, the giant component spans more than 99% of the network. Age and proximity based homophily guarantees the intended internal cohesion of single age-groups and a positive assortativity, yet the transitivity remains weak. By introducing a Lognormally distributed sociability we obtain the often desired heavy-tailed degree distribution. However, when long edges are strongly penalized, sociable hubs tend to concentrate in densely populated areas, thus intensifying the correlation between favorable positioning and degree. Nevertheless, the variability of the degree internal to a tile is preserved, being entirely controlled by the social fitness. A weak penalization of long edges, on the other hand, makes it more difficult to partition the network into well-defined communities. Almost regardless of their size, the communities tend to have a large spatial extension, even though the average distance of their members is small.

Some of the above properties are reflected in the outcomes of a SIR epidemic simulation on the network. The penalization imposed to long edges has an impact on the speed and pervasiveness – both quantitatively and geographically – of the contagion, whereas the age-based social mixing determines which age-groups drive the infection to a greater extent than the prevalence of different age-groups. Regardless of the specific epidemic use case considered in this paper, our model appears well suited to support the analysis of dynamic processes occurring within a urban population, thanks to its adherence to real data and its flexibility that allow easily evaluating the impact of socio-demographic and geographic features.

The intrinsic ambiguity in the concept of "friendship" leaves a few issues, somehow, open to further investigations. We plan to explore the integration of explicit preferential attachment mechanisms, and to verify whether a fine-grained age-stratification or age-specific β's do foster triangles within certain (*e.g.*, school age) groups. Further, we will consider a density-aware dependence on the distance [20], to gain more control on the social advantage associated to high density areas – which may still be desirable if the goal is predicting physical interactions, *e.g.*, for use in computational epidemiology. Finally, closed results binding the distribution of the social fitness to the obtained degree distribution may significantly improve the usability of our simulator. That said, we believe that our model presents unique features that make it a valuable resource for computational social scientists. Since the simulator is fully parametric and available as open source software, any potential user may adjust it to her/his needs and possibly contribute to its further development.

Acknowledgment. The authors thank the municipality of Florence for the kind support provided and Francesca Colaiori for useful discussions.

References

1. Amblard, F., Bouadjio-Boulic, A., Gutiérrez, C.S., Gaudou, B.: Which models are used in social simulation to generate social networks? A review of 17 years of publications in jasss. In: 2015 Winter Simulation Conference (WSC), pp. 4021–4032. IEEE (2015)
2. de Arruda, G.F., Rodrigues, F.A., Moreno, Y.: Fundamentals of spreading processes in single and multilayer complex networks. Phys. Rep. **756**, 1–59 (2018)
3. Barrett, C.L., et al.: Generation and analysis of large synthetic social contact networks. In: Proceedings of the 2009 Winter Simulation Conference (WSC), pp. 1003–1014. IEEE (2009)
4. Barthélemy, M.: Spatial networks. Phys. Rep. **499**(1–3), 1–101 (2011)
5. Beckman, R.J., Baggerly, K.A., McKay, M.D.: Creating synthetic baseline populations. Transp. Res. Part A: Policy Practice **30**(6), 415–429 (1996)
6. Bernaschi, M., Celestini, A., Guarino, S., Lombardi, F., Mastrostefano, E.: Spiders like onions: on the network of tor hidden services. In: The World Wide Web Conference, pp. 105–115 (2019)
7. Büchel, K., Ehrlich, M.V.: Cities and the structure of social interactions: evidence from mobile phone data. J. Urban Econ. **119**, 103276 (2020)
8. Caldarelli, G., Capocci, A., De Los Rios, P., Munoz, M.A.: Scale-free networks from varying vertex intrinsic fitness. Phys. Revi. Lett. **89**(25), 258702 (2002)
9. Cauchemez, S., et al.: Role of social networks in shaping disease transmission during a community outbreak of 2009 h1n1 pandemic influenza. In: Proceedings of the National Academy of Sciences, vol. 108(7), pp. 2825–2830 (2011)
10. Del Valle, S.Y., Hyman, J.M., Hethcote, H.W., Eubank, S.G.: Mixing patterns between age groups in social networks. Soc. Netw. **29**(4), 539–554 (2007)
11. Eagle, N., Pentland, A.S., Lazer, D.: Inferring friendship network structure by using mobile phone data. Proc. Natl. Acad. Sci. **106**(36), 15274–15278 (2009)
12. Goldenberg, J., Levy, M.: Distance is not dead: social interaction and geographical distance in the internet era. arXiv:0906.3202 (2009)
13. Guarino, S., et al.: Inferring urban social networks from publicly available data. Future Internet **13**(5), (2021). https://doi.org/10.3390/fi13050108
14. Guarino, S., Trino, N., Celestini, A., Chessa, A., Riotta, G.: Characterizing networks of propaganda on Twitter: a case study. Appl. Netw. Sci. **5**(1), 1–22 (2020)
15. Herrera-Yagüe, C., et al.: The anatomy of urban social networks and its implications in the searchability problem. Sci. Rep. **5**, 10265 (2015)

16. Illenberger, J., Nagel, K., Flötteröd, G.: The role of spatial interaction in social networks. Netw. Spat. Econ. **13**(3), 255–282 (2013)
17. Klepac, P., et al.: Contacts in context: large-scale setting-specific social mixing matrices from the BBC pandemic project. medRxiv (2020)
18. Krackhardt, D.: The strength of strong ties: The importance of philos in organizations. Networks and Organizations: Structure, Form, and Action, 216–239 (1992)
19. Lambiotte, R., et al.: Geographical dispersal of mobile communication networks. Phys. A **387**(21), 5317–5325 (2008)
20. Liben-Nowell, D., Novak, J., Kumar, R., Raghavan, P., Tomkins, A.: Geographic routing in social networks. Proc. Natl. Acad. Sci. **102**(33), 11623–11628 (2005)
21. Mastrandrea, R., Fournet, J., Barrat, A.: Contact patterns in a high school: a comparison between data collected using wearable sensors, contact diaries and friendship surveys. PLOS ONE **10**(9), 1–26 (2015)
22. Mistry, D., et al.: Inferring high-resolution human mixing patterns for disease modeling. arXiv:2003.01214 (2020)
23. Mitzenmacher, M.: A brief history of generative models for power law and lognormal distributions. Internet Math. **1**(2), 226–251 (2004)
24. Mossong, J., et al.: Social contacts and mixing patterns relevant to the spread of infectious diseases. PLOS Med. **5**(3), 1 (2008)
25. Newman, M.: Networks: An Introduction. OUP Oxford, Oxford (2010)
26. Onnela, J.P., Arbesman, S., González, M.C., Barabási, A.L., Christakis, N.A.: Geographic constraints on social network groups. PLoS one **6**(4), e16939 (2011)
27. Onnela, J.P., et al.: Analysis of a large-scale weighted network of one-to-one human communication. New J. Phys. **9**(6), 179 (2007)
28. Read, J.M., et al.: Social mixing patterns in rural and urban areas of southern china. Proc. Royal Soc. B: Biol. Sci. **281**(1785), 20140268 (2014)
29. Robins, G., Snijders, T., Wang, P., Handcock, M., Pattison, P.: Recent developments in exponential random graph (p*) models for social networks. Soc. Netw. **29**(2), 192–215 (2007)
30. Ryan, J., Maoh, H., Kanaroglou, P.: Population synthesis: comparing the major techniques using a small, complete population of firms. Geogr. Anal. **41**(2), 181–203 (2009)
31. Scellato, S., Noulas, A., Lambiotte, R., Mascolo, C.: Socio-spatial properties of online location-based social networks. ICWSM **11**, 329–336 (2011)
32. Squazzoni, F., et al.: Computational models that matter during a global pandemic outbreak: A call to action. J. Artif. Soc. Soc. Simul. **23**(2), 10 (2020)
33. Voas, D., Williamson, P.: An evaluation of the combinatorial optimisation approach to the creation of synthetic microdata. Int. J. Popul. Geogr. **6**, 349–366 (2000)
34. Walsh, F., Pozdnoukhov, A.: Spatial structure and dynamics of urban communities (2011)
35. Willem, L., Van Hoang, T., Funk, S., Coletti, P., Beutels, P., Hens, N.: SOCRATES: an online tool leveraging a social contact data sharing initiative to assess mitigation strategies for COVID-19. medRxiv (2020)
36. Wong, L.H., Pattison, P., Robins, G.: A spatial model for social networks. Phys. A **360**(1), 99–120 (2006)

Three-State Opinion Q-Voter Model
with Bounded Confidence

Wojciech Radosz$^{(\boxtimes)}$ and Maciej Doniec

Department of Theoretical Physics, Wroclaw University of Science and Technology,
Wybrzeze Wyspianskiego 27, 50–370 Wroclaw, Poland
wojciech.radosz@pwr.edu.pl

Abstract. We study the q-voter model with bounded confidence on the
complete graph. Agents can be in one of three states. Two types of agents
behaviour are investigated: conformity and independence. We analyze
whether this system is qualitatively different from a corresponding model
without bounded confidence. The key result of this paper is that the
system has two phase transitions: one between order-order phases and
another between order-disorder phases.

Keywords: q-voter · Complex systems · Monte Carlo simulations

1 Introduction

The q-voter model is widely used in the area of opinion dynamics [1–3]. Within
the q-voter model opinions are usually binary dynamical variables. Only recently,
a new version of the model with multi-state opinions was introduced [4,5]. In [5]
agents can change opinions without any limitations: all opinions are equivalent.
The situation when opinions are not equivalent was analyzed by Stauffer for the
Sznajd model [6]: one agent can convince another to its opinion only if they
share similar opinions, i.e. not too distant from each other (this rule is known
as a bounded confidence [7]). The simplest multi-state model with bounded
confidence is a model with three opinions. In this case two opinions are considered
as extreme and agents do not change their opinion from one extreme to another
due to the bounded confidence in a single update. Agents with the middle opinion
can change it to any other. One can think about this simple realisation of multi-
state opinion model with bounded confidence in terms of political parties: left-
and right-wing extreme parties and centrist party. There is empirical evidence
that agents opinion has multidimensional nature [8] and cannot be reduced to
simple *yes-no* case. The model with three-state opinion and bounded confidence
is a step to make it more realistic. In the future even more states can be added.

In this paper we analyze to what extent the q-voter model with three-state
opinion and bounded confidence is different from the one without bounded
confidence.

Research is supported by the National Science Center (NCN, Poland) through grant
no. 2019/35/B/HS6/02530.

M. Paszynski et al. (Eds.): ICCS 2021, LNCS 12744, pp. 295–301, 2021.
https://doi.org/10.1007/978-3-030-77967-2_24

2   Methodology

Considered system consists of N agents. Each agent is characterized by the dynamical variable named opinion $o_i(t)$, $i = 1, ..., N$, where t denotes simulation time measured in Monte Carlo steps (MCS). Opinion takes one of three possible values: $o_i(t) = k \in \{1, 2, 3\}$ and only the following transitions between states are allowed:

$$1 \; \underset{\curvearrowleft}{\overset{\curvearrowright}{} } \; 2 \; \underset{\curvearrowleft}{\overset{\curvearrowright}{} } \; 3. \tag{1}$$

Transitions between states 1 and 3 are forbidden, so opinions 1 and 3 have only opinion 2 as a neighbouring one, whereas for opinion 2 both 1 and 3 are the neighbouring states.

Number of agents with opinion k at time t is denoted by $N_k(t)$, and their concentration is $c_k(t) = N_k(t)/N$. Agents are placed in the vertices of a complete graph (CG). We distinguish two types of behaviour: independence and conformity [2,5]. Agents behave independently with probability p and conform to others with probability $1 - p$. For the latter case agent is influenced by the q-panel of unique neighbours (chosen without repetition). Changes of opinion are limited by restrictions from Eq. (1). In simulations we use random sequential updating, which means that in a single update only one agent can change its opinion. Pseudocod of a single update is presented below.

Algorithm 1: Pseudocod

$i \leftarrow random_{int}(1, N)$;
if $p > random(0, 1)$ **then**
| $i.state \leftarrow state\,(random_{int}\,(i.state - 1, i.state + 1))$;
else
| Find q neighbours;
| **if** *states of neighbours are equal* **then**
| | **if** *neighbour.state* $\in (state\,(i.state - 1), state\,(i.state + 1))$ **then**
| | | $i.state \leftarrow neighbour.state$;

---

3   Mean-Field Approach

We investigate presented model on the CG - mainly because it allows for an exact theoretical calculations as CG is equivalent with the mean-field approach. Such an approach enables verification of the model by comparison of the Monte Carlo (MC) results with analytical ones. The dynamics of the system can be in general described as the flow of agents from one opinion to another. Opinions 1 and 3 have different dynamics than opinion 2 because of different number of neighbours (see Eq. (1)).

We want to calculate the flow between opinions for certain values of parameters and define the stationary state for each of those. Let us define concentration of a given state in the form of:

$$c_k(t) = \frac{N_k(t)}{N} = \frac{1}{N} \sum_{i=1}^{N} \delta\left(o_i(t), k\right), \tag{2}$$

where $\delta\left(i, j\right)$ is the Kronecker delta function.

In a random sequential updating the elementary change of concentration $c_k(t)$ for all k in a single update is $\Delta c = 1/N$. Concentration $c_k(t)$ can increase or decrease with corresponding probabilities

$$\gamma_k^+ = P(c_k \to c_k + \Delta c), \quad \gamma_k^- = P(c_k \to c_k - \Delta c). \tag{3}$$

Up to this moment we have been dealing with random variable $c_k(t)$. We can also write the evolution equation of the corresponding expected values. For $N \to \infty$ we assume that random variable $c_k(t)$ localizes to the expected value. The time evolution of the expected value of c_k is

$$c_k(t + \Delta t) = c_k(t) + \frac{1}{N}(\gamma_k^+ - \gamma_k^-). \tag{4}$$

Since there is N agents and one MCS means N individual updates then $N\Delta t = 1$ and $\Delta t = \frac{1}{N}$. If the system is large enough and $N \to \infty$ then Eq. (4) simplifies to

$$\frac{\partial c_k}{\partial t} = \gamma_k^+ - \gamma_k^-. \tag{5}$$

Now we write explicitly $\gamma_k^{\pm}$. With probability $(1-p)$ agent is a conformist. There are $N_{k'}$ agents with opinions in states different than k, but achievable for agent in state k. We randomly choose q neighbours and they all need to share opinion k. For the first neighbour there are N_k available agents out of $N-1$. For every next neighbour there is one less unique agent. Finally the conformism part γ_{con}^+ and the outflow γ_{con}^- can be written as

$$\gamma_{k,con}^+ = \frac{N_{k'}}{N}(1-p)\left(\prod_{i=0}^{q} \frac{N_k - i}{N-1-i}\right), \quad \gamma_{k,con}^- = \frac{N_k}{N}(1-p)\left(\prod_{i=0}^{q} \frac{N_{k'} - i}{N-1-i}\right). \tag{6}$$

With probability p agent is independent. There are $N_{k'}$ agents in different state than k and achievable for agent in state k. Due to the bounded confidence the independence term is different for $k = 1, 3$ and $k = 2$. Random choice of state for $k = 1, 3$ means that with probability $\frac{1}{2}$ we can change to $k' = 2$ or stay in the same state. For state $k = 2$ there are three options: $k' = 1, 3$ and preserving opinion, so each term has probability $\frac{1}{3}$. So for example $\gamma_{1,ind}$ yields

$$\gamma_{1,ind}^+ = \frac{N_2}{N}\frac{p}{3}, \quad \gamma_{1,ind}^- = \frac{N_1}{N}\frac{p}{2}. \tag{7}$$

Combining Eqs. (6–7) for state $k = 1$ we obtain

$$\gamma_1^+ = \frac{N_2}{N}\left((1-p)\left(\prod_{i=0}^{q} \frac{N_1 - i}{N-1-i}\right) + \frac{p}{3}\right), \tag{8}$$

and

$$\gamma_1^- = \frac{N_1}{N} \left((1-p) \left(\prod_{i=0}^{q} \frac{N_2 - i}{N - 1 - i} \right) + \frac{p}{2} \right). \tag{9}$$

When $N \to \infty$, the above equations simplify to

$$\gamma_1^+ = c_2 \left((1-p)c_1^q + \frac{p}{3} \right), \quad \gamma_1^- = c_1 \left((1-p)c_2^q + \frac{p}{2} \right). \tag{10}$$

Analogic formulas were derived for $k = 2, 3$. The time evolution of the system can be described via three equations:

$$\frac{\partial c_1}{\partial t} = c_2 \left((1-p)(c_1^q) + \frac{p}{3} \right) - c_1 \left((1-p)(c_2^q) + \frac{p}{2} \right), \tag{11}$$

$$\frac{\partial c_2}{\partial t} = (c_1 + c_3) \left((1-p)(c_2^q) + \frac{p}{2} \right) - c_2 \left((1-p)(c_1^q + c_3^q) + \frac{2p}{3} \right), \tag{12}$$

$$\frac{\partial c_3}{\partial t} = c_2 \left((1-p)(c_3^q) + \frac{p}{3} \right) - c_3 \left((1-p)(c_2^q) + \frac{p}{2} \right). \tag{13}$$

In the next Section we compare results obtained from MC simulations with numerical results from Eqs. (11–13).

4   Simulations

For the simulations we use the system of size $N = 25\,000$ and simulation time $t = 5\,000$ MCS. Results were averaged over 64 independent realisations.

Figure 1 shows the plot of concentration c_1 against p for different values of q (see legend). Solid lines represent theoretical solutions for $c_1(p)$, whereas symbols denote the outcome of MC simulations. Coloured and empty symbols stand for different initial conditions: the former for $c_1(0) = 1$ and $c_2(0) = c_3(0) = 0$, the latter for $c_2(0) = 1$ and $c_1(0) = c_3(0) = 0$. It can be seen in more detail in the inset of Fig. 1, that those data sets are different but they tend to overlap for certain values of p. At first we focus on $c_1(p)$ for $q = 2$. Data maintains order up to $p \approx 0.19$ when c_1 drops from ~ 0.7 to ~ 0.1 suggesting some kind of phase transition in the system. For higher values of p, c_1 slowly and smoothly grows, through the inflection point to equilibrium value ~ 0.27. Final value of c_1 is not equal to $\frac{1}{3}$ because, as mentioned, opinions $k = 1, 3$ have only one neighbour while central opinion $k = 2$ has two neighbours. Data for $c_1(0) = 0$ (empty squares) slowly growths up to $p \approx 0.2$. For higher p both data sets overlap.

The data $c_1(p)$ for $q = 3$ (blue and empty circles) has very similar character to $q = 2$. The drop in c_1 value takes place earlier: for $p \approx 0.17$, after which coloured and empty circles overlap. Later growth is more rapid than for $q = 2$.

This character repeats for $q = 4$ (green and empty triangles). The only noticeable difference is that growth after $p \approx 0.13$ is much faster and an eye-inspection reveals the position of inflection point.

The data for $q = 5$ (magenta and empty diamonds) displays qualitatively different character. After the initial drop of c_1 for $p \approx 0.1$ we notice short, rapid

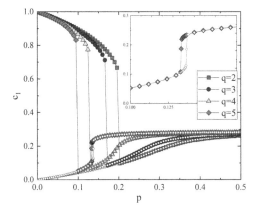

Fig. 1. Dependence between the stationary concentration c_1 of opinion 1 and probability of independence p for several values of q obtained from MC simulations (symbols) and numerical solutions of Eqs.(11–13) (solid lines). Coloured symbols stand for the initial configuration $c_1(0) = 1$. Empty symbols stand for $c_1(0) = 0$ (inset shows this in detail). Inset: hysteresis loop for $q = 5$, scaled up.

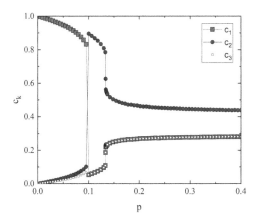

Fig. 2. Dependence between the stationary values of all opinions and the probability of independence p for $q = 5$.

growth and another jump in value for $p \approx 0.14$. The second jump is discontinuous and displays hysteresis loop that can be seen via discrepancy between the two data sets: coloured and empty symbols. This part of the graph can be seen in more detail in the inset. This is the first case when we see double discontinuous transition: from one order to another (dominant $k = 1$ into dominant $k = 2$ or 3) and from order into disorder (dominant $k = 2$ or 3 into disorder).

The next key issue is what happens with the system after the first and the second transition. To explain this we plot the concentration of all states $c_{1,2,3}(p)$ for $q = 5$ - see Fig. 2. Initial condition was $c_1(0) = 1$ and $c_2(0) = c_3(0) = 0$.

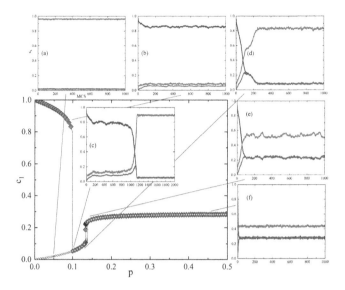

Fig. 3. Trajectories of $c_k(t)$ against MC time for $q = 5$ and various p. Each trajectory graph is connected to red point showing its position on $c_1(p)$ graph. Black line corresponds to c_1, red to c_2 and blue to c_3. All the trajectory plots have the same axis labels as plot (a). (Color figure online)

Up to $p = 0.1$ the concentration of states is the following: dominating state is $k = 1$, while states $k = 2, 3$ are in minority. Afterwards there is a rapid drop of c_1 and a rapid growth of c_2. For rather narrow region $p \in [0.1, 0.14]$ state $k = 2$ dominates while states $k = 1, 3$ are equal. Then there is a second rapid drop of value: this time c_2 drops while $c_{1,3}$ grow. For further increase of p there is a smooth, rather slight change of values of all concentrations reaching final value around $p \approx 0.3$.

Figure 2 clearly shows that the first transition for $p \approx 0.1$ means the change of domination from state $k = 1$ into $k = 2$. The second transition for $p \approx 0.14$ means that state $k = 2$ looses decisive domination and for $p > p^* \approx 0.14$ almost disordered phase is reached with only slight dominance of the central opinion c_2 over the extremes $c_1 = c_3$. This results is in agreement with theoretically calculated values of stationary concentrations c_k for high p, namely $c_{1,3,st} = \frac{2}{7}$ and $c_{2,st} = \frac{3}{7}$.

To gain deeper insight into the behaviour of the system we present the trajectories – plots of time evolution of concentration $c_k(t)$ (see Fig. 3). We have chosen data for $q = 5$ as for this value we obtained the most intriguing results. Trajectories are presented together with data $c_1(p)$ for better understanding of their position on the phase diagram. Each red point in graph $c_1(p)$ corresponds to the given inset trajectory plot. For $p = 0.05$ and $c_2(0) = 1$ (panel (a) in Fig. 3) opinion $k = 2$ dominates for the whole simulation time. Trajectory (b) for $p = 0.09$ and $c_1(0) = 1$ shows similar character with domination of $k = 1$.

Trajectory (c) for $p = 0.1$ and $c_1(0) = 1$ is qualitatively different from both previous plots. State $k = 1$ dominates for about 1000 MCS and then rapidly looses domination in favour of $k = 2$. Opinion $k = 3$ initially slowly increases, then the sudden rapid growth is observed but after very short time it drops to the final steady value. This surprising non-monotonic behaviour originates from the growth of neighbouring opinion $k = 2$. Later the system stabilizes with domination of $k = 2$. Trajectory (d) for $p = 0.125$ and $c_1(0) = 1$ has similar character to plot (c) but the change of domination happens much faster (~ 100 MCS). Trajectory (e) for $p = 0.142$ and $c_1(0) = 1$ has similar character to (d) but $k = 2$ becomes dominant much faster. Its final concentration is considerably smaller than in all previous cases. Plot (f) for $p = 0.45$ and $c_1(0) = 1$ shows very fast transition into domination of $k = 2$ on the lowest value of all cases: $c_2 \approx 0.43$.

Those results indicate once more that the first transition localized at $p \approx 0.1$ corresponds to the change of domination: from $k = 1$ to $k = 2$. The second transition at $p \approx 0.14$ corresponds to the loss of the overwhelming domination of any state. From this point with growing p the system evolves towards mixed state with equally numerous states $k = 1, 3$ and slightly more numerous middle state $k = 2$.

5   Discussion

When opinion is multi-state people are rather unlikely to change their opinion dramatically [6,7]. In the model we express this in terms of bounded confidence. We analyzed the q-voter model with three-state opinions and bounded confidence on a complete graph. We took into account two types of the social response: conformity and independence. There are two phase transitions: the first from a certain order into a different order, and the second from an order into disorder. The first transition is discontinuous in all analyzed cases, while the second transition is discontinuous only for $q \geq 5$. The system shows dynamics that is qualitatively different than the corresponding system without bounded confidence [5].

References

1. Castellano, C, Munoz, M.A., Pastor-Satorras, R.: Phys. Rev. E **80**, 041129 (2009)
2. Jedrzejewski, A., Sznajd-Weron, K.: Comptes Rendus Physique **20**, 244–261 (2019)
3. Chmiel, A., Sznajd-Weron, K.: Phys. Rev. E **92**, 052812 (2015)
4. Martins, A.C.R.: Eur. Phys. J. B **93**, 1 (2020)
5. Nowak, B., Ston, B., Sznajd-Weron, K.: Sci. Rep. **11**, 6098 (2021)
6. Stauffer, D.: Advances in Complex Systems **5**, 97–100 (2002)
7. Hegselmann, R., Krause, U.: JASSS **5** (2002)
8. Alos-Ferrer, C., Granic, D.-G.: Elect. Stud. **39**, 56–71 (2015)

The Evolution of Political Views Within the Model with Two Binary Opinions

Magdalena Gołębiowska[ID] and Katarzyna Sznajd-Weron(✉)[ID]

Department of Theoretical Physics, Wrocław University of Science and Technology,
Wybrzeże Wyspiańskiego 27, 50-370 Wrocław, Poland
katarzyna.weron@pwr.edu.pl

Abstract. We study a model aimed to describe political views within two-dimensional approach, known as the Nolan chart or the political compass, which distinguish between opinions related to economic and personal freedom. We conduct Monte Carlo simulations and show that in the lack of noise, i.e. at social temperature $T = 0$, the consensus is impossible if there is a coupling between opinions related to economic and personal freedom. Moreover, for $T > 0$ we show how the strength of the coupling between these opinions can hamper or facilitate the consensus.

Keywords: Complex social systems · Agent-based simulation · Opinion dynamics · Political compass

1 Introduction

The temptation to model human behavior appeared numerous times not only in the history of science [5] but also in the fiction science. Probably the most obvious example of such a temptation is psycho-history, introduced by Asimov in his famous Foundation cycle.

One of the approaches to understand the human behavior is agent-based modeling (ABM) [6,12], which allows to describe the macroscopic behaviors based on the microscopic rules that define how individuals interact with each other. Among many different subjects studied within agent-based (i.e. microscopic) approach, one of the most popular and interdisciplinary one is the opinion dynamics [6,11,12,14–16,18].

When it comes to modeling opinions related to political views, binary variables seem to be particularly natural choice being a discretization of the traditional left–right/progressive-conservative division [10,13,17]. However, numerous empirical studies show that such a one-dimensional description may not be sufficient enough [1,8].

Placing political views along two axes, representing economic and personal freedom, is known presently as the Nolan chart or the political compass. To our

Supported by the National Science Center (NCN, Poland) through Grant No. 2016/21/B/HS6/01256.

ⓒ Springer Nature Switzerland AG 2021
M. Paszynski et al. (Eds.): ICCS 2021, LNCS 12744, pp. 302–308, 2021.
https://doi.org/10.1007/978-3-030-77967-2_25

best knowledge, such a two-dimensional description of political views was used for the first time within ABM in [21]. Originally it was introduced as a non-equilibrium model described by the set of dynamical rules. However, this year, the model was reformulated in the spirit of statistical physics equilibrium model and studied analytically within the renormalization group technique [20].

Despite the obvious advantages of such an analytical method, there are several disadvantages when looking from the social point of view. Firstly, it allows to study only infinite systems, which do not exist in reality. Secondly, it does not allow to follow the temporal evolution of the system, which is particularly interesting from the social point of view. Therefore, we decided to analyze the model within computer simulations and focus mainly on the temporal behavior of the small systems.

2 Model

We consider a one dimensional lattice of size N with periodic boundary conditions. Each site $i = 1, \ldots, N$ of the lattice is occupied by exactly one agent, who is described by two binary dynamical variables (S_i, σ_i), where $S_i = \pm 1$ denotes the view related to the economic freedom and $\sigma_i = \pm 1$ the view related to the private freedom [20,21]. According to the political compass we define: (S)ocialists $S_i = -1, \sigma_i = +1$, they want the strong government in the economic area but they value the freedom in the personal life, (L)ibertarians $S_i = +1, \sigma_i = +1$, they value the freedom in both areas, (A)uthoritarians $S_i = -1, \sigma_i = -1$ they are against any freedom and (C)onservatives $S_i = +1, \sigma_i = -1$ they want the economic freedom but strict rules in the private area.

The model, as usually in the equilibrium statistical physics, is defined by the Hamiltonian [20]:

$$
H = - J_1 \sum_{i=1}^{N} S_i S_{i+1} - J_2 \sum_{i=1}^{N} S_i S_{i+2} - K_1 \sum_{i=1}^{N} \sigma_i \sigma_{i+1} - K_2 \sum_{i=1}^{N} \sigma_i \sigma_{i+2}
$$
$$
- M_0 \sum_{i=1}^{N} \sigma_i S_i, \tag{1}
$$

where J_1, J_2, K_1, K_2 are the coupling constants between agents, and M_0 describes the interaction between views related to the private and to the economic freedom.

In this paper we investigate the system within Monte Carlo simulations and we use the standard Metropolis algorithm. It means that the transition rate from one state $r \equiv (S_1, \ldots, S_i, \ldots, S_N)$ to the new one $r' \equiv (S_1', \ldots, S_i', \ldots, S_N')$:

$$
P(r \to r') = min\left(1, e^{-(H(r')-H(r))/T}\right), \tag{2}
$$

where T is so called *social temperature* introduced within micro-sociology by Bahr and Passerini [4].

We use a random sequential updating, which mimics the continuous time. It means that within a single update we choose randomly only one agent and we try to update its state. In case of a single binary opinion it is straightforward – we try to change the state of an agent to the opposite one $S_i \rightarrow -S_i$. However, in our model each agent is described by two opinions and thus it is less obvious how we should update the system.

We can use many different updating schemes (US), analogously as it was done for the Ashkin-Teller two-spin model [7]: (US1) choose randomly agent i, update its opinion S_i and then σ_i, (US2) choose randomly agent i, update its opinion σ_i and then S_i, (US3) first update opinions S_i in the random order for all i and then do the same for σ_i, (US4) first update opinions σ_i in the random order for all i and then S_i. We conducted simulations within all above schemes and we obtained the same results for all of them. Therefore, here we present the algorithm and results only for US1.

As usually, we count the time in Monte Carlo steps (MCS) and one MCS consists of N elementary updates given by the following algorithm:

1. Choose randomly agent i from all N agents, $i \sim U\{1, N\}$
2. Update opinion related to the economic freedom:
   (a) Calculate the change $\Delta H = H(r') - H(r)$ caused by the potential change $S_i \rightarrow -S_i$
   (b) If $\Delta H \leq 0$ then update the state $S_i \rightarrow -S_i$ else
   (c) Choose a random number $r \sim U(0,1)$. If $r < e^{-\Delta H/T}$ then update the state $S_i \rightarrow -S_i$.
3. Update opinion related to the private freedom:
   (a) Calculate the change $\Delta H = H(r') - H(r)$ caused by the potential change $\sigma_i \rightarrow -\sigma_i$
   (b) If $\Delta H \leq 0$ then update the state $\sigma_i \rightarrow -\sigma_i$ else
   (c) Choose a random number $r \sim U(0,1)$. If $r < e^{-\Delta H/T}$ then update the state $\sigma_i \rightarrow -\sigma_i$.

3   Results

The model consists of 6 parameters – 5 coupling constants and the social temperature T. For $M_0 = 0$ there are no interactions between chains $(\sigma_1, \ldots, \sigma_N)$ and $(S_1, \ldots, S_N)$, and thus the model reduces to two independent next-nearest neighbor Ising (ANNNI) models [9]. Because of that we focus mainly on the role of M_0. We focus on the very particular case, $J_1 = K_2 = 0$ and $J_2 = K_1 > 0$, which means that in the private area we try to follow the nearest neighbors (like in the basic Ising model), whereas in the economic one the next-nearest neighbors (like in the basic Sznajd model) [22]. Such a choice was inspired by the original paper on the political compass within ABS [21].

Let us start with the case of $M_0 = 0$ as the reference one. In such a case for $T = 0$ the system evolves towards an absorbing state and its configuration can be easily predicted for all combinations of parameters J_1, J_2, K_1, K_2. For

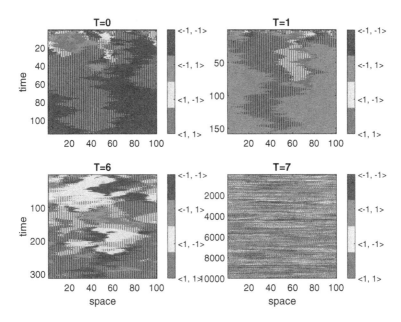

Fig. 1. The time evolution of the system of size $N = 100$ for the coupling constant $M_0 = 0$ and four values of the social temperature T, as indicated in the title of each subplot. Vertical axis represents the time and horizontal one the space, which means that each horizontal line is a visualization of the system's configuration at a given time. Individual states $< S_i, \sigma_i >$ are indicated by the color-bar.

$J_2 = K_1 = 0$ and $J_1, K_2 > 0$, even if initially $\forall_i S_i(0) = \sigma_i(0)$ like in Fig. 1, during the evolution all four political views appear. However, eventually the system reaches one of possible states: (1) consensus, i.e. all agents have the same political attitude (S,L,A or C) or (2) the war between two states (two-party system).

All these absorbing states can be identified by measuring four quantities: two magnetizations

$$m_S = \frac{1}{N} \sum_{i=1}^{N} S_i, \qquad m_\sigma = \frac{1}{N} \sum_{i=1}^{N} \sigma_i, \tag{3}$$

and two densities of active bonds

$$\rho_S = \frac{1}{2N} \sum_{i=1}^{N} (1 - S_i S_{i+1}), \qquad \rho_\sigma = \frac{1}{2N} \sum_{i=1}^{N} (1 - \sigma_i \sigma_{i+1}). \tag{4}$$

The consensus within a chain is reached if $|m_\alpha| = 1$, where $\alpha = \{S, \sigma\}$, whereas the war corresponds to $\rho_\alpha = 1$.

For $0 < T < T^*$ all these states can still be reached but they are not absorbing anymore, because due to the noise the system evolves for ever. It is not seen on presented figures, because we stop the simulation, once the consensus or the war,

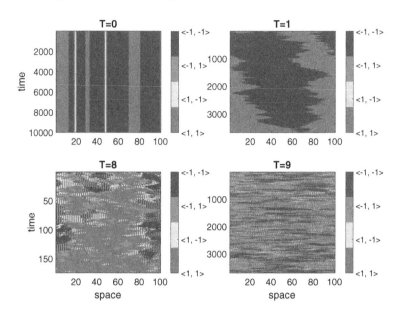

Fig. 2. The time evolution of the system of size $N = 100$ for the coupling constant $M_0 = 1, J_2 = K_1 = 10$ and four values of the social temperature T, as indicated in the title of each subplot. Vertical axis represents the time and horizontal one the space, which means that each horizontal line is a visualization of the system's configuration. Individual states $< S_i, \sigma_i >$ are indicated by the color-bar.

is reached, i.e., ($|m_S| = 1$ or $\rho_S = 1$) and ($|m_\sigma| = 1$ or $\rho_\sigma = 1$). For $T > T^*$ the we do not observe growth of any consensus or war domains, as shown in the bottom right panel of Fig. 1.

What changes if we allow for the coupling between the economic and private area, e.g., $M_0 > 0$? For $M_0 > min(2|J_2|, 2|K_1|)$ the system almost immediately blocks and no time evolution is seen. Independently on the initial state, in the blocked state $S_i = \sigma_i$ for all agents. It means that Libertarians and Authoritarians coexist but consensus is never reached. Analogous discussion can be provided for $M_0 < 0$, but in such a case Socialists coexist with Conservatives.

The most interesting behavior is observed for intermediate values $M \in (0, min(2|J_2|, 2|K_1|))$. Sample time evolution is shown in Fig. 2. In this case for $T = 0$ the system blocks after several time steps and coexistence of A and L (for $M_0 > 0$) or S and C (for $M_0 < 0$) is observed. On the other hand, for relatively small social temperature $0 < T < T_0^*$ it evolves very slowly towards consensus and all agents reach the state $S_i = \sigma_i$ (for $M_0 > 0$) or $S_i = -\sigma_i$ (for $M_0 < 0$). For $T_0^* < T < T^*$, all four political attitudes can appear and both types of clusters (consensus and war) are observed. However, only the consensus can spread over the whole system and it spreads relatively fast, what is clearly seen in Fig. 3. We do not give here any particular values of T_0^*, T^* because they depend on the particular values of the interaction constants, as seen in the right panel of Fig. 3.

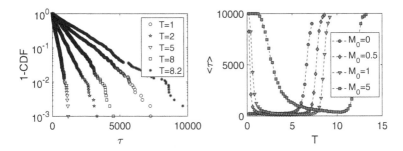

Fig. 3. Waiting time τ to reach consensus for the system of size $N = 100$ obtained within 10^4 samples for interaction coefficients $J_1 = 10, J_2 = 0, K_1 = 0, K_2 = 10$. **Left panel:** The tail of the empirical cumulative distribution function of the waiting times for $M_0 = 1$ and several values of social temperature T. **Right panel:** The average waiting time to reach consensus as a function of social temperature T for several values of M_0.

4 Discussion

In Fall 2011 Reason-Rupe Poll reported that 24 % of Americans are Economically Conservative and Socially Liberal, 28 % Liberal, 28 % Conservative, and 20 % Communitarian (check on $https : //reason.com/poll/$). This is one of numerous studies, which confirms that one-dimensional description of political views is not sufficient [8]. Here, we presented briefly results of the simple equilibrium model with two binary opinions.

We would like to stress that there is a significant difference between a multi-state but one-dimensional description, such as in the Potts model [23], and the multi-dimensional description, such as in the Axelrod model [3] or in the one discussed here. The natural question is why not to use the Axelrod model to represent political views, when it allows to describe many dimensions and to measure each of them by a multi-state variable? The answer, and simultaneously the justification of our approach, is that within the Axelrod model all dimensions are equivalent and there are no direct interactions between them.

We are aware of many limitations of our study, which should be treated as a zero-level approach to the real political system. First of all, a one-dimensional lattice with periodic boundary conditions is an appropriate structure for a round-table discussion but not for the large real-life social networks. In the future, one should rather consider heterogeneous graphs with a small-world property. Moreover, it has been suggested, based on the empirical research, that even a two-dimensional space may be not sufficient to describe political landscapes [2].

However, even a two-dimensional description is a step forward in relation to a single left-right representation of political attitudes, that has been mostly used so far [10,13,19].

References

1. Albanese, F., Tessone, C.J., Semeshenko, V., Balenzuela, P.: Data-driven model for mass media influence in electoral context, pp. 3–9 (2019). arXiv:1909.10554v2
2. Alós-Ferrer, C., Granić, T.G.: Political space representations with approval data. Electoral Stud. **39**, 56–71 (2015)
3. Axelrod, R.: The dissemination of culture: a model with local convergence and global polarization. J. Conf. Resolut. **41**(2), 203–226 (1997)
4. Bahr, D., Passerini, E.: Statistical mechanics of opinion formation and collective behavior: micro-sociology. J. Math. Sociol. **23**(1), 1–27 (1998)
5. Ball, P.: The physical modelling of society: a historical perspective. Physica A Stat. Mech. Appl. **314**(1–4), 1–14 (2002)
6. Bianchi, F., Squazzoni, F.: Agent-based models in sociology. Wiley Interdisc. Rev. Comput. Stat. **7**(4), 284–306 (2015)
7. Ditzian, R., Banavar, J., Grest, G., Kadanoff, L.: Phase diagram for the Ashkin-Teller model in three dimensions. Phys. Rev. B **22**(5), 2542–2553 (1980)
8. Eysenck, H.: The Psychology of Politics. Routledge, London (1998)
9. Fisher, M., Selke, W.: Infinitely many commensurate phases in a simple using model. Phys. Rev. Lett. **44**(23), 1502–1505 (1980)
10. Galam, S.: Sociophysics: A Physicist's Modeling of Psycho-Political Phenomena. Springer, Heidelberg (2012). https://doi.org/10.1007/978-1-4614-2032-3
11. Grabisch, M., Rusinowska, A.: A survey on nonstrategic models of opinion dynamics. Games **11**(4), 1–28 (2020)
12. Jackson, J., Rand, D., Lewis, K., Norton, M., Gray, K.: Agent-based modeling: a guide for social psychologists. Social Psychol. Pers. Sci. **8**(4), 387–395 (2017)
13. Kononovicius, A.: Empirical analysis and agent-based modeling of the Lithuanian parliamentary elections. Complexity **2017** (2017)
14. Noorazar, Hossein: recent advances in opinion propagation dynamics: a 2020 survey. Eur. Phys. J. Plus **135**(6), 1–20 (2020). https://doi.org/10.1140/epjp/s13360-020-00541-2
15. Proskurnikov, A., Tempo, R.: A tutorial on modeling and analysis of dynamic social networks. Part i. Ann. Rev. Control **43**, 65–79 (2017)
16. Proskurnikov, A., Tempo, R.: A tutorial on modeling and analysis of dynamic social networks. Part ii. Ann. Rev. Control **45**, 166–190 (2018)
17. Qiu, L., Phang, R.: Agent-based modeling in political decision making. In: Thompson, W.R. (ed.) Oxford Research Encyclopedia, Politics. Oxford University Press, USA (2020)
18. Sobkowicz, P.: Whither now, opinion modelers? Front. Phys. **8** (2020)
19. Stauffer, D.: Better being third than second in a search for a majority opinion. Adv. Complex Syst. **05**(01), 97–100 (2002)
20. Sznajd, J.: From modeling of political opinion formation to two-spin statistical physics model. J. Stat. Mech. Theory Exp. **2021**(1), 013210 (2021)
21. Sznajd-Weron, K., Sznajd, J.: Who is left, who is right? Physica A Stat. Mech. Appl. **351**(2–4), 593–604 (2005)
22. Sznajd-Weron, K., Sznajd, J., Weron, T.: A review on the Sznajd model–20 years after. Physica A Stat. Mech. Appl. **565**, 125537 (2021)
23. Wu, F.Y.: The Potts model. Rev. Mod. Phys. **54**(1), 253–268 (1982)

How to Reach Consensus? Better Disagree with Your Neighbor

Tomasz Weron[1]([✉]) [iD] and Katarzyna Sznajd-Weron[2] [iD]

[1] Department of Applied Mathematics, Wrocław University of Science and
Technology, 50-370 Wrocław, Poland
tomasz.weron@pwr.edu.pl
[2] Department of Theoretical Physics, Wrocław University of Science and Technology,
50-370 Wrocław, Poland
katarzyna.weron@pwr.edu.pl

Abstract. We study the basic first passage properties of a discrete one-dimensional mathematical model of opinion dynamics. The model studied here is a generalization of the original Sznajd model, which consists of introducing a new parameter p, being the probability of disagreement with the nearest neighbor in case of uncertainty. We study the model via Monte Carlo simulations and show that the exit probability does not change with the size of the system N, whereas the average exit time τ scales with N as $\tau \sim N^\alpha$. Moreover, we show that generally the consensus is reached more rapidly if agents disagree more often with their nearest neighbors in case of uncertainty.

Keywords: Mathematical sociology · Complex social systems · Agent-based simulation · Opinion dynamics

1 Introduction

Agent based models (ABMs) are known as a tool which builds a bridge between micro and macro scale. Moreover, the macro outcome of the microscopic rules can be surprising and sometimes contra-intuitive, which is one of the key features of the complex systems, known as emergence [1].

In this short paper, we discuss one of such a non-intuitive result that is obtained within a simple opinion dynamics model, recently introduced in the review paper on the Sznajd model (SM) [12]. The model we study here can be treated as a generalization of the original one-dimensional SM and reduces in limiting cases to two versions of SM, which were usually studied in the literature [12].

We focus on one of the most studied issues in a field of opinion dynamics, namely reaching the consensus [2,4–6,11]. Our intuition says that the more you

---

T. W. was supported by the Ministry of Science and Higher Education, Poland within the "Diamond Grant" Program through grant no. DI2019 0008 49 and K. S-W. by the National Science Center (NCN, Poland) through grant no. 2016/21/B/HS6/01256.

M. Paszynski et al. (Eds.): ICCS 2021, LNCS 12744, pp. 309–315, 2021.
https://doi.org/10.1007/978-3-030-77967-2_26

agree with your neighbors the easier consensus should be reached. Here, we will show that this is not necessarily true.

2 The Model

We study a system of N agents placed in the cells of the one-dimensional lattice with periodic boundary conditions. Each agent can be in one of two states, $S_i(t) = \pm 1$, which represent alternative opinions (yes/no, agree/disagree, etc.), that changes in time t due to interactions between agents. Each cell is occupied by an exactly one agent and agents cannot move. Therefore, equivalently we could describe the system as the one-dimensional grid of binary cells.

At each elementary update, a pair of neighboring cells $(i, i+1)$ is chosen at random and it influences two neighboring cells: one on the left side of a pair, i.e. $i - 1$, and one on the right side, i.e. $i + 2$. If pair $(i, i+1)$ is unanimous, i.e. $S_i(t) = S_{i+1}(t)$, then two neighbors take the same state as a pair: $S_{i-1}(t+\Delta t) = S_i(t)$, $S_{i+2}(t+\Delta t) = S_i(t)$, where Δt is a time period needed for a single update. Otherwise, if $S_i(t) = -S_{i+1}(t)$, cells $(i-1, i+2)$ take the opposite states to their nearest neighbors with probability p: $S_{i-1}(t + \Delta t) = -S_i(t)$, $S_{i+2}(t + \Delta t) = -S_{i+1}(t)$. It means that in case of uncertainty, agents disagree with their nearest neighbors with probability p: $p = 1$ corresponds to the original"United we stand, divided we fall" rule, whereas $p = 0$ to the rule, which has been mostly used in the literature within the Sznajd model [12]. As usual in this type of models, a time unit consists of N elementary updates, i.e. $N\Delta t = 1$.

We realize that the assumptions we made here, including one-dimensional structure and the lack of movement, describe only a limited number of real-life scenarios, such as opinion formation during the round-table discussions. In many other cases, such as discussion in various social media, the assumption of the lack of the movement would be still valid. However, the structure of the social network in such a case should be modeled by some heterogeneous, even multiplex graphs [11]. To model face-to-face contacts, allowing for agents' movement would be also an interesting idea but this would require an additional parameter describing the density of occupied nodes.

We study the model within Monte Carlo simulations from different initial conditions, parametrized by the concentration $c_0 \equiv c(0)$ of agents with positive opinion at time $t = 0$:

$$c(t) = \frac{N_+(t)}{N} = \frac{1}{2N} \sum_{i=1}^{N} (S_i(t) + 1), \tag{1}$$

where $N_+(t)$ is the number of agents with opinion $+1$ at time t. To define precisely initial conditions, besides the concentration of positive argents c_0, we have to decide on their spacial distribution. We consider two limiting cases, as in [12]:

- **Random**: choose randomly $N_+(0)$ out of N cells for agents with positive opinions. For large systems it is almost identical with a much simpler rule:

for $i = 1$ to N, with probability $c_0 = N_+(0)/N$, set $S_i(0) = 1$ and, with complementary probability $1 - c_0$, set $S_i(0) = -1$.
- **Sorted**: for $i = 1$ to $N_+(0)$ set $S_i(0) = 1$ and for $i = N_+(0)$ to N set $S_i(0) = -1$.

We average results over $L = 10^3$ independent samples, which means that for each set of parameters (p, c_0, N) we run L independent simulations. We stop simulation once the absorbing state is reached and then, we collect data:

- The final (absorbing) configuration, which for this model is one of the following [12]: (1) positive consensus $(+ + + + + \cdots)$, (2) negative consensus $(- - - - - \cdots)$ or (3) a stalemate state $(+ - + - + \cdots)$. This allows us to calculate the probability of each absorbing state: P_+, P_-, P_{+-}, so called **exit probability** [7,11].
- The time needed to reach the absorbing state, so called **exit time**, which allows us to calculate the average exit time τ [7].

3   Results

It was already shown in [12] that only for $p = 1$ the consensus or stalemate state can be reached, whereas for $p < 1$ only consensus is possible. For all values of p both the consensus, as well as the stalemate state are the fixed points, i.e. once the system is in one of these states it will never leave them. However, for $p < 1$ the stalemate state is unstable, so it can never be reached from other states.

In [12] the exit probability was measured within the Monte Carlo simulations for $N = 100$. Here, we checked it more systematically for different $N \geq 100$. It occurs that for any $p < 1$ the exit probability of the positive consensus can be approximated by:

- for random initial conditions (see Fig. 1):

$$P_+ = \frac{c_0^2}{c_0^2 + (1 - c_0)^2}, \tag{2}$$

- for sorted initial conditions $P_+ = c_0$.

Exactly the same results have been obtained previously for the original Sznajd model without disagreement rule, which corresponds to $p = 0$ [3,9].

The second important first passage property, namely the exit time, was never measured so far for such a generalized model, but only for $p = 0$ [9]. It is obvious that $\tau = 0$ for $c_0 = 0$ or $c_0 = 1$ because the initial state is already an absorbing one. Moreover, we expect that τ has the maximum value for $c_0 = 0.5$ [8,9]. Finally, we expect that the exit time for the random initial conditions should be a bit shorter than for the ordered ones [9].

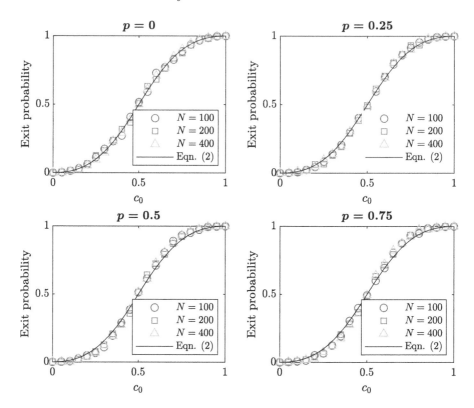

Fig. 1. Exit probability from random initial conditions for several system sizes N and four different values of p.

What we do not know, is the role of the parameter p. As discussed above, the exit probability does not depend on p, as long as $p \neq 1$, for which additional absorbing stalemate state appears. It could indicate that p is an unnecessary parameter, which could be omitted. However, it turns out that it affects the exit time in a non-trivial way, as shown in Fig. 2.

For each c_0 there is an optimal value $p = p^* = p^*(c_0)$, for which consensus is the most rapid, i.e., the exit time is the shortest. This optimal value p^* is surprisingly high, as shown in Fig. 2. For example, if initially the number of positive and negative opinions is more or less the same, $c_0 \approx 0.5$ the consensus is reached the most rapidly for $p = p^* \approx 0.9$. It means that consensus is reached faster if agents disagree with their nearest neighbors more often.

On contrary to the exit probability, the average exit time depends on the system size, what is rather expected, shown in Fig. 3. For $p = 0$ it scales with an exponent 2, i.e. $\tau \sim L^2$, what has been already shown in [9]. For other values of p the scaling exponent $\alpha \approx 2$, but it is not exactly equal to 2.

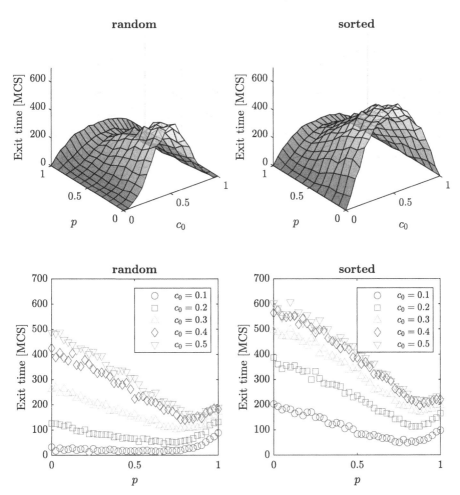

Fig. 2. Dependence between the average exit time to reach consensus and two model's parameters, for the system of size $N = 100$: initial concentration of positive opinions c_0 and the probability of disagreement p. Results from two types of initial conditions are shown: random (left panels) and sorted (right panels).

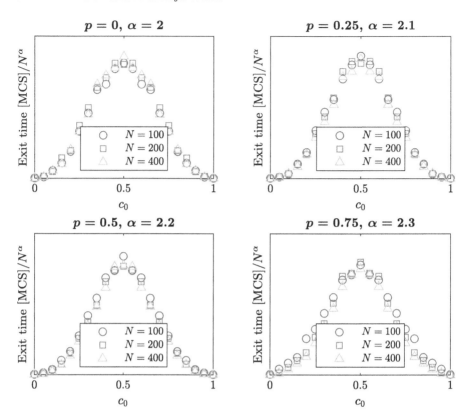

Fig. 3. Rescaled exit time from random initial conditions for several system sizes N and four different values of p.

4   Discussion

It is often claimed that the key lesson from agent-based modeling can be summarized by the sentence verbalized by Epstein [1,10]: "We get macro-surprises despite complete micro-level knowledge". In this paper, we show one of such surprises. We expected that for $p = 0$ the time evolution would be long, because then the change of the state is possible only if the source pair is unanimous. However, we did not expect that for $p > 0$ the exit time will decrease with an increasing p, and definitely not that the optimal value for the most rapid consensus will be so high.

The second surprise came with the finite size scaling. We expected the exit time to scale as the power law with an exponent $\alpha = 2$, but it occurs that it scales with $\alpha = 2$ only for the limiting values of $p = 0$ and $p = 1$. This result is far from being obvious and definitely needs deeper insight, what is a planned task for the future.

We are aware that Epstein's notion on macro-surprises can be criticized by saying that the level of surprise depends on the perceptiveness and experience of the researcher, the ability to find cause-effect relationships, etc. We cannot argue with that, just as we cannot state whether our intuition about what facilitates reaching consensus is the same as the intuition of the reader of this work. However, we hope that our short paper will inspire at least some readers to explore the model deeper, for example on more realistic social structure.

References

1. Epstein, J.: Generative Social Science: Studies in Agent-Based Computational Modeling. Princeton University Press, Princeton (2020)
2. Grabisch, M., Rusinowska, A.: A survey on nonstrategic models of opinion dynamics. Games **11**(4), 1–28 (2020)
3. Lambiotte, R., Redner, S.: Dynamics of non-conservative voters. EPL **82**(1) (2008)
4. Noorazar, H.: Recent advances in opinion propagation dynamics: a 2020 survey. Eur. Phys. J. Plus **135**(6), 1–20 (2020). https://doi.org/10.1140/epjp/s13360-020-00541-2
5. Proskurnikov, A., Tempo, R.: A tutorial on modeling and analysis of dynamic social networks. Part i. Ann. Rev. Control **43**, 65–79 (2017)
6. Proskurnikov, A., Tempo, R.: A tutorial on modeling and analysis of dynamic social networks. Part ii. Ann. Rev. Control **45**, 166–190 (2018)
7. Redner, S.: A Guide to First Passage Processes. Cambridge University Press, Cambridge (2001)
8. Slanina, F., Lavicka, H.: Analytical results for the Sznajd model of opinion formation. Eur. Phys. J. B **35**(2), 279–288 (2003)
9. Slanina, F., Sznajd-Weron, K., Przybyła, P.: Some new results on one-dimensional outflow dynamics. EPL **82**(1) (2008)
10. Smith, E., Conrey, F.: Agent-based modeling: a new approach for theory building in social psychology. Pers. Social Psychol. Rev. **11**(1), 87–104 (2007)
11. Sood, V., Redner, S.: Voter model on heterogeneous graphs. Phys. Rev. Lett. **94**(17), 178701 (2005)
12. Sznajd-Weron, K., Sznajd, J., Weron, T.: A review on the sznajd model—20 years after. Physica A Stat. Mech. Appli. **565**, 125537 (2021)

Efficient Calibration of a Financial Agent-Based Model Using the Method of Simulated Moments

Piotr Zegadło[(⊠)] [iD]

Kozminski University, Warsaw, Poland
pzegadlo@kozminski.edu.pl

Abstract. We propose a new efficient method of calibrating agent-based models using the Method of Simulated Moments. It utilizes the Filtered Neighborhoods optimization algorithm, gradually narrowing down the search area by examining local neighborhoods of promising solutions. The new method obtains better calibration accuracy for a benchmark financial agent-based model in comparison to a broad selection of other methods, while using just a tiny fraction of their computational budget.

Keywords: Agent-based models · Heterogeneous agent models · Model calibration

1 Introduction

Agent-based models (ABMs) have been extensively used to investigate the rich dynamics of financial time series. They can easily incorporate investor heterogeneity and bounded rationality, replicating the most important stylized facts of the financial markets as a result [10]. Such models have been also found to assist in asset pricing [13]. The vast body of literature on this topic can be generally divided into two main strands. One involves large, multi-agent computational models emulating financial market structure [20]. The other focuses on simpler models, with usually only a few groups of interacting agents guided by certain behavioral rules [14]. Such models, sometimes referred to as heterogeneous agent models (HAMs), sacrifice a certain amount of flexibility and complexity for the sake of greater interpretability and an insight into the underlying mechanism of market price dynamics [8].

Although many theoretical HAMs have been shown to be useful in analyzing financial market dynamics, they have also come under criticism for their apparent subjectivity. The need to reconcile theoretical underpinnings with empirical observations has spurred the growth of HAM calibration and validation literature in the recent years [21]. In that context, calibration can be defined as matching the output of a HAM employing a specific set of parameter values to the

ⓒ Springer Nature Switzerland AG 2021
M. Paszynski et al. (Eds.): ICCS 2021, LNCS 12744, pp. 316–329, 2021.
https://doi.org/10.1007/978-3-030-77967-2_27

characteristics of a reference dataset[1]. The quality of that match and the parameter values can then be taken under scrutiny, potentially vindicating a specific model – or, conversely, suggesting its modification or abandonment altogether.

A canon of best practices for agent-based models' calibration has not yet been firmly established. This is partially because the modern models are often presented in the form of an analytically intractable set of stochastic difference equations. This spurs most researchers tackling the task of finding the parameters of such models to take the approach of numerical optimization based on a certain heuristic. The Method of Simulated Moments (MSM) is widespread, where parameter calibration is performed through minimizing the distance between a set of statistical moments for the actual observations and the simulated series [7].

Due to the complex dynamics inherent in HAMs, assuring the convergence of a distance metric to the global minimum is difficult. It necessitates evaluating the objective function numerous times through simulations, which may generate a substantial computational cost [22]. What is more, even finding the global minimum of a loss function may not bring the correct parameter set values, due to the stochastic component present in typical financial HAMs. Figure 1 exemplifies this problem using two stylized loss functions - noiseless and noisy. The global minimum at $X = 1000$ is clearly visible for the noiseless function. After adding noise to the loss function, the actual minimum appears in the vicinity of $X = 750$, although we would prefer our optimizer to still uncover $X = 1000$ as the optimum here. The fact of the loss function being noisy should clearly be taken into account in the HAM calibration process.

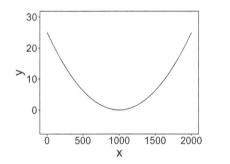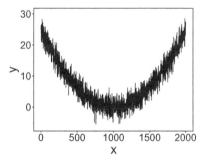

Fig. 1. Example of stylized noiseless (LHS) and noisy (RHS) loss functions.

This work proposes a new algorithm for calibrating HAM parameter values within the Method of Simulated Moments paradigm, called Filtered Neighborhoods. The algorithm evaluates the distance function for different parameter sets chosen with random search. The search area is gradually narrowed down through

---

[1] The notions of "calibration" and "validation" of HAMs do not have unambiguous definitions. Here they are used broadly, designating a process of arriving at HAM parameter values that provide the best fit of model output to some reference series.

the examination of the local neighborhoods of promising solutions. Parameter sets associated with large errors are filtered out from the process. In each iteration, new model calls are allotted to searching the areas picked with the help of an acquisition function often employed in Bayesian optimization.

The proposed Filtered Neighborhoods method is applied to calibrate a classic benchmark heterogeneous agent model [4]. Referencing the broad set of methods tested by [22] to solve the same task, it is shown that the new algorithm achieves better results than several competing HAM calibration methods, employing at most ca. 0.2% of the model calls allotted for the competition. The algorithm allows the usage of any distance metrics and objective functions. It can potentially be used for other noisy optimization problems.

2 Construction of the Proposed Calibration Approach

2.1 The Method of Simulated Moments

The family of methods applied in the calibration of agent-based models is broad. Recent surveys suggest different breakdowns of this family into multiple groups [11,21,22]. The Method of Simulated Moments is nevertheless prominently featured in the literature, as shown by [7]. It is also considered to be the most straightforward way for calibrating complex models [21].

MSM is a simulation-based method, where first a set of statistical moments for the actual data is picked. For financial HAMs, that actual data is typically a vector (time series) of prices or returns. Then multiple runs of the time series are simulated within the HAM, using different sets of parameter values. The resulting moments for the simulated data are then compared against the moments of the actual dataset. The parameter set best calibrated to the real data is chosen based on the value of the distance function between the moments of the simulated and the real samples.

However, the rather intuitive concept of MSM is not without its issues. Let us consider the following general representation of MSM, where we assume that the reference data at our disposal was generated by a process described perfectly by a given Agent-Based Model:

$$\hat{\theta} = arg \left\{ min\, \mathcal{D} \left(f(\theta, \varepsilon_r), f(\hat{\theta}, \varepsilon_s) \right) \right\} \tag{1}$$

where $\hat{\theta}$ denotes the calibrated set of parameter values based on the simulations and θ is the actual (unobservable) set of parameter values for the reference data. The value of function f is the vector of moments to be used in the calibration process. In accordance with the specification of the ABM being calibrated, function f takes a set of parameter values as an argument, alongside a noise term ε_r for the reference data and ε_s for the simulations. This notation clearly shows that even perfectly matching $\hat{\theta}$ to θ may not drive down the value of the distance function $\mathcal{D}$ to 0. What is more, it may not result in achieving the global minimum of the distance function for a given realization of the noise term.

This stems from the fact that $\mathcal{D}\left(f(\theta, \varepsilon_r), f(\hat{\theta}, \varepsilon_s)\right)$ is essentially a random variable with a (usually) unknown distribution. Conversely, finding a global minimum of the distance function does not guarantee minimizing the fitness function $\mathcal{F}$ which measures the distance between the true and the calibrated parameter set.

This dichotomy of working with two objective functions in the process of noisy calibration needs to be stressed to avoid confusion. In this paper, the distance function $\mathcal{D}$ denotes the distance between the true set of moments and the simulated set of moments. Its values are observable and it serves as the driver of the calibration process. However, the final goodness of fit of the calibration is evaluated using the fitness function $\mathcal{F}$, gauging the distance between the calibrated set of parameters and the true set of parameters. The actual goal in the process is to minimize the value of $\mathcal{F}$, and not necessarily $\mathcal{D}$. Importantly, in the simulation exercises, the true set of parameters is known, which enables calculating the actual value of $\mathcal{F}$. In empirical work, the value of the fitness function remains unobservable.

Chen, Chang and Du [6] remarked that this indirect inference about θ (using the observable $\mathcal{D}$ in place of the unobservable $\mathcal{F}$) can be impacted by a number of decisions a researcher has to make, including:

- the number of the moments,
- the selection of the moments,
- the (sub)set of model parameters being calibrated,
- the form of the distance function $\mathcal{D}$,
- the optimization algorithm used to arrive at $\hat{\theta}$,
- the computational budget, i.e. the number of model calls available during simulations.

This broad degree of subjectivity remains the main cause for criticism of MSM [11,22]. Therefore in this work we concentrate on developing just a single element of the process - the optimization algorithm for searching of $\hat{\theta}$ - considered to be an important research topic [21]. Other design choices for the MSM specification used in this paper will be made in accordance with common choices made in the literature.

2.2   Details of the Filtered Neighborhoods Optimization Algorithm

The proposed algorithm entails a modified random search procedure through a gradually narrowed search space. The narrowing is performed with the help of an acquisition function typically used in Bayesian optimization.

Random search is a widespread approach to hyper-parameter optimization in machine learning [2]. It can be applicable in this case, too, as loss functions for both machine learning models and HAMs are essentially nonlinear, noisy functions of multiple parameters. The first step of the proposed algorithm entails drawing an initial sample of n HAM parameter values to obtain an initial position for the later search. In order to ensure even sampling across the dimensions of the parameter space, quasi-random low-discrepancy Sobol sequences with Owen and Faure-Tezuka scrambling are employed [9,15]. Bergstra and Bengio [2] demonstrated that the use of the Sobol sequence in random search can be advantageous, especially if the generated number of observation points amounts to several hundred.

The second step of the algorithm filters out all parameter value sets with a high value of the distance function $\mathcal{D}$, leaving only β points with the lowest values of the distance metric for further examination. This is done for two reasons. Firstly, the noisiness of the objective function may be very complex. The relationship between specific parameter values and the distance metric can vary with the value of the objective function. As the goal of the optimization process is searching for minima, that relationship is crucial above all for the low values of the objective function. Secondly, limiting the sample to the points which have the potential of having the most informational value can be beneficial for computational time, without adversely affecting the results [16].

The third step entails picking the most promising areas of the parameter space in order to perform the next batch of simulations concentrated on such areas. Acquisition functions such as probability of improvement, expected improvement or lower confidence bound are often employed for a similar purpose in Bayesian optimization [25]. However, typically they are set up to pick only a single point from the parameter space before proceeding with the simulation. Also, the process depends strongly on the chosen prior [3]. Instead of assuming a prior distribution, we find α nearest neighbors of each set of parameter values in the sample[2] (including itself). In this way, β neighborhoods including α points each are created. Such local neighborhoods can then be characterized by the mean value of the distance metric and its standard error. Minimum and maximum parameter values for each of the points constitute the borders of the neighborhood's area. This setup allows the use of an acquisition function to rank the neighborhoods from the most to the least promising in the fourth step of the algorithm. Using a portfolio of multiple acquisition functions can also be beneficial [24]. It is not considered in this paper in order to keep the number of parameters in the algorithm as low as possible. For the centroid of the single most promising neighborhood in each optimization round, the value of the distance function $\mathcal{D}$ is recorded (a single additional model call per round).

Then, in the fifth step, the budget of n simulations is distributed among γ most promising neighborhoods, resulting in n/γ additional model calls per neighborhood.

The sixth step of the algorithm entails randomly picking the allocated number of new parameter set values for each of the neighborhoods within the designated borders, using the continuation of the Sobol sequence initialized in the first step. This approach is analogous to implicit averaging, advocated as one of the optimization strategies in noisy settings within the field of evolutionary optimization [23]. It assumes that gathering the values of the distance function in the neighborhood of previous parameter sets with a good fit can implicitly compensate for the noisiness of the function's values. This enables us to search a

---

[2] Search ranges for all parameters need to be scaled in the same way first.

broader range of parameter values and save model calls. Alternatively, averaging out the noise could be achieved by calculating the distance metrics for a single parameter set using different realizations of the noise term. We consider this alternative inefficient, which is supported by the results in Sect. 3.2.

Steps 2–6 can be repeated until the computational budget is fully exhausted. After the budget is finished, from the most promising neighborhoods we pick the centroid with the lowest value of the distance function $\mathcal{D}$ as the final set of calibrated parameters $\hat{\theta}$. Importantly, this is not the point that minimizes $\mathcal{D}$. Our calibration process may have uncovered parameter sets with a lower distance metric. However, the chosen parameter set offers a low value of $\mathcal{D}$ within a neighborhood of other low values of the distance function. We find that concentration purely on minimizing $\mathcal{D}$ may be counterproductive, as it does not necessarily lead to a lower value of the fitness function $\mathcal{F}$. Simulations in Sect. 3.2 provide an illustration of that issue.

The full proposed procedure is summarized as Algorithm 1. Algorithm parameters include:

- n: number of available model calls per 1 iteration of the algorithm,
- β: number of the parameter sets corresponding with the lowest loss function values to be left for further processing in each iteration,
- α: number of nearest neighbors used to create local neighborhoods,
- γ: number of the most promising neighborhoods to be allocated additional model calls.

Algorithm 1: Filtered Neighborhoods optimization algorithm

Result: calibrated parameter set $\hat{\theta}$
1   draw the initial sample of n HAM parameter sets;
2   **repeat**
3   leave only β parameter sets with the lowest values of the distance metric;
4   create β local neighborhoods, taking α nearest neighbors for each parameter set in the sample;
5   use the acquisition function to rank the neighborhoods;
6   **if** *performed model calls* $+n >=$ *computational budget* **then**
7   set the output parameter vector to the values associated with the neighborhood centroid having the lowest value of the loss function;
8   **break**
9   **end**
10   distribute the next-round budget of n simulations among γ most promising neighborhoods;
11   draw the next sample of n HAM parameter sets within the chosen neighborhood areas;
12 **until** *break*;

From a functional perspective, parameter n regulates the computational cost of the optimization process. β governs filtering out the least promising parameter sets, with lower values meaning stricter filtering. α and γ are directly related to the exploration-exploitation trade-off - a crucial issue in optimization [5]. Higher α increases the size of the neighborhoods, thus promoting exploration over exploitation. Similarly, higher γ spreads the available model calls over more neighborhoods, also promoting exploration.

3   Calibration Results of a Benchmark Agent-Based Model

3.1   The Brock and Hommes 1998 Model

The Brock and Hommes 1998 model (abbreviated further as BH98) is one of the most important benchmark financial ABMs [11]. It served as a platform for testing calibration methods in the recent years [18,22]. This HAM models the interaction of traders following h strategies, each having the following forecast $f_{h,t}$ for the market's price deviation x_t from the fundamental price p^*:

$$f_{h,t} = g_h x_{t-1} + b_h \tag{2}$$

g_h constitutes a trend parameter and b_h is a bias parameter. Market price dynamics within this model is characterized by the following 3 equations:

$$(1 + r)x_t = \sum_{h=1}^{H} n_{h,t}(g_h x_{t-1} + b_h) + \epsilon_t \tag{3}$$

$$n_{h,t} = \frac{\exp(\beta U_{h,t-1})}{\sum_{h=1}^{H} \exp(\beta U_{h,t-1})} \tag{4}$$

$$U_{h,t-1} = (x_{t-1} - Rx_{t-2})\left(\frac{g_h x_{t-3} + b_h - Rx_{t-2}}{a\sigma^2}\right) \tag{5}$$

Symbol r denotes the risk-free rate, with $R = 1 + r$. $U_{h,t-1}$ is the fitness measure of strategy h assessed at the beginning of period t. The noise term ϵ_t depicts the uncertainty about the economic fundamentals. a is the risk aversion parameter and β is the intensity of choice parameter. σ^2 denotes the variance of excess return. $n_{h,t}$ is the fraction of market participants following strategy h.

Figure 2 depicts the noisiness of BH98, presenting two time series output by the model with the same parameter values, but different random seeds.

3.2   Calibration Process and Results

The recent comparison of calibration methods by Platt [22] becomes our benchmark, as it tests multiple ABM calibration methods (MSM among them) on a few ABMs, including BH98. In the calibration process we focus on analyzing the performance of our optimization algorithm, relying on Platt for other choices necessary in the calibration process.

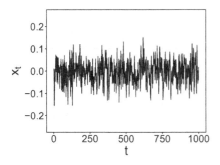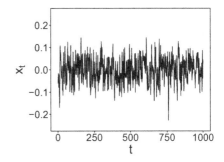

Fig. 2. Brock and Hommes 1998 model output with the same parameter values and different random seeds.

We follow the most complex case involving the BH98 model presented in [22], choosing:

- 4 strategies interacting in the model;
- 4 out of 11 model parameters to be calibrated;
- 7 moments used in calibration: the variance of the raw series, kurtosis of the raw series, autocorrelation coefficients of the raw series, absolute value series and squared series at lag 1, and the autocorrelation coefficients of the absolute value series and squared series at lag 5;

As pointed out by [6], calibrating only a subset of parameters is common practice due to the size of the search space growing exponentially with additional parameters. The chosen moments are supposed to constitute a measure of stylized facts observable in financial time series [21].

The distance function $\mathcal{D}$ used to evaluate the distance between the set of simulated moments and the moments for the "true" dataset is chosen to be the weighted sum of squared errors. Following [22], the weighting matrix is the inverse of the covariance matrix of the "true" moments (Newey-West estimator).

The final calibrated set of parameters obtained through the optimization algorithm is compared with the "true" parameter set using the following fitness function:

$$\mathcal{F}\left(\theta, \hat{\theta}\right) = ||\theta - \hat{\theta}||_2 \tag{6}$$

The experimental process assumes first generating 100 time series from BH98, 1,000 observations each. They are generated using a single set of model parameters and 100 different seeds for the random number generator. These will be treated as the "true" data and the optimization algorithm will be used to uncover the values of model parameters used when generating the data. Using 100 "true" datasets, differing just because of the random noise terms and not the model parameters, enables us to calculate confidence intervals for various algorithm performance statistics. This is a key difference between the experimental procedure used in this paper and [22], who uses only a single run of the BH98 as his "true" dataset.

We opt for a rather modest computational budget, assuming 5,500 to 6,000 model calls per 1 "true" dataset: 1 initialization round and 10 optimization rounds with 500 calls per round. Some algorithm setups use 1,000 model calls in the initialization round only. As in [22], each model call generates a simulated run of 1,000 observations, for which the 7 aforementioned moments are calculated and compared with the moments of a "true" dataset. In accordance with the suggestions of the algorithm, each call uses a different model parameter set and a different realization of the noise term. This differs from Platt [22], who generally uses 10 optimization rounds with 1,000 different parameter sets (compared to 500 in this paper), but also performs 250 Monte Carlo trials per each parameter set to average out the random noise. His technique drives the total budget to 2,500,000 model calls.

The "true" parameter set used in the calibration process follows [22] and is depicted in Table 1. Out of these parameters, the calibration process seeks the values of g_2, b_2, g_3 and b_3, treating the other parameters as known.

Table 1. True model parameters.

Parameter	g_1	b_1	g_2	b_2	g_3	b_3	g_4	b_4	r	β	σ
Value	0	0	0.6	0.2	0.7	−0.2	1.01	0	0.01	10	0.04

Within the experiment, 10 sets of Filtered Neighborhoods (FN) algorithm parameters are tested, keeping the computational budget stable with $n = 500$. Some algorithm setups utilize 1,000 instead of 500 model calls in the initial round only. The acquisition function employed in the algorithm is the well-known Probability of Improvement [3]. The parameter space is shown in Table 2.

Table 2. Filtered Neighborhoods algorithm parameter sets used in the calibration process.

Parameter	n	β	α	γ
Set 1	500	250	30	20
Set 2	500	250	10	20
Set 3	500	250	50	20
Set 4	500	100	30	20
Set 5	500	500	30	20
Set 6	500	250	30	10
Set 7	500	250	30	50
Set 8	500 (1,000 for initial draw)	250	30	20
Set 9	500 (1,000 for initial draw)	200	50	50
Set 10	500 (1,000 for initial draw)	100	50	50

Competing HAM calibration methods include:

- Method of Simulated Moments (MSM),
- Generalized Subtracted L-divergence, which measures the distance between distributions of patterns in the "true" and simulated time series [19] (GSL),
- Markovian Information Criterion (MIC), based on the Kullback-Leibler distance between the "true" and simulated data [1],
- Bayesian estimation approach presented by [12] (BE).

MSM, GSL and MIC are all optimized using either Particle Swarm Optimization (MSM/PS, GSL/PS, MIC/PS) or the approach of Knysh and Korkolis [17] (MSM/KK, GSL/KK, MIC/KK). Additionally, the newly proposed Filtered Neighborhoods algorithm is used within the Method of Simulated Moments (MSM/FN).

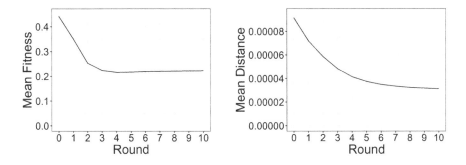

Fig. 3. Comparison of the mean fitness function $\mathcal{F}$ value and mean distance function value $\mathcal{D}$ per subsequent optimization round for all tested Filtered Neighborhoods parameter sets.

Figure 3 shows the mean fitness function $\mathcal{F}$ value for all 10 tested Filtered Neighborhoods parameter sets in each optimization round. Fitness for MSM/FN seems to converge already in round 4 at the value of 0.22. This is less than half of the error value achieved by the direct competitors, MSM/PS (0.45) and MSM/KK (0.54). What is more, instead of using 2,500,000 model calls, MSM/FN uses ca. 5,000–5,500 calls for 10 rounds - and only 2,000–2,500 calls if stopped after 4 rounds. Interestingly, continuing the optimization beyond round 4 does decrease the value of the distance function $\mathcal{D}$, but does not improve HAM parameter calibration accuracy.

The best discovered MSM/FN parameter set is Set 10 in Table 2. It is characterized by the increased number of model calls in the initial optimization round (1,000), prominent filtering of the dataset ($\beta = 100$), and leaning towards exploration ($\alpha = 50$, $\gamma = 50$). The average achieved fitness function value across 100 different realizations of the "true" time series was 0.172, with the 99% confidence interval being 0.152–0.193. This is a significantly better result than obtained by any of the competing methods, with the exception of Bayesian Estimation. However, it is not significantly worse than the performance of Bayesian Estimation, notwithstanding the lack of confidence intervals provided for the results presented in [22].

Table 3. Calibration results.

Method	g_2	b_2	g_3	b_3	$\mathcal{F}$
True θ	0.6	0.2	0.7	−0.2	−
MSM/FN	0.6096	0.1609	0.7037	−0.1598	**0.0570**
MSM/KK	0.8742	0.2424	0.2424	−0.2040	0.5352
MSM/PS	1	0.1941	0.5015	−0.2174	0.4469
GSL-div/PS	0.6047	0	0.5277	0	0.3312
GSL-div/KK	0.5575	0	0.5529	0	0.3216
MIC/KK	0.3663	0.2431	0.8766	−0.2387	0.2985
MIC/PS	0.8105	0.2278	0.7430	−0.2186	0.2173
BE	0.4651	0.2468	0.6251	−0.2388	0.1658

What is more, mean calibrated HAM parameter values for MSM/FN are very close to the "true" ones and compare favorably with the competition - as evident in Table 3. Fitness value for the mean parameters calibrated with MSM/FN is only 0.057. Due to the noisiness of the HAM model, in 12 out of 100 cases the value of the distance function $\mathcal{D}$ for the calibrated parameter set turns out to be lower than for the "true" parameter set.

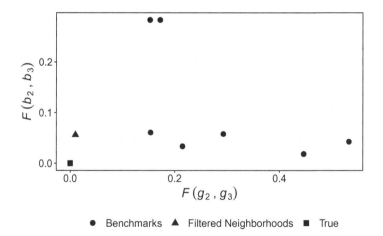

Fig. 4. Fitness metrics for the pairs of calibrated b and g parameters.

Figure 4 presents the fitness value of the examined models decomposed into two fitness metrics for both pairs of calibrated b and g parameters. On average, the Filtered Neighborhoods algorithm finds both g parameters with almost perfect accuracy. Most of the algorithm error comes from missing the "true" value of b parameters.

4   Conclusions and Further Research Directions

The newly proposed Filtered Neighborhoods optimization method for the Method of Simulated Moments clearly outperforms both Particle Swarm Optimization and the approach of Knysh and Korkolis within the context of the BH98 model. The quality of the calibration is superior to most of the other benchmarks as well, and similar to that of Bayesian Estimation. All that is achieved while using roughly 500 times less model calls, which saves a considerable amount of computational time.

The advantage of the new method holds for a range of different algorithm parameter values. Stronger filtering and pursuing exploration seem to be beneficial for calibration accuracy.

In this paper, the efficiency of our method has been proved only for the BH98 model and for a specific calibration issue involving 4 parameters. Still, as postulated in [21], the best choice of a calibration framework for an agent-based model might be model-specific. Our method seems to be well-suited to the problem at hand, and the conclusions seem rather robust to MSM/FN parameter changes. However, more work is needed to examine the optimal choice of the FN algorithms parameter values. Other acquisition functions (including function portfolios) should also be tested. A much larger effort altogether (and considerable computation costs) would be necessary to prove the Filtered Neighborhoods algorithm's reliability over a broader range of parameter sets of BH98, as well as other important HAMs - or any other noisy optimization problems. The results for BH98 definitely make this avenue worth pursuing.

On a more general note, the presented calibration results indicate how a long pursuit of the minimum of the observable distance function may not contribute to finding the minimum of the unobservable fitness function - the true purpose of the whole calibration process. The relationship between the distance metric and the fitness metric should be deeply examined and taken into account in the optimization process. This may not only improve calibration accuracy, but lessen the necessary computational workload.

References

1. Barde, S.: A practical, accurate, information criterion for Nth order Markov processes. Comput. Econ. **50**(2), 281–324 (2016). https://doi.org/10.1007/s10614-016-9617-9
2. Bergstra, J., Bengio, Y.: Random search for hyper-parameter optimization. J. Mach. Learn. Res. **13**, 281–305 (2012)

3. Brochu, E., Cora, V.M., de Freitas, N.: A tutorial on Bayesian optimization of expensive cost functions, with application to active user modeling and hierarchical reinforcement learning. CoRR abs/1012.2599 (2010)
4. Brock, W.A., Hommes, C.H.: Heterogeneous beliefs and routes to chaos in a simple asset pricing model. J. Econ. Dyn. Control **22**, 1235–1274 (1998). https://doi.org/10.1016/S0165-1889(98)00011-6
5. Chen, J., Xin, B., Peng, Z., Dou, L., Zhang, J.: Optimal contraction theorem for exploration-exploitation tradeoff in search and optimization. IEEE Trans. Syst. Man Cybern. Part A Syst. Hum. **39**(3), 680–691 (2009). https://doi.org/10.1109/TSMCA.2009.2012436
6. Chen, S.H., Chang, C.L., Du, Y.R.: Agent-based economic models and econometrics. Knowl. Eng. Rev. **27**(2), 187–219 (2012). https://doi.org/10.1017/S0269888912000136
7. Chen, Z., Lux, T.: Estimation of sentiment effects in financial markets: a simulated method of moments approach. Comput. Econ. **52**(3), 711–744 (2016). https://doi.org/10.1007/s10614-016-9638-4
8. Dieci, R., He, X.Z.: Heterogeneous agent models in finance. In: Hommes, C., LeBaron, B. (eds.) Handbook of Computational Economics, vol. 4, chap. 5, pp. 257–328. Handbooks in Economics, Elsevier (2018). https://doi.org/10.1016/bs.hescom.2018.03.002
9. Dutang, C., Savicky, P.: randtoolbox: Generating and Testing Random Numbers (2020). R package version 1.30.1
10. Ellen, S., Verschoor, W.F.C.: Heterogeneous beliefs and asset price dynamics: a survey of recent evidence. In: Jawadi, F. (ed.) Uncertainty, Expectations and Asset Price Dynamics. DMEEF, vol. 24, pp. 53–79. Springer, Cham (2018). https://doi.org/10.1007/978-3-319-98714-9_3
11. Fagiolo, G., Guerini, M., Lamperti, F., Moneta, A., Roventini, A.: Validation of agent-based models in economics and finance. In: Beisbart, C., Saam, N.J. (eds.) Computer Simulation Validation. SFMA, pp. 763–787. Springer, Cham (2019). https://doi.org/10.1007/978-3-319-70766-2_31
12. Grazzini, J., Richiardi, M.G., Tsionas, M.: Bayesian estimation of agent-based models. J. Econ. Dyn. Control **77**, 26–47 (2017). https://doi.org/10.1016/j.jedc.2017.01.014
13. He, X.Z.: Recent developments in asset pricing with heterogeneous beliefs and adaptive behaviour of financial markets. In: Bischi, G., Chiarella, C., Sushko, I. (eds.) Global Analysis of Dynamic Models in Economics and Finance: Essays in Honour of Laura Gardini, pp. 3–34. Springer, Berlin (2013). https://doi.org/10.1007/978-3-642-29503-4_1
14. Hommes, C.H.: Heterogeneous agent models in economics and finance. In: Tesfatsion, L., Judd, K.L. (eds.) Handbook of Computational Economics, vol. 2, chap. 23, pp. 1109–1186. Handbooks in Economics, Elsevier (2006). https://doi.org/10.1016/S1574-0021(05)02023-X
15. Hong, H.S., Hickernell, F.J.: Algorithm 823: implementing scrambled digital sequences. ACM Trans. Math. Softw. **29**(2), 95–109 (2003). https://doi.org/10.1145/779359.779360
16. Klein, A., Falkner, S., Bartels, S., Hennig, P., Hutter, F.: Fast Bayesian Optimization of Machine Learning Hyperparameters on Large Datasets. In: Singh, A., Zhu, J. (eds.) Proceedings of the 20th International Conference on Artificial Intelligence and Statistics. Proceedings of Machine Learning Research, vol. 54, pp. 528–536. PMLR, Fort Lauderdale (2017)

17. Knysh, P., Korkolis, Y.: Blackbox: a procedure for parallel optimization of expensive black-box functions. CoRR abs/1605.00998 (2016)
18. Kukacka, J., Barunik, J.: Estimation of financial agent-based models with simulated maximum likelihood. J. Econ. Dyn. Control **85**, 21–45 (2017). https://doi.org/10.1016/j.jedc.2017.09.006
19. Lamperti, F.: An information theoretic criterion for empirical validation of simulation models. Econom. Stat. **5**, 83–106 (2018). https://doi.org/10.1016/j.ecosta.2017.01.006
20. LeBaron, B.: Agent-based computational finance. In: Tesfatsion, L., Judd, K.L. (eds.) Handbook of Computational Economics, vol. 2, chap. 24, pp. 1187–1233. Handbooks in Economics, Elsevier (2006). https://doi.org/10.1016/S1574-0021(05)02024-1
21. Lux, T., Zwinkels, R.C.J.: Empirical validation of agent-based models. In: Hommes, C., LeBaron, B. (eds.) Handbook of Computational Economics, vol. 4, chap. 8, pp. 437–488. Handbooks in Economics, Elsevier (2018). https://doi.org/10.1016/bs.hescom.2018.02.003
22. Platt, D.: A comparison of economic agent-based model calibration methods. J. Econ. Dyn. Control **113** (2020). https://doi.org/10.1016/j.jedc.2020.103859
23. Rakshit, P., Konar, A., Das, S.: Swarm and evolutionary computation. J. Am. Stat. Assoc. **33**, 18–45 (2017). https://doi.org/10.1016/j.swevo.2016.09.002
24. Shahriari, B., Swersky, K., Wang, Z., Adams, R.P., de Freitas, N.: Taking the human out of the loop: a review of Bayesian optimization. Proc. IEEE **104**(1), 148–175 (2016). https://doi.org/10.1109/JPROC.2015.2494218
25. Snoek, J., Larochelle, H., Adams, R.P.: Practical Bayesian optimization of machine learning algorithms. In: Pereira, F., Burges, C.J.C., Bottou, L., Weinberger, K.Q. (eds.) Advances in Neural Information Processing Systems, vol. 25, pp. 2951–2959. Curran Associates, Inc. (2012)

Computational Collective Intelligence

A Method for Improving Word Representation Using Synonym Information

Huyen Trang Phan[1] ⓘ, Ngoc Thanh Nguyen[2] ⓘ, Javokhir Musaev[1] ⓘ,
and Dosam Hwang[1](✉) ⓘ

[1] Department of Computer Engineering, Yeungnam University,
Gyeongsan, South Korea
{javokhirmuso,dshwang}@yu.ac.kr
huyentrangtin@ynu.ac.k
[2] Department of Applied Informatics, Wroclaw University of Science and Technology,
Wroclaw, Poland
Ngoc-Thanh.Nguyen@pwr.edu.pl

Abstract. The emergence of word embeddings has created good conditions for natural language processing used in an increasing number of applications related to machine translation and language understanding. Several word-embedding models have been developed and applied, achieving considerably good performance. In addition, several enriching word embedding methods have been provided by handling various information such as polysemous, subwords, temporal, and spatial. However, prior popular vector representations of words ignored the knowledge of synonyms. This is a drawback, particularly for languages with large vocabularies and numerous synonym words. In this study, we introduce an approach to enrich the vector representation of words by considering the synonym information based on the vectors' extraction and presentation from their context words. Our proposal includes three main steps: First, the context words of the synonym candidates are extracted using a context window to scan the entire corpus; second, these context words are grouped into small clusters using the latent Dirichlet allocation method; and finally, synonyms are extracted and converted into vectors from the synonym candidates based on their context words. In comparison to recent word representation methods, we demonstrate that our proposal achieves considerably good performance in terms of word similarity.

Keywords: Synonym words · Word embeddings · Synonym vector

1 Introduction

Embeddings, similar to vector models that capture relational meaning, are more fine-grained than just a string or index; in particular, embeddings are good at modeling similarities/analogies. To apply embeddings, we only need to download

M. Paszynski et al. (Eds.): ICCS 2021, LNCS 12744, pp. 333–346, 2021.
https://doi.org/10.1007/978-3-030-77967-2_28

and use them. They are useful tools in practice and are more popular in several fields, specifically, word embeddings in the natural language processing area. Word embeddings represent word meanings from corpus statistics.

The use of word embeddings has been widely increasing in applications related to natural language processing, such as sequence tagging [15], machine translation [21], language understanding [18], text classification [13] and sentiment analysis [20]. In addition, word embeddings are useful in machine learning and deep learning algorithms. Therefore, several pre-trained word embeddings exhibit state-of-the-art performance. These methods are applied in several studies, such as Word2Vec [16], FastText [12], GloVe [17], and BERT [4]. Word2Vec[1] includes two models: skip-gram and continuous bag-of-words. The first model predicts the surrounding words for the current word, meanwhile the second model uses the context words to predict the current word [28,29]. This model produces the same vector for a word irrespective of its meaning and context. GloVe[2], used for word representations, leverages the statistics of word occurrences in the corpus and uses a neural network to represent the meaning of such statistics [28,29]. The idea of this model is similar to that of latent semantic analysis. It captures the global and local contexts of the word. FastText[3] is improved from the Word2Vec skip-gram model by considering the subword information. A word is represented as a sum of character n-gram embeddings that appeared in the word. The FastText model outperforms skip-gram model in most scenarios and datasets when dealing with syntactic tasks [29]. However, for semantic tasks, the FastText model is less accurate than the skip-gram model [29]. It can generate out-of-vocabulary word embeddings. BERT[4] combines several tasks. It predicts masked words in a sentence and indicates whether sentence A is followed by sentence B, as embedding combines several hidden layers of the network [28,29]. In addition, BERT learns relationships between sentences and predicts whether sentence B is the actual sentence that follows sentence A or whether it is a random sentence. However, the above methods have the same limitation that ignores the impact of synonyms when representing words [28].

According to WordNet, synonym words are introduced as *"words that denote the same concept and are interchangeable in many contexts"*. To clearly understand the importance of synonyms in the tweet sentiment analysis task, we consider the following small example. Assume that there are two tweets as follows: Tweet 1: *"The color of this phone is outdated"*. Tweet 2: *"The color of this phone is outmoded"*. It can be seen that the words "outdated" and "outmoded" have the same meaning, but they have entirely different characters in words. These words are called synonyms. However, most of the previous methods used different vectors to represent them. This leads to a misunderstanding of the word meaning by the computer. Therefore, the quality of word representations is decreased, and the performance of applications is also affected. To address the aforementioned synonyms problem, we introduce a model to enrich the

---

[1] https://code.google.com/archive/p/word2vec/.
[2] https://nlp.stanford.edu/projects/glove/.
[3] https://fasttext.cc/docs/en/crawl-vectors.html.
[4] https://mccormickml.com/2019/05/14/BERT-word-embeddings-tutorial/.

vector representation of words by adding the synonym information. This proposal focuses on the extraction and presentation of synonyms based on their context words by considering three main steps: First, the context words of the synonym candidates are extracted; second, these context words are grouped into small clusters using the latent Dirichlet allocation (LDA) method; and finally, synonyms are extracted and converted into vectors from the synonym candidates based on their context words. In comparison to recent word representation methods, we demonstrate that our proposed method achieves state-of-the-art performance on a given task in terms of word similarity. Our proposal is motivated by the distributional hypothesis [9] that says: *"words that occur in the same contexts tend to have similar meanings"* and the basic hypothesis investigated by Rubenstein *et al.* [24]: *"there is a positive relationship between the degree of synonymy (semantic similarity) existing between a pair of words and the degree to which their contexts are similar"*.

The remainder of this paper is organized as follows: The literature regarding sentiment analysis methods is summarized in Sect. 2. We describe the research problem in Sect. 3 and introduce the proposed method in Sect. 4. The information related to experimental results, such as data acquisition, evaluation method, and result discussion, is provided in Sect. 5. The conclusions and future work are discussed in the final section.

2 Related Works

Recently, several methods have been published to enrich word embeddings. In this section, we represent certain recent and outstanding methods by discussing their processes, advantages, and disadvantages.

Svoboda *et al.* introduced two approaches to improve the quality of vector representations. In [27], the authors enriched word embeddings by considering global information. In [26], the authors improved word meaning representations using Wikipedia categories. Jianqiang *et al.* [11] provided enriching word embedding approaches by using the word vectors in the GloVe collection, n-grams of words, and the sentiment score of words. The model obtained good results. However, the authors did not compare the performance with other studies on the same datasets. Meanwhile, Hassan *et al.* [14] converted words into feature vectors of real values. These features include the semantic and syntactic information. However, this method only considered the word's surface features ignoring the impact of the in-depth features. Therefore, in [1], the authors improved word embeddings by adding the information related to the features, such as generic words and sentiment specific words. Rezaeinia *et al.* [22] increased the performance of available word embeddings by considering the information of words, such as the part-of-speech (POS) tag, lexicon, and position. Nevertheless, the authors ignored the subword information, global information, and temporal and spatial information. Therefore, Bojanowski *et al.* [3] proposed a new approach based on the skip-gram model. Each word was represented as a bag of character n-grams to overcome the limitation of ignoring the morphology of words.

A vector representation was associated with each character n-gram, and words were represented as the sum of these representations. This method could quickly train models on large corpora and compute word representations for words that did not appear in the training data. Besides, Gong *et al.* [6] proved that the meaning of a word is closely linked to sociocultural factors that can change over time and location, resulting in corresponding meaning changes. Therefore, they presented a model for learning word representation conditioned on time and location to solve the problem of ignoring the previous methods' temporal or spatial information. In addition, to capture meaning changes over time and location, the authors required that the resulting word embeddings retained salient semantic and geometric properties. This model was trained on time- and location-stamped corpora and used both quantitative and qualitative evaluations to capture semantics across time and locations. Whatever, the discussed methods did not consider the impact of the polysemous words in word embeddings. To solve this limitation, Gou *et al.* [7] presented an approach to convert the polysemous into vectors by clustering the context words. This method is the basis of our improvement method. The difference is selecting the word embedding model, extracting and clustering context words, and determining the parameters' value to predict the synonyms.

Notably, the prior methods did not consider the impact of synonyms when representing words in the vector space. Therefore, we chose to study this problem.

3   Model

Let $T = \{t_1, t_2, ..., t_n\}$ represent a set of tweets, where $W_t = \{w_1, w_2, ..., w_h\}$ represent a set of words existing in tweet t. Let $W = \{w_1, w_2, ..., w_m\}, (m > h)$ represent a set of words in the vocabulary, where $W = \cup_{t \in T}\{W_t\}$. Let $V = \{V_{w_1}, V_{w_2}, ..., V_{w_m}\}$, where $V_{w_j} = \{v_{w_j}^1, v_{w_j}^2, ..., v_{w_j}^q\}$, represent a set of context words that surround the words in set W.

3.1   General Model

In this section, we briefly review the skip-gram model introduced by Mikolov *et al.* [16]. Given a word vocabulary W, the goal of the skip-gram model is to learn a vector representation for each word w. In other words, the aim of this model is to maximize the average log probability as follows:

$$\frac{1}{m}\sum_{i=1}^{m}\sum_{j=1}^{q} log\, p(v_{w_i}^j | w_i) \tag{1}$$

The probability of observing a context word $v_{w_i}^j$, given w_i, is parameterized. Let Sc denote a scoring function that maps pairs of (word, context) to scores in $\mathbb{R}$. The problem is to predict context words. For the word w_i, all context words as positive examples and sample negatives from vocabulary are considered [3]. For a target context word $v_{w_i}^j$, using the binary logistic loss, the negative log-likelihood is shown as follows:

$$log(1 + e^{-Sc(w_i, v_{w_i}^j)}) + \sum\nolimits_{n \in \mathcal{N}_{i,j}} log(1 + e^{Sc(w_i, n)}) \qquad (2)$$

where $\mathcal{N}_{i,j}$ represents a set of negative examples sampled from the vocabulary. By denoting the logistic loss function $\ell : \chi \rightarrow log(1 + e^{-\chi})$, Eq. 1 is rewritten as follows:

$$\sum_{i=1}^n \left[\sum_{j=1}^k \ell(Sc(w_i, v_{w_i}^j)) + \sum_{n \in \mathcal{N}_{i,j}} \ell(-Sc(w_i, n)) \right] \qquad (3)$$

Assume two vectors z_{w_i} and $z_{v_{w_i}^j}$, corresponding to word w_i and context word $v_{w_i}^j$, respectively. Then the score Sc is computed as the scalar product between word w_i and context word $v_{w_i}^j$ using the following equation:

$$Sc(w_i, v_{w_i}^j) = z_{w_i}^\top z_{v_{w_i}^j} \qquad (4)$$

3.2   Synonym Representation Model

Given a word $w \in \mathcal{W}$, let $\mathcal{S}_w = \{s_w^1, s_w^2, ..., s_w^g\}$ represents a set of synonym words of word w, where each synonym word $s_w^g \in \mathcal{S}_w$ is associated to a vector representation $z_{s_w^g}$. A synonym word is represented by the sum of the vector representations of its contexts. Thus, the scoring function of our model is shown as follows:

$$Sc(s_w, v_w) = \sum\nolimits_{v_w \in \mathcal{V}_w} z_{v_w}^\top z_{s_w} \qquad (5)$$

Similar to the Word2Vec skip-gram model, our model decides whether the target word is a synonym word or not by using the context words of the target word. In addition, to identify the synonym word, we use a context window of size between one and five words to decide the context words of the synonym candidate word. Next, we formally define the problems related to the enrichment of word embeddings by considering synonym information. As a computational problem, the improvement of word embeddings assumes that the input is a set of tweets $\mathcal{T} = \{t_1, t_2, ..., t_n\}$.

For $t \in \mathcal{T}$: let $\mathcal{W}_t = \{w_1, w_2, ..., w_h\}$ represent a set of words appearing in t. Let $\mathcal{W} = \{w_1, w_2, ..., w_m\}, (m > h)$ represent a set of words in the vocabulary, where $\mathcal{W} = \cup_{t \in \mathcal{T}} \{\mathcal{W}_t\}$. For $w \in \mathcal{W}$: let $\mathcal{U} = \{u_{w_1}, u_{w_2}, ..., u_{w_m}\}$ be a set of baseline word embeddings of words in $\mathcal{W}$, and u_{w_j} $(j = 1, ..., m)$ denote the vector of word w_j. For $w \in \mathcal{W}_t$ and $u_w \in \mathcal{U}$: let $\mathcal{M}$ represent a word embeddings mapping table, $\mathcal{M} = \{[w_1, u_{w_1}], [w_2, u_{w_2}], ..., [w_m, u_{w_m}]\}$. For $w \in \mathcal{W}_t$ and $\mathcal{V}_w \in \mathcal{V}$: let $\mathcal{P}$ represent a context words mapping table of words, $\mathcal{P} = \{[w_1, \mathcal{V}_{w_1}], [w_2, \mathcal{V}_{w_2}], ..., [w_m, \mathcal{V}_{w_m}]\}$. From $\mathcal{P}$ and $\mathcal{M}$: let $\mathcal{G} = \{\mathcal{G}_{w_1}, \mathcal{G}_{w_2}, ..., \mathcal{G}_{w_m}\}$ represent a set of clusters of context words, in which $\mathcal{G}_{w_j} = \{g_{w_j}^1, g_{w_j}^2, ..., g_{w_j}^k\}$ $(j = 1, ..., m)$; $g_{w_j}^i = \{v_{w_j}^1, v_{w_j}^2, ..., v_{w_j}^h\}$, $(i = 1, ..., k)$ represents a set of context words in the i-th cluster of the word w_j.

Definition 1. w_i and w_j $(i \neq j)$ are called synonyms if:

- $w_i \in t_x$ and $w_j \in t_y$, $(x \neq y)$
  And
- $\mathcal{G}_{w_i} = \mathcal{G}_{w_j}$

Definition 2. The word embedding of synonym word s, denoted by z_s, is a transform of synonym s into a d-dimensional vector by calculating the average vector of context words of this synonym. The synonym word embedding z_s is defined as follows:

$$z_s = \mathcal{AVG}(vector(\mathcal{M}, \mathcal{G}_s)) \tag{6}$$

where $\mathcal{AVG}$ is a function to calculate the average vector, vector is a mapping function that is used to map each context word into one word embedding.

3.3   Research Question

In this study, we attempt to answer the main question: *How to determine the vector representations of synonym words?* This question is divided into the following two questions:

1. *How to identify the synonym words by using their context words?*
2. *How to convert the synonym words into numerical vectors?*

4   Proposed Method

In this section, we present a methodology to enrich word embeddings by adding synonym information. The workflow of our method is illustrated in Fig. 1.

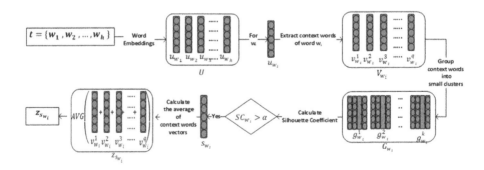

Fig. 1. Workflow of the proposed method.

Our proposed method consists of three main steps. First, the context words of the synonym candidates are extracted using a context window to scan the entire corpus. Second, these context words are grouped into small clusters using the LDA method. Finally, synonyms are extracted and converted into vectors from the synonym candidates based on their context words.

4.1   Word Embeddings

Word embeddings are created using the available text representation models that are used to convert words in a corpus into a vector space. Here, we used the model introduced in [20] to obtain word embeddings for our corpus. The detailed steps to create these vectors are presented in Fig. 2. Its components are described as follows.

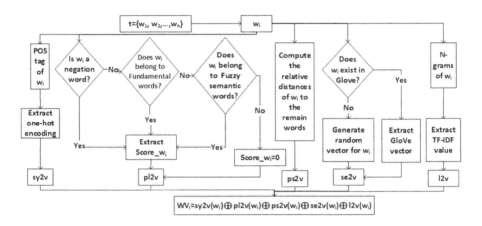

Fig. 2. The word embeddings model architecture.

Parameter *l2v* denotes the lexicon vector. To create the lexicon vector of a word, first, n-grams starting from this word, such as 1-gram, 2-grams, and 3-grams, are extracted. Then, the term frequency-inverse document frequency (TF-IDF) value of these n-grams is calculated. Finally, the TF-IDF values of n-grams are concatenated into one vector.

Parameter *sy2v* denotes the word-type vector of a word. This vector is built as follows: First, the POS tag of this word is identified. Then, this word is converted into a one-hot encoding vector based on the position of the corresponding POS tag.

Parameter *se2v* denotes the semantic vector. This vector is created based on the GloVe embeddings [17]. If this word exists in the GloVe dataset, the word vector of this word is extracted and assigned to the semantic vector. If not, a random vector of this word is created and assigned to the semantic vector.

Parameter *pl2v* denotes the polarity sentiment vector of a word. To create this vector, first, the kind of word of this word is determined. Then, the sentiment score of this word is calculated. Finally, the polarity sentiment vector is created based on this sentiment score.

Parameter *ps2v* denotes the position vector. To build the position vector of the word, first, the position of this word is extracted by calculating the distances from this word to the remaining words in a tweet. Then, the position vector is created based on these distances.

4.2 Context Words Extraction

For each w : let $region(w, d)$ represent a context region of word w with the region length as $2 \times r + 1$, where r denotes the window size for a region. In this study, synonym words are identified via the position of their context words. The aim of this phase is to find set $\mathcal{V} = \{\mathcal{V}_{w_1}, \mathcal{V}_{w_2}, ..., \mathcal{V}_{w_m}\}$. Therefore, any word in a region of text can become the context word of a target word. The context words set is determined according to the following equation.

$$\mathcal{V}_{w_i} = \bigcup_{r=1}^{2 \times r + 1} region(w_i, r) \tag{7}$$

Hence,

$$\mathcal{V} = \bigcup_{i=1}^{h} \mathcal{V}_{w_i} \tag{8}$$

The positions of context words of word w are determined determined according to Algorithm 1.

Algorithm 1. Context words extraction

Input: $\mathcal{M}$;
Output: $\mathcal{P}$;
1: **for** $i = 1$ to m **do**
2:    **for** each w_i **do**
3:       $w_c := w_i$;
4:       **for** $\theta = 1$ to $2 \times r + 1$ **do**
5:          if $w_c \in region(w_i, \theta)$; then insert w_c into $\mathcal{V}_{w_i}$;
6:       **end for**;
7:    **end for**;
8:    insert $\mathcal{V}_{w_i}$ into $\mathcal{V}$;
9: **end for**;
10: **for** $i = 1$ to m **do**
11:    **for** $\mathcal{V}_{w_i} \in \mathcal{V}$ **do**
12:       $\mathcal{P} = \{[w_i, \mathcal{V}_{w_i}]\}$
13:    **end for**;
14: **end for**;
15: **return** $\mathcal{P} = \{[w_1, \mathcal{V}_{w_1}], [w_2, \mathcal{V}_{w_2}], ..., [w_m, \mathcal{V}_{w_m}]\}$

4.3 Clustering Context Words

Given a set of context vocabularies denoted by $\mathcal{V} = \{\mathcal{V}_{w_1}, \mathcal{V}_{w_2}, ..., \mathcal{V}_{w_m}\}$, assuming that we have a target word w and its context vocabulary $\mathcal{V}_w = \{v_w^1, v_w^2, ..., v_w^q\}$ in the mapping table $\mathcal{P}$. In this phase, we use the LDA model [2] to group the context words of the target word w into small clusters, denoted by $\mathcal{G}_w = \{g_w^1, g_w^2, ..., g_w^k\}$, where $g_w^i = \{v_w^1, v_w^2, ..., v_w^h\}, (h < q$ and $i = 1, .., k)$ represent a set of words that are grouped into the i-th class of the context vocabulary of the word w. The LDA model can be generated using the following process

[8,19]: Let m denote the size of the vocabulary and n the total number of context vocabularies in $\mathcal{V}$. A statistical topic model represents the words in a collection of tweets as mixtures of k topics, words within context vocabularies $w_{e,q}, (e = 1, ..., m; q = 1, ..., n)$ are observed variables while the probabilistic distribution over words of each latent topic $\varphi_l (l = 1, ..., k)$ with hyper parameter γ, the topic distribution per tweet $\theta_e, (e = 1, ..., m)$ with hyperparameter δ and the perword topic assignment $z_{e,q}$ are hidden variables. For each tweet, the words are created by the following steps: First, a distribution over topics is randomly selected. A topic is randomly selected for each word in the tweet based on the distribution over topics. Second, the hidden random variables (φ_l and θ_e) are not observed that could be learned through Gibbs sampling method[5] via maximizing the probability $p(\mathcal{V}|\delta, \gamma)$ as the following equation:

$$p(\mathcal{V}|\delta, \gamma) = \prod_{e=1}^{m} \int p(\theta_e|\delta)(\prod_{q=1}^{n} \sum_{z_{e,q}} p(z_{e,q}|\theta_e)p(w_{e,q}|z_{e,q}, \gamma))d\theta_e \qquad (9)$$

The LDA model provides the outputs including k sub-clusters of context words that belong to the given cluster $G_w = \{g_w^1, g_w^2, ..., g_w^k\}$, the word distribution per topic $\varphi_l, (l = 1, ..., k)$ and the topic distribution per the context vocabulary $\theta_e, (e = 1, ..., m)$. The steps to cluster the context words are presented as the following Algorithm 2.

Algorithm 2. Clustering context words

Input: $\mathcal{P}$;
Output: $G_{w_l}, \varphi_{w_l}, \theta_{w_l}$;
1: **for** $l = 1$ to k **do**
2:     **for** $e = 1$ to m **do**
3:         $g_{w_l}^l, \varphi_l, \theta_e = \text{LDA}(\mathcal{P})$;
4:         insert $g_{w_l}^l$ into G_{w_l};
5:     **end for**;
6:     assign φ_l to φ_{w_l};
7:     assign θ_e to θ_{w_l};
8: **end for**;
9: **return** $G_{w_l}, \varphi_{w_l}, \theta_{w_l}$

4.4   Synonym Words Extraction

In this study, the synonym words are extracted by calculating the Silhouette Coefficient [23]. Thus, for a cluster of context words of a target word $g_{w_l}^l, (e = 1, ..., m; l = 1, ..., k)$, we have to calculate the Silhouette coefficient of

---

[5] https://gist.github.com/mblondel/542786#file-lda_gibbs-py.

each $w_e \in g_{w_e}^l$. Let SC_{w_e} denote the Silhouette coefficient of word w_e in cluster $g_{w_e}^l$.

$$SC_{w_e} = \frac{1}{m} \sum_{e=1}^{m} (Sw_{g_{w_e}^l}) \tag{10}$$

where

$$Sw_{g_{w_e}^l} = \frac{b_{w_e} - a_{w_e}}{max(a_{w_e}, b_{w_e})} \tag{11}$$

where

$$a_{w_e} = \frac{1}{n_l - 1} \sum_{w_f \in g_{w_e}^l (f \neq e)} \mathcal{D}_{ef} \tag{12}$$

$$b_{w_e} = min_{h \neq l} \left(\frac{1}{n_h} \sum_{w_f \in g_{w_e}^h} \mathcal{D}_{ef} \right) \tag{13}$$

where

$$\mathcal{D}_{ef} = \sqrt{\sum_{e,f=1}^{m} (Q_{w_e} - Q_{w_f})^2} \tag{14}$$

$$Q_{w_e} = \varphi_{w_e} \times \theta_{w_e}; Q_{w_f} = \varphi_{w_f} \times \theta_{w_f} \tag{15}$$

where $\mathcal{D}_{ij}$ is the Euclidean distances for all pairs of the words in cluster $g_{w_e}^l$; $n_l (n_h)$ is the number of words in the l-th (h-th) cluster. The value of the Silhouette Coefficient is in the interval $[-1,1]$. A higher value implies a better assignment of words into clusters. Therefore, in this study, a word is decided as a synonym word when the value of the Silhouette coefficient is equal to 1. The steps to determine the synonym words are illustrated in Algorithm 3:

Algorithm 3. Synonym words extraction

Input: $G_{w_e}, \varphi_{w_e}, \theta_{w_e}$;
Output: S_{w_e};
1: **for** $l = 1$ to k **do**
2:    **for** $e = 1$ to m **do**
3:       $SC_{w_e} = Silhouette\_Coefficient(g_{w_e}^l)$;
4:    **end for**;
5:    **if** $SC_{w_e} > \alpha$ **then**
6:       w_e is determined as a synonym word;
7:       $s_{w_e} := w_e$;
8:    **end if**
9:    insert s_{w_e} into S_{w_e};
10: **end for**;
11: **return** S_{w_e};

4.5 Synonym Words Representation

Using the aforementioned steps, the synonym words are extracted from the tweets. Next, we have to determine a way to convert these synonym words into numerical vectors. In this study, the synonym words are represented as vectors by calculating their context words' average vectors. The equation to calculate the context words' average vectors is described in Definition 2. The overall algorithm of the synonym word representation is presented in the following Algorithm 4.

Algorithm 4. Synonym words representation

Input: a set of synonym words S_{w_ℓ};
     a set of context words of synonym word s_{w_ℓ}, denoted by $\mathcal{G}_{s_{w_\ell}}$, where $s_{w_\ell} \in S_{w_\ell}$;
Output: $z_{s_{w_\ell}}$;
1: **for** $e = 1$ to m **do**
2:    $z_{s_{w_\ell}} = \mathcal{AVG}(vector(\mathcal{M}, \mathcal{G}_{s_{w_\ell}}))$;
3: **end for**;
4: **return** $z_{s_{w_\ell}}$;

---

5 Experiment

5.1 Data Acquisition

The proposed method was applied to tweet data. The tweets in Semeval-2013[6] were used to train our proposal. Then, the unnecessary factors in tweets, such as punctuation, retweet marks, URLs, hashtags, and query terms were discarded. The Python emoji package[7] was used to replace each emoji with descriptive text. Tweets often include acronyms, spelling errors, and symbols. It is necessary to correct them. We fixed these spellings using the Python-based Aspell library[8]. In addition, to evaluate the performance of our method, we experimented with three English word datasets, namely, WordSim-353 [5], RG-65 [24], and SimLex-999 [10]. These datasets were obtained from Svoboda et al. [27][9]. WordSim-353 includes 353 word pairs, including both concepts and named entities. RG-65 includes 65 word pair similarities. The SimLex-999 dataset is composed of 999 word pairs, 666 of which are noun pairs.

5.2 Evaluation Method

We evaluated the performance of our proposal for the task of word similarity (relatedness). This evaluation method was implemented by computing Spearman's rank correlation coefficient [25] between annotators and the obtained vectors of our system. Furthermore, to prove the quality of our approach, we compared our synonym word embedding model with corresponding state-of-the-art models by implementing the following baseline methods: Baseline 1: A Word2Vec model that is trained on the entire corpus without considering the synonym information. Baseline 2: A GloVe model that is also trained on the entire corpus without considering the synonym information. Baseline 3: A text representation model regarding tweets containing fuzzy sentiment that considers elements such as lexicon, word-type, semantic, position, and sentiment polarity of words [20].

---

[6] https://www.kaggle.com/azzouza2018/semevaldatadets?select=semeval-2013-train.csv.

[7] https://pypi.org/project/emoji/.

[8] https://pypi.org/project/aspell-python-py2/.

[9] https://github.com/Svobikl/global_context/tree/master/AnalogyTester/evaluation_data.

5.3 Training Setup

In this study, the tweets in the dataset were tokenized into separate words using the NLTK package[10]. This model was trained on 10 iterations. After training, the vector dimension for our model was set to $d = 300$. In addition, we used a context window of size 5 to the left and 5 to the right from the target word to extract the context words. Additionally, the threshold α is chosen by 0.7. All above parameters were set manually following the experiments. We conducted an exhaustive search for d from 50 to 400, α from 0.5 to 1. In each trial, we adjusted the thresholds with increments of 50 and 0.1 for d and α, respectively. An evaluation measure was necessary to select the highly reliable instances of d and α in an exhaustive search. Therefore, in this study, the basis of choosing the above parameters' value is based on the value of the word-similar score. The highest word-similar score was obtained for the threshold $d = 300$ and $\alpha = 0.7$. Selecting a higher or lower values for d and α would result in a misprediction of more synonyms. The baseline methods were also trained with the same dataset.

5.4 Result and Discussion

The results for our method and the baseline methods are presented in Table 1. Notably, some words in testing datasets were not included in our training data. Therefore, we could not obtain the vector representation of these words. Hence, for these words, we created random vectors to provide comparable results.

Table 1. Word similarity results (%).

Method	WordSim-353	RG-65	SimLex-999
Baseline 1	76.12	69.31	42.26
Baseline 2	73.64	65.12	41.25
Baseline 3	78.73	69.32	40.57
Our approach	79.54	72.32	41.20

According to Table 1, it can be seen that for the WordSim-353 dataset, our approach can improve the performance of the Baselines 1,2, and 3 by 3.42%, 5.9%, and 0.81%, respectively. Besides, for the RG-65 dataset, our method can increase the word similarity accuracy of the baseline methods by up to 7.2% (for Baseline 2), by at least 3% (for Baseline 3). However, for the SimLex-999 dataset, although the performance of our proposal was higher than Baseline 3, but it was lower than the remaining methods by up to 1.06% comparison with Baseline 1, by at least 0.05% comparison with Baseline 2.

As our assessment, our approach outperformed the baselines on the RG-65 and WordSim-353 datasets, but not on the SimLex-999 dataset. In this dataset,

---

[10] https://www.nltk.org/_modules/nltk/tokenize/api.html.

the performance of our method was lower than that of Baseline 1 and Baseline 2 methods. This regard was because words in the SimLex-999 dataset were common words for good vectors obtained without exploiting synonym information. When evaluating less frequent words, we noted that using the context words of a target word helped to learn good word vectors.

In general, our method proved the role of synonym information when enriching word representations. Our approach improved the performance of the prior techniques, but not always, owing to the imbalance of synonym frequency in the datasets.

6 Conclusion and Future Work

We improved the vector representation of words by adding the synonym information. This improvement focuses on the extraction and presentation of synonyms based on their context words. We show that our method has extracted synonym words based on grouping their context words. The synonym words have been represented to vectors by computing the average vector from vectors of their context words. By comparing to recent word representation methods, we proved that our proposal achieved a quite good performance on a given task in terms of word similarity. The main limitation is that we have not compared this proposal's performance to other methods of synonym representations because of the difficulty in determining similar methods for comparison. We will open-source our implementation to make easy the comparison of future work on learning synonym vectors.

References

1. Al-Twairesh, N., Al-Negheimish, H.: Surface and deep features ensemble for sentiment analysis of Arabic tweets. IEEE Access. **7**, 84122–84131 (2019)
2. Blei, D.M., Ng, A.Y., Jordan, M.I.: Latent Dirichlet allocation. J. Mach. Learn. Res. **3**(Jan), 993–1022 (2003)
3. Bojanowski, P., Grave, E., Joulin, A., Mikolov, T.: Enriching word vectors with subword information. Trans. Assoc. Comput. Linguist. **5**, 135–146 (2017)
4. Devlin, J., Chang, M.W., Lee, K., Toutanova, K.: Bert: pre-training of deep bidirectional transformers for language understanding. arXiv preprint arXiv:1810.04805 (2018)
5. Finkelstein, L., et al.: Placing search in context: the concept revisited. In: Proceedings of the 10th International Conference on World Wide Web, pp. 406–414 (2001)
6. Gong, H., Bhat, S., Viswanath, P.: Enriching word embeddings with temporal and spatial information. arXiv preprint arXiv:2010.00761 (2020)
7. Guo, S., Yao, N.: Polyseme-aware vector representation for text classification. IEEE Access. **8**, 135686–135699 (2020)
8. Hamzehei, A., Wong, R.K., Koutra, D., Chen, F.: Collaborative topic regression for predicting topic-based social influence. Mach. Learn. **108**(10), 1831–1850 (2019). https://doi.org/10.1007/s10994-018-05776-w

9. Harris, Z.S.: Distributional structure. Word. **10**(2–3), 146–162 (1954)
10. Hill, F., Reichart, R., Korhonen, A.: Simlex-999: evaluating semantic models with (genuine) similarity estimation. Comput. Linguist. **41**(4), 665–695 (2015)
11. Jianqiang, Z., Xiaolin, G., Xuejun, Z.: Deep convolution neural networks for Twitter sentiment analysis. IEEE Access. **6**, 23253–23260 (2018)
12. Joulin, A., Grave, E., Bojanowski, P., Douze, M., Jégou, H., Mikolov, T.: Fasttext. zip: Compressing text classification models. arXiv preprint arXiv:1612.03651 (2016)
13. Kim, Y.: Convolutional neural networks for sentence classification. arXiv preprint arXiv:1408.5882 (2014)
14. Kundi, F.M., Ahmad, S., Khan, A., Asghar, M.Z.: Detection and scoring of internet slangs for sentiment analysis using sentiwordnet. Life Sci. J. **11**(9), 66–72 (2014)
15. Lample, G., Ballesteros, M., Subramanian, S., Kawakami, K., Dyer, C.: Neural architectures for named entity recognition. arXiv preprint arXiv:1603.01360 (2016)
16. Mikolov, T., Chen, K., Corrado, G., Dean, J.: Efficient estimation of word representations in vector space. arXiv preprint arXiv:1301.3781 (2013)
17. Pennington, J., Socher, R., Manning, C.D.: Glove: global vectors for word representation. In: Proceedings of the 2014 Conference on Empirical Methods in Natural Language Processing (EMNLP), pp. 1532–1543 (2014)
18. Peters, M.E., et al.: Deep contextualized word representations. arXiv preprint arXiv:1802.05365 (2018)
19. Phan, H.T., Nguyen, N.T., Tran, V.C., Hwang, D.: An approach for a decision-making support system based on measuring the user satisfaction level on Twitter. Inf. Sci. (2021). https://doi.org/10.1016/j.ins.2021.01.008
20. Phan, H.T., Tran, V.C., Nguyen, N.T., Hwang, D.: Improving the performance of sentiment analysis of tweets containing fuzzy sentiment using the feature ensemble model. IEEE Access. **8**, 14630–14641 (2020)
21. Qi, Y., Sachan, D.S., Felix, M., Padmanabhan, S.J., Neubig, G.: When and why are pre-trained word embeddings useful for neural machine translation? arXiv preprint arXiv:1804.06323 (2018)
22. Rezaeinia, S.M., Rahmani, R., Ghodsi, A., Veisi, H.: Sentiment analysis based on improved pre-trained word embeddings. Expert Syst. Appl. **117**, 139–147 (2019)
23. Řezanková, H.: Different approaches to the silhouette coefficient calculation in cluster evaluation. In: 21st International Scientific Conference AMSE Applications of Mathematics and Statistics in Economics 2018, pp. 1–10 (2018)
24. Rubenstein, H., Goodenough, J.B.: Contextual correlates of synonymy. Commun. ACM. **8**(10), 627–633 (1965)
25. Sedgwick, P.: Spearman's rank correlation coefficient. Bmj. **349** (2014)
26. Svoboda, L., Brychcın, T.: Improving word meaning representations using wikipedia categories. Neural Netw. World. **523**, 534 (2018)
27. Svoboda, L., Brychcín, T.: Enriching word embeddings with global information and testing on highly inflected language. Computación y Sistemas. **23**(3) (2019)
28. Ulčar, M., Robnik-Šikonja, M.: High quality ELMo embeddings for seven less-resourced languages. arXiv preprint arXiv:1911.10049 (2019)
29. Wang, B., Wang, A., Chen, F., Wang, Y., Kuo, C.C.J.: Evaluating word embedding models: methods and experimental results. APSIPA Trans. Signal Inf. Process. **8** (2019)

Fast Approximate String Search for Wikification

Szymon Olewniczak$^{(\boxtimes)}$ and Julian Szymański

Faculty of Electronics, Telecommunications and Informatics,
Gdańsk University of Technology, Gdańsk, Poland
{szymon.olewniczak,julian.szymanski}@eti.pg.edu.pl

Abstract. The paper presents a novel method for fast approximate string search based on neural distance metrics embeddings. Our research is focused primarily on applying the proposed method for entity retrieval in the Wikification process, which is similar to edit distance-based similarity search on the typical dictionary. The proposed method has been compared with symmetric delete spelling correction algorithm and proven to be more efficient for longer stings and higher distance values, which is a typical case in the Wikification task.

Keywords: Information retrieval · Neural embeddings · Edit distance · Convolutional neural networks · Approximate matching

1 Introduction

The goal of our research is to build an efficient method for Wikification [12], a process of creating links between arbitrary textual content and Wikipedia articles. The links must be relevant to the context of the processed content what is a non-trivial task. Wikification might be considered as a variant of the more general task of Entity Linking.

The Entity Linking systems usually split the process into two stages [10]: entity retrieval and entity disambiguation. In the first stage, the goal is to extract all candidates from content i.e. spans of text that might be linked to our knowledge-base (Wikipedia article's base in case of Wikification). In the second stage among all candidates only the relevant ones, that make sense in the context of the processed text, are selected.

Wikipedia is composed of many articles and covers a large number of topics from different disciplines. Carefully implemented Wikification tool can support many crucial information retrieval tasks such as text representation that is the basis of achieving good results of text classification. It makes Wikification a very important problem among other NLP tasks.

The first stage of Wikification - an entity retrieval may be considered as a specific variant of the Named Entity Recognition (NER) problem. We called it extended NER because we consider here not only named entities but also some common nouns and phrases. One of the possible approaches to this problem is

© Springer Nature Switzerland AG 2021
M. Paszynski et al. (Eds.): ICCS 2021, LNCS 12744, pp. 347–361, 2021.
https://doi.org/10.1007/978-3-030-77967-2_29

to extract all links' labels from Wikipedia articles and use them as a dictionary of possible entities.

To further improve the quality of entity retrieval, we can adopt an approximate string matching for our dictionary of possible entities. The goal of approximate string search is for a given query string q, we retrieve a dictionary element or elements that is the most similar to q according to some metric.

There are many reasons to adopt this strategy for entity retrieval. First, the terms in source text might be misspelled what is even more common for rare named entities. Second, the words in a phrase may be compounded (some spaces might be omitted). Third, the words in phrases can have slightly different variants regarding their position in a sentence, which is common in many languages. Fourth, the words in phrase might be reordered, for example, name and surname may be swapped in an entity describing a person, without changing the meaning of the entity.

2   Approximate String Search

Formally, approximate string search is defined as a task of retrieving elements from dictionary D that are similar to query string q, according to a given metric $dist(.)$. We will denote the set of retrieved elements as X. The dictionary elements and query string are sequences constructed from some finite alphabet A.

There are two variants of approximate string search. The first is called a radius based nearest neighbors search (rNN). In this variant we receive all the terms X from the D which satisfy the condition for some predefined r:

$$\forall x \in X : dist(q, x) \leq r \tag{1}$$

Second, called k-nearest neighbours search (kNN) retrieves k nearest neighbours for a query q such as:

$$\forall d \in D \setminus X : dist(q, d) \geq max(dist(q, X)) \tag{2}$$

In the entity retrieval task, we are rather interested in the best match rather than all possible matches for two reasons. First, we don't want to create too many possibilities for the entity disambiguation stage. Second, there is usually one correct match that the user really meant. Thus in our research, we use kNN variant for approximate string search.

The most common distance metric for approximate string search is edit distance. Its simplest variant is called Levenshtein distance, where we count the smallest possible number of basic transformations that are required in order to transform one string into another. There are three kinds of transformation defined in Levenshtein distance. The first is insertion which means inserting an additional character on a selected position. The second is deletion which means deleting a selected charter from a string and the last one is substitution which replaces one character with the other.

For misspellings correction, an extended variant of Levenshtein distance called Damerau-Levenshtein is commonly used [3]. In this method, we allow the additional transformation of text called transposition which means swapping two adjacent characters in a string. This modified metric is known to better represent misspellings occurring on real data.

The problem with edit distance metrics is that calculating them between two strings is computationally costly. The best currently known algorithm of computing the distance between two strings has $O(n^2)/log(n)$ complexity [8]. In addition, if the strong exponential time hypothesis (SETH) is true, the distance cannot be calculated in time lower than $O(D^{2-\gamma})$ for any fixed $\gamma > 0$.

Another problem with edit distance metrics is that they are not perfect when it comes to real data. For example, in our experiments on Wikipedia's List of common misspellings dataset [16] and the dictionary from SymSpell project [4], the Damerau-Levenshtein resulted in 86% of correct matches, while the Levenshtein metric over the same data achieved 79% accuracy. There are several reasons for the mismatches, where among others there is the fact that similarly sounding phones are more easily mistaken.

Additionally, from the entity retrieval point of view, the edit distance metrics are useless when it comes to word rearranging in a phrase or synonyms detection. Nevertheless, they might be a good approximation in many use cases.

3   Related Work

Calculating the edit distance between two strings is a computationally costly task. It causes that the most straightforward approach to approximate string search, that is iterating over the entire dictionary D and comparing each element with q is usually too slow. To speed up that process we may use an auxiliary data structure, called index, that is intended to reduce the number of actual commissions that we conduct.

There are many indexing methods proposed both for rNN and kNN approximate searches [9,17]. Most generally we split them into two main categories: exact indexes and approximate indexes. Exact indexes guarantee that if there exists a record satisfying the search criteria, it will be returned. On the other hand, an approximate index might sometimes fail but by relaxing the conditions, it might also work faster.

For exact indexes, most common are solutions based on inverted indexes or trees. In an inverted index approach, we create a data structure that allows us to narrow the set of possible results, before conducting actual edit distance calculations. As an example of this approach, we can give the DevideSkip [7] algorithm or AppGram [15]. The main disadvantage of this approach is that we still need to do the manual purification of candidates, which can make them impractical when the dictionary elements are long or a high edit distance threshold is required.

Another class of exact indexes is trees-based methods. Most generally these methods reduce the time complexity of a dictionary search from linear to logarithmic. As an example for kNN approximate search, we can give a classical

BK-Tree data structure or HS-Topk algorithm [14] which uses hierarchical segment tree index together with preliminary purifying. The disadvantage of these approaches is that the time complexity is also dependant on the dictionary size and the length of a query string.

Another interesting class of solutions is indexes based on all potential misspellings generation. The solution has $O(1)$ time complexity but their spatial complexity is tremendous and grows really fast with the maximum supported edit distance between strings. However, in the case of Levenshtein distance (or Damerau-Levenshtein) the spatial complexity of these methods can be highly improved by generating not all the possible misspellings but only the deletions. This method is called symmetric delete spelling correction algorithm and was utilized by FastSS [13], which is currently state-of-the-art for small edit distance values. The method was future improved by Wolf Garbe in SymSpell [4] where we the all deletions generation can be reduced only to the prefix of selected length. This allows us to establish a compromise between the space and time complexity of the method.

Another category of index structures is approximate indexes. The main idea behind this class of solutions is embedding a costly edit distance metric space in another metric space (for example Euclidean) that will retain the properties of the original metric. After the projection, we can use locality sensitive hashing function to quickly compute kNN for our query term.

The main challenge for approximate indexes is to find a good embedding function. There were several proposed embedding functions for edit distance, for example, [11] or more recently CGK [1]. The main drawback of these embedding functions is that they are data independent. It means they work exactly the same regardless of the dictionary used in the search, which reduces the accuracy of the method. To mitigate the problem, the approach of training embedding functions from data using neural networks (a.k.a. learning to hash) was proposed recently: [2,18]. These approaches turned out to have much better properties than previously used functions, and what is also very important, they offer a more general framework, that can be used to embed many different metrics, not only the edit distance based ones.

4   SimpleWiki Labels Dataset

Our study aimed to create an efficient index for Wikification. To test the relevance of a proposed method we decided to test it with Wikipedia in Simple English. We called our test dictionary: SimpleWiki labels dataset.

The SimpleWiki labels dataset was created by parsing all the SimpleWiki articles and extracting all the links from them. Then, for each link we got its anchor text (which we called label) and added it to our dictionary. Before storing in dictionary we also removed all the charters from the label that are not English letters, digits or space character.

Generated dataset consists of 227,575 unique labels. Comparing to the dataset of 82,767 English words[1] it has different characteristics which are summarized in Table 1.

In the Fig. 1 we have also presented the differences in distribution of Damerau-Levenshtein metric between elements in both datasets. It is visible that in case of English words the distribution is much more concentrated than in the SimpleWiki labels. This differences in datasets might cause that the methods that were designed to work optimally for English words, might not be optimal for entity retrieval in Wikification process.

Table 1. Comparison of the SimpleWiki labels and english words datasets.

Dataset	SimpleWiki labels	English words
Dataset size	227,575	82,767
Avg. len.	14.13	8.1
Std. dev.	7.66	2.55
Min. len.	1	1
Max. len.	164	28

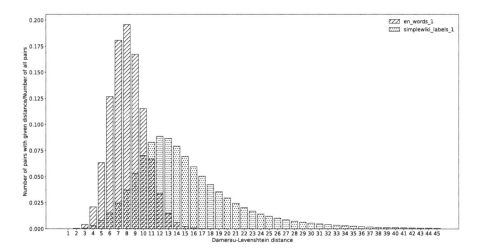

Fig. 1. A compression between distributions of Damerau-Levenshtein distances in english dictionary and SimpleWiki labels dataset.

5   Our Method

Proposed index structure for approximate entity retrieval for Wikification uses approximate index with embedding function. Our solution is inspired by CNN-ED [2] but it was improved to better fit the task. First, our embedding function

---

[1] We refer here to English words dictionary from SymSpell project [4].

is trained with approximate string search in mind, not the approximate string join as in the original paper. Second, we trained our function for Damerau-Levenshtein distance, not the Levenshtein, which better reflects the actual misspellings made by people. Finally, we provided a complete end-to-end solution, not only the embedding function itself, which can be compared with other approximate string search methods.

Our index structure consists of three main components. The first is an embedding function, that is trained to map our Damerau-Levenshtein metric space to Euclidean metric space. The second is a training component of the embedding function. The third is an efficient kNN index for Euclidean metric space, which we use to perform the actual search.

5.1   Embedding Function

Our embedding function is neural network with one convolutional layer, one pooling layer, and the final dense layer. In the input, the function receives one-hot encoded strings. The maximum length of the input string is $M = 167$ characters (the maximum label length of SimpleWiki labels is 164). The alphabet consists of 37 characters which are: "qwertyuiopasdfghjklzxcvbnm1234567890" and space.

One-hot encoding means that we transform each input string S of length L to the matrix X of size: $|A| \times L$ where $|A|$ is the size of our alphabet. Then for each character in a string:

$$\sum_{i=1}^{|A|} \sum_{j=1}^{L} X_{ij} = \begin{cases} 1 & \text{if } S_j = A_i \\ 0 & \text{if } S_j \neq A_i \end{cases} \tag{3}$$

If the string length is lower than the maximum string input, we fill the rest of the input matrix with zeros.

The convolutional layer uses 1D convolutional with 64 output channels, the kernel of size 3, stride 1, and padding 1, without a bias. As a result the network transforms the input of size $N \times |A| \times M$ (N is a batch size) to matrix of size $N \times 64 \times M$. The results of convolution are further passed to ReLU for non-linearity. The convolutional step is crucial in our function because it detects the local modifications in the input, without being sensitive to the modification position, which would be a case in deeply connected layers.

After the convolutional layer comes the pooling layer. We decided to use a max-pooling function with kernel size 2, which was inspired by the CNN-ED model. We tested our model both with and without the pooling layer and it turned out that the pooling significantly reduces the size of the network without a negative impact on the predictions. The max-pooling reduces the convolutional layer output from $N \times 64 \times M$ to $N \times 64 \times M/2$.

The output is constructed from the dense layer that maps its input to the vector of floats of size 100. This vector forms our final embeddings. The network has 538,404 trainable parameters and takes 2,05 MB of memory. Figure 2 shows the network architecture.

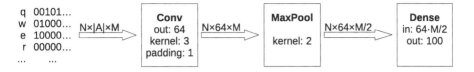

Fig. 2. The architecture of neural embedding function.

5.2 Training Component

Loss Function. The proposed embedding function was trained using triplet loss [5], together with mean square error:

$$\mathcal{L}(s_{acr}, s_{pos}, s_{neg}) = \mathcal{L}_t(s_{acr}, s_{pos}, s_{neg}) + \alpha \mathcal{L}_m(s_{acr}, s_{pos}, s_{neg}) \tag{4}$$

where α is the scaling factor, set to 0.1.

In the triplet loss approach, we train our network by sampling and comparing triplets from our training set. The first element of triplet is called anchor, the second is a positive example and the third is a negative example. In Damerau-Levenshtein metric space the distance between s_{acr} and s_{pos} is smaller than the distance between s_{acr} and s_{neg}. In a triplet loss we want to move the relation from Damerau-Levenshtein metric space to Euclidean space using a vector representations from embedding function: $y_{acr}, y_{pos}, y_{neg}$. The triplet loss is formally defined as follows:

$$\mathcal{L}_t(s_{acr}, s_{pos}, s_{neg}) = \max(0, \|y_{acr} - y_{pos}\| - \|y_{acr} - y_{neg}\| + \eta) \tag{5}$$

where $\eta = \overline{dist}(s_{acr}, s_{neg}) - \overline{dist}(s_{acr}, s_{pos})$ and $\overline{dist}(.)$ is a function that returns a Damerau-Levenshtein distance between its arguments, divided by the average distance between all pairs in the dictionary:

$$\overline{dist}(s_1, s_2) = \frac{dist(s_1, s_2)}{\frac{1}{|D|^2} \sum_{i=1}^{|D|} \sum_{j=1}^{|D|} dist(d_i, d_j)} \tag{6}$$

The triplet loss function pushes $\|y_{acr} - y_{pos}\|$ to 0 and $\|y_{acr} - y_{neg}\|$ to be greater than $\|y_{acr} - y_{pos}\| + \eta$. Where the η is in fact the actual distance between the negative and the positive examples.

The triplet loss itself is relative positioning loss function, which means that it only positions the learning set elements in a correct order, without preserving the absolute distance values between them. To mitigate the issue the additional mean square error loss is introduced. The $\mathcal{L}_m$ is formally defined as:

$$\mathcal{L}_m(s_{acr}, s_{pos}, s_{neg}) = (\|y_{acr} - y_{pos}\| - \overline{dist}(s_{acr}, s_{pos}))^2$$
$$+ (\|y_{acr} - y_{neg}\| - \overline{dist}(s_{acr}, s_{neg}))^2 \tag{7}$$
$$+ (\|y_{pos} - y_{neg}\| - \overline{dist}(s_{pos}, s_{neg}))^2$$

This loss component aims to make the Euclidean distance between embedding vectors the same as the averaged Damerau-Levenshtein distance between strings. Figure 3 presents a visual summary of the proposed learning architecture.

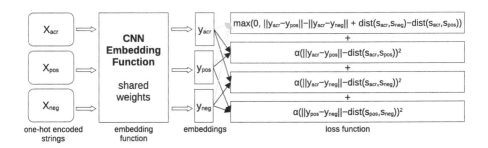

Fig. 3. Triplet loss network architecture.

Training Samples. During each epoch of the training process, we iterate over all elements of the dictionary in random order, considering them our anchor elements. Then for each anchor, we select one positive and one negative example. The examples are selected either from the other dictionary elements or generated by corrupting the anchor word. In the first case, we consider the top 100 kNN of the anchor word and choose the positive and negative examples at random from them. In the second case, the positive and the negative examples are generated by corrupting the anchor, such that the positive example is at Damerau-Levenshtein distance 1 from the anchor and the negative at distance 2.

We train our Triplet Loss Network with a batch size of 64 and a learning rate of 0.001 until the epoch loss stabilizes.

5.3   Index

In order to achieve the full potential of our solution, in addition to a good hashing function, we need an efficient method to retrieve near neighbors from the dictionary. In our solution, we decided to use a faiss library [6], which is currently state-of-the-art for kNN search in Euclidean distance, using GPU.

We construct our index by creating a hash for every element in our dictionary. These hashes are then used to create the faiss index structure. Then for every incoming query string, we calculating the hash for it and look up its nearest neighbors in the index structure. Figure 4 summarizes this process.

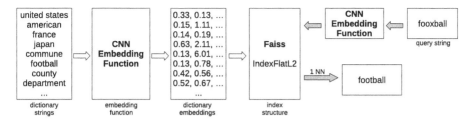

Fig. 4. The index construction process (white arrows) and approximate string search using index structure (gray arrows).

6   Results

To test our method for different dictionary elements' lengths and edit distances, we prepared three different test cases. For the first test case, we took only the SimpleWiki labels that are no longer than 10 characters. For the second test case, we took the labels that are longer than 20 characters. For the last one we took all the SimpleWiki labels. The distances' distribution of the first and second test case labels is presented in Fig. 5. The distribution of distances over all SimpleWiki labels was already presented in Fig. 1.

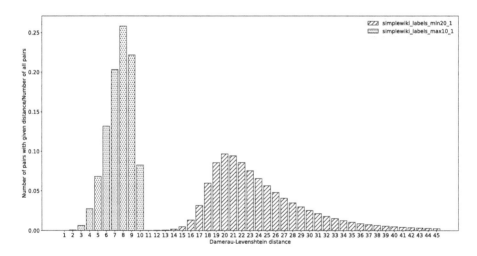

Fig. 5. Comparison of distributions of Damerau-Levenshtein distances between SimpleWiki labels of maximum length 10 and minimum length 20.

To test the performance of our method, we decided to compare it with Sym-Spell 6.7, which is currently state-of-the-art for small edit distances and short dictionary elements [4]. All of our tests were 1 NN searches. When there were several possibilities (several words with an identical distance to the query), any of them was considered a correct result.

Additionally, we tested the neural hashing method in two variants. In the first valiant, we retrieved the 1 NN for the query string hash and returned it as the result. In the second variant, we retrieved 10 NN for the query hash and returned the dictionary element with the smallest Damerau-Levenshtein distance to the query string.

All benchmarks were performed on Ubuntu 20.04, with Intel Core i7-8700 CPU, 16 GB RAM, and NVIDIA Geforce GTX 1660 GPU. For running Sym-Spell, we used .NET Core SDK 2.1.811 and for distance embeddings, Python 3.8.5, PyTorch 1.6.0, and Faiss 1.6.3.

For the first test case, we prepared three test sets. Every set contained 10 misspellings per word in the dictionary, which gave us 766,370 examples per set. The first set contained misspellings generated at Damerau-Levenshtein distance 1 from the original word. The second contained misspellings at distance 2. The third contained misspellings at distance 3.

Our hashing function was separately trained for the SimpleWiki labels sub-set used in this test case. In order to better fit the new dictionary statistics, we changed the input size of the function to 15 and increased the number of con-volution output channels to 4096. Table 2 summarizes the running time of the SymSpell and both variants of the neural hashing method, for processing all test sets. Figure 6 shows the dependency between the maximum allowed Damerau-Levenshtein distance between the correct and misspelled word and running time of each procedure. As we can observe, the execution time of the SymSpell method grows within the maximum allowed distance between a misspelled and correct version of a label, while the execution time of the neural hashing method remains constant.

Table 2. The execution times (in seconds) of processing test sets for the first test case. "Ed" is the maximum allowed Damerau-Levenshtein distance between the correct and misspelled label for the test set.

Ed	Symspell, prefixLength = 7	Neural embeddings, 1 NN	Neural embeddings, 10 NN
1	2.508	74.784	74.264
2	11.834	75.909	75.913
3	48.143	74.146	80.055

Table 3 shows the percent of correct results for the neural hashing method in its 1 NN and 10 NN variants. We consider the result correct, when the returned dictionary element is the correct form of the misspelled word or any other dic-tionary element with the same distance to the query string as the misspelled word.

Our second test case also contained the three test sets. Same as in the pre-vious case, each set contained 10 misspellings per every word in the dictionary, which gave us 344,250 examples per test set. Because the second test case was

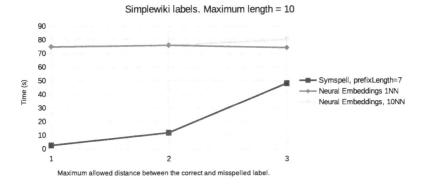

Fig. 6. The execution time compared to growing maximum allowed distance between correct and misspelled labels.

Table 3. The percent of correct results for the two variants of the neural hashing method, for the first test case. "Ed" is the maximum allowed Damerau-Levenshtein distance between the correct and misspelled label for the test set.

Ed	Neural embeddings, 1NN	Neural embeddings, 10NN
1	94.993%	99.700%
2	84.169%	97.133%
3	77.515%	94.329%

Table 4. The execution times (in seconds) of processing test sets for the second test case. "Ed" is the maximum allowed Damerau-Levenshtein distance between the correct and misspelled label for the test set.

Ed	Symspell, prefixLength = 7	Neural embeddings, $k = 1$	Neural embeddings, $k = 10$
3	45.727	70.367	72.145
4	193.047	72.668	74.922
5	1057.782	74.066	76.962

built from longer labels, we also decided to test it against higher Damerau-Levenshtein distances. The first test set in the case contained the misspelled labels with Damerau-Levenshtein distance 3 to the correct form. The second set contained the misspelled labels with Damerau-Levenshtein distance 4 and the final set with the distance 5.

For the neural hashing method, we used here the convolutional network with the same architecture as for the full labels set but trained only on the subset examples. Table 4 shows the execution time for SymSpell and both variants of the neural method for the test sets. Figure 7 plots the execution time of all the methods according to the growing maximum distance of misspelled words. As we can see, the neural method outperforms the SymSpell starting from edit distance = 4 and is much faster for the edit distance = 5.

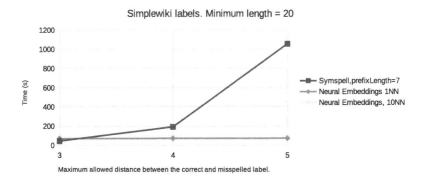

Fig. 7. The execution time compared to growing maximum allowed distance between correct and misspelled labels.

Table 5 shows the precision of the neural method for both variants. As we can see the results are very precise, even for 1NN. This shows that the neural method is well suited for the dictionaries with more sparse distance distributions.

Table 5. The percent of correct results for the two variants of the neural hashing method, for the second test case. "Ed" is the maximum allowed Damerau-Levenshtein distance between the correct and misspelled label for the test set.

Ed	Neural embeddings, 1 NN	Neural embeddings, 10 NN
3	97.599%	98.865%
4	96.403%	98.05%
5	95.395%	97.423%

Our final test case had only one test set that contained 10 misspellings per each label in the SimpleWiki labels dataset, which gave us 2,275,750 examples. The misspellings were introduced here in a progressive manner, which means that the maximum allowed Damerau-Levenshtein distance between the correct and the misspelled word grew with the label's length. We were calculating the Maximum Damerau-Levenshtein distance between the correct word w of length L and the misspelled word using the following formula:

$$max_{ed} = \begin{cases} \lceil \frac{L}{5} \rceil & \text{if } L <= 40 \\ 8 & \text{if } L > 40 \end{cases} \tag{8}$$

The progressive error rate was meant to reflect the real-world cases, where the probability of the typo, grows accordingly to the phrase length. The execution times for the test set and the accuracy of the neural algorithm are presented in Table 6. As we can see here, the performance of the neural hashing method is superior to the SymSpell algorithm, while the correctness is still on a high level.

Table 6. The execution times (in seconds) of processing test set and the percent of correct results for the two variants of the neural hashing method, for the third test case.

Method	Symspell, prefixLength = 9	Neural emb.,1NN	Neural emb., 10 NN
Execution time (s)	4933.292	508.678	556.225
Correctness	100%	88.851%	96.933%

7 Conclusions and Future Works

The paper presented a novelty solution for an approximate string matching index, that is based on the latest research in the field of neural network metrics embeddings. The work has two important contributions. First is the convolution neural network-based embedding function optimized directly for Wikification. Secondly, according to our knowledge, this is the first attempt to combine the embedding function with an efficient Euclidean space search algorithm for the approximate string search task.

Our results show that the neural hashing method might be a good alternative to the other approximate string matching indexes, for the entity retrieval in the Wikification process. However, there are still many areas that might be further explored.

Firstly, we want to test our method for different distance metrics, e.g. weighted edit distance or phonetic algorithms. Thanks to the generality of our solution, it can be used with any distance metric. Particularly, we want to test the method for learning metrics from the data approach, where the metric is learned from the real user misspellings.

Secondly, we want to find the correlation between the dictionary and the actual size of the embedding neural network. We should consider here not only the dictionary size but also the element lengths and used metric. The correlation would be very important in the adaptation of the method for different use cases.

Finally, we want to future investigate the possibilities of joining the neural method with the traditional ones, to get the best from both worlds. We think that using the SymSpell for the shorter strings and the neural hashing method for longer ones might be the best solution for practical applications.

Acknowledgments. The work was supported by funds of Department of Computer Architecture, Faculty of Electronics, Telecommunications and Informatics, Gdańsk University of Technology.

References

1. Chakraborty, D., Goldenberg, E., Koucký, M.: Streaming algorithms for embedding and computing edit distance in the low distance regime. In: Proceedings of the Forty-Eighth Annual ACM Symposium on Theory of Computing, STOC 2016, pp. 712–725. Association for Computing Machinery, New York (2016). https://doi.org/10.1145/2897518.2897577
2. Dai, X., Yan, X., Zhou, K., Wang, Y., Yang, H., Cheng, J.: Convolutional embedding for edit distance. In: Proceedings of the 43rd International ACM SIGIR Conference on Research and Development in Information Retrieval (2020). https://doi.org/10.1145/3397271.3401045
3. Damerau, F.J.: A technique for computer detection and correction of spelling errors. Commun. ACM **7**(3), 171–176 (1964). https://doi.org/10.1145/363958.363994
4. Garbe, W.: Symspell. https://github.com/wolfgarbe/symspell. Accessed 18 Dec 2020
5. Hermans, A., Beyer, L., Leibe, B.: In defense of the triplet loss for person re-identification. arXiv preprint arXiv:1703.07737 (2017)
6. Johnson, J., Douze, M., Jégou, H.: Billion-scale similarity search with gpus. IEEE Trans. Big Data. p. 1–1 (2019). https://doi.org/10.1109/TBDATA.2019.2921572
7. Li, C., Lu, J., Lu, Y.: Efficient merging and filtering algorithms for approximate string searches. In: Alonso, G., Blakeley, J.A., Chen, A.L.P. (eds.) Proceedings of the 24th International Conference on Data Engineering, ICDE 2008, 7–12 April 2008, Cancún, Mexico, pp. 257–266. IEEE Computer Society (2008). https://doi.org/10.1109/ICDE.2008.4497434
8. Masek, W.J., Paterson, M.S.: A faster algorithm computing string edit distances. J. Comput. Syst. Sci. **20**(1), 18–31 (1980). https://doi.org/10.1016/0022-0000(80)90002-1
9. Rachkovskij, D.: Index structures for fast similarity search for symbol strings. Cybern. Syst. Anal. **55**(5), 860–878 (2019)
10. Sevgili, O., Shelmanov, A., Arkhipov, M., Panchenko, A., Biemann, C.: Neural entity linking: a survey of models based on deep learning (2020)
11. Sokolov, A.: Vector representations for efficient comparison and search for similar strings. Cybern. Syst. Anal. **43**(4), 484–498 (2007)
12. Szymański, J., Naruszewicz, M.: Review on wikification methods. AI Commun. **32**(3), 235–251 (2019)
13. Bocek, T.E., Hunt, B.S.: Fast similarity search in large dictionaries. Technical Report ifi-2007.02, Department of Informatics, University of Zurich (April 2007)
14. Wang, J., Li, G., Deng, D., Zhang, Y., Feng, J.: Two birds with one stone: an efficient hierarchical framework for top-k and threshold-based string similarity search. In: Gehrke, J., Lehner, W., Shim, K., Cha, S.K., Lohman, G.M. (eds.) 31st IEEE International Conference on Data Engineering, ICDE 2015, Seoul, South Korea, 13–17 April 2015, pp. 519–530. IEEE Computer Society (2015). https://doi.org/10.1109/ICDE.2015.7113311

15. Wang, X., Ding, X., Tung, A.K.H., Zhang, Z.: Efficient and effective knn sequence search with approximate n-grams. In: Proceedings of VLDB Endow, vol. 7, no. 1, 1–12 September 2013 (2013). https://doi.org/10.14778/2732219.2732220
16. Wikipedia: Lists of common misspellings/For machines. https://en.wikipedia.org/wiki/Wikipedia:Lists_of_common_misspellings/For_machines. Accessed 18 Dec 2020
17. Yu, M., Li, G., Deng, D., Feng, J.: String similarity search and join: a survey. Front. Comput. Sci. **10**(3), 399–417 (2016)
18. Zhang, X., Yuan, Y., Indyk, P.: Neural embeddings for nearest neighbor search under edit distance (2020). https://openreview.net/forum?id=HJlWIANtPH

ASH: A New Tool for Automated and Full-Text Search in Systematic Literature Reviews

Marek Sośnicki[✉][iD] and Lech Madeyski[iD]

Department of Applied Informatics, Wroclaw University of Science and Technology, Wyb.Wyspianskiego 27, 50-370 Wroclaw, Poland

Abstract. Context: Although there are many tools for performing Systematic Literature Reviews (SLRs), none allows searching for articles using their full text across multiple digital libraries. **Goal**: This study aimed to show that searching the full text of articles is important for SLRs, and to provide a way to perform such searches in an automated and unified way. **Method**: The authors created a tool that allows users to download the full text of articles and perform a full-text search. **Results**: The tool, named ASH, provides a meta-search interface that allows users to obtain much higher search completeness, unifies the search process across all digital libraries, and can overcome the limitations of individual search engines. We use a practical example to identify the potential value of the tool and the limitations of some of the existing digital library search facilities. **Conclusions**: Our example confirms both that it is important to create such tools and how they can potentially improve the SLR search process. Although the tool does not support all stages of SLR, our example confirms its value for supporting the SLR search process.

Keywords: Systematic literature review · Systematic review · SLR · Knowledge synthesis · Automated search · Meta-search

1 Introduction

Systematic literature review (SLR) is a commonly and increasingly used process to systematically analyze all research related to some specific area. One of the most difficult parts of preparing SLR is the search and selection phase, where researchers must gather a base set of papers to be analyzed. It is necessary to obtain very high levels of search completeness—which means getting possibly all available literature on the specific topic of interest. Researchers have developed several approaches which aid in this task. The basic method is an automated search, which is involves running search queries across different search engines [4]. However, with the increasing amount of research, it gets harder to perform a complete search of potentially relevant computer science and software engineering articles. The SLR process recommends using multiple search engines,

© Springer Nature Switzerland AG 2021
M. Paszynski et al. (Eds.): ICCS 2021, LNCS 12744, pp. 362–369, 2021.
https://doi.org/10.1007/978-3-030-77967-2_30

however, each search engine has slightly different features, which may bias the search process and increase the amount of work needed in order to perform a comprehensive search [1].

As SLRs are effort-intensive, researchers have developed different tools to support the process [12]. Such tools can assist in the SLR process by automating some phases or by providing a unified interface for navigation across different literature. Unfortunately, although the concept of meta-search tools is understood [17], there is currently no tool that would gather data from all important software engineering search engines and provide a unified interface to facilitate the search and selection phase [13].

Hence, the main contribution of this article is an Automated Search Helper (ASH)—a tool which automates the process of downloading articles and allows the user to perform an initial search and selection phase for the SLR process. ASH takes a list of paper identifiers, downloads their bibliographic information along with their full text, and saves the data in a unified format, allowing users to run searches on the collected data. Additionally, the tool's interface presents papers in a way that accelerates the decision making in the selection process for SLR. In this paper, we present the first version of the application, which focuses on the download and unification of the article data. This shows the potential of the tool, and is intended to encourage future users to use it in their SLR routine. Currently, ASH supports six main digital libraries that catalog Computer Science research.

This article is divided into the following sections: Sect. 2 presents the current state of the art, Sect. 3 explains the motivation to create such a tool along with an example problem, Sect. 4 shows the workflow of the tool, Sect. 5 explains how ASH was used to address the example problem, and Sect. 6 draws conclusions and proposes future improvements.

2 State-of-the-Art

SLRs were introduced to software engineering by Kitchencham et al. [6,10]. The SLR process consists of three main phases: planning, conducting, and reporting. In the planning phase, researchers identify the need for a review and prepare a protocol that contains research questions, search strategy, study selection criteria and procedures, quality assessment checklists, data extraction and synthesis strategy, and the project's timetable. In the second phase, researchers conduct the review according to the protocol. The first part of this phase is a search for all studies, next, there are one or more selection phases, e.g., initially, studies are excluded based on a title, abstract, and keywords and only the remaining ones have their full text analyzed. The next parts are the quality assessment of the selected studies and data extraction and analysis. The final phase is about reporting the review which is preparing a paper or technical report [2].

2.1   Search Process

One of the most difficult and critical parts of the review is the search for, and the selection of, primary studies. The search process must be defined in a way that provides the highest possible level completeness, and should also support the reproducibility of the research process. Researchers have developed many techniques that can help to achieve it. The first and most commonly used technique to gather research is an automated search. The main source of studies for the search are digital libraries. They allow researchers to find all articles related to the topic using their search engines. A well-prepared search string may allow users to obtain most of the research on a given topic, on the other hand, a poorly-constructed string may miss relevant papers or provide a large number of irrelevant articles. Preparing an effective search string is not an easy task and it requires extensive knowledge of the topic and often requires considerable iteration to get the best possible outcome [4]. Digital libraries differ in their query languages, so researchers usually have to prepare different strings for different libraries [7]. Moreover, some digital libraries do not allow for some exclusions to be expressed in a search string, or do not allow searches related to the full text of the paper [9] which can make it difficult to find all relevant studies [1].

2.2   Current Search Problems

Al Zubidy et al. [1] identified five main groups of problems: problems with search strings—building, manipulation, etc., problems with digital libraries front-end and back-end functionalities, lack of tool support, and problems related to researchers. Some of those problems can be solved by researchers, others can be solved with tools, but many can only be solved by digital library providers.

The main source of problems is the lack of support for performing SLRs provided by the digital libraries. Each digital library works differently, the search strings are built with slightly different syntax, results are returned in a different format. Each digital library allows for different search exclusions (e.g., limits to the scope of the search). Some libraries do not allow to search in full text, and sometimes the full text is missing or cannot be downloaded. Moreover, when the search is performed many duplicates and irrelevant results are returned—especially when the search is done on multiple databases [1]. Even if the above problems are mitigated, some issues can be efficiently solved with proper tools. Although many stages of performing SLR can be automated (sometimes only partially), the search and selection process often requires substantial manual effort. Manual search and selection are prone to human errors, which can reduce the validity of the process. However, with proper tooling, the risk of misclassification of articles can be reduced and the whole process can be accelerated.

2.3   Tools Supporting SLR Process

Many tools designed for SLRs were designed to support medical reviews [15], although some studies summarized existing tool support for SLRs in software engineering [1,12,13].

One of the most recent tools for SLR is Thoth [11] that tries to facilitate all phases of SLR. The developers' goal was to allow the user to perform the whole SLR process using their tool, but it focuses mainly on the selection phase. It allows user to perform automatic searches but only in Scopus. Other recent tools that support all phases of SLR are SLuRp [3] and Buhos [15]. Both have multiple useful features, but their main focus is to improve communication between reviewers working on the same SLR project.

Some researchers are content to use multipurpose tools like Excel or Jabref [1] in their SLRs. Such tools allow data storage (both of citation information and extracted data), but they were not explicitly designed to support the SLR process, thus, their interfaces and facility do not properly support the aggregation and organization of results from all SLR phases. Al. Zubidy [1] analyzed also other tools which are used for SLRs including, e.g., SLRTool [5], SESRA [14]. Each of them has multiple useful features, but none has all the features desired by reviewers [1].

There currently is no tool that would fully automate the whole search and selection process. Some tools try to apply some automation (e.g., SLuRp can perform semi-automated downloads of articles), but their main focus is on later stages of SLR—which do not depend on external factors such as digital library facilities.

3   Motivation and Significance

In this section, we present a motivational example based on a problem we encountered in our research. The problem occurred when we were undertaking an SLR about mutation testing in C++ programming language. The goal was to gather all papers related to mutation testing in which authors used C++—either to create a tool for mutation testing, or to review the quality of software in this language.

The problem appeared during the search phase. First, we prepared a list of articles which would be considered as a checklist for validation of the search [8]. Due to a large number of articles in this area, we decided that the search phrase would be performed as an automated search in most well-known Computer Science related digital libraries . The following query was used to find all relevant articles: ("mutation testing" OR "mutation analysis" OR "mutant analysis") AND "C++". If it was possible for a specific digital library, we limited the search to the Computer Science area. In Scopus, this query produces about 80 results. Unfortunately, it is the only major search engine that can perform a search of the phrase "C++", other libraries omit "++" which produces hundreds of results most of them relating to "C" instead of "C++". Moreover, even this large set of articles did not cover all the articles from the checklist. The reason for this was that some articles (e.g., [16]) provided information about the programming language used for mutation testing within the full article text, e.g., in sections about the evaluation of new approaches.

In this example, we had two options. Firstly, we could modify the queries to include searching the full text of a paper. Unfortunately, some sources (including Scopus) do not offer full-text search. Secondly, if full-text search was not available in a given digital library, we could use the following query: "mutation testing" OR "mutation analysis" OR "mutant analysis" and then search the full text of the identified articles manually. The second approach yields thousands of results, where more than 95% are irrelevant. Assessment of all the papers would require substantial manual effort. It would mean downloading and opening the text of the full article, searching for usages of "C++" phrase, analyzing the context of its use, and finally accepting or rejecting the article. Having such large numbers of papers, it would be easy to misclassify some of them which would undermine the goal of maximising search completeness. Additionally, there is no consistent standard across digital libraries—as mentioned above, different queries would be used for different libraries which is not a desirable situation [1].

4   Tool

To solve the problem stated in previous section a tool (Automated Search Helper) was created. ASH downloads articles from across multiple digital libraries, extracts the text from them, transform them into a common format and removes duplicates among them. Moreover, ASH provides a way to search inside full text of the downloaded articles. Results of queries are displayed as web pages, served by the web server included in the tool, they allow user to quickly analyze the query hits.

ASH is still under development, but current version is already functional and can be efficiently used in the SLR process. The tool is an application written in Python and was tested on Linux, Mac, and Windows devices, but it can run in any system that can fulfil the requirements described in the documentation [18]. Current version of application supports papers from the following digital libraries: IEEE, ACM, Science Direct, Springer, Wiley and Scopus. More digital libraries will be supported in the future. The tool can obtain full text of the article directly from the publisher website or, using optical character recognition, parse the full text from PDF of the paper. Detailed description of ASH architecture and usage manual can be found in the ASH Documentation [18].

ASH fulfils many of the SLR tool requirements described in [1], but because the solution already integrates the search from across the different libraries, its possible to add more features in order to satisfy other requirements and overcome the barriers Al-Zubidy et al. mentioned. The tool proposed here covers only the search phase of SLR, it is not intended to support the later stages of SLR, for which there are already useful tools available, e.g., Buhos [15]. This means that our tool allows a user to easily generate the input required by others tools that can be used to support the rest of the SLR process.

5   Evaluation

The tool was evaluated on the problem described in motivation, see Sect. 3. The tool was tasked to collect all the papers related to "Mutation testing" and in the obtained set to find all papers that mentioned "Mutation testing in C++". ASH managed to successfully download full text of more then 95% of articles from the search results of "Mutation testing" across all supported digital libraries (for all other references only the abstract with bibliographic information was downloaded). It managed to remove the duplicates among the papers (about 20% of all results from all digital libraries). This means that, even though some articles full text could not be downloaded, the set of papers which was obtained by the tool as a base for further search was not worse than the one provided by digital libraries.

The tool managed to find all the relevant articles, even these which did not mention "C++" in title, abstract or keywords. For most of the digital libraries, the search results of "Mutation testing in C++" were close to the results obtained by digital libraries search engine. However, for a few of the digital libraries, ASH obtained much better results, either by providing more relevant results due to searching inside full text, or by filtering the false positive search results (e.g., when digital library search engine treated "C++" as "C").

The detailed results and description of the evaluation can be found in Evaluation section (https://github.com/LechMadeyski/AutomatedSearchHelper/wiki/Evaluation) of the ASH documentation [18] due to the imposed paper length limit. The results are very promising, although ASH does not support all digital libraries yet, it can already be used to reduce the amount of manual work during the SLR process.

6   Conclusions

In this article, we explained how important it can be to analyze the full text of articles during the initial search in the SLR process. Our contribution is a tool (ASH) that automatically downloads and searches the full text of articles, allowing users to obtain higher levels of completeness for their SLR searches. The tool does not yet support all digital libraries, but we demonstrated, that even without this, ASH is already able to obtain the full text of about 90% articles, for searches related to Computer Science topics. Moreover, the results obtained by the tool can be much more accurate for specific research questions than standard approaches.

The tool is available for use, and further features of tool are currently under development. We aim to increase the number of supported Digital Libraries, including also ones not related to Computer Science. Additionally, the user interface of the tool can be improved depending on the future user needs and the feedback received in further testing stages.

Acknowledgements. The authors thank Prof. Barbara Kitchenham for reviewing this paper before its submission.

References

1. Al-Zubidy, A., Carver, J.C.: Identification and prioritization of SLR search tool requirements: an SLR and a survey. Empir. Softw. Eng. **24**(1), 139–169 (2019)
2. BA, K., Charters, S.: Guidelines for performing Systematic Literature Reviews in Software Engineering. Technical report (2007), version 2.3
3. Bowes, D., Hall, T., Beecham, S.: SLuRp: a tool to help large complex systematic literature reviews deliver valid and rigorous results. In: Proceedings of the 2nd International Workshop on Evidential Assessment of Software Technologies, pp. 33–36. EAST 2012. ACM, New York, NY, USA (2012)
4. Dieste, O., Grimán, A., Juristo, N.: Developing search strategies for detecting relevant experiments. Empir. Softw. Eng. **14**(5), 513–539 (2008)
5. Fernández-Sáez, A.M., Genero Bocco, M., Romero, F.P.: SLR-TOOL - a tool for performing systematic literature reviews. In: Proceedings of the 5th International Conference on Software and Data Technologies, vol. 2: ICSOFT, pp. 157–166. INSTICC, SciTePress (2010). https://doi.org/10.5220/0003003601570166
6. Kitchenham, B.: Procedures for performing systematic reviews. Technical report (08 2004)
7. Kitchenham, B., Brereton, P., Budgen, D.: The educational value of mapping studies of software engineering literature. In: Proceedings of the 32nd ACM/IEEE International Conference on Software Engineering, vol. 1, pp. 589–598. ACM, New York, NY, USA (2010)
8. Kitchenham, B., Budgen, D., Brereton, P.: Evidence-Based Software Engineering and Systematic Reviews. CRC Press, Boca Raton (2016)
9. Kitchenham, B.A., et al.: Refining the systematic literature review process-two participant-observer case studies. Empir. Softw. Eng. **15**(6), 618–653 (2010)
10. Kitchenham, B.A., Dybå, T., Jørgensen, M.: Evidence-based software engineering. In: ICSE 2004: International Conference on Software Engineering, pp. 273–281 (2004)
11. Marchezan, L., Bolfe, G., Rodrigues, E., Bernardino, M., Basso, F.P.: Thoth: a web-based tool to support systematic reviews. In: ACM/IEEE International Symposium on Empirical Software Engineering and Measurement (ESEM). pp. 1–6 (2019)
12. Marshall, C., Brereton, P.: Tools to support systematic literature reviews in software engineering: a mapping study. In: ACM/IEEE International Symposium on Empirical Software Engineering and Measurement, pp. 296–299 (2013)
13. Marshall, C., Kitchenham, B., Brereton, P.: Tool features to support systematic reviews in software engineering - a cross domain study. e-Informatica Softw. Eng. J. **12**(1), 79–115 (2018). https://doi.org/10.5277/e-Inf180104
14. Molléri, J.S., Benitti, F.B.V.: SESRA: a web-based automated tool to support the systematic literature review process. In: Proceedings of the 19th International Conference on Evaluation and Assessment in Software Engineering. EASE 2015, ACM, New York, NY, USA (2015)
15. Navarrete, C.B., Malverde, M.G.M., Lagos, P.S., Mujica, A.D.: Buhos: a web-based systematic literature review management software. SoftwareX **7**, 360–372 (2018)

16. Petrović, G., Ivanković, M.: State of mutation testing at Google. In: Proceedings of the 40th International Conference on Software Engineering Software Engineering in Practice - ICSE-SEIP 2018. ACM Press (2018)
17. Ramampiaro, H., Cruzes, D., Conradi, R., Mendona, M.: Supporting evidence-based software engineering with collaborative information retrieval. In: 6th International Conference on Collaborative Computing: Networking, Applications and Worksharing (CollaborateCom 2010), pp. 1–5 (2010). https://doi.org/10.4108/icst. collaboratecom.2010.9
18. Sośnicki, M.: (2020). https://github.com/LechMadeyski/AutomatedSear chHelper/wiki

A Voice-Based Travel Recommendation System Using Linked Open Data

Krzysztof Kutt[1(✉)] ⓘ, Sebastian Skoczeń[2] ⓘ, and Grzegorz J. Nalepa[1,2] ⓘ

[1] Jagiellonian Human-Centered Artificial Intelligence Laboratory (JAHCAI)
and Institute of Applied Computer Science, Jagiellonian University,
31-007 Kraków, Poland
`krzysztof.kutt@uj.edu.pl, gjn@gjn.re`
[2] AGH University of Science and Technology, 30-059 Kraków, Poland

Abstract. We introduce J.A.N.E. – a proof-of-concept voice-based travel assistant. It is an attempt to show how to handle increasingly complex user queries against the web while balancing between an intuitive user interface and a proper knowledge quality level. As the use case, the search for travel directions based on user preferences regarding cuisine, art and activities was chosen. The system integrates knowledge from several sources, including Wikidata, LinkedGeoData and OpenWeatherMap. The voice interaction with the user is built on the Amazon Alexa platform. A system architecture description is supplemented by the discussion about the motivation and requirements for such complex assistants.

Keywords: Voice assistant · Human-computer interaction · Linked Open Data

1 Introduction and Motivation

Apple's Siri, Amazon's Alexa, Microsoft's Cortana, and Google's Assistant – are the names of the most popular voice assistants. They are software agents that provide an interface between the user and the device, e.g., a smartphone, a PC or a stand-alone home appliance. With their help, one can send text messages, set calendar entries or control media playback using only own voice [5]. As they are becoming more popular, more problems appear. The failure of Microsoft's *Tay* showed us that as a society we are not ready yet for loose conversations with an electronic agent. Both from the design point of view, where unpredictable user behaviour has to be predicted, as well as unrealistic user expectations for such systems [9]. Additionally, it is important to pay attention to the problems of automatic processing of knowledge in general-purpose assistants. An example is Google's Assistant, which believes that the horse has 6 legs [10].

To overcome problems related to the quality of knowledge, it is better to focus on creating modules responsible for smaller, but more complicated tasks. Data available in the Semantic Web and the whole stack of Semantic Web

© Springer Nature Switzerland AG 2021
M. Paszynski et al. (Eds.): ICCS 2021, LNCS 12744, pp. 370–377, 2021.
https://doi.org/10.1007/978-3-030-77967-2_31

technologies may provide a reliable architecture for knowledge representation and reasoning [1]. In particular, the projects grouped in the Linked Open Data (LOD)[1] community may be useful. They are publicly accessible for free, without any licensing, that encourages collaboration between people and boost the development of knowledge graphs [3]. The strength of the LOD is the use of URIs as identifiers, thanks to which we can easily combine and transform data from many different sets, resulting in a so-called *semantic mashup* [7]. Such mashups were also created for travel-related tasks (e.g., [4]), but they were not combined with other data sources nor with a voice interface.

In this paper, the J.A.N.E. voice assistant is introduced. In contrast to general-purpose assistants, J.A.N.E. is prepared for specific applications using reliable sources of knowledge and the interaction is conducted in a loose but very controlled manner. Our second motivation was to show the promising capabilities of a hybrid system that combines (a) standard methods of search, (b) powerful semantic queries, and (c) voice-based user interface to process complex time-consuming queries in a natural way. This is achieved by creating a solution based on Amazon's Alexa platform that will be capable of recommending the best travel destination based on user's preferences.

The rest of paper is organized as follows. In Sect. 2 we discuss the specification and the workflow of the tool. Then, four datasets used by J.A.N.E. to perform its tasks are described in more detail in Sect. 3. The evaluation is provided in Sect. 4. The paper is concluded in Sect. 5.

2   J.A.N.E. Overview

J.A.N.E. (Just Another Neat Expeditionist) name has been chosen to explicitly show the focus area of this voice assistant and to make the whole system more "human". It was also taken into consideration that female voice assistants are better perceived by users and are considered more trustworthy [8]. Finally, this approach encourages end-users to interact with the system in a more natural way.

The goal of J.A.N.E. is to seek travel destinations based on user preferences. The system takes into consideration the initial point of the journey, all the destinations available via a direct flight, the weather in the final location, on top of which the alignment with user preferences is checked. Three specific use cases are defined as a workbench:

Cuisine: query all of the possible locations for restaurants, cafés and bars that aligns with the requested type of cuisine, e.g., Italian, Asian, French,

Art: choose a destination based on art collection that can be found in local galleries and museums, and

Activities: search for possible sport activities in the destination location, e.g., riding a bike, swimming, ice skating.

---

[1] See: https://lod-cloud.net/.

J.A.N.E. is built on Alexa, as it is a part of well-established stack of Amazon Web Services (AWS) technologies. The system consists of three main parts (see Fig. 1):

1. The Alexa Echo Dot Gen. 2 device, a hardware interface for the user, supported by the voice front-end interface developed with Alexa Voice Services (AVS),
2. The AWS Lambda back-end, responsible for processing the request and querying the data sources,
3. Four services (DynamoDB, Wikidata, LinkedGeoData, OpenWeatherMap) providing the actual data (see Sect. 3).

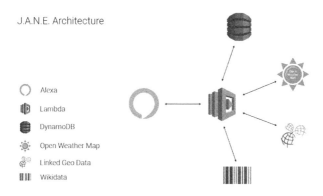

Fig. 1. J.A.N.E. architecture overview

After the user asks a question, J.A.N.E. takes a few steps to generate the answer:

1. First, the user voice request is parsed by the AVS accordingly to the defined interaction model (see Listing 1.1). It consists of distinguishable utterances that are used as schemes to process the voice.
2. Then, generated JSON is sent to AWS Lambda back-end, and the user is informed about the current state (e.g., *Give me a while... I'll do a small research for you*).
3. All available connections from the current location are checked using DynamoDB database (see Sect. 3.1).
4. The user preferences are transformed into SPARQL queries to search for matches in all of the previously discovered locations (see Sect. 3.2–3.3).
5. The locations that are fulfilling users requirements along with the number of matches are then passed through the forecast check (see Sect. 3.4).
6. Finally, the destination with the highest score is chosen and delivered as a voice response to the user.

3 Data Sources

3.1 Amazon DynamoDB Database

As there are no reliable LOD services with flight connections data, two database dumps were used: one containing airport details and the other containing information about all routes operated by the airlines in January 2012. The final dataset containing four columns: *Airport IATA Code, Municipality, Coordinates* and *Destinations* was stored in the DynamoDB.

The current location of the user is used for a simple database lookup. It is given either as the city name (specified by the user) or as the GPS coordinates from the device.

3.2 Wikidata Query Service

Wikidata[2] is a database that contains structured data aimed at support for other wiki projects such as Wikipedia and Wikivoyage. Access to the data is free of charge and the dataset is developed by both human editors and automated bots [12].

In Listing 1.1 one can notice that the interaction model concerning art includes questions (specified in `samples` element) about specific painters or artistic movements. When such a statement is recognized, the corresponding excerpt from the user's input is stored in one of the defined slots: `artMovement` or `painter`. For both types, separate Wikidata queries are prepared to obtain relevant information (see Listing 1.2). Both are processed using the https:// query.wikidata.org/sparql?query= endpoint.

3.3 LinkedGeoData SPARQL Endpoint

LinkedGeoData(See footnote 4) is a project that uses the information collected by the OpenStreetMap and transforms it into RDF triples [11]. J.A.N.E. uses this resource particularly to handle cuisine and activities use cases. A sample query for counting all gastronomy locals in a particular location can be seen in Listing 1.3. Activities are pre-processed before being placed in the corresponding placeholder in a query, e.g., if a user is searching for "swim" or "dive", `SwimmingPool`, `WaterPark`, `DiveCenter` and `Beach` are added to the activity list.

3.4 OpenWeatherMap API

OpenWeatherMap[3] is a free web service that provides weather forecast. It can be queried using the endpoint: http://api.openweathermap.org/data/2.5/forecast? lat=_LAT_&lon=_LON_&appid=_APIKEY_. The result consists of 40 weather

---

[2] See: https://www.wikidata.org.
[3] See: https://openweathermap.org/.

records that represent the weather state every 3 h for the upcoming 5 days. Every record is evaluated and the points are cumulated into a weather score, in the manner presented in Table 1. The final score is then calculated as: $(1 + weatherScore) \cdot matchScore$.

4 Evaluation

Because one of the aims of this project is to provide the user with the answer in less than 120 s, the following 3 approaches were tested, using a "cuisine" intent query launched on 10 biggest cities in Europe, to find the quickest solution:

Querying All Destinations with a Single Query: the query held all 10 cities in a form of an array consisting of the city name, latitude and longitude: VALUES (?city ?latt ?long) {("saint petersburg" 59.9343 30.3351), ...}. This query took 56 s on average to complete the task.

Querying All Destinations One After Another: the queries were launched one after another. Surprisingly, the task was finished in average time of 49 s, which is 7 s on average faster than performing a joint query.

Querying All Destinations in Parallel: the queries were launched in parallel (respecting the endpoint limitations), not waiting for the previous one to finish. In this case the task was completed in 9 s on average.

Furthermore, a side effect was observed. While querying endpoints with the same query multiple times, the performance boost was visible and the query was executed in the lower amount of time until it reached the lowest possible time. This fact might be used to establish the connection with endpoints before they are actually queried.

To compare J.A.N.E. performance with the performance of real users, a simplified use case has been made based on "artist" intent. All of the subjects were asked to manually find a destination that fulfills two requirements: 1) It is possible to fly there directly from Cracow, Poland, 2) There is the biggest collection of Vincent Van Gogh paintings in the world. The participants could

```
1  { "interactionModel": {
2      "languageModel": {
3          "invocationName": "jane",
4          "intents": [ {
5              "name": "art",
6              "slots": [
7                  { "name": "artMovement", "type": "AMAZON.SearchQuery" },
8                  { "name": "painter", "type": "AMAZON.Author" }
9              ],
10             "samples": [
11                 "where should I travel to discover {painter} works",
12                 ...
13                 "where I should travel to discover {artMovement} movement",
14                 "where I should travel to learn more about {artMovement}"
15             ]
16         }, ... ], } } }
```

Listing 1.1. An excerpt from the Alexa interaction model concerning art.

```
 1  PREFIX wd: <http://www.wikidata.org/entity/>
 2  PREFIX wdt: <https://www.wikidata.org/wiki/Property:>
 3  PREFIX wikibase: <http://wikiba.se/ontology#>
 4  PREFIX bd: <http://www.bigdata.com/rdf#>
 5
 6  SELECT ?coord (COUNT(*) as ?count)
 7  WHERE {
 8      ?painting wdt:P31 wd:Q3305213 ;        ## instance of (P31) painting (Q3305213)
 9                wdt:P276 ?location ;         ## location (P276)
10                wdt:P135 wd:_USERINPUT_ . ## movement (P135)
11      ?location wdt:P625 ?coord .            ## coordinate location (P625)
12      SERVICE wikibase:label { bd:serviceParam wikibase:language "en" }
13  }
14  GROUP BY ?coord
15  ORDER BY DESC(?count)
```

Listing 1.2. SPARQL query against Wikidata to search and count all places related to art movement processed from user input (_USERINPUT_ placeholder). The creator property (P170), instead of the movement (P135), is used for **painter**-related query.

```
 1  PREFIX rdfs: <http://www.w3.org/2000/01/rdf-schema#>
 2  PREFIX lgdo: <http://linkedgeodata.org/ontology/>
 3  PREFIX ogc: <http://www.opengis.net/ont/geosparql#>
 4  PREFIX geom: <http://geovocab.org/geometry#>
 5
 6  SELECT ?city ?latt ?long (COUNT(?name) AS ?count)
 7  WHERE { VALUES (?city ?latt ?long) { ( _MUNICIPALITY_ _LAT_ _LON_ ) }.
 8      ?instance a ?gastronomy_local ;
 9                rdfs:label ?name ;
10                geom:geometry [ ogc:asWKT ?entityGeo ] .
11      FILTER (bif:st_intersects (?entityGeo, bif:st_point(?long, ?latt), 20)).
12      FILTER (?gastronomy_local = lgdo:Restaurant ||
13              ?gastronomy_local = lgdo:Cafe ||
14              ?gastronomy_local=lgdo:Bar).
15  }
```

Listing 1.3. GeoSPARQL query against LinkedGeoData to count all gastronomical locals in the radius of 20 km from specific location. _MUNICIPALITY_, _LAT_ and _LON_ are placeholders replaced with details from DynamoDB.

Table 1. Weather records evaluation scheme.

Weather name:	Clear	Cloudy	Foggy	Drizzle	Snowy	Rainy	Thunderstorm
Points:	+0.02	+0.01	−0.002	−0.004	−0.006	−0.0125	−0.02

use any website they wanted along with the device that they felt the most comfortable with to simulate normal conditions of such research (e.g. google search engine, chrome web browser and windows operating system). In this experiment participants weren't asked to check the weather, check other possible locations, calculate how many matching paintings are there and then calculate the final score picking up the most promising location. It is worth noting, that these are tasks J.A.N.E. is performing every time when handling a user's request.

The total of 30 participants (19–56 years old) took part in the experiment. It took 153 s on average for the participants to search the query[4], whereas J.A.N.E. needed 13 s to find the solution performing all of the additional processing. The results of this experiment are very promising showing that such system is almost 12 times faster than a regular user performing manual research. Moreover, the superiority of such travel recommendation system over manual research not only lies in the speed but also in the precision of the results giving here an exact number of the matching entities and providing a recommendation that also takes into account the latest weather forecast.

5   Discussion

J.A.N.E.—the voice assistant presented in this paper—provides an intermediary layer between user voice requests (gathered with Alexa platform) and the data stored in the Semantic Web. The system is a hybrid solution using data from both standard databases and Open Linked Data, gathering user preferences, interpreting their intents and returning the desired response using a synthesised voice.

The presented solution is in line with current trends in the e-Tourism, where attention is drawn to the need for voice communication in natural language and the integration of various sources of knowledge. Other use cases developed in this area include, e.g., room booking and conversation about local attractions (see [2] for overview on current issues in knowledge-based assistants in e-Tourism).

This study has some potential limitations. First of all, only three use cases of seeking flights based on user preferences were introduced as a proof-of-concept to show the possibility of using a vocal assistant with heterogeneous domains. We assume, that the tool has the potential to be integrated (and extended) using different external knowledge sources. Secondly, currently used voice-to-SPARQL mechanism may be improved with the use of recent automatic state-of-the-art methods [6,13]. However, it should be noted that, unlike these methods, we also query various APIs and classic SQL databases, and not just SPARQL endpoints. Finally, the workflow and the queries could be optimized so that the answer for the user will be generated faster.

We assume that the presented approach can have a wide range of applications everywhere, where a user-friendly endpoint is needed to access well-defined data. Considering the use cases implemented so far, the system could be extended with modules checking available accommodation, the price of the ticket or even if there are no other overlapping events in users' personal calendar. The system can also be enriched with the ability to learn the user priorities, e.g., weather over the number of museums, providing more precise results over time. Finally, we plan to examine how people are using a voice-based travel assistants and identify the limitations in current solutions that can be addressed via knowledge-based

---

[4] There were some participants that knew the answer and they had significantly lower result from the average.

reasoning. One of these deficiencies may be a mechanism to clarify ambiguities, e.g., to determine whether Venice in Italy or Poland[5] is in question.

References

1. Allemang, D., Hendler, J.A.: Semantic Web for the Working Ontologist - Effective Modeling in RDFS and OWL, 2nd edn. Morgan Kaufmann, Burlington (2011)
2. Angele, K., et al.: Semantic Web Empowered E-Tourism, pp. 1–46. Springer International Publishing, Cham (2020). https://doi.org/10.1007/978-3-030-05324-6_22-1
3. Bizer, C., Heath, T., Berners-Lee, T.: Linked data - the story so far. Int. J. Semantic Web Inf. Syst. **5**(3), 1–22 (2009). https://doi.org/10.4018/jswis.2009081901
4. Cano, A.E., Dadzie, A.S., Ciravegna, F.: Travel mashups. In: Endres-Niggemeyer, B. (ed.) Semantic Mashups: Intelligent Reuse of Web Resources, pp. 321–347. Springer, Heidelberg (2013). https://doi.org/10.1007/978-3-642-36403-7_11
5. Hoy, M.B.: Alexa, Siri, Cortana, and more: an introduction to voice assistants. Med. Ref. Serv. Q. **37**(1), 81–88 (2018). https://doi.org/10.1080/02763869.2018. 1404391, pMID: 29327988
6. Jung, H., Kim, W.: Automated conversion from natural language query to SPARQL query. J. Intell. Inf. Syst. **55**(3), 501–520 (2020). https://doi.org/10. 1007/s10844-019-00589-2
7. Malki, A., Benslimane, S.M.: Building semantic mashup. In: ICWIT, pp. 40–49 (2012)
8. Mitchell, W.J., Ho, C.C., Patel, H., MacDorman, K.F.: Does social desirability bias favor humans? Explicit–implicit evaluations of synthesized speech support a new HCI model of impression management. Comput. Hum. Behav. **27**(1), 402–412 (2011). https://doi.org/10.1016/j.chb.2010.09.002
9. Neff, G., Nagy, P.: Talking to bots: Symbiotic agency and the case of Tay. Int. J. Commun. **10**, 4915–4931 (2016)
10. Schwartz, B.: Google works to fix how many legs horses & snakes have, April 2019. https://www.seroundtable.com/google-how-many-legs-horses-snakes-27491.html
11. Stadler, C., Lehmann, J., Höffner, K., Auer, S.: Linkedgeodata: a core for a web of spatial open data. Semant. Web **3**(4), 333–354 (2012). https://doi.org/10.3233/SW-2011-0052
12. Vrandečić, D., Krötzsch, M.: Wikidata: a free collaborative knowledge base. Commun. ACM **57**, 78–85 (2014). http://cacm.acm.org/magazines/2014/10/178785-wikidata/fulltext
13. Yin, X., Gromann, D., Rudolph, S.: Neural machine translating from natural language to sparql. Future Gener. Comput. Syst. **117**, 510–519 (2021). https://doi.org/10.1016/j.future.2020.12.013

---

[5] See https://en.wikipedia.org/wiki/Wenecja for the polish village called Venice.

Learning from Imbalanced Data Streams Based on Over-Sampling and Instance Selection

Ireneusz Czarnowski$^{(\boxtimes)}$ ⓘ

Department of Information Systems, Gdynia Maritime University,
Morska 83, 81-225 Gdynia, Poland
i.czarnowski@umg.edu.pl

Abstract. Learning from imbalanced data streams is one of the challenges for classification algorithms and learning classifiers. The goal of the paper is to propose and validate a new approach for learning from data streams. However, the paper references a problem of class-imbalanced data. In this paper, a hybrid approach for changing the class distribution towards a more balanced data using the over-sampling and instance selection techniques is discussed. The proposed approach assumes that classifiers are induced from incoming blocks of instances, called data chunks. These data chunks consist of incoming instances from different classes and a balance between them is obtained through the hybrid approach. These data chunks are next used to induce classifier ensembles. The proposed approach is validated experimentally using several selected benchmark datasets and the computational experiment results are presented and discussed. The results of the computational experiment show that the proposed approach for eliminating class imbalance in data streams can help increase the performance of online learning algorithms.

Keywords: Classification · Learning from data streams · Imbalanced data · Over-sampling · Instance selection

1 Introduction

Data analysis is an area of intense research because data analysis is important from the perspective of potential business, medical, social, and industrial applications (see for example [1–4]). Much attention has been paid to data analysis because data analysis tools can be used to support decision-making processes. On the other hand, there are many real applications (including in human activities) which result in the growing volume of data as well as evolving their characteristics. Today, the commonly used term Big data emphasises both the aspect of data growth and the importance of data and its analysis, i.e. the volume and value of data are emphasized.

Big data is also defined in terms of velocity, which refers to data properties that are changed over time. It is another important dimension of the current data trend. In many real implementations, the data are accumulated with high speeds and flows in from different sources like, for example, machines, networks, social media, mobile phones,

© Springer Nature Switzerland AG 2021
M. Paszynski et al. (Eds.): ICCS 2021, LNCS 12744, pp. 378–391, 2021.
https://doi.org/10.1007/978-3-030-77967-2_32

etc. Examples include Google or Facebook, on which 3.5 billion searches are made every day, and where the number of users has been seen to increase by approximately 22% from year to year. This also implies that there is a massive and continuous flow of data. Such a change of data properties over time is referred to as data drift, which has a relation with the dynamic character of the data source [5, 6]. Such changes are also referred to as concept drift [7] or dataset shift [8]. Data streams are also referred to when new instances of the data are continuously provided.

When data analysis is a process of seeking out some important information in raw data as well as organising raw data to determine the usefulness of information, the process of extracting important patterns from such datasets is carried out under the umbrella of data mining tasks. Among these tasks is a classification, where assigning data into predefined classes is core. A classifier, which a role in the assigning of data to classes, is a model produced under the machine learning process. The aim of the process is to find a function that describes and distinguishes data classes. The process is called learning from data (learning from examples or, shortly, learning classifier from data) [9], where machine learning tools are used as learner models. From the implementation point of view, the process consists of a training and testing phase.

It should be noted that if a dataset with examples is categorical, then learning from examples is based on the existence of certain real-world concepts which might or might not be stable during the process of learning [9]. In the case of data streams, data which change over time together with the concept, involve changes which are difficult to predict in advance. Indeed, standard machine learning algorithms do not work with such data when their properties change over time. In other words, the algorithms cannot efficiently handle changes and the concept drift. Thus, learning classifiers from data streams are one of the recent important challenges in data mining. The challenge is to find the best way to automatically detect the concept drift and to adapt to the drift, as well as the associated algorithms, which will enable the fulfilment of all the data streaming requirements such as constraints of memory usage or restricted processing times. A review of different approaches for concept drift detection and their discussion is included in [14].

Data streams can be provided online, instance by instance or in block. It means that the learning algorithm can process instances appearing one by one over time, or in sets called data chunks. When the latter case, all learning processes are performed when all instances from the data chunk are available.

To deal with the limitations deriving from the stream character of data, so-called summarisation techniques can be used for data stream mining. Sampling or window models are proposed for such summarisations. From an implementation point of view, this means that a relatively small subset of data is processed. Of course, the size of the subset must be pre-set, but the core of such an approach is based on updating such a subset of data after a new chunk of data has arrived, and the removal of some instances from the subset with a given probability instead of periodically selecting them [10]. Other techniques are also available, including the weighted sampling [11] method or sampling within the sliding window model [12]. For example, the sliding window approach assumes that the analysis of the data stream is limited to the most recent instances. In a simple approach, sliding windows with a fixed size include the most recent instances and each new instance replaces the oldest instance. Window-based approaches are also useful due to their ability to quickly react to data changes. However, the size of the window is crucial. When the size is small the reaction can be relatively quick, although to the small size of the window may also lead to a loss of classification accuracy. Data summarisation techniques are also promising because they integrate well with drift detection techniques. Thus, when changes in the concept are detected a learner is updated or rebuilt. The main idea is based on keeping informative instances in the window frame, forgetting instances at a constant rate, and using only a window of the latest instances to train the classifier [13]. Of course, the question is which strategy should be implemented to the updating, including the forgetting, of the instances in the data window.

Another problem merging with learning from data streams concerns the process of classifier induction. A basic approach is based on so-called incremental learning. Incremental learning predicts a class label of the incoming instance and afterwards, information about whether the prediction was correct or not becomes available. The information may then be used for updating a classifier. However, a decision to modify or induce a new one classifier depends on the implemented adaptation mechanism. There are several different approaches to incrementally build new classifiers (see for example [13, 15]).

An approach supporting incremental learning from data streams can be also based on the decomposition of a multi-class classification problem into a finite number of the one-class classification problems [16]. This approach allows an independent analysis of the instances of each considered class, as well as the process of drift change monitoring. However, the results of such an independent analysis must be finally merged to obtain a classification model which will readily predict class labels for new instances following into the system.

Data stream mining requires the monitoring of drift detection as well as class distributions. Both problems belong to a challenging task itself. When the concept drift is detected it results in the data becoming unbalanced. Class imbalance is typical of streams of data and diametrically increases the difficulties associated with the learning of data process. Class imbalance may also negatively influence the learning process and decrease its accuracy. This phenomenon is important because these changes can be very dynamic. As has been underlined in [14], classes may switch roles, and minority classes may become the majority one, and vice versa. Such phenomena may have a dynamic character with a high frequency. The problem of class imbalance in learning from data streams is the main topic of this paper. This paper deals with a problem eliminating this phenomenon.

The aim of this work is to show that extending the functionality of the online learning approach, previously proposed in [17], by adding methods to balance the minority and majority classes, i.e., for supporting the analysis of the imbalanced data within the stream, increases the performance of the online learning algorithm. The proposed approach is based on over-sampling and under-sampling (i.e., instance selection) techniques that are implemented to form data chunks which are then used to induce the ensemble of classifiers. The main contribution of the paper is therefore to propose an over-sampling approach and extend the online learning framework presented in [17] through new computational functionalities.

This paper is organised as follows. The following section includes problem formulation. Then, a framework for learning from data streams is presented. In Subsect. 3.1, a proposed approach for changing the class distribution towards a more balance form are described; this subsection presents details of the proposed over-sampling approach. A detailed description of the computational experiment setup and the discussion of the experimental results is then included in Sect. 4. Lastly, the final section contains conclusions and directions for future research.

2   Learning from Data Streams – Problem Formulation

A data stream can be considered as a sequence $X_1, .., X_t, ...,$, where X_i may be defined as a single instance or a set of instances, when instances appear not one by one in time but form sets, called data chunks. So, in the case of online learning from data streams, the sequence appears in a form $\{x_1, x_2, ..., x_t, ...\}$, where x_t is the t-th example's feature vector and t is a step of learning.

Considering the training phase, where the learner is produced, the sequence has a form of pairs $\{(x, c)_1, (x, c)_2, ..., (x, c)_t, ...\}$, where c is a class label associated with x, and is taken from a finite set of decision classes $C = \{c_i : i = 1, ..., d\}$. d is the number of classes established for the current step (the time step). During the training phase, such sequence pairs are provided to the learning algorithm as the training instances T. The role of the machine learning algorithm is to use these data pairs to find the best possible approximation f' of the unknown function f such that $f(x) = c$. After that, based on f' a class $c = f'(x) = c$ for x, where $(x, c) \notin T$ can be predicted.

In online learning, feedback on the actual label can be obtained after a prediction is made. In the case of an incorrect class assigned to x in step t, the feedback information can be used to update the function f' for the next steps.

When the classified data stream is given in a form of data chunks, it can be denoted by S_t meaning that it is the t-th data chunk. In such a case, the training set is formed from such data chunks ($S_t \subset T$ and $|S_t| < |T|$) or, of course, may form such a training set itself ($S_t = T$ and $|S_t| = |T|$).

Set T can be also noted as a sum of subsets to whom the instances from different classes belong, i.e., as $T = T^1 \cup T^2 \cup \cdots \cup T^d$. When set T is analysed with regards to the number of different classes to whom the instances forming this set belong, the problem of within-class imbalance can be observed. The problem of imbalanced data exists, when $\exists_{i,j \in \{1,...,d\}} |T^i| \neq |T^j|$, where $i \neq j$. In such a case $T^i_{minority} (i \in [1, ..., d])$ contains the minority class dataset, which means that the cardinality of $T^i_{minority}$ is smaller than the cardinality of each of the remaining subsets of T representing the remaining classes. On the other hand, among these remaining subsets, there is the majority class subset containing the majority class instances.

In this paper, it is assumed that a training set is formed by one data chunk, so it is a case of ($S_t = T$).

3   An Approach for Learning from Imbalanced Data Streams

3.1   Online Classifier for Data Streams

In this paper, the problem of online learning from data streams is solved using the framework that had previously been proposed in [17]. The framework is based on three components which focus on data summarisation, learning, and classification.

The framework involves the processing of data chunks which consist of prototypes formed from a sequence of incoming instances for which predictions were incorrect. Data chunks are formed by the data summarisation component. This component consists of methods responsible for extracting instances incoming from a classification component. The aim of the component is to permanently update data chunks by selecting adequate instances (called prototypes) from incoming data, memorising them and forgetting/removing other (non-informative) instances. The role of this component is to then pass the created or updated data chunk to the learning component. Such an updating mechanism is intended to help the system adapt to outside changes viewed in the data. Forming data chunks from incoming instances for which predictions were incorrect is also an approach for concept drift detection within data streams.

The idea implemented within the discussed approach is also based on the integration of online learning by sequentially inducing the prediction model. The existing feedback is assumed from a comparison of predicted class labels with the true labels. It means, in the proposed case, that a new classifier is constructed whenever a new data chunk becomes available.

The discussed framework is also based on a decomposition of the single multi-class classification problem into a set of one-class classification problems. This implies that a multi-class classification problem is solved using an ensemble of single one-class classifiers, i.e., one for each target class. It thus means that the incoming learning component data chunks are partitioned into subsets with so-called positive and negative instances for each considered decision class. Based on such a preparation of training data, a pool of the simple base classifiers is induced. These classifiers are represented by the matrix Φ consisting of $d \times \tau$ elements, i.e., K one-class classifiers, one for each target class. The approach is also based on the remembering of earlier-induced classifiers and τ represents the earlier steps with respect to data chunks that do not already exist in the system and which have been forgotten. This also means that the ensemble consists of the fixed-size set of classifiers, depending on the value of τ. The ensemble is updated when a new data chunk arrives. Then, a new induced classifier replaces the worst component in the ensemble. However, the process is associated with weights assigned to each of the base classifiers, based on the WAE approach (see [18]), where the value of weight increases if the classifier has been taking the correct decisions.

Finally, the aim of the classification component is to classify new incoming instances using ensemble classifiers. Because the classifier is constructed from K one-class classifiers, the prediction result produced by the ensemble classifier is determined through the weighted majority vote.

In summary, online learning from data streams based on the proposed framework is the following:

- three components are integrated with feedback from a comparison of predicted class labels with the true labels as a core to adapting the system to changes within data,
- the learning classifier is carried out using data chunks,
- a training data set consists of one data chunk,
- data chunks are formed from incoming instances for which former predictions were incorrect,
- a data chunk is updated to represent the current and representative instances,
- the learning classifiers are based on a decomposition of the single multi-class classification problem into a set of one-class classification problems,
- data classification is carried out using a weighted ensemble for one-class classification,
- the system is based on updating existing ensemble classifiers according to new incoming data.

The structure of the framework is shown in Fig. 1. The next subsection provides more details on the process forming the data chunks, which is core to this paper.

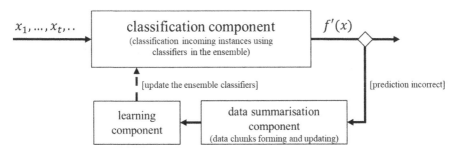

Fig. 1. Data processing by the proposed online learning approach

3.2   Training Data Set Forming and Updating

This subsection addresses the problem of forming data chunks from incoming instances. These data chunks are next used as a training data set. The proposed approach assumes that the size of the data chunk cannot be greater than the acceptable threshold. However, the following cases must be considered during the forming data chunks:

- the size of the data chunk is determined by the defined threshold, that respectively means that $\sum_{i=1}^{d} |T^i| = \alpha$, where α is the value of the threshold,
- when the size of the data chunk is smaller than the threshold size, incoming instances are being added to data chunk. However, each new instance is allocated to the corresponding subset of T, depending on the decision class of these instances,
- when the size of the data chunk is equal to the threshold, the chunk is updated and a new incoming instance replaces the other instance included in the current data chunk,
- the process of forming and updating the data chunk is guided in such a way as to maintain a balance between instances belonging to the considered classes, meaning that the sum of the sizes of the subsets for instances representing considered decision classes is not greater than the threshold,

The balance between instances belonging to minority and majority classes within data chunks is carried out using so-called data level methods. Data level methods aim to transform an imbalanced dataset into a better-balanced one by reducing the imbalance ratio between the majority and minority classes. The reduction can be carried out by over-sampling or under-sampling.

The aim of the under-sampling approaches is to balance the distribution of data classes. In practice, under-sampling techniques just remove instances from the majority class. The strength of this approach depends on the kind of rules that have been implemented for instance removal. Many methods belonging to this group are based on clustering and instance selection [19].

The proposed ensemble classifier for mining data streams assures a balanced distribution between minority and majority class instances using an approach based on instance selection. Instance selection aims to decide which from available and incoming instances should be finally retained and used during the learning process. So, the data chunk can be updated by replacing an older instance with a new one. Of course, the instance selection process can also decide whether or not to update the data chunk. The process of instance selection is carried out only on this part of instances belonging to the same decision class.

To decide whether the instances can be added to the current set, two well-known instance selection techniques, i.e., the Condensed Nearest Neighbour (CNN) algorithm and the Edited Nearest Neighbour (ENN) algorithm - both adopted to the considered one-class classification problem through applying the one-class k Nearest Neighbour method – called the Nearest Neighbour Description (NN-d) [20] – have been used. The adaptation assures instance selection processing for instances independently from considered decision classes. The pseudocode of the updating methods, denoted CNN-d and ENN-d respectively, are included in [17].

When in the current step of learning the number of instances in the subsets of T is not equal to the assumed threshold, the over-sampling procedure is activated on these subsets of instances to obtain a more balanced distribution of instances belonging to all classes.

The over-sampling procedure starts with identifying reference instances for two clusters of instances which do not belong to the minority class. In the presented clustering procedure, a k-means algorithm is used. The centres of the produced clusters are used for representing the reference instances. Next, for the reference instances, the procedure finds their neighbours belonging to the minority class. The closeness of neighbours is measured using the Euclidean measure and the number of neighbours is a parameter of the procedure. Next, an artificial instance located between the identified neighbours is generated randomly.

The pseudo-code explaining how an artificial instance is generated is shown as Algorithm 1. The algorithm is also illustrated in Fig. 2. Algorithm 2 shows the pseudo-code of the proposed over-sampling procedure.

Algorithm 1 Generation of an artificial instance (GAI)

Input: x_1, x_2 – reference instances for the minority class; S – a subset of instances; k - number of neighbours;
Output: x_a – an artificial instance;

Begin

For x_1 and x_2 find its k-nearest neighbour instances, which belong to S and where N contains the neighbour instances;
Generate randomly an artificial instance x_a located between instances from N;
Return x_a;
End

Algorithm 2 Over-sampling procedure for the minority set of instances

Input: T - training set; k - number of neighbours; d - the number of classes; α is the value of the threshold
Output: $T = T^1 \cup T^2 \cup ... \cup T^d$ - sets of balanced instances forming a training set.

Begin

$\beta \cong \frac{\alpha}{d}$;
For $i:=1,...,d$ **do**
If $|T^i| < \beta$ **then**
For subset $\bigcup_{j:j\in\{1,..,d\}\setminus\{i\}} T^j$ and run k-means algorithm for $k=2$ and for each obtained cluster return their centres as reference instances x_1^r and x_2^r;
$T^i = T^i \cup \{GAI(x_1^r, x_2^r, k, T^i)\}$;
End If
Return $T^1, T^2, ..., T^d$;
End

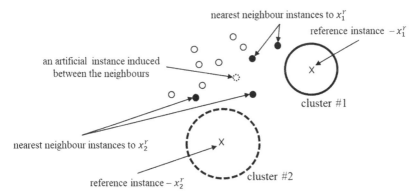

Fig. 2. Generation of an artificial instance

4   Computational Experiment

The computational experiment results are provided in this section. The aim of the experiment computation was to evaluate the performance of the approach discussed in this paper. The performance of the approach has been measured based on classification accuracy.

Based on this simple measure, the reported experiment aimed to answer the following question: whether the proposed approach, assuring a uniform class distribution in the training data set, is competitive with the other approaches dedicated to solving online learning from data streams.

In this paper, the proposed approach has been denoted as WECOI (Weighted Ensemble with one-class Classification and with Over-sampling and Instance selection). The WECOI approach has been implemented in two versions using the CNN-d and ENN-d algorithms for instance selection.

The WECOI has been also compared with its earlier versions denoted as WECU (Weighted Ensemble with one-class Classification based on data chunk Updating) and OLP (Online Learning based on Prototypes) (see [17]). The WECU has been implemented as an ensemble model updated based on the weights assigned to each one-class classifier. The WECU has also been implemented in a version based on a simple ensemble model with a simple majority voting to combine member decisions; such versions of the algorithm are denoted as WECUs. The OLP uses a simple ensemble model in which ensembles are updated by removing the oldest classifier. For comparison, WECOI has been also implemented in a version based on a simple ensemble model, denoted as WECOIs.

Both algorithms, i.e., WECU and OLP use CNN or ENN (or CNN-d and ENN-d in the case of WECU) to update data chunks but without balancing the class distribution within a training data set.

The aim has also been to compare the obtained results with others i.e., the Accuracy Weighted Ensemble (AWE) [22], the Hoeffding Option Tree (HOT), and the iOVFDT (Incrementally Optimized Very Fast Decision Tree) [5, 21], which are implemented as extensions of the Massive Online Analysis package within the WEKA environment. In the case of the WECOI and WECU approaches, the POSC4.5 algorithm has been used as a learning tool to induce base classifiers. In the case of OLP, the C4.5 algorithm has been applied to induce all the base models for ensemble classifiers.

The computational experiments have been carried out with parameter settings presented in Table 1. The values of the parameters have been set arbitrarily based on the trial-and-error procedure.

All algorithms have been applied to solve the respective problems using several benchmark datasets obtained from the UCI Machine Learning Repository [23] and IDA repositories [24]. The used datasets are shown in Table 2.

An experiment plan was based on 30 repetitions of the proposed schema. The instances for the initial training set were selected randomly from each considered dataset. The number of selected instances was limited to the set threshold. The mean values of the classification accuracy obtained by the WECOI model based on an assumed experiment

Table 1. Parameter setting in the reported computational results

Parameter	
τ– number of base classifiers	5
α– size of the data chunk	The value is shown in Table 2
Number of neighbors for GAI	2
Number of neighbors for ENN	3
Metric distance for GAI	Euclidean
Metric distance for ENN and CNN	Euclidean

Table 2. Characteristics of the dataset used in the experiment

Dataset	Source	#instances	#attributes	#classes	Best reported results classification accuracy	Threshold (as % of the data set)
Heart	UCI	303	13	2	83.8%[a]	10%
Diabetes	UCI	768	8	2	80.12%[b]	5%
WBC	UCI	699	9	2	99.3%[b]	5%
Australian credit (ACredit)	UCI	690	15	2	92.1%[b]	9%
German credit (GCredit)	UCI	1000	20	2 .	80.3%[b]	10%
Sonar	UCI	208	60	2	97.1%[d]	10%
Satellite	UCI	6435	36	6	–	10%
Banana	IDA	5300	2	2	89.26%[a]	20%
Image	UCI	2310	18	2	80.3%[c]	20%
Thyroid	IDA	215	5	2	95.87%[a]	10%
Spambase	UCI	4610	57	2	82.37%[e]	20%
Twonorm	IDA	7400	20	2	97.6%[a]	20%

Sources:
[a][25], [b][26], [c][27], [d][23], [e][28]

plan are shown in Table 3[1]. In the table the performances of WECU, OLP, AWE, HOT, as well as iOVFDT are also presented.

The results presented in Table 3 show that the WECOI model can be considered as a competitive algorithm. When the results are analysed for the approach used for instance

---

[1] The best solution obtained by the compared algorithms is indicated in bold. The underline indicates the best solution obtained by the WECOI or WECU algorithm.

Table 3. Average classification accuracy (in %) obtained by the compared algorithms

Algorithm	Heart	Diabetes	WBC	ACredit	GCredit	Sonar	Satellite	Banana	Image	Thyroid	Spambase	Twonorm
WECOI-ENN	84,01	79,8	72,58	83,4	74,61	84,11	81,34	86,57	92,57	94,15	79,3	**97,73**
WECOIs-ENN	83,2	78,15	72,4	83,4	73,21	82,6	80,5	84,6	91,84	93,25	78,62	96,52
WECOI-CNN	80,81	78,15	71,89	82,06	74,03	82,32	80,78	87,02	91,8	94,06	77,86	96,43
WECOIs-CNN	81,24	77,58	71,64	81,52	73,25	82,2	79,1	86,4	91,42	93,52	77,05	95,62
WECU- ENN	**84,14**	79,62	72,54	83,74	**75,4**	**84,21**	79,14	88,1	91,47	94,01	78,5	97,7
WECUs-ENN	81,1	76,75	71,4	83,48	73,04	81,63	80,75	86,45	91,08	93,08	78,5	96,81
WECU- CNN	80,4	77,8	72,4	82,24	74,35	81,24	78,4	87,12	90,47	93,4	77,54	97,51
WECUs-CNN	81,24	76,57	72,11	80,84	73,91	82,2	76,54	86,02	91,21	91,47	76,34	94,24
OLP-ENN	80,4	72,82	71,21	82,4	71,84	75,44	78,25	85,72	90,07	91,47	78,17	96,4
OLP-CNN	78,14	73,2	70,1	81,52	70,06	76,81	80,45	84,1	90,32	93,14	77,61	95,1
AWE	78,01	72,5	**72,81**	84,5	73,5	77,02	82,4	87,4	91,61	93,1	75,4	76,8
HOT	81,4	**80,42**	72,67	82,41	72,87	76,05	**83,4**	86,77	92,2	94,21	77,4	96,06
iOVFDT	81,7	77,4	71,04	**84,5**	75,21	78,38	81,54	**89,21**	**95,07**	**94,63**	**80,2**	97

selection, the general conclusion is that using ENN is superior to CNN. The observation holds true for all considered datasets.

Another observation is that using a simple majority voting does not guarantee competitive results like when using weighted majority votes.

WECOI can be also considered as competitive with the other algorithms for which results have also been obtained, i.e., AWE, HOT, and iOVFDT.

However, the main conclusion should be formulated for the results obtained by the approach aiming for a balance of instances distribution belonging to different decision classes. When analysing the results obtained by the WECOI and WECU models (results for these algorithms have been underlined in the table), better results were obtained with WECOI. Thus, the WECOI model is superior to the WECU one. On the other hand, both algorithms outperform OLP.

5   Conclusions

This paper contributes to our understanding of how data streams can work with imbalanced data, through extending the presented framework. Here, an approach for balancing class distributions within data chunks based on over-sampling and instance selection is proposed. These techniques have been implemented within the data summarisation component which aims to prepare training data to be used for learning classifiers. Over-sampling has been used as a tool for instance duplication in the minority class when instance selection has been used as a procedure for reducing the number of instances in the majority class.

The proposed approach has been evaluated and compared with other approaches and the computational experiment results show that the approach is competitive with others. Specifically, the proposed algorithm for balancing class distribution within the data stream overcomes its previous version where the imbalanced data problem was ignored.

Future research will focus on studying the influence of the size of data chunks as well as the number of neighbours in the generation of an artificial instance (GAI) on the accuracy of the proposed online learning approach. The future direction of the research will also allow the validation of different over-sampling techniques and the selection of the best one to be made.

References

1. Kaplan, A.M., Haenlein, M.: Users of the world, unite! the challenges and opportunities of social media. Bus. Horiz. **53**(1), 59–68 (2010). https://doi.org/10.1016/j.bushor.2009.09.003
2. Chan, J.F., et al.: A familial cluster of pneumonia associated with the 2019 novel coronavirus indicating person-to-person transmission: a study of a family cluster. Lancet **395**(10223), 514–523 (2020). https://doi.org/10.1016/S0140-6736(20)30154-9
3. Phan, H.T., Nguyen, N.T., Tran, V.C., Hwang, D.: A sentiment analysis method of objects by integrating sentiments from tweets. J. Intell. Fuzzy Syst. **37**(6), 7251–7263 (2019). https://doi.org/10.3233/JIFS-179336

4. Wang, Y., Zheng, L., Wang, Y.: Event-driven tool condition monitoring methodology considering tool life prediction based on industrial internet. J. Manuf. Syst. **58**, 205–222 (2021). https://doi.org/10.1016/j.jmsy.2020.11.019
5. Bifet, A.: Adaptive learning and mining for data streams and frequent patterns. PhD thesis, Universitat Politecnica de Catalunya (2009)
6. Sahel, Z., Bouchachia, A., Gabrys, B., Rogers, P.: Adaptive mechanisms for classification problems with drifting data. In: Apolloni, B., Howlett, R.J., Jain, L. (eds.) Knowledge-Based Intelligent Information and Engineering Systems. LNCS (LNAI), vol. 4693, pp. 419–426. Springer, Heidelberg (2007). https://doi.org/10.1007/978-3-540-74827-4_53
7. Widmer, G., Kubat, M.: Learning in the presence of concept drift and hidden contexts. Mach. Learn. **23**(1), 69–101 (1996)
8. Tsymbal, A.: The problem of concept drift: definitions and related work. Technical Report. TCD-CS-2004–15, Department of Computer Science, Trinity College Dublin, Dublin, Ireland (2004)
9. Mitchell, T.: Machine Learning. McGraw-Hill, New York (1997)
10. Vitter, J.S.: Random sampling with a reservoir. ACM Trans. Math. Softw. **11**(1), 37–57 (1985)
11. Chaudhuri, S., Motwani, R., Narasayya, V.R. On random sampling over joins. In: Delis, A., Faloutsos, C., Ghandeharizadeh, S. (eds.) SIGMOD 1999, pp. 263–274. ACM Press (1999)
12. Guha, S., Mishra, N., Motwani, R., O'Callaghan, L.: Clustering data streams. In: Proceedings of the 41st Annual Symposium on Foundations of Computer Science (FOCS), pp. 359–366. IEEE Computer Society, Washington (2000)
13. Kuncheva, L.I.: Classifier ensembles for changing environments. In: Roli, F., Kittler, J., Windeatt, T. (eds.) Multiple Classifier Systems. LNCS, vol. 3077, pp. 1–15. Springer, Heidelberg (2004). https://doi.org/10.1007/978-3-540-25966-4_1
14. Fernández, A., García, S., Galar, M., Prati, R.C., Krawczyk, B., Herrera, F.: Learning from imbalanced data streams. In: Learning from Imbalanced Data Sets, pp. 279–303. Springer, Cham (2018). https://doi.org/10.1007/978-3-319-98074-4_11
15. Stefanowski, J.: Multiple and hybrid classifiers. In: Polkowski L. (ed.) Formal Methods and Intelligent Techniques in Control, Decision Making. Multimedia and Robotics, pp. 174–188. Warszawa (2001)
16. Zhu, X., Ding, W., Yu, P.S.: One-class learning and concept summarization for data streams. Knowl. Inf. Syst. **28**, 523–553 (2011)
17. Czarnowski, I., Jędrzejowicz, P.: Ensemble online classifier based on the one-class base classifiers for mining data streams. Cybern. Syst. **46**(1–2), 51–68 (2015). https://doi.org/10.1080/01969722.2015.1007736
18. Woźniak, M., Cal, P., Cyganek, B.: The influence of a classifiers' diversity on the quality of weighted aging ensemble. In: Nguyen, N.T., Attachoo, B., Trawiński, B., Somboonviwat, K. (eds.) ACIIDS 2014. LNCS (LNAI), vol. 8398, pp. 90–99. Springer, Cham (2014). https://doi.org/10.1007/978-3-319-05458-2_10
19. Tsai, C.-F., Lin, W.-C., Hu, Y.-H., Ya, G.-T.: Under-sampling class imbalanced datasets by combining clustering analysis and instance selection. Inf. Sci. **477**, 47–54 (2019). https://doi.org/10.1016/j.ins.2018.10.029
20. Khan, S., Madden, M.G.: One-class classification: taxonomy of study and review of techniques. Knowl. Eng. Rev. **29**(3), 345–374 (2014)
21. Bifet, A., Holmes, G., Kirkby, R., Pfahhringer, B.: MOA: Massive online analysis. J. Mach. Learn. Res. **11**, 1601–1604 (2010)
22. Wang, H., Fan, W., Yu, P.S., Han, J.: Mining concept-drifting data streams using ensemble classifiers. In: Proceedings of 9th ACM SIGKDD International Conference on Knowledge Discovery and Data Mining, 226–235 (2003). https://doi.org/10.1145/956750.956778

23. Asuncion, A., Newman, D.J.: UCI machine learning repository. University of California, School of Information and Computer Science, Irvine, CA (2007). http://www.ics.uci.edu/~mlearn/MLRepository.html

24. IDA Benchmark Repository (2014). https://mldata.org/

25. Wang, L., Hong-Bing, J., Jin, Y.: Fuzzy passive-aggressive classification: a robust and efficient algorithm for online classification problems. Inf. Sci. **220**, 46–63 (2013)

26. Jędrzejowicz, J., Jędrzejowicz, P.: Rotation forest with GEP-induced expression trees. In: Shea, J.O., et al. (eds.) Systems: Technologies and Applications, LNAI, vol. 6682, pp. 495–503. Springer, Heidelberg (2011)

27. Jędrzejowicz, J., Jędrzejowicz, P.: A family of the online distance-based classifiers. In: Nguyen, N.T., Attachoo, B., Trawiński, B., Somboonviwat, K. (eds.) Intelligent Information and Database Systems. LNCS (LNAI), vol. 8398, pp. 177–186. Springer, Cham (2014). https://doi.org/10.1007/978-3-319-05458-2_19

28. Bertini, J.B., Zhao, L., Lopes, A.A.: An incremental learning algorithm based on the K-associated graph for non-stationary data classification. Inf. Sci. **246**, 52–68 (2013)

Computational Intelligence Techniques for Assessing Data Quality: Towards Knowledge-Driven Processing

Nunik Afriliana[1,2]([✉]) [iD], Dariusz Król[1] [iD], and Ford Lumban Gaol[3] [iD]

[1] Wroclaw University of Science and Technology, Wroclaw, Poland
{nunik.afriliana, dariusz.krol}@pwr.edu.pl
[2] Universitas Mutimedia Nusantara, Jakarta, Indonesia
[3] Computer Science Department, Bina Nusantara University, Jakarta, Indonesia
fgaol@binus.edu

Abstract. Since the right decision is made from the correct data, assessing data quality is an important process in computational science when working in a data-driven environment. Appropriate data quality ensures the validity of decisions made by any decision-maker. A very promising area to overcome common data quality issues is computational intelligence. This paper examines from past to current intelligence techniques used for assessing data quality, reflecting the trend for the last two decades. Results of a bibliometric analysis are derived and summarized based on the embedded clustered themes in the data quality field. In addition, a network visualization map and strategic diagrams based on keyword co-occurrence are presented. These reports demonstrate that computational intelligence, such as machine and deep learning, fuzzy set theory, evolutionary computing is essential for uncovering and solving data quality issues.

Keywords: Big data · Data quality · Computational intelligence · Knowledge engineering · SciMAT · Uncertainty processing · VOSviewer

1 Introduction

Data plays a significant role in every organization. High-quality data are those that can be quickly analyzed to reveal valuable information [19]. Therefore, data quality assessment is essential to be performed to ensure the quality of data. Improving data quality is the most crucial process of a today's zettabyte era [4].

Nowadays, data shows explosive growth, with collaborative, heterogeneous, and multi-source characters, which increases the complexity and difficulty of data assessment. Assessing data quality is a challenging task. Data quality (DQ) has been investigated extensively, however, only a few research looks at the actual data quality level within organizations [22]. Along with the rapid development of computer science, computational intelligence techniques become noticeably promising.

© Springer Nature Switzerland AG 2021
M. Paszynski et al. (Eds.): ICCS 2021, LNCS 12744, pp. 392–405, 2021.
https://doi.org/10.1007/978-3-030-77967-2_33

In this paper, a complete bibliometric analysis is developed, by retrieving articles from the following databases: IEEE Xplore, ProQuest, ScienceDirect, Scopus, Springer Link. The time window to be analyzed is 2001–2020. The keywords for this searching were "data quality" or "data quality assessment" or "data quality improvement" or "assessing data quality". The number of documents retrieved was 354 articles including 351 unique documents. These articles were screened and analyzed based on their title and abstract following the PRISMA flow diagram guideline [18] shown in Fig. 1. According to the screening process, 190 articles were excluded based on the title. We then evaluate the abstract of 161 articles. 42 articles were excluded based on the abstract resulting in 119 articles eligible to be reviewed. Eventually, 93 articles were included in this paper based on the full-text review.

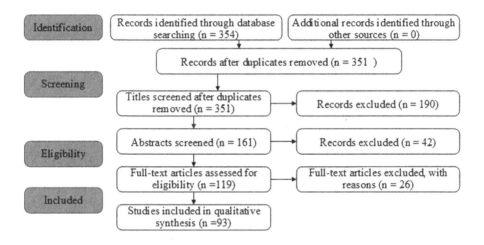

Fig. 1. The article screening process flow based on PRISMA guideline.

An analysis was performed using two bibliometric software tools, namely VOSviewer ver. 1.6.16 and SciMAT ver. 1.1.04. VOSviewer is a freely available computer program for constructing and viewing bibliometric maps [10]. VOSviewer was utilized to generate a network and density visualization. SciMAT is a performance analysis and science mapping software tool for detecting and visualizing conceptual subdomains [7]. In this research, the SciMAT was utilized to build a strategic map for data quality assessment trends. Both analyses were conducted based on the article's keywords in order to show growing patterns in DQ-related techniques.

2 Performance Analysis

The mapping analysis based on the co-occurrence of keywords from articles was carried out using VOSviewer. There were 1430 keywords identified from those articles. We used the threshold of at least three co-occurrence of the keywords to be represented in a node. It has clustered the research trend into 7 clusters. These clusters consist of 91 items and 851 links. Every cluster is represented by different colors. Therefore, it can be easily investigated, see Fig. 2. The total items from cluster 1 to cluster 7 are 21, 15, 15, 13, 10, 9, and 8, respectively. The density of every item is visualized by the size of every node in the network. In cluster 1, the densest item is data quality with 59 occurrences, followed by data quality assessment and data reduction with 30 and 17, respectively. The distribution of the four densest items and the related techniques or algorithms that occurred in every cluster is presented in Table 1.

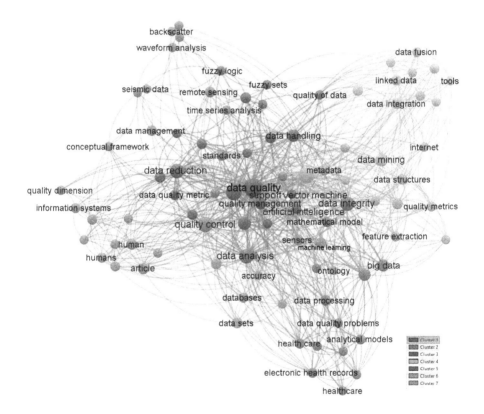

Fig. 2. Network map based on keywords co-occurrence generated by VOSviewer.

The second analysis was performed using SciMAT science mapping software tool. It generated the strategic diagram in the selected period. We divided the whole period (2001–2020) into four consecutive slides listed in Table 2.

A strategic diagram is a two-dimensional space built by plotting themes according to their centrality and density rank value [7]. It consists of four quadrants with the classification as follows:

Table 1. Four densest items on the network.

Cluster	Four densest items	Techniques
Cluster 1	Data quality, data quality assessment, data reduction, decision making	Fuzzy set theory, fuzzy logic, artificial intelligence, support vector machine
Cluster 2	Quality control, article, standard, quality dimensions	Qualitative analysis
Cluster 3	Data analysis, data integrity, data model, big data	Analytic hierarchy process, mathematical model
Cluster 4	Quality assessment, data mining, data integration, linked data	
Cluster 5	Data management, conceptual framework, remote sensing, time series analysis	Time series analysis
Cluster 6	Measurement, data structure, meta data, feature extraction	Feature extraction
Cluster 7	Sensor, machine learning, quality, data sets	Machine learning

Table 2. Performance distribution divided into four consecutive slides.

Period	Number of documents	Number of citations
2001–2005	5	28
2006–2010	15	802
2011–2015	30	620
2016–2020	43	159

- The upper-right quadrant is a motor theme with strong centrality and high density. It is important for structuring a research.
- The upper-left quadrant includes highly developed internal ties, very specialized and isolated themes.
- The lower-left quadrant shows emerging or declining themes, weakly developed (low density) and marginal (low centrality).
- The lower-right quadrant presents transversal and general themes. These themes are important for the research field but are not developed.

There are 4 strategic diagrams have been generated by the SciMAT that represent 4 periods from 2001 to 2020. Due to limitation of the space we only presented one strategic diagram from 2016–2020 period shown in Fig. 3, however Fig. 4 gives a more straightforward overview and summarizes the essential themes for structuring research in data quality assessment from all slides in two decades. We considered only those themes that resided in two quadrants: motor themes and basic and transversal themes.

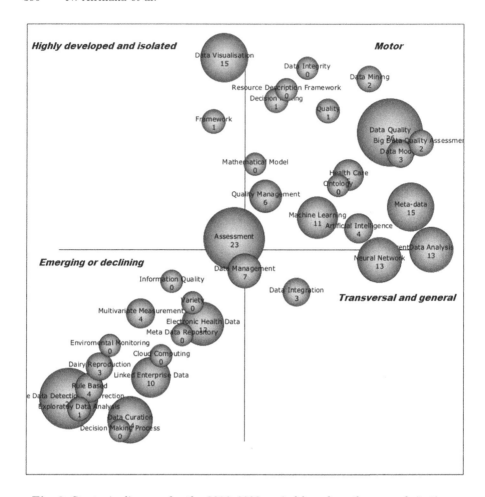

Fig. 3. Strategic diagram for the 2016–2020 period based on the sum of citations.

Based on the strategic diagram for the first period (2001–2005), the research motors are mathematical themes and some themes that are not specified in any particular field. Quality control theme is transversal and general, which means it is essential for the study but not well developed. The highest citation is quality control themes with 28 citations from 3 documents. Therefore, it can be concluded that there was no sufficient research concerning the intelligence techniques for assessing data quality during this period.

Quality assessment, quality improvement, unsupervised classification, medical audit are mainly the motor themes within the 2006–2010 period. In this period, data quality assessment gained more attention along with the intelligence techniques such as unsupervised classification. These themes were playing essential roles in the research field as they were in the motor theme quadrant. The most cited theme with 622 citations from 8 documents are data quality,

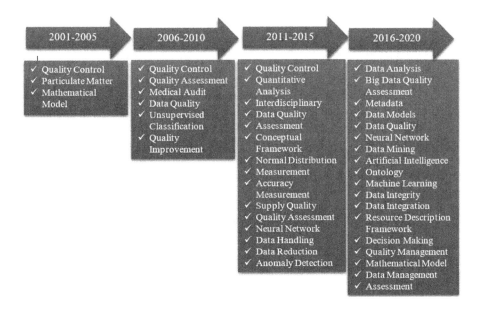

Fig. 4. Essential themes for structuring research in the whole period.

which belonged to the transversal and general quadrant. In this period, data quality and quality control were important for structuring research, although they are general themes.

Quality assessment, anomaly detection, measurement, qualitative analysis, conceptual framework, and neural network are mainly the most cited themes from the motor quadrant within 2011–2015 period. The number of citations of these themes is 460. Artificial intelligence and machine learning themes are placed in the highly developed and isolated quadrant. In the transversal and general quadrant, there are data quality, assessment, quality control, accuracy measurement, and data reductions. These themes were also crucial but not well developed as they belong to the transversal and general quadrant.

Figure 3 shows the recent themes for the 2016–2020 period. They are mainly distributed into two quadrants: motor and emerging or declining quadrant. Only a few themes were plotted in the highly developed and isolated, or transversal and general quadrant. Thus, it would be easier to conclude what are the trends and important themes. From this strategic diagram, we observe that intelligence techniques represented by artificial intelligence, machine learning, and neural network are important for structuring the research field. Artificial intelligence and machine learning are both plotted in motor quadrant, which is well-developed and important for a research, while neural network is at the border between transversal and general quadrant.

3   Top 3 Most Common Data Quality Issues

3.1   Data Quality Dimensions

Data quality dimension is intended to categorize types of data quality measurement. It consists of accuracy, completeness, concordance, consistency, currency, redundancy, or any other dimension [28,34]. For instance, the World Health Organization identifies the completeness and timeliness of the data, the internal coherence of data, the external coherence of data, and data comparisons on the entire population as data quality indicators [24]. Therefore, stating a set of requirements for data quality is crucial in establishing the quality of the data despite standards and methodology used [25]. The authors of [6] concluded that the most common data quality dimensions are completeness, timeliness, and accuracy, followed by consistency and accessibility.

3.2   Outlier Detection

Outlier detection is a process to identify objects that are different than the majority of data, resulting from contamination, error, or fraud [15]. Outliers are either errors or mistakes that are counterproductive. Identifying outliers can lead to better anomaly detection performance [36]. Detecting outliers can also be utilized to give insight into the data quality [26]. In [17], an outlier is used to measure the consistency of climate change data, while in [20], an outlier detection method based on time-relevant k-means was used to detect the voltage, curve, and power data quality issues in electricity data.

In the strategic diagram presented in Fig. 5, outlier or anomaly detection was a motor theme for research on data quality within the 2011–2015 period. Therefore, it is crucial to identify the outlier in order to assess and improve the data quality.

3.3   Data Cleaning and Data Integration

Data cleaning is the backbone of data quality assessment. It purposes to clean the raw data into new data that meet the quality criteria. It includes cleaning the missing values, typos, mixed formats, replicated entries, outliers, and business rules violations [5].

When we take a look at the strategic diagram in Fig. 3, data integration was plotted in the transversal and general quadrant. This theme has the highest citation among others. It means that data integration has an essential role in data quality research although it was classified as not developed. The study in [23] gives an example of a framework for data cleaning and integration in the clinical domain of interest. The framework includes data standardization to finally enable data integration.

4 Computational Intelligence Techniques

4.1 Neural Networks for Assessing Data Quality

The neural network was plotted at the border between the motor and transversal quadrant (Fig. 3), while its bibliometric network is shown in Fig. 5(a). To be able to display all the networks, the analysis was made slightly different from Fig. 2. We changed the network reduction parameter from 3 to 1. The neural network has relations with items such as artificial neural network, autoencoders, data set, fuzzy logic, etc. It has also links with quality assessment and quality control.

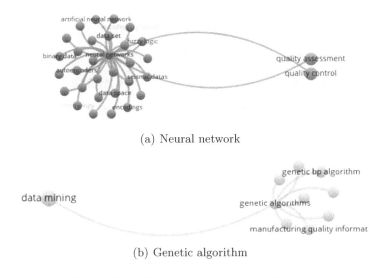

(a) Neural network

(b) Genetic algorithm

Fig. 5. The bibliometric network visualization.

Based on the literature, here are some examples of neural network implementation for assessing data quality. The VGG-16 and the Bidirectional Encoder Representations from Transformers are employed in a task-oriented data quality assessment framework proposed in [16]. Another usage of a neural network model is presented in [8]. Multi-Layer Perceptron was utilized to evaluate the framework of data cleaning in the regression model. The result shows that the models give a better result after the cleaning process. Long Short-Term Memory Autoencoder based on RNNs architecture was used in the DQ assessment model for semantic data in [14]. This framework presents the web contextual data quality assessment model with enhanced classification metric parameters. This contextual data contains metadata with various threshold values for different types of data.

4.2   Fuzzy Logic for Assessing Data Quality

The network of fuzzy set theory in the data quality research is shown in Fig. 6(a). It is a part of the network created by VOSviewer in Fig. 2. It has 16 links, including the link to the data quality, data quality assessment, data integrity, decision making, and data models. Some fuzzy set theories were identified during this study. The Choquet integral in a fuzzy logic principle is utilized by data, and information quality assessment from the framework proposed in [3]. It is used in the context of fuzzy logic and as part of the multi-criteria decision aid system.

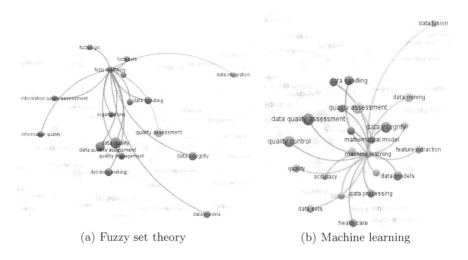

(a) Fuzzy set theory   (b) Machine learning

Fig. 6. The bibliometric network visualization.

Fuzzy TOPSIS was used to measure the results of the systematic literature review approach to identify critical and challenging factors in DevOps data quality assessment [27]. It was implemented to prioritize the investigated challenging factors concerning the DevOps data quality assessment.

4.3   Evolutionary Computation for Assessing Data Quality

Two works in our set have been identified which use a genetic algorithm. The first study was using the genetic algorithm to measure the data quality in dimensions of accuracy, comprehensibility, interestingness, and completeness [21]. The second study [9] employed the association rule for the purpose of quality measurement. A multi-objective genetic algorithm approach for data quality with categorical attributes was utilized. The bibliometric network of the genetic algorithm is presented in Fig. 5(b). It has eight links, one of them for data mining.

4.4   Computational Learning Theory for Assessing Data Quality

The computational learning theory is a fundamental building block of a mathematical formal representation of a cognitive process widely adapted by machine learning. The overview of machine learning's bibliometric map is presented in Fig. 6(b). It has 22 links connected to some items such as data quality assessment, data integrity, data fusion, mathematical model, etc. Random forests were utilized in [13] to predict the accuracy of an early-stage data cleaning, which was used to assess the quality of fertility data stored in dairy herd management software. Time-relevant k-means was employed to measure data accuracy by detecting the electricity data outlier [20].

Local Outlier Factor (LOF), an algorithm for identifying distance-based local outliers, was used in [8]. With LOF, the local density of a particular point is compared with its neighbors. An outlier is suggested if the point is in a sparser region than its neighbors. Another outlier detection used with respect to DQ is Density-Based Spatial Clustering of Applications with Noise (DBSCAN). It chooses an arbitrary unassigned object from the dataset. If it is a core object, DBSCAN finds all connected objects, and these objects are assigned to a new cluster. If it is not a core object, then it is considered as an outlier object [8].

Two clustering techniques, namely t-SNE and PCA, were utilized in a framework that aims to evaluate the precision and accuracy of experimental data [30]. Using t-SNE and PCA gives the dimensional reduction while retaining the most of the information to detect outlier eventually. For unstructured data, the study in [31] suggests to combine techniques such as machine learning, natural language processing, information retrieval, and knowledge management to map the data into a schema.

4.5   Probabilistic and Statistical Methods for Assessing Data Quality

The probabilistic and statistical approach was rarely being included in the abstract or articles' keywords, therefore its appearance in the network map or strategic diagram was not identified. However, this approach was identified varying from many methods and purposes, therefore human-oriented analysis on this approach was made. Univariate and multivariate methods for outlier detection are utilized in [23]. The framework of DQ Assessment for Smart Sensor Network presented in [1] used the interquartile statistical approach to detect outliers. If the data received is lower or greater than the boundaries, it is considered as an outlier. To find a mislabeled data, Shannon Index was utilized in [33]. Shannon Index is a quantitative measure that reflects how many different data there are in a dataset. If the Shannon index is lower than the threshold, it can be predicted as a mislabeled data.

In [2], the completeness of data is measured using logistic regression, while the timeline is measured using binomial regression. Chi-square of patient's age and sex in the medical health record was utilized in [32]. It computes mean and median age by sex for the database population and compares to an external/standard population. Another utilization of the statistical approach is shown in [29]. An ANOVA test was performed to validate the selected characteristics and dimensions for assessing data quality.

In [11,12], a probability-based metric for semantic consistency and assessing data currency is performed using a set of uncertain rules. This metric would allow conditions that are supposed to be satisfied with unique probabilities to be considered. The last approach was found in a medical big data quality evaluation model based on credibility analysis and analytic hierarchy process (AHP) in [35]. Firstly, data credibility is evaluated. It calculates the data quality dimensions after excluding the inaccurate data. Then by combining all dimensions with AHP, it obtains the data quality assessment outcome.

5   Final Remarks

Regarding uncertainty and a vast amount of data, computational intelligence such as machine learning, deep learning, fuzzy set theory, etc., are potent approaches for DQ assessment problems. The use of these methods has been identified in this work. However, the best practice still varies depending on the characteristics and goals of the assessment. The increasing number of successful implementations of these approaches demonstrates the versatility of computational intelligence techniques in assessing DQ.

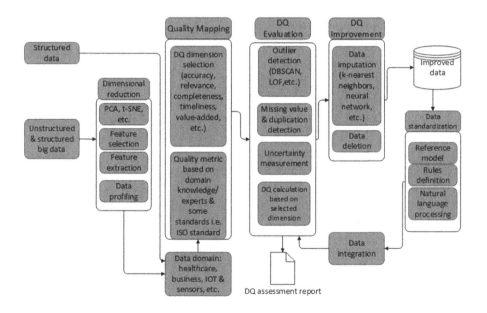

Fig. 7. Proposed data quality assessment framework.

As a final note, we propose a conceptual framework for data quality evaluation, improvement, and possibly data integration, as shown in Fig. 7. This proposed framework will mainly consist of the data model, quality mapping, DQ evaluation, DQ improvement, data standardization, and data integration. In the future, we will evaluate the suitable computational intelligence techniques to be implemented into this framework. From our point of view, developing an intelligent data quality framework would require further advancement in computational science.

Acknowledgments. Part of the work presented in this paper was received financial support from the statutory funds at the Wrocław University of Science and Technology.

References

1. de Aquino, G.R.C., de Farias, C.M., Pirmez, L.: Hygieia: data quality assessment for smart sensor network. In: Proceedings of the 34th ACM/SIGAPP Symposium on Applied Computing, SAC 2019, pp. 889–891. Association for Computing Machinery, New York (2019)
2. BoesBoes, L., et al.: Evaluation of the German surveillance system for hepatitis B regarding timeliness, data quality, and simplicity, from 2005 to 2014. Public Health **180**, 141 (2020)
3. Bouhamed, S.A., Dardouri, H., Kallel, I.K., Bossé, E., Solaiman, B.: Data and information quality assessment in a possibilistic framework based on the Choquet Integral. In: 5th International Conference on Advanced Technologies for Signal and Image Processing (ATSIP), pp. 1–6 (2020)
4. Cai, L., Zhu, Y.: The challenges of data quality and data quality assessment in the big data era. Data Sci. J. **14**, 1–10 (2015)
5. Chu, X., Ilyas, I.F., Krishnan, S., Wang, J.: Data cleaning: overview and emerging challenges. In: Proceedings of the ACM SIGMOD International Conference on Management of Data, pp. 2201–2206 (2016)
6. Cichy, C., Rass, S.: An overview of data quality frameworks. IEEE Access **7**, 24634–24648 (2019)
7. Cobo, M.J., López-Herrera, A.G., Herrera-Viedma, E., Herrera, F.: An approach for detecting, quantifying, and visualizing the evolution of a research field: a practical application to the fuzzy sets theory field. J. Inf. **5**(1), 146–166 (2011)
8. Corrales, D.C., Corrales, J.C., Ledezma, A.: How to address the data quality issues in regression models: a guided process for data cleaning. Symmetry **10**(4), 99 (2018)
9. Das, S., Saha, B.: Data quality mining using genetic algorithm. Int. J. Comput. Sci. Secur. IJCSS **3**(2), 105–112 (2009)
10. van Eck, N.J., Waltman, L.: Software survey: VOS viewer, a computer program for bibliometric mapping. Scientometrics **84**(2), 523–538 (2010)
11. Heinrich, B., Klier, M.: Metric-based data quality assessment - developing and evaluating a probability-based currency metric. Decis. Support Syst. **72**, 82–96 (2015)
12. Heinrich, B., Klier, M., Schiller, A., Wagner, G.: Assessing data quality - a probability-based metric for semantic consistency. Decis. Support Syst. **110**, 95–106 (2018)
13. Hermans, K., et al.: Novel approaches to assess the quality of fertility data stored in dairy herd management software. J. Dairy Sci. **100**(5), 4078–4089 (2017)

14. Jarwar, M.A., Chong, I.: Web objects based contextual data quality assessment model for semantic data application. Appl. Sci. **10**(6), 2181 (2020)
15. Larson, S., et al.: Outlier detection for improved data quality and diversity in dialog systems. In: Proceedings of the 2019 Conference of the North American Chapter of the Association for Computational Linguistics: Human Language Technologies, vol. 1, pp. 517–527 (2019)
16. Li, A., et al.: TODQA: efficient task-oriented data quality assessment. In: 15th International Conference on Mobile Ad-Hoc and Sensor Networks, pp. 81–88. IEEE (2019)
17. Li, J.S., Hamann, A., Beaubien, E.: Outlier detection methods to improve the quality of citizen science data. Int. J. Biometeorol. **64**, 1825–1833 (2020)
18. Liberati, A., et al.: The PRISMA statement for reporting systematic reviews and meta-analyses of studies that evaluate health care interventions: explanation and elaboration, J. Clin. Epidemiol. **62**(10), e1–e34 (2009)
19. Liu, H., Ashwin Kumar, T.K., Thomas, J.P.: Cleaning framework for big data - object identification and linkage. In: 2015 IEEE International Congress on Big Data, pp. 215–221 (2015)
20. Liu, H., Wang, X., Lei, S., Zhang, X., Liu, W., Qin, M.: A rule based data quality assessment architecture and application for electrical data. In: Proceedings of the International Conference on Artificial Intelligence, Information Processing and Cloud Computing. Association for Computing Machinery (2019)
21. Malar Vizhi, J.: Data quality measurement on categorical data using genetic algorithm. Int. J. Data Min. Knowl. Manage. Process. **2**(1), 33–42 (2012)
22. Nagle, T., Redman, T., Sammon, D.: Assessing data quality: A managerial call to action. Bus. Horiz. **63**(3), 325–337 (2020)
23. Pezoulas, V.C., et al.: Medical data quality assessment: on the development of an automated framework for medical data curation. Comput. Biol. Med. **107**, 270–283 (2019)
24. Pietro Biancone, P., Secinaro, S., Brescia, V., Calandra, D.: Data quality methods and applications in health care system: a systematic literature review. Int. J. Bus. Manage. **14**(4), 35 (2019)
25. Plotkin, D.: Important Roles of Data Stewards. In: Data Stewardship, pp. 127–162. Morgan Kaufmann (2014)
26. Pucher, S., Król, D.: A quality assessment tool for Koblenz datasets using metrics-driven approach. In: Fujita, H., Fournier-Viger, P., Ali, M., Sasaki, J. (eds.) IEA/AIE 2020. LNCS (LNAI), vol. 12144, pp. 747–758. Springer, Cham (2020). https://doi.org/10.1007/978-3-030-55789-8_64
27. Rafi, S., Yu, W., Akbar, M.A., Alsanad, A., Gumaei, A.: Multicriteria based decision making of DevOps data quality assessment challenges using fuzzy TOPSIS. IEEE Access **8**, 46958–46980 (2020)
28. Rajan, N.S., Gouripeddi, R., Mo, P., Madsen, R.K., Facelli, J.C.: Towards a content agnostic computable knowledge repository for data quality assessment. Comput. Methods Programs Biomed. **177**, 193–201 (2019)
29. Simard, V., Rönnqvist, M., Lebel, L., Lehoux, N.: A general framework for data uncertainty and quality classification. IFAC PapersOnLine **52**(13), 277–282 (2019)
30. Symoens, S.H., et al.: QUANTIS: data quality assessment tool by clustering analysis. Int. J. Chem. Kinet. **51**(11), 872–885 (2019)
31. Taleb, I., Serhani, M.A., Dssouli, R.: Big Data Quality Assessment Model for Unstructured Data. In: Proceedings of the 13th International Conference on Innovations in Information Technology, pp. 69–74 (2018)

32. Terry, A.L., Stewart, M., Cejic, S., et al.: A basic model for assessing primary health care electronic medical record data quality. BMC Med. Inform. Decis. Mak. **19**, 30 (2019)

33. Udeshi, S., Jiang, X., Chattopadhyay, S.: Callisto: entropy-based test generation and data quality assessment for machine learning systems. In: 13th International Conference on Software Testing, Verification and Validation, pp. 448–453 (2020)

34. Valencia-Parra, Á., Parody, L., Varela-Vaca, Á.J., Caballero, I., Gómez-López, M.T.: DMN4DQ: when data quality meets DMN. Decis. Support Syst. **141**, 113450 (2021)

35. Zan, S., Zhang, X.: Medical data quality assessment model based on credibility analysis. In: 4th Information Technology and Mechatronics Engineering Conference, pp. 940–944 (2018)

36. Zimek, A., Schubert, E.: Outlier detection. In: Liu, L., Öszu, T. (eds.) Encyclopedia of Database Systems. Springer, Cham (2017) https://doi.org/10.1007/978-1-4899-7993-3_80719-1

The Power of a Collective: Team of Agents Solving Instances of the Flow Shop and Job Shop Problems

Piotr Jedrzejowicz[ID] and Izabela Wierzbowska[(✉)][ID]

Gdynia Maritime University, Gdynia, Poland
p.jedrzejowicz@umg.edu.pl, i.wierzbowska@wpit.umg.edu.pl

Abstract. The paper proposes an approach for solving difficult combinatorial optimization problems integrating the mushroom picking population-based metaheuristic, a collective of asynchronous agents, and a parallel processing environment, in the form of the MPF framework designed for the Apache Spark computing environment. To evaluate the MPF performance we solve instances of two well-known NP-hard problems – job shop scheduling and flow shop scheduling. In MPF a collective of simple agents works in parallel communicating indirectly through the access to the common memory. Each agent receives a solution from this memory and writes it back after a successful improvement. Computational experiment results confirm that the proposed MPF framework can offer competitive results as compared with other recently published approaches.

Keywords: Collective of agents · Metaheuristics · Parallel computations · Computationally hard combinatorial optimization problems

1 Introduction

Computational collective intelligence (CCI) techniques use computer-based models, algorithms, and tools that take advantage of the synergetic effects of interactions between agents acting in parallel to reach a common goal. In the field of optimization, applications of the CCI techniques usually involve the integration of multiple agent systems with the population-based metaheuristics including the cooperative co-evolutionary algorithms.

Population-based metaheuristics are used to deal with computationally difficult optimization problems like, for example, combinatorial optimization, global optimization in complex systems, multi-criteria optimization as well as optimization and control in dynamic systems. Population in a population-based metaheuristic represents solutions or some constructs that can be easily transformed into solutions. Population-based algorithms reach their final solutions after having carried out various operations transforming populations, sub-populations, or population members to find the best solution. Advantages of the population-based algorithms can be attributed to their following abilities:

© Springer Nature Switzerland AG 2021
M. Paszynski et al. (Eds.): ICCS 2021, LNCS 12744, pp. 406–419, 2021.
https://doi.org/10.1007/978-3-030-77967-2_34

- Reviewing in a reasonable time a big number of possible solutions from the search space.
- Directing search processes towards more promising areas of the search space.
- Increasing computation effectiveness through implicit or explicit cooperation between population members and thus achieving a synergetic effect.
- Performing a search for the optimum solution in parallel and a distributed environment.

More details on the population-based metaheuristics can be found in reviews of [6], [12] and [20].

An important tool for increasing computation effectiveness in solving difficult optimization problems is the decentralization of efforts and cooperation between decentralized computational units. To achieve full advantages of such a cooperation, multiple agent frameworks have been proposed and implemented. Agent-based implementation of metaheuristics allows autonomous agents to communicate and cooperate through information exchange synchronously or asynchronously. Besides, there might be some kind of learning implemented in agents. This feature enables the agent to assimilate knowledge about the environment and other agents' actions and use it to improve the consequences of their actions. The review of frameworks for the hybrid metaheuristics and multi-agent systems for solving optimization problems can be found in [19]. Example frameworks used for developing multi-agent systems and implementing population-based metaheuristic algorithms include AMAM - a multi-agent framework applied for solving routing and scheduling problems [18] and JABAT, a middleware for implementing JADE-based and population-based A-Teams [3].

Effects of integrating population-based metaheuristics and multi-agent technology for solving difficult computational problems, especially combinatorial optimization problems, are constrained by the available computation technologies. Recent developments in the field of parallel and distributed computing make it possible to alleviate some of these constraints. Several years ago parallel and distributed computing were a promising, but rather a complex way of programming. At present every programmer should have a working knowledge of these paradigms, to exploit current computing architectures [8].

This paper aims to show that integrating an approach involving a population-based metaheuristic, a collective of asynchronous agents, and a parallel processing environment, may benefit the search for a solution in case of difficult combinatorial optimization problems. To demonstrate that the above statement holds we show the results of a computational experiment involving parallel implementation of the Mushroom Picking Algorithm (MPA) with asynchronous agents. Our test-bed consists of two well-known NP-hard scheduling problems – flow shop (PFSP) and job shop (JSSP), and the MPF framework designed to enable MPA implementation using the Apache Spark, an open-source data-processing engine for large data sets. It is designed to deliver the computational speed, scalability, and programmability required for Big Data [22].

The rest of the paper is constructed as follows. Section 2 contains a brief description of the considered scheduling problems. Section 3 reviews currently published algorithms for solving instances of PFSP and JSSP. Section 4 gives details of the implementation of the parallel, agent-based, using the MPF framework. Section 5 contains the results of the computational experiment. Section 6 includes conclusions and suggestions for future research.

2   Scheduling Problems

Job Shop Scheduling Problem (JSSP)   consists of a set of n jobs $(j_1, \ldots, j_n)$ to be scheduled on m machines $(m_1, \ldots, m_m)$. Each job consists of operations (tasks) that have to be processed in the given order. Each operation within a job must be processed on a specific machine, only after all preceding operations of this job are completed.

Further constraints include:

- Operations cannot be interrupted.
- Each machine can handle only one job at a time.

The goal is to find the job sequences on machines minimizing the makespan. A single solution may be represented as the ordered list of the numbers of the jobs. The length of the list is $n \times m$. There are m occurrences of each job in such a list. When examining the list from the left to the right, the ith occurrence of job j refers to the ith operation (task) of this job.

The problem was proven to be NP-hard in [16].

Permutation Flow Shop Scheduling Problem (PFSP)   consists of a set of different machines that carry out operations (tasks) of jobs. All jobs have identical processing order of their operations. Following [5], assume that the order of processing a set of jobs J on m different machines is described by the machine sequence $P_1, \ldots, P_m$. Hence, job $J_j \in J$ consists of m operations $O_{1j}, \ldots, O_{mj}$ with processing times p_{ij}, $i = 1, \ldots, m$, $j = 1, \ldots, n$ where n is the number of jobs in J.

The following constraints on jobs and machines have to be met:

- Operations cannot be interrupted.
- Each machine can handle only one job at a time.

While the machine sequence of all jobs is identical, the problem is to find the job sequence minimizing the makespan (maximum of the completion times of all tasks). A single solution may be represented as the ordered list of the numbers of the jobs of the length n.

The problem was proven to be NP-hard in [10].

3   Current Approaches for Solving PFSP and JSSP

3.1   Algorithms and Approaches for Solving JSSP Instances

Population-based algorithms including swarm intelligence and evolutionary systems have proven successful in tackling JSSP, one of the hard optimization problems considered in this study. A state of the art review on the application of the

AI techniques for solving the JSSP as of 2013 can be found in [29]. Recently, several interesting swarm intelligence solutions for JSSP were published. In [11] the local search mechanism of the PSA and large-span search principle of the cuckoo search algorithm are combined into an improved cuckoo search algorithm (ICSA). A hybrid algorithm for solving JSSP integrating PSO and neural network was proposed in [36]. An improved whale optimization algorithm (IWOA) based on quantum computing for solving JSSP instances was proposed by [37]. An improved GA for JSSP [7] offers good performance. Their niche adaptive genetic algorithm (NAGA) involves several rules to increase population diversity and adjust the crossover rate and mutation rate according to the performance of the genetic operators. Niche is seen as the environment permitting species with similar features to compete for survival in the elimination process. According to the authors, the niche technique prevents premature convergence and improves population diversity. Recently, a well-performing GA for solving JSSP instances was proposed in [14]. The authors suggest a feasibility preserving solution representation, initialization, and operators for solving job-shop scheduling problems. Another genetic algorithm combined with the local search was proposed in [30]. The approach features the use of a local search strategy in the traditional mutation operator; and a new multi-crossover operator.

A novel two-level metaheuristic algorithm was suggested in [21]. The lower-level algorithm is a local search algorithm searching for an optimal JSSP solution within a hybrid neighborhood structure. The upper-level algorithm is a population-based search algorithm developed for controlling the input parameters of the lower-level algorithm.

A discrete wolf pack algorithm (DWPA) for job shop scheduling problems was proposed in [32]. DWPA involves 3 phases: initialization, scouting, and summoning. During initialization heuristic rules are used to generate a good quality initial population. The scouting phase is devoted to the exploration while summoning takes care of the intensification. In [31] a novel biomimicry hybrid bacterial foraging optimization algorithm (HBFOA) was developed. HBFOA is inspired by the behavior of E. coli bacteria in its search for food. The algorithm is hybridized with simulated annealing. Additionally, the algorithm was enhanced by a local search method. Evaluation of the performance of several PSO-based algorithms for solving the JSSP can be found in [1].

As in the case of other computationally difficult optimization problems, an emerging technology supported development of parallel and distributed algorithms for solving JSSP instances. A scheduling algorithm, called MapReduce coral reef (MRCR) for JSSP instances was proposed in [28]. The basic idea of the proposed algorithm is to apply the MapReduce platform and the Spark Apache environment to implement the coral reef optimization algorithm to speed up its response time. More recently, a large-scale flexible JSSP optimization by a distributed evolutionary algorithm was proposed in [25]. The algorithm belongs to the distributed cooperative evolutionary algorithms class and is implemented on Apache Spark.

3.2   Algorithms and Approaches for Solving PFSP Instances

There exist several heuristic approaches for solving PFSP. Selected heuristics, namely CDS, Palmer's slope index, Gupta's algorithm, and concurrent heuristic algorithm for minimizing the makespan in permutation flow shop scheduling problem were studied in [24]. An improved heuristic algorithm for solving the flow shop scheduling problem was proposed in [23]. In [4] the adapted Nawaz-Enscore-Ham (NEH) heuristic and two metaheuristics based on the exploration of the neighborhood are studied. Another modification of NEH heuristic was suggested in [17] where a novel tie-breaking rule was developed by minimizing partial system idle time without increasing the computational complexity of the NEH heuristic.

A Tabu Search with the intensive concentric exploration over non-explored areas was proposed in [9] as an alternative solution to the simplest Tabu Search with the random shifting of two jobs indexes operation for Permutation Flow Shop Problem (PFSP) with the makespan minimization criterion.

Recently, several metaheuristics have proven effective in solving PFSP instances. In [33] the authors propose two water wave optimization (WWO) algorithms for PFSP. The first algorithm adapts the original evolutionary operators of the basic WWO. The second further improves the first algorithm with a self-adaptive local search procedure. Application of the cuckoo search metaheuristic for PFSP was suggested in [35]. The approach shows good performance in solving the permutation flow shop scheduling problem. Modified Teaching-Learning-Based Optimization with Opposite-Based-Learning algorithm was applied to solve the Permutation Flow-Shop-Scheduling Problem under the criterion of minimizing the makespan was proposed in [2]. To deal with the complex PFSPs, the paper of [34] proposed an improved simulated annealing (SA) algorithm based on the residual network. First, this paper defines the neighborhood of the PFSP and divides its key blocks. Second, the residual network algorithm is used to extract and train the features of key blocks. Next, the trained parameters are used in the SA algorithm to improve its performance.

4   An Approach for Solving JSSP and PFSP

4.1   The MPF Framework

To deal with the considered combinatorial optimization problems we use the Mushroom Picking Framework (MPF). The MPF is based on the Mushroom Picking Algorithm (MPA) originally proposed in [13] for solving instances of the Traveling Salesman Problem and job shop scheduling. The metaphor of MPA refers to a situation where many mushroom pickers, with different preferences as to the collected mushroom kinds, explore the woods in parallel pursuing individual or random, or mixed strategies and trying to increase the current crop. Pickers exchange information indirectly by observing traces left by others and modifying their strategies accordingly. In case of finding interesting species, they intensify search in the vicinity hoping to find more specimens. In the MPA a set of simple, dedicated, agents, metaphorically mushroom pickers, explore in parallel the search space. Agents differ between themselves by performing different operations on the encountered solutions. They may have also different computational complexities. Agents explore a search space randomly intensifying their efforts after having found an improved solution.

MPF differs from MPA in being only a framework, allowing the user to define the internal algorithm controlling a solution improvement processes performed by an agent. There are no constraints on the number of agents with different internal algorithms used. There are also no constraints on the overall number of agents employed for solving a particular instance of the problem at hand. The user is also responsible for generating the initial population of solutions and for storing it in the common memory. MPF provides the capability of reading one or more solutions from the common memory and the capability of writing an improved by an agent solution in the common memory.

Agents in the MPF work in parallel, in threads, and cycles. Each cycle involves the following steps:

- Solutions in the common memory are randomly shuffled.
- The population of solutions in the common memory is divided into several subpopulations of roughly equal size. Observe that shuffling at stage I assures that subpopulations do not consist of the same solutions in different cycles.
- Each subpopulation is processed by a set of agents in a separate thread. The same composition of agent kinds and numbers is used in each thread. Each agent receives a solution or solutions (depending on the number of arguments of the agent) and runs its internal algorithm which could be, for example, a local search algorithm, to produce an improved solution. If such a solution is found, it replaces the solution drawn from the subpopulation in the case of the single argument agents. Otherwise, it replaces the worst one, out of all solutions processed by an agent.
- The cycle ends after a predefined number of trials to improve the subpopulations have been applied in all threads.
- At the end of a cycle all current subpopulations are appended into the common memory.

The overall stopping criterion is defined as no improvement of the best result (fitness) after the predefined number of cycles has elapsed.

4.2   The MPF Framework Implementation for Scheduling Problems

The general scheme of the MPF implementation for scheduling problems is shown in a pseudo-code as Algorithm 1.

In Algorithm 1, *applyOptimization* is responsible for improving solutions in each subpopulation in all threads. In the first cycle, *ApplyOptimizations* receives solutions not yet initialized as for the sequence of jobs, and it starts with filling these solutions with randomly generated sequences of jobs.

For the proposed implementation of the MPF for solving PFSP and JSSP instances in each thread we use the following set of agents:

Algorithm 1: MPA

$n \leftarrow$ the number of parallel threads
$solutions \leftarrow$ a set of solutions with empty sequence of jobs
while $!stoppingCriterion$ **do**
  $populations \leftarrow$ solutions randomly split into n subsets of equal size
  $populationsRDD \leftarrow populations$ parallelized in ApacheSpark
  $populationsRDD \leftarrow populationsRDD.map(p => p.applyOptimizations)$
  $solutions = populationsRDD.flatMap(identity).collect()$
         // thanks to $flatMap$, *collect* returns list
         // of solutions, not list of populations
  $bestSolution \leftarrow$ a solution from $solutions$ with the best fitness
return $bestSolution$

- randomReverse—takes a random slice of the list of jobs and reverses the order of its elements;
- randomMove – takes one random job from the list of jobs and moves it to another, random position,
- randomSwap – replaces jobs on two random positions in the list of jobs,
- crossover – requires two solutions. A slice from the first solution is extended with the missing jobs in the order as in the second solution.

During computations, solutions in each subpopulation may, with time, become similar or even the same. To assure the required level of diversification of the solutions, two measures are introduced:

- The crossover agent is chosen by the $ApplyOptimization$ procedure twice less often than each of the one-argument agents (in each thread there is only one such agent, while the other agents come in pairs).
- If two solutions drawn for the crossover agent have the same fitness, or fitness differing by 1, the worse solution is replaced by a new random one.

$ApplyOptimization$ is shown as Algorithm 2.

The implementation for both considered problems that is PFSP and JSSP differs mainly in how solutions are represented as explained in Subsect. 2 and Subsect. 2. In both cases, a solution is represented by a list of numbers and such solutions are processed in the same way in all subpopulations, and by the same agents, as described earlier.

If a method of calculating the length of the makespan is defined for the JSSP, then it may be also used for PFSP, however first the solution of PFSP must be transformed to represent the sequence of operations as in the JSSP case. The solution $(j_1, j_2, \ldots, j_n)$ is mapped to $(j_1, j_1, \ldots, j_1, j_2, j_2, \ldots, j_2, \ldots, j_n, j_n, \ldots, j_n)$. Thus the same code with very few changes (including the mapping procedure) has been used for both problems.

Algorithm 2: applyOptimizations

$solutions \leftarrow$ solutions in the subpopulation
foreach $s \in solutions$ **do**
    if $s.jobs == null$ **then**   // s has empty sequence of jobs
      | $s.jobs \leftarrow$ random sequence of jobs
for $k \leftarrow 1$ **to** given number of iterations **do**
    $A \leftarrow$ random agent from the available agents
    if A is two argument agent **then**
        $s1, s2 \leftarrow$ two solutions drawn from $solutions$
        $sw \leftarrow s1 \max s2$ // solution with the bigger makespan
        if $abs(s1.makespan - -s2.makespan) < 2$ **then**
        | in solutions replace sw with a random solution
        else
            $newSolution \leftarrow A(s1, s2)$
            if $newSolution.makespan < sw.makespan$ **then**
            | in $solutions$ replace sw with $newSolution$
    else
        $s \leftarrow$ draw one solution from $solutions$
        $newSolution \leftarrow A(s)$
        if $newSolution.makespan < s.makespan$ **then**
        | in solutions replace s with newSolution
return $solutions$

---

5   Computational Experiment Results

To validate the proposed approach, we have carried out several computational experiments. Experiments were based on two widely used benchmark datasets: the Lawrence dataset for JSSP [15], and the Taillard dataset for PFSP [27]. Both datasets contain instances with known optimal solutions for the minimum makespan criterion. All computations have been run on Spark cluster consisting of 8 nodes with 32 virtual central processing units at the Academic Computer Center in Gdansk. Performance measures included errors calculated as a percentage deviation from the optimal solution value and computation time in seconds.

In [13] it has been shown, that in the MPA the choice of agents that are used to improve solutions may lead to significant differences in the produced results. For the current MPF implementation, agents have been redesigned and changed as described in Subection 4.2. In Table 1 the performance of the proposed approach denoted as MPF is compared with results from [13] and performances of other recently published algorithms for solving JSSP on Lawrence benchmark instances. The errors for [14] have been calculated based on the results from their paper. The results for MPF have been averaged over 30 runs for each problem instance. For solving the JSSP instances by the MPF, the following parameter settings have been used:

– for instances from la01 to la15–200 subpopulations, each consisting of 3 solutions, 3000 iterations in each cycle and stopping criterion as no change in the best solution for two consecutive cycles;

– for instances from la16 to la40–400 subpopulations, each consisting of 3 solutions, 6000 iterations in each cycle, and stopping criterion as no change in the best solution for five consecutive cycles.

From Table 1 it can be observed that MPF outperforms in terms of both measures - average error and computation time - MPA, GA of [14], enhanced GA of [30]. The enhanced two-level metaheuristic (MUPLA) of [21] offers smaller average errors at the cost of exceedingly high computation times.

To gain better insight into factors influencing the performance of the proposed approach we have run several variants of MPF with different components using a sample of instances from the Lawrence benchmark dataset as shown in Table 2. These experiments were run with the same parameter settings as in the case of results shown in Table 1.

From the results shown in Table 2, it can be observed that both mechanisms introduced within the proposed approach, that is shuffling of solutions in the common memory, and diversification by introducing random solutions, enhance the performance of the MPF. Shuffling stands behind the indirect cooperation between agents and both – diversification and shuffling help getting out of local optima. It should be also noted that results produced by MPF are fairly stable in terms of the average standard deviation of errors.

The PFSP problem experiment has been based on the Taillard benchmark dataset consisting of 10 instances for each considered problem size. Best known values for Taillard instances can be found online [26]. In the experiment the following settings for the proposed MPF have been used:

– for sizes 200×20 and 500×20–112 three-solution subpopulations, 100 iterations in each cycle and stopping criterion as no change in the best solution for 10 and 5 consecutive cycles respectively;
– for 50×20, 100×10, 100×20, 200×10–200 three-solution subpopulations, 1500 iterations in each cycle and stopping criterion as no change in the best solution for 5 consecutive cycles;
– for all other instances – 200 three-solution subpopulations, 3000 iterations in each cycle, and stopping criterion as no change in the best solution for 5 consecutive cycles;

In Table 3 the performance of MPF is compared with results of other, recently published, approaches.

From the results shown in Table 3, it can be observed that in terms of average error outperforms other approaches except for the water wave optimization algorithm implementation of [33]. Unfortunately, information as to computation times is not available for other approaches. The average standard deviation of errors in the case of the MPF is fairly stable for smaller instances, growing with the problem size.

Table 1. Comparison of results for the JSSP problem

Data-set	Make-span	MPF Error %	Time s	SD %	MPA [13] Error %	Time s	GA[14] Error %	mXLSGA [30] Error %	Time s	MUPLA [21] Error %	Time s
la01	666	0.00%	1	0.00%	0.00%	1	0.00%	0.00%	n.a	0.00%	1
la02	655	0.00%	1	0.00%	0.00%	1	1.22%	0.00%	n.a	0.00%	2
la03	597	0.41%	2	0.54%	0.84%	2	1.01%	0.00%	34	0.00%	10
la04	590	0.00%	1	0.00%	0.24%	1	2.37%	0.00%	n.a	0.00%	2
la05	593	0.00%	1	0.00%	0.00%	1	0.00%	0.00%	n.a	0.00%	0
la06	926	0.00%	1	0.00%	0.00%	1	0.00%	0.00%	n.a	0.00%	2
la07	890	0.00%	1	0.00%	0.00%	1	0.00%	0.00%	n.a	0.00%	2
la08	863	0.00%	1	0.00%	0.00%	1	3.23%	0.00%	n.a	0.00%	2
la09	951	0.00%	1	0.00%	0.00%	1	0.00%	0.00%	n.a	0.00%	2
la10	958	0.00%	1	0.00%	0.00%	1	0.00%	0.00%	n.a	0.00%	2
la11	1222	0.00%	2	0.00%	0.00%	1	0.00%	0.00%	n.a	0.00%	4
la12	1039	0.00%	2	0.00%	0.00%	1	0.00%	0.00%	n.a	0.00%	8
la13	1150	0.00%	2	0.00%	0.00%	1	0.00%	0.00%	n.a	0.00%	6
la14	1292	0.00%	2	0.00%	0.00%	1	0.00%	0.00%	n.a	0.00%	6
la15	1207	0.00%	3	0.00%	0.00%	31	0.75%	0.00%	n.a	0.00%	7
la16	945	0.07%	21	0.05%	0.31%	41	2.96%	0.00%	n.a	0.00%	294
la17	784	0.01%	16	0.07%	0.06%	40	1.66%	0.00%	70	0.00%	33
la18	848	0.00%	21	0.00%	0.20%	41	2.48%	0.00%	n.a	0.00%	24
la19	842	0.20%	26	0.35%	1.01%	42	4.87%	0.00%	n.a	0.00%	149
la20	901	0.46%	16	0.21%	0.50%	33	1.77%	0.00%	n.a	0.00%	1073
la21	1046	1.55%	64	0.58%	3.03%	79	10.07%	1.24%	n.a	0.06%	30668
la22	927	1.23%	63	0.42%	2.08%	75	11.00%	0.86%	n.a	0.00%	1439
la23	1032	0.00%	26	0.00%	0.00%	45	5.72%	0.00%	n.a	0.00%	25
la24	935	1.68%	57	0.78%	3.50%	65	10.37%	1.17%	n.a	0.26%	21350
la25	977	1.60%	67	0.74%	3.52%	78	8.50%	0.92%	n.a	0.00%	15827
la26	1218	0.13%	88	0.27%	1.61%	106	11.17%	0.00%	n.a	0.00%	82
la27	1235	3.21%	85	0.56%	4.49%	121	13.52%	2.75%	n.a	0.03%	194427
la28	1216	1.90%	117	0.65%	3.15%	114	13.65%	1.89%	n.a	0.00%	1972
la29	1152	5.69%	114	1.05%	7.77%	121	16.58%	4.26%	n.a	1.02%	130059
la30	1355	0.01%	68	0.05%	0.61%	112	9.30%	0.00%	236	0.00%	123
la31	1784	0.00%	64	0.00%	0.00%	64	3.25%	0.00%	n.a	0.00%	306
la32	1850	0.00%	76	0.00%	0.00%	85	5.46%	0.00%	n.a	0.00%	172
la33	1719	0.00%	60	0.00%	0.00%	74	4.07%	0.00%	n.a	0.00%	313
la34	1721	0.00%	87	0.00%	0.36%	172	7.50%	0.00%	n.a	0.00%	448
la35	1888	0.00%	71	0.00%	0.03%	95	3.92%	0.00%	n.a	0.00%	393
la36	1268	3.23%	87	0.72%	4.53%	80	10.02%	2.12%	n.a	0.00%	85418
la37	1397	3.84%	87	0.96%	4.94%	94	13.10%	1.28%	n.a	0.00%	60481
la38	1196	4.27%	116	1.17%	7.02%	106	17.56%	4.18%	n.a	0.25%	169974
la39	1233	2.39%	108	0.45%	4.25%	99	12.08%	2.02%	n.a	0.00%	18057
la40	1222	2.61%	102	0.99%	3.69%	109	13.26%	1.71%	n.a	0.16%	119463
Avg		0.86%	43	0.00%	1.44%	55	5.56%	0.61%	n.a	0.04%	21316

Table 2. MPF performance with different variants of components used

Dataset	Random, Shuffling			No random, Shuffling			Random, No shuffling			No random, No shuffling		
	Error	Time	SD	Error	Time	SD	Error	Time	SD	Error	Time	SD
	%	s	%	%	s	%	%	s	%	%	s	%
la20	0.46%	16	0.2%	0.48%	14	0.2%	0.46%	14	0.2%	0.63%	15	0.2%
la21	1.55%	64	0.6%	2.37%	43	0.8%	3.52%	54	0.8%	4.65%	39	0.7%
la22	1.23%	63	0.4%	1.70%	47	0.4%	2.44%	64	0.6%	3.32%	42	0.9%
la23	0.00%	26	0.0%	0.00%	26	0.0%	0.00%	34	0.0%	0.00%	32	0.0%
la24	1.68%	57	0.8%	2.74%	53	0.7%	3.28%	57	0.7%	4.72%	46	0.8%
la25	1.60%	67	0.7%	2.81%	57	1.0%	3.22%	59	1.0%	5.36%	42	0.7%
la26	0.13%	88	0.3%	0.35%	89	0.5%	1.42%	113	0.9%	2.61%	88	0.9%
la27	3.21%	85	0.6%	3.58%	84	0.6%	4.96%	95	0.6%	5.66%	78	0.8%
la28	1.90%	117	0.6%	2.20%	96	0.5%	3.75%	106	0.6%	4.90%	83	1.0%
la29	5.69%	114	1.1%	6.56%	106	1.3%	8.45%	90	0.5%	9.67%	83	0.9%
la30	0.01%	68	0.1%	0.11%	91	0.3%	0.52%	106	0.5%	1.92%	77	0.8%
Avg	1.59%	70	0.5%	2.08%	64	0.6%	2.91%	72	0.6%	3.95%	57	0.7%

Table 3. Performance of the MPF versus other approaches

Size	MPF			CH [24]	[23]	NEHLJP1 [17]	WWO [33]
	Error	Time	SD	Error	Error	Error	Error
	%	s	%	%	%	%	%
20 × 5	0.04%	5	0.00%	5.94%	1.99%	2.16%	0.00%
20 × 10	0.03%	15	0.03%	8.77%	3.97%	3.68%	0.01%
20 × 20	0.02%	27	0.02%	9.46%	3.26%	3.06%	0.02%
50 × 5	0.03%	25	0.02%	5.10%	0.57%	0.64%	0.00%
50 × 10	0.82%	94	0.17%	7.04%	4.24%	4.25%	0.19%
50 × 20	1.29%	164	0.30%	8.78%	5.29%	6.15%	0.28%
100 × 5	0.06%	70	0.02%	3.57%	0.36%	0.36%	0.00%
100 × 10	0.55%	162	0.17%	6.92%	1.50%	1.72%	0.21%
100 × 20	1.96%	830	0.28%	8.28%	4.68%	4.81%	0.86%
200 × 10	0.49%	861	0.13%	5.60%	0.96%	0.89%	0.08%
200 × 20	3.04%	881	0.38%	7.60%	4.14%	3.65%	2.36%
500 × 20	3.05%	3138	0.42%	5.50%	1.89%	1.62%	2.08%
Avg	0.95%	523	0.16%	6.88%	2.74%	2.75%	2.74%

6   Conclusions

The paper proposes a framework for solving combinatorial optimization problems using Apache Spark computation environment and a collective of simple optimization agents. The proposed framework denoted as MPF is flexible and can be used for solving a variety of combinatorial optimization problems. In the current paper, we demonstrate the MPF application for solving instances of job shop and flow shop scheduling problems. The idea of the MPF is based on recently proposed by the authors mushroom picking metaheuristic, where many agents explore randomly the solution space intensifying their search around promising solutions with diversification mechanism enabling escape from local optima. The approach assumes indirect cooperation between the collective members sharing access to the common memory containing a population of solutions. The computational experiment carried out, and comparisons with several recently published approaches to solving both considered scheduling problems, show that the proposed MPF implementation can obtain competitive results in a reasonable time.

Future research will focus on designing and testing a wider library of optimization agents allowing for the effortless implementation of the approach for solving a more extensive range of difficult optimization problems. Also, the framework may be extended by some new features, like for example online adjustments in the intensity of usage of the available agents. At the current version the number of agents and the frequency with which they are called is predefined. Both values could be automatically adapted during computations.

References

1. Anuar, N.I., Fauadi, M.H.F.M., Saptari, A.: Performance evaluation of continuous and discrete particle swarm optimization in job-shop scheduling problems. In: Materials Science and Engineering Conference Series. Materials Science and Engineering Conference Series, vol. 530, p. 012044 (June 2019). https://doi.org/10.1088/1757-899X/530/1/012044
2. Balande, U., Shrimankar, D.: A modified teaching learning metaheuristic algorithm with opposite-based learning for permutation flow-shop scheduling problem. Evol. Intell. 1–23 (2020). https://doi.org/10.1007/s12065-020-00487-5
3. Barbucha, D., Czarnowski, I., Jedrzejowicz, P., Ratajczak, E., Wierzbowska, I.: Jade-based a-team as a tool for implementing population-based algorithms. In: Sixth International Conference on Intelligent Systems Design and Applications, vol. 3, pp. 144–149 (2006). https://doi.org/10.1109/ISDA.2006.31
4. Belabid, J., Aqil, S., Allali, K.: Solving permutation flow shop scheduling problem with sequence-independent setup time. J. Appl. Math. **2020**, 1–11 (2020). https://doi.org/10.1155/2020/7132469
5. Blazewicz, J., Ecker, K., Pesch, E., Schmidt, G., Weglarz, J.: Scheduling computer and manufacturing processes, Springer, Berlin (1996). https://doi.org/10.1007/978-3-662-03217-6
6. Boussaïd, I., Lepagnot, J., Siarry, P.: A survey on optimization metaheuristics. Inf. Sci. **237**, 82–117 (2013). https://doi.org/10.1016/j.ins.2013.02.041
7. Chen, X., Zhang, B., Gao, D.: Algorithm based on improved genetic algorithm for job shop scheduling problem, pp. 951–956 (2019). https://doi.org/10.1109/ICMA.2019.8816334
8. Danovaro, E., Clematis, A., Galizia, A., Ripepi, G., Quarati, A., D'Agostino, D.: Heterogeneous architectures for computational intensive applications: a cost-effectiveness analysis. J. Comput. Appl. Math. **270**, 63–77 (2014). https://doi.org/10.1016/j.cam.2014.02.022
9. Dodu, C., Ancau, M.: A tabu search approach for permutation flow shop scheduling. Studia Universitatis Babes-Bolyai Informatica **65**, 104–115 (2020). https://doi.org/10.24193/subbi.2020.1.08

10. Garey, M.R., Johnson, D.S., Sethi, R.: The complexity of flowshop and jobshop scheduling. Math. Oper. Res. **1**(2), 117–129 (1976). http://www.jstor.org/stable/3689278
11. Hu, H., Lei, W., Gao, X., Zhang, Y.: Job-shop scheduling problem based on improved cuckoo search algorithm. Int. J. Simul. Model. **17**, 337–346 (2018). https://doi.org/10.2507/IJSIMM17(2)CO8
12. Jedrzejowicz, P.: Current trends in the population-based optimization. In: Nguyen, N.T., Chbeir, R., Exposito, E., Aniorté, P., Trawiński, B. (eds.) ICCCI 2019. LNCS (LNAI), vol. 11683, pp. 523–534. Springer, Cham (2019). https://doi.org/10.1007/978-3-030-28377-3_43
13. Jedrzejowicz, P., Wierzbowska, I.: Parallelized swarm intelligence approach for solving tsp and jssp problems. Algorithms **13**(6), 142 (2020). https://doi.org/10.3390/a13060142
14. Kalshetty, Y., Adamuthe, A., Kumar, S.: Genetic algorithms with feasible operators for solving job shop scheduling problem. J. Sci. Res. **64**, 310–321 (2020). https://doi.org/10.37398/JSR.2020.640157
15. Lawrence, S.: Resource constrained project scheduling - technical report (1984)
16. Lenstra, J., Rinnooy Kan, A., Brucker, P.: Complexity of machine scheduling problems. Ann. Discrete Math. **1**, 343–362 (1977). https://doi.org/10.1016/S0167-5060(08)70743-X
17. Liu, W., Jin, Y., Price, M.: A new improved NEH heuristic for permutation flowshop scheduling problems. Int. J. Prod. Econ. **193**, 21–30 (2017). https://doi.org/10.1016/j.ijpe.2017.06.026
18. Lopes Silva, M.A., de Souza, S.R., Freitas Souza, M.J., Bazzan, A.L.C.: A reinforcement learning-based multi-agent framework applied for solving routing and scheduling problems. Exp. Syst. Appl. **131**, 148–171 (2019). https://doi.org/10.1016/j.eswa.2019.04.056
19. Lopes Silva, M.A., de Souza, S.R., Freitas Souza, M.J., de França Filho, M.F.: Hybrid metaheuristics and multi-agent systems for solving optimization problems: A review of frameworks and a comparative analysis. Appl. Soft. Comput. **71**, 433–459 (2018). https://doi.org/10.1016/j.asoc.2018.06.050
20. Ma, X., et al.: A survey on cooperative co-evolutionary algorithms. IEEE Trans. Evol. Comput. **23**(3), 421–441 (2019). https://doi.org/10.1109/TEVC.2018.2868770
21. Pongchairerks, P.: An enhanced two-level metaheuristic algorithm with adaptive hybrid neighborhood structures for the job-shop scheduling problem. Complexity **2020**, 1–15 (2020). https://doi.org/10.1155/2020/3489209
22. Salloum, S., Dautov, R., Chen, X., Peng, P., Huang, J.: Big data analytics on apache spark. Int. J. Data Sci. Anal. **1** (2016). https://doi.org/10.1007/s41060-016-0027-9
23. Sharma, S., Jeet, K., Nailwal, K., Gupta, D.: An improvement heuristic for permutation flow shop scheduling. Int. J. Process. Manage. Benchmarking **9**, 124 (2019). https://doi.org/10.1504/IJPMB.2019.10019077
24. Soltysova, Z., Semanco, P., Modrak, J.: Exploring heuristic techniques for flow shop scheduling. Manage. Prod. Eng. Rev. **10**(3), 54–60 (2019) https://doi.org/10.24425/mper.2019.129598
25. Sun, L., Lin, L., Li, H., Gen, M.: Large scale flexible scheduling optimization by a distributed evolutionary algorithm. Comput. Indus. Eng. **128**, 894–904 (2019). https://doi.org/10.1016/j.cie.2018.09.025

26. Taillard, E.: Summary of best known lower and upper bounds of Taillard's instances. http://mistic.heig-vd.ch/taillard/problemes.dir/ordonnancement.dir/ordonnancement.html (2015). Accessed 23 Nov 2020
27. Taillard, E.: Benchmarks for basic scheduling problems. Eur. J. Oper. Res. **64**(2), 278–285 (1993). https://doi.org/10.1016/0377-2217(93)90182-M, project Management anf Scheduling
28. Tsai, C.W., Chang, H.C., Hu, K.C., Chiang, M.C.: Parallel coral reef algorithm for solving jsp on spark. In: 2016 IEEE International Conference on Systems, Man, and Cybernetics, SMC 2016, Budapest, Hungary, 9–12 October 2016, pp. 1872–1877. IEEE (2016). https://doi.org/10.1109/SMC.2016.7844511
29. Çaliş Uslu, B., Bulkan, S.: A research survey: review of ai solution strategies of job shop scheduling problem. J. Intell. Manuf. **26** (2013). https://doi.org/10.1007/s10845-013-0837-8
30. Viana, M.S., Morandin Junior, O., Contreras, R.C.: A modified genetic algorithm with local search strategies and multi-crossover operator for job shop scheduling problem. Sensors **20**(18), 5440 (2020). https://doi.org/10.3390/s20185440
31. Vital-Soto, A., Azab, A., Baki, M.F.: Mathematical modeling and a hybridized bacterial foraging optimization algorithm for the flexible job-shop scheduling problem with sequencing flexibility. J. Manuf. Syst. **54**, 74–93 (2020). https://doi.org/10.1016/j.jmsy.2019.11.010
32. Wang, F., Tian, Y., Wang, X.: A discrete wolf pack algorithm for job shop scheduling problem. In: Proceedings of the 2019 5th International Conference on Control, Automation and Robotics (ICCAR), Beijing, China, pp. 19–22 (April 2019)
33. Wu, J.-Y., Wu, X., Lu, X.-Q., Du, Y.-C., Zhang, M.-X.: Water wave optimization for flow-shop scheduling. In: Huang, D.-S., Huang, Z.-K., Hussain, A. (eds.) ICIC 2019. LNCS (LNAI), vol. 11645, pp. 771–783. Springer, Cham (2019). https://doi.org/10.1007/978-3-030-26766-7_70
34. Yang, L., Wang, C., Gao, L., Song, Y., Li, X.: An improved simulated annealing algorithm based on residual network for permutation flow shop scheduling. Complex Intell. Syst. 1–11 (2020). https://doi.org/10.1007/s40747-020-00205-9
35. Zhang, L., Yu, Y., Luo, Y., Zhang, S.: Improved cuckoo search algorithm and its application to permutation flow shop scheduling problem. J. Algorithms Comput. Technol. **14**, 1748302620962403 (2020). https://doi.org/10.1177/1748302620962403
36. Zhang, Z., Guan, Z., Zhang, J., Xie, X.: A novel job-shop scheduling strategy based on particle swarm optimization and neural network. Int. J. Simul. Model. **18**, 699–707 (2019). https://doi.org/10.2507/IJSIMM18(4)CO18
37. Zhu, J., Shao, Z., Chen, C.: An improved whale optimization algorithm for job-shop scheduling based on quantum computing. Int. J. Simul. Model. **18**, 521–530 (2019). https://doi.org/10.2507/IJSIMM18(3)CO13

Bagging and Single Decision Tree Approaches to Dispersed Data

Małgorzata Przybyła-Kasperek$^{(\boxtimes)}$ and Samuel Aning

Institute of Computer Science, University of Silesia in Katowice,
Będzińska 39, 41-200 Sosnowiec, Poland
{malgorzata.przybyla-kasperek,samuel.aning}@us.edu.pl
https://us.edu.pl/

Abstract. The article is dedicated to the issue of classification based on independent data sources. A new approach proposed in the paper is a classification method for independent local decision tables that is based on the bagging method. For each local decision table, sub-tables are generated with the bagging method, based on which the decision trees are built. Such decision trees classify the test object, and a probability vector is defined over the decision classes for each local table. For each vector decision classes with the maximum value of the coordinates are selected and the final joint decision for all local tables is made by majority voting. The results were compared with the baseline method of generating one decision tree based on one local table. It cannot be clearly stated that more bootstrap replicates guarantee better classification quality. However, it was shown that the bagging classification trees produces more unambiguous results which are in many cases better than for the baseline method.

Keywords: Ensemble of classifiers · Dispersed data · Bagging method · Classification trees · Independent data sources

1 Introduction

Classification based on data provided by independent sources is a current and challenging issue. Data can be provided by different sensors, collected in separate data sets or provided by various devices/units. Methods for classification based on independent sources were used for example in the streaming domain [3], transfer learning [9], medicine [8], land cover identification [1] and others. In this paper, independent data sources are decision tables that are collected independently by various units/entities/agents. They must relate to the same domain and have a common decision attribute, besides these requirements, there are no other restrictions as to their form. Decision trees are one of the most popular methods used in classification problems. The best-known algorithm for building decision trees are ID3, C4.5, CART and CHAID [2]. Bagging, boosting and random forests [7] methods are the next stage in the development of decision trees.

© Springer Nature Switzerland AG 2021
M. Paszynski et al. (Eds.): ICCS 2021, LNCS 12744, pp. 420–427, 2021.
https://doi.org/10.1007/978-3-030-77967-2_35

The novelty of the work is the use of the bagging method and classification trees for independent data sources stored in local decision tables. We use the bagging method as it is simple, often produces very good results, and works well with diverse data (so appropriate for independent data). We are dealing with a two-level division of the data in the method. The first level concerns the independent way of collecting data. The second level concerns the using of the bagging method. Based on each local table, we generate sub-tables from which decision trees are constructed to define prediction vectors. Based on these vectors, the votes for the decisions with the maximum vector's coordinates are calculated. At the end, the majority voting method combines the classification results for local tables.

The structure of the paper is organised as follows. Section 2 is dedicated to the classification tree and the bagging method. Section 3 contains a description of the proposed approach. Section 4 addresses the data sets that are used. Section 5 presents the conducted experiments and comments on the obtained results. Section 6 gives conclusions and future research plans.

2   Classification Tree and Bagging Method

ID3 and CART, were the first algorithms to be proposed independently by Quinlan and by the team Breiman, Friedman, Stone and Olshen [2]. Both algorithms are greedy. At first, the full training set of objects is considered, and the division of the set defined by the conditional attributes is optimized to a specific measure. The two most popular measures are information gain and the Gini index. The Gini index measures the purity of the set $Gini(X) = 1 - \sum_{i=1}^{m} p_i^2$, where m is the number of decision classes, $p_i = \frac{|C_{i,X}|}{|X|}$, $|X|$ is the size of the training set X and $|C_{i,X}|$ is the number of objects from the i-th decision class. The Gini index of division X_1, X_2, that is defined based on the attribute a is calculated as follows $Gini_a(X) = \frac{|X_1|}{|X|} Gini(X_1) + \frac{|X_2|}{|X|} Gini(X_2)$. The tree induction algorithm selects a conditional attribute that minimizes the Gini index. The minimum number of objects in a given node or the maximum tree height is used as a stop condition - we stop splitting the nodes and defined leaves. In an ensemble of classifiers approach, each new test object is classified by a set of classifiers, and the final decision is made by majority voting. The most popular ensemble methods proposed are: bagging, boosting and random forests methods [7]. In the bagging method K bootstrap samples $X_1, \ldots, X_K$ are created based on the training set X. Each X_i is defined by drawing with replacement from the set X to create diversity in data set. A decision tree is built based on each set X_i. For the test object decisions trees make decisions and the most common decision is chosen. Usually this approach improves the quality of the classification and is more resistant to outliers and overfitting.

3   Classification Tree Applied for Dispersed Data

Dispersed data is a set of local decision tables that share the same decision attribute. $S_i = (U_i, A_i, d)$, $i \in \{1, \ldots, N\}$ is called a i-th decision table, where

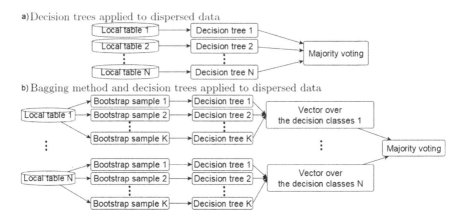

Fig. 1. Graphical representation of the proposed approaches

U_i is the universe, a set of objects of i−th decision table, A_i is a set of condition attributes of i−th decision table and $d \notin \bigcup_{i=1}^{N} A_i$ is a decision attribute. Both the conditional attribute sets and the sets of objects in such local tables have no restrictions or constrains.

Research on the use of dispersed data was conducted in earlier papers [4–6]. These articles mainly used the k-nearest neighbors classifier to generate decisions based on dispersed data. In this paper, for the first time, a decision tree-based classifier has been used for dispersed data. Two approaches are used in this article. The first, simpler, approach consist of building one decision tree based on each local decision table. For this purpose, the Gini index and the stop criterion determined by two objects in the node were used. When a test object is classified, each decision tree votes for one decision value, the final decision is made by majority voting. A graphic illustration of this approach is presented on Fig. 1.

In the second approach, the bagging method was used for each local decision table. Based on the bootstrap samples the decision trees are built in analogous way to that described above. It should be noted that a double dispersion of data occurs in this approach. Firstly, data is in dispersed form because it is collected by independent units. Secondly, a set of bootstrap samples is generated based on each local decision table. Since we have a two-level process of dispersion in this approach, a two-step process of aggregation of results is also used. The results of classification obtained based on the bootstrap samples and decision trees is aggregated into a vector over the decision classes. Each vector coordinate corresponds to one decision class and represents the number of decision trees that indicated such a decision for the test object. In this study, the majority voting method is used to aggregate such vectors. Votes are calculated based on each vector, each decision class with the maximum value of the coordinates is given a voice. Then, the final decisions are the classes that received the maximum number of votes defined based on all vectors. This aggregation method can generate ties. A graphic illustration of this approach is presented on Fig. 1.

4   Description of the Data and Experiments

In the experimental part of the paper, three data sets were used: the Vehicle Silhouettes, the Soybean (Large) and the Lymphography data sets. All of these data are available in a non-dispersed version (single decision table) at the UC Irvine Machine Learning Repository. The quality of the classification for the proposed approaches was evaluated by the train and test method. The analysed data sets are multidimensional (from 18 to 35 conditional attributes) and have several decision classes (from 4 to 19 decision classes). Dimensionality is important because a set of local decision tables are created based on each training sets. The dispersion on 3, 5, 7, 9 and 11 local tables were created such that the lesser number of local tables, the greater the number of attributes in the tables. All local tables contain the full set of objects. The large number of decision classes in the analysed data sets is important because we allow that ties occur. The application of the proposed approaches to data with missing values is also analysed in this article. Missing values occur in the Soybean data set. Four different approaches to dealing with such data were analysed: the global dominant method (Gl) - for each attribute the dominant value based on all values in the table is selected and objects with missing values are supplemented with this dominant value; the dominant in relation to the decision classes method (Dec) - the dominant values are determined separately for each decision class and each attribute; objects with more than 50% of conditional attributes with missing values are removed (50%); all objects, no matter how many missing values they have, are used (all).

4.1   Approach with Single Decision Tree Created Based on Local Table

For all data sets based on each local table, a decision tree was built using the Python language and the function sklearn.tree.DecisionTreeClassifier. The final decision was made using majority voting - ties may occur. Therefore, the following two measures are used: the classification error e – the fraction of the number of misclassified objects by the total number of objects in the test set; the average number of generated decisions sets $\bar{d}$. Results of classifications are considered unambiguous if $\bar{d} = 1$ and otherwise when $\bar{d} > 1$. The results are given in Table 1. Based on the presented results, it can be concluded that for the Lymphography data set, unambiguous results are mostly obtained. There is ambiguity for the Vehicle data set, however, it is small and acceptable. Greater ambiguity occurs for the Soybean data set. However, this data set has 19 decision classes, so such ambiguity is acceptable. A graph (Fig. 2) was created. It can be seen that for a smaller number of local tables, rather better classification quality were achieved. This is due to the fact that in the case of greater dispersion, the number of conditional attributes in local tables is smaller. Decision trees that are built based on a very small number of conditional attributes do not provide good classification quality. When we consider different methods of completing the missing values in the Soybean data set, it is observed that the

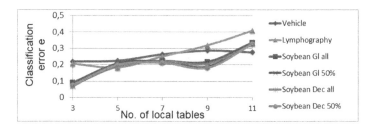

Fig. 2. Classification error e for decision trees created directly based on local tables

Table 1. Classification error e and the average number of generated decisions $\bar{d}$ for decision trees created directly based on local tables.

No. of local tables	Data sets										
	Vehicle		Lymphography		Soybean Gl all		Soybean Gl 50%		Soybean Dec all		Soybean Dec 50%
	e	$\bar{d}$	e	$\bar{d}$	e	$\bar{d}$	e	$\bar{d}$	e	$\bar{d}$	e	$\bar{d}$
3	0.220	1.260	0.205	1	0.090	1.457	0.068	1.450	0.074	1.388	0.075	1.352
5	0.224	1.169	0.182	1.045	0.205	1.258	0.208	1.283	0.189	1.274	0.199	1.303
7	0.264	1.134	0.250	1.045	0.223	1.189	0.218	1.235	0.221	1.152	0.212	1.218
9	0.287	1.154	0.318	1	0.218	1.231	0.189	1.235	0.207	1.234	0.182	1.241
11	0.276	1.146	0.409	1	0.335	1.225	0.332	1.215	0.324	1.197	0.326	1.215

method with the global dominance (Gl) and leaving all objects (all) produces the worst results.

4.2 Approach with Bagging Method and Decision Trees

One of the experimental goals was to check the impact of the number of bootstrap samples on the quality of classification. Therefore different number of samples were tested for the bagging method: 10, 20, 30, 40, 50. All evaluations were performed five times due to the indeterminism of generating bootstrap samples. The results that are given in Table 2 are the average value of the results obtained from these five runs. In Table 2, the best results, in terms of classification error e, within each data set and the number of local tables are shown in blue. If for two different numbers of bootstrap samples the same values of classification error were obtained, the result with a smaller number of the $\bar{d}$ measure was selected. As before, better results are obtained for a smaller number of local tables.

From Table 2 it can be noticed that the higher the number of bootstrap samples is, the lesser the $\bar{d}$ measure occur. In order to compare the obtained results for two proposed approaches Table 3 was created. In this table, for each dispersed data set, the lowest values of the classification error obtained for both approaches are given. As can be seen for the Vehicle data set the approach (2) – the bagging method produces better results. Unlikely results of Lymphography is because decision trees built from bagging method is not able to learn well from its small number of objects. For the Soybean data set, in the case of 5

Table 2. Classification error e and the average number of generated decisions $\bar{d}$ for bagging method, decision trees and the two-step process of aggregation results.

No. of local tables	No. of bootstrap samples	Data sets											
		Vehicle		Lympho-graphy		Soybean Gl all		Soybean Gl 50%		Soybean Dec all		Soybean Dec 50%	
		e	$\bar{d}$	e	$\bar{d}$	e	$\bar{d}$	e	$\bar{d}$	e	$\bar{d}$	e	$\bar{d}$
3	10	0.226	1.030	0.214	1.014	0.121	1.021	0.113	1.015	0.122	1.026	0.110	1.016
	20	0.220	1.017	0.205	1.005	0.120	1.016	0.117	1.005	0.123	1.016	0.110	1.010
	30	0.225	1.010	0.205	1.009	0.132	1.009	0.114	1.003	0.131	1.003	0.116	1.005
	40	0.219	1.010	0.209	1.005	0.124	1.009	0.114	1.004	0.135	1.006	0.117	1.005
	50	0.227	1.010	0.205	1	0.131	1.005	0.117	1.006	0.132	1.005	0.114	1.004
5	10	0.214	1.021	0.232	1.009	0.180	1.043	0.183	1.029	0.200	1.047	0.170	1.026
	20	0.203	1.013	0.250	1	0.176	1.023	0.178	1.011	0.192	1.029	0.170	1.016
	30	0.216	1.006	0.245	1.014	0.188	1.015	0.175	1.006	0.188	1.014	0.186	1.013
	40	0.204	1.003	0.245	1	0.178	1.006	0.177	1.010	0.190	1.013	0.183	1.009
	50	0.205	1.005	0.241	1	0.175	1.004	0.177	1.045	0.192	1.015	0.180	1.007
7	10	0.251	1.017	0.323	1.023	0.220	1.014	0.207	1.023	0.216	1.023	0.213	1.016
	20	0.254	1.010	0.295	1	0.207	1.010	0.206	1.008	0.218	1.012	0.219	1.008
	30	0.255	1.008	0.318	1.023	0.214	1.008	0.208	1.005	0.215	1.007	0.207	1.005
	40	0.258	1.007	0.332	1.005	0.209	1.006	0.208	1.007	0.212	1.003	0.217	1.004
	50	0.255	1.008	0.323	1	0.216	1.004	0.215	1.006	0.212	1.004	0.218	1.005
9	10	0.277	1.017	0.354	1.018	0.279	1.013	0.267	1.020	0.278	1.018	0.256	1.020
	20	0.286	1.007	0.359	1.005	0.287	1.010	0.270	1.012	0.275	1.011	0.262	1.018
	30	0.288	1.006	0.355	1	0.279	1.003	0.273	1.005	0.282	1.007	0.256	1.004
	40	0.270	1.006	0.359	1.005	0.283	1.005	0.267	1.005	0.285	1.005	0.252	1.005
	50	0.277	1.009	0.350	1	0.284	1.002	0.263	1.007	0.283	1.002	0.266	1.005
11	10	0.280	1.019	0.373	1.009	0.352	1.015	0.349	1.018	0.360	1.014	0.339	1.012
	20	0.288	1.005	0.386	1.014	0.349	1.010	0.350	1.005	0.358	1.007	0.338	1.007
	30	0.291	1.002	0.368	1	0.348	1.006	0.345	1.004	0.361	1.004	0.341	1.007
	40	0.286	1.003	0.364	1.009	0.346	1.005	0.353	1.003	0.363	1.004	0.341	1.004
	50	0.294	1.006	0.386	1	0.349	1.003	0.347	1.006	0.361	1.001	0.343	1.005

and 7 local decision tables the approach (2) provides better results, for the remaining versions of dispersion the approach (1) gives better results. However, it should be noted that for the Soybean data set, better results for the approach (1) are generated with greater ambiguity. Therefore, in applications where the unambiguity of the generated decisions matters, the bagging method should be used. Based on Table 3 it can also be concluded that for almost all cases the best quality of classification is obtained with using the dominant value in relation to the decision class and removing objects with more than half of attributes with the missing values. Paper [4] presents the results obtained by direct aggregation of the predictions generated by the k-nearest neighbors classifier instead of the decision trees. It can be concluded that when using decision trees and bagging method, better results were obtained in most cases.

Table 3. Comparison of classification error (e) for approaches: (1) single decision tree created based on one local table vs. (2) bagging method with decision trees

No. of local tables	Data sets											
	Vehicle		Lympho-graphy		Soybean Gl all		Soybean Gl 50%		Soybean Dec all		Soybean Dec 50%	
	(1)	(2)	(1)	(2)	(1)	(2)	(1)	(2)	(1)	(2)	(1)	(2)
	e	e	e	e	e	e	e	e	e	e	e	e
3	0.220	0.219	0.205	0.205	0.090	0.120	0.068	0.113	0.074	0.122	0.075	0.110
5	0.224	0.203	0.182	0.232	0.205	0.175	0.208	0.175	0.189	0.188	0.199	0.170
7	0.264	0.251	0.250	0.295	0.223	0.207	0.218	0.206	0.221	0.212	0.212	0.207
9	0.287	0.270	0.318	0.350	0.218	0.279	0.189	0.263	0.207	0.275	0.182	0.252
11	0.276	0.280	0.409	0.364	0.335	0.346	0.332	0.345	0.324	0.358	0.326	0.338

5 Conclusions

In this paper, two new approaches on applying decision trees to dispersed data were presented: the approach with decision trees directly generated based on local tables and the approach with the bagging method and decision trees. It was found that the bagging method gives more unambiguous results than the method based on the direct generation of decision trees based on local tables. Moreover, it was noticed that the higher the number of bootstrap samples is, the lesser the $\bar{d}$ measure occur. When dealing with missing data, it was found that the method with the dominant value in relation to the decision class and removing objects with more than half of attributes with the missing values provide the best results. In future work, it is planned to analyse various parameters when building decision trees (different stop conditions and applying information gain). It is also planned to use other fusion methods to combine the predictions of decision trees.

References

1. Elmannai, H., et al.: Rule-based classification framework for remote sensing data. J. Appl. Remote Sens. **13**(1), 014514 (2019)
2. Kotsiantis, S.B.: Decision trees: a recent overview. Artif. Intell. Rev. **39**(4), 261–283 (2013)
3. Li, Y., Li, H.: Online transferable representation with heterogeneous sources. Appl. Intell. **50**(6), 1674–1686 (2020). https://doi.org/10.1007/s10489-019-01620-3
4. Przybyła-Kasperek, M.: Generalized objects in the system with dispersed knowledge. Expert Syst. Appl. **162**, 113773 (2020)
5. Przybyła-Kasperek, M.: Coalitions weights in a dispersed system with Pawlak conflict model. Group Decis. Negot. **29**, 549–591 (2020). https://doi.org/10.1007/s10726-020-09667-1
6. Przybyła-Kasperek, M., Wakulicz-Deja, A.: Dispersed decision-making system with fusion methods from the rank level and the measurement level - a comparative study. Inf. Syst. **69**, 124–154 (2017)
7. Sagi, O., Rokach, L.: Ensemble learning: a survey. Wiley Interdiscip. Rev. Data Min. Knowl. Discov. **8**(4), e1249 (2018)

8. Giger, M.L.: Machine learning in medical imaging. J. Am. Coll. Radiol. **15**(3), 512–520 (2018)
9. Wu, Q., et al.: Online transfer learning with multiple homogeneous or heterogeneous sources. IEEE Trans. Knowl. Data Eng. **29**(7), 1494–1507 (2017)

An Intelligent Social Collective with Facebook-Based Communication

Marcin Maleszka[(✉)]

Wroclaw University of Science and Technology, St. Wyspianskiego 27,
50-370 Wroclaw, Poland
marcin.maleszka@pwr.edu.pl

Abstract. This paper describes the model of an intelligent social collective based on the Facebook social network. It consists of three main elements: social agents with a specified knowledge structure, a list of communication modes describing how agents send outbound messages, and a list of integration strategies describing how agents react to incoming messages. The model is described in detail, with examples given for important subalgorithms. The model is then analyzed in comparison to epidemic SI models in knowledge diffusion tasks and tested in a simulated environment. The tests show that it behaves according to the expectations for real world groups.

Keywords: Collective intelligence · Group modeling · Multi-agent simulation · Knowledge diffusion · Collective model

1 Introduction

Computational Collective Intelligence is a current research area that tackles multiple problems of modern computing. One of those is modelling groups of intelligent agents that work towards some common purpose or exchange knowledge to solve specific problems. Such intelligent collectives are a method used in the area to work with problems of group dynamics, knowledge diffusion, opinion formation, or problems from other research areas, e.g., influence maximization in social network research. A properly constructed model of an intelligent collective should be usable to work with at least one of those tasks.

In our previous research, we first focused on asynchronous communication between agents, but as more and more parameters were added to the developing model of an intelligent social collective, it became similar to how communication works in the Twitter social network [10]. In this paper, we translate the same model to represent the Facebook social network, in most part by changing the modes of communication between agents to reflect it. We also improve the formal description of the intelligent social collective, which allows us to perform an analytical evaluation of some aspects of the model, instead of only using simulations.

© Springer Nature Switzerland AG 2021
M. Paszynski et al. (Eds.): ICCS 2021, LNCS 12744, pp. 428–439, 2021.
https://doi.org/10.1007/978-3-030-77967-2_36

This paper is organized as follows: in Sect. 2 we describe the research works that were key to creating the proposed social collective model and some models that describe the problem from the point of view of other research areas; Sect. 3 contains details of the Facebook-based model, with some examples of its functioning; in Sect. 4 we present a short analysis of some properties of the model and describe simulation experiments testing the model; finally in Sect. 5 we give some concluding remarks and present our intended further work on intelligence social collectives.

2   Related Works

In order to develop the presented model of social collective, we applied theoretical approaches from multiple varying fields of research, mostly focused on sociology and collective intelligence research.

The research in the area of social influence is the main motivation for using multiple possible modes of communication and multiple possible responses to communication in the model. The focus of the area is describing how people are influenced by others in social settings. Following some classical research in the area, [5] discusses different levels of influence in a collective depending on individual and group factors, e.g. subjective norms, group norms, or social identity. Other works discuss compliance with and internalization of group rules to better fit the group [8]. Both approaches allow different levels of response to outside information, from no resistance to full resistance. A divergent area of research called social proof takes additional elements into account, including the competence or the amount of knowledge of the receiver – the less they know, the more likely they are to learn [14].

A parallel area in the field of computer science are influence maximization models in social network research. They are often built as a type of predictive models, that is, they are created with the aim of predicting the network behavior when a new node is added to the network. Approaches bearing the most similarity to ours would be those focusing on calculating the probability of a person sending outgoing messages [12] and those observing changes in the distribution of discussed topics in the network [1]. There is also a subgroup of linear threshold models, where the most similarity to our model can be found in [4]. In that paper, the authors consider a threshold number of received messages before a person acknowledges incoming information. There are also explanatory influence models, where the aim is to determine the node or community with the most influence on others in the existing network. In this group, the ones most similar to our research work with classifying agent behavior to specific groups, e.g., active, subject leader [3].

As stated, our research was also influenced by different aspects of collective intelligence research, especially in its computational aspects. There is some criticism about its applicability to realistic groups [15], but we consider the typical approaches from the position of a single agent. This allows us to assume that the messages received by an agent fit the Surowiecki postulates [16], which cannot be said about the whole group. Instead of focusing on the whole group, we derive the collective behavior from changes occurring in such single agents. The specific methods we use are mostly derived from consensus theory [11], that states several requirements for algorithms, but in practical terms often requires only to calculate the centroid of a group as a median or average.

3   Model of the Social Collective

Following our previous research, where we introduced a social collective based on the Twitter social network [10], the model of the Facebook-based collective proposed in this paper comprises of three main parts: the individual social agents, the methods they use to initiate communication and the methods they use to integrate any received knowledge or opinions.

Each social agent $a_i \in A$ has some internal knowledge or opinion k_i, at least one associated communication mode Mo_i^a that he uses, and at least one Integration Strategy S_j^a that he uses on received messages. Additionally, the agent may follow some context-based rules (e.g., in context A they use integration strategy Mo_i^1, and in context B they use integration strategy Mo_i^2).

The knowledge of agents is represented in the form of "vector of statements" with associated weights (sentiment represented as numerical values): $k_i = \{<$ $k_i^a, w_i^a >\}$. The use of simple statements (e.g. "light bulbs require electricity") is based on the observation that social network messages often cover only a single issue, but are augmented with multiple emotional descriptors. In turn, the messages that are generated for such representation of knowledge consist of single pairs from the internal knowledge base of an agent, selected at random with probability proportional to the associated weight (more precisely to the absolute value of the weight). For simplicity of implementation, we consider only positive and negative weights (generic positive and negative sentiment) in range $[-W, W]$.

There are multiple allowed approaches to communication in the model. While in our previous Twitter-based model, only asynchronous communication was possible, in Facebook-based model we also allow a specific type of synchronous and symmetric communication (simulating real-time chat). Each communication mode has a probability P_i^c of using this mode, which sums up to 1 for all modes. The communication modes allowed in this model are based on the Facebook social network and there are multiple possible levels of relation and communication between agents (user accounts): synchronous bi-directional chat, asynchronous posting, liking and replying to wall messages, and derived approaches defaulting to the previous (e.g. interest groups). Following, each social agent in the model has a list of *friends* (bi-directional relation) and we define four communication modes:

– At any moment an agent may send a message consisting of part of their knowledge (opinion) to any one or more agents from their list of *friends*. The receivers are selected at random, with possibility of the same agent being selected multiple times. This represents posts on Facebook *wall*, where people may skip a message or read it several times.

- Each agent may also, instead of initiating communication on some new topic, make a public agreement with a previous message from some other agent. In effect, the agent copies a previously received message, then uses the same approach to determine a new list of receivers, and sends the same message again. This represents people using the *Like* function of some Facebook wall posts. Again, people may see the *Like* many times, or not see it at all. There is an additional parameter determining the additional probability of not using this mode.
- Similarly to the previous option, the agent may make an own statement based on the message that they received. In this case, the agent also uses a previously received message as a template, but instead of copying it, they instead create a message with their own knowledge (opinion) on the same topic. Then they determine the list of receivers, using the same approach as in the two previous communication modes, and send the message. This represents people commenting on posts on Facebook *wall* (including commenting on other comments). Again, people may see the comment many times, or not see it at all. There is an additional parameter determining the additional probability of not using this mode.
- Each agent may also initiate a bi-directional communication with any single agent from his list of *friends*. In this case, the agent selects the part of their knowledge and sends it as a message to the selected agent and the selected agent sends back a message containing their state of knowledge (opinion) of the same topic. As processing incoming messages occurs independently of sending them, the returning message may either already consider the message from the first agent, or be the original knowledge of the second agent. This type of communication represents discussions via chat option between two different people.

The reaction of a social agent to incoming messages is one of the integration strategies S_j, which are knowledge integration algorithms we have based on various research in the areas of sociology and collective intelligence. It is also possible for incoming messages to be aggregated and the integration to be done at a later time. The input of the algorithms are the incoming message (messages) and agents own knowledge on the specific topic (two or more pairs $< k_i^a, w_i^a >$, where $\forall_{j1 \neq j2} k_{j1}^a = k_{j2}^a$), and the output is the new knowledge of the agent (a single such pair). In the model, we use the following possible integration strategies, but more can be introduced as needed:

- Substitution – this integration strategy is based on sociological works in the area of social influence (mainly [8]) and follows the concept of *no resistance to induction*. This can be understood as a person accepting any outside knowledge and immediately adding it to their internal knowledge base (colloquially: *is very naive*). This basic integration strategy uses the same approach: upon receiving any incoming knowledge (opinion), the agent immediately adds it to his own knowledge base and, if necessary, substitutes his previous knowledge on the topic with the newly received one.

Example 1:
- agent previous knowledge : $\{< A, 3 >, < B, -2 >, < C, 7 >\}$
- new message : $\{< C, -1 >\}$
- agent new knowledge : $\{< A, 3 >, < B, -2 >, < C, -1 >\}$

Example 2:
- agent previous knowledge : $\{< A, 3 >, < B, -2 >, < C, 7 >\}$
- new message : $\{< D, -1 >\}$
- agent new knowledge : $\{< A, 3 >, < B, -2 >, < C, 7 >, < D, -1 >\}$

– Discard – following the same sources, the opposite reaction is called *full resistance to induction*. In such situation, a person does not internalize any received knowledge. As an integration strategy, this means no change to the internal knowledge base and is only used for some specific experiments.

Example:
- agent previous knowledge : $\{< A, 3 >, < B, -2 >, < C, 7 >\}$
- new message : $\{< C, -1 >\}$
- agent new knowledge : $\{< A, 3 >, < B, -2 >, < C, 7 >\}$

– Delayed voting – this integration strategy is based on research in the area of collective intelligence, specifically consensus theory [11]. It requires buffering several (T_i^{dv}, consensus theory favors odd numbers) messages on the same knowledge statement before the integration occurs (this includes own knowledge on the topic, if it exists). The new knowledge-weight pair is selected based on plurality vote on the weight. In case of ties, a random pair is selected among the winners.

Example 1:
- agent previous knowledge : $\{< A, 3 >, < B, -2 >, < C, 7 >\}$
- full buffer for $T_i^{dv} = 7$: $\{< C, -1 >, < C, -1 >, < C, 2 >, < C, 2 >, < C, 2 >, < C, 7 >\}$
- agent new knowledge : $\{< A, 3 >, < B, -2 >, < C, 2 >\}$

Example 2:
- agent previous knowledge : $\{< A, 3 >, < B, -2 >, < C, 7 >\}$
- full buffer for $T_i^{dv} = 7$: $\{< D, -1 >, < D, -1 >, < D, -1 >, < D, 2 >, < D, 2 >, < D, 2 >, < D, 7 >\}$
- agent new knowledge : $\{< A, 3 >, < B, -2 >, < C, 7 >, < D, -1 >\}$

– Delayed weighted average consensus – this integration strategy is based on similar research as the previous one, and it also requires a buffer of messages. The integration is done by determining the average weight of the pairs used in calculation.

Example 1:
- agent previous knowledge : $\{< A, 3 >, < B, -2 >, < C, 7 >\}$
- full buffer for $T_i^{dv} = 7$: $\{< C, -1 >, < C, -1 >, < C, 2 >, < C, 2 >, < C, 2 >, < C, 7 >\}$
- agent new knowledge : $\{< A, 3 >, < B, -2 >, < C, 2.6 >\}$

Example 2:
- agent previous knowledge : $\{< A, 3 >, < B, -2 >, < C, 7 >\}$

- full buffer for $T_i^{dv} = 7$: $\{< D, -1 >, < D, -1 >, < D, -1 >, < D, 2 >,$ $< D, 2 >, < D, 2 >, < D, 7 >\}$
- agent new knowledge : $\{< A, 3 >, < B, -2 >, < C, 7 >, < D, 1.4 >\}$
- Polarization – this integration strategy is based on a different approach to social influence called Social Judgment Theory [2]. It is based on the notion that a person hearing an opinion opposed to theirs, will further distance themselves from it, while when hearing a similar opinion, they will take it into account to increase the similarity. In the strategy, we calculate the distance to weights in incoming messages ($\delta(w_1, w_2) = |w_1 - w_2|$), if it smaller than the threshold D_0 then the weight associated with agents internal knowledge changes towards it by $\delta(w_1, w_2) \cdot d_1$, otherwise it changes to increase the distance by $\delta(w_1, w_2) \cdot d_2$. Our initial assumption based on sociological literature study was that $d_1 > d_2$ and our experiments have shown that it is close to $d_1 = 10 \cdot d_2$. If the knowledge is previously unknown to the agent, the strategy defaults to *Substitution*.

Example 1:
- agent previous knowledge : $\{< A, 3 >, < B, -2 >, < C, 7 >\}$
- parameters $D_0 = 5, d_1 = 0.5, d_2 = 0.05$
- new message : $\{< C, -1 >\}$
- agent new knowledge : $\{< A, 3 >, < B, -2 >, < C, 7.4 >\}$
Example 2:
- agent previous knowledge : $\{< A, 3 >, < B, -2 >, < C, 7 >\}$
- parameters $D_0 = 5, d_1 = 0.5, d_2 = 0.05$
- new message : $\{< C, 3 >\}$
- agent new knowledge : $\{< A, 3 >, < B, -2 >, < C, 5 >\}$
Example 3:
- agent previous knowledge : $\{< A, 3 >, < B, -2 >, < C, 7 >\}$
- parameters $D_0 = 5, d_1 = 0.5, d_2 = 0.05$
- new message : $\{< D, -1 >\}$
- agent new knowledge : $\{< A, 3 >, < B, -2 >, < C, 7 >, < D, -1 >\}$

4   Evaluation of the Model

4.1   General Properties

Assuming that the network is the *small world* type (at most L connection in the graph between any two nodes) and agents do not change their internal knowledge, time to communicate the statement $k_1^{a'}$ from agent a_1 to a random agent in a network (a_L) is:

$$T(a_1, a_L) = \tau \Pi_{i \in \{1, \dots, L-1\}} \left(\frac{w_i^{a'}}{\sum_a w_i^a} \cdot \frac{r_i}{card(R_i)} \right) \tag{1}$$

where τ is the base delay between subsequent communications, r_i is the (expected value of) number of other agents that receive each message and $card(R_i)$ is the total number of agents that can be communicated with (i.e. in the agents *friend* list). Similar equations would also hold for other types of networks, with a more complex estimation of the number of required intermediate steps. In a more practical situation, where other statements are also communicated and lead to changes in internal knowledge, the equation needs to be modified by making the weights a function of time.

Based on the above, we can estimate how the proposed social collective behaves in terms of an epidemic SI model of a social network [13]. This type of model divides the group into those Susceptible and Infected ($S(t)$ and $I(t)$, respectively, with $S(t) + I(t) = N$ as the entire group) and it can be described in terms of equations determining the increase of the number of Infected over time. Assuming that we have one person with some knowledge $I(0) = 1$, the time until it is spread to the entire collective depends on the time required to spread to the furthest member, and using the equation from [13] for change of the number of Infected:

$$\frac{d(\frac{I(t)}{N})}{dt} = \lambda \frac{I(t)}{N}(1 - \frac{I(t)}{N}) \qquad (2)$$

The parameter λ that describes the network can be calculated using Eq. 1 as follows:

$$\lambda = \frac{N-1}{\tau \Pi_{i \in \{1,...,L-1\}}(\frac{w_i^{a'}}{\sum_a w_i^a} \cdot \frac{r_i}{card(R_i)})} \cdot \frac{1}{\frac{1}{N}(1 - \frac{1}{N})} \qquad (3)$$

The proposed model is not inherently of the SI class, so Eq. 3 describes only the spread of the knowledge statement throughout the network, not the weight of it (i.e. if it is positive or negative statement).

On the other hand, the proposed model does not conform to some proposals of general collective intelligence in real-world groups, as described in [6,9], in large part because of the lack of deeper intelligence present in any individual agent. The former paper defines a measure of collective intelligence based on specific abilities of the group: generating tasks, choosing tasks, executing tasks, remembering tasks, and sensing tasks. The proposed model can be slightly expanded to allow choosing or remembering tasks by changing the knowledge representation and adding memory (forgetting parts of knowledge), but other abilities would require a complete rebuild of the model towards a more objective-oriented one. The latter paper describes real-world collectives where members exchange information on multiple topics. The authors have observed a specific type of peak of interest (increase of communication frequency) on some topics and described it in mathematical terms. The main parameters of such dynamics are growth by imitation, self-inhibiting saturation, and competition with other topics. While our proposed model contains elements of both growth and competition, the frequency of communication on each topic does not tend to peak in the same manner. It would be possible to modify the model to introduce an element of saturation, but without a thorough rebuilding of the whole model, it would lead to limiting its applicability in knowledge diffusion.

4.2   Simulation of Social Collectives

To further evaluate the proposed model, we implemented it in the simulation environment we have been using in our previous research [10], which uses discrete time moments (iterations) during which the agents may communicate. The simulation setups a group of agents with uniformly random distribution of knowledge and opinions and lets them interact for some time. Meanwhile we can observe how the knowledge (opinion) of selected agents or the whole group changes. We can also add agents with atypical knowledge or atypical behaviour.

For the purposes of testing and comparing the models, we have adapted the notion of *drift* from sociological research [7] and defined it as follows:

Definition 1. k_i-**drift** is the absolute value of the average change of weights describing the knowledge statement k_i in the whole collective over one iteration.

Definition 2. Collective Drift is the average of k_i-drifts about every knowledge statement k_i possible in the Closed World interpretation.

Definition 3. ϵ, τ-**stability**. A collective is ϵ, τ-stable, if the average weights describing knowledge of its members change over time τ by no more than ϵ.

Definition 4. Stable collective. A collective is called stable if it is drift is no larger than 0.1. Otherwise, it is called unstable.

Following the sociological research, a proper group with initially uniform distribution of opinions (here: weights describing knowledge statements) should be stable, but any introduced heterogeneity should make it unstable. Therefore, in the experiments we conduct a large-scale simulation of a collective of agents over a long period of time and compare the initial and final knowledge (weights) to calculate the Collective Drift.

The overall model has multiple parameters that we have initially determined in our previous research and further tuned for purposes of the Facebook model. Additionally, some specific parameters were adjusted for different experimental runs (and are provided in their descriptions). The common parameters are:

- Number of agents in simulation : 1000,
- Number of possible knowledge statements: 100,
- Initial number of statements for each agent : 20,
- Range of allowed weights : [-10,10],
- Weight distribution : uniform in the entire range,
- Number of agents in *friend* relation : 10,
- Length of simulation τ : 1000 (discrete time moments),
- Probability of starting communication by an agent in each time moment : 0.2,
- Maximum number of receivers for a message : 5,
- Size of message buffer for delayed strategies T^{dv} : 11,
- Polarization threshold D_0 : 5,
- Polarization weights $d_1 = 10 \cdot d_2$: 0.5.

The general aim of the simulations was to determine the model parameters, where a homogeneous collective is stable, but introducing even one agent using Discard strategy makes it unstable. We have conducted multiple runs and compiled the results of the most interesting and realistic configurations (averaged over multiple simulation runs for each combination of parameters) in Table 1.

The gathered results show that Substitution, Delayed voting and Delayed weighted average consensus for all realistic and most overall combinations of tuned parameters behave as expected from a real group, that is, they are stable when homogeneous and unstable when an outside disturbance (Discard agent) is introduced. Polarization is a different class of integration strategies as it leads to the gradual evolution of two opposed subgroups for every issue. Introducing an outside interference leads to faster (and larger) distancing between the groups. Otherwise, it behaves like any polarizing group [2].

Table 1. Collective opinion on a single knowledge statement for different integration strategies, results averaged over several simulation runs. Selected interesting configurations only. The value of *Collective Drift*, which is calculated as average change of opinion over time. Homogeneous and heterogeneous (one Discard agent) collectives for different configurations of the experimental environment. Probabilities given for modes: bi-directional messages / wall posts / wall likes / comments, as well as probability of not responding to post/comment.

Configuration (prob.)	Sub	D.vote	D.avg	Pol	D.+Sub	D+D.vote	D+D.avg	D+Pol
0.1/0.4/0.3/0.2, 0.2/0.5	0.17	0.07	0.08	0.93	2.55	1.12	0.89	4.12
0.3/0.3/0.3/0.1, 0.2/0.5	0.75	0.07	0.1	1.27	3.01	1.11	1.2	4.89
0.5/0.2/0.2/0.1, 0.2/0.5	0.6	0.1	0.08	2.14	2.97	1.32	1.33	5.4
0.5/0.2/0.2/0.1, 0.0/0.0	0.33	0.11	0.08	1.89	2.63	1.25	1.57	4.66
0.5/0.2/0.2/0.1, 0.5/0.5	0.47	0.05	0.09	1.74	2.73	0.97	1.64	4.39
0.5/0.2/0.2/0.1, 0.7/0.7	0.59	0.11	0.12	1.61	2.89	1.22	1.66	4.17
0.5/0.2/0.2/0.1, 0.9/0.9	0.71	0.08	0.07	1.67	3.15	1.18	1.89	4.26
0.1/0.1/0.7/0.1, 0.7/0.5	0.28	0.04	0.03	0.99	2.49	1.01	0.97	4.23

To better present the results, we have also organized them by increasing the probability of using specific communication modes (with other modes having equal probability), as shown in Fig. 1. Here one may observe that the changing probabilities of communication modes do not have a significant influence on the drift values, but in several cases such influence can be found. Polarization integration strategy is slightly influenced by the increasing probability of bi-directional chat communication, less influenced by the probabilities of post or like communication modes, and not influenced by the comment type of communication. Substitution integration strategy is also slightly influenced by bi-directional chat, post and like communication and not by comment communication mode. The remaining integration strategies are not influenced by these probabilities in a statistically significant way.

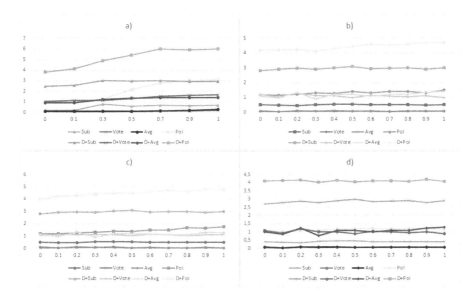

Fig. 1. Drift of specific homogeneous (Sub = Substitution, D.vote = Delayed voting, D.avg = Delayed weighted average consensus, Pol = Polarization) and heterogeneous (D.+ = one Discard agent added to homogeneous group) collectives. Graphs represent varying probabilities of using specific communication modes (a) bi-directional chat b) posting c) likes d) comments) with other modes having equal probability.

The parameters of the communication modes, that are based on the notion of *"skipping post/comment"* in the Facebook social network, do not change the behavior of the strategies, but rather lower the drift by extending the effects of communication in time.

5   Conclusions

This paper details a model of an intelligent social collective with communication based on Facebook social network. It operates on a simple structure of knowledge and uses multiple methods for internalization of knowledge by agents. We have evaluated the model in terms of drift – a measure proposed by us in some previous papers, as well as by translating it to a standard epidemic model.

In some of our other papers, we have developed more integration strategies that can easily be applied to this model. We have also used the notion of agents forgetting part of their knowledge and it is also possible to use this method for Facebook-based model. Both those enhancements do not provide a novel research point, so were not included in this paper.

We find that the largest step required for the completion of our model is the further development of the use of more complex knowledge structures, e.g., ontologies. Such a structure may better reflect the internal organization of knowledge in real people, but it requires the development of advanced knowledge integration algorithms that could later be translated into functional integration strategies.

While the group of models we develop in our overall research works well for tasks related to computational collective intelligence, it can be also applied to social network research. There is however a difficulty in obtaining sufficient knowledge about the social network to receive any substantial gains from the social collective model. The main problem are the integration strategies, which are internal processes occurring when social network users internalize knowledge. We can only approximate them based on how their expressed knowledge changes. To gain this knowledge for a fully comprehensive test of the model, we would need to conduct an experiment on a working social network with explicit questionnaires given after each new message is read. In fact we are taking preliminary steps to prepare such experiment, with input from researchers in the sociology area.

References

1. Barbieri, N., Bonchi, F., Manco, G.: Topic-aware social influence propagation models. Knowl Inf. Syst. **37**, 555–584 (2012)
2. Cameron, K.A.: A practitioner's guide to persuasion: an overview of 15 selected persuasion theories, models and frameworks. Patient Educ. Couns. **74**(3), 309–317 (2009)
3. Chen, B., Tang, X., Yu, L., Liu, Y.: Identifying method for opinion leaders in social network based on competency model. J. Commun. **35**, 12–22 (2014)
4. Chen, H., Wang, Y.T.: Threshold-based heuristic algorithm for influence maximization. J. Comput. Res. Dev. **49**, 2181–2188 (2012)
5. Cheung, C.M.K., Lee, M.K.O.: A theoretical model of intentional social action in online social networks. Decis. Support Syst. **49**, 24–30 (2010)
6. Engel, D., Woolley, A.W., Jing, L.X., Chabris, C.F., Malone, T.W.: Reading the Mind in the Eyes or Reading between the Lines? Theory of Mind Predicts Collective Intelligence Equally Well Online and Face-To-Face. PLoS ONE **9**(12), (2014). https://doi.org/10.1371/journal.pone.0115212
7. Gardner, W.L., Garr-Schultz, A.: Understanding our groups, understanding ourselves: the importance of collective identity clarity and collective coherence to the self. In: Lodi-Smith, J., DeMarree, K.G. (eds.) Self-Concept Clarity, pp. 125–143. Springer, Cham (2017). https://doi.org/10.1007/978-3-319-71547-6_7
8. Kelman, H.C.: Interests, relationships, identities: three central issues for individuals and groups in negotiating their social environment. Annu. Rev. Psychol. **57**, 1–26 (2006)
9. Lorenz-Spreen, P., Mønsted, B.M., Hövel, P., et al.: Accelerating dynamics of collective attention. Nat. Commun. **10**, 1759 (2019). https://doi.org/10.1038/s41467-019-09311-w
10. Maleszka, M.: Application of collective knowledge diffusion in a social network environment. Enterp. Inf. Syst. **13**(7–8), 1120–1142 (2019)

11. Nguyen, N.T.: Inconsistency of knowledge and collective intelligence. Cybern. Syst. Int. J. **39**(6), 542–562 (2008)
12. Saito, K., Nakano, R., Kimura, M.: Prediction of information diffusion probabilities for independent cascade model. In: Lovrek, I., Howlett, R.J., Jain, L.C. (eds.) KES 2008. LNCS (LNAI), vol. 5179, pp. 67–75. Springer, Heidelberg (2008). https://doi.org/10.1007/978-3-540-85567-5_9
13. Pastorsatorras, R.: Epidemic spreading in scale-free networks. Phys. Rev. Lett. **86**, 3200–3203 (2001)
14. Pratkanis, A.R.: Social influence analysis: An index of tactics. In: Frontiers of social psychology: The Science of Social Influence: Advances and Future Progress (1st ed.), pp. 17–82 (2007)
15. Søilen, K.S.: Making sense of the collective intelligence field: a review. J. Intell. Stud. Bus. **9**(2), 6–18 (2019)
16. Surowiecki, J.: The wisdom of crowds: why the many are smarter than the few and how collective wisdom shapes business, economies, societies, and nations. 1st Doubleday Books, New York (2004)

Multi-agent Spatial SIR-Based Modeling and Simulation of Infection Spread Management

Amelia Bădică[1], Costin Bădică[1(✉)], Maria Ganzha[2], Mirjana Ivanović[3], and Marcin Paprzycki[4]

[1] University of Craiova, Craiova, Romania
costin.badica@software.ucv.ro
[2] Warsaw University of Technology, Warsaw, Poland
[3] University of Novi Sad, Novi Sad, Serbia
[4] Polish Academy of Sciences, Warsaw, Poland

Abstract. This paper proposes a multi-agent system for modeling and simulation of epidemics spread management strategies. The core of the proposed approach is a generic spatial Susceptible-Infected-Recovered stochastic discrete system. Our model aims at evaluating the effect of prophylactic and mobility limitation measures on the impact and magnitude of the epidemics spread. The paper introduces the modeling approach and, next, it proceeds to the development of a multi-agent simulation system. The proposed system is implemented and evaluated using the GAMA multi-agent platform, using several simulation scenarios, while the experimental results are discussed in detail. Our model is abstract and well defined, making it very suitable as a starting point for extension and application to more detailed models of the specific problems.

Keywords: SIR model · Multi-agent system · Epidemics spread mitigation

1 Introduction

Research concerning mathematical modeling of the spread of infectious and contagious diseases has started almost one century ago [16]. The goal was the development of dynamic models that are able to capture and explain the various factors that can influence the intensity and span of the disease, in a given population of individuals.

A well known case that threatened humanity at the beginning of the twentieth century is the "1918 Flu Pandemic", also known as the "Spanish Flu". It lasted for more than 2 years causing the death of more than 500 million people [19]. Another very actual case, experienced nowadays by the whole humanity is "Coronavirus disease 2019" or "COVID 19" contagious respiratory and vascular disease, also known as "Severe Acute Respiratory Syndrome Coronavirus 2" or "SARS-CoV-2" [14]. It triggered a huge research interest in biomedical and computing sciences, and the need of development of detailed simulation models of the dynamics of the spread of the disease, including the design and analysis of various mitigation strategies and policies [2, 10].

© Springer Nature Switzerland AG 2021
M. Paszynski et al. (Eds.): ICCS 2021, LNCS 12744, pp. 440–453, 2021.
https://doi.org/10.1007/978-3-030-77967-2_37

Agent-based modeling and simulation (ABMS), is a well established approach that provides multiple application opportunities in natural systems including, among others, the simulation of stochastic diffusion processes in continuous and nonlinear environments [3]. ABMS enables the domain experts to choose between various levels of granularity of modeling at the macroscopic, mesoscopic or microscopic level. This flexibility allows focusing on various elements of the target system, by trading-off capturing of details and exploiting of efficiency of the modeling. It worth noting that the ABMS techniques and tools played a crucial role in the development of virus spread simulations using novel ABMS platforms, like GAMA, MESA and JS-son [2, 8, 10, 12, 13, 15].

The core of our proposal is a generic spatial Susceptible-Infected-Recovered (SIR) model of epidemics spread in a population. In particular we exploit our initial results in ABMS of core spatial SIR models for epidemics spread in a population, presented in [5]. Here we expand our work by adding an infection spread mitigation control agent and modeling the effects of its mitigation strategies and policies. Our proposal follows a control system approach representing the "intelligence" of the system. Various decisions and strategies, for limiting the epidemics, are conceptualized as dynamic controls that are defined based on the status information, acquired from the system using available social sensors. The upgraded multi-agent model is simulated using the GAMA multi-agent platform [20], following the methodology proposed in [5].

The paper is structured as follows. Section 2 contains an overview of relevant related works. Section 3 generalizes the generic SIR model from [5] by adding a spatial attenuation function to the infection propagation part. We propose and model the strategy of infection mitigation and management, modeling the behavior of the control agent supporting this strategy. Section 4 contains experimental results for the baseline system (without infection mitigation and management, considering a scenario with 15000 rather than 10000 agents, as in [5]), as well as for the same system including the control part. Section 5 presents the conclusions and points to future works.

2   Related Works

Compartmental models originate from the early work of Kermack and McKendrick [16]. They are used by epidemiologists and mathematicians to abstract the spreading process of infectious diseases, and to simplify their mathematical analysis. These models propose to structure the spreading process of the epidemics within the population by compartments, depending on the state of the population individuals. Examples of compartments are: Susceptible, Infectious, and Recovered, usually labeled as S, I and R. These models are defined as differential or difference equations that capture the dynamics of the population transfer between compartments, according to the degree, or stage, of the infection. Such a model can be utilized to theoretically, or experimentally, investigate the spread of the epidemic both in time and in (geo)space [21].

Compartmental models can be classified according to several criteria: i) depending on spatial and temporal characteristics of the model: temporal and spatial-temporal; ii) depending on the nature of modeling: discrete and continuous; iii) depending on model uncertainty: deterministic and stochastic; iv) depending on granularity of modeling: microscopic, mesoscopic and macroscopic.

Related models of diffusion processes were investigated using multi-agent simulation, including agent-based emotion contagion [4], information propagation in large teams [11] and BDI models for predator-prey systems [6].

Agent technologies include tools and platforms for software development, modeling and simulation [7,17,18]. Multi-agent systems can model emergent phenomena, and provide a natural representation of the system under observation. They are flexible, allowing the development of dynamic models and simulations. ABMS were traditionally employed for modeling and simulation of diverse human and social systems, combining strategic thinking with flow and diffusion processes, in many areas of business, society, biology, and physics [3]. A recent survey of modern ABMS tools is given in [1].

In the context of the recent COVID pandemic, several multi-agent simulations have been proposed, some of them based on GAMA platform [2,10,22]. Moreover, control system approaches have been found useful to theoretically investigate the control dynamic of a computer virus [9].

3   Multi-agent SIR Model

3.1   Core Spatial SIR Model

Let us consider the general core spatial SIR model introduced in [5] that contains:

- S compartment, denoting the set of susceptible individuals. They are not infected. When they enter in contact with infectious individuals, they can become infected. The population of susceptible individuals, at time t, is denoted by S_t.
- I compartment, denoting the set of infected individuals. They are infected and capable of infecting other susceptible individuals. The population of infected individuals, at time t, is denoted by I_t.
- R compartment, denoting the set of "removed" individuals. This set includes all individuals that either recovered, became resistant, or died. The population of removed individuals at time t is denoted by R_t.

The aim of the model is to capture the population dynamics in space and time. Here:

$$\Delta S = S_{t+\Delta t} - S_t$$
$$\Delta I = I_{t+\Delta t} - I_t \qquad (1)$$
$$\Delta R = R_{t+\Delta t} - R_t$$

Moreover, the total population of individuals at time t, denoted with N_t, is defined by:

$$N_t = S_t + I_t + R_t \qquad (2)$$

The spread of the disease is modeled by function F, representing the "force of infection". In each step, a fraction F of susceptible individuals, from set S_t, gets infected. Obviously, the set F depends on the current population of susceptible individuals S_t, as well as on the current set of infections I_t. Hence, the amount of susceptible individuals that get infected, in the current step, can be written as $F(S_t, I_t)$. The infected individuals will be "transferred" from the S compartment to the I compartment. Hence:

$$\Delta S = -F(S_t, I_t) \tag{3}$$

Let us also assume that a proportion $\gamma \in (0, 1)$ of infected individuals (denoted in what follows with $\gamma \cdot I_t$) is removed (either because they became resistant or they died). Adding Eq. (3) to this system, it follows that:

$$\begin{aligned} \Delta S &= -F(S_t, I_t) \\ \Delta I &= F(S_t, I_t) - \gamma \cdot I_t \\ \Delta R &= \gamma \cdot I_t \end{aligned} \tag{4}$$

This system is discrete, so time has a discrete variation $t_0 = 0, t_1 = h, \ldots, t_n = n \cdot h, \ldots$, where $h > 0$ is the constant time step. In what follows, we denote the values of model variables at time t_n simply with subscript n, where n is the current step number. For example, the number of infected individuals at step n is I_n.

3.2 Modeling Infection Spread

The specification of F depends on the considered model. We have chosen a spatial model, in which each individual has a certain physical location. This allows capturing the local spread of the disease, in which a certain amount of individuals, in the spatial vicinity of an infected individual, will get infected. Moreover, our individuals are "mobile", i.e. they are allowed to change their location in each step. Thus, "locality" of infection propagation, and "mobility" of individuals, enable our simple model to capture both the spatial and the temporal dimensions of the spread of the disease.

Let us denote with $i.x$ the physical coordinate of an individual i and let d be a distance function, in the space of physical locations (for example the Euclidean distance). We denote with $V(x)$ a vicinity of x, i.e. the set of physical locations that are "close" to x, according to some specific rule. Here, one possibility is to define $V(x)$ as the set of locations inside the ball of radius ϵ, centered at x, i.e.:

$$B(x, \epsilon) = \{y | d(x, y) \le \epsilon\} \tag{5}$$

Let us now consider an individual j and let $I(j)$ be the set of all infected individuals from the vicinity of j, defined as:

$$I(j) = \{i | i.x \in B(j.x, \epsilon)\} \tag{6}$$

Let $p_i, q_i \in (0, 1)$ be the probabilities that a certain individual i will transmit the infection in its vicinity and, respectively, will be infected if is present nearby. Let us assume that the transmission of the disease from an infected individual i to an individual j located at a distance x from i is also affected by a spatial attenuation function $\phi(x)$ defined as:

$$\begin{aligned} &\phi(x) : \mathbb{R}^+ \longrightarrow [0, 1] \\ &\phi(0) = 1 \\ &\lim_{x \to \infty} \phi(x) = 0 \\ &x \le y \Rightarrow \phi(x) \ge \phi(y) \end{aligned} \tag{7}$$

Let us denote by d_{ij} the distance between individuals i and j, defined as:

$$d_{ij} = d(i.x, j.x) \tag{8}$$

Following simple probabilistic reasoning, it is easy to see that the probability that a susceptible individual j will get infected is defined by:

$$q_j \cdot \left(1 - \prod_{i \in I(j)} (1 - p_i \cdot \phi(d_{ij}))\right) \tag{9}$$

Note that Eq. (9) generalizes the simple interaction model introduced in [5], where it was considered that the probability of infection of a susceptible individual, in the vicinity of an infected individual, is constant.

Observe that (9) is a well-defined probability value, in the interval $[0, q_i]$, and that this value is 0 when $I(j) = \emptyset$. This corresponds to the intuition that there is no infection propagation to a susceptible individual when there are no infected peers in its vicinity.

3.3 Discussion

Let us denote with β the average infection probability, determined by Eq. (9). This value depends on the average number of individuals in the vicinity of the considered individual, as well as on probabilities p and q. Moreover, the infection will affect only the susceptible individuals from the population. Hence, β should be weighted with the fraction $\frac{S_t}{N_t}$. Then we could estimate $F(S_t, I_t)$ as:

$$F(S_t, I_t) = \beta \cdot \frac{S_t}{N_t} \cdot I_t \tag{10}$$

Replacing (10) in the second equation of (4) it can be established that:

$$\Delta I = \left(\frac{\beta}{\gamma} \cdot \frac{S_t}{N_t} - 1\right) \cdot \gamma \cdot I_t \tag{11}$$

A simple mathematical analysis of Eq. (11) shows that the ratio $\frac{\beta}{\gamma}$ is the "reproduction rate" (i.e. $R_0 = \beta/\gamma$) from the classical SIR model, defined using differential equations [16]. The difference between the two models is that, here, β is a random variable that is determined by the spatial definition of what is meant by the "vicinity" of an individual. Moreover, the dynamic model is discrete and stochastic, rather than continuous and deterministic.

Note also that the value of parameter γ determines the average duration of the infection of an individual (removal time), i.e. the number of steps from the moment it became infected until it is removed. The expected value of this random variable is:

$$T_r = 1 \cdot \gamma + 2 \cdot \gamma \cdot (1 - \gamma) + 3 \cdot \gamma \cdot (1 - \gamma)^2 + \dots$$
$$= \gamma \cdot (1 + 2 \cdot (1 - \gamma) + 3 \cdot (1 - \gamma)^2 + \dots) = \gamma \cdot Q$$

However:

$$Q = (1 + (1 - \gamma) + (1 - \gamma)^2 + \dots) + (1 - \gamma) \cdot Q = 1/\gamma + (1 - \gamma) \cdot Q$$

Solving the equation we obtain $Q = 1/\gamma^2$ and substituting it in expression of T_r results:

$$T_r = 1/\gamma \tag{12}$$

3.4 Infection Management

The proposed approach for infection spread management assumes three steps: situation assessment, detection and mitigation. The system block diagram is shown in Fig. 1.

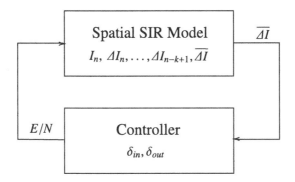

Fig. 1. Block diagram of infection management system.

Situation assessment is based on the information obtained from social sensors. Basically, individuals with symptoms and their peers, determined for example using techniques of contact tracing, are tested. Hence, the system monitors the variation of the number of infections, and is able to estimate the rate of infection spread. Obviously, the accuracy of obtained estimators depends on the capacity and precision of the testing.

In the stochastic system, the moving average of the last $k \geq 1$ steps can be used to estimate the infection rate, i.e.:

$$\Delta I_n = I_n - I_{n-1} \quad \text{for } n \geq 1$$
$$\overline{\Delta I_n} = \frac{\sum_{i=1}^{k} \Delta I_{n+i-1}}{k} \quad \text{for a given } k \geq 1 \tag{13}$$

Note that $\overline{\Delta I_n}$ can be computed iteratively as follows:

$$\overline{\Delta I_{n+1}} = \overline{\Delta I_n} + \frac{\Delta I_{n+1} - \Delta I_{n-k+1}}{k} \tag{14}$$

Equation (14) can be implemented in $O(1)$ time, and with $O(k)$ memory, using a circular queue represented as an array Q of size k, and its pointer l to store the last k values of ΔI; as shown in (15). The new value of ΔI, determined in the current simulation step, is ΔI_{new}. The update of Q is done as follows (here, $\leftarrow$ denotes the programming assignment):

$$m \leftarrow (l+1) \mod k$$
$$\overline{\Delta I} \leftarrow \overline{\Delta I} + \frac{\Delta I_{new} - Q_m}{k}$$
$$Q_m \leftarrow \Delta I_{new} \tag{15}$$
$$l \leftarrow m$$

The value of k is set as proportional to T_r, using a constant $k^{mem} > 0$, as follows:

$$k = k^{mem} \cdot T_r \qquad (16)$$

Detection is achieved by comparing, in each simulation step, the infection rate (per 1000 individuals) with a given threshold. Initially, the threshold $\delta_{in} > 0$ is used. So, if the condition $\overline{\Delta I} > \delta_{in}$ holds, the process enters the emergency (mitigation) stage.

Mitigation is achieved by declaring an "emergency situation" that assumes adoption of special measures, involving the self-protection of individuals, and limitation of their mobility (up to its prohibition). The situation is also continuously monitored in the mitigation stage, by comparing, in each step, the infection rate (per 1000 individuals) with the second threshold value δ_{out}. If the condition $\overline{\Delta I} < \delta_{out}$ holds, the process reenters the normal stage. Note that, in order to be sure that the infection is decreasing when reentering the normal state, a negative threshold is used, i.e. $\delta_{out} < 0$. Additionally, the "emergency situation" is kept for a minimum number of steps T_e that is defined as proportional (using given parameter $k_e^1 > 1$) with removal time T_r defined by (12):

$$T_e = k_e^1 \cdot T_r \qquad (17)$$

Note that the values of the thresholds are updated each time the system leaves the "emergency situation", using a preset coefficient $0 < k_e^2 < 1$ as follows:

$$\begin{aligned} \delta_{in} &\leftarrow \delta_{in} \cdot k_e^2 \\ \delta_{out} &\leftarrow \delta_{out} / k_e^2 \end{aligned} \qquad (18)$$

The initial values of δ_{in} and δ_{out} are defined as follows (k_{in} and k_{out} are two positive constants such that k_{out} is smaller than k_{in}):

$$\begin{aligned} \delta_{in} &= k_{in} \cdot (N/1000) \\ \delta_{out} &= -k_{out} \cdot (N/1000) \end{aligned} \qquad (19)$$

3.5 Multi-agent Representation

The mapping of the model onto a multi-agent representation is straightforward. Basically, each individual is represented as an agent. The compartment, to which each individual is assigned, is captured as the agent state, i.e. S, I or R. Initially most of the agents are in state S, few of them are in state I, and neither is in state R. At the end of the simulation, when the process stabilizes, each agent is either in state S or in state R. So state I is a "transitory state" of the agents. Moreover, if after the stabilization process all the agents are in state R, it means that the epidemic reached the whole population.

The state transition diagram of each individual is modeled as a finite state automaton (see, left part of Fig. 2). Transitions are triggered synchronously for all the agents, at each simulation step. Note that state transitions are stochastic, with exact probabilities determined by the parameters of the model. Hence, in fact, the behavior of the agents can be captured by a Markov chain.

Infection propagation (from an agent representing an infected individual, to an agent representing a susceptible individual) is achieved by agent interaction. Basically, each

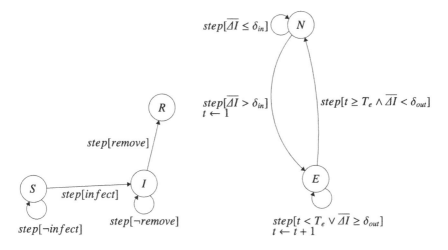

Fig. 2. State transition diagram of individual (left). State transition diagram of controller (right)

agent representing an infected individual sends an "infection message" to all susceptible agents in its vicinity, with the computed probability, according to Eq. (9).

The transition of an agent, from state I to state R, is achieved by letting the agent itself update its state from I to R with probability γ.

Agents in the system are mobile, as they represent also mobility of the individuals in the population. Hence, by design, they are endowed with a spatial mobility behavior.

Similarly, the state transition diagram of the controller is modeled as a finite state automaton (see, right part of Fig. 2). State N models the normal situation of the system, while E models the "emergency situation". The transitioning back and forth between these two states can take place at each simulation step, being triggered by conditions $\overline{\Delta I} > \delta_{out}$ and, respectively, $t \geq T_e \wedge \overline{\Delta I} < \delta_{out}$. Note that variable t represents the time (in number of steps) elapsed since the triggering of the "emergency situation". Variable t is initialized to 1, during the transition from state N to state E, and it is incremented for each step performed while system is in state E.

4   Results and Discussion

4.1   GAMA Implementation

In GAMA [20], models are specifications of simulations that can be executed and controlled using "experiments", defined by experiment plans. Experiments facilitate the intuitive visualization, as well as the extraction of useful simulation results.

Implementation of the, above proposed, pandemic model contains:

- The global section representing the agent environment. Here the definitions of all the model parameters, as well as other implementation-dependent parameters that were needed for the model visualization, like for example visual attributes used for agents visualization, have been included.

- The initialization section for setting up the initial population of agents, representing the initial populations of susceptible and infected individuals.
- The class defining the agents that instantiate the individuals of our model, together with their internal state and behavior.
- The section defining our experiment, visualizing the results, and extracting the simulation information. Here, a "gui" experiment has been implemented, -to allow the graphical visualization and animation of the simulation.

Each individual is modeled by an agent of *Host* species that contains the following attributes and behaviors:

- Boolean attributes *is_susceptible*, *is_infected*, and *is_immune* such that exactly one of them is *true*, for representing agent states S, I and R. Susceptible individuals have *is_susceptible=true*, while infected individuals have *is_infected=true*.
- Mobility is achieved by endowing agents with *moving* skill. It defines the required actions for spatially mobile agents. In particular, the *wander* action, for moving the agent forward to a random location at the distance computed based on its *speed* in a random direction (an angle, in degrees) determined using its *amplitude* attribute, has been used. Here, if the current value of the direction is h and the value of the *amplitude* is a, the new value of h is chosen from the interval $[h - a/2, h + a/2]$. Moreover, $a = 360$ was used, to allow full coverage of all directions.
- State transitions are achieved by the user-defined skills *infect* and *remove* that define the transition of an agent from state S to I, and from state I to R. Agents vulnerable for getting the infection were determined by enumerating all agents located at a given maximum distance, from an infected agent, using the *at_distance* operator.
- Initially, agents representing susceptible individuals are randomly distributed on a square-bounded 2D space.

Controller parameters are maintained as global attributes of the simulation environment, as follows:

- Boolean attribute *is_emergency* is used to distinguish between the normal (N) and the emergency state (E) of the controller.
- Values of k_{in} and k_{out} to compute initial values of thresholds δ_{in} and δ_{out} using Eqs. (19) are defined as: $k_{in} = 0.40$ and $k_{out} = 0.02$.
- Values of k_e^1 and k_e^2 used in Eqs. (17) and (18) that control the span of the emergency situation are defined as: $k_e^1 = 1.1$ and $k_e^2 = 0.9$.
- The value of parameter k^{mem} that defines the memory of the moving average $\overline{\Delta I}$ using Eq. (16) was set to $k^{mem} = 0.5$. This produces a value $k = 10$ for the "memory" of computing the moving average $\overline{\Delta I}$.
- The queue representing the "memory" of computing the moving average $\overline{\Delta I}$, implemented as a list of k values.

Overall, three visualizations of the simulation results are provided (see, Fig. 3):

- 2D spatial visualization of the agents. Susceptible agents are shown in green, infected agents are shown in red, while removed agents are shown in blue.
- The dynamics of susceptible, infected and removed individuals, the simulation timeline using the same color code.
- The dynamics of the instantaneous and moving averages of the infection rate.

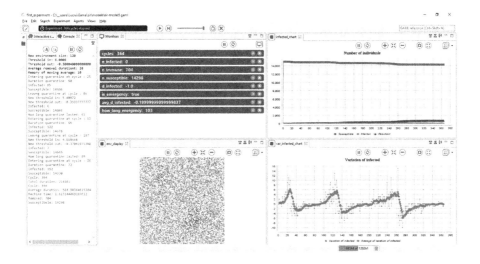

Fig. 3. System GUI.

4.2   Experimental Setup

The simulation is controlled by the following parameters that dictate the speed and magnitude of the epidemics spread:

- Size d of the agents' environment (a $d \times d$ rectangle); $d = 120$ was used in experiments.
- Population size N, including initial number of susceptible individuals S_0, and infected individuals I_0. Note that $N = S_0 + I_0$ and $I_0 \ll S_0$. To keep things simple and observable $I_0 = 2$ was set. This made possible to observe the epidemics spread starting from two distinct source points that define two independent infection outbreaks. Note that the size of the 2D space was kept constant. Hence, the population size dictates the population density that is expected to directly influence the spatial-temporal dynamics of the epidemics. In all experiments $S_0 = 15000$ was used.
- Since the considered model is a stochastic system, each simulation was repeated multiple times, while keeping fixed the initial locations of the infection outbreaks, at: $(0.25 \cdot d, 0.25 \cdot d)$ and $(0.75 \cdot d, 0.75 \cdot d)$.
- The radius ϵ defining "active vicinity" of an infected agent was set to $\epsilon = 0.8$.
- The probabilities that an agent will transmit, respectively will get the infection (see Eq. (9)), p_i and q_i were constant for all individuals. However, note that: i) before the outbreak these probabilities have high values $p^H = q^H = 0.90$; ii) during the emergency situation these probabilities have low values $P^L = q^L = 0.30$, and iii) after the first emergency situation, as people got to know about the danger of the infection, these probabilities have average values $p^M = q^M = 0.75$ (many people are likely to proactively take into account protection measures).

- The speed v of the wandering process was set to $v^H = 1.0$ during the normal situation and to $v^L = 0.1$ during the emergency situation. Note that the angle that gives the range of the direction of the wandering was kept constant and equal to 360°.
- Coefficient γ, defining number of infected individuals removed at each iteration; was set to $\gamma = 0.05$.
- Spatial attenuation function $\phi(x)$ (see Eq. (7)) was chosen as follows:

$$\phi(x) = \begin{cases} 1 - \frac{\log(1+\alpha \cdot x)}{\log(1+\alpha \cdot \epsilon)} & 0 \le x \le \epsilon \\ 0 & x > \epsilon \end{cases}$$

with constant parameter α defined as $\alpha = 10.0$.

4.3 Experimental Results

Due to space limitation we report results of a representative experiment evaluating the effect of managing the infection spread, using the proposed mechanisms. Initially we ran the simulation without the controls. Here, probabilities for transmitting and getting the infection were set to $p^H = q^H = 0.90$, and then to $p^L = q^L = 0.75$. The results for the first case are shown in Fig. 4. Quite similar results have been observed also in the second case. In both cases, the infection propagated to almost entire population. Only a few individuals remained untouched (456 in the first case and 1003 in the second case; from the total of 15002 individuals).

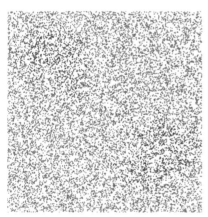

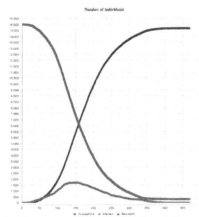

a. Agent population after 104 cycles.

b. Dynamics of population size at the end of the simulation after 469 cycles.

Fig. 4. Results of simulation without control system.

In the second part of the experiment, the control subsystem was turned on. The simulation was repeated 6 times to observe the effect of the controls under the uncertainties of the model. Note that among multiple simulations that were executed, we have retained 6 simulations that are characterized by 2 outbreaks (not all runs generated two outbreaks, corresponding to the 2 initially infected individuals).

The results of a single simulation run that included the control system are presented in Fig. 5. Here, 5 emergency situations were determined between cycles 32–92, 147–213, 278–350, 429–510 and 661–762. These situations can be observed as a decreasing slope of the thick curve in Fig. 5d, representing the variation of the moving average $\overline{\Delta I}$. The number of susceptible individuals, that were not touched by the infection was 12899 for this simulation, and this value can be roughly observed in Fig. 5c.

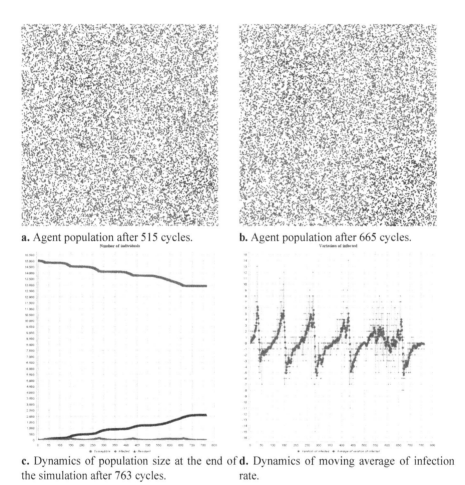

a. Agent population after 515 cycles.

b. Agent population after 665 cycles.

c. Dynamics of population size at the end of the simulation after 763 cycles.

d. Dynamics of moving average of infection rate.

Fig. 5. Results of simulation with control system.

Figures 5a and 5b depict the agent population immediately after the end of the 4th emergency situation and immediately after the start of the 5th emergency situation. Interestingly, one of the outbreaks was extinguished, while the second remains active.

The simulation finished after an of 619.66 cycles (minimum 444 and maximum 874) with an average number of 13465 susceptible individuals (minimum 12080 and maximum 14156), not touched by the infection. The average number of "emergency situations" that were determined by the control system, during the simulation, was 4.16 (minimum 3 and maximum 6).

5   Concluding Remarks

We have developed a multi-agent simulation of the spatial SIR model for epidemics spread management in a population, using the GAMA agent-platform. We have used the simulation system to experimentally investigate a scenario, in which the spread of the infection is mitigated by mobility limitation and by enforcement of protection. While the simulation revealed the usefulness of the approach, we also noticed that it is very sensible to the settings of the parameters of the control system. The approach also depends on the accuracy of the social sensors, used for monitoring and detecting the speed of the infection. We plan to extend our simulation by relaxing the rationality and caution assumptions of the agents and adding different agent profiles, as well as by refining the parameters that control the virus transmission.

Acknowledgement. This work has been supported by the joint research project "Novel methods for development of distributed systems" under the agreement on scientific cooperation between the Polish Academy of Sciences and Romanian Academy for years 2019–2021.

References

1. Abar, S., Theodoropoulos, G.K., Lemarinier, P., O'Hare, G.M.: Agent based modelling and simulation tools: a review of the state-of-art software. Comput. Sci. Rev. **24**, 13–33 (2017). https://doi.org/10.1016/j.cosrev.2017.03.001
2. Ban, T.Q., Duong, P.L., Son, N.H., Dinh, T.V.: Covid-19 disease simulation using gama platform. In: 2020 International Conference on Computational Intelligence (ICCI), pp. 246–251 (2020). https://doi.org/10.1109/ICCI51257.2020.9247632
3. Bonabeau, E.: Agent-based modeling: methods and techniques for simulating human systems. Proc. National Acad. Sci. **99**(suppl 3), 7280–7287 (2002). https://doi.org/10.1073/pnas.082080899
4. Bosse, T., Duell, R., Memon, Z.A., Treur, J., van der Wal, C.N.: Agent-based modeling of emotion contagion in groups. Cogn. Comput. **7**(1), 111–136 (2014). https://doi.org/10.1007/s12559-014-9277-9
5. Bădică, A., Bădică, C., Ganzha, M., Ivanović, M., Paprzycki, M.: Multi-agent simulation of core spatial sir models for epidemics spread in a population. In: 5th International Conference and Workshops on Recent Advances and Innovations in Engineering (ICRAIE 2020), IEEE (2020)
6. Bădică, A., Bădică, C., Ivanović, M., Dănciulescu, D.: Multi-agent modelling and simulation of graph-based predator-prey dynamic systems: a BDI approach. Expert Syst. J. Knowl. Eng. **35**(5), (2018). https://doi.org/10.1111/exsy.12263
7. Bădică, C., Budimac, Z., Burkhard, H., Ivanović, M.: Software agents: languages, tools, platforms. Comput. Sci. Inf. Syst. **8**(2), 255–298 (2011). https://doi.org/10.2298/CSIS110214013B

8. Cuevas, E.: An agent-based model to evaluate the covid-19 transmission risks in facilities. Comput. Biol. Med. **121**, (2020). https://doi.org/10.1016/j.compbiomed.2020.103827

9. Dang, Q.A., Hoang, M.T., Tran, D.H.: Global dynamics of a computer virus propagation model with feedback controls. J. Comput. Sci. Cybern. **36**(4), 295–304 (2020). https://doi.org/10.15625/1813-9663/36/4/15555

10. Gaudou, B., et al.: Comokit: a modeling kit to understand, analyze, and compare the impacts of mitigation policies against the covid-19 epidemic at the scale of a city. Front. Public Health **8**, 587 (2020). https://doi.org/10.3389/fpubh.2020.563247

11. Glinton, R., Scerri, P., Sycara, K.: Exploiting scale invariant dynamics for efficient information propagation in large teams. In: Proceedings of the 9th International Conference on Autonomous Agents and Multiagent Systems, AAMAS 2010, pp. 21–30. International Foundation for Autonomous Agents and Multiagent Systems, Richland, SC (2010), https://doi.org/10.5555/1838206.1838210

12. Patlolla, P., Gunupudi, V., Mikler, A.R., Jacob, R.T.: Agent-based simulation tools in computational epidemiology. In: Böhme, T., Larios Rosillo, V.M., Unger, H., Unger, H. (eds.) IICS 2004. LNCS, vol. 3473, pp. 212–223. Springer, Heidelberg (2006). https://doi.org/10.1007/11553762_21

13. Kampik, T., Nieves, J.C.: JS-son - a lean, extensible JavaScript agent programming library. In: Dennis, L.A., Bordini, R.H., Lespérance, Y. (eds.) EMAS 2019. LNCS (LNAI), vol. 12058, pp. 215–234. Springer, Cham (2020). https://doi.org/10.1007/978-3-030-51417-4_11

14. Karia, R., Gupta, I., Khandait, H., Yadav, A., Yadav, A.: COVID-19 and its modes of transmission. SN Compr. Clin. Med. **2**(10), 1798–1801 (2020). https://doi.org/10.1007/s42399-020-00498-4

15. Kazil, J., Masad, D., Crooks, A.: Utilizing python for agent-based modeling: the mesa framework. In: Thomson, R., Bisgin, H., Dancy, C., Hyder, A., Hussain, M. (eds.) SBP-BRiMS 2020. LNCS, vol. 12268, pp. 308–317. Springer, Cham (2020). https://doi.org/10.1007/978-3-030-61255-9_30

16. Kermack, W.O., McKendrick, A.G., Walker, G.T.: A contribution to the mathematical theory of epidemics. In: Proceedings of the Royal Society of London. Series A, Containing Papers of a Mathematical and Physical Character, vol. 115 no. 772, pp. 700–721 (1927). https://doi.org/10.1098/rspa.1927.0118

17. Kravari, K., Bassiliades, N.: A survey of agent platforms. J. Artif. Soc. Soc. Simul. **18**(1), 11 (2015). https://doi.org/10.18564/jasss.2661

18. Pal, C.V., Leon, F., Paprzycki, M., Ganzha, M.: A review of platforms for the development of agent systems. arXiv.org (2020). https://arxiv.org/abs/2007.08961

19. Spreeuwenberg, P., Kroneman, M., Paget, J.: Reassessing the Global Mortality Burden of the 1918 Influenza Pandemic. Am. J. Epidemiol. **187**(12), 2561–2567 (2018). https://doi.org/10.1093/aje/kwy191

20. Taillandier, P., et al.: Building, composing and experimenting complex spatial models with the GAMA platform. Geo Informatica **23**(2), 299–322 (2018). https://doi.org/10.1007/s10707-018-00339-6

21. Tolles, J., Luong, T.: Modeling epidemics with compartmental models. JAMA Guide Stat. Methods **323**(24), 2515–2516 (2020). https://doi.org/10.1001/jama.2020.8420

22. te Vrugt, M., Bickmann, J., Wittkowski, R.: Effects of social distancing and isolation on epidemic spreading modeled via dynamical density functional theory. Nature Commun. **11**(1), 5576 (2020). https://doi.org/10.1038/s41467-020-19024-0

Multi-criteria Seed Selection for Targeted Influence Maximization Within Social Networks

Artur Karczmarczyk[1]([✉])(iD), Jarosław Jankowski[1](iD), and Jarosław Wątrobski[2](iD)

[1] Faculty of Computer Science and Information Technology, West Pomeranian University of Technology in Szczecin, Żołnierska 49, 71-210 Szczecin, Poland
{artur.karczmarczyk,jaroslaw.jankowski}@zut.edu.pl
[2] The Faculty of Economics, Finance and Management of the University of Szczecin, Mickewicza 64, 71-101 Szczecin, Poland
jwatrobski@usz.edu.pl

Abstract. Information spreading and influence maximization in social networks attracts attention from researchers from various disciplines. Majority of the existing studies focus on maximizing global coverage in the social network through initial seeds selection. In reality, networks are heterogeneous and different nodes can be a goal depending on campaign objectives. In this paper a novel approach with multi-attribute targeted influence maximization is proposed. The approach uses the multi-attribute nature of the network nodes (age, gender etc.) to better target specified groups of users. The proposed approach is verified on a real network and compared to the classic approaches delivers 7.14% coverage increase.

Keywords: Seed selection · Targeted influence maximization · MCDA

1 Introduction

Social media are used for maintaining connections with relatives, friends and to access information sources. Virtual marketing within social media is strategized to reach people with specific interests. It results in a better engagement of the potential client thereof [4] and makes possible to avoid targeting users not interested in products or services. While most of the research focused on influence maximization and global coverage, social networking platforms deliver the ability to pick multiple choice parameters for an exact target class. The need to better address the real specifics of campaigns is visible, but the targeted approaches are introduced in a limited number of studies and are focused mainly on single node attributes [7,15].

The approach presented in this paper deals with the selection of nodes for seeding the social platform on the basis of manifold criteria, as well as diverse attributes within agent based computational environment. The MCDA foundations of the proposed approach enable to adjust the gravity of each touchstone

© Springer Nature Switzerland AG 2021
M. Paszynski et al. (Eds.): ICCS 2021, LNCS 12744, pp. 454–461, 2021.
https://doi.org/10.1007/978-3-030-77967-2_38

to be computed for selection purpose, in order to meet the requirements of the advertiser. Moreover, the relevant MCDA tools and computations enable to gauge the impact of nodes seeding individually on the viral marketing strategy to hit the target groups. The paper comprises of five sections. The Introduction is followed by the Literature review Sect. 2. Next, the methodology discussion is presented 3. After that, experimental results are showed 4 and followed by concluding statements 5.

2   Literature Review

In the area of information spreading within social platforms, it was supposed in the early stages of research that all the nodes of a network carry the same level of inclination towards a promulgated product or service or any other content [6]. However, in reality more result-oriented campaigns allow multiple node behaviors to be taken into consideration and better nodes allocation [7]. Recent studies used the cost assignment to the user of the network combined with the user interest benefits [8]. The goal of nodes selection can be also avoidance of intense campaign with unnecessarily repeated messages [1]. Pasumarthi et al. identified a targeted influence maximization problem, introducing an objective functionality and a penalizing criterion for adopting non targeted nodes [9].

Recently, initial studies are held discussing the application of MCDA techniques in the areas related to social networking. TOPSIS [1] method is used by Yang et al., in SIR (Susceptible Infected Recovered) model for identification of influential nodes in complex network [13]. Entropy weight method is used to measure and set up the weight values [14]. For maximizing the coverage and reducing the overlap, TOPSIS method is used by Zarei et al., while a social network is being influenced [16]. PROMETHEE[2] method was used by Karczmarczyk et al., to evaluate the responsiveness of viral marketing campaigns within social networking portals and also for providing decision support in order to plan these campaigns [5].

Review of studies in the area of information spreading and influence maximization has shown that among large number of studies only a small chunk is targeting the most common problem such as reaching out the specific user with multiple characteristics. Most of the existing approaches behave mono-trait by addressing nodes as a single attribute. However, social networks generally identify the target groups relying on multiple parameters, such as gender, localization or age. This identifies a research gap for seed selection based on a multi-characteristic computation in order to target specific multi-attribute network nodes, which this paper addresses.

3   Methodology

The proposed methodology complements the widely-used Independent Cascade (IC) model for modeling the spread within the complex networks [6], by taking

---

[1] Technique for Order Preference by Similarity to Ideal Solution.
[2] Preferences Ranking Organization METHod for Enrichment of Evaluations.

into account the problem of reaching targeted multi-attribute nodes in social networks by the information propagation processes. In the proposed approach, it is assumed that the network nodes are characterized not only by the centrality relations between them and other nodes, but also by a set of custom attributes $C_1, C_2, \ldots, C_n$. The nodes can also be characterized by the computed attributes derived from the network characteristics and measures, such as degree. Last, but not least, additional attributes can be derived as a composite of the two aforementioned types of attributes, by computing centrality measures based on limited subsets of the nodes' neighbors. For example, if attribute C_i represented the degree of a node, i.e. the total count of its neighbors, the C_{i_1} could represent the count of its male neighbors.

The aim of the proposed methodological framework is to maximize the influence within the targeted group of multi-attribute network nodes. While other approaches focus on generating the ranking of seeds based on a single centrality measure, in the authors' proposed methodological framework, the seeds are selected based on multiple attributes. This allows to select seeds which might be worse at maximizing global influence in the network, but which are better at maximizing influence in the targeted group of multi-attribute network nodes.

The approach presented in this paper is based on the MCDA methodology foundations [11]. The assumed modeling goal is to reach only the targeted set of multi-attribute nodes, instead of maximixing global influence in the network. Based on the guidelines provided by [3], it was decided that the PROMETHEE II method is most suitable for the proposed approach. It is an MCDA method that uses pairwise comparison and outranking flows to produce a ranking of the best decision variants. In the proposed approach, PROMETHEE II is used to produce a multi-criteria ranking of the nodes in the network with the aim to shortlist the ones which have the best chances to maximize influence in the targeted group of multi-attribute nodes. A detailed description of the PROMETHEE methods can be found in [2]. The MCDA foundations of the proposed approach help maximizing influence in the targeted group of multi-attribute nodes by selecting the seeds which have the highest, according to the marketer, potential to reach the targeted nodes in the social network. Moreover, the use of tools such as GAIA visual aid allows to understand the preferences backing the actual seed selection, and provide feedback which allows to further iteratively improve the obtained solution.

4   Empirical Study

In order to illustrate the proposed approach, the empirical study with the use of agent based simulations was performed on a relatively small real network [10] with 143 vertices and 623 edges giving the ability of detailed multi-criteria analysis. The proposed approach is intended for networks whose nodes are described with multiple attributes. However, the publicly available network datasets predominantly consist only of information on their nodes and edges, without information on the node attributes. To overcome this problem, the node attributes

Table 1. Criteria used in the empirical research.

Criterion	Values	Criterion	Values
C1 Degree	Integer [1-42]	C5 Age	1: 0-29, 2: 30-59, 3: over 60
C2 Gender	1: male, 2: female	C6 Deg. younger	Integer [0-18]
C3 Deg. male	Integer [0-20]	C7 Deg. medium	Integer [0-15]
C4 Deg. female	Integer [0-22]	C8 Deg. older	Integer [0-9]

Table 2. Top 7 network nodes used as seeds in the empirical research. A - degree; B - betweenness, C - closeness, D - eigen centrality, E-G - the proposed multi-attribute approach

A	105	17	95	48	132	43	91	E	17	95	48	132	50	105	20
B	107	17	48	91	32	95	141	F	19	95	48	50	132	91	105
C	105	17	95	37	74	48	91	G	132	20	136	19	50	122	3
D	105	31	136	132	20	19	69								

were artificially overlaid over the network, following the attributes' distribution from demographic data. Two demographic attributes were overlaid on the network – gender and age. For illustrative purposes, the target for the viral marketing campaign was chosen for the empirical research. In this experiment, the male users from the youngest age group were targeted, which translates to 28 of all the 143 users of the network. In the proposed approach, the seeds are selected from the network based on multiple criteria. In the empirical research, apart from the two aforementioned demographic attributes, also the degree measure was taken into account, as well as 5 criteria based on a mix of the degree and the demographic measures. This resulted in a total of 8 seed evaluation criteria, which are presented in Table 1.

Initially, the classic single-metric approaches were tested on the network, to provide a benchmark for the proposed approach. Four centrality metrics (degree, closeness, betweenness and eigen centrality) were used individually to first rank all vertices in the network, and then select the top nodes as seeds. It was decided for the seeding fraction to be set to 0.05 (seven seeding nodes) and propagation probability to 0.10. Moreover, in order to allow repeatability of the experiment for seeds selected by each approach, 10 pre-defined scenarios were created, in which each node was assigned a pre-drawn weight. The seeds obtained from rankings based on each centrality measure, i.e. degree, betweenness, closeness and eigen centrality, are presented in Table 2A–D respectively. The averaged simulation results are presented in Table 3A–D.

Table 3. Aggregated results from the empirical study simulations

	Iterations	Infected	Coverage	Infected targeted	Coverage
A	6.6	41.2	0.2881	7.7	0.2750
B	6.1	33.7	0.2357	5.5	0.1964
C	6.2	39.2	0.2741	6.2	0.2214
D	6.5	34.3	0.2399	9.0	0.3214
E	5.9	40.6	0.2839	9.2	0.3286
F	6.0	40.7	0.2846	9.5	0.3393
G	6.4	30.1	0.2105	9.7	0.3464

Table 4. Utilized PROMETHEE II parameters

	Criteria	C1	C2	C3	C4	C5	C6	C7	C8
E	Weight	1	1	1	1	1	1	1	1
	Preference function	Usual	Usual	Usual	Usual	Usual	Usual	Usual	Usual
F	Weight	1	1	1	1	1	1	1	1
	Preference function	Linear	Usual	Linear	Linear	Usual	Linear	Linear	Linear
	q; p	3; 9	1; 2	1; 4	1; 2	1; 2	1; 4	1; 3	1; 2
G	Weight	8.2	25.4	12.6	3.8	28.4	14	3.8	3.8
	Preference function	Linear	Usual	Linear	Linear	Usual	Linear	Linear	Linear
	q; p	3; 9	1; 2	1; 4	1; 2	1; 2	1; 4	1; 3	1; 2

In the next step of the empirical study, the authors' proposed approach was used to choose the seeds based on a multi-criteria ranking produced by the PROMETHEE II method. All eight criteria were taken into account. Initially, the usual preference function was used for comparing each vertex under all criteria. Also, all criteria were given an equal preference weight (see Table 4E). As a result, seven seeds were selected (see Table 2E). It can be noticed that the produced seed set is considerably different than the ones produced by the classic approaches (compare with Table 2A-2D).

After the simulations were executed with the newly selected seeds, it was observed that averagely 40.6 network nodes were infected (0.2839 coverage, see Table 3E). It is a worse result than for the degree-based approach. What is important to note, however, is that averagely 9.2 targeted nodes were infected, i.e. 0.3286 targeted coverage, which was the best result so far.

One of the benefits of using the PROMETHEE methods is the possibility to adjust the preference function used in pairwise comparisons of the nodes under individual criteria. While the usual preference function provides a simple boolean answer for the pairwise comparison of gender (C2) and age (C5) criteria, in case of the criteria based on degree, usage of a linear preference function with indifference and preference thresholds can yield better results. Therefore, in the subsequent step of the empirical research, a linear preference function was applied to all degree-based criteria (see Table 4F). The change in the preference function resulted in a different set of seeds selected for simulations (see Table 2F). The averaged results from the simulations are presented in Table 3F. It can be observed, that both global and targeted coverage values improved slightly.

Depending on the target group, the marketer can decide that some criteria can better help to reach the target group than the other criteria. Therefore, the marketer can adjust the preference weights of each criterion. Before the last set of simulations in this empirical research, an expert knowledge was elicited from the marketer with the use of the Analytical Hierarchy Process (AHP) [12], to adjust the preference weights of all criteria. The elicited weights used in the final set of simulations is presented in Table 4G. The adjusted preference weights resulted in a significantly different set of nodes used as seeds in the campaign (see Table 2G). The averaged simulation results are presented in Table 3G. The approach resulted in the best coverage in the targeted group (0.3464, compared to 0.2750 for the degree-based approach, 0.0714 difference). In the final step of the research, the GAIA visual analysis aid was used to study the criteria preference relations in the seed selection decision model (see Fig. 1). The analysis of Fig. 1 allows to observe that criteria C2 and C5 are not related to each other in terms of preference. This is quite straightforward, because these criteria represent the gender and age respectively. On the other hand, the remaining criteria are similar in terms of preference, possibly because they are all partially based on the degree measure.

Fig. 1. GAIA visual analysis

5   Conclusions

The existing research in the area of information spreading focuses mainly on influence maximisation. Only limited number of studies discuss targeting nodes with specific characteristics with main focus on their single attributes. This paper proposes a novel approach to multi-attribute targeted influence maximization in social networks, focused on a multi-attribute seed selection. In the proposed approach, the seeds for initializing the campaign are chosen based on a ranking obtained with an MCDA method. The weights of individual criteria can be adjusted, as well as criteria values' comparison preference functions can be chosen to best fit the marketer's needs. In the experimental research, the proposed approach resulted in target nodes' coverage superior by as much as 7.14% compared to traditional degree-based approaches. The research opens some possible future directions. It would be beneficial to further broaden the research scope by studying how the changes in seeding fraction and propagation probability affect the efficiency of the proposed approach. Moreover, this research was performed on a network with attributes superimposed artificially. A research project can be run in order to collect knowledge about a real multi-attribute social network.

Acknowledgments. This work was supported by the National Science Centre of Poland, the decision no. 2017/27/B/HS4/01216 (AK, JJ) and within the framework of the program of the Minister of Science and Higher Education under the name "Regional Excellence Initiative" in the years 2019–2022, project number 001/RID/2018/19, the amount of financing PLN 10,684,000.00 (JW).

References

1. Abebe, R., Adamic, L., Kleinberg, J.: Mitigating overexposure in viral marketing. In: Proceedings of the AAAI Conference on Artificial Intelligence, vol. 32 (2018)
2. Brans, J.-P., Mareschal, B.: Promethee methods. Multiple Criteria Decision Analysis: State of the Art Surveys. ISORMS, vol. 78, pp. 163–186. Springer, New York (2005). https://doi.org/10.1007/0-387-23081-5_5
3. Cinelli, M., Kadziński, M., Gonzalez, M., Słowiński, R.: How to support the application of multiple criteria decision analysis? Let us start with a comprehensive taxonomy. Omega **96**, 102261 (2020)
4. Iribarren, J.L., Moro, E.: Impact of human activity patterns on the dynamics of information diffusion. Phys. Rev. Lett. **103** (2009)
5. Karczmarczyk, A., Jankowski, J., Wątróbski, J.: Multi-criteria decision support for planning and evaluation of performance of viral marketing campaigns in social networks. PloS one **13**(12), 1–32 (2018)
6. Kempe, D., Kleinberg, J., Tardos, É.: Maximizing the spread of influence through a social network. In: Proceedings of the Ninth ACM SIGKDD International Conference on Knowledge Discovery and Data Mining, pp. 137–146 (2003)
7. Mochalova, A., Nanopoulos, A.: A targeted approach to viral marketing. Electron. Commer. Res. Appl. **13**(4), 283–294 (2014)
8. Nguyen, H.T., Dinh, T.N., Thai, M.T.: Cost-aware targeted viral marketing in billion-scale networks. In: IEEE INFOCOM 2016-The 35th Annual IEEE International Conference on Computer Communications, pp. 1–9. IEEE (2016)
9. Pasumarthi, R., Narayanam, R., Ravindran, B.: Near optimal strategies for targeted marketing in social networks. In: Proceedings of the 2015 International Conference on Autonomous Agents and Multiagent Systems, pp. 1679–1680 (2015)
10. Rossi, R.A., Ahmed, N.K.: The network data repository with interactive graph analytics and visualization. In: AAAI (2015). http://networkrepository.com/email-enron-only.php
11. Roy, B., Vanderpooten, D.: The European school of MCDA: emergence, basic features and current works **5**(1), 22–38
12. Saaty, T.L.: Decision-making with the AHP: why is the principal eigenvector necessary. Eur. J. Oper. Res. **145**(1), 85–91 (2003)
13. Yang, P., Liu, X., Xu, G.: A dynamic weighted TOPSIS method for identifying influential nodes in complex networks. Mod. Phys. Lett. B **32**(19), 1850216 (2018)

14. Yang, Y., Yu, L., Zhou, Z., Chen, Y., Kou, T.: Node importance ranking in complex networks based on multicriteria decision making. Math. Probl. Eng. (2019)
15. Zareie, A., Sheikhahmadi, A., Jalili, M.: Identification of influential users in social networks based on users' interest. Inf. Sci. **493**, 217–231 (2019)
16. Zareie, A., Sheikhahmadi, A., Khamforoosh, K.: Influence maximization in social networks based on TOPSIS. Expert Syst. Appl. **108**, 96–107 (2018)

An Adaptive Network Model of Attachment Theory

Audrey Hermans[1], Selma Muhammad[2], and Jan Treur[2(✉)]

[1] Athena Institute, Vrije Universiteit Amsterdam, Amsterdam, The Netherlands
a.m.m.hermans@student.vu.nl
[2] Social AI Group, Department of Computer Science, Vrije Universiteit Amsterdam, Amsterdam, The Netherlands
s.muhammad@student.vu.nl, j.treur@vu.nl

Abstract. The interactions between a person and his or her primary caregiver shape the attachment pattern blueprint of how this person behaves in intimate relationships later in life. This attachment pattern has a lifelong effect on an individual, but also evolves throughout a person's life. In this paper, an adaptive network was designed and simulated to provide insights into how an attachment pattern is created and how this pattern then has its effects and evolves as the person develops new intimate relationships at older age.

Keywords: Attachment theory · Adaptive temporal causal network

1 Introduction

For many adults, establishing and maintaining emotionally intimate relationships with loved ones, friends or family members may raise difficulties such as problems with intimacy, dependence or abnormal emotional reactivity. The root of establishing healthy relationships can be found in the attachment theory introduced by Mary D. Salter Ainsworth and John Bowlby [5–6, 15–20], which provides a framework on how distorted personal bonds between an individual and his primary caregiver leads to a faulty development of intimate relationships later on in life.

Attachment theory has had a profound impact on changing institutional care provided for children (e.g., in care homes and in hospitals). However, the theory's wider application to adult mental health has not been investigated extensively. While the attachment patterns originating from interactions with the primary caregiver have been studied thoroughly, little attention has been paid to how these attachment styles are expressed and evolve at an older age. Modelling the expression ad evolution of adult attachment patterns may fill part of this gap by providing insight in how the concepts of attachment theory affect the individual's interactions when this person meets a securely attached person in later life and how these adult experiences may change attachment patterns.

In this study, an adaptive network model was designed to model, firstly how attachment styles are established, and secondly, how these patterns then affect the individual in later life when trying to build an intimate relationship with a securely attached person.

© Springer Nature Switzerland AG 2021
M. Paszynski et al. (Eds.): ICCS 2021, LNCS 12744, pp. 462–475, 2021.
https://doi.org/10.1007/978-3-030-77967-2_39

For the latter situation, this model also shows how the individual's attachment style is adapted at adult age by learning from such a new relationship.

These insights behind the model and the patterns generated by it may be an helpful instrument in therapeutic settings. The model can be used in the therapeutic setting to help patients understand how the process continues, e.g., with virtual role play sessions using avatars such as pointed out in [23]. In this way, this paper provides a new perspective on the use of computational modeling [9] in a psychological context.

This paper is divided into five sections. After the current section, the second section introduces attachment theory and its main concepts. In the third section, the design of the network model will be discussed, which consists of the main attachment concepts, the individual's inner working model and the interactions with the securely attached person. The resulting simulations of the scenarios will then be laid out in the fourth section. The final section presents the conclusions for and discussion of this study.

2 Attachment Theory

The Attachment Theory concerning the relationships between humans was developed from the 1940s and 1950s on mainly by developmental psychologist Mary D. Salter Ainsworth and psychologist and psychiatrist John Bowlby [5–6, 15–20] as a successor of security Theorey developed by William E. Blatz and Mary D. Salter Ainsworth [4, 15]. The Attachment Theory explains an important evolutionary function of the relationship between the child and caregiver. This has been supported by empirical research in various settings. For example, Salter Ainsworth did research on mother-child relationships for two years from 1954 on in Uganda [16] and also Bowlby has investigated the empirical basis of the theory among humans and non-human primates [6]. The theory is often applied in therapeutical contexts; e.g., [8, 10–11].

According to Attachment theory, the first attachment relationship is between a child and its primary caregiver (PC), which has a significant effect on the child's cognitive and socio-emotional development. Research has shown that early attachment is correlated with the PC's sensitivity, reliability and responsiveness [7]. These behaviours of the PC lead to the child's development of the three principles of attachment theory: bonding as an intrinsic human survival strategy, regulation of emotion and fear to enhance resilience and vitality, and flexible adaptiveness and growth [10]. In this way, the early experiences with the PC form a main input for the child to develop the 'internal working model of social relationships' (including a 'model of self' and a 'model of other'), which continues to change with age and experience [12]. For example, if a caretaker shows high levels of parental sensitivity, reliability and responsiveness, the child will develop positive models of self and other [7].With these internal models, children predict the PC's behaviour and plan their own behaviour accordingly [7]. Important relationships later in life may be built on the quality of early attachment [20]. Accordingly, the formed relationship with the PC can be seen as a 'blueprint' for future relationships [3]; see Fig. 1 for an overview of this. The attachment behaviour of children has been classified in four patterns or styles: secure, anxious-avoidant, anxious-ambivalent, and disorganized [6, 13]. Adult attachment behaviour corresponds to these categories, but is named differently. The classification of attachment is based on the balance between intimacy and independence

[3]. It is measured in levels of avoidance and dependence, where avoidance can be related to the 'model of other' and dependence can be related to the 'model of self'. Table 1 gives an overview of the classifications, internal working models, behaviours and parental styles.

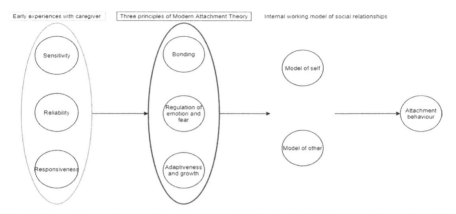

Fig. 1. How early experiences with a primary caregiver lead to the development of internal models of self and other that in turn lead to attachment behaviour.

Table 1. Overview of classifications, internal working models, behaviours and parental styles.

Infant attachment category	Adult attachment category	Internal working model	Avoidance and dependence	Parental style	Behaviour
Secure	Secure	Positive model of self	Low dependence	Sensitive Reliable Responsive	Self is worthy of love and support Others are trustworthy and reliable Seek proximity
		Positive model of other	Low avoidance		
Anxious-avoidant	Dismissing	Positive model of self	Low dependence	Insensitive Rejecting	Self is worthy of love and support Others are unreliable and rejecting Avoid proximity to protect self against disappointment
		Negative model of other	High avoidance		
Anxious-ambivalent	Pre-occu-pied	Negative model of self	High dependence	Unreliable Unresponsive	Self is not worthy of love and support Others are trustworthy and reliable Hesitant in seeking proximity, but strive for self-acceptance by gaining acceptance of valued others
		Positive model of other	Low avoidance		
Disorganized	Fearful	Negative model of self	High dependence	Parental abuse and neglect	Self is not worthy of love and support Others are unreliable and rejecting Avoid proximity to protect self against anticipated rejection by others
		Negative model of other	High avoidance		

Secure attachment indicates a positive model of both self and other and thus low levels of dependence and avoidance. This is the most prevalent attachment pattern [6]. Anxious-avoidant attachment, also referred to as 'dismissing', indicates a positive 'model of self', resulting in low dependence, and a negative 'model of other', resulting in high avoidance. Anxious-avoidant is the second most prevalent attachment style [6]. Anxious-ambivalent attachment, also referred to as 'preoccupied', indicates a negative 'model of self', resulting in high dependence, and a positive 'model of other', resulting in low avoidance. Finally, disorganized attachment is also referred to as 'fearful' and indicates a negative model of both self and other and thus high levels of dependence and avoidance.

3 The Modeling Approach Used

For this study, the Network-Oriented Modelling approach, as described in [21, 22], was used to design a model based on a network structure which can be used to simulate and analyse attachment behaviour. The elements of such a network model are:

- the *states* Y of the network
- the *connections* from states X to Y, with their *connection weights* $\omega_{X,Y}$ specifying different strengths for these connections
- a *speed factor* η_Y for each state Y to express how fast state Y can change
- a *combination function* c_Y for each state Y to indicate how all incoming connections for each state combine to impact that state

The designed attachment behaviour model comprises the main attachment concepts as states and relations between these concepts as connections. The numerical representation created by the available dedicated software environment is based on the following equations (where $X_1, \ldots, X_k$ are the states from which state Y gets incoming connections):

$$\mathbf{impact}_{X,Y}(t) = \omega_{X,Y}X(t) \tag{1}$$

$$\mathbf{aggimpact}_Y(t) = \mathbf{c}_Y(\mathbf{impact}_{X_1,Y}(t), \ldots, \mathbf{impact}_{X_k,Y}(t)) = \mathbf{c}_Y(\omega_{X_1,Y}X_1(t), \ldots, \omega_{X_k,Y}X_k(t)) \tag{2}$$

$$
\begin{aligned}
Y(t + \Delta t) = Y(t) &= \eta_Y\big[\mathbf{aggimpact}_Y(t) - Y(t)\big]\Delta t \\
&= Y(t) + \eta_Y\big[\mathbf{c}_Y\big(\omega_{X_1,Y}X_1(t), \ldots, \omega_{X_k,Y}X_k(t)\big) - Y(t)\big]\Delta t \tag{3}
\end{aligned}
$$

The combination functions from the library used are shown in Table 2.

For the learning, the modeling approach provides the possibility to include *self-models* in a network model. This idea is inspired by the idea of self-referencing or 'Mise en abyme' in art, sometimes also called 'the Droste-effect' after the famous Dutch chocolate brand who uses this effect in packaging and advertising of its products already since 1904[1]. This effect occurs in art when within an artwork a small copy of the same artwork is included. For Network-Oriented Modeling, this idea leads to *self-modeling networks*, also called reified networks [22]. These are networks that represent some of the network structure characteristics by self-model states within the network. As an example used here, the weight $\omega_{X,Y}$ of a connection from (base) state X to Y can be represented by a (first-order) self-model state $\mathbf{W}_{X,Y}$. Such a first-order self-model state is depicted

---

[1] E.g.,   https://en.wikipedia.org/wiki/Mise_en_abyme,   https://en.wikipedia.org/wiki/Droste_effect.

in a 3D format (as in Fig. 2) in a separate (blue) plane above the (pink) plane for the base network. Like any other state, such a self-model state $\mathbf{W}_{X,Y}$ has an activation value that changes over time, based on its incoming connections from other states. Through a downward connection from $\mathbf{W}_{X,Y}$ to Y (indicated by pink arrows in Fig. 2), the weight $\omega_{X,Y}$ of the related connection from state X to state Y within the base network will adapt accordingly, which creates a form of learning for that connection. In the current paper *plasticity* is modeled by Hebbian learning and *metaplasticity* to control this learning [1, 14]. For more details, see [2].

Table 2. Combination functions from the library used in the presented model

	Notation	Formula	Parameters
Steponce	$\text{steponce}_{\alpha,\beta}(V)$	1 if $\alpha \leq t \leq \beta$, else 0	α begin, β end time
Scaled sum	$\text{ssum}_{\lambda}(V_1, ..., V_k)$	$\frac{V_1 + \cdots + V_k}{\lambda}$	Scaling factor $\lambda > 0$
Advanced logistic sum	$\text{alogistic}_{\sigma,\tau}(V_1, ..., V_k)$	$[\frac{1}{1+e^{-\sigma(V_1+\cdots+V_k-\tau)}} - \frac{1}{1+e^{\sigma\tau}}](1 + e^{-\sigma\tau})$	Steepness $\sigma > 0$ Excitability threshold τ
Scaled minimum	$\text{smin}_{\lambda}(V_1,V_2)$	$\frac{\min(V_1,V_2)}{\lambda}$	Scaling factor $\lambda > 0$
Scaled maximum	$\text{smax}_{\lambda}(V_1, ..., V_k)$	$\frac{\max(V_1,...,V_k)}{\lambda}$	Scaling factor $\lambda > 0$
Min advanced logistic	$\text{minalogistic}_{\sigma,\tau}(V_1,...,V_{k+1})$	$\min(\text{alogistic}_{\sigma,\tau}(V_1, \ldots, V_k, V_{k+1})$	Steepness $\sigma > 0$ Excitability threshold τ

4   Design of the Adaptive Network Model for Attachment Theory

This section describes the adaptive temporal-causal network model designed to investigate the formation of an attachment style and its impact on the individual's relationship at older age. Figure 2 provides an overview of all the states and their connections for a person A who develops a 'blueprint' from his/her primary caregiver C and then interacts with a new person B. The connection weights, speed factors, and combination functions are labels for the nodes and arrows. These are not presented in the figure, but can be found for each scenario in [2] in the form of role matrices. Table 3 presents the nomenclature and explanation for all 47 states of the model.

Note that, triggered by when the other person is there, base states X_{10} to X_{15} are the states that become activated according to person A's model of the other, and base states X_{16} to X_{21} according to person A's model of self. If there is no person, these states are not activated. To achieve this, the persistent models of the other and of the self are represented by first-order self-model **W**-states X_{35} to X_{40} and X_{41} to X_{46}, respectively. These **W**-states represent the weights of the connections from X_3 (presence of a person) to the respective base states X_{10} to X_{15} and X_{16} to X_{21}. The specific combinations of values of these **W**-states define the attachment style, and when a person is present (i.e., X_3 is activated), the base states as mentioned get their (temporary) values accordingly. The different types of states were assigned different combination functions to ensure that incoming connections impact the state activations in a proper manner. The six different combination functions shown in Table 2 have been used as follows:

- The *steponce* combination function. The two initiating states X_1 for primary caregiver C and X_2 for other person B were assigned the steponce function. This function was used to ensure that each of them initiates the process at the specified time interval.
- The *scaled sum* combination function. This function was assigned to state X_3 and the sensor and sensory representation states X_{22} and X_{23}.
- The *alogistic* combination function. The states for the early experiences with the primary caregiver C (X_4, X_5, X_6), the three principles of modern attachment theory (X_7, X_8, X_9) and the action, reaction and evaluation states (X_{26} to X_{31}) were assigned this function. The learning control state (X_{47}) was also assigned the alogistic combination function. By this, person A processes the arrival of a person and the possible adaptation of the attachment blueprint for the person.
- The *maximum* combination function. For the states of the internal self and other working models (X_{10} to X_{21}) this function was used.
- The *min-alogistic* combination function. The preparation states (X_{24}, X_{25}) were given a composition of two functions: first, the alogistics function is applied and then the minimum function to select the outcome with the lowest value.
- The *minimum* combination function. All **W**-states (X_{32} to X_{46}) were assigned this function.

5 Simulation Scenarios

This section describes two scenarios that were simulated by the adaptive network model and the resulting simulation graphs. The first scenario concerns the secure attachment style and the second scenario the anxious-avoidant attachment style. The latter scenario was differentiated into a scenario for an attachment style that is highly adaptive to the interactions at older age and one for a pattern that is more rigid. For both scenarios the first peak illustrates the interactions of the individual with the primary caregiver C, which then leads to a persistent 'blueprint' in A for the level of bonding, emotion regulation, adaptivity, and the model of self and other. The second peak illustrates the interaction with the other person B, which either continues the learnt pattern or leads to an adaptation of person A's attachment pattern to B.

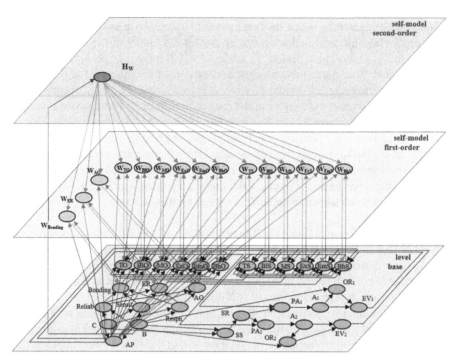

Fig. 2. Graphical representation of the connectivity of the adaptive network model for attachment behaviour.

Scenario 1: Secure Attachment Style

The simulation results from the first scenario are displayed in Fig. 3.

In this scenario, the main character has a primary caregiver C who is sensitive, reliable and responsive. Therefore, the main character develops a secure attachment pattern. Next, at an older age, the main character encounters person B who is also sensitive, reliable and responsive. In Fig. 3 for t from 30 to 60, A interacts with C and these interactions lead to the formation of A's attachment style, represented by the persistent W-states. This comprises developing bonding, emotion regulation, adaptiveness and the construction of the 'model of self' and 'model of other'. After these interactions with C, the attachment blueprint is carried over through this person's life through the persistent W-states, which can be seen in the graph for t from 60 to 150, where the W-states remain at a constant level. Finally, for t from 150 to 200, the individual encounters B with whom he or she interacts. Since A had developed a secure attachment with C in this scenario, the blueprint, as represented by the W-states, does not actually adapt when meeting another securely attached person. The degree of flexibility to adapt later in life, incorporated in the model by the connection weight of X_2 to X_{47}, was given different levels. This connection weight was set to 0.2 to express a relatively low flexibility (a more rigid evolving attachment style) as shown in the upper graph and to 0.5 to simulate a higher flexibility shown in the lower graph. Note that there is a dip immediately after encountering B, which reflects how A becomes open to a certain extent to adapt his/her blueprint when meeting B. On the longer term, eventually the attachment blueprint

Table 3. Nomenclature and explanation of the states in the network model

Nr	Abb.	Full name	Definition
X_1	C	Primary caregiver	A being in contact with caregiver C
X_2	B	Other person	A being in contact with other person B at adult age
X_3	AP	Abstract Person	A being in contact with any person; all learned attachment patterns and personal characteristics are related to this state
X_4	Reliab	Reliability	Early experiences with reliability of C in relation to A
X_5	Sensit	Sensitivity	Early experiences with sensitivity of C in relation to A
X_6	Respn	Responsiveness	Early experiences with responsiveness of C in relation to A
X_7	Bond	Bonding	First principle of Modern Attachment Theory
X_8	ER	Emotion regulation	Second principle of Modern Attachment Theory
X_9	AG	Adaptiveness and growth	Third principle of Modern Attachment Theory
X_{10}	TO	Thoughts about other	First part of A's internal other-model of attachment
X_{11}	BlfO	Beliefs about other	Second part A's of internal other-model of attachment
X_{12}	MO	Memories about other	Third part A's of internal other-model of attachment
X_{13}	ExO	Expectations about other	Fourth part A's of internal other-model of attachment
X_{14}	EmO	Emotions about other	Fifth part of A's internal other-model of attachment
X_{15}	BhvO	Behaviours about other	Sixth part of A's internal other-model of attachment
X_{16}	TS	Thoughts about self	First part of A's internal self-model of attachment
X_{17}	BlfS	Beliefs about self	Second part of A's internal self-model of attachment
X_{18}	MS	Memories about self	Third part of A's internal self-model of attachment
X_{19}	ExS	Expectations about self	Fourth part of A's internal self-model of attachment
X_{20}	EmS	Emotions about self	Fifth part of A's internal self-model of attachment
X_{21}	BhvS	Behaviours about self	Sixth part of A's internal self-model of attachment
X_{22}	SS	Sensor state (feelings for person B)	Feelings developed by A through perceiving B.
X_{23}	SR	Sensory representation	A's sensory representation of feelings for B
X_{24}	PA1	Preparation action 1	A prepares for A's first attachment behaviour
X_{25}	PA2	Preparation action 2	A prepares for A's second attachment behaviour
X_{26}	A1	Action 1: Dependence	Degree of dependence: A's first attachment behaviour
X_{27}	A2	Action 2: Proximity seeking	Degree of proximity seeking: A's second attachment behaviour
X_{28}	OR1	Other person's reaction to action 1	Reaction of B to degree of shown dependence of A
X_{29}	OR2	Other person's reaction to action 2	Reaction of B to degree of expressed proximity seeking of A
X_{30}	EV1	Evaluation of action 1	A's evaluation of A's first attachment behaviour
X_{31}	EV2	Evaluation of action 2	A's evaluation of A's second attachment behaviour
X_{32}	$\mathbf{W}_{bond}$	W-state of bonding	A's first-order self-model state for connection weight ω_{bond}
X_{33}	$\mathbf{W}_{ER}$	W-state of emotion regulation	A's first-order self-model state for connection weight ω_{ER}
X_{34}	$\mathbf{W}_{AG}$	W-state of adaptiveness and growth	A's first-order self-model state for connection weight ω_{AG}
X_{35}	$\mathbf{W}_{TO}$	W-state of thoughts about other	A's first-order self-model state for connection weight ω_{TO}
X_{36}	$\mathbf{W}_{BlfO}$	W-state of beliefs about other	A's first-order self-model state for connection weight ω_{BlfO}
X_{37}	$\mathbf{W}_{MO}$	W-state of memories about other	A's first-order self-model state for connection weight ω_{MO}
X_{38}	$\mathbf{W}_{ExO}$	W-state of expectations about other	A's first-order self-model state for connection weight ω_{ExO}
X_{39}	$\mathbf{W}_{EmO}$	W-state of emotions about other	A's first-order self-model state for connection weight ω_{EmO}
X_{40}	$\mathbf{W}_{BhvO}$	W-state of behaviours about other	A's first-order self-model state for connection weight ω_{BhvO}
X_{41}	$\mathbf{W}_{TS}$	W-state of thoughts about self	A's first-order self-model state for connection weight ω_{TS}
X_{42}	$\mathbf{W}_{BlfS}$	W-state of beliefs about self	A's first-order self-model state for connection weight ω_{BlfS}
X_{43}	$\mathbf{W}_{MS}$	W-state of memories about self	A's first-order self-model state for connection weight ω_{MS}
X_{44}	$\mathbf{W}_{ExS}$	W-state of expectations about self	A's first-order self-model state for connection weight ω_{ExS}
X_{45}	$\mathbf{W}_{EmS}$	W-state of emotions about self	A's first-order self-model state for connection weight ω_{EmS}
X_{46}	$\mathbf{W}_{BhvS}$	W-state of behaviours about self	A's first-order self-model state for connection weight ω_{BhvS}
X_{47}	Hw	Learning control state	A's second-order self-model state: control state for learning of attachment patterns, representing the learning speed

returns to the previous values during the interactions with B, because the new person B has the same secure attachment style as the primary caregiver C.

Scenario 2: Anxious-Avoidant Attachment Style

The simulations of the second scenario are displayed in Fig. 4. In this scenario, A has a primary caregiver C who is insensitive and rejecting. To express this in the network characteristics, the connection weight for the connection from X_1 (for C) to X_4, X_5

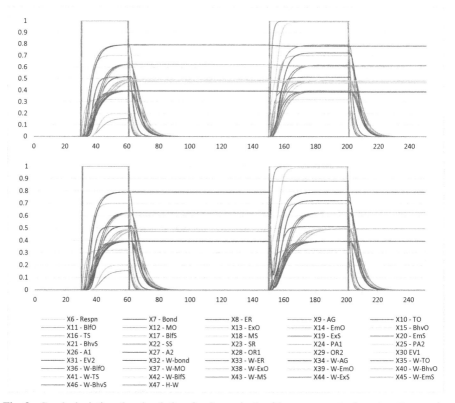

Fig. 3. Graph depicting the simulation for Scenario 1 with a secure attachment pattern and a relatively low flexibility (upper graph, 0.2) and high flexibility (lower graph, 0.5) to adapt later in life.

and X_6 (sensitivity, reliability and responsiveness) was lowered from 0.4 to respectively 0.1, 0.2 and 0.2. These values represent a different ratio which would lead the main character to develop an anxious-avoidant attachment pattern. In this way, the impact from the initiating state X_1 is now much lower than in the secure attachment pattern described in Scenario 1 above, leading the main character A to develop an anxious-avoidant attachment pattern. Furthermore, again the degree of flexibility to adapt later in life, incorporated in the model by the connection weight of X_2 to X_{47}, was given different levels. This time this connection weight was set to 0.1 to simulate a very low, rigid evolving attachment style for the upper graph and to 0.9 to express a very high flexibility for the lower graph. Additionally, the threshold for X_{27} (Action 2: proximity seeking) was increased from 0.5 to 1 to ensure that the outcome of this action was lower, in accordance with this attachment style. As described in Sect. 2, the anxious-avoidant attachment pattern leads to more avoidance, which results in decreased proximity seeking.

Later in life, the now adult main character encounters person B who is sensitive, reliable and responsive at time point 150. Since person A's attachment blueprint from the interactions with C expresses an anxious-avoidant attachment style, A will conduct actions in accordance with this attachment style. However, A's continuous evaluations of A's actions, both based on the reaction of B and from A's own 'model of self' and 'model of other' will lead him to adapt his/her behaviour when interacting with B by selecting a different action. By doing so, the attachment blueprint will adapt and converge more towards a secure attachment pattern. The upper graph in Fig. 4 shows a simulation of the anxious-avoidant primary caregiver with A who expresses a low ability to adapt his attachment pattern, whereas the lower graph shows a simulation of A who has a high flexibility to change. Note that the extent of flexibility by which the attachment pattern is adapted may vary, as many other factors besides the attachment pattern may increase or suppress the ability to learn from the interactions at older age.

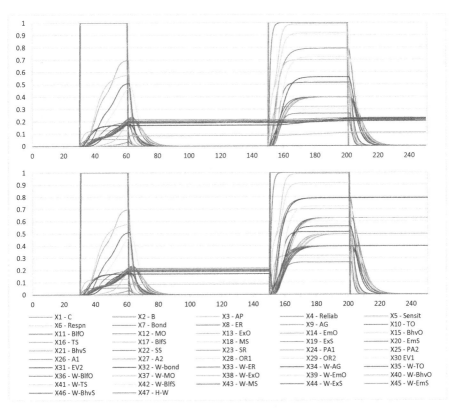

Fig. 4. Graph depicting the simulation for Scenario 2 with an anxious-avoidant attachment pattern which adapts with a low flexibility (upper graph, 0.1) and a high flexibility (lower graph, 0.9) later in life.

In the graphs in Fig. 4, this adaptation of the blueprint is visible for t between 150 and 200. A dip similar to Scenario 1 is visible at timepoint 150. Due to lower flexibility to adapt later in life, the upper graph shows less adaptation than the lower graph. A only slightly adapts their behavior after encountering the securely attached B, which can be seen as the **W**-states are much more rigid and only differ slightly from their previous values that originated from the interactions with the primery caregiver.

Comparing the Two Scenarios. To compare the two scenarios and assess the impact an insensitive and rejecting parent can have on a child's development, this part provides a closer look into the **W**-states and the action-states for each scenario. In Fig. 5 (the double secure case) the two actions A1 for dependence (X_{26}) and A2 for proximity seeking (X_{27}) overlap (indicated in red). As found in literature, in secure attachment, the two actions are in an equilibrium. This equilibrium implies that the main character is self-confident and seeks out others. The same graph shows that the values for X_7 to X_{21} learnt from C remain at the same value when encountering a B who also gives input for a secure attachment. There is a dip shortly after meeting B, which can be explained by the fact that A needs to process encountering B and to possibly adapt some of the learnt patterns. In Fig. 6 adressing the anxious-avoidant case, action A1 for dependence (X_{26}) remains at the same level as in Scenario 1. However, action A2 for proximity seeking (X_{27}) is now at a much lower value. This is in line with literature stating that for anxious-avoidant attachment there is more avoiding behaviour and thus lower proximity seeking. Additionally, the graph shows very low levels for the learnt bonding, emotion regulation and adaptiveness (X_{32}, X_{33}, X_{34}) which increase after meeting B who is sensitive, reliable and responsive.

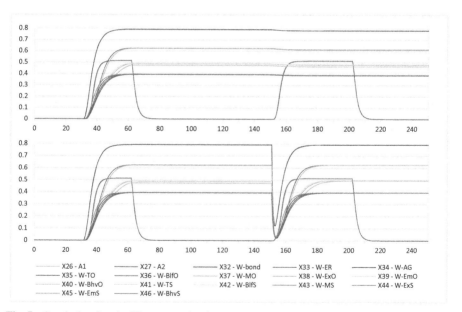

Fig. 5. Graph showing the **W**-states and action states A1 and A2 (in red) for Scenario 1 for secure attachment with a lower (upper graph, 0.2) and higher (lower graph, 0.5) adaptation flexibility.

The **W**-states for the 'model of other' states (X_{35} to X_{40}) are also lower. The **W**-states for the 'model of self' states (X_{41} to X_{46}) are also lower, but less deviant from the secure attachment scenario. When encountering B, values closer to the 'healthy' ones for the internal models are learnt, which results in values similar to Scenario 1. Lastly, the upper graph in Fig. 7 displays the **W**-states and action states where A's attachment pattern remains unopen to change. Here, the **W**-states, in contrast to the lower graph in Fig. 6, only change slightly after meeting B, which implies that A is less affected by the new interactions. This simulation shows the difficulty that A experiences in evolving his/her attachment style to form healthy intimate relationships at older age.

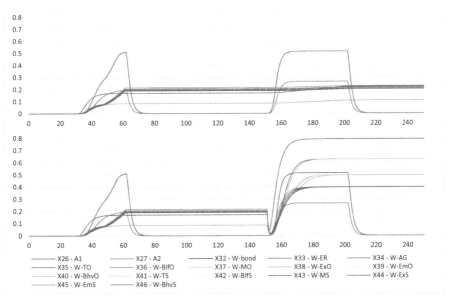

Fig. 6. Graph showing the **W**-states and action states A1 and A2 (in red and purple) for Scenario 2 anxious-avoidant attachment with a low (upper graph, 0.1) and high (lower graph, 0.9) adaptation flexibility.

6 Discussion

This paper presented an adaptive network model of attachment theory [5–6, 15–20] which simulates how an individual can develop his or her attachment pattern and adapt this through social interactions later in life. The model was built according to the Network-Oriented Modelling approach for adaptive networks from [21, 22]. Such a computational model of attachment patterns has not been created before as far as the authors are aware of. Additionally, the role of attachment patterns in adult relationships has not been extensively investigated.

A literature review on attachment theory was conducted to understand which factors influence the formation and adaptation of attachment styles and which internal processes play a role in this development [5–6, 15–20]. This literature study resulted in a total of 47 states which play a causal role in attachment patterns and which were used for the model.

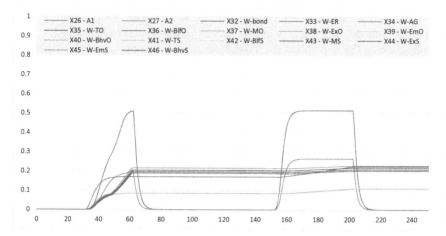

Fig. 7. Graph showing the **W**-states and action states A1 and A2 (in red and purple) for Scenario 2 anxious-avoidant attachment with a low adaptation flexibility.

Two scenarios were explored; secure attachment to the primary caregiver as a child, and anxious-avoidant attachment to the primary caregiver as a child. In both scenarios the main character encounters a securely attached other person at an adult age. The distinction between the two scenarios was made by adjusting the connection weights for the connections from the primary caregiver to the characteristics of the primary caregiver (i.e., sensitivity, reliability and responsiveness). Additionally, the extent to which an attachment pattern can evolve later in life was also differentiated for Scenario 2, leading to personal differences from highly flexible and rigid adjustment of the attachment pattern.

While the simulations of the model do correspond with what is described in the literature, it should be noted that simulating social interactions remains complex. The model does not and could not include all the elements that may influence social interaction and which might play a role in forming and adapting attachment patterns. This means that the model is still not reality but a simplified version of reality, as any model necessarily is. Furthermore, it takes time to adapt one's attachment style. In the presented scenarios, for the sake of simplicity of these scenarios, the main character quickly adapts when meeting one other person. In reality, this adaptation requires more time and may need multiple interactions with different persons. As the model has certain personal characteristics that determine the strength of adaptation, weaker settings for these characteristics enable the model to simulate these slower forms of adaptation as well.

This model can serve as a basis for further investigation of the role of attachment patterns and how they can be altered through adaptive learning. Future research could explore the other two attachment patterns that have not been simulated here (i.e., anxious-ambivalent and disorganized attachment). Based on the literature review as described in Sect. 2, it is hypothesized that in the anxious-ambivalent scenario simulation the behaviour would show heightened levels of dependence and balanced levels of proximity seeking. In the disorganized scenario, the simulation would show behaviour with heightened levels of dependence and lowered levels of proximity seeking. Finally, in future studies, scenarios could be explored where the other person who is encountered later in life is not securely attached; the interactions between two insecurely attached persons has not been investigated yet in the current study.

Reference

1. Abraham, W.C., Bear, M.F.: Metaplasticity: the plasticity of synaptic plasticity. Trends Neurosci. **19**(4), 126–130 (1996)
2. Appendix (2021). https://www.researchgate.net/publication/348402591
3. Bartholomew, K., Horowitz, L.M.: Attachment styles among young adults: a test of a four-category model. J. Pers. Soc. Psychol. **61**(2), 226–244 (1991)
4. Blatz, W.E.: Human Security: Some Reflections. University of Toronto Press, Toronto (1966)
5. Bowlby, J.: Forty-four juvenile thieves: their characters and their home life. Int. J. Psycho-Anal. **25**, 19–52, 107–127 (1944)
6. Bowlby, J.: Attachment. Basic books, New York (2008)
7. Bretherton, I.: The origins of attachment theory: John Bowlby and Mary Ainsworth. Dev. Psychol. **28**, 759–775 (1992)
8. Fonagy, P.: Attachment theory and psychoanalysis. Random House, New York (2001)
9. ICCM: International Conferences on Cognitive Modeling (2020). https://iccm-conference.github.io/previous.html
10. Johnson, S.M.: Attachment theory in Practice: Emotionally Focused Therapy with Individuals, Couples, and Families. Guilford, New York (2019)
11. Marmarosh, C., Markin, R., Spiegel, E.: Attachment in Group Psychotherapy. The American Psychological Association, Washington, D.C. (2013)
12. Mercer, J.: Understanding Attachment: Parenting, Child Care, and Emotional Development. Greenwood Publishing Group, Westport (2006)
13. Parkes, C.M., Stevenson-Hinde, J., Marris, P. (eds.): Attachment Across the Life Cycle. Routledge, London (2006)
14. Robinson, B.L., Harper, N.S., McAlpine, D.: Meta-adaptation in the auditory midbrain under cortical influence. Nat. Commun. **7**, e13442 (2016)
15. Salter, M.D.: An Evaluation of Adjustment Based on the Concept of Security. Ph.D. Thesis, University of Toronto Studies, Child Development Series, vol. 18, 72 (1940)
16. Salter Ainsworth, M.D.: Infancy in Uganda. Johns Hopkins, Baltimore (1967)
17. Salter Ainsworth, M.D.: Security and attachment. In: Volpe, R. (ed.) The Secure Child: Timeless Lessons in Parenting and Childhood Education, pp. 43–53. Information Age Publishing, Charlotte, NC (2010)
18. Salter Ainsworth, M.D., Blehar, M., Waters, E., Wall, S.: Patterns of Attachment. Erlbaum, Hillsdale (1978)
19. Salter Ainsworth, M.D., Bowlby, J.: Child Care and the Growth of Love. Penguin Books, London (1965)
20. Salter Ainsworth, M.D., Bowlby, J.: An ethological approach to personality development. Am. Psychol. **46**(4), 333–341 (1991)
21. Treur, J.: Network-Oriented Modeling: Addressing Complexity of Cognitive, Affective and Social Interactions. Springer Nature, Cham (2016). https://doi.org/10.1007/978-3-319-45213-5
22. Treur, J.: Network-Oriented Modeling for Adaptive Networks: Designing Higher-Order Adaptive Biological, Mental and Social Network Models. Springer Nature, Cham (2020). https://doi.org/10.1007/978-3-030-31445-3
23. Treur, R.M., Treur, J., Koole, S.L.: From Natural Humans to Artificial Humans and Back: An Integrative Neuroscience-AI Perspective on Confluence. Proc. of the First International Conference on Being One and Many: Faces of the Human in the 21st Century (2021).

Computational Health

Hybrid Predictive Modelling for Finding Optimal Multipurpose Multicomponent Therapy

Vladislav V. Pavlovskii⬤, Ilia V. Derevitskii$^{(\boxtimes)}$ ⬤, and Sergey V. Kovalchuk⬤

ITMO University, Saint Petersburg, Russia
ivderevitckii@itmo.ru

Abstract. This study presents a new hybrid approach to predictive modelling of disease dynamics for finding optimal therapy. We use existing methods, such as expert-based modelling methods, models of system dynamics and ML methods in compositions together with our proposed modelling methods for simulating treatment process and predicting treatment outcomes depending on the different therapy types. Treatment outcomes include a set of treatment-goal values, therapy types include a combination of drugs and treatment procedures. Personal therapy recommendation by this approach is optimal in terms of achieving the best treatment multipurpose outcomes. We use this approach in the task of creating a practical tool for finding optimal therapy for T2DM disease. The proposed tool was validated using surveys of experts, clinical recommendations [1], and classic metrics for predictive task. All these validations have shown that the proposed tool is high-quality, interpretable and usability, therefore it can be used as part of the Decision Support System for medical specialists who work with T2DM patients.

Keywords: Optimal therapy · Predictive modeling · Expert-based modeling · Hybrid approach · Diabetes mellitus · Machine learning

1 Introduction

In modern practical medicine, there are many approaches to personalize recommending therapy to a patient. Medical experts without experience selecting therapy based on clinical guidelines [1] for the treatment of a specific disease. However, clinical guidelines cannot consider the whole variety of combinations of patient conditions and combinations of drugs. More experienced specialists select therapy based on their own experience, studies, and fundamental knowledge of the particular disease course. However, the combinations space of indicators with individual patient's treatment history is multidimensional and multicomponent. The experience of experts may be insufficiently to make a decision in each specific case from this space. Therefore, special methods are required for making decisions in selecting therapy tasks. In this work, we present a new hybrid approach for finding the optimal therapy based on statistical modeling and modeling of the dynamics of the course of the disease.

© Springer Nature Switzerland AG 2021
M. Paszynski et al. (Eds.): ICCS 2021, LNCS 12744, pp. 479–493, 2021.
https://doi.org/10.1007/978-3-030-77967-2_40

2   Related Works and Problem Definition

Methods for solving the problem of finding optimal therapy for chronic disease are widely described in the literature. These methods include 3 approaches.

Articles in the first approach describe methods for identifying patterns of the effect of a particular therapy on specific treatment targets. Patterns are identified using statistical modeling tools such as statistical hypothesis testing and correlations. In work [2] authors discusses the principles of rational using of antibiotics for sepsis and septic shock and presents scientifically based recommendations for optimal antibiotic therapy. In this work [3], experts studied the effect of two diabetic drugs on blood composition and calculated the coefficients of the effectiveness of these drugs. Jason K. et al. show the advantages of personalizing selection of cancer therapy in the work [4]. Burgmaier et al. have reviewed the potential action drugs on cardiovascular disease and summarize the potential role of present glucagon-like peptide-1-based therapies from a cardiologist's point of view [5]. The methods from this approach are not applicable for personalized selection of the optimal therapy for a particular case. However, using these methods, it is possible to identify patterns that can be a basis for creating methods for finding the optimal therapy.

The second approach includes articles describing methods of identifying linear or non-linear patterns between particular drugs and treatment-goal indicators. Tools for identifying patterns includes Machine Learning methods [6, 7] including Deep Learnings using neural networks [8]. Menden M. and al. predict the response of cancer cell lines to drug treatment. Models predicted IC50 values by 8-fold cross-validation and an independent blind test with coefficients of determination R2 0.72 and 0.64, respectively [6]. In this study [7], Khaledi A. and al. sequenced the genomes and transcriptomes of 414 drug-resistant clinical Pseudomonas aeruginosa isolates. Researchers generated predictive models and identified biomarkers of resistance to four commonly administered antimicrobial drugs. In the work [8] Barbieri C. and al. use feedforward artificial neural network for predicting the response to anemia treatment. Using this approach experts can predict effectiveness of particular drugs. However, a lot of this articles haven't proposed method for finding the most effective therapy in drugs combinations form.

In contrast to the second approach, the third approach includes methods for predicting the synergy of new drug combinations. This approach is based on special methods, such as Tree Combo [9], Deep Synergy [10], and also Machine Learning methods [11]. Janizek J. and al. introduce new extreme gradient boosted tree-based approach to predict synergy of novel drug combinations, using chemical and physical properties of drugs and gene expression levels of cell lines as features [10]. The second examples of this approach is work [10], that describes method of predicting drugs synergy based on Deep Learning. This method was compared with other machine learning methods, DeepSynergy significantly outperformed the other methods with an improvement of 7.2% over the second-best method at the prediction of novel drug combinations within the space of explored drugs and cell lines. In the work [12] Kuenzi B. and al. developed DrugCell, an interpretable deep learning model of human cancer cells trained on the responses of 1,235 tumor cell lines to 684 drugs. Analysis of the DrugCell results leads to the development of synergistic drug combinations that are validated using combinatorial CRISPR, in vitro drug-drug screening, and patient-derived xenografts. DrugCell provides a blueprint for constructing interpretable predictive medicine models. Using models from this approach we can predict the synergy of drugs combinations. However, the space of drugs combinations and goal-treatment is multidimensional and multicomponent.

Also, there are a lot of novelties in the pharmacological sphere, and it is necessary to consider those in our task. Butler at all. described new knowledge and new developments in the pharmacological sphere for diabetes treatment [13].

Special methods are needed to find the optimal combination of drugs based on the assessment of the effectiveness of a particular combination (which this approach can predict). Therefore, a new method of searching for optimal therapy is needed. It should include all the advantages of the above methods. In this work, we propose a hybrid method that includes methods of the first and second approaches to identify the relationship between a drug and a target indicator of treatment, methods of the third approach to assesses the synergy of drug combinations, and new methods to find the optimal drug combination for a multi-component treatment goal.

3   Hybrid Predictive Modelling for Finding Optimal Therapy

In the previous items, we explained necessarily to create an approach for personalized finding optimal therapy.

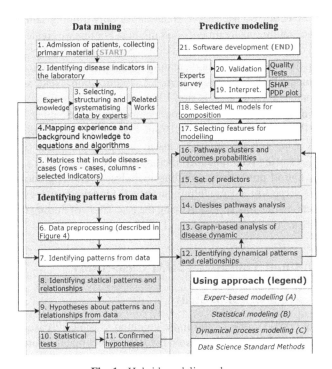

Fig. 1. Hybrid modeling scheme

In this work we propose new hybrid approach for this goal. Scheme of creating this approach is shown in Fig. 1.

The first stage is data mining. We collaborate with medical experts to transform real-word medical processes into digital form. Primary information includes data from patient's survey, laboratory results, records from electronic medical cards, medical images and other. Medical experts structuring, selecting, and aggregating information for hypothesis and modelling. Our team studying related works and mapping experience and background knowledge to equations, algorithms, and digital patterns. Next, we create special scripts for transforming information by medical experts to matrices. Rows are cases of diseases; columns are important indicators of course of the disease. In this stage of method, we use only expert-based modelling. This stage includes 1th–5th steps.

The second stage is identification of patterns from data. This stage includes steps 6th–10th from scheme on Fig. 1 (further just a scheme). In the first, data are pre-processed, it is 6th steps. This step includes the following: noise and emission processing, removing/replacing data gaps, coding categorical features using one-hot-label-encoding/dummy encoding methods, scaling, and logging. Next, in the 8th step, we are identifying statical patterns from data. In parallel, we are using dynamical process modelling for identifying dynamical patterns and relationships in the 12th steps.

The third stage is predictive modelling. This stage includes steps 11th–17th from scheme. We use dynamical and statical patterns for selecting indicators for final features set. All samples are divided into training and testing parts. Next, we create ML predictive models. Models' selection do use cross-validation for only training samples.

Next, we create optimizing treatment goals method. For this we create set of synthetic therapies in form of random combination of drugs. Set include a lot of variants of drugs. Then, we find top-100 best variants and creating only one drugs combination using this combination. This method describes in item 4.3 Optimization. Next, we upgrade this method using dynamical patterns, it is step 13th.

Next step is interpretation. Interpretation methods include Shapley Additive explanations (SHAP), Partial Depends Plots (PDP) and expert-based interpretation. Then, we validate methods using expert's surveys, classic metric of predictive task quality and comparing results of using recommended therapy with real results.

To summary, hybrid predictive modelling includes several approaches for finding optimal therapy – expert-based modeling, statistical modeling, and dynamic process modeling. We demonstrate this approach using the case study of finding optimal therapy for treatment diabetes mellitus of two type (T2DM).

4 Finding Optimal Therapy: T2DM Case-Study

4.1 Problem Description: T2DM-Study

Diabetes mellitus (DM) is one of the most common chronic diseases in the world. Experts from the World Diabetes Federation predict, predict that the number of patients with diabetes by 2030 will increase 1.5 times and reach 552 million people, mainly due to patients with type 2 diabetes (T2DM). For public health, this type of diabetes is one of the most priority problems, since this disease is associated with a large number of concomitant diseases, leading to early disability, and increased cardiovascular risk. Therefore, it is especially important for patients with this disease to prevent the development of serious complications. The risks of complications can be reduced by right selected therapy, but it is important to choose the right drugs, considering the synergy of drugs and the patient's personal characteristics. There are many brands to treat this disease. Treatment targets are multicomponent and include carbohydrates metabolism compensation (glycated hemoglobin), lipid control (total cholesterol), and optimization of systolic and diastolic pressure-level. The space of drug combinations and treatment targets is multidimensional and multi-component, special methods are required for personalized searching drug combinations, that are optimal in terms of better treatment multi-component outcomes. Therefore, the task of creating a method for a personalized search for optimal therapy for patients with type 2 diabetes is relevant and suitable for demonstrating the proposed hybrid approach.

4.2 Data Mining

The study was based on dataset including 189 671 medical records for patients who were treated for diabetes type 2 in Almazov National Medical Research Centre or in Pavlov First Saint Petersburg State Medical University, St. Petersburg, Russia in 2008–2018. There are several entry and exclusionary criteria's for including treatment case in study. Criteria showed in Table 1.

Table 1. Entry and exclusionary criteria for including treatment case in dataset.

Entry criteria	Exclusionary criteria
1. Diabetes type 2	1. Early stages of diabetes, prediabetes, impaired glucose tolerance
2. At least 2 measurements of glycated hemoglobin	2. The presence of a large number of gaps in key-indicators
3. Age between 18 and 80 years	3. The observation period is less than half a year

Each treatment case describes using set of indicators. This is shown in Table 2.

Table 2. Medical indicators for treatment cases

Feature's group	Features
Measurements	Group includes height, weight, age, gender, SBP, DBP, pulse, body mass index, body surface area;
Hypertensions	Group includes i10, i11, i12, i13, i15 ICD codes;

(continued)

Table 2. (*continued*)

Feature's group	Features
Heart complications	Group includes chronic heart failure, chronic obstructive pulmonary disease, atherosclerosis, myocardial infarction, acute coronary syndrome, and others;
Diabetic complications	Group includes retinopathy, angiopathy, nephropathy, neuropathy, foot ulcer, diabetic coma, osteoarthropathy and others;
Other nosologies	Group includes anemia, hypothyroidism, acute pulmonary complications (pneumonia, bronchitis, other types of pneumonia);
Insulin	Group includes 117 insulin preparations, short, medium, long-acting insulins, genetically engineered insulins, and many others:
Sugar-lowering drugs	Group includes 213 types of drugs. These drugs are most often used in the treatment of diabetes in the Russian Federation. This drug includes metformin, diabetalong, lixumia, and others;
Other drugs	Group includes 211 different types of diuretics (e.g., Aldacton Saltucin), 108 different types of statins (e.g., Anvistat), 193 types of betablockers (e.g., Normoglaucon)

These medical indicators include information about diseases in anamnesis, analysis' values, physical measurements, lifestyle, a lot of drug types.

4.3 Optimization

Data Mining. The data were presented as a time series of treatment of all patients. However, this data was not suitable for processing by traditional methods like ARIMA or LSTM network. It was caused by availability of a large number of data gaps. The second reason for not using these methods was the presence of various distances between visits. Due to these reasons series for all patients were compressed into one-dimensional vector, which represents series with its statistical characteristics. The compression scheme showed in Fig. 2.

This transformation was applied for visits which were between any two measurements between target values. As targets for predicting were chosen 4 features, they are glycated hemoglobin, total cholesterol, systolic and diastolic pressure. For targets were applied other transformations, for glycated hemoglobin and total cholesterol the difference between the end and the start values in the series was calculated. However, for systolic and diastolic pressure this operation should not be applied, since these features change throughout the day, then the use of one value will be incorrect. Therefore, for pressures were calculated mean values within six months after the treatment.

Predictive Modeling. Since the main goal is to reduce key indicators, first of all it is necessary to be able to predict their values by using treatment as a predictor. Thus, the problem becomes building a regression model.

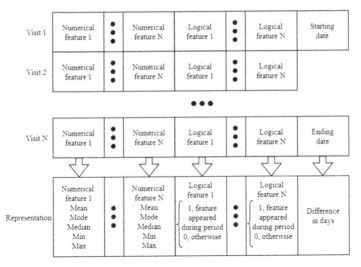

Fig. 2. Time series transformation scheme

Machine learning models were created for all key indicators. The following algorithms were used: Decision Tree, Random Forest, XGBoost, SGD, CatBoost. Mean squared error and coefficient of determination were used as validation metrics. Metrics of trained models showed in the Tables 3, 4, 5 and 6.

Table 3. Metrics of glycated hemoglobin predicting model.

Model	Mean squared error (CI = 95%)	R^2 (CI = 95%)
Decision tree	(0.2220–0.5290) 0.3755	(0.4270–0.7616) 0.5943
Random forest	(0.1518–0.3134) 0.2326	(0.6792–0.8338) 0.7565
XGBoost	(0.1312–0.2784) 0.2048	(0.7042–0.8594) 0.7818
SGD	(0.5104–0.7710) 0.6407	(0.2412–0.4118) 0.3265
CatBoost	(0.1330–0.2546) 0.1938	(0.7312–0.8594) 0.7953

Table 4. Metrics of total cholesterol predicting model.

Model	Mean squared error (CI = 95%)	R^2 (CI = 95%)
Decision tree	(0.6806–0.9546) 0.8176	(0.3278–0.5294) 0.4286
Random forest	(0.4086–0.5642) 0.4864	(0.6160–0.7076) 0.6618
XGBoost	(0.4106–0.5690) 0.4898	(0.6080–0.7094) 0.6587
SGD	(1.0048–1.3264) 1.1656	(0.1420–0.2414) 0.1917
CatBoost	(0.4214–0.5712) 0.4963	(0.6092–0.6986) 0.6539

Table 5. Metrics of systolic pressure predicting model.

Model	Mean squared error (CI = 95%)	R^2 (CI = 95%)
Decision tree	(24.4212–29.8376) 27.1294	(0.9072–0.9236) 0.9154
Random forest	(15.4982–18.5108) 17.0045	(0.9424–0.9516) 0.9470
XGBoost	(31.6220–35.1730) 33.3975	(0.8908–0.9010) 0.8959
SGD	(190.2478–203.8474) 197.0476	(0.3756–0.3962) 0.3859
CatBoost	(32.2876–35.7460) 34.0168	(0.8892–0.8990) 0.8941

Table 6. Metrics of diastolic pressure predicting model.

Model	Mean squared error (CI = 95%)	R^2 (CI = 95%)
Decision tree	(17.8134–21.7346) 19.7740	(0.9068–0.9232) 0.9150
Random forest	(10.9792–13.0486) 12.0139	(0.9438–0.9528) 0.9483
XGBoost	(23.7260–26.0768) 24.9014	(0.8886–0.8972) 0.8929
SGD	(161.9712–171.8524) 166.9118	(0.2708–0.2918) 0.2813
CatBoost	(24.5408–26.8876) 25.7142	(0.8850–0.8938) 0.8894

Since trained models could predict our target values it is possible to change medicine, which was used in treatment and to see how key indicator changed. For choosing best medicine combination it is necessary to apply models for all drugs combinations. However, there are 87 different medicaments in the dataset, thus 2^{87} combinations should be checked, which is time-consuming task.

Treatment selection becomes an optimization problem, where our target to find vector of medicines, which gives most acceptable values of key indicators. There are several difficulties with solving this problem. First problem is that our target function is a black box function, which means that we do not know its behavior. The second one is that it has high evaluation cost, consequently using metaheuristic optimization algorithms will be time consuming. The third problem is that it is necessary to pick up only integer values, because vector of medicines consists of zeros and ones.

Since no suitable method was found to solve the optimization problem, the following approach was suggested. Scheme of this approach showed in Fig. 3.

For a patient randomly generates 100000 different combinations of drugs. Next step is to predict key indicators values with ML models. Then the difference between the predicted value and the value that the patient needs to obtain to get within the acceptable interval is calculated. Next it is necessary to normalize these differences, multiply by the coefficient of importance and summarize them. Then we need to select the top 100 obtained values with their corresponding medicines combinations and use them to calculate the probability of taking the drug into treatment. For each drug we calculate its frequency of occurrence in top 100 combinations and divide it by 100. The last step

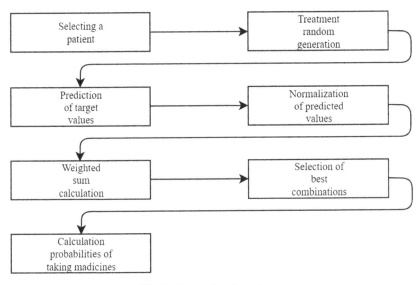

Fig. 3. Drug selection scheme

is the selection of drugs according to a random number generator based on the obtained probabilities.

Interpreting. With using SHAP values it is possible to interpret model output, which means that an importance of concrete drug and its influence could be found out. SHAP values of glycated hemoglobin prediction model showed in Fig. 4.

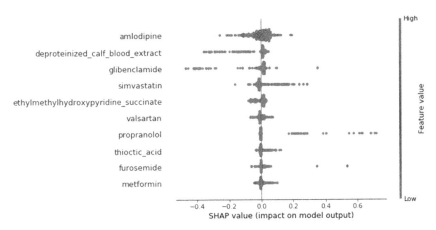

Fig. 4. SHAP values of glycated hemoglobin prediction model. (Color figure online)

The graph shows the 10 drugs most affecting glycated hemoglobin. Red color means that the medicine was included in the treatment, blue is the opposite. According to a model deprotenized calf blood extract and glibenclamide are most effective drugs in terms of the lowering the indicator. On the other hand, simvastatin and propranolol raise this value most effectively. Also, there is interesting case with metformin, it does not decrease target value, but without using it this value is increasing, which mean that it could be used for keeping the indicator in appropriate range.

SHAP values of systolic pressure prediction model showed in Fig. 5.

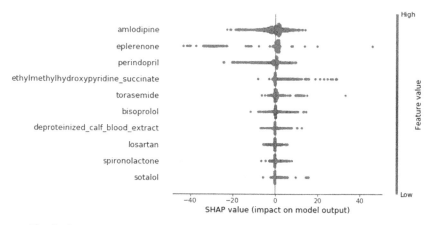

Fig. 5. SHAP values of systolic pressure prediction model. (Color figure online)

According to this plot ethylmethyldroxypytidine succinate is most systolic pressure affective in terms of increasing this value. However, this medicine also in the list of most glycated hemoglobin affective drugs, but it lowers this value. This is a good example of why it is important to choose the right treatment, because bad combination could optimize only one target to the detriment of another. Therefore, patients taking this drug should also be prescribed medication to compensate for the increase in systolic pressure.

SHAP values of diastolic pressure prediction model showed in Fig. 6.

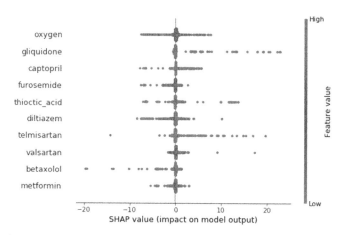

Fig. 6. SHAP values of diastolic pressure prediction model. (Color figure online)

Graph above shows good example of drug, which can be used to maintain acceptable values of key indicators. This is metformin, when using this drug, there is no need to compensate for the change in any indicator.
SHAP values of total cholesterol prediction model showed in Fig. 7.

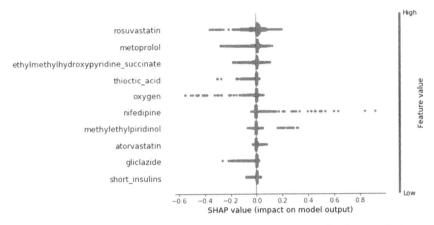

Fig. 7. SHAP values of total cholesterol prediction model. (Color figure online)

From this graph we can see that using medical oxygen greatly decreases total cholesterol value, but there is a problem with using this drug. SHAP values of this drug from diastolic pressure table are hard to interpret. It means that if this medication will be included into a treatment, it will be hard to compensate increasing or decreasing of diastolic pressure, since changes in target value can be in any direction.

Validating. The optimization scheme was applied to patients for whom it was possible to calculate all 4 key indicators. Results showed in Table 7.

Table 7. Optimization scheme validation results.

Target value	Percentage of cases, when target value brought back to normal range	Percentage of cases, when target value became better that in real treatment
Glycated hemoglobin	67.64	**91.17**
Total cholesterol	20.58	**26.47**
Systolic pressure	47.05	**82.35**
Diastolic pressure	64.70	**79.41**
GH + TC	20.58	26.47
GH + SP	26.47	73.52

(continued)

Table 7. (*continued*)

Target value	Percentage of cases, when target value brought back to normal range	Percentage of cases, when target value became better that in real treatment
GH + DP	50.00	70.58
TC + SP	17.64	20.58
TC + DP	14.70	17.64
SP + DP	26.47	70.58
GH + TC + SP	17.64	20.58
GH + TC + DP	14.70	17.64
GH + SP + DP	20.58	61.74
TC + SP + DP	11.76	17.64
GH + TC + SP + DP	11.76	17.64

According to given results, most important target, which is glycated hemoglobin, becomes much better than in real cases, systolic and diastolic pressure also becomes better in big number of cases, however total cholesterol value is poorly improved. To increase the number of cases of improvement in total cholesterol, it is necessary to improve the model.

We improve this model using dynamic patterns of the treatment process. For this, we analyze space of therapy trajectories (Fig. 8 shows this space). For each patient, its treatment trajectory is identified in the form of a sequence of diagnosed certain complications from the past history of the disease. Then, using the Tanimoto coefficient [14], we find similar trajectories (patients with similar medical past histories). Further, for these patients, we determine a set of pharmacological groups of drugs that show the best results for optimizing a multipurpose result. To determine this set of pharmacological groups, we use our developed algorithm based on the analysis of precedents. Further, we apply the above method, choosing drugs not from 87 drugs, but from a reduced number of drugs from individually selected pharmacological groups of drugs. Table 8 shows the change in the quality indicators of the model from Table 7.

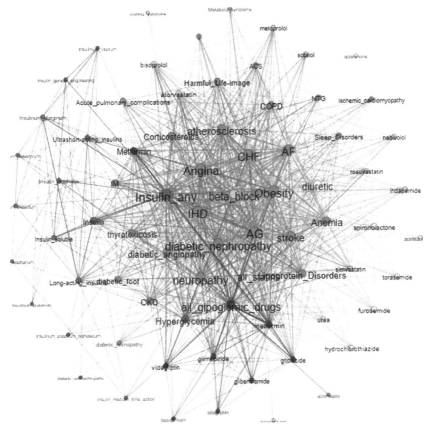

Fig. 8. Graph of therapy trajectories. Colors are marks for clusters by modularity maximization. (Color figure online)

Table 8. Result of improve model for increasing number improving total cholesterol.

Target value	Percentage of cases when target value brought back to normal range	Percentage of cases when target value became better that in real treatment
Total cholesterol	–	44.11 (**+17,64%**)
GH + TC	94.11 (**+73,53%**)	41.17 (**+14,7%**)
TC + SP	26.47 (**+9%**)	29.41 (**+9%**)
TC + DP	–	20.58 (**+2,94%**)

5   Conclusion and Future Work

This paper proposes a hybrid approach for creating a method for finding optimal therapy in the terms of optimizing the multipurpose outcome of patient treatment. This approach is based on identifying statistical and dynamic patterns from a course of the disease using expert-based modelling methods, machine learning methods, and predictive modelling methods. This method was demonstrated on the practical task of finding the optimal therapy for type 2 diabetes mellitus in terms of achieving 4 treatment goals - compensation of carbohydrate metabolism (target value is glycated hemoglobin), compensation of lipid metabolism (target value is total cholesterol), optimization of half-year indicators of arterial pressure (target value are systolic and diastolic). This method was validated using a survey of experts-endocrinologists, classical metrics of predictive modelling tasks. Also, we have validated this method using real-treatments cases. We have found the optimal therapy for each case and have predicted results of the optimal therapy for each case using developed predictive models. As a result, the proposed hybrid method improves the target indicators of carbohydrate metabolism compensation in 91% of cases, the target of lipid metabolism in 44% of cases, the average semi-annual systolic pressure in 82% of cases, and average semi-annual diastolic pressure in 80% of cases, compared with real therapy for the selected patients. In summary, the method is of high quality, it can be applied as part of a support and decision-making system for medical specialists working with T2DM patients.

Acknowledgments. The reported study was funded by RFBR according to the research project № 20–31-70001. Participation in the ICCS conference was supported by the NWO Science Diplomacy Fund project # 483.20.038 "Russian-Dutch Collaboration in Computational Science".

References

1. Dedov, I.I., et al.: Standards of specialized diabetes care. In: Dedov, I.I., Shestakova M.V., Mayorov A.Yu Diabetes Mellit, 9th ed. (2019)
2. Kumar, A.: Optimizing antimicrobial therapy in sepsis and septic shock. Crit. Care Clin. **25**, 733–751 (2009)
3. Srividya, P., Devi, T.S.R., Gunasekaran, S.: Ftir spectral study on diabetic blood samples – monotherapy and combination therapy. Ojp **25**, 744–750 (2012)
4. Sicklick, J.K., et al.: Molecular profiling of cancer patients enables personalized combination therapy: the I-PREDICT study. Nat. Med. **30**, 289–299 (2019)
5. Burgmaier, M., Heinrich, C., Marx, N.: Cardiovascular effects of GLP-1 and GLP-1-based therapies: implications for the cardiovascular continuum in diabetes? Diab. Med. **30**, 289–299 (2013)
6. Menden, M.P., et al.: Machine Learning Prediction of Cancer Cell Sensitivity to Drugs Based on Genomic and Chemical Properties. PLoS One **8**, e61318 (2013)
7. Khaledi, A., et al.: Predicting antimicrobial resistance in Pseudomonas aeruginosa with machine learning-enabled molecular diagnostics. EMBO Mol. Med. **12**, e10264 (2020)
8. Barbieri, C., et al.: A new machine learning approach for predicting the response to anemia treatment in a large cohort of End Stage Renal Disease patients undergoing dialysis. Comput. Biol. Med. **61**, 56–61 (2015)
9. Janizek, J.D., Celik, S., Lee, S.I.: Explainable machine learning prediction of synergistic drug combinations for precision cancer medicine. bioRxiv **8**, 1-115 (2018)
10. Preuer, K., Lewis, R., Hochreiter, S., Bender, A., Bulusu, K., Klambauer, G.: DeepSynergy: predicting anti-cancer drug synergy with Deep Learning. Bioinformatics **34**(9), 1538–1546 (2018)

11. Tabl, A., Alkhateeb, A., Pham, H., Rueda, L., ElMaraghy, W., Ngom, A.: A novel approach for identifying relevant genes for breast cancer survivability on specific therapies. Evol. Bioinf. **14**, 117693431879026 (2018)

12. Kuenzi, B., et al.: Predicting drug response and synergy using a deep learning model of human cancer cells. Cancer Cell **38**(5), 672-684.e6 (2020). https://doi.org/10.1016/j.ccell. 2020.09.014

13. Butler, J., Januzzi, J.L., Rosenstock, J.: Management of heart failure and type 2 diabetes mellitus: Maximizing complementary drug therapy. Diab. Obes. Metab. **22**, 1243–1262 (2020)

14. Bajusz, D., Rácz, A., Héberger, K.: Why is Tanimoto index an appropriate choice for fingerprint-based similarity calculations? J. Cheminf. **7**(1), 1–13 (2015)

Towards Cost-Effective Treatment of Periprosthetic Joint Infection: From Statistical Analysis to Markov Models

Yulia E. Kaliberda[1], Vasiliy N. Leonenko[1(✉)] (iD), and Vasiliy A. Artyukh[2]

[1] ITMO University, Kronverksky Pr. 49A, 197101 St. Petersburg, Russia
[2] Russian Scientific Research Institute of Traumatology and Orthopedics named after R.R. Vreden, Akademika Baykova St., 8, 195427 St. Petersburg, Russia

Abstract. The aim of the research is to perform statistical analysis and to build probabilistic models for the treatment of periprosthetic joint infection (PJI) based on available data. We assessed and compared the effectiveness of different treatment procedures from the terms of the objective result (successful PJI treatment without relapse) and the subjective assessment of the condition of the patients (Harris Hip Score). The ways to create prognostic models and analyze cost-effectiveness of treatment strategies are discussed based on the results obtained.

Keywords: Periprosthetic joint infection · Statistical analysis · Markov models · Python

1 Introduction

Periprosthetic joint infection (PJI) is a serious complication which may occur after arthroplasty [1]. It is associated with high morbidity and requires complex treatment strategies including multiple surgical revisions and long-term antimicrobial treatment, because the implant as a foreign body increases the pathogenicity of bacteria and the presence of biofilm makes the diagnosis and treatment problematic [2]. The investigations related to the creation of cost-effective approaches to PJI treatment [3, 4] mention the issues connected with the corresponding analysis due to lack of quality studies.

Beside the problem of finding the best PJI treatment method in general, relying on both direct (percentage of successful PJI elimination) and indirect treatment outcomes (such as resulting increase in quality of life of the patients who underwent the treatment), there is an arising challenge of finding an optimal treatment strategy in advance for the particular patient based on his individual characteristics [5]. This challenge became actual as a part of the personalized medicine concept and requires the research based on multidisciplinary approach and relying on statistical analysis, mathematical modeling and machine learning [6]. The ultimate aim of that direction of research consists in developing a computational tool to predict the consequences of a fixed treatment strategy for a given patient. Such a tool, when put into use by the healthcare professionals, will

© Springer Nature Switzerland AG 2021
M. Paszynski et al. (Eds.): ICCS 2021, LNCS 12744, pp. 494–505, 2021.
https://doi.org/10.1007/978-3-030-77967-2_41

help enhance the quality of PJI treatment both in terms of cost-effectiveness and the quality of life of individuals undergoing treatment.

The aim of the research described in the presented paper was to perform statistical analysis of the PJI treatment methods effectiveness used in clinical practice of Russian Scientific Research Institute of Traumatology and Orthopedics named after R.R. Vreden (St Petersburg, Russia) and to formulate a discrete-event stochastic model for the clinical paths of the patients exposed to PJI, taking into account both the treatment procedure itself and its long-term treatment consequences, including the relapse of PJI and the related interventions. We assessed and compared the effectiveness of different treatment procedures from the terms of the objective result (successful PJI treatment without relapse) and the subjective assessment of the condition of the patients (Harris Hip Score). The ways to create prognostic models and analyze cost-effectiveness of treatment strategies are discussed based on the results obtained.

2 Data

The dataset used in the analysis contains the records of 609 patients who received treatment for PJI infection in the period of 2000–2020. The patient records are divided into two observation groups – the retrospective and the prospective one. The retrospective group was composed of the disease histories taken from the archives, whereas the prospective group was filled by constantly adding the data of the patients currently undergoing treatment starting from 2014. The structure of the dataset with respect to treatment methods is shown in Fig. 1.

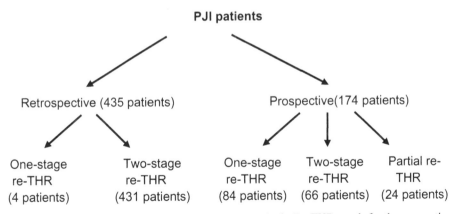

Fig. 1. The dataset structure related to the treatment methods. Re-THR stands for the reoperation of total hip replacement

In retrospective group, the applied treatment methods included one-stage re-endoprosthesis and two-stage re-endoprosthesis, also known as reoperations of total hip replacement (re-THR). Since the set with one-stage re-THR patients contained only 4 records, we excluded it from further observation. In the prospective group, a new method was presented in addition to the two named before, which was partial re- THR.

The two consecutive interventions of two-stage re-THR in the retrospective group were separated by more than 2 months, while in the prospective group waiting time between surgeries did not exceed 2 months. The two-stage re-THR prospective group was divided into 2–3 weeks waiting time and 6–8 weeks waiting time. All people were being observed till the end of 2020.

In patients' data for each patient there are defined features:

- Name, birthdate
- Operation dates (the total number of operations varied between 1 and 10)
- Operation types
- Two sets of Harris Hip Scores (HHS) [7], - before and after treatment, - used to measure the quality of life of the patient
- The resulting state of the patient related to PJI, measured during his last attendance to healthcare services (PJI relapse or no PJI)

In the ideal situation, the prospected number of operations performed on each patient is fixed and defined solely by the PJI treatment method, but in many cases additional operations were required due to the relapse of PJI or other issues (postoperative wound hematomas, spacer dislocations, etc.). There exists 15 different types of operations in overall, which could be divided into three groups, depending on how the particular operation type was connected with PJI:

- Operations which have no connection with PJI

  - Endoprosthesis (EP) installation + spacer removal
  - EP installation (no spacer)
  - Non-infectious: spacer dislocation
  - Other (suturing, etc.)
  - Non-infectious: periprosthetic fracture case
  - Unknown

- First case of PJI or PJI relapse

  - Debridement (DAIR)
  - Debridement + spacer installation
  - EP components replacement + debridement
  - Debridement + full EP replacement
  - Joint drainage + long-term suppressive antibiotic therapy (ABT)

- PJI relapse

  - Debridement + spacer reinstallation
  - Disarticulation
  - Spacer removal + support osteotomy
  - Debridement + support osteotomy + Girdlestone resection arthroplasty
  - Joint drainage

Further on, we do not distinguish the types of operations and consider them equal (see more comments on the matter in Sect. 4).

3 Statistical Analysis

3.1 Comparing the Treatment Outcomes

The first aim of the statistical analysis was to compare the effectiveness of different treatment methods in terms of two indicators available from the data:

- The ratio of PJI relapse cases among the patients (reflects the objective result of the treatment method application)
- The improvement of Harris Hip Score according to the answers of the patients (reflects the increase in the quality of life as a result of performing the treatment procedures).

The corresponding proportions and confidence intervals were calculated using the algorithm implemented in Python 3.6. The results are presented in Table 1 and Fig. 2.

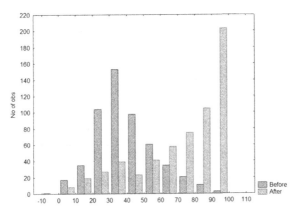

Fig. 2. The frequency of particular Harris Hip Scores among the patients before (blue) and after the treatment (red). (Color figure online)

Table 1. Harris score mean confidence intervals and PJI relapse ratio for each treatment type. First questioning was made when a patient was admitted for a treatment, second questioning corresponds to current patients' score.

Group name	HHS before treatment	HHS after treatment	Proportion of people with PJI
Two-stage > 2 months	(37.3685, 40.8680)	(66.4771, 71.6734)	0.092
One-stage	(33.2760, 40.8489)	(73.6527, 84.4723)	0.110
Two-stage 2–3 weeks	(32.3068, 64.4205)	(56.6777, 104.0496)	0.077
Two-stage 6–8 weeks	(39.5684, 50.8899)	(79.4123, 89.7127)	0.060
Partial	(37.5556, 55.5277)	(60.6142, 83.3858)	0.091

Figure 2 demonstrates that the distribution of Harris scores after the treatment compared to the one before confirms the increase of functional capacity, since it is shifted to the right by its median and is left-skewed. The difference was proved to be statistically significant by the Sign test ($p < 0.05$), applied to the whole sample and to the subsamples corresponding to each of the five treatment methods. As it can be seen from Table 1, the lowest PJI proportion (0.06) is shown by a two-stage 6–8 weeks treatment method. This method may also boast the second highest median Harris score (see Fig. 3). The best median HHS and the second best proportion of PJI-free patients corresponds to two-stage 2–3 weeks treatment method, however more data might needed to verify this result since the sample of patients for the method is not big enough.

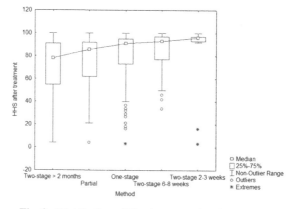

Fig. 3. Distribution of Harris scores after the treatment

3.2 Quantifying the Role of Multiple Operations on the Well-Being of Patients

As it was mentioned earlier, in the course of treatment additional operations might be required, which are caused by the necessity to deal with PJI relapse and other unexpected situations. Since every additional surgical intervention might negatively influence both the PJI treatment outcome and the functional capacity of the patients, it is important to analyze the dynamics of the corresponding indicators. Thus we counted the dependence of PJI ratio on the number of operations for different treatment methods - Table 2 shows the example for the retrospective group (Two-stage re-THR, > 2 months).

Table 2. Operation connections with PJI stage in retrospective group

Number of operations in the group	Number of patients in the group	Number of patients without PJI	PJI relapse proportion
1	211	202	0.04265
2	121	109	0.09917

(continued)

Table 2. (*continued*)

Number of operations in the group	Number of patients in the group	Number of patients without PJI	PJI relapse proportion
3	58	48	0.17241
4	25	21	0.16000
5	13	10	0.23077
6	5	3	0.40000
7	2	2	0.00000
8	2	2	0.00000
9	0	0	N/D
10	3	1	0.66667

Such transformation serves to a future target of an overall work - creating a probabilistic model for each group of patients.

The dependence between the number of operations and mean Harris score is shown in Table 3, while Fig. 4 demonstrates the distribution of Harris scores. The average statistic was calculated in both cases without division on treatment methods due to the small sample sizes.

Table 3. Average Harris scores depending on the number of operations

Number of operations in the group	Average HHS after the treatment	Average HHS improvement	Number of patients in the group
1	74.432	33.382	273
2	75.383	36.605	187
3	69.375	27.344	75
4	60.851	29.000	30
5	43.800	9.100	14
6	45.500	9.750	5
7	3.000	−18.000	3
8	N/D	N/D	2
9	N/D	N/D	0
10	39.666	−16.334	3

It is seen from the Tables 2 and 3 that PJI ratio is an increasing function of the resulting operation number while Harris score tends to decrease. In other words, according to the data, the more operations a particular patient had, the bigger risk of PJI he has and the lower functional capacity he retains.

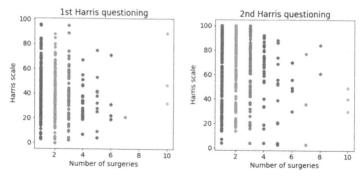

Fig. 4. Harris scale results before and after treatment by patient groups distinguished by overall resulting number of operations

It is important to note that the number of patients who has 5 operations and more is rather small (see Tables 2 and 3), which might alter the correctness of the conclusions. Another issue that might potentially decrease the quality of comparison of the methods (particularly, the comparison of final PJI ratios) and the subsequent model calibration on data was the difference in the time periods when different treatment methods were first introduced. As a result, the older methods might have records with longer observation time compared to the newer ones. If a patient in the record is listed as one without PJI relapse, having had few operations, it might mean that he was indeed effectively treated by the method, or, alternatively, that the sufficient time after the treatment has not passed to register the subsequent relapse and start treating it with the help of surgical manipulations. As a result of analysis of observation times in each treatment group (Fig. 5), we have concluded that although their distributions are indeed different, the mean observation times do not differ dramatically (with a two-stage > 2 months group being somewhat an exception).

4 Model

After performing the statistical analysis of the treatment outcomes, the next step of the study was to formulate a Markov model which could be used in predicting the individual patient trajectories in case of PJI treatment. There is a number of the corresponding models published [8–11] which differ by their structure. The additional data analysis was performed to assess the applicability of different concepts to our situation. The following model features were analyzed.

Generalized States [8, 9] vs Explicit States [10]. In some of the proposed Markov models there are no explicit states taking into the account consecutive numerous operations after the first surgical intervention. For instance, in [8] and other similar studies all the operations after the first unsuccessful revision are generalized in one "Re-revision" state which might lead to death or recovery. In our case, since we have found a dependence of a number of operations on the patients' well-being and we wanted to have a model capable of predicting those indicators, we decided to explicitly state all the operations one by one along with the probabilities that after operation N an operation N + 1

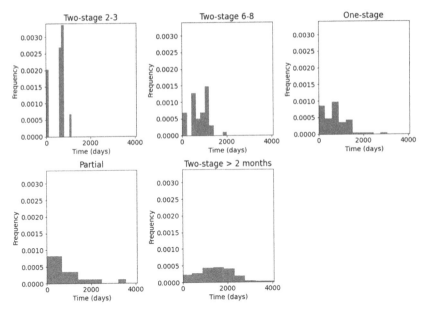

Fig. 5. Observation time distribution plots

might be required. At the same time, we have not distinguished the operation types as it was made in [10]. Particularly, we have not used the information on whether the operation was PJI-related or not (see Sect. 2).

The selected approach has several advantages and disadvantages:

- Since the majority of considered operations imply aggressive surgery, we assumed that they equally might affect patients' condition – this was supported by the analysis of Harris scores and PJI ratios in the previous section.
- To expand the study with the cost-effectiveness assessment, which is planned in the future, it is necessary to take into account all the operations performed – so aggregating the states would cause issues with the cost calculation.
- On the other hand, there exist a mandatory number of surgical interventions connected to the treatment methods, which is not taken into account and might affect the method comparison results and the accuracy of models (for instance, two-stage re-THR has a planned additional operation by default compared to one stage re-THR, with the transition probability between the stages close to 1).

Time-Dependent [11] vs Time-Independent State Transitions. The transition probabilities between the model states might be calculated in two ways:

- The probability of transition per a fixed time period (essentially a rate)
- The probability of transition *per sé* without regarding time

In the former case the model explicitly includes time, whereas in the latter the time "jumps" according with the state transitions, so those jumps are not equal in length. This might affect the accuracy of treatment method comparison in case when the average time period between the states is substantially different for different methods. For instance, the method which in average requires PJI relapse treatment not earlier than in five years after the first PJI treatment should be considered more effective compared with the method which causes a PJI relapse after five months. At the same time, it is more beneficial to perform all the necessary operations within the minimal time interval, because it enhances the quality of life of a patient. In our case we decided to disregard these nuances, because the distribution of time between the operations (Table 4, Fig. 6) seem to be quite similar among the prospective methods. In case of the retrospective method (two-stage > 2 months) it is twice as big, which additionally discourages from using this method.

Table 4. Confidence intervals in days for the average time between the operation for each treatment type

Group name	Confidence interval	Mean	Size
Two-stage 2–3 weeks re-THR	(402.613, 802.617)	602.615	13
Two-stage 6–8 weeks re-THR	(658.482, 895.910)	777.196	81
One-stage re-THR	(583.224, 826.704)	704.964	53
Two-stage > 2 months	(1402.159, 1567.156)	1484.657	435
Partial re-THR	(476.210, 1186.707)	831.458	24

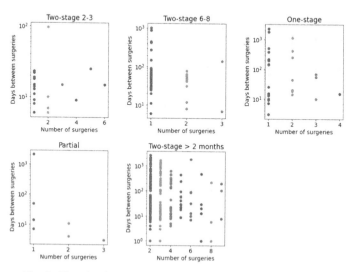

Fig. 6. Plots for times between surgeries for each treatment group

As a result of the mentioned considerations, we have chosen to develop a Markov model which complies to the following principles:

- The possible model states are "Operation 1", "Operation 2",...., "Operation N", "PJI" and "No PJI".
- Each patient proceeds through the sequence of operations until he dies or his monitoring stops for other reasons
- The state "PJI"/ "No PJI" is final and reflects the condition of the patient before his death or at the moment of final monitoring event

Probabilistic trees which correspond to modeling the transitions of patients between the Markov model states for three different treatment methods are demonstrated in Figs. 7, 8. These trees are used to predict the trajectory of a fixed patient undergoing a certain treatment, with the help of Monte Carlo methods. Model implementation is performed via Python programming language. We have selected the trees for "one-stage re-THR" and "two-stage re-THR with 6–8 weeks" for the demonstration purposes, because the corresponding samples are big enough to hope for the correct estimation of the transition probabilities (apart from "Two-stage 2–3 weeks re-THR" and "Partial re-THR") and also they are in active use these days (apart from "Two-stage > 2 months"). The model structure is not dependent on the treatment method type and can be further verified when the new data will become available.

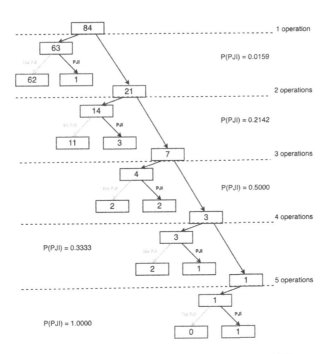

Fig. 7. Results of the treatment with one-stage re-THR

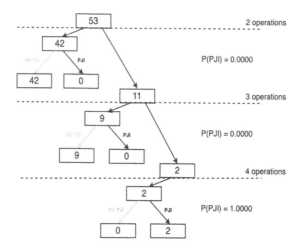

Fig. 8. Results of the treatment with two-stage re-THR with 6–8 weeks between operations

5 Discussion

In the current study, we have analyzed the data on PJI treatment and proposed a Markov model to reflect the transition between the treatment stages for the particular patient, with transition probabilities depending on the treatment method. The size of the sample under study somewhat limits the possibility to make explicit conclusions from the analysis as to comparative effectiveness of the treatment methods. It was however shown that the treatment itself enhances the functional capacity of the patients, and that this capacity is badly affected by repetitive operations, not depending on treatment method. Also the PJI relapse chance correlates positively with the number of operations performed (although this result should be considered *cum grano salis* due to necessity of providing a meaningful interpretation to it).

The Markov model presented in the study serves as a first step towards the prediction of patient trajectories. The absence of a timer in our model might affect the perspectives of cost-effectiveness analysis (in case of evaluating characteristics such as QALY gained per year). Also, calculating time between the model states might be necessary for the application of the model in the decision support system, so in the future the model structure might be reconsidered.

The issue worth noticing is that the transitions between the operations have a complex nature and are not always directly connected with treatment method effectiveness. For instance, repetitive operations due to PJI relapse are the example of the process connected with the disease course, and more operations mean less efficient treatment. At the same time, some operations follow each other in a regular fashion (like two stages in PJI treatment), they are planned according to the schedule and there should not be any probabilistic transitions between the corresponding model states. Finally, some of the operations are caused by accidents, which are random but not connected to PJI (for instance, spacer dislocation). They should be considered an external factor. The model which will distinguish these three groups of situations will be more correct in terms of describing the patient treatment dynamics.

In addition to the mentioned model improvement, future directions of the study will include cost-effectiveness analysis based on the cost of particular operations, and alongside it the consideration of the model with explicit operation states. The obtained study results related to changes in the quality of life of patients might be enhanced with the usage of EQ-5D scale beside Harris score. Finally, the classification of clinical pathways is to be done using CLIPIX software [12].

Acknowledgement. This research is financially supported by The Russian Science Foundation, Agreement #19-11-00326.

References

1. Assmann, G., et al.: Comparison of health care costs between aseptic and two stage septic hip revision. J. Arthroplasty **29**(10), 1925–1931 (2014)
2. Masters, E.A., et al.: Evolving concepts in bone infection: redefining "biofilm", "acute vs. chronic osteomyelitis", "the immune proteome" and "local antibiotic therapy". Bone Res. **7**(1), 1–18 (2019)
3. Li, C., Renz, N., Trampuz, A.: Management of periprosthetic joint infection. Hip Pelvis **30**, 138–146 (2018)
4. Hernández-Vaquero, D., et al.: Treatment of periprosthetic infections: an economic analysis. Sci. World J. **2013**, 1–6 (2013). https://doi.org/10.1155/2013/821650
5. Kovalchuk, S.V., Funkner, A.A., Metsker, O.G., Yakovlev, A.N.: Simulation of patient flow in multiple healthcare units using process and data mining techniques for model identification. J. Biomed. Inform. **82**, 128–142 (2018)
6. Kovalchuk, S.V., Knyazkov, K.V., Syomov, I.I., Yakovlev, A.N., Boukhanovsky, A.V.: Personalized clinical decision support with complex hospital-level modelling. Procedia Comput. Sci. **66**, 392–401 (2015)
7. Harris, W.H.: Traumatic arthritis of the hip after dislocation and acetabular fractures: treatment by mold arthroplasty: an end-result study using a new method of result evaluation. J. Bone Joint Surg. Am. **51**, 737–755 (1969)
8. Tack, P., Victor, J., Gemmel, P., Annemans, L.: Do custom 3D-printed revision acetabular implants provide enough value to justify the additional costs? the health-economic comparison of a new porous 3D-printed hip implant for revision arthroplasty of Paprosky type 3B acetabular defects and its closest alternative. Orthop. Traumatol. Surg. Res. **107**(1), 102600 (2021)
9. Parisi, T.J., Konopka, J.F., Bedair, H.S.: What is the long-term economic societal effect of periprosthetic infections after THA? a Markov analysis. Clin. Orthop. Related Res.®, **475**(7), 1891–1900 (2017)
10. Wolf, C.F., Gu, N.Y., Doctor, J.N., Manner, P.A., Leopold, S.S.: Comparison of one and two-stage revision of total hip arthroplasty complicated by infection: a Markov expected-utility decision analysis. JBJS **93**(7), 631–639 (2011)
11. Fisman, D.N., Reilly, D.T., Karchmer, A.W., Goldie, S.J.: Clinical effectiveness and cost-effectiveness of 2 management strategies for infected total hip arthroplasty in the elderly. Clin. Infect. Dis. **32**(3), 419–430 (2001)
12. Funkner, A.A., Yakovlev, A.N., Kovalchuk, S.V.: Towards evolutionary discovery of typical clinical pathways in electronic health records. Procedia Comput. Sci. **119**, 234–244 (2017)

Optimization of Selection of Tests in Diagnosing the Patient by General Practitioner

Jan Magott[1]([⊠]) [iD] and Irena Wikiera-Magott[2] [iD]

[1] Faculty of Electronics, Technical Informatics Chair,
Wrocław University of Science and Technology, Wrocław, Poland
jan.magott@pwr.edu.pl
[2] Faculty of Medicine, Clinic of Pediatric Nephrology, Medical University in Wrocław,
Wrocław, Poland

Abstract. In General Practitioner's work the fundamental problem is the accuracy of the diagnosis under time constraints and health care cost limitations. The General Practitioner (GP) after an interview and a physical examination makes a preliminary diagnosis. The goal of the paper is to find the set of tests with such total diagnostic potential in verification of this diagnosis that is not smaller than a threshold value and with minimal total cost of tests. In proposed solution method, the set of preliminary diagnoses after the interview and the physical examination is given. For each preliminary diagnosis, for each test, diagnostic potential of the test in verification of the diagnosis is determined using Analytic Hierarchy Process based method with medical expert participation. Then binary linear programming problem with constraint imposed on total diagnostic potential of tests but with criterion function of minimal total test cost is solved for each diagnosis. For the case study when the patient with lumbal pain is coming to the GP, for each of six preliminary diagnoses, for each test, the diagnostic potentials of tests have been estimated. Then for each diagnosis, the solution of the binary linear programming problem has been found. A limitation of the case study is the estimation of diagnostic potential of tests by one expert only.

Keywords: Diagnostic potential · Analytic Hierarchy Process · Binary linear programming problem

1 Introduction

In General Practitioner's work the fundamental problem is the accuracy of the diagnosis under time constraints and cost limitations on health care. The General Practitioner (GP) after interview with the patient or his/her guardian and physical examination makes a preliminary diagnosis. In order to check this diagnosis, GP can order the tests (examinations) from the admitted set of tests. Under-testing may result in delayed or missed diagnosis, while over-testing can cause a cascade of unnecessary activities, and costs [3, 8]. Medical diagnosis with medical tests cost is studied in the papers based on the following approaches: naive Bayes classification [2], decision trees [6], genetic algorithms and fuzzy logic [4], rough sets [5], Analytic Hierarchy Process [1].

© Springer Nature Switzerland AG 2021
M. Paszynski et al. (Eds.): ICCS 2021, LNCS 12744, pp. 506–513, 2021.
https://doi.org/10.1007/978-3-030-77967-2_42

The goal of the paper is to help General Practitioner (GP) to find the sets of tests with sufficiently great diagnostic potential but with minimal total cost of tests.

In the case study when the patient with lumbal pain is coming to the GP, for each of six assumed preliminary diagnoses, the BLPP is solved. In this case study, the AHP pairwise comparisons of diagnostic potentials of tests has been based on expert's opinion. In general in these comparisons, statistical data about tests can be taken into account too.

The structure of the paper is as follows. The research problem and its solution method are outlined in Sect. 2. In next section the case study about patient with lumbal pain is presented. Finally there are summary and conclusions.

2 Research Problem and Research Method

The following problem is studied in the paper. The patient with health problem is coming to GP. The GP should decide what to do: start the treatment by himself or direct the patient to a hospital or direct the patient to a specialist. The proposed solution method is as follows. The set of preliminary diagnoses after the interview and physical examination is given. For each diagnosis, for each test, e.g.: urine, blood, ultrasound, X-ray, the diagnostic potential of the test in verification of the diagnosis is determined. In order to define the diagnostic potential of these tests when verifying a diagnosis, Analytic Hierarchy Process (AHP) [7, 9] based method with medical expert participation is applied. These tests are diagnosis dependent, e.g., the extent of blood test depends on diagnosis. For given diagnosis the costs of tests are calculated. The selected set of tests need to have the total diagnostic potential which is greater or equal to the given threshold. The criterion is the minimal cost of selected tests. Hence, for each diagnosis, the binary linear programming problem (BLPP) is defined. The analysis is executed for different values of threshold of total diagnostic potential of tests.

Now the proposed method will be presented in details.

Let

$D = \{D_1, \ldots, D_k, \ldots, D_s\}$ – the set of preliminary (hypothetical) diagnoses for the interview and the physical examination,

$T = \{T_1, \ldots, T_l, \ldots, T_t\}$ – the set of tests (examinations) that the GP can order.

Diagnostic potential of test T_l, where $l \in \{1, \ldots, t\}$, for diagnosis D_k, where $k \in \{1, \ldots, s\}$ is the number $0 \le p_{T_l} \le 1$ such that $\sum_{l=1}^{t} p_{T_l} = 1$. The diagnostic potential expresses a capacity of this test in verification of the diagnosis. Greater the capacity is greater the number is.

Selection algorithm of tests for diagnosis D_k

1. For each test from the set T define its extent (set of elementary tests) and estimate the diagnostic potential of the test in verification of the diagnosis D_k using AHP.
2. For each test calculate the total cost of all elements in the extent using a table.
3. Find such subset S of T that total diagnostic potential in verification of diagnosis D_k of elements of S is not smaller than required threshold diagnostic potential but the total cost is minimal using BLPP.

AHP is linear algebra based method that is used in expressing the expert opinion in multi-step multi-criteria decision process. In this paper diagnostic potential estimation process is one-step. In this paper in AHP based approach the comparison between pairs of test diagnostic potentials in diagnosis D_k verification is performed. This comparison is done by domain (medical) expert. Scale of relative importance when test T_i is not weaker than test T_j is given in Table 1.

Table 1. Scale of relative importance when test T_i is not weaker than test T_j

Value a_{ij}	Estimation of diagnostic potential (importance) of test T_i in regard to T_j
9	T_i is extremely preferred (absolutely more important) in regard to T_j
7	T_i is very strongly preferred (definitely more important)
5	T_i is strongly preferred (clearly more important)
3	T_i is moderately preferred (slightly more important)
1	T_i is equivalent (equally important) with T_j

When intermediate values between those in Table 1. are required, then the values from the set {2, 4, 6, 8} can be assumed. For a_{ji} (estimation of diagnostic potential of test T_j in regard to T_i), the following condition $a_{ji} = 1/a_{ij}$ need to be satisfied. Hence, if T_i is extremely preferred (absolutely more important) in regard to T_j, then T_j is absolutely less important than T_i, i.e. $a_{ji} = 1/9$. Sample values of relative comparisons are contained in Table 2. Elements $a_{ii} = 1$, because they concern the comparison of importance of T_i with itself. Having all pairwise comparisons of all test diagnostic potentials, the diagnostic potential of all tests is calculated according to AHP approach. It will be illustrated in case study (Sect. 3). Then *Consistency ratio* [7, 9] is calculated. It should be smaller than 0,1. Sometime, value 0,15 is accepted as the upper bound. If the requirement imposed on Consistency ratio is not satisfied then the above pair-wise comparisons need to be done once more.

In order to present BLPP for diagnosis D_k, the following notation will be introduced.

p_l – diagnostic potential of test T_l in verification of diagnosis D_k obtained in point 1 of the above algorithm,
x_l – binary decision variable; $x_l = 1$ if test T_l should be executed for diagnosis D_k, $x_l = 0$ otherwise,
$P_{min} \in\ < 0, 1]$ - required minimal total diagnostic potential (sum of potentials) of selected tests; the same value have been accepted for all diagnoses,
c_l – test T_l cost.

BLPP for selection of tests for verification of diagnosis D_k.
Constraint: $\sum_{l=1}^{t} p_l \cdot x_l \geq P_{min}$
Criterion: $min \sum_{l=1}^{t} c_l \cdot x_l$

The constraint imposes that decision variables satisfy the threshold of total diagnostic potential of selected tests requirements, while the criterion requires the minimal total cost value of these tests. In order to solve the BLPP, SIMPLEX method will be applied and supported with software addition SOLVER for Excel.

In [1], diagnosis ability, cost of testing and other criteria are submitted to subjective AHP evaluation. In our approach the test cost is not subjectively weighted in comparing with the diagnostic potential. It is difficult to compare the test cost with diagnostic potential of the test.

3   Case Study

The case of the patient with lumbal pain will be examined. Let the sample set of diagnoses D contain the following elements: Spine disease (SD), Urolithiasis (U), Aortic dissecting aneurysm (ADA), Oncological disease (OD), Pancreatic disease (PD), Acute pyelonephritis (AP). The abbreviations in the parentheses will be used further in the paper. The set of tests that the GP can order contain the elements: Urine (UR), Blood (B), Ultrasound examination (US), X-ray examination (X).

3.1   Analytic Hierarchy Process (AHP) in Estimation of Diagnostic Potential of Tests in Verification of the Diagnoses

Let us take urolithiasis as an example of the estimation of diagnostic potential using AHP. In Table 2 in rows and columns that are labelled by symbols p_{UR}, p_B, p_{US}, p_X of diagnostic potential of tests UR, B, US, X there are pair-wise comparison values a_{ij} of the relative importance of test T_i in regard to T_j when urolithiasis (U) is the preliminary diagnosis. The disease symbol U is put in the element of first row and first column. Then the sums of elements in columns are calculated. These sums are in the last row of this table.

Table 2. Values of pair-wise comparisons of diagnostic potentials p_{UR}, p_B, p_{US}, p_X of tests UR, B, US, X with sums of entries in columns for diagnosis U.

U	p_{UR}	p_B	p_{US}	p_X
p_{UR}	1	2	0,142857	1
p_B	0,5	1	0,111111	1
p_{US}	7	9	1	7
p_X	1	1	0,142857	1
Sum of entries in column	9,5	13	1,396825	10

In order to calculate normalized values of pair-wise comparisons of diagnostic potentials in Table 3, elements a_{ij} from Table 2 are divided by sums of entries in columns.

Table 3. Normalized values of pair-wise comparisons of diagnostic potential of tests UR, B, US, X with means of entries in the rows that are diagnostic potentials of tests for diagnosis U

U	p_{UR}	p_B	p_{US}	p_X	Mean value of the row (diagnostic potential for U)
p_{UR}	0,105263	0,153846	0,102273	0,1	0,11534551
p_B	0,052632	0,076923	0,079545	0,1	0,077275028
p_{US}	0,736842	0,692308	0,715909	0,7	0,711264722
p_X	0,105263	0,076923	0,102273	0,1	0,096114741

Then mean values of entries in the rows are calculated (last column in Table 3). These means are diagnostic potentials of tests for diagnosis U.

The diagnostic potentials of tests UR, B, US, X in verification of the other diagnoses SD, ADA, OD, PD, AP are given in Table 4.

Table 4. Diagnostic potentials of tests UR, B, US, X in verification of the other diagnoses SD, ADA, OD, PD, AP

	Diagnostic potential for SD	Diagnostic potential for ADA	Diagnostic potential for OD	Diagnostic potential for PD	Diagnostic potential for AP
p_{UR}	0,06575919	0,065678508	0,067565247	0,059815141	0,506360018
p_B	0,296573726	0,162151802	0,251287775	0,443705986	0,263267192
p_{US}	0,057469716	0,706491183	0,613581731	0,443705986	0,195249668
p_X	0,580197368	0,065678508	0,067565247	0,052772887	0,035123122

For each diagnosis for the set of diagnostic potentials of the tests, the Consistency ratios have been calculated. In all above cases these ratios are smaller than 0,1.

3.2 Test Costs for Diagnoses

Now for each diagnosis the test costs will be given. Urine, blood, ultrasound, X-ray tests are complex tests. They consist of elementary tests, e.g. urine test for diagnosis U consists of general urine test and urine culture. Costs of tests will be given in polish zloty (PLN). Costs of elementary tests will be assumed according to service price list of University Clinical Hospital in Wrocław [10], provided there is such information in this list. Otherwise the prices of the following elementary tests are taken from the sources: blood count with smear [12], phosphorus [13], Vitamin D concentration 25OHD3 [11], calcium [14]. Because of many medical terms the calculations of costs of tests will be omitted, and final results only are given in Table 5.

Table 5. Test costs for all preliminary diagnosis

Diagnosis	Urine test cost	Blood test cost	Ultrasound examination cost	X-ray examination cost
SD	10	112,00	100,00	50,00
U	30	126,00	100,00	35,00
ADA	10	35,00	100,00	35,00
OD	10	69,00	100,00	35,00
PD	10	71,00	100,00	35,00
AP	30	40,00	100,00	35,00

3.3 Binary Linear Programming Problems for Finding the Sets of Tests for Diagnoses

The BLPP for diagnosis U for total diagnostic potential threshold equal to 0,8 is defined as follows.

Constraint:

$$0,11534551 \cdot x_1 + 0,077275028 \cdot x_2 + 0,711264722 \cdot x_3 + 0,096114741 \cdot x_4 \geq 0,8$$

Criterion:

$$\min (30 \cdot x_1 + 126 \cdot x_2 + 100 \cdot x_3 + 35 \cdot x_4)$$

where

x_1, x_2, x_3, x_4, respectively, are binary decision variables for urine, blood, ultrasound examination, X-ray examination tests, respectively.

For all six diagnosis, the solutions of BLPP, values of total diagnostic potential, and total tests costs for the constraint: total diagnostic potential threshold equal to 0,8 are given in Table 6.

Table 6. For all diagnoses, the solutions of BLPP, values of total diagnostic potential, and total tests costs for the total diagnostic potential threshold equal to 0,8

	x_1	x_2	x_3	x_4	Total diagnostic potential	Total tests cost
SD	0	1	0	1	0,876771	162
U	1	0	1	0	0,82661	130
ADA	0	1	1	0	0,868643	135
OD	0	1	1	0	0,86487	169
PD	0	1	1	0	0,887412	171
AP	1	1	0	1	0,80475	105

The same results as in Table 6, however, for total diagnostic potential threshold equal to 0,9 are given in Table 7.

Table 7. For all diagnoses, the solutions of BLPP, values of total diagnostic potential, and total tests costs for the total diagnostic potential threshold equal to 0,9

	x_1	x_2	x_3	x_4	Total diagnostic potential	Total test cost
SD	1	1	0	1	0,94253	172
U	1	0	1	1	0,922725	165
ADA	1	1	1	0	0,934321	145
OD	1	1	1	0	0,932435	179
PD	1	1	1	0	0,947227	181
AP	1	1	1	0	0,964877	170

For the total diagnostic potential threshold equal to 1,0, all four tests are required.

4 Summary and Conclusions

The selection method of medical tests that General Practitioner can order for specific health problem, interview, and physical examination has been presented. The method of estimating the diagnostic potentials of tests is Analytic Hierarchy Process based. Having these potentials and test costs, binary linear programming problem is solved in order to find the set of tests with sufficiently great total diagnostic potential but with minimal cost. The method is different when comparing with typical usage of AHP where diagnostic potentials of tests and test costs are subjectively weighted. The approach has been applied for case study of patient with lumbal pain and six diagnostic hypotheses verified by medical tests. A limitation of the case study is the estimation of diagnostic potential of tests by one expert only. In the similar way, the method can be applied to a wider range of diagnostics of diseases.

References

1. Castro, F., et al.: Sequential test selection in the analysis of abdominal pain. Med. Decis. Making **16**(2), 178–183 (1996)
2. Chai, X., Deng, L., Yang, Q., Ling, C.X.: Test-cost sensitive naive Bayes classification. In: 4th IEEE International Conference on Data Mining, pp. 51–58 (2004)
3. Duddy, C., Wong, G.: Explaining variations in test ordering in primary care: protocol for a realist review. BMJ Open (2018). https://doi.org/10.1136/bmjopen-2018-023117
4. Ephzibah, E.P.: Cost effective approach on feature selection using genetic algorithms and fuzzy logic for diabetes diagnosis. Int. J. Soft Comput. **2**(1), 1–10 (2011)
5. Fakih, S.J., Das, T.K.: LEAD: a methodology for learning efficient approaches to medical diagnosis. IEEE Trans. Inf Technol. Biomed. **10**(2), 220–228 (2006)

6. Ling, C.X., Sheng, V.S., Yang, Q.: Test strategies for cost sensitive decision trees. IEEE Trans. Knowl. Data Eng. **8**(8), 1055–1067 (2006)
7. Manoy, M.: Multicriteria decision making, Analytic hierarchy process (AHP) method. https://www.youtube.com/watch?v=J4T70o8gjlk, Accessed 21 Nov 2020
8. Morgan, S., Coleman, J.: We live in testing times, teaching rational test ordering in general practice. Aust. Fam. Phys. **43**(5), 273–276 (2014)
9. Saaty, T.L.: The Analytic Hierarchy Process: Planning, Priority Setting, Resource Allocation. McGraw-Hill, New York (1980)
10. Service price list of J. Mikulicz-Radecki University Clinical Hospital in Wrocław. (in Polish)
11. https://www.medicover.pl/badania/witamina-d/, Accessed 22 Dec 2020. (in Polish)
12. https://www.medonet.pl/zdrowie,ile-kosztuje-morfologia--ceny-podstawowych-badan-krwi,artykul,1734721.html, Accessed 22 Dec 2020. (in Polish)
13. https://www.synevo.pl/fosfor/, Accessed 22 Dec 2020. (in Polish)
14. https://www.synevo.pl/wapn-calkowity/, Accessed 22 Dec 2020. (in Polish)

Simulation of Burnout Processes
by a Multi-order Adaptive Network Model

Louis Weyland[1], Wiebe Jelsma[1], and Jan Treur[2(✉)]

[1] Computational Science, University of Amsterdam, Amsterdam, The Netherlands
{l.f.weyland,w.a.jelsma}@student.vu.nl
[2] Social AI Group, Vrije Universiteit Amsterdam, Amsterdam, The Netherlands
j.treur@vu.nl

Abstract. In this paper, an adaptive network model for the development of and recovery from burnout was designed and analysed. The current literature lacks adequate adaptive models to describe the processes involved in burnout. In this research, the informal conceptual models from Golembiewski and Leiter-Maslach were combined with additional first- and second-order adaptive components and used to design a computational network model based on them. Four different scenarios were simulated and compared, where the importance of the therapy and the ability to learn from it was emphasised. The results show that if there was no therapy present, the emotion regulation was too poor to have effect. However, at the moment therapy was applied, the emotion regulation would start to help against a burnout. Another finding was that one long therapy session has more effect than several shorter sessions. Lastly, therapy only had a significant long-lasting effect when adaquate neuro-plasticity occurred.

Keywords: Burnout · Second-order adaptive · Network model

1 Introduction

Burnout is defined as a syndrome which occurs after being exposed for a longer time to chronic stress and frustration. The symptoms are mental (emotional) exhaustion, lack of personal accomplishment and depersonalization. Emotional exhaustion corresponds to persisting fatigue with which the person has to deal in his/her daily life. The lack of personal accomplishment represents the depletion of one's productivity. It is linked with being overwhelmed with the slightest task/stress which in turn will impact the effectiveness of efforts made. The depersonalization process characterises the detachment of oneself to his/her work and ambition [9]. Once the burnout stage is reached, the consequence is a psychological trauma which often complicates the integration back into the working-based society.

Given its complex dynamics based on a number of interacting factors, a thorough understanding of the syndrome and its underlying nature is challenging and important. This in return can offer effective intervention and avoid psychological trauma. However,

© Springer Nature Switzerland AG 2021
M. Paszynski et al. (Eds.): ICCS 2021, LNCS 12744, pp. 514–527, 2021.
https://doi.org/10.1007/978-3-030-77967-2_43

not all models provide a consistent and straight forward intervention method. Furthermore, it has been noticed that the current literature is lacking adequate adaptive network models for burnout. In this paper, the revised informal conceptual Leiter-Maslach model is used as inspiration but extended by introducing second-order adaptive network components. It is shown that the adaptive attributes have a non-negligible impact on the simulation of the development of burnout. Furthermore, the adaptive components will play a crucial role in recovery by an effective therapy against burnout.

2 Background Theory

The aforementioned symptoms (exhaustion, lack of personal accomplishment and depersonalization) correspond to the sub-scales of the Maslach Burnout Inventory (MBI) [6]. Many studies base their research on the MBI instrument. However, the order in which the symptoms occur is subject to variation. In literature, well-studied models vary on the order of personal development. In the informal Golembiewski model [1], burnout is described in eight progressive phases which represent a between-subject phase model, starting with depersonalization which impacts the personal accomplishment and leads to emotional exhaustion [1]. According to the model, personal trauma or chronic stress at work could trigger the depersonalization process as a defence mechanism. This leads to low personal accomplishment. Eventually, the emotional exhaustion stage is reached and a full-blown burnout is diagnosed [9]. Another well-known informal model is the Leiter-Maslach model which is a within-subject path model [3].

In the latter model, the process of burnout starts with the emotional exhaustion stage due to the influence of the workload. Then, the repercussion of the emotional exhaustion is the depersonalization phase as a defence strategy (Fig. 1).

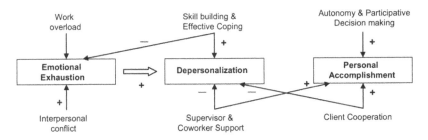

Fig. 1. Leiter-Maslach's informal model [9].

However, in the revised model, the personal accomplishment evolves independent and in parallel to the two MBI factors. Thereby, the three main MBI components are all influenced by external factors such as workload, moral support or internal conflict. In the Golembiewski model, the order of the progressive phases is subject to individual cases, which makes it diffcult to formulate a concrete intervention method. On the other hand, the Leiter-Maslach model formulates that the development of burnout can be prevented if the factors affecting directly the emotional exhaustion are being handled appropriately

[7]. These factors are foremost the workload and personal conflicts. However, it is also mentioned that in advanced stages of a burnout it is advised to undertake organisational changes such as increasing external moral and work support or allowing more autonomy.

As a mental condition, burnout is part of the cognitive neuroscience domain. Hence, it has been shown in the literature that many processes including adaptive behaviour, are reflecting a more accurate image of the studied process [11]. Existing studies proposed an adaptive model for the Maslach model or the job demands-resources model [5, 12]. However, no relevant study has been published on an adaptive Leiter-Maslach model. Additionally, no research has been found to introduce a multi-order adaptive mechanism.

3   The Designed Adaptive Computational Network Model

The designed adaptive network model is based on the revised informal Leiter-Maslach model as shown in Fig. 1. However, a few nodes and connections have been added to the network model (see Fig. 2). First of all, the node Personal accomplishment (which actually concerns lack thereof) has been renamed to Personal non-achievement to better fit its actual meaning and avoid any ambiguity. A connection between depersonalization and personal non-achievement has been created as it is originally in the Maslach model [6]. We believe that a detachment of one's personality affects personal ambitions and objectives. Additionally, a connection between personal non-achievement and workload has been made. An increasing lack of accomplishments will provoke an accumulation of the workload on the other end. An emotion regulation component has been added to the emotional exhaustion. It is known that, based on activity of certain parts of the prefrontal cortex in the brain, a person may control his/her emotional responses and feelings. However, for some people this mechanism may function poorly so that the effect is not sufficient to handle the regulation well.

Furthermore, therapy is added which focuses on improving the emotion regulation explicitly. In the proposed model, for three nodes adaptive characteristics have been included in the network model: the workload, the depersonalization and the emotion regulation. The workload is influenced by personal non-achievement due to which it changes in a non-linear way. The higher the unfulfillment, the more overwhelmed the person will be. Thus the productivity will decrease in a non-linear way. Consequently, the workload will increase in a dynamical fashion. Emotional exhaustion has also underlying adaptive behaviour. Burnout is a slow process, thus the emotion regulation can be seen as an adaptive component where neural networks reorganise based on previous experiences with external stress factors. For the same reasons, the depersonalization has also been equipped with adaptive characteristics.

To represent the network model with its adaptive characteristics well, a multi-order reified temporal-causal network (also called *self-modeling network*) was designed. For its numerical representation, the model was specified by the following *network characteristics*:

- **Nodes** Y (also called states) with state value $Y(t)$ at each time point t.
- **Connectivity** defined by connections from node X to Y and described by their *connection weight* $\omega_{X,Y}$.

- **Aggregation** defined by *combination functions* c_Y determining how multiple incoming single impacts $\omega_{X,Y} Y(t)$ on one node Y are aggregated.
- **Timing** defined by *speed factors* η_Y specifying how fast the state of a node Y changes based on the incoming aggregated impact.

Based on the above network characteristics, the underlying standard difference equation used for all nodes Y is

$$Y(t + \Delta t) = Y(t) + \eta_Y[c_Y(\omega_{X_1,Y} X_1(t), \ldots, \omega_{X_k,Y} X_k(t)) - Y(t)]\Delta t \qquad (1)$$

where $X_1, \ldots, X_k$ are the nodes from which Y gets incoming connections. Note that when modeling a specific person and/or context, these network characteristcs are used to represent personal and/or contextual characteristics, including individual differences. In this way such a network model can be used for a wide variety of specific circumstances.

Equations (1) are already given in the available dedicated software environment; see [10], Ch 9. Within the software environment described there, around 40 useful basic combination functions are included in a combination function library; see Table 1 for the ones used in this paper. The selected ones for a model are assigned to states Y by specifying combination function weights $\gamma_{i,Y}$ and their parameters used by $\pi_{i,j,Y}$.

Note that 'network characteristics' and 'network states' are two distinct concepts for a network. Self-modeling (or reification) is a way to relate these distinct concepts to each other in an interesting and useful way:

- A *self-model* is making the implicit network characteristics (such as connection weights ω or excitability thresholds τ) explicit by adding states for these characteristics; thus the network gets a self-model of part of the network structure; this can easily be used to obtain an *adaptive network*.
- In this way, multiple self-modeling levels can be created where network characteristics from one level relate to states at a next level. This can cover *second-order* or *higher-order adaptive networks*; see, for example, [10, 11].

Adding a self-model for a given temporal-causal (base) network is done in the way that for some of the states Y of this base network and some of its related network structure characteristics for connectivity, aggregation and timing (in particular, some from $\omega_{X,Y}$, $\gamma_{i,Y}$, $\pi_{i,j,Y}$, η_Y), additional network states $\mathbf{W}_{X,Y}$, $\mathbf{C}_{i,Y}$, $\mathbf{P}_{i,j,Y}$, $\mathbf{H}_Y$ (*self-model states*) are introduced:

a) **Connectivity self-model**
   Self-model states $\mathbf{W}_{Xi,Y}$ are added representing connectivity characteristics, in particular connection weights $\omega_{Xi,Y}$
b) **Aggregation self-model**
   Self-model states $\mathbf{C}_{j,Y}$ are added representing aggregation characteristics, in particular combination function weights $\gamma_{i,Y}$
   Self-model states $\mathbf{P}_{i,j,Y}$ are added representing aggregation characteristics, in particular combination function parameters $\pi_{i,j,Y}$

c) **Timing self-model**

Self-model states $\mathbf{H}_Y$ are added representing timing characteristics, in particular speed factors η_Y

The notations $\mathbf{W}_{X,Y}$, $\mathbf{C}_{i,Y}$, $\mathbf{P}_{i,j,Y}$, $\mathbf{H}_Y$ for the self-model states indicate the referencing relation with respect to the characteristics $\omega_{X,Y}$, $\gamma_{i,Y}$, $\pi_{i,j,Y}$, η_Y: here $\mathbf{W}$ refers to ω, $\mathbf{C}$ refers to γ, $\mathbf{P}$ refers to π, and $\mathbf{H}$ refers to η, respectively. For the processing, these self-model states define the dynamics of state Y in a canonical manner according to Eqs. (1) whereby $\omega_{X,Y}$, $\gamma_{i,Y}$, $\pi_{i,j,Y}$, η_Y are replaced by the state values of $\mathbf{W}_{X,Y}$, $\mathbf{C}_{i,Y}$, $\mathbf{P}_{i,j,Y}$, $\mathbf{H}_Y$ at time t, respectively. As the outcome of the addition of a self-model is also a temporal-causal network model itself, as has been shown in detail in [10], Ch 10, this construction can easily be applied iteratively to obtain multiple levels of self-models.

For the designed model, four different combination functions were used; see Table 1. For nodes with one incoming connection, the identity function was used. For nodes with multiple incoming connections, the advanced logistic sum combination function was used see the second row in the table. Table 2 provides an overview of all nodes of the network model; note that when using these functions in (1), the variables V, V_i are applied to single impacts $\omega_{X,Y} Y(t)$.

Table 1. Basic combination functions from the library used in the presented model

	Notation	Formula	Parameters
Identity	$\mathbf{id}(V)$	V	-
Advanced logistic sum	$\mathbf{alogistic}_{\sigma,\tau}(V_1, ...,V_k)$	$[\frac{1}{1+e^{-\sigma(V_1+\cdots+V_k-\tau)}} - \frac{1}{1+e^{\sigma\tau}}](1 + e^{-\sigma\tau})$	Steepness $\sigma > 0$ Threshold τ
Hebbian learning	$\mathbf{hebb}_\mu(V_1, V_2, W)$	$V_1 V_2(1 - W) + \mu W$	Persistence factor $\mu > 0$
Stepmod	$\mathbf{stepmod}_{\rho,\delta}(V_1, ...,V_k)$	0 if $t \bmod \rho < \delta$, else 1	Repetition ρ Duration δ

The first-order self-model states are: Workoverload adaptation, Depersonalisation adaptation and Emotion regulation adaptation. In terms of the above explanation and terminology, these three states are $\mathbf{W}$-states representing weights ω of connections between base states: respectively, from state Personal nonachievement to Workoverload, and from state Emotional exhaustion both to Depersonalisation and Emotion regulation control. The adaptation mechanisms for them were based on Hebbian learning. In the literature, Hebbian learning has shown adequate results [2]. It reinforces connections of nodes that are activated simultaneously:

'Neurons that fire together, wire together'

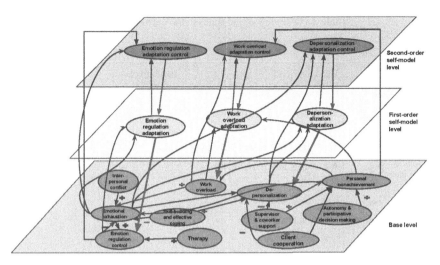

Fig. 2. Extended Leiter-Maslach model by introduction of first- and second-order adaptation. Furthermore, a direct connection from depersonalization to personal non-achievements is made. Finally, an emotion regulation and therapy component is added. The red connections represent the effects of the self-models on the adaptive network characteristics. (Color figure online)

The formula shown in the third row of Table 1 models this Hebbian learning principle where W corresponds to the value of the first-order connection weight self-model **W**-state and V_1, V_2 represent the activation levels $X(t)$ and $Y(t)$ of the connected base states X and Y. The second-order adaptation nodes are Workoverload adaptation control, Depersonalisation adaptation control, and Emotion regulation adaptation control. Given the above explanation and terminology, these three states are **H**-states representing the speed factors η of the related first-order **W**-states. They control the three types of adaptation modeled by the first-order self model **W**-states. For the second-order self-model states, the advanced logistic sum combination function was used for the aggregation [11]. These second-order nodes model that the speed factors η of the related first-order nodes are adaptive. Through this, the learning process was equipped with an adaptive learning rate which changes over time depending on influences of the external factors. The second-order adaptation (also called *metaplasticity*) principle modeled here is:

'Adaptation accelerates with increasing stimulus exposure' ([8], p. 2).

The fourth combination function was used specifically for the therapy node and is referred as steppmod function. It creates a repeated occurrence of an external factor (here the therapy) with some time duration and time of resting.

The model was implemented using the available dedicated software environment in Matlab, using a library for combination functions; see [10], Ch 9. The role matrices fully specifying the adaptive network model are shown in the Appendix [13].

Table 2. Overview of the states in the network model

State	Name	Level
X_1	Workoverload	Base level
X_2	Interpersonal conflict	
X_3	Skill building and effective coping	
X_4	Supervisor and coworker support	
X_5	Autonomy and participative decision	
X_6	Client cooperation	
X_7	Emotional exhaustion	
X_8	Depersonalisation	
X_9	Personal nonachievement	
X_{12}	Emotion regulation control	
X_{10}	Workoverload adaptation	First-order self-model
X_{11}	Depersonalisation adaptation	
X_{13}	Emotion regulation adaptation	
X_{14}	Workoverload adaptation control	Second-order self-model
X_{15}	Depersonalisation adaptation control	
X_{16}	Emotion regulation adaptation control	
X_{17}	Therapy	Base level

4   Simulation Results

Four different scenarios were simulated. The first scenario shows the simulation of developing a burnout, without therapy. The second scenario simulates a developed burnout with short and frequent therapy sessions. The third one simulates a burnout with longer and less frequent therapy sessions. The fourth scenario shows a therapy session where the higher-order adaptation of emotion regulation is deactivated.

For each scenario, a fictive character with a tendency to develop a burnout was addressed. This was achieved by setting a higher interpersonal conflict and low support from the external factors, such as the Client cooperation or Supervisor/Co-worker support. According to the complex causal relationships as modeled, the external factors influence the development. To trigger a burnout, for the first three scenarios, a high workload was set. The initial value for the connection weight representation for emotion regulation was set to a very low value, thus representing a person with very poor emotion regulation capabilities: 0.0005. The external factors are set to relatively low values. The connections from the first-order states to the second-orders were set to -0.2 to avoid an exaggerated learning rate. Additionally, the speed factor for the workload was decreasing. The base nodes emotional exhaustion, depersonalization and personal non-achievement were all set to a low initial value of 0.03 to represent a very healthy starting point for the person. The second-order speed factors were set to 0.001. Concerning the advanced logistic sum combination function, the threshold for the depersonalization and the second-order depersonalization was set slightly higher than for the emotional exhaustion and personal non-achievement due to a higher number of incoming connections. By doing so, the excitation of the node is delayed and an early triggering is circumvented. For the Hebbian function, the persistence factor μ was set to 0.98 for all first-order adaptation nodes. Therefore the learning is effective, most what is learned will be remembered (only 2% loss per time unit). The simulation was set to a duration of 20000 time units (but the graphs shown in Figs. 3 and 7 only display a focus on a shorter time interval). The step size Δt was set to 0.5. The external factors are not shown in the figures due to the fact that they are constant and don't significantly contribute to the understanding of the figures.

Scenario 1: Development and Persistence of Burnout Without Therapy

Figure 3 shows the simulation with an initial high workload and no therapy. One can see a rapid increase in the emotional exhaustion. For the depersonalization and personal non-achievement, a short dip is noticeable before they increases rapidly and go to an equilibrium. The dip is caused by the second-order dip at the beginning.

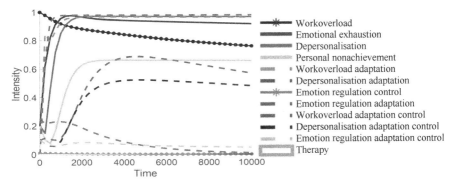

Fig. 3. Running the simulation with a duration of 10000 and a step size Δt of 0.5. The simulation starts with a high workload, no therapy is applied.

The emotion regulation and corresponding adaptive nodes stay largely unaffected in this simulation. The therapy here shown by the rainbow coloured line is set to 0.

Scenario 2: Short Therapy Sessions

Figure 4 (upper graph) shows the simulation with short and repetitive therapy sessions. As expected, the simulation has the same initial evolution as in Scenario 1. However, due to influences of the therapy a more complex pattern emerges. The therapy starts at $t = 3000$ and repeats itself in short intervals of 1500 time units. First of all, a constant decrease of the workload can be observed. Emotional exhaustion, depersonalization and personal non-achievement fluctuate in a periodic manner. During the therapy, the three Maslach sub-scales decrease. In between, the sub-scales underwent a small increase. However, in overall, the sub-scales decreased. All the first-order nodes increase immediately and stay around an equilibrium.

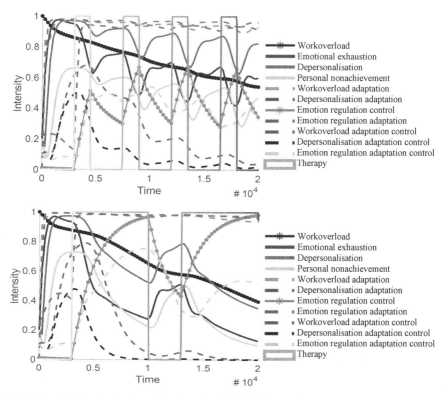

Fig. 4. Running the simulation with a duration of 20000 time units and a step size Δt of 0.5. The upper graph (Scenario 2) starts with high workload, while over time short and frequent therapy sessions are applied. The lower graph (Scenario 3) starts with high workload, while over time long therapy sessions are applied.

Once the therapy was activated, an instantaneous increase of the first-order adaptation states for emotional regulation and the emotional regulation itself were provoked. However, between the sessions, the emotion regulation decreased. The second-order adaptation of emotion regulation had a steady fluctuation before the therapy. During the therapy and shortly after, the second-order adaptation states increased. In the middle of the in-between sessions, the second-order adaptation states seemed to decrease. The second-order adaptation states for workload and for depersonalization nodes underwent a decrease after the first therapy.

Scenario 3: Long Therapy Sessions

Scenario 3 (Fig. 4 lower graph) used the same parameters as in Scenario 2. However, the duration of the therapy was set to a duration of 7000 time units. Before the therapy, the same pattern as described before can be observed. During the therapy, an important increase of the emotion regulation is noticeable. Furthermore, a significant decrease of the emotional exhaustion, depersonalization and personal non-achievement can be seen during the therapy. At $t = 10000$, the three sub-scales had a lower value compared to the simulation in Scenario 2. The second-order adaptation state for emotion regulation has reached overall higher values.

Scenario 4: Short Therapy and No Second-Order Neuro-Plasticity

Scenario 4 (see Fig. 5) was build upon Scenario 3. However, the emotion regulation was not made adaptive. Thus, one can see that the second-order adaptation state for emotion regulation is not active (constant) during the simulation. The first-order adaptation state for emotion regulation starts to increase once the first therapy starts but doesn't increase further after the consecutive therapies. A fast-growing trend can be observed for the Maslach sub-scales. During the therapy sessions, the emotional exhaustion, depersonalization and the personal non-achievement decrease slowly. In the end, these nodes seem to undergo periodical fluctuations.

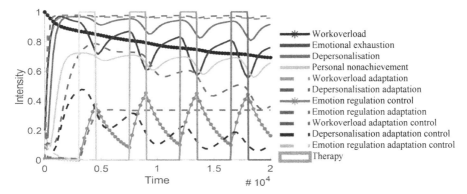

Fig. 5. A simulation (Scenario 4) with a duration of 20000 time units and a step size Δt of 0.5. The simulation is run with short therapies and no adaptive behaviour for the emotional regulation.

5 Verification by Mathematical Analysis of Stationary Points

In scientific computing, truncating error and rounding error induce inaccurate results. Additionally, the time steps Δt used for the integration can produce unstable algorithms with important deviation to the true results. Therefore, it is important to verify the exactness of the implemented model. The verification of the model can be done by comparing the theoretically determined values based on the high-level specification of the model to the ones computed in a simulation generated by the implemented model.

This verification was done by selecting randomly points at times t where a considered node Y of the model is stationary, i.e., $dY(t)/dt = 0$. According to difference Eq. (1), if $\eta_Y > 0$, being stationary is equivalent to the standard criterion

$$\mathbf{c}_Y(\omega_{X_1,Y}X_1(t), \ldots, \omega_{X_k,Y}X_k(t)) = Y(t) \tag{2}$$

where $X_1, \ldots, X_k$ are the nodes from which Y gets incoming connections. Here, the left hand side is also denoted by **aggimpact**$_Y(t)$; this is what we call here theoretical value, whereas for the state value at the right hand side the simulation value is used. Thus, the deviation between the theoretical and simulation value indicates the accuracy of the model. In this case, the verification was applied to the model in Scenario 1. From each level, a random node was selected for the analysis. Thus for the picked nodes Y the theoretical values computed by the formula from Table 1 were compared to the value $Y(t)$ from the simulation. The deviations found are small (see Table 3), which provides evidence that the implemented model is correct with respect to the high-level model specification.

Table 3. Verification results.

State X_i	Personal non-achievement	First-order adaptation for workload	Second-order adaptation for emotion regulation
Time point t	200000	200000	200000
$X_i(t)$	0.9584	0.9638	0.0350
aggimpact$_{X_i(t)}$	0.9432	0.9712	0.0354
Deviation	0.0152	0.0074	0.0004

6 Model Validation

The challenge in burnout research is the quantification of the different stages. In [4], empirical data was collected using a latent profile analysis with 5-class scale (see Fig. 6). Each stage is characterised by the combination of the different scores for exhaustion, cynicism (depersonalization) and efficiency. For the fine-tuning, the scale was normalised to a maximum of 1 instead of 5. Thus, by doing so, the score can be expressed in percentage where 100% means a score of 1 out of 1.

The burnout stage as it is described in [4] would have a score of 100% for exhaustion, 84% for cynicism and 60% (see the empirical values in Table 4). For the validation, the two stages burnout and ineffective were chosen.

An ineffective profile means that the personal non-achievement is high without compromising the mental health of a person. The simulation without therapy was chosen to be fine-tuned. The validation process entailed time-consuming computations. Thus the maximum time t was set to 5000 and the step size Δt to 0.5. Furthermore, the number of parameters fine-tuned was limited to 3. From these, one corresponded to the initial value of emotional exhaustion. The other two parameters fine-tuned were the speed factors of the depersonalization and personal non-achievement. The empirical data to which the model was fine-tuned were chosen to arbitrarily with the end value corresponding to the empirical values found in [4]. The parameter tuning was done using a simulated annealing algorithm from Matlab's Optimization Tool. The root means square error (RMSE) was chosen as an error function. The results are shown in Table 4. An example simulation after the fine-tuning in comparison to the data points used is shown in Fig. 7.

Class	N	Exhaustion	Cynicism	Inefficacy
Burnout	48 (4%)	5.04	4.19	2.98
Disengaged	136 (12%)	4.67	3.78	1.22
Overextended	201 (17%)	4.18	1.62	0.92
Ineffective	235 (20%)	2.54	1.99	2.43
Engagement	546 (47%)	1.81	0.87	0.82
Overall	1166	2.83	1.70	1.30
SD		1.46	1.30	0.92

Fig. 6. Visualisation (left) of 5 different progressive stages included the final burnout stage with their respective score determined in [4]. The table (right) contains the score of the Maslach Subscales for the different stages. The column N corresponds to the percentages of patients diagnosed with the respective stage.

Table 4. Results from fine-tuning the model to the different stages from [4] using Matlab's Optimization Tool

	RMSE	Emotional exhaustion		Depersonalization		Personal non-achievement	
		Empirical values	Computed values	Empirical values	Computed values	Empirical values	Computed values
Burnout	0.17	100%	91%	84%	94%	60%	70%
Ineffective	0.35	50%	40%	40%	30%	49%	21%

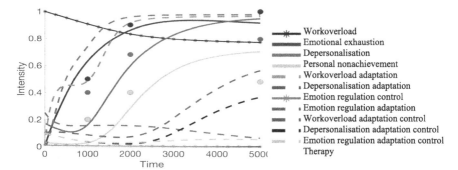

Fig. 7. Fine-tuning result for the model to the burnout stage determined in [4]. The numerical results are shown in Table 4.

7 Discussion and Conclusion

The simulations have shown promising results. In accordance with the literature, a burnout is triggered by having high emotional exhaustion, depersonalization and personal non-achievement (Fig. 4). However, the interesting results are shown in Fig. 5 where the importance of the first and second-order components of the model are shown. Especially, the impact on the emotion regulation is of great interest. Before the therapy, when the symptoms increased rapidly, the ability to cope with the situation were poor. The emotion regulation had no impact on the situation. Furthermore, the learning rate was constant. However, once the therapy started, an instant increase in the emotion regulation emerged. The learning capacity improved fast due to an adaptive learning rate that became higher. Despite a small decrease of the emotion regulation between the therapy sessions, the learning ability is retained over time, which allows to further increase the emotion regulation for each therapy. Comparing to Fig. 5 (lower), one can see that one longer therapy shows to be more effective than having multiple shorter sessions. The speed of learning how to regulate the emotions increased further.

In Scenario 4, a simulation was conducted where no second-order neuro-plasticity occurs. In this case, the previous experiences and therapy have no long-term effect since the ability to learn from it is limited. Thus, one can see that after each therapy, the emotion regulation regress to earlier stages. The second-order adaptation state for emotion regulation node had no effect over the whole simulation. The first-order adaptation state increases only after the first therapy. In other words, after the first therapy, the first-order adaptation state was not further stimulated. Thus the results show that without second-order plasticity, the therapy has limited effect. Since we know that in practice the therapy is effective and essential, second-order neuro-plasticity must occur. Thus, (second-order) adaptive behaviour and adaptive learning from previous experiences play an important role and need to be taken into account.

In [5] also an adaptive network model was shown for burnout. A main difference compared to the current model is that in [5] no second-order adaptation is modelled. A similar difference applies to [12]. Another difference with [5] is that their focus is on the role of dreaming whereas for the current model it is on the informal models from [1] and [4], which were not considered in [5].

The study could be taken further were the environmental nodes such as personal conflict, client cooperation, skill-building and such vary over time. By creating a feedback loop from the Maslach sub-scales to the environmental nodes, a more refined model can be created.

References

1. Golembiewski, R.T., Munzenrider, R., Carter, D.: Phases of progressive burnout and their work site covariants: critical issues in OD research and praxis. J. Appl. Behav. Sci. **19**(4), 461–481 (1983)
2. Hebb, D.O.: The Organization of Behavior: A Neuropsychological Theory. Chapman & Hall, London (1949)
3. Leiter, M.P.: Burnout as a developmental process: consideration of models. In: Schaufeli, W., Maslach, S., Marek, T. (eds.) Professional burnout: Recent developments in theory and research, pp. 237–250. Taylor and Francis, Washington (1993)
4. Leiter, M.P., Maslach, C.: Latent burnout pro les: a new approach to understanding the burnout experience. Burn. Res. **3**(4), 89–100 (2016)
5. Maijer, M., Solak, E., Treur, J.: An adaptive network model for burnout and dreaming. In: Krzhizhanovskaya, V.V., et al. (eds.) ICCS 2020. LNCS, vol. 12137, pp. 342–356. Springer, Cham (2020). https://doi.org/10.1007/978-3-030-50371-0_25
6. Maslach, C., Jackson, S.E., Leiter, M.P.: The Maslach Burnout Inventory. Consulting Psychologists Press, Palo Alto (1981)
7. Richardsen, A.M., Burke, R.J.: Models of burnout: Implications for interventions. Int. J. Stress Manag. **2**(1), 31–43 (1995)
8. Robinson, B.L., Harper, N.S., McAlpine, D.: Meta-adaptation in the auditory midbrain under cortical influence. Nat. Commun. **7**, e13442 (2016)
9. Sharma, R.R., Cooper, C.L.: Executive Burnout: Eastern and Western Concepts, Models and Approaches for Mitigation. Emerald Group Publishing, Bradford (2016)
10. Treur, J.: Network-Oriented Modeling for Adaptive Networks: Designing Higher-Order Adaptive Biological, Mental and Social Network Models. Springer, Cham (2020). https://doi.org/10.1007/978-3-030-31445-3
11. Treur, J.: Network-Oriented Modeling: Addressing Complexity of Cognitive, Affective and Social Interactions. Springer, Cham (2016). https://doi.org/10.1007/978-3-319-45213-5
12. Veldhuis, G., Sluijs, T., van Zwieten, M., Bouwman, J., Wiezer, N., Wortelboer, H.: A proof-of-concept system dynamics simulation model of the development of burnout and recovery using retrospective case data. Int. J. Environ. Res. Public Health **17**(16), 5964 (2020). https://doi.org/10.3390/ijerph17165964
13. Appendix (2021). https://www.researchgate.net/publication/348405068

Reversed Correlation-Based Pairwised EEG Channel Selection in Emotional State Recognition

Aleksandra Dura⬤, Agnieszka Wosiak$^{(\boxtimes)}$⬤, Bartłomiej Stasiak⬤,
Adam Wojciechowski⬤, and Jan Rogowski⬤

Institute of Information Technology, Lodz University of Technology,
Wólczańska 215, 90-924 Łódź, Poland
aleksandra.dura@dokt.p.lodz.pl,
{agnieszka.wosiak,bartlomiej.stasiak,adam.wojciechowski,
jan.rogowski}@p.lodz.pl

Abstract. Emotions play an important role in everyday life and contribute to physical and emotional well-being. They can be identified by verbal or non-verbal signs. Emotional states can be also detected by electroencephalography (EEG signals). However, efficient information retrieval from the EEG sensors is a difficult and complex task due to noise from the internal and external artifacts and overlapping signals from different electrodes. Therefore, the appropriate electrode selection and discovering the brain parts and electrode locations that are most or least correlated with different emotional states is of great importance. We propose using reversed correlation-based algorithm for intra-user electrode selection, and the inter-subject subset analysis to establish electrodes least correlated with emotions for all users. Moreover, we identified subsets of electrodes most correlated with emotional states. The proposed method has been verified by experiments done on the DEAP dataset. The obtained results have been evaluated regarding the recognition of two emotions: valence and arousal. The experiments showed that the appropriate reduction of electrodes has no negative influence on emotion recognition. The differences between errors for recognition based on all electrodes and the selected subsets were not statistically significant. Therefore, where appropriate, reducing the number of electrodes may be beneficial in terms of collecting less data, simplifying the EEG analysis, and improving interaction problems without recognition loss.

Keywords: Emotion recognition · Feature selection · EEG analysis · Data analysis · Machine learning · EEG electrodes selection

1 Introduction

In everyday life and interpersonal communication, emotions play an essential role. Positive emotions promote well-being, while negative emotions can

M. Paszynski et al. (Eds.): ICCS 2021, LNCS 12744, pp. 528–541, 2021.
https://doi.org/10.1007/978-3-030-77967-2_44

lead to the worsening of health problems. Neurobiological and psychological research confirms that emotions are a vital factor that influences rational behavior. Besides, patients with emotional disorders have trouble conducting daily activities [1].

Emotions can be recognized by verbal (e.g., emotional vocabulary) or non-verbal signs such as intonation, expressions, and gestures. Moreover, emotional states can also be detected by infrared thermal images or EEG signals [2–4]. EEG examination is non-invasive and inexpensive when compared to other types of signal acquisition. However, the accuracy of semantic EEG signal analysis in mental activity classification is still a significant problem. Hence, research on identifying emotions is of great importance in the EEG signal analysis [5,6].

During EEG examination, the electrical patterns of the brain are recorded. Sensors (electrodes and wires) placed on the patient's scalp monitor the potentials of the synchronous work of groups of cerebral cortical neurons. Therefore, the correct positions of measuring electrodes are one of the positive key factors that influence the analysis of EEG recordings. On the other hand, the noise from internal (physiological) and external (hardware-related) artifacts and overlapping signals from different electrodes may have a negative impact on the EEG analysis [7].

The optimal selection of measuring electrodes is the first and a key stage of works related to the automatic selection of EEG signals [8]. It also leads to improved brain activity analysis. However, due to the complexity of the emotion recognition process, the attempts made so far have not clearly indicated the number and the locations of the best-suited electrodes. For example, Nakisa et al. in [9] suggest using nine electrodes. In contrast, the authors of the paper [10] indicate 3 to 8 electrodes depending on the experiment. An additional difficulty is that the smaller sets of electrodes do not constitute subsets of the covering sets of electrodes in the cited papers.

We propose to reverse the problem of electrode selection and discover the electrodes and their locations that are least correlated with one another. Consequently, our research can contribute to classification advancement due to redundancy reduction and better recognition of brain regions responsible for emotional state conditions. These regions, in turn, may be stimulated to improve human health state and well-being [11]. As recognizing cross-subject emotions based on brain EEG data has always been perceived as difficult due to the poor generalizability of features across subjects [12], we propose a two-phased approach. First an intra-subject electrode selection is performed. Then, the inter-subject subset analysis is proceeded to sum individual subsets and establish parts of brain and electrode locations least correlated with emotions for all users, which makes our solution suitable for unseen subjects as well. We used a reversed correlation-based algorithm (RCA) to select electrodes in conjunction with bands of frequencies automatically. The RCA was described in [13]. The method has already been used in emotion recognition and yielded promising results [14]. The experimental research was carried out on the reference DEAP set [5], which is the primary data source in research on emotions' classification. We applied inter-subject analysis to identify band-channel combinations most correlated with emotional states and intra-subject analysis to make the results universal. This two-phases procedure revealed that keeping only 12.5% of the initial number of electrodes does not negatively influence the process of emotion recognition, which contributes to EEG signal analysis in terms of emotional state recognition.

The remainder of the paper is organized as follows. In the next section, the emotion classification problem in terms of human well-being and the EEG test is introduced, and the whole methodology is described. Next, the experiments carried out are presented, and the obtained results are discussed. The final section presents the study's conclusions and delineates future research.

2   Materials and Methods

2.1   EEG Analysis in Emotional State Recognition and Brain Stimulation for Improved Health and Well-Being

Many psychological studies undertake the problem of emotion classification, as emotions play a vital role in human communication and interactions. Various scales of emotion categorization were proposed [15–17]. One of the widely used scales is Russell's valence-arousal scale [18] which is often applied in the automated interpretation of emotions [5]. In this scale, each psychological state is placed on a two-dimensional plane with arousal and valence as the vertical and horizontal axes (Fig. 1).

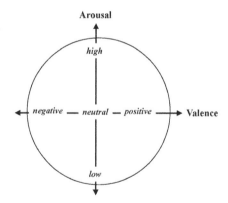

Fig. 1. Russell's circumplex model of affect.

Positive emotions and feelings are fundamentals in cultivating resilience, vitality, happiness, and life satisfaction, which contribute to physical and emotional well-being [19]. Advancing knowledge on how the nervous system implements positive emotions and feelings is critical in developing strategies that enhance the experience of healthy positive emotions and associated well-being outcomes.

According to neurophysiological and clinical research, electroencephalogram (EEG) reflects the electric activity of the brain and the functional condition of mind. The electrodes gather signals from the four lobes of the cerebral cortex and each of them is perceived as responsible for different activities [20]:

1. the frontal lobe, responsible for thinking, memory, evaluation of emotions and situations,
2. the parietal lobe, located behind the frontal lobe and responsible for movement, recognition, a sensation of temperature, touch, and pain,
3. the occipital lobe, causative for seeing and analyzing colors and shapes, and
4. the temporal lobe, located in the lateral parts, and responsible for speech and recognition of objects.

As the human brain activity also reflects the emotional state, in the last several years EEG analysis have been slowly introduced to emotion recognition [21]. However, as the process usually involves gathering EEG signals and comparing them with expert classification or self-assessment, it is a challenging task due to its subjectivity. Moreover, recent EEG research has also tended to show that during the performance of cognitive tasks many different parts of the brain are activated and communicate with one another, thus making it difficult to isolate one or two regions where the activity will take place [22].

There is no doubt the successful analysis of the user's state can contribute to medical and psychology-related applications, either to identify possible pathology or to assist in developing well-being tools and healthcare-related solutions. Such solutions would provide treatments to improve cognitive performance, enhance mental focus attention, and promote a feeling of well-being and relaxation. Therefore, a considerable amount of research focused on the neural correlations of positive or negative emotions and brain stimulation contributing to good health [23–26]. The EEG analysis can also be beneficial in reflecting the real emotional state of people with neuropsychiatric disorders and deficits in processing emotions, e.g. Parkinson's disease (PD) [27].

EEG signals are gathered from many locations across the brain and frequently entail more than 100 electrodes. However, the huge number of electrodes negatively influence the computational complexity while assessing EEG signals. It also increases the risk of overlapping signals and causes interactions problems. Hence, the efficient channel selection is of great importance. The literature describes that similar or even the same performance could be accomplished using a more compact group of channels. Furthermore, electrode selection may be crucial for overcoming problems of complex and high-intrusive devices [28]. However, researchers argue about the definite number and locations of the EEG electrodes [9,10].

Therefore, we propose an approach that reverses the problem of EEG channel classification and indicates the electrodes least correlated with emotional states. In addition, we also identify subsequent, inclusive subsets of the electrodes most correlated with emotions.

2.2   The Method Overview

The considered method for reducing the number of EEG electrodes by excluding least correlated with emotional state recognition can be presented in the following steps:

1. Selecting bands of frequencies based on calculating average frequencies for each second of a trial.
2. Selecting band-electrode combinations based on a statistical analysis of correlation coefficients using intra-subject approach.
3. Building an inter-subject subset of electrodes by eliminating all the electrodes that did not appear in any user's subset and summing up the occurrences.
4. Evaluation of emotions' classification results in terms of valence and arousal.

The description of EEG data, as well as the main steps of the methodology, are presented in Subsects. 2.3, 2.4, 2.5 and 2.6.

2.3   EEG Dataset

The experimental study was carried out on the reference DEAP dataset [5], created by researchers at Queen Mary University of London. It is among the most frequently used open-source datasets in research studies on emotional state classification. It contains multiple physiological signals with the psychological evaluation. The data comes from 32 subjects. For every subject, 32 electrodes were placed on the scalp, and 13 physiological electrodes were put on the examined person's fingers and face and recorded bioelectrical signals while the subject was watching 40 trials of music movies with different psychological tendencies.

The distribution of channels is shown in Fig. 2. In our research we consider only 32 EEG electrodes placed on the scalp, marked as gray in Fig. 2.

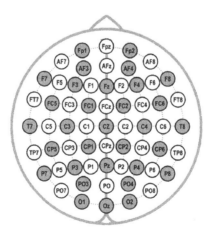

Fig. 2. Names of channels.

Four psychophysical states were assessed (valence, arousal, dominance, liking) while viewing 1-min movies in the DEAP dataset. Each video lasted 60 s, and 3 s were reserved for preparing the person for the next trial. Each movie has been ranked on a scale of 1–9. The smallest values indicate negative and higher – positive "polarity" of a given emotional quality.

Our research focused on the two most objective feelings: valence and arousal. The valence reflects what the person feels while viewing the film (happy or sad), while arousal demonstrates the degree of the impression the film makes (calm, enthusiastic).

The data was preprocessed by downsampling 128 Hz and the eye movement (EOG) artifacts removal.

2.4 Band Selection

There are five widely recognized frequency bands in the EEG signal:

- delta (δ) with frequencies in a range of 0.5–4 Hz, specific for deep sleep,
- theta (θ) with 4–8 Hz frequency range typical for mental relaxation,
- alpha (α) of 8–14 Hz frequencies characteristic for clear-headed and passive attention,
- beta (β), with frequencies ranged in 14–30 Hz typical for brain awakening alertness, and active, external attention, and
- gamma (γ) in a range of 30–60 Hz, specific for high concentration.

For valence and arousal classification, the delta band can be excluded from analysis. As a result, after preprocessing our data, we have obtained 128 band-channel combinations: 32 electrodes x 4 bands.

In our approach we proposed choosing the mean statistics of frequencies for every second of the recordings.

2.5 Automated Selection of Band-Electrode Combinations

Feature selection algorithms enhance computation time and classification accuracy and identify the most appropriate channels for a specific application or task. As it had been introduced in Sect. 2.1, the process of channel selection might be of great importance. Hence, the researchers develop new techniques for selecting the optimal number and location of electrodes.

We propose using the Reversed Correlation Algorithm (RCA). The algorithm belongs to unsupervised methods of machine learning and uses reversed correlation coefficients between all the parameters. It means that the RCA suggests features that are the least connected with all their predecessors. For that reason, the RCA might be beneficial in reduction of overlapping EEG signals.

The algorithm was proposed by Wosiak and Zakrzewska in [13] and has already been successfully applied in EEG signal analysis [14]. First, we start building a subset of selected elements with the band-channel combination that is the least correlated with the others. Then, correlation coefficients between the chosen element and the rest of the combinations are calculated. The element with the lowest correlation value is indicated as the second component. The obtained subset of two elements is further extended by adding the band-channel combination of the correlation coefficient with the lowest value between the subset and the rest of the items. The process of appending the components of the lowest correlation values is repeated unless the number of elements in the subset R is equal to the initially determined number N. The whole procedure is presented in Algorithm 1.

Algorithm 1. Reversed Correlation Algorithm

1: **function** RCA(N)   ▷ N is a desired number of elements from R subset
2:     $//Ch = ch_1, ch_2, ..., ch_{128}$ ▷ set of all elements
3:     $R_1 \leftarrow$ take the first element with the min correlation
4:     **while** $i < N$ **do**
5:         **while** $j <$ length of Ch **do**
6:             **while** $k <$ length of R **do**
7:                 $value \leftarrow$ compute correlation between elements
8:                 $sum \leftarrow sum + value$
9:                 $k \leftarrow k + 1$
10:             **end while**
11:             $R_i \leftarrow sum/len(R)$
12:             $sum \leftarrow 0$
13:             $j \leftarrow j + 1$
14:         **end while**
15:         $R_i \leftarrow$ choose element with the lowest sum of value
16:         $k \leftarrow k + 1$
17:     **end while**
18:     **return** R_i ▷ selected subset of elements
19: **end function**

We propose using RCA approach for building individual subsets of electrodes adapted to each user [12,29] and as a result we get an intra-subject (within-subject) subsets of electrodes.

However, such an approach cannot be used for unseen subjects as it only uses data related to the particular individual. Therefore, we propose the inter-subject generalization step. First, we summed up occurrences of every channel. Then, we built subsets of electrodes, starting from the most frequent channels and adding subsequent electrodes with fewer and fewer occurrences until at least one appeared.

2.6 Evaluation Criteria

Limiting the number of EEG channels and selecting an optimum subset of electrodes requires criteria of evaluation. Concerning emotion recognition, classifier-based measures are usually used. However, considering the fact, that emotional state in the DEAP dataset is self-assessed by real values ranging in 1–9, linear regression was applied. The regression analysis estimates relationships between a dependent variable (in our case users rating) and independent variables (values for selected subset of channels) using linear predictor functions. For regression evaluation we calculated two measures: Mean Absolute Error (MAE) and Root Mean Square Error (RMSE). Both measures express average model prediction error. However, the RMSE gives a relatively high weight to large errors, since the errors are squared before they are averaged. As a result, the RMSE would be observed as large if our predictions were significantly different from the target participant's rating value.

3 Results and Discussion

The purpose of the experiments was to indicate a subset of EEG channels, which reduces the complexity of the analysis and signal noise and positively influences classifying emotions.

The experiments were conducted according to the methods introduced in Sect. 2 on the dataset described in Sect. 2.3. Only valence and arousal were analyzed. The procedures were repeated ten times and the mean values were calculated. The experimental environment was based on Python programming language and its libraries.

The proposed steps of the experiments can be presented as follows:

1. The RCA analysis performed for every user to choose intra-subject electrode-band combinations, including validation of results.
2. The analysis of occurrences of channels and building the most frequent subsets of channels - the inter-subject generalization step.
3. Applying the subsets from the previous step on randomly picked users for final evaluation.

According to the description in Sect. 2.4, first, every channel was divided into four main frequencies: theta, alpha, beta, gamma. Therefore, as an input, we have 128 features (32 channels x 4 frequencies), which represent Ch subset. Then, the RCA algorithm was applied to every user. We set the number of channels selected to three, according to other studies and data limitations [14]. Every selection was validated according to the evaluation criteria introduced in Sect. 2.6 to make sure that the selection does not worsen the valence and arousal classification. The aggregated preferences are illustrated in Fig. 3 as a heat-map.

The horizontal axis of the map represents the electrode index (as defined in the original DEAP dataset) and the vertical axis of the map - the frequency subbands. The color intensity highlights the oftenness of a given combination indicated by the RCA algorithm, and varies from 0 to 4 (the occurrence range).

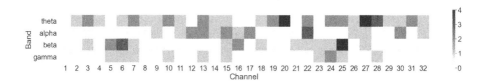

Fig. 3. The heat-map for channel-subband frequency selections

In the second step of the experiments, we summarized the occurrences of the channels and built the most frequent subsets of channels. Moreover, we identified four channels not selected by the RCA algorithm for any user and built a subset for all channels selected at least once. The results of occurrence frequencies are highlighted in Fig. 4 and presented in Table 1. Table 2 depicts the most frequent subsets of channels and their elements.

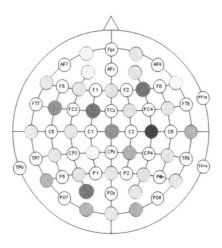

Fig. 4. The channel occurrence frequency illustration

We can observe in Fig. 4 that the central and right areas of the brain are favored. It can be justified by the psychological factor showing that negative emotions are generally believed to elicit more reactivity, and they are reflected by higher right-brain activity [31].

The authors of [30] carried out the research considering temporal channels with features from eye movements. In that way, the researchers achieved accuracy 72%. Our research shows that temporal channels are less important than other lobes in terms of emotional state recognition. Higher accuracy results could be gained by additional EOG signal analysis.

Medical research [32] confirms that neurotransmitter serotonin, also known as the happiness hormone, which is responsible for depression detection [33], is produced in the frontal lobe. Therefore, it might explain that positive emotions are correlated with channels in the front of the scalp. However, our research shows that these channels have a small impact on emotion recognition, which confirms that temporary feelings are not related to happiness hormone.

Table 1. The summarized channels' occurrences

Channel identifiers	No. of occurrences
C4	6
FC1, PO3, F4	4
FC5, Cz	3
P7, O1, T8, CP2, O2	2

Table 2. The most frequent subsets of channels

Subset ID	Channel identifiers
A	C4, FC1, PO3, F4
B	C4, FC1, PO3, F4, FC5, Cz
C	C4, FC1, PO3, F4, FC5, Cz, P7, O1, T8, CP2, O2
D	28 channels (FP1, AF3, CP1 and F8 were excluded from the initial set)

In the final step, we applied the selected subsets of electrodes on six randomly picked users to verify if our approach, aiming at the reduction of the number of electrodes and - as a consequence - simplification of data gathering, does not worsen the valence and the arousal classification. The evaluation was performed according to the methodology described in Sect. 2.6. The values of RMSE and MAE are presented in Fig. 5. Each chart refers to one of the randomly picked users, whereas each group of bars presents emotion-error combination types, and each series represents a subset of electrodes. The first series illustrates errors for all 32 channels.

One can notice that the error values slightly differ between subjects. It can be justified by the fact that experiencing emotions by people has a different impact on the brain. Moreover, we may assume that emotion evaluation by the users can be unstable. A similar feeling can be overestimated or underestimated by different subjects.

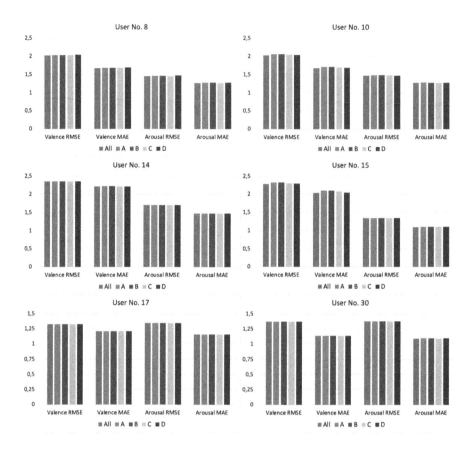

Fig. 5. The values of RMSE and MAE for valence and arousal recognition and random subjects

One can notice that our proposed reduction of electrodes has no negative influence on emotion recognition correctness. Even the most radical reduction, resulting in limiting the number of electrodes to four most often selected, enabled recognition of a similar level. The differences between errors for recognition based on all electrodes and the selected subsets were not statistically significant (p-value $\geq$ 0.68). Therefore, where appropriate, reducing the number of electrodes may be beneficial in terms of collecting less data, simplifying the EEG analysis, and improving interaction problems without recognition loss.

4 Conclusions

The research presented in the paper addresses the problem of building reduced and, therefore, more optimal EEG channel set for human emotion recognition. Optimal EEG channel selection and location is a challenging task due to the interconnections between individual channels.

The proposed method incorporating the RCA algorithm finds combinations of electrodes and their frequency bands, which are least correlated with one another. It, therefore, enables noise reduction and classification advancement in the intra-subject approach. The generalization inter-subject step makes our conclusions more universal.

Our method is simple in use. At the same time, it may significantly reduce the number of data needed for the analysis when appropriate. The experiments revealed that we can reduce number of electrodes from 32 channels to 4 channels. It means that keeping only 12.5% of the initial number of electrodes does not negatively influence the process of emotion recognition.

Nonetheless, there is still a need for additional investigations. Therefore, further research is planned to investigate multiple datasets. Moreover, deeper insight into denoising methods is planned. Future research may also explore neural networks and deep learning approaches to find unknown attribute correlations.

References

1. Nesse, R.M., Ellsworth, P.C.: Evolution, emotions, and emotional disorders. Am. Psychol. **64**(2), 129–139 (2009)
2. Petrantonakis, P., Hadjileontiadis, L.: Emotion recognition from EEG using higher order crossings. IEEE Trans. Inf Technol. Biomed. **14**(2), 186–197 (2010)
3. Masruroh, A.H., Imah, E.M., Rahmawati, E.: Classification of emotional state based on EEG signal using AMGLVQ. Procedia Comput. Sci. **157**, 552–559 (2019). https://doi.org/10.1016/j.procs.2019.09.013
4. Topalidou, A., Ali, N.: Infrared emotions and behaviours: thermal imaging in psychology. Int. J. Prenat. Life Sci. **1**(01), 65–70 (2017). https://doi.org/10.24946/IJPLS
5. Koelstra, S., et al.: DEAP: a database for emotion analysis; using physiological signals. IEEE Trans. Affect. Comput. **3**(1), 18–31 (2012)
6. Shu, L., Xie, J., Yang, M., Li, Z., Liao, D., et al.: A review of emotion recognition using physiological signals. Sensors **18**(7), 2074 (2018)
7. Maswanganyi, C., Owolawi, Ch., Tu, P., Du, S.: overview of artifacts detection and elimination methods for BCI using EEG. In: 3rd IEEE International Conference on Image, Vision and Computing (2018)
8. Baig, M.Z., Aslam, N., Shum, H.P.H.: Filtering techniques for channel selection in motor imagery EEG applications: a survey. Artif. Intell. Rev. **53**(2), 1207–1232 (2019). https://doi.org/10.1007/s10462-019-09694-8
9. Nakisa, B., Rastgoo, M.N., Tjondronegoro, D., Chandran, V.: Evolutionary computation algorithms for feature selection of EEG-based emotion recognition using mobile sensors. Expert Syst. Appl. **93**, 143–155 (2018)
10. Lahiri, R., Rakshit, P., Konar, A.: Evolutionary perspective for optimal selection of EEG electrodes and features. Biomed. Signal Process. Control **36**, 113–137 (2017)
11. Fidalgo, T.M., Morales-Quezada, L., Muzy, G.S., Chiavetta, N.M., Mendonca, M.E., Santana, M.V., et al.: Biological markers in non-invasive brain stimulation trials in major depressive disorder: a systematic review. Nat. Inst. Health J. ECT **30**(1), 47 (2014)

12. Li, X., Song, D., Zhang, P., Zhang, Y., Hou, Y., Hu, B.: Exploring EEG features in cross-subject emotion recognition. Front. Neurosci. **12**(162), (2018)
13. Wosiak, A., Zakrzewska, D.: Integrating correlation-based feature selection and clustering for improved cardiovascular disease diagnosis. Complexity **250706**, (2018). https://doi.org/10.1155/2018/2520706
14. Wosiak, A., Dura, A.: Hybrid method of automated EEG signals selection using reversed correlation algorithm for improved classification of emotions. Sensors **20**, 7083 (2020)
15. Ekman, P., et al.: Universals and cultural differences in the judgments of facial expressions of emotion. J. Pers. Soc. Psychol. **53**(4), 712–717 (1987)
16. Parrott, W.G.: Emotions in Social Psychology: Essential Readings. Psychology Press, Amsterdam (2001)
17. Plutchik, R.: The nature of Emotions: Human emotions have deep evolutionary roots, a fact that may explain their complexity and provide tools for clinical practice. Am. Sci. **89**, 344–350 (2001)
18. Russell, J.A.: A circumplex model of affect. J. Pers. Soc. Psychol. **39**(6), 1161–1178 (1980)
19. Silton, R.L., Kahrilas, I.J., Skymba, H.V., Smith, J., Bryant, F.B., Heller, W.: Regulating positive emotions: implications for promoting well-being in individuals with depression. Emotion **20**(1), 93–97 (2020). https://doi.org/10.1037/emo0000675
20. Jaušovec, N., Jaušovec, K.: EEG activity during the performance of complex mental problems. Int. J. Psychophysiol. **36**(1), 73–88 (2000)
21. Tong, L., Zhao, J., Wenli, F.: Emotion recognition and channel selection based on EEG Signal. In: Proceedings of the 11th International Conference on Intelligent Computation Technology and Automation, Changsha, China, pp. 101–105 (2018)
22. Curran, E.A., Stokes, M.J.: Learning to control brain activity: a review of the production and control of EEG components for driving brain-computer interface (BCI) systems. Brain Cogn. **51**(3), 326–336 (2003). https://doi.org/10.1016/S0278-2626(03)00036-8
23. Alexander, R., Aragón, O.R., Bookwala, J., Cherbuin, N., Gatt, J.M., Kahrilas, I.J., et al.: The neuroscience of positive emotions and affect: implications for cultivating happiness and wellbeing. Neurosci. Biobehav. Rev. **121**, 220–249 (2021)
24. Cromheeke, S., Mueller, S.C.: Probing emotional influences on cognitive control: an ALE meta-analysis of cognition emotion interactions. Brain Struct. Funct. **219**(3), 995–1008 (2013). https://doi.org/10.1007/s00429-013-0549-z
25. Kelley, N.J., Gallucci, A., Riva, P., Romero Lauro, L.J., Schmeichel, B.J.: Stimulating self-regulation: a review of non-invasive brain stimulation studies of goal-directed behavior. Front. Behav. Neurosci, 12, 337 (2019)
26. Anchieta da Silva, P., Dantas Alves Silva Ciaccia F.R.: Brain stimulation system and method to provide a sense of wellbeing. U.S. Patent Application No. 16/332,173 (2019)
27. Yuvaraj, R., Murugappan, M., Ibrahim, N.M., Omar, M.I., Sundaraj, K., Mohamad, K., et al.: Emotion classification in Parkinson's disease by higher-order spectra and power spectrum features using EEG signals: a comparative study. J. Integr. Neurosci. **13**(01), 89–120 (2014)
28. Garcia-Moreno, F.M., Bermudez-Edo, M., Garrido, J.L., Rodriguez-Fortiz, M.J.: Reducing response time in motor imagery using a headband and deep learning. Sensors **20**, 6730 (2020). https://doi.org/10.3390/s20236730
29. Arevalillo-Herráez, M., Cobos, M., Roger, S., García-Pineda, M.: Combining inter-subject modeling with a subject-based data transformation to improve affect recognition from EEG signals. Sensors **19**(13), 2999 (2019)

30. Zheng, W.L., Liu, W., Lu, Y., Lu, B.L., Cichocki, A.: Emotion meter: a multimodal framework for recognizing human emotions. IEEE Trans. Cybern. (2018). https://doi.org/10.1109/TCYB.2018.2797176
31. Coan, J.A., Allen, J.J.: Frontal EEG asymmetry as a moderator and mediator of emotion. Biol. Psychol. **67**(1–2), 7–50 (2004)
32. Lu, H., Liu, Q.S.: Serotonin in the frontal cortex: a potential therapeutic target for neurological disorders. Biochem. Pharmacol. (2016). https://doi.org/10.4172/2167-0501.1000e184
33. Li, X., Hu, B., Sun, S., Cai, H.: EEG-based mild depressive detection using feature selection methods and classifiers. Elsevier (2016). https://doi.org/10.1016/j.cmpb.2016.08.010

Theory of Mind Helps to Predict Neurodegenerative Processes in Parkinson's Disease

Andrzej W. Przybyszewski[1,2(✉)]

[1] Polish-Japanese Academy of Information Technology, 02-008 Warsaw, Poland
przy@pjwstk.edu.pl, Andrzej.Przybyszewski@umassmed.edu
[2] Department of Neurology, University of Massachusetts Medical School, Worcester, MA 01655, USA

Abstract. Normally, it takes many years of theoretical and clinical training for a physician to be the movement disorder specialist. It takes additional multiple years of the clinical practice to handle various "non-typical" cases. The purpose of our study was to predict neurodegenerative disease development by abstract rules learned from experienced neurologists. Theory of mind (ToM) is human's ability to represent mental states such as emotions, intensions or knowledge of others. ToM is crucial not only in human social interactions but also is used by neurologists to find an optimal treatment for patients with neurodegenerative pathologies such as Parkinson's disease (PD). On the basis of doctors' expertise, we have used supervised learning to build AI system that consists of abstract granules representing ToM of several movement disorders neurologists (their knowledge and intuitions). We were looking for similarities between granules of patients in different disease stages to granules of more advanced PD patients. We have compared group of 23 PD with attributes measured three times every half of the year (G1V1, G1V2, G1V3) to other group of 24 more advanced PD (G2V1). By means of the supervised learning and rough set theory we have found rules describing symptoms of G2V1 and applied them to G1V1, G1V2, and G1V3. We have obtained the following accuracies for all/speed/emotion/cognition attributes: G1V1: 68/59/53/72%; G1V2: 72/70/79/79%; G1V3: 82/92/71/74%. These results support our hypothesis that divergent sets of granules were characteristic for different brain's parts that might degenerate in non-uniform ways with Parkinson's disease progression.

Keywords: Granular computing · Rough set · Rules · Cognition

1 Introduction

We are interested in the mechanisms related to the neural death with related compensation and reorganization mechanisms in different brain's neural circuits.

The majority of the reorganization mechanisms are related to human's learning and adaptation inspired by our rich environment, and they are the biological basis of our intelligence. In the consequence, the most patients are not able to notice significant

© Springer Nature Switzerland AG 2021
M. Paszynski et al. (Eds.): ICCS 2021, LNCS 12744, pp. 542–555, 2021.
https://doi.org/10.1007/978-3-030-77967-2_45

cognitive, emotional and behavioral changes related to their brain neurodegeneration processes for over 20 years before their first noticed symptoms.

Another side of this long period of individual compensatory learning processes is that "there is no two PDs with exactly same symptoms". It is not effective and possible to observe every single neuron (there is about $8.6 * 10^{10}$ neurons and just as many non-neuronal cells, which actively participate in the neurodegeneration, in the human brain) and its connections (about 10^4 for each neuron) during neurodegenerative process, so we will observe meta-learning by recording attributes related to changes in the different brain structures. These processes are related to many different neuronal changes that are principally compensated by two major kinds of learning. The first one is the supervised learning based on the "teacher's feedback" (beginning with our Mothers), and the second kind of compensation is related to the reinforced learning (RL) [1]. The RL is based on the selection of such activity (behaviour) that gives reward. As the reward is associated with the pleasure and release of the neurotransmitter - dopamine, the RL mechanisms might change in Parkinson's disease (PD). PD is primary caused by dopamine depletion related to the neurodegeneration of the substantia nigra. PD has characteristic dominating motor symptoms (bradykinesia) with emotional and cognitive dysfunctions. There are also related subtle adaptation (RL) problems e.g., responses to sudden changes in patient's environment, as one PD patient said: *"when my husband went to hospital for three days, I became crazy"*. The reliable (supportive) part of the environment has changed; therefore, patient has problems to adapt that caused emotional instabilities (another role of the dopamine).

In order to understand the complex interactions between different mechanisms related to the neurodegeneration (loss of neurons) and also to the compensatory learning, we have introduced the Theory of Mind (ToM). Generally, in the literature the ToM was used in the domain of cognitive and motor related (verbal fluency) social cognition [2]. There are also findings related to deficits of the cognitive components of ToM in early stages, and affective parts of ToM in the late stages of PD patients [3]. But in order to find patients' ToM abilities we need to follow neurologists' ToM to "get inside" of the patient's changes in the brain. For example, the social emotional thinking is based on the mirroring [4] of movements and emotions introduced by others' facial expressions (movements) [5] that might be formalized by rough set theory [6].

In summary, our purpose was twofold, not only to look into the ToM ability in different PD patients, but also to propose the machine's ToM that will mirror neurologists reasoning and make it more universal by introducing the *abstract rules*. We wanted to check the following hypothesis: if our abstract rules related to the disease progression of different patients are appropriate, then they should be more similar to rules describing disease symptoms of the more advanced patients. This postulate is evident for the most neurologists but notice that each patient has different mechanisms and rates of the disease progression. It follows by another more detailed question: are different structures, such as related predominantly to the movement, cognition, and emotion have similar rates of the disease progression or not?

The structure of our paper is the following: in the Methods section we have described four different tests that involve: all parameters (general test), movements related parameters (movements test), emotion related parameters (emotional test), and cognition related parameters (cognition test), in addition we have review our method based on rough set theory (RST). In the Result section we have performed statistical evaluation of all our parameters and in the following paragraphs we have evaluated four different tests mentioned above. For each set of tests, the discretization and parameters reduction were performed with help of the RSES software. The RST rules for the more advanced patient's group were found and applied to other groups. The prediction accuracy and coverage for each group and each test were found and compared between different groups and different tests. In the Discussion and Conclusion sections the meaning of our findings and the practical consequences were discussed, as well as our future plans.

2   Methods

2.1   Review Stage

All 47 PD patients were divided into two groups: the first group of 23 patients was tested three times every half of the year (visits were numbered as G1V1, G1V2, G1V3), and the second group (G2V1) of more advanced 24 patients was a reference model of disease progression to the first group. All patients were tested in two sessions: with- Ses $= 2$ or without-medication Ses $= 1$. The neurologists in Brodno Hospital, Department of Neurology, Faculty of Health Science, Medical University Warsaw, Poland performed all tests [7]. In the present work, in addition to standard neurological tests, we have measured the fast eye movements: reflexive saccades (RS) by means of saccadometer (Ober Consulting) using methodology as described in [8]. In short, every subject was sitting in a stable position without head movements and watching a computer screen before him/her. At the beginning he/she has to fixate in the center of the screen, and to keep on moving light spot. This spot was jumping randomly, ten degrees to the right or ten degrees to the left. Patient has to follow movements of the light spot during 20 trails. The following parameters were measured: the latency - *RSLat* as time difference between beginning of spot and eyes movements, the saccade duration - *RSDur;* the saccade amplitude - *RSAmp* and the saccade velocity - *RSVel.*

In addition to the general test (**General ToM**) where all 12 attributes were used, all PD patients have three distinct groups of tests related to functions of different systems in the brain:

1. **Movements ToM - Speed and accuracy of movements**: reflexive eye movements parameters, *Epworth* (quality of sleep) and *Trail A* (speed and precision of connecting circled numbers) results
2. **Emotional ToM - Emotional stage** of patients estimated by the *PDQ 39* (quality of life), *Beck* depression tests, and eye movements parameters.
3. **Cognitive ToM - Cognitive processes** tested by *FAS* test (test od language fluency) and *Epworth* (sleep quality test that is related to memory consolidation) *Trail B* (speed and precision of connecting circled numbers and letters), and eye movements measures.

We have analyzed all attributes together or alternatively in three separated mentioned above tests in order to predict developments of the diseases progression that is estimated by the standard PD test: the UPDRS (Unified Parkinson's Disease Rating Scale) that has parts related to behaviour and mood, activities of daily living, motor symptoms and estimation of patient's stage of the disease; or by the UPDRS III that is a part of the UPDRS limited to only motor symptoms. The UPDRS scale has 42 items and it is a 'golden standard' for estimation of the progression in Parkinson's disease.

2.2 Rough Set Theory

Our data mining analysis follows rough set theory (RST) discovered by Prof. Zdzislaw Pawlak [9]. He has considered the problem of the boundaries after the philosophical approach of Frege that "concepts must have sharp boundaries". Prof. Pawlak solution of the vague concept of boundaries is to approximate them by sharp sets of the upper and lower approximations (Fig. 1).

It was demonstrated previously that RST gave the best results in the PD symptoms classifications in comparison to other methodologies [10]. Our data are represented as a decision table where rows represented different measurements (from the same or different patients) and columns were related to different attributes. An information system [9] is as a pair $S = (U, A)$, where U, A are nonempty, finite sets: U is the universe of objects; and A is the set of attributes. The value $a(u)$ is a unique element of V (where V is a value set) for $a \in A$ and $u \in U$. The RST *indiscernibility relation* is defined as: $(x, y) \in IND(B)$ or $xI(B)y$ iff $a(x) = a(y)$ for every $a \in B$ where the value of $a(x) \in V$. It is an equivalence relation $[u]_B$ that we understand as a *B-elementary granule*.

Fig. 1. Rough set concept explanation. Interrupted curve represents properties of the complex object S. Squares represent elementary granules (atoms); squares in black are related to the lower approximation of S, grey and black squares represent the upper approximation of S, and white squares are placed outside of S.

A *lower approximation* of set $X \subseteq U$ in relation to an attribute B is defined as:

$$\underline{B}X = \{u \in U : [u]_B \subseteq X\}$$

The *upper approximation* of X is defined as:

$$\overline{B}X = \{u \in U : [u]_B \cap X \neq \phi\}$$

The difference of $\underline{B}X$ and $\overline{B}X$ is the boundary region of X that we denote as $BN_B(X)$. If $BN_B(X)$ is empty then set than X is *exact* with respect to B; otherwise, if $BN_B(X)$ is not empty and X is not *rough* with respect to B [9, 11]. A decision table for S is the triplet: $S = (U, C, D)$ where: C, D are condition and decision attributes. Each row of the information table gives a particular rule that connects condition and decision attributes for a single measurement of a particular patient. As there are many rows related to different patients and sessions, they gave many particular rules. Rough set approach allows generalizing these rules into universal hypotheses that may determine optimal treatment options for an individual PD patient. However, an important difference to other classification system is that RST by using rules (with explicit meanings) is easy understand, also gives better accuracies than most AI but does not cover all cases (coverage is smaller than 1). Fuzzy RST gives coverage = 1 but has lower accuracies for the same data e.g. [12].

We have used Rough Set Exploration System RSES 2.2 as a toolset for analyzing data with rough set methods [13].

3   Results

3.1   Statistics

For the first group of PD patients we have performed three tests, every half-year, whereas the second group of more advanced PD we have measured only one time. The mean age of the first group (G1) was 57.8 ± 13 (SD) years with disease duration 7.1 ± 3.5 years. It is very strong and significant influence of medication, but only UPDRS and eye movements parameters are measured in without/with medication (MedOff/On).

UPDRS MedOff/On was 48.3 ± 17.9 and 23.6 ± 10.3 for the first visit (V1); 57.3 ± 16.8 and 27.8 ± 10.8 for the second visit (V2), 62.2 ± 18.2 and 25 ± 11.6 for the third visit (V3). The second group (G2) of patients was more advanced with mean age 53.7 ± 9.3 years, and disease duration 10.25 ± 3.9 years; UPDRS MedOff/On was 62.1 ± 16.1 and 29.9 ± 13.3 measured one time only. In all cases influences of medications on UPDRS were stat. sig. ($p < 0.001$).

The eye movements parameters were the following for G1V1 MedOff/On: RSLat 257 ± 78 ms/220 ± 112 ms; RSDur: 50.3 ± 5.1 ms/46 ± 16 ms; RSAmp: 10.5 ± 2.4/8.6 ± 7.0; RSVel 409 ± 104/471 ± 354.

For G2V1 MedOff/On: RSLat 247 ± 69 ms/250 ± 60 ms; RSDur: 49.3 ± 5.7 ms/48 ± 5 ms; RSAmp: 9.6 ± 2.4/7.6 ± 3.9; RSVel 402 ± 104/453 ± 101.

The quality of sleep measured by the Epworth score was the following for G1V1/G2V1: 7.9 ± 4.9/9.1 ± 5.5; for Trail A: 55.5 ± 29.7/50.0 ± 13.0; for Trail B: 141 ± 99/108.7 ± 68.5; FAS: 43.9 ± 12.9 / 39.9 ± 14.5; the Beck depression inventory (Beck test): 14.2 ± 9.8/14.8 ± 10.1; the quality-of-life score (PDQ 39): 48.3 ± 29.3/56.5 ± 22.8.

As states above mean values between groups are different, but because large variabilities between patients not all parameters are stat. different.

There were several stat. sig. differences between G1 and G2 patients: UPDRS III (characteristic for Parkinson's movement disorders); MedOff G1V1: 29.4 ± 16.1; G2: 35.8 ± 9.9 ($p < 0.04$), AIMS (abnormal involuntary movements score) G1V1: 2.3 ± 4.0; G2: 9.1 ± 5.7; ($p < 0.0001$), and significantly different for all G1 visits.

The learning Slope (CVLT – California Verbal Learning Test) G1V1: 2.9 ± 1.36; G2: 2.0 ± 0.8 ($p < 0.016$), and significantly different for all G1 visits. There were other significantly different parameters between both groups like the means time of dyskinesia and mean OFF time [7], also many other cognitive parameters were recently statistically analyzed [14].

Data were placed in four information tables: G1V1, G1V2, G1V3, and G2V1.

3.2 General ToM

We have used rough set theory [9] in order to obtain rules connecting decision and condition attributes for the advanced group of PD patients: G2V1. We have placed all data in the information table (as described above) that had 48 rows: 24 patients measured in two sessions (see above) each. Columns of this table were related to different 12 attributes and rows to results of different patients testing.

There were 12 columns related to the condition attributes: patient number id: *#P*, *Ses:* session number, *dur:* disease duration, *PDQ39* – quality of life, *Epworth* – quality of sleep, *Beck* depression test, RS (reflexive saccade): *RSLat, RSDur, RSAmp, RSVel* (as explained above), *Trail A* and *B* (as described above). The last column was related to the decision attribute: *UPDRS* (Unified Parkinson's Disease Rating Scale) (as above). In the next step, by using RST algorithms (RSES 2.2) we *have discretized (found optimal bin width) and reduced number of attributes.* As the results of the reduction the following attributes: RSDur, RSAmp, RSVel, Trail A, B were discarded. UPDRS was optimally divided by RSES into 4 ranges: *"(−Inf, 18.5)", "(18.5, 43.0)", "(43.0, 54.0)", "(54.0, Inf)".* We have divided G2V1 data (48 objects) into 6 groups and predictions were performed by rules learned from 5 groups in order to predict UPDRS of 6th group then it was performed 6 times for different groups (6-fold). We have used LEM 2 [15] algorithm with its parameters: *coverage 0.8 and with a simple voting.* These tests with different algorithms and parameters were performed in this and other cases in order to find maximum prediction accuracy.

From above data we have obtained 71 rules from which after filtering for removing single matches we have got the following 7 rules:

$$(Ses = 1)\&(Beck = ''(12.5, Inf)'')\&(dur = ''(8.5, Inf)'') => (UPDRS = ''(54, Inf)''[8]) \tag{1}$$

$$(Ses = 1)\& (Beck = '' (12.5, Inf)'')\&(RSLat = '' (219, Inf)'') => (UPDRS = ''(54, Inf)''[6]) \tag{2}$$

$$(Ses = 1)\&(PDQ39 = '' (58.5, Inf)'')\&(dur = ''(8.5, Inf)'') => (UPDRS = ''(54, Inf)''[5]) \tag{3}$$

$$(Ses = 2)\&(Beck = '' (12.5, Inf)'')\&(PDQ39 = '' (58.5, Inf)'')\&RSLat = '' (-Inf, 219)'') => (UPDRS = '' (18.5, 43)'' [5]) \tag{4}$$

$$(dur = '' (8.5, Inf)'')\&(PDQ39 = '' (58.5, Inf)'')\&(RSLat = '' (Inf, 219)'') => (UPDRS = '' (18.5, 43)'' [3]) \tag{5}$$

$$(Ses = 2)\&(Beck = '' (12.5, Inf)'')\&(dur = '' (8.5, Inf)'')\&(RSLat = '' (219, Inf)'') => (UPDRS = '' (18.43)''[2]) \tag{6}$$

$$(Ses = 2)\&(Beck = ''(12.5, Inf)'')\&(dur = '' (-Inf, 8.5)'')\&RSLat = ''(-Inf, 219)'' => (UPDRS = '' (18.5, 43)'' [2]) \tag{7}$$

Equations (1–3) were for the *Ses = 1* (patient without medication) and they were fulfilled by 8 (1), 6 (2) and 5 (3) cases, whereas Eqs. (4, 6, 7) were for the *Ses = 2* (patients on medication) and they were fulfilled by 5 (4), and by 2 (6, 7) cases. Equation (5) was session independent. We read Eq. (1) as **if** the patient is without medication (*Ses = 1*) and has the *Beck* depression score larger than 12.5 and with the *disease duration* longer than 8.5 years **then** his/her UPDRS will be above 54.

On the basis of above rules, we have estimated similarities between the UPDRS values obtained during three visits (every half-year) of G1 patients (less advanced group of patients) to symptoms of more advanced group of patients (G2). If the effect of the disease progression is that granules from group G2V1 become more similar to granules from PD group, it would suggest that G2V1 is a good model *M* of the disease progression. On the basis of above rules, we have predicted that UPDRS values of G1V1 group can be predicted from above rules (1–7) with global accuracy 0.68, and global coverage 0.48. Whereas G1V2 UPDRS, on the basis of the same rules, can be predicted with the global accuracy 0.72 and coverage 0.39; G1V3 UPDRS with the global accuracy 0.82 and coverage 0.37.

UPDRS
In summary, application of G2V1 rules to less advanced PD patients have demonstrated that all used significant attributes predicted disease progression as accuracy of the UPDRS estimation was increasing, in agreement with doctors' expectations, from 0.68 (G1V1), to 0.72 (G1V2), and 0.82 (G1V3).

In the next step, we were looking for the more elementary granules that were associated with the disease progression related to different parts of the brain. We have analyzed three sets of attributes related to properties of movements, emotions and cognitive changes of patients. The first what neurologists specialized in Parkinson's disease (doctors ToM) are looking for is the slowness of patients' movements. On this basis they normally estimate disease stage.

3.3   Movements ToM

Deficits in movements such as speed or precision are primary PD symptoms; they are like light spots in the visual system.

We took the following six condition attributes as patient number id: *P#*, *Ses*: session number, *dur* – disease duration, *RSLat* - reflexive saccade latency, and *Trail A*: speed of circled numbers connection, *Epworth* score (quality of sleep). The decision attribute was the *UPDRS*. The *UPDRS* was optimally divided by RSES into 4 ranges: *"(−Inf, 33.5)"*, *"(33.5, 43.0)"*, *"(43.0, 63.0)"*, *"(63.0, Inf)"*. As above, we have divided G2V1 data (48 objects) into 4 groups and predictions were performed by rules learned from 3 groups in order to predict *UPDRS* of 4th group then it was performed 4 time for different groups (4-fold).

We have obtained the following three rules:

$$(Ses = 2) \& (TrailA ='' (-Inf, 42)'') => (UPDRS ='' (-Inf, 33.5)'' \ [5]) \tag{8}$$

$$(Ses = 1) \& (RSLat ='' (264.0, Inf)'') \& (Eworth ='' (Inf, 14.0)'') => UPDRS ='' (63.0, Inf)''[3]) \tag{9}$$

$$(Ses = 2) \& (dur ='' 5.695, Inf)'') \& ((RSLat ='' (-Inf, 264.0)'')) \& (Epworth ='' (14.0, Inf)'')$$
$$=> (UPDRS ='' (63.0, Inf)''[2]) \tag{10}$$

Equation (8) is relatively simple and describe UPDRS predictions as a function of the session number (*MedOn*) and *Trail A* tests only. Equations (9,10) are more complex as they depend on the *Ses* number, *RSLat* and *Epworth* (quality of the sleep) and also Eq. (10) depends on the *dur* – disease duration.

It is interesting that there are only 3 relatively simple equations, but disadvantage is that they are only estimation of two ranges of the UPDRS. All other ranges are patients specific so there are no universal rules for their estimation.

On the basis of above rules, we have estimated similarities between G1V1, G1V2, G1V3 and G2V1 groups. We have used LEM 2 [15] algorithm with a simple voting and with 4-fold that gave the highest accuracy.

We have applied above rules to speed-related attributes of G1V1 group and obtained TPR: True positive rates for decision classes were {(0.29, 0.0, 0.0, 0.8)}, ACC: Accuracy for decision classes were {(1.0, 0.0, 0.0, 0.89)}, the global accuracy was 0.59 and global coverage was 0.37. As you may notice only the 1st and the 4th rages of UPDRS were predicted with the high accuracy.

We have estimated similarities between G1V2 and G2V1 groups:

We have applied above rules to speed-related attributes of G1V2 group and obtained TPR: True positive rates for decision classes were {(0.43, 0.0, 0.0, 1.0)}, ACC: Accuracy for decision classes were {(1.0, 0.0, 0.0, 0.843)}, the global accuracy was 0.70 and global coverage was 0.435.

We have estimated similarities between G1V3 and G2V1 groups:

We have applied above rules to speed-related attributes of G1V3 group and obtained TPR: True positive rates for decision classes were {(1.0, 0.0, 0.0, 0.9)}, ACC: Accuracy for decision classes were {(1.0, 0.0, 0.0, 1.0)}, the global accuracy was 0.923 (great) and global coverage was only 0.3.

In summary, application of the G2V1 rules to less advanced PD patients' groups have demonstrated that such elementary attribute as speed of eyes and hands movements can predict disease progression at accuracy of the UPDRS estimation was increasing from 0.59 (PDV1), to 0.7 (PDV2), and 0.92 (PDV3). These results, with the high accuracy of 4 UPDRS ranges, confirm doctors' ToM intuitions.

3.4 Emotional ToM

As in PD is lack of the dopamine; there are related emotional self-problems that projects to the social interactions (one of the major social problem of PD leading to the isolation). Emotions are like higher visual areas integrating all parts together. Movements evoke the pleasure and emotions are also visible in movements.

We took the following five condition attributes: patient number id: *P#*, *Ses*: session number, *PDQ39* (quality of life test), *Beck* (depression test), and *RSLat*: saccade latency. As the decision attribute was the *UPDRS*. The *UPDRS* was optimally divided by RSES into 3 ranges: "$(-Inf; 43.0)$", "$(43.0; 63.0)$", "$(63.0; Inf)$". We have divided G2V1 data (48 objects) into 5 groups and predictions were performed by rules learned from 4 groups in order to predict UPDRS of 5th group then it was performed 5 time for different groups (5-fold). We have used LEM 2 algorithm [15] with coverage 0.8 and with a standard voting. TPR: True positive rates for decision classes were (0.6, 0.8, 0.0), ACC: Accuracy for decision classes were (0.6, 0.75, 0.0), the global accuracy was 0.7 and global coverage was 0.29 As you may notice only the first and the second rages of UPDRS were predicted with the high accuracy. We have obtained the following six rules:

$$(Ses = 2) \& (RSLat ='' (208.0, 244, 5)'') => (UPDRS ='' (-Inf, 43.0)'' [7]) \tag{11}$$

$$(Ses = 2)\&(RSLat =" (194.5, 20.0)")\&(Bec =" (9.5, Inf)")$$
$$=> (UPDRS =" (-Inf; 43.0)"[4]) \tag{12}$$

$$(Ses = 1)\&(RSLat =" (244.5, 342.0)")=> (UPDRS =" (63.0, Inf)"[4]) \tag{13}$$

$$(Ses = 2)\&(RSLat =" (342.0, Inf)")=> (UPDRS =" (-Inf, 43.0)"[3]) \tag{14}$$

$$(Ses = 1)\&(RSLat =" (194.5, 208.0)")\&(Beck =" (9.5, Inf)")$$
$$=> (UPDRS =" (43.0, 63.0)"[2]) \tag{15}$$

$$(Ses = 1)\&(RSLat =" (34.0, Inf)")\&(Beck =" (9.5; Inf)")$$
$$=> (UPDRS =" (63.0, Inf)"[2]) \tag{16}$$

Equations (11–16) describe precisely UPDRS changes as dependent on emotional progressions. Notice that the parameters of eye movements play here a significant role as they are in all equations (see in the Discussion section). In three Eqs. (12, 15, 16) there is the attribute related to the depression (Beck test score), but only with higher values that indicates the emotional problems.

On their basis we have estimated similarities between G1V1 and G2V1 groups:

We have applied above rules to emotion-related attributes of G1V1 group and obtained TPR: True positive rates for decision classes were (0.2, 0.9, 0,0), ACC: Accuracy for decision classes were (1.0, 0.8, 0.0), the global accuracy was 0.53 and global coverage was 0.41. As you may notice only the first and the second rages of UPDRS were predicted with the high accuracy.

We have estimated similarities between G1V2 and G2V1 groups:

We have applied above rules to emotion-related attributes of G1V2 group and obtained TPR: True positive rates for decision classes were (0.56, 0.9, 1.0), ACC: Accuracy for decision classes were (1.0, 0.9, 0.2), the global accuracy was 0.74 and global coverage was 0.41. As you may notice in this case all three rages of UPDRS were predicted and two of them very high accuracy.

We have estimated similarities between G1V3 and G2V1 groups:

We have applied above rules to emotion-related attributes of G1V3 group and obtained TPR: True positive rates for decision classes were (0.5, 0.9, 0.5), ACC: Accuracy for decision classes were (0.7, 1.0, 25), the global accuracy was 0.71 and global coverage was 0.30. As you may notice in this case all three rages of UPDRS were predicted and two of them very high accuracy.

It is the first surprising result that emotions were not progressing with the disease development. In the early phase progressions were more significant and later they have stabilized: accuracy for G1V1 was 0.53, for G1V2 was 0.79, but later for G1V3 went down to 0.71.

3.5 Cognitive ToM

Cognitive processes play the role of the integrator of different neurological systems. There are related to the consciousness of self: influencing movements like e.g., equilibrium makes us aware of subliminal emotions, as well as related emotions.

We took the following seven condition attributes: patient number id: *P#; Ses*: session number; results of *FAS* test (it is related to the speech fluency); *Epworth* score (quality of sleep test related to memory consolidation); *Trail B* results (speed and precision of connecting circled numbers and letters), and parameters of the eye movements: *RSLat*, *RSDur*. As the decision attribute was the *UPDRS*. After discretization and parameter reduction by RSES, the *UPDRS* was optimally divided by RSES into 4 ranges: $a =$ "(−Inf; 43,0)", $b =$ "(43,0; 47,5)", $c =$ "(47.5, 63.0)", $d =$ "(63,0; Inf)". Only four condition attributes are left: *Ses* number *(MedOff/On)*, *FAS* (speech fluency), parameters of saccadic eye movements: *delay (RSLat)* and *saccade duration (RSDur)*.

From G2V1 group, we have obtained 25 rules that gave seven rules after removing rules fullfield in single cases, below are all seven rules as a basis for prediction of the possible longitudinal cognitive problems in G1 group:

$$(Ses = 2)\&(RSLat =" (213.5, Inf)")\&(RSDur =" (-Inf, 48.5)")$$
$$=> (UPDRS\_T =" (-Inf, 43.0)"[7]) \tag{17}$$

$$(Ses = 2)\&(FAS =" (-Inf, 46.5)")\&(RSDur =" (48.5, Inf)")\&(RSLat =" (Inf, 213.5)")$$
$$= > (UPDRS =" (-Inf, 43.0)"[5]) \tag{18}$$

$$(Ses = 1)\&(RSLat =" (-Inf, 213.5)")\&(RSDur =" (48.5, Inf)")\&(FAS =" (Inf, 46.5)")$$
$$=> (UPDRS =" (47.5, 63.0)"[3]) \tag{19}$$

$$(Ses = 1)\&(FAS =" (-Inf, 46.5)")\&(RSDur =" (-Inf, 48.5)")\&(RSLat =" (Inf, 213.5)")$$
$$=> (UPDRS =" (63.0, Inf)"[2]) \tag{20}$$

$$(Ses = 2)\&(RSLat =" (-Inf, 213.5)")\&(RSDur =" (-Inf, 48.5)")$$
$$=>(UPDRS =" (-Inf, 43.0)") [2] \tag{21}$$

$$(FAS =" (46.5, Inf)")\&(RSLat =" (-Inf, 213.5)")\&(Pat = 76)$$
$$=>(UPDRS =" (Inf, 43.0)"[2]) \tag{22}$$

$$(Ses = 1)\&(RSDur =" (48.5, Inf")\&(RSLat =" (-Inf, 213.5)"))\&(FAS =" (46.5, Inf)")$$
$$=> (UPDRS =" (43.0, 47.5)"[2]) \tag{23}$$

Equations (17–20) describe the *UPDRS* changes as function of cognitive changes in 7, 5, 3, 2 cases. From statistics: the *UPDRS* range *(−Inf, 43.0)* was used in four rules, other ranges were used each one in one rule. Parameters of the eye movements (EM) are in all rules, so the EM plays an important role in the cognition. Only one rules Eq. (22) is not depend on medication, but it is the patient's dependent. Also, only two rules are not dependent on (speed fluency) attribute.

We have divided G2V1 data (48 objects) into 10 groups (10-fold) as described above. We have used the Exhaustive algorithm [11] with standard voting (RSES). For G2V1 population we have obtained global accuracy of 0.73 with coverage 0.67.

On basis of above rules (17–13) we have estimated similarities between G1V1, G1V2, G1V3 and G2V1 groups.

Table 1, 2 and 3 demonstrate changes of cognitive symptoms with the disease progression in comparison to the more advanced PD group (G2V1). The cognitive accuracy is not changing so dramatically like the movements or even less than the emotional symptoms. The values of this attribute are significant different between Alzheimer's and Parkinson's diseases that also means that neurodegeneration processes with many similarities are basically different, mainly related to the different structures. There are

Table 1. Confusion matrix for UPDRS of G2V1 patients base on rules (17–23), TPR: True positive rates for decision classes; ACC: Accuracy for decision classes; Coverage for decision classes: (0.33, 0.47, 0.125, 0.4); the global accuracy was 0.72 and global coverage was 0.39.

Actual	Predicted				
	"(63.0, Inf)"	"(−Inf, 43.0)"	"(47.5, 63.0)"	"(43.0, 47.5)"	ACC
"(63.0, Inf)"	0. 0	0.0	1.0	0.0	0.0
"(-Inf, 43.0)"	0.0	12.0	1.0	0.0	0.86
"(47.5, 63.0)"	2.0	0.0	0.0	1.0	0.0
"(43.0, 47.5)"	0.0	0.0	1.0	1.0	0.5
TPR	0.0	1.0	0.0	0.5	

Table 2. Confusion matrix for UPDRS of G2V2 patients base on rules (17–23), TPR: True positive rates for decision classes; ACC: Accuracy for decision classes: Coverage for decision classes: (0.14, 0.5, 0.36, 0.5); the global accuracy was 0.79 and global coverage was 0.41.

Actual	Predicted				
	"(63.0, Inf)"	"(−Inf, 43.0)"	"(47.5, 63.0)"	"(43.0, 47.5)"	ACC
"(63.0, Inf)"	0. 0	0.0	0.0	1.0	0.0
"(-Inf, 43.0)"	1.0	11.0	0.0	0.0	0.92
"(47.5, 63.0)"	1.0	0.0	0.0	1.0	0.5
"(43.0, 47.5)"	0.0	0.0	1.0	2.0	1.0
TPR	0.0	1.0	0.0	0.5	

PD patients with cognitive problems but their influence in these group of 47 patients is not dominant.

In our longitudinal study, comparison of the cognition with more advanced patients did not show large changes with time. The accuracy for G1V1 was 0.72, for G1V2 was 0.79, but later for G1V3 went down to 0.74.

What is interesting that the coverage with time was increasing, from 0.39, 0.41 to 0.50. It means that with the time patients from G1 group become more similar to cases in G2 group but not necessarily that their cognitions are significantly deteriorating.

4   Discussion

We have used the granular computing to estimate disease progression in our longitudinal study of patients with Parkinson's disease (PD). We've applied granular computing with RST (rough set theory [9]) that looks into "crisp" granules (in the contrast to Fuzzy RST [12]) and estimate objects/symptoms by upper and lower approximations that determine precision of the description as dependent from properties of granules [9]. As we are able

Table 3. Confusion matrix for UPDRS of G2V3 patients base on rules (17–23), TPR: True positive rates for decision classes; ACC: Accuracy for decision classes: Coverage for decision classes: (0.33, 0.47, 0.125, 0.4); the global accuracy was 0.74 and global coverage was 0.5.

Actual	Predicted "(63.0, Inf)"	"(−Inf, 43.0)"	"(47.5, 63.0)"	"(43.0, 47.5)"	ACC
"(63.0, Inf)"	1. 0	0.0	1.0	2.0	0.25
"(-Inf, 43.0)"	0.0	13.0	0.0	0.0	1.0
"(47.5, 63.0)"	0.0	1.0	2.0	2.0	0.4
"(43.0, 47.5)"	0.0	0.0	0.0	1.0	1.0
TPR	1.0	0.93	0.67	0.2	

to precisely classify a complex, unknown objects as we are tuning and comparing their particular attributes in many different levels (with help of rough set theory).

This approach has similarities to other works using intelligent classification methods in order to test influences of different systems (like dissimilar object's properties – [16]) e.g., elementary granules related to the speed, cognition or depression [17] in their variable influences on the PD progression (object recognition).

Our results generally support intuitions of the neurologists that even if every patient is different, the most of PD patients' attributes become, with time development, similar to symptoms of more advanced group (G2V1) of patients. These intuitions are probably based on patients' movement changes that were confirmed in our study. However, we found that disease progression is not directly related to the emotional changes (even if depression might be advanced before PD [17]) and cognitive changes are decaying very slowly.

There is a significant number of papers that studied ToM in Parkinson's disease, e.g., see review in [3]. Generally, they assumed that the abilities to understand, and recognize mental state and intends of others are deteriorated during the course of Parkinson's disease [18]. They suggested that the cognitive impact may influence affective ToM in PD, and it is related to the involvement of the visual spatial abilities (VSA) [18]. It is in agreement with our findings that the saccade latencies are important to estimate movements and cognition ToM and saccade latencies and durations are important in the cognitive ToM. As ToM is the basic skill for development of the social relationships, PD patients showed impairments in ToM connected to the working memory and executive functions that were related to the white volume matter and grey matter decreases. These changes are mainly related to the frontal cortex and inferior frontal gyrus [19] and are associated largely with the cognitive changes. Another cause of PD patient's poor performance on tests of ToM, might be explained by the deficits in the inhibitory mechanisms [20]. Inhibitory mechanisms are important in the executive functions such as Trial B that was a significant parameter in our Cognitive ToM.

In summary, we have demonstrated that there are many mechanisms related to the Social Brain that are affected in Parkinson's disease (PD) patients, and they can also estimate PD progression. The first practical meaning for neurologists is to pay attention not only to movements deficits, but also to emotional and cognitive changes. In our work, we have estimated changes in different deficits (brain structures) related to neurodegenerations in Parkinson's disease progression by our abstract rules, and we found that they are not changing uniform with the disease progression (the second practical meaning). The third practical consequence of our study is that the eye movements (EM) parameters is the very important attribute that helps to estimate not only the peripheral movements symptoms, but also emotional and cognitive related disorders and their progressions.

5 Conclusions

Our different Machine's ToM follows changes in the human brain, and they use abstract rules based on the visual brain mechanisms. We have used the principle of object recognition as a comparison of the actual sensory input with the Model of the object saved in the higher visual areas [16]. Our 'Model' is related to attributes of advanced PD patients and object 'recognition' is related to similarities between attributes of PD patients progressing in time and the Model. We have demonstrated that PD disease progression is generally not uniform in relationship to the movements, emotions and cognitions changes, even if an individual patient may be more or less affected by depression or cognitive problems. Our rules gave very good prediction accuracy, but not very good coverage. It is mostly related to small groups of patients. Therefore, in order to get more subjects, our future projects will be related on on-line testing that should significantly increase the number of subjects. As an automatic evaluation of different test results is relatively easy, but it is a problem with precise and automatic estimation of the EM. We are actually working on it by using the OpenCV approach with the real-time computer vision and their hardware implemented AI libraries.

References

1. Sutton, R.S., Barto, A.G.: Reinforcement Learning: An Introduction, 2nd edn. MIT Press, Cambridge (2018)
2. Nobis, L., Schindlbeck, K., Ehlen, F., Tiedt, H., Rewitzer, C., Duits, A.A., Klostermann, F.: Theory of mind performance in Parkinson's disease is associated with motor and cognitive functions, but not with symptom lateralization. J. Neural Transm. **124**, 1067–1072 (2017)
3. Poletti, M., Enrici, I., Bonuccelli, U., Adenzato, M.: Theory of mind in Parkinson's disease. Behav. Brain Res. **219**, 342–350 (2011)
4. Rizzolatti, G., Fabbri-Destro, M.: The mirror system and its role in social cognition. Curr. Opinion Neurobiol. **5**(1), 24–34 (2009)
5. Ekman, P.: Emotion in the Human Face. Malor Books, Los Angeles (2015). ISBN-10: 1933779829, 13:978-1933779829
6. Przybyszewski, A.W., Polkowski, L.T.: Theory of mind and empathy. Part I - model of social emotional thinking. Fundamenta Informaticae **150**, 221–230 (2017)
7. Szlufik, S., et al.: Evaluating reflexive saccades and UDPRS as markers of deep brain stimulation and best medical treatment improvements in Parkinson's disease patients: a prospective controlled study. Pol. J. Neurol. Neurosurg. **53**(5), 341–347 (2019)
8. Szymański, A., Szlufik, S., Koziorowski, D.M., Przybyszewski, A.W.: Building classifiers for Parkinson's disease using new eye tribe tracking method. In: Nguyen, N.T., Tojo, S., Nguyen, L.M., Trawiński, B. (eds.) ACIIDS 2017. LNCS (LNAI), vol. 10192, pp. 351–358. Springer, Cham (2017). https://doi.org/10.1007/978-3-319-54430-4_34

9. Pawlak, Z.: Rough Sets - Theoretical Aspects of Reasoning Abot Data. Kluwer Academic Publisher, Boston (1991)
10. Przybyszewski, A.W., et al.: Multimodal learning and intelligent prediction of symptom development in individual Parkinson's Patients. Sensors **16**(9), 1498 (2016). https://doi.org/10. 3390/s16091498
11. Bazan, J.G., Szczuka, M.: The rough set exploration system. In: Peters, J.F., Skowron, A. (eds.) Transactions on Rough Sets III. LNCS, vol. 3400, pp. 37–56. Springer, Heidelberg (2005). https://doi.org/10.1007/11427834_2
12. Przybyszewski, A.W.: Fuzzy RST and RST rules can predict effects of different therapies in Parkinson's disease patients. In: Ceci, M., Japkowicz, N., Liu, J., Papadopoulos, G.A., Raś, Z.W. (eds.) ISMIS 2018. LNCS (LNAI), vol. 11177, pp. 409–416. Springer, Cham (2018). https://doi.org/10.1007/978-3-030-01851-1_39
13. Bazan, J.G., Szczuka, M.: RSES and RSESlib - a collection of tools for rough set computations. In: Ziarko, W., Yao, Y. (eds.) RSCTC 2000. LNCS (LNAI), vol. 2005, pp. 106–113. Springer, Heidelberg (2001). https://doi.org/10.1007/3-540-45554-X_12
14. Szlufik, S., et al.: The potential neuromodulatory impact of subthalamic nucleus deep brain stimulation on Parkinson's disease progression. J. Clin. Neurosci. **73**, 150–154 (2020)
15. Grzymala-Busse, J.: A new version of the rule induction system LERS. Fund. Inform. **31**(1), 27–39 (1997)
16. Przybyszewski, A.W.: SI: SCA measures - fuzzy rough set features of cognitive computations in the visual system. J. Intell. Fuzzy Syst. **36**, 3155–3167 (2019). https://doi.org/10.3233/JIFS-18401
17. Przybyszewski, A., Nowacki, J., Drabik, A., Szlufik, S., Habela, P., Koziorowski, D.: Granular computing (GC) demonstrates interactions between depression and symptoms development in parkinson's disease patients. In: Nguyen, Ngoc Thanh, Gaol, Ford Lumban, Hong, Tzung-Pei., Trawiński, Bogdan (eds.) ACIIDS 2019. LNCS (LNAI), vol. 11432, pp. 591–601. Springer, Cham (2019). https://doi.org/10.1007/978-3-030-14802-7_51
18. Romosan, A.-M., Romosan, R.-S., Bredicean, A.C., Simu, M.A.: Affective theory of mind in Parkinson's disease: the effect of cognitive performance. Neuropsychiatr Dis. Treat. **15**, 2521–2535 (2019)
19. Díez-Cirarda, M., Ojeda, N., Peña, J.,Cabrera-Zubizarreta, A.,Gómez-Beldarrain, M.Á., Gómez-Esteban, J.C., et al.: Neuroanatomical correlates of theory of mind deficit in Parkinson's disease: a multimodal imaging study. PLoS One **10** (2015). Article no. e0142234
20. Foley, J.A., Lancaster, C., Poznyak, E., et al.: Impairment in theory of mind in Parkinson's disease is explained by deficits in inhibition. Parkinson's Dis. **2019** (2019). Article no. 5480913, 8 pages

Regaining Cognitive Control: An Adaptive Computational Model Involving Neural Correlates of Stress, Control and Intervention

Nimat Ullah[(✉)] [iD] and Jan Treur[iD]

Social AI Group, Vrije Universiteit Amsterdam, Amsterdam, The Netherlands
nimatullah09@gmail.com, {nimat.ullah,j.treur}@vu.nl

Abstract. Apart from various other neural and hormonal changes caused by stress, frequent and long-term activation of the hypothalamus–pituitary–adrenal (HPA) axis in response to stress leads in an adaptive manner to the inadequacy of the stress response system. This leads to a cognitive dysfunction where the subject is no more able to downregulate his or her stress due to the atrophy in the hippocampus and hypertrophy in the amygdala. These atrophies can be dealt with by antidepressant treatment or psychological treatments like cognitive and behavioural therapies. In this paper, an adaptive neuroscience-based computational network model is introduced which demonstrates such a cognitive dysfunction due to a long-term stressor and regaining of the cognitive abilities through a cognitive-behavioural therapy: Mindfulness-Based Cognitive Therapy (MBCT). Simulation results are reported for the model which demonstrates the adaptivity as well as the dynamic interaction of the involved brain areas in the phenomenon.

Keywords: Stress induced neural anatomy · Negative metaplasticity · Mindfulness · Adaptive causal modeling · Cognition · Positive metaplasticity · Therapy

1 Introduction

Alteration in cognitive abilities can, potentially, be caused by the various ups and downs in humans' life and body. For instance, although termed to vary person to person, decline in cognitive abilities with increasing age and long-term stress have been confirmed by [1, 2]. Similarly, another discrepancy in the cognitive abilities is the lack of flexibility with age [3] which is considered very essential by many, specifically in changing situations. Taking the potentially negative consequences of long-term stress into account, various studies have reported similar findings regarding its effects in the long run [2, 4, 5], i.e., cognitive decline. At the cellular level, according to [6, 7], the cell loss and, therefore, changes in the synaptic plasticity take place because of the decrease in the brain-derived neurotrophic factor (BDNF) caused by the increase in the glucocorticoids.

To handle this severe problem in cognition, various studies, for instance [2], suggest antidepressant treatment but on the other hand, [8, 9] come up with Mindfulness-Based Cognitive Therapy (MBCT) [10] as an effective treatment for similar problems in general

© Springer Nature Switzerland AG 2021
M. Paszynski et al. (Eds.): ICCS 2021, LNCS 12744, pp. 556–569, 2021.
https://doi.org/10.1007/978-3-030-77967-2_46

and cognitive impairments caused by long-term stress. In MBCT, the subject is trained to focus on the present moment, gain awareness of himself and accept reality. Cognitive Behaviour Therapy (CBT) is another, almost similar therapy but according to [11] MBCT was found more effective when compared to CBT and that's also the reason why the study presented here considers MBCT. The reason may lie in the fact that the later combines techniques from the former with a mindfulness training program which provides added value.

Moreover, to combine these concepts into a single model, this study considers an adaptive network modeling approach [12] because of its efficacy and suitability for the adaptive and cyclic processes, as demonstrated in [13, 14]. In rest of the paper, Sect. 2 gives brief account of the literature on the subject, Sect. 3 presents the adaptive network model, which is explained by simulation results in detail in Sect. 4. Finally, the paper is concluded in Sect. 5.

2 Related Work

The alteration in cognitive abilities caused by long-term stress are attributed to the neuronal losses at the cellular level caused by stress. These changes are considered similar to those caused by depression [2]. For instance [6] links such cellular changes in the hippocampus to the increased level of glucocorticoid hormones, i.e., cortisol. Similarly, at the molecular level too, these cellular paucities were found in the hippocampus which are, most of the time, caused by the decrease expression of BDNF and resultant increased level of glucocorticoid/cortisol [6, 7, 15]. The down-regulating role of the increased level of glucocorticoids in the hippocampal expression has also been reported by [16]. BDNF is considered essential for neuronal survival, but [17] attributes reduction of BDNF to the potential mediating action of glucocorticoid on the hippocampus.

The effect of the boost of glucocorticoids is referred to as negative metaplasticity as it downregulates adaptivity of the hippocampal synaptic connectivity. In contrast, the boost in the expression of BDNF is referred to as positive metaplasticity as it strengthens connectivity in the hippocampus. These changes in the background, at the neural level, cause lack of control at the forefront or what we know as cognitive loss whereby the subject lacks the ability to regulate his or her emotions in an adaptive manner. Having said this, it is also possible that the same process is reversed by adequate means (antidepressant treatment for instance [2]), increasing the expression of BDNF. Synapses process and transmit neural information with some efficacy. Alteration in the synapsis is called synaptic plasticity or (first-order) synaptic adaptation. As mentioned above, synaptic plasticity itself can also change which is referred to as second-order adaptation or metaplasticity. According to [2], if metaplasticity improves the adaptive cognitive function, it's considered positive metaplasticity but on the contrary, if it brings impairment to the aforementioned adaptive cognitive function then it's called negative metaplasticity. This kind of cognitive impairment has been observed in both humans [18] and animals [19] as a result of long-term stress [20, 21].

MBCT, that is modeled here as a treatment for the above cognitive deficit, is considered a very effective approach [8, 9]. This therapy improves psychological health by increasing mindfulness. It combines Kabat-Zinn's [10] mindfulness-based stress reduction program with the techniques used in CBT. MBCT, therefore, promotes acceptance of feelings without judgement, focusing on the present moment and awareness of self [22]. Acceptance enables the person to disintegrate him or herself from the negative thoughts and consider emotions as a non-permanent event [23]. After this disengagement from negative thoughts, the mindfulness training helps the person in positive reappraisal [24]. Similarly, the focus on the present moment helps the person get insight of his or her own feelings and sensations for successful reappraisal of his thoughts. Generally, there are various brain areas involved in all these processes of MBCT but the most responsible parts that are considered essential for successful MBCT are the anterior cingulate cortex (ACC), insula, temporo-parietal junction, posterior cingulate cortex and prefrontal cortex (PFC) [15]. Activation of ACC helps enhance attention regulation by sustaining attention on a chosen object. Insula and temporo-parietal junction enhance body awareness by focusing on the internal experience like emotions, breathing and body sensation. PFC is responsible for the control of emotion regulation. Moreover, PFC together with posterior cingulate cortex, insula, temporo-parietal junction also helps the person change his perspective on himself [15].

Currently, there are various modeling techniques used in the field of artificial intelligence, specifically for modeling and simulating brain processes as summarized in [25, 26] but this study uses [12] because of its suitability for the model presented in this paper. This modeling approach comes under the umbrella of causal modeling which has a long history in Artificial Intelligence, e.g., [27, 28]. The dynamic and adaptive perspective on causal relations makes this technique unique among other similar approaches. Here, causal effects are exerted over time. Interestingly, the causal relations themselves are adaptive and can change over time too. Moreover, this type of adaptation can itself be adaptive too, leading to second-order adaptivity as occurs in metaplasticity; e.g., [2]. The network model introduced here is a *second-order adaptive temporal-causal network model* whereby adding dynamics and adaptation makes the model capable of application that would otherwise be out of scope of the causal modeling. This provides us with a useful opportunity to transform qualitative processes as described in empirical literature into adaptive causal network models. Simulations then can show that the underlying neural mechanisms that according to the assumptions made in this empirical literature explain certain observed emerging phenomena are indeed able to generate the phenomena computationally.

3   Multilevel Adaptive Cognitive Modeling

The multilevel adaptive causal network modeling approach [12, 29] has been used as a tool for the development and simulation of the adaptive causal model. The conceptual and numerical representation of the network characteristics used are summarized below in Table 1. Currently, this technique provides a dedicated software environment with a library of over 40 combination functions, publically available at https://www.researchg ate.net/publication/336681331, for combining the incoming causal impacts to a network state. The library also includes facilities to compose the existing functions into new functions by mathematical function composition. Moreover, self-defined functions can also be added to the library easily as per need of the model and phenomenon which makes this technique very feasible and flexible. The combination functions used in the current paper are shown in Table 2.

Table 1. Conceptual and numerical representations of the network characteristics used

Concept	Conceptual representation	Explanation
Connectivity characteristics		
States and connections	$X, Y, X \rightarrow Y$	Describes the nodes (representing state variables, shortly called *states*) and links (representing causal *connections* between states) of the network
Connection weight	$\omega_{X,Y}$	A *connection weight* $\omega_{X,Y}$ (usually in $[-1, 1]$) represents the strength of the causal impact of state X on state Y through connection $X \rightarrow Y$
Aggregation characteristics		
Aggregating multiple impacts on a state	$\mathbf{c}_Y(..)$	For each state Y (a reference to) a *combination function* $\mathbf{c}_Y(..)$ is chosen to combine the causal impacts of other states on state Y
Timing characteristics		
Timing of the effect of causal impact	η_Y	For each state Y a *speed factor* $\eta_Y \geq 0$ is used to represent how fast a state is changing upon causal impact
Concept	**Numerical representation**	**Explanation**
State values over time t	$Y(t)$	At each time point t each state Y in the model has a real number value, usually in $[0, 1]$
Single causal impact	$\mathbf{impact}_{X,Y}(t)$ $= \omega_{X,Y}X(t)$	At t state X with a connection to state Y has an impact on Y, using connection weight $\omega_{X,Y}$
Aggregating multiple causal impacts	$\mathbf{aggimpact}_Y(t)$ $= \mathbf{c}_Y(\mathbf{impact}_{X1,Y}(t),...,$ $\mathbf{impact}_{Xk,Y}(t))$ $= \mathbf{c}_Y(\omega_{X1,Y}X_1(t), ...,$ $\omega_{Xk,Y}X_k(t))$	The aggregated causal impact of multiple states X_i on Y at t, is determined using combination function $\mathbf{c}_Y(..)$
Timing of the causal effect	$Y(t + \Delta t) = Y(t) +$ $\eta_Y [\mathbf{aggimpact}_Y(t) - Y(t)] \Delta t$ $= Y(t) + \eta_Y [\mathbf{c}_Y(\omega_{X1,Y}X_1(t), ...,$ $\omega_{Xk,Y}X_k(t)) - Y(t)] \Delta t$	The causal impact on Y is exerted over time gradually, using speed factor η_Y; here the X_i are all states with outgoing connections to state Y

Using this technique, we propose an adaptive causal network model with connectivity as given in Fig. 1. A description of the various states of the model is provided in Table 3 where the background colors differentiate between the different levels of the model. The base level refers to the basic functioning of the model, involving the regulation of the negative emotions.

Table 2. Basic combination functions from the library used in the presented model

	Notation	Formula	Parameters
Advanced logistic sum	$\mathbf{alogistic}_{\sigma,\tau}(V_1,...,V_k)$	$[\frac{1}{1+e^{-\sigma(V_1+\cdots+V_k-\tau)}} - \frac{1}{1+e^{\sigma\tau}}](1+e^{-\sigma\tau})$	Steepness $\sigma > 0$ Excitability threshold τ
Hebbian learning	$\mathbf{hebb}_\mu(V_1, V_2, W)$	$V_1 V_2 (1-W) + W$	Persistence factor $\mu > 0$
Identity	$\mathbf{id}(V)$	$\mathbf{id}(V) = V$	

The first-order adaptation levels of the model explicitly represent weights $\omega_{X,Y}$ of some of the connections in the base model by *first-order self-model* states $\mathbf{W}_{X,Y}$ (also called *reification states*). For instance, X_{13} and X_{14} are first-order self-model states representing the adaptive connection weights $\omega_{adrenalcortex,hippocampus}$ and $\omega_{adrenalcortex,PFC}$, i.e., the connections represented by the two outgoing light-blue colored arrows from X_6, in the base model, respectively. The persistence μ and speed factors η of these connections' adaptation states X_{13} and X_{14} are represented by *second-order self-model* states X_{15} ($\mathbf{M}_{cortisol-feedback}$), X_{16} ($\mathbf{H}_{cortisol-feedback}$) and X_{17} ($\mathbf{M}_{cortisol}$), X_{18}($\mathbf{H}_{cortisol}$), respectively. The impact of these self-modeling states on their respective states in the lower order is represented by the red downward connections from the upper levels to the lower levels.

Table 3. States and their explanation

States		Role in the model	Level
X_1	stimulus	Anything causing stress in the real world	
X_2	thalamus	Processing of sensory information	
X_3	amygdala	Detects negative emotions and informs HPA to respond [15]	
X_4	hypothalamus	Part of autonomic stress response system which releases cortisol in the body to handle the situation [30].	
X_5	anterior-pituitary	Also called HPA axis	
X_6	adrenal-cortex		
X_7	hippocampus	Memory formation [15]	Base Level
X_8	PFC	Regulator of the emotions [15]	
X_9	ACC	Activated by MBCT where:	
X_{10}	insula	- ACC regulates attention,	
X_{11}	temporo-parietal-junction	- Insula together with temporo-parietal-junction gives body awareness,	
X_{12}	posterior-cingu-late-cortex	- PFC, posterior cingulate cortex, insula and temporo-parietal junction helps in changing one's perspective on the self [15].	
X_{13}	$\mathbf{W}_{cortisol-feedback}$	First-order self-model states for hebbian learning representing connection weights $\omega_{adrenalcortex,hippocampus}$ and $\omega_{adrenalcortex,PFC}$	First-Order Self-Model Level
X_{14}	$\mathbf{W}_{cortisol}$		
X_{15}	$\mathbf{M}_{cortisol-feedback}$	These states represent the adaptive control of plasticity, also called metaplasticity as described for instance in [2, 4, 5]. The hormones released by HPA which can cause negative as well as positive metaplasticity in different brain parts [30]	Second-Order Self-Model Level
X_{16}	$\mathbf{H}_{cortisol-feedback}$		
X_{17}	$\mathbf{M}_{cortisol}$		
X_{18}	$\mathbf{H}_{cortisol}$		

Generally, there are various adaptive connections in the brain, the plasticity and metaplasticity of which are subject to various factors, for instance reward is one of those factors to be mentioned [31]. This model is however motivated by the psychological computational model presented in [32] but the network in this model is modeled based on anatomical knowledge in the light of the findings from neurosciences as presented in Sect. 2. This model, therefore, only considers the aforementioned two adaptive connections out of the many adaptive connections in the brain. It demonstrates the phenomenon of negative and positive metaplasticity at a neural level where long-term stress causes cognitive loss through negative metaplasticity whereby the person loses control on regulation capabilities. As a treatment, MBCT has been used in the model which enables the person to regain his or her cognitive control through positive metaplasticity. The base model is a network of main parts of human brain and body involved in the stress experiences and the MBCT. The first-order adaptation represents the hormonal changes taking place as a result of stress and its treatment i.e. MBCT. The first-order adaptation uses a Hebbian learning principle [33]. The second-order adaptation represents the adaptation of the first-order adaptation to control the adaptation.

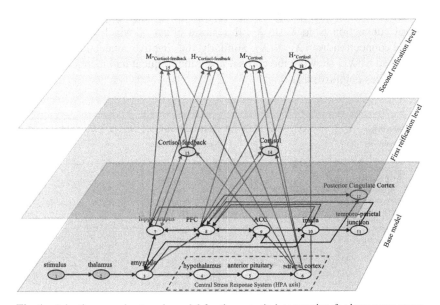

Fig. 1. Adaptive causal network model for therapeutic intervention for long-term stress

In the base model when the person faces some negative stressing stimulus, it's detected by the amygdala through the thalamus. Detection of stress by the amygdala automatically activates the Stress Response System which means activation of the Hypothalamic-pituitary-adrenal (HPA) axis as a result [30, 34]. The HPA releases cortisol to handle the situation. This works fine if this is not very frequent but repeated and prolonged activation of the HPA axis and hence prolonged release of cortisol blunts the stress response system; this is where the problem begins. In the model, the connections to PFC and hippocampus from the HPA model the hormonal effect of HPA on the two, which impairs the function of the PFC and hippocampus leading to the lack of cognitive control called negative metaplasticity, as mentioned.

The MBCT practice, on the other hand, activates the ACC, insula, temporo-parietal junction and posterior cingulate cortex which helps the person decrease activation of the HPA and hence less release of cortisol over time [15]. At the neural level these changes are considered as positive metaplasticity as the person regains control over his cognitive abilities. The **M-** and **H-**states represent the persistence and speed factor of the learning taking place at the respective base level connections.

In Box 1 and Box 2, the full specification of the network characteristics needed for reproduction of the model results are given. These specifications are not only essential for the reproduction of the results demonstrated in Fig. 2, 3, 4 and 5 but also qualitatively validates the model against the relevant literature in the sense that they show that personal characteristics exist by which indeed the assumed neural mechanisms lead to the overall patterns reported in the literature. Box 1 contains the connectivity role matrices called **mb** and **mcw**. Here **mb** gives all those incoming connection to a state which are either at the same level or from a lower level. The downward connections are indicated in role matrix **mcw** wherein they are used as indicator of their respective adaptive connection. For instance, in the model in Fig. 1, state X_{13} (i.e., a **W**-state) represents the adaptive base level connection from X_6 to X_7, the causal effect of which is modeled by the downward connection from X_{13} to X_7. Similarly, the adaptive connection from X_6 to X_8 is represented by X_{14} showing the cortisol level, the frequent and increased expression of which causes cognitive loss.

mb connectivity: base connectivity		1	2	3	4	5	6	7	mcw connectivity: connection weights		1	2	3	4	5	6	7
X_1	stimulus	X_1							X_1	stimulus	1						
X_2	thalamus	X_1							X_2	thalamus	1						
X_3	amygdala	X_2	X_7	X_8	X_{10}				X_3	amygdala	.7	.1	-.8	.1			
X_4	hypothalamus	X_3							X_4	hypothalamus	1						
X_5	anterior-pitui-tary	X_4							X_5	anterior-pitui-tary	1						
X_6	adrenal-cortex	X_5							X_6	adrenal-cortex	1						
X_7	hippocampus	X_3	X_6	X_8	X_{10}				X_7	hippocampus	.15	X_{13}	.4	.22			
X_8	PFC	X_3	X_6	X_7	X_9	X_{10}	X_{11}	X_{12}	X_8	PFC	.15	X_{14}	.22	.2	.2	.2	.2
X_9	ACC	X_3	X_8	X_{12}					X_9	ACC	.74	.01	1				
X_{10}	insula	X_3	X_9						X_{10}	insula	.45	1					
X_{11}	temporo-parie-tal-junction	X_{10}							X_{11}	temporo-parie-tal-junction	1						
X_{12}	posterior-cingu-late-cortex	X_8	X_9						X_{12}	posterior-cin-gulate-cortex	.15	1					
X_{13}	$\mathbf{W}_{cortisol-feedba}$	X_6	X_7	X_{13}					X_{13}	$\mathbf{W}_{cortisol-feedb}$	1	1	1				
X_{14}	$\mathbf{W}_{cortisol}$	X_6	X_8	X_{14}					X_{14}	$\mathbf{W}_{cortisol}$	1	1	1				
X_{15}	$\mathbf{M}_{cortisol-feedba}$	X_6	X_7	X_{13}	X_{15}				X_{15}	$\mathbf{M}_{cortisol-feedb}$	-1	1	1	1			
X_{16}	$\mathbf{H}_{cortisol-feedba}$	X_6	X_7	X_{13}	X_{16}				X_{16}	$\mathbf{H}_{cortisol-feedb}$	-1	1	1	1			
X_{17}	$\mathbf{M}$	X_6	X_8	X_{14}	X_{17}				X_{17}	$\mathbf{M}$	-1	1	1	1			

Box 1. Role matrices for connectivity characteristics.

Similarly, role matrices **mcfw, mcfp** for the aggregation characteristics and **ms** for the timing characteristics are given in **Box 2**. Matrix **mcfw** contains selection of the combination functions used for aggregation of the incoming causal impact at a state X_i.

For instance, state X_8 uses **alogistic(..)** and state X_{14} uses **hebb(..)** combination function as given in Table 2. Moreover, the first-order adaptation state X_{18} uses the Hebbian learning combination function **hebb(..)** from the same table. Role matrix **mcfp** specifies the parameter values for each of the combination function as indicated in **mcfw**. Note here that the red cells with numbered state names X_i in it, indicate the downward connections from these states in all the matrices except **mb**. Role matrix **ms** carries all the speed factor values of the states. In role matrix **ms**, the rows with red cells represent the state with adaptive speed factors i.e. X_{13} and X_{14}.

mcfw — Combination function weights	aggregation:	1 alogistic	2 hebb	3 Id
X_1	stimulus			1
X_2	thalamus			1
X_3	amygdala	1		
X_4	hypothalamus			1
X_5	anterior-pituitary			1
X_6	adrenal-cortex			1
X_7	hippocampus	1		
X_8	PFC	1		
X_9	ACC	1		
X_{10}	insula	1		
X_{11}	temporo-parietal-junction	1		
X_{12}	posterior-cingulate-cortex	1		
X_{13}	$W_{cortisol-feedback}$		1	
X_{14}	$W_{cortisol}$		1	
X_{15}	$M_{cortisol-feedback}$	1		
X_{16}	$H_{cortisol-feedback}$	1		
X_{17}	$M_{cortisol}$	1		
X_{18}	$H_{cortisol}$	1		

mcfp — Combination function parameters	aggregation	1 Alogistic σ	1 Alogistic τ	2 Hebb μ	3 id
X_1	stimulus				1
X_2	thalamus				1
X_3	amygdala	8	.4		
X_4	hypothalamus				1
X_5	anterior-pituitary				1
X_6	adrenal-cortex				1
X_7	hippocampus	8	.52		
X_8	PFC	8	.56		
X_9	ACC	18	.69		
X_{10}	insula	18	.64		
X_{11}	temporo-parietal-junction	18	.6		
X_{12}	posterior-cingulate-cortex	18	.4		
X_{13}	$W_{cortisol-feedback}$			X_{15}	
X_{14}	$W_{cortisol}$			X_{17}	
X_{15}	$M_{cortisol-feedback}$	10	.91		
X_{16}	$H_{cortisol-feedback}$	10	1.05		
X_{17}	$M_{cortisol}$	10	.75		
X_{18}	$H_{cortisol}$	10	.75		

ms — timing:		1 Speed factors
X_1	stimulus	0
X_2	thalamus	1
X_3	amygdala	.2
X_4	hypothalamus	.3
X_5	anterior-pituitary	.3
X_6	adrenal-cortex	.3
X_7	hippocampus	.3
X_8	PFC	.2
X_9	ACC	.01
X_{10}	insula	.015
X_{11}	temporo-parietal-junction	.01
X_{12}	posterior-cingulate-cortex	.015
X_{13}	$W_{cortisol-feedback}$	X_{16}
X_{14}	$W_{cortisol}$	X_{18}
X_{15}	$M_{cortisol-feedback}$	0.01
X_{16}	$H_{cortisol-feedback}$	0.01
X_{17}	$M_{cortisol}$	0.01
X_{18}	$H_{cortisol}$	0.01

Box 2. Role matrices for aggregation and timing characteristics

4 Simulation Results

Simulation results for an example scenario are provided here with and without MBCT, which shows how a person can go into a complete loss of cognitive abilities (caused by long-term stress) contrary to recovery from the cognitive loss. The results can be obtained by providing the values given in Box 1 and Box 2 to the dedicated software as mentioned above with the initial values of the states as shown in Table 4.

Figure 2 demonstrates the effect of long-term stress at the neural level where frequent and long-term expression of the cortisol by HPA blunts the autonomic stress response system. It can be seen that initially when the amygdala gets activated by some kind of

Table 4. Initial values of the states

State	Stimulus	All other base states	Cortisol-feedback (W)	Cortisol (W)	$M_{Cortisol-feedback}$	$H_{Cortisol-feedback}$	$M_{Cortisol}$	$H_{Cortisol}$
Value	1	0	0.3	0.3	0.5	0.9	0.5	0.9

stressful event, the hippocampus and PFC also gets activated which helps in activating the associated memory and handling of the stress respectively. But as this goes longer, the person's hippocampus and PFC are no longer activated despite the fact that the amygdala and the HPA are still very high.

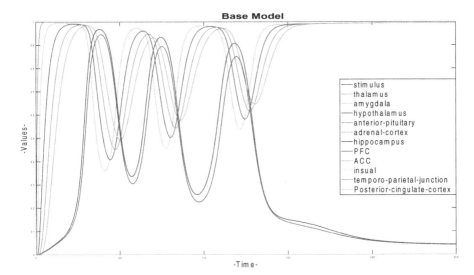

Fig. 2. Base model without therapy

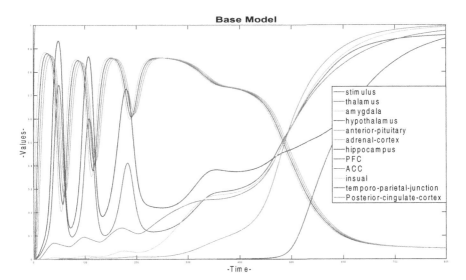

Fig. 3. Base model with therapy

Contrary to Fig. 2, in Fig. 3 it can be seen that although the person's cognitive abilities go down for some period, this doesn't remain like this for longer. It's because the person undergoes the proposed therapy which helps the person slowly regain his cognitive abilities. The therapy, on one hand makes the person not get stressed so easily and on the other hand it decreases the activation of HPA and hence expression of cortisol which has positive plastic and metaplastic effects on the Hippocampus and PFC. Therefore, both of these important parts of the brain start functioning as normal and regulate the negative stress the person is facing. At the neural level, this happens because in the MBCT, the person activates his or her other brain parts like ACC, insula, temporo-parietal-junction and posterior-cingulate-cortex which helps the person regulate his attention, get awareness of himself and change his perspective about himself, respectively.

In connection to Fig. 2 above, Fig. 4 shows the first- and second-order adaptation. Cortisol-feedback shows the Hebbian learning taking place at the connection in the base level between the HPA and hippocampus wherein impairment takes place at the hippocampus due to the increase level of cortisol. These states in the first-order adaptation level are the **W**-states. Similarly, the cortisol represents the second **W**-state which represent the learning taking place at the connection in base level between HPA and PFC. Moreover, the two **M**- and **H**-states represent the persistence and speed factor of the negative plasticity here, for metaplasticity. As this figure only shows the negative plasticity, therefore these connections only decrease, representing cognitive loss.

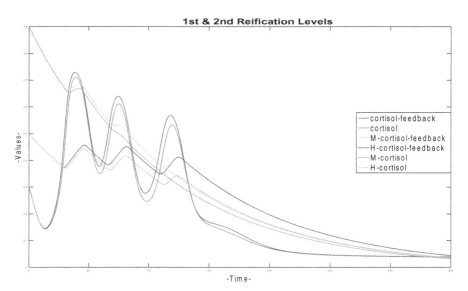

Fig. 4. First and second-order self-model states indicating negative plasticity and metaplasticity

Figure 5 in connection to Fig. 3 shows negative as well as positive metaplasticity. As already explained above, initially negative plasticity is taking place because of the excessive expression of the cortisol but when the person starts MBCT training, the situation starts getting reversed. Initially the person reverses the learning as can be seen that the cortisol-feedback and cortisol (the learning taking place at the HPA to hippocampus and PFC connections respectively) starts getting increasing. While the **M**- and **H**-states increasing slowly representing the persistence and speed factor of the

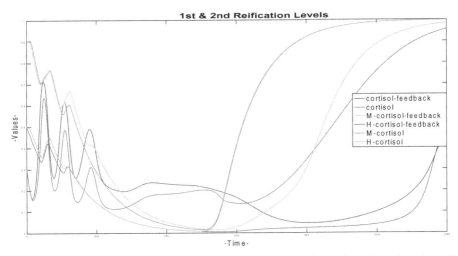

Fig. 5. First and second-order self-model states indicating negative and positive plasticity and metaplasticity

learning taking place called positive metaplasticity. These changes show it effect in the form of normal activation of the hippocampus and PFC in response to stress as discussed in Fig. 3 above.

5 Conclusion

The introduced adaptive network model is based on the neural correlates of stress response system and MBCT. It was designed using a multilevel adaptive network-oriented causal modeling approach in such a way that the anatomy of stress and MBCT induced brain parts were incorporated. The concepts of plasticity and metaplasticity have a long history in neuroscience. The model demonstrates the processes through simulations, showing how negative and positive metaplasticity occur with their effects on health. These results can be made as close to available empirical data as possible. This can also prove as a base for virtual training agent for therapies. The implementation of these techniques in the way done in this paper through the multilevel adaptive causal network model makes these processes easily understandable but also makes it an easy choice for implementation in the form of a complex artificially intelligent systems to work in a human-like manner.

During this study, it was learnt that, although quite a lot of work has been done in these areas of neuroscience, the anatomy of these processes, specifically in case of the aforementioned therapy are still not fully clear. Therefore, a temporal anatomy of the brain parts activated by such therapies would be a valuable contribution. This will not only make it easier to understand the flow of these complex processes going on in the brain but also make its implementation feasible in a more realistic way.

Apart from the added values of the model to neuroscience research, this paper also acknowledges the scope of causal modeling e.g., [27, 28] which has gotten even wider with the dynamicity brought by the multi-order adaptation [12, 29] as it has enabled this modeling approach to model phenomenon that would otherwise be not possible. In the future, the authors aim at developing a virtual agent system for training based on this model where the agent would collect data from body sensors of the patient and help him in undergoing therapies accordingly.

References

1. Verhaeghen, P.: Cognitive processes and ageing. In: Stuart-Hamilton, I. (ed.) An Introduction to Gerontology, pp. 159–193. Cambridge University Press, Cambridge (2011)
2. Garcia, R.: Stress, Metaplasticity, and Antidepressants. Curr. Mol. Med. 2(7), 629–638 (2002). https://doi.org/10.2174/1566524023362023
3. Charles, S.T.: Strength and vulnerability integration: a model of emotional well-being across adulthood. Psychol. Bull. 136(6), 1068–1091 (2010). https://doi.org/10.1037/a0021232
4. Mazure, C.M., Maciejewski, P.K., Jacobs, S.C., Bruce, M.L.: Stressful life events interacting with cognitive/personality styles to predict late-onset major depression. Am. J. Geriatr. Psychiatry 10(3), 297–304 (2002)
5. Tennant, C.: Work-related stress and depressive disorders. J. Psychosom. Res. 51(5), 697–704 (2001). https://doi.org/10.1016/s0022-3999(01)00255-0
6. Sapolsky, R.M.: Glucocorticoids, stress, and their adverse neurological effects: relevance to aging. Exp. Gerontol. 34(6), 721–732 (1999). https://doi.org/10.1016/s0531-5565(99)00047-9
7. Fuchs, E., Gould, E.: In vivo neurogenesis in the adult brain: regulation and functional implications. Eur. J. Neurosci. 12(7), 2211–2214 (2000). https://doi.org/10.1046/j.1460-9568.2000.00130.x
8. Garland, E., Gaylord, S., Park, J.: The role of mindfulness in positive reappraisal. Explore 5(1), 37–44 (2009). https://doi.org/10.1016/j.explore.2008.10.001
9. Garland, E.L., Gaylord, S.A., Fredrickson, B.L.: Positive reappraisal mediates the stress-reductive effects of mindfulness: an upward spiral process. Mindfulness 2(1), 59–67 (2011). https://doi.org/10.1007/s12671-011-0043-8
10. Kabat-Zinn, J.: Full catastrophe living; using the wisdom of your body and mind to face stress, pain and illness. Delta, New York (1990)
11. Troy, A.S., Shallcross, A.J., Davis, T.S., Mauss, I.B.: History of mindfulness-based cognitive therapy is associated with increased cognitive reappraisal ability. Mindfulness 4(3), 213–222 (2013). https://doi.org/10.1007/s12671-012-0114-5
12. Treur, J.: Network-Oriented Modeling for Adaptive Networks: Designing Higher-Order Adaptive Biological, Mental and Social Network Models. SSDC, vol. 251. Springer, Cham (2020). https://doi.org/10.1007/978-3-030-31445-3
13. Ullah, N., Treur, J.: Better late than never: a multilayer network model using metaplasticity for emotion regulation strategies. In: Cherifi, H., Gaito, S., Mendes, J.F., Moro, E., Rocha, L.M. (eds.) COMPLEX NETWORKS 2019. SCI, vol. 882, pp. 697–708. Springer, Cham (2020). https://doi.org/10.1007/978-3-030-36683-4_56
14. Ullah, N., Treur, J.: The choice between bad and worse: a cognitive agent model for desire regulation under stress. In: Baldoni, M., Dastani, M., Liao, B., Sakurai, Y., Zalila Wenkstern, R. (eds.) PRIMA 2019. LNCS (LNAI), vol. 11873, pp. 496–504. Springer, Cham (2019). https://doi.org/10.1007/978-3-030-33792-6_34
15. Hölzel, B.K., Lazar, S.W., Gard, T., Schuman-Olivier, Z., Vago, D.R., Ott, U.: How does mindfulness meditation work? proposing mechanisms of action from a conceptual and neural perspective. Perspect. Psychol. Sci. 6(6), 537–559 (2011). https://doi.org/10.1177/1745691611419671

16. Smith, M.A., Makino, S., Kvetnansky, R., Post, R.M.: Stress and glucocorticoids affect the expression of brain-derived neurotrophic factor and neurotrophin-3 mRNAs in the hippocampus. J. Neurosci. **15**(3), 1768–1777 (1995)
17. Mocchetti, I., Spiga, G., Hayes, V., Isackson, P., Colangelo, A.: Glucocorticoids differentially increase nerve growth factor and basic fibroblast growth factor expression in the rat brain. J. Neurosci. **16**(6), 2141–2148 (1996). https://doi.org/10.1523/JNEUROSCI.16-06-02141.1996
18. Lupien, S.J., et al.: Stress-induced declarative memory impairment in healthy elderly subjects: relationship to cortisol reactivity. J. Clin. Endocrinol. Metab. **82**(7), 2070–2075 (1997). https://doi.org/10.1210/jcem.82.7.4075
19. Mizoguchi, K., Yuzurihara, M., Ishige, A., Sasaki, H., Chui, D.H., Tabira, T.: Chronic stress induces impairment of spatial working memory because of prefrontal dopaminergic dysfunction. J. Neurosci. **20**(4), 1568–1574 (2000)
20. Kim, J.J., Yoon, K.S.: Stress: metaplastic effects in the hippocampus. Trends Neurosci. **21**(12), 505–509 (1998). https://doi.org/10.1016/s0166-2236(98)01322-8
21. Foster, T.C.: Involvement of hippocampal synaptic plasticity in age-related memory decline. Brain Res. Rev. **30**(3), 236–249 (1999). https://doi.org/10.1016/S0165-0173(99)00017-X
22. Coffey, K.A., Hartman, M., Fredrickson, B.L.: Deconstructing mindfulness and constructing mental health: understanding mindfulness and its mechanisms of action. Mindfulness **1**(4), 235–253 (2010). https://doi.org/10.1007/s12671-010-0033-2
23. Allen, N.B., Blashki, G., Gullone, E.: Mindfulness-based psychotherapies: a review of conceptual foundations, empirical evidence and practical considerations. Aust. N. Z. J. Psychiatry **40**(4), 285–294 (2006). https://doi.org/10.1111/j.1440-1614.2006.01794.x
24. Jha, A.P., Krompinger, J., Baime, M.J.: Mindfulness training modifies subsystems of attention. Cogn. Affect. Behav. Neurosci. **7**(2), 109–119 (2007). https://doi.org/10.3758/CABN.7.2.109
25. Moustafa, A.A.: Computational Models of Brain and Behavior. John Wiley & Sons Ltd., Chichester (2017)
26. Pastur-Romay, L., Cedrón, F., Pazos, A., Porto-Pazos, A.: Computational models of the brain. In: MOL2NET, International Conference on Multidisciplinary Sciences, pp. 1–10. MDPI, Basel (2015)
27. Kuipers, B., Kassirer, J.P.: How to discover a knowledge representation for casual reasoning by studying an expert physician. In: Proceedings IJCAI 1983 (1983)
28. Kuipers, B.: Commonsense reasoning about causality: deriving behavior from structure. Artif. Intell. **24**(1–3), 169–203 (1984). https://doi.org/10.1016/0004-3702(84)90039-0
29. Treur, J.: Network-Oriented Modeling: Addressing Complexity of Cognitive, Affective and Social Interactions. Springer International Publishing, Cham (2016). https://doi.org/10.1007/978-3-319-45213-5
30. Bezdek, K.G., Telzer, E.H.: Have no fear, the brain is here! how your brain responds to stress. Frontiers for Young Minds **5**(December), 1–8 (2017). https://doi.org/10.3389/frym.2017.00071
31. Ullah, N., Treur, J.: The older the better: a fourth-order adaptive network model for reward-driven choices of emotion regulation strategies over time. Appl. Netw. Sci. **5**(1), 1–15 (2020). https://doi.org/10.1007/s41109-020-00267-1
32. Ullah, N., Treur, J.: Know yourself: an adaptive causal network model for therapeutic intervention for regaining cognitive control. In: Maglogiannis, I., Iliadis, L., Pimenidis, E. (eds.) AIAI 2020. IAICT, vol. 584, pp. 334–346. Springer, Cham (2020). https://doi.org/10.1007/978-3-030-49186-4_28
33. Hebb, D.O.: The Organization of Behavior: A Neuropsychological Theory. Chapman & Hall limited, London (1949)
34. Pagliaccio, D., et al.: Stress-system genes and life stress predict cortisol levels and amygdala and hippocampal volumes in children. Neuropsychopharmacology **39**(5), 1245–1253 (2014). https://doi.org/10.1038/npp.2013.327

MAM: A Metaphor-Based Approach for Mental Illness Detection

Dongyu Zhang[1], Nan Shi[1], Ciyuan Peng[2], Abdul Aziz[1], Wenhong Zhao[3], and Feng Xia[2(✉)] (iD)

[1] School of Software, Dalian University of Technology, Dalian 116620, China
[2] School of Engineering, IT and Physical Sciences, Federation University Australia, Ballarat, VIC 3353, Australia
f.xia@ieee.org
[3] Ultraprecison Machining Center, Zhejiang University of Technology, Hangzhou 310014, China

Abstract. Among the most disabling disorders, mental illness is one that affects millions of people across the world. Although a great deal of research has been done to prevent mental disorders, detecting mental illness in potential patients remains a considerable challenge. This paper proposes a novel metaphor-based approach (MAM) to determine whether a social media user has a mental disorder or not by classifying social media texts. We observe that the social media texts posted by people with mental illness often contain many implicit emotions that metaphors can express. Therefore, we extract these texts' metaphor features as the primary indicator for the text classification task. Our approach firstly proposes a CNN-RNN (Convolution Neural Network - Recurrent Neural Network) framework to enable the representations of long texts. The metaphor features are then applied to the attention mechanism for achieving the metaphorical emotions-based mental illness detection. Subsequently, compared with other works, our approach achieves creative results in the detection of mental illnesses. The recall scores of MAM on depression, anorexia, and suicide detection are the highest, with 0.50, 0.70, and 0.65, respectively. Furthermore, MAM has the best F1 scores on depression and anorexia detection tasks, with 0.51 and 0.71.

Keywords: Mental illness · Metaphor · Attention model · Text classification

1 Introduction

According to the WHO survey, about 13% of people worldwide have a mental illness[1], and approximately 800,000 people suicide every year, with almost one death every $40\,s$[2]. Most people with mental illness are reluctant to share their

---

[1] https://www.who.int/health-topics/mental-health.
[2] http://www.who.int/data/gho/data/themes/mental-health.

© Springer Nature Switzerland AG 2021
M. Paszynski et al. (Eds.): ICCS 2021, LNCS 12744, pp. 570–583, 2021.
https://doi.org/10.1007/978-3-030-77967-2_47

feelings with others in real life and have a prevalent tendency to be alone. However, with the emergence of social media platforms, such as Facebook, WeChat and Twitter, etc., many mental disorder patients are willing to talk about their illnesses and share their feelings with strangers online. Thus, many reliable texts that reflect people's emotions and sentiments can be obtained from social media. Likewise, it becomes easy and possible to detect one's mental state by analyzing the text content posted by them.

In particular, many patients with mental illness tend to hide their feelings and emotions in the text through implicit means of expression, such as metaphor, rather than expressing them directly [7,13,15]. While sharing the feelings and sentiments, many people with mental illness often express their emotions through many different metaphorical languages, by which they intend to describe something other than what has been written on social media. For example, some people post the metaphorical expressions such as "My father is a monster" or "My life is a prison" frequently, these kind of expressions can help in detecting their hidden mental illnesses.

In general, there are many ongoing studies about mental illness detection via social media [17,19]. Typically, text classification algorithms have broadly been applied for prediction of people's mental health state based on their social status in the form of text posted on social media platform. To improve text classification models' performance, additional information extracted from the texts needs to be as the learning features [23]. However, current research on mental health issues only focuses on extracting some explicit features such as topics to indicate the text classification task. The implicit emotional feelings expressed by metaphors and inherently related to mental illnesses are ignored. This paper proposes a metaphor-based approach to fill this gap, which takes advantage of metaphorical expressions in the texts to detect mental illnesses. We extract the metaphor features which reflect some implicit emotions of potential mental illness patients as the primary indicator of the text classification task. However, it should be noted that there are also some patients who do not use metaphors to express their emotions. Therefore, our approach only focuses on detecting the mental illness of the patients who write in a metaphorical way to express their feelings. Our approach is divided into three main processes: long text representation based on CNN-RNN (Convolution Neural Network - Recurrent Neural Network) framework, metaphor feature extraction, and attention mechanism based text classification.

Initially, we take the advantage of CNN-RNN framework to represent the textual data. Some studies on deep learning-based text classification have been done in this regard [3,28]. Nevertheless, most of them only focus on short texts which contain hundreds of words. Social media texts contain thousands or millions of words, and it is hard for the ordinary models to achieve good performance on the social media text classification task. We apply the CNN-RNN framework to process long text data by extracting the information from words and sentences, respectively. Similar to TextCNN (the convolutional neural network for text) [6], We firstly apply CNN to extract word features from sentences and then obtain sentence representations. All the sentence representations in a text are processed by the bidirectional Long Short Term Memory (Bi-LSTM) model to obtain text representation. Conspicuously, the CNN-RNN framework's application makes our approach more efficient on long text extraction tasks with lower memory usage.

Because of the inherent connection between metaphorical expression and mental health, we extract the metaphor features as the text classification task's primary indicator. To extract metaphor features, we apply the token-level metaphor identification method RNN_MHCA (Recurrent Neural Network Multi-Head Contextual Attention) [5,16] to identify the metaphors from text. Significantly, for better performance of our approach, we extract sentence-metaphors and text-metaphor features respectively. Then, an attention mechanism is used to calculate attention weights based on the sentence-metaphor and text-metaphor features. Consequently, the attention mechanism integrates the text representation obtained from the CNN-RNN framework with the attention weights to get the text vectors for the text classification.

Finally, we test our model on the datasets eRisk [14] and CLPsych (The Computational Linguistics and Clinical Psychology) [1,30]. The experimental results demonstrated that our model performs well.

The main contributions of this study are as follows:

- A novel metaphor-based approach is proposed to detect users' mental illness status with the help of social media platforms. Specifically, our approach classifies social media texts based on mental illness patients' metaphorical expressions expressed over social media platform.
- Basically the initial aim of our approach is to represent the long texts based on the CNN-RNN framework. Remarkably, our model gains the competitive results with processing long textual data from social media.

The remainder of this paper is organized as follows. Section 2 introduces related work on text classification and metaphor recognition. Section 3 presents details about the proposed metaphor-based attention model. Experiments and results are discussed in Sect. 4. Finally, Sect. 5 concludes the paper.

2   Related Work

Research on textual data, specifically text classification, has gained much attention in recent years. In particular, the most popular text classification methods focus on semantic feature extraction, and inner logic in context [27].

Bi-LSTM has been used by [28] to obtain the contextual representation of the words in the sentence. Then, in their model, CNN is employed on the encoder's output to obtain text features. However, CNN's output vector is treated as the calculation benchmark of the attention mechanism instead of the representation for classification. The contextual representation of each position is connected with the last hidden layer state in the recurrent network. Moreover, the text representation is the weighted sum of contextual representation accumulated through the attention mechanism. Likewise, Yao et al. [24] combined the graph convolutional neural network with the text classification task. By converting documents and words into graph nodes, the graph neural network algorithm is successfully applied to text classification. They introduced the graph's adjacency matrix and degree matrix with stacked multiple graph convolutional network layers for multi-level extraction of information about the target node and surrounding neighbors.

Many text classification methods have been applied to mental illness detection tasks [17,29]. In particular, a large number of them focus on social media text-based mental illness detection [19]. Hierarchical recurrent neural networks with attention mechanism are used to extract the specific expressions at word-level and sentence-level (e.g., I have depression) [9,20]. Then, their model classifies the texts for mental illness detection based on these specific expressions. Considering the connection between words and categories of text [2,12] learned and integrated the correlation information for text classification. Compared with methods based on deep learning, their method has a stronger interpretability and supports in incremental learning instead of retraining when adding new samples.

Yao et al. [23] directly trained patterns to capture useful phrases with word embedding technique. They combined these word embeddings with entity embeddings of the Unified Medical Language System and Concept Unique Identifiers as external knowledge. These embeddings are independently condensed through CNN and pooling layer, then connected for text classification after multi-layer perceptron. According to [10] texts' sentiment and topic are the leading indicators for mental illness detection. They obtained the emotional features by counting emotional words and topics obtained from the LDA (Latent Dirichlet Allocation) topic model. The contextual vectors, encoded by the RNN module, are concatenated with representations of sentiment or topic. Then the contextual vectors are inputted into relation networks to calculate relation vectors that are used in the attention mechanism to get text representation.

Metaphor identification is a linguistic metaphor processing task with concrete words [4,25]. There are many methods for metaphor identification, such as Selectional Preference Violation (SPV) [4,22] and Metaphor Identification Procedure (MIP) [21]. SPV depends on the semantic contrast between the target word and its context. MIP focuses on the contrast between the literal meaning and the contextual meaning of the target word. Nevertheless, metaphor identification has been treated as a sequence tagging task recently [16,26]. Therefore, the deep learning methods, similar kind of approaches and the attention network have widely been applied for the tasks of metaphor identification [18].

Although metaphor identification research is extending in terms of development, few studies directly use metaphors as the text feature to achieve text classification, specifically, the text classification for mental illness detection. This paper considers the metaphor features in the texts as the primary indicators of text classification for mental illness detection.

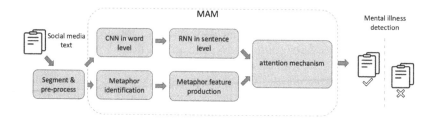

Fig. 1. Data processing in MAM.

3   Metaphor-Based Attention Model

This section highlights the details of our approach MAM that achieves creative results on mental illness detection. There are three main processes of our approach MAM namely, the CNN-RNN framework based long text representation, metaphor feature extraction, and attention mechanism based text classification. The overall data processing of MAM approach is illustrated in Fig. 1. As it is clearly shown in Fig. 1, we first segment and pre-processed the social media textual data. Then, CNN is applied to process the words in the texts and passes the obtained sentence representation to RNN to process the sentence-level data. Therefore, the text representation can be obtained after the processing of RNN. On the other hand, we also extracted metaphor features by identifying metaphors and then producing the features from them. Lastly, an attention mechanism calculates attention weights based on the metaphor features and integrates the text representation with the attention weights to get the text vectors for text classification. In our approach, we classify the texts by calculating the texts' association with the labels (0: health control group, 1: positive group).

The entire architecture of MAM is shown in the Fig. 2. Each sentence is composed of words $\{x_1, x_2, ..., x_m\}$. Firstly, the word features in each sentence are extracted in the convolutional layer, and then we obtain sentence representations $\{c_1, c_2, ..., c_n\}$ in the average pooling layer. Afterward, the sentence representations are encoded by bidirectional long short-term memory that captures the sequential contextual information in each sentence. Consequently, the text is represented as $\{h_1, h_2, ..., h_n\}$. Then, in the hidden layer, the attention weights $\{w_1, w_2, ..., w_n\}$ are calculated based on sentence-metaphor feature $\{m_1, m_2, ..., m_n\}$ and text-metaphor feature t. Finally, the text representation is integrated with the attention weights to obtain the text vectors for text classification.

3.1   CNN-RNN Framework for Long Text Representation

We first apply CNN to capture word features since CNN can notice the collocation between words in a sentence, i.e. abnormal word pair [11]. Taking a single $sample = (text, label)$ as an example of our model, the words $\{x_1, x_2, ..., x_m\}$ in each sentence are embedded by 300d Glove vector. Each sentence's input is a matrix of $m \times 300$, where m is the number of words in the sentence. The convolutional kernel k performs the convolution operation with a window $x_{i:i+h-1}$ on the input matrix and produces the representation of each sentence as formulas:

$$c_{ki} = f(w_k \cdot x_{i:i+h-1} + b_k)$$

$$c_k = ave\_pool([c_{k1}, c_{k2}, ..., c_{k(m-h+1)}])$$

$$c_j = [c_1, c_2, ..., c_k].$$

Here, $x_{i:i+h-1}$ denotes a window of size $h \times 300$ consisting of row i to row $i+h-1$ of the input matrix. w_k is the weight matrix, and b indicates the bias. f is the activation function of the convolutional kernel k which performs sequential scanning on the input matrix and stitches results to obtain $[c_{k1}, c_{k2}, ..., c_{k(m-h+1)}]$ as the feature maps. These feature maps are inputted to the average pooling layer that obtain the average value c_k for the overall position collocation information. Then, the sentence representation c_j is obtained by using various convolution kernels in different kernel size to convolve the sentence. Here, j denotes the j-th sentence in the text.

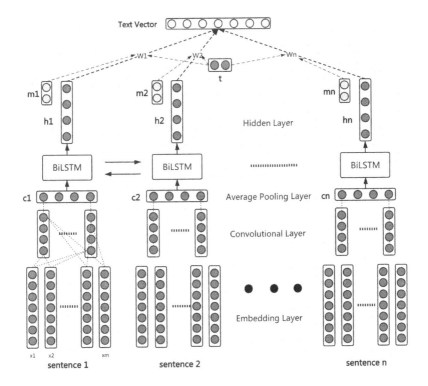

Fig. 2. The entire architecture of MAM.

Because the normal RNN does not perform well when dealing with the long sequences. Moreover, the long short term memory (LSTM) improves the performance of RNN by introducing the gate mechanism and the cell state [8]. Therefore, we apply the Bi-LSTM to encode the sentence vectors based on the sequential and positional information between sentences in the text. The formulas of LSTM are shown as follows:

$$forget_j = \sigma(W_{forget} \cdot [h_{j-1}, c_j] + b_{forget})$$

$$input_j = \sigma(W_{input} \cdot [h_{j-1}, c_j] + b_{input})$$

$$\widetilde{C_j} = tanh(W_c \cdot [h_{j-1}, c_j] + b_c)$$

$$C_j = forget_j * C_{j-1} + input_j * \widetilde{C_j}$$

$$output_j = \sigma(W_{output} \cdot [h_{j-1}, c_j] + b_{output})$$

$$h_j = tanh(C_j) * output_j,$$

where, W denotes the weight matrix, b indicates the bias. c_j is the j-th sentence representation obtained from the CNN model, and the hidden representation of the j-th sentence h_j participates in the calculation as a part of the input. The cell state C_j is used for recording the sequential state information in the j-th sentence, and the temporary cell state $\widetilde{C_j}$ is used to record the sequential state information temporarily. Gates $forget_j, input_j, output_j$ control what is forgotten, inputted and produced, respectively, and they are calculated by input vector c_j and h_{j-1}, here, h_{j-1} denotes the representation of the $(j-1)$th sentence. The new hidden representation h_j and cell state C_j will be employed in the next hidden layer.

Though the single direction structure of LSTM can not wholly extract all context in actual usage. Thus, we choose bidirectional LSTM, which can capture sequential information more effectively to process the data. In Bi-LSTM, the hidden sentence representation consists of the representations with two different directions. The formulas for calculating the sentence representation with different directions are as follows:

$$\overrightarrow{h_j} = hidden\_layer(c_j, h_{j-1})$$

$$\overleftarrow{h_j} = hidden\_layer(c_j, h_{j+1})$$

$$h_j = [\overrightarrow{h_j}, \overleftarrow{h_j}],$$

where, $\overrightarrow{h_j}$ indicates the hidden sentence representation with direction from left to right, and $\overleftarrow{h_j}$ is the hidden sentence representation with direction from right to left. Afterward, the sentence representation of j-th sentence is h_j which consist of $\overrightarrow{h_j}$ and $\overleftarrow{h_j}$.

Finally, the text can be represented by the list of sentence representations $\{h_1, h_2, ..., h_n\}$. Here, n is the number of sentences in the text.

3.2   Metaphor Feature Extraction

People with mental illnesses use metaphors separately with various people, we extract metaphorical information as a selection standard of the contextual sentence representation. In terms of the extraction of metaphors, we use the token-level metaphor identification method RNN_MHCA to identify the metaphors. Although it cannot guarantee that all metaphors in the text can be exactly detected, because the identified metaphors directly affect the performance of our model, the accuracy of metaphor identification is high if our model performs well. RNN_MHCA compares a representation of the target word with its context that is captured by a bidirectional attention network. Then, according to the difference between the context and the target word, we can judge whether the target word has metaphorical meanings. Then, the target words with metaphorical meanings are considered as the metaphorical words in the text, called metaphors.

According to the identified metaphors, our model extracts sentence-metaphor and text-metaphor features, respectively, as the text classification task's primary indicator. We consider that people with mental illness use metaphors in different frequency, so the number of metaphors in a text is one of the important metaphor features. Furthermore, patients with different mental illnesses use metaphors in different forms. Therefore, we also consider that the POS (Parts-Of-Speech) tag of metaphors is one of the metaphor features. Because of the different ways, people with a mental health condition using metaphor are based on many aspects. We extract metaphor features based on the following factors:

- The number and proportion of metaphors
- The POS tag of metaphors, including noun, verb, and adjective, etc.
- The length and number of sentences

The number and proportion of obtained metaphors are considered the most important factors for extracting metaphor features from a text. Also, we consider the POS tags of metaphors obtained by NLTK[3] as one of the factors of metaphor features. In particular, for the extraction of text-metaphor features, the length and the number of total sentences in a text are also considered as the leading factors. Therefore, the sentence-metaphor features are mainly extracted based on the number and proportion of metaphors in one sentence and the metaphors' POS tags. For the text-metaphor features, they are not only extracted based on the number and the proportion of metaphors and the POS tags, but also based on the length and number of the sentences in the text.

3.3   Attention Mechanism for Text Classification

Compared with the methods adding metaphorical information as an additional feature directly, we add metaphor features to the representation and use the metaphor features as the text information selection benchmark. It makes the

---

[3] http://www.nltk.org.

approach pay more attention to the sentence information, which has some characteristics in using the metaphors. In attention mechanism, the text-metaphor vector and the sentence-metaphor vector are transformed into 20 dimensional metaphor representations $t \in R^{1 \times d}$ and $m_i \in R^{1 \times d}$, respectively, through a fully connected layer. Here, m_i means the metaphor vector of the i-th sentence in the text. d is the dimension of the metaphor representation. Afterwards, the attention mechanism calculates the attention weights w_i of the i-th sentence based on the metaphor representations. The calculation is as follows:

$$w_i = Softmax(t \cdot m_i^T).$$

Then, the hidden representation of the i-th sentence h_i and the i-th sentence-metaphor vector m_i are concatenated by a a fully-connected layer L. w_i multiplies the sentence vector which is transformed from the concatenation of them. Consequently, the text vector r can be obtained:

$$r = \sum_{i=1}^{n} w_i \cdot (L \cdot [h_i, m_i] + b_L),$$

here, n is the number of sentences, b_L denotes the bias. Lastly, the text vector r is connected with output layer for classification:

$$p(la\hat{b}el|r, t) = \sigma(W_r \cdot [r, t] + b_r),$$

where, the *label* indicates the labels of the training data. Typically, the labels are denoted as 0 (the health control group) and 1 (the positive group). W_r is the weight matrix, and b_r denotes the bias.

4   Experiments

4.1   Datasets and Implementation

We have chosen two datasets for testing our model: eRisk and CLPsych. These two datasets are both constructed from social media texts and applied in shared tasks. We use the second edition of eRisk that contains two tasks - depression and anorexia and the suicide assessment dataset in the 2019 CLPsych shared task. For more details, please refer to [14,30].

The dataset of eRisk serves as a collection of writings (posts or comments) from a set of social media users. The newest edition includes three types of characteristics for mental illnesses: depression, anorexia and self-harm. The eRisk dataset considers time sequence information by grouping social media text in chronological order. However, in our experiment, we do not consider the time factor and treat each chunk of a user as an individual sample with the same label to extend the contribution in experimental datasets. The CLPsych is a shared risk detection task for mental illnesses based on social media texts. It retains the texts' time information while the eRisk dataset groups the texts by time to meet the early detection tasks requirement. Based on the considerations of CLPsych data's reliability, we employ *Expert* dataset that was annotated by experts for 245 users who posted on SuicideWatch and 245 control users who did not. The annotation scale includes a - no risk, b - low risk, c - moderate risk and d - severe risk. According to the explanation of low risk: annotators do not think this person has a high risk of suicide, we divide c and d into positive data, and the rest is negative data. The *Expert* dataset is not divided into the training set and the test set. We use ten-fold cross-validation in the experiment.

For the validity of data, we remove the sentences with less than three words and the texts with less than two sentences from the datasets. The statistic of the datasets is shown in Table 1 (de: depression, an: anorexia).

Table 1. The statistics of the experimental dataset.

	Train_de	Test_de	Train_an	Test_an	Suicide
The number of samples	8810	8116	1513	3165	479
Average tokens of sentence	16.78	16.69	16.73	17.04	25.51
Average sentence of sample	132.96	169.17	128.08	118.62	298.69

During the experimental process, we set the batch size of 4, kernel size in the CNN layer of 2 and 3, and the number of filters is 200. The output dimension of Bi-LSTM and the fully-connected layer is 200. The RNN_MHCA is pre-trained on VUA dataset [21] for the process of metaphor feature extraction. Due to the data imbalance, the positive samples in all experiments have two times the negative samples' weight in the calculation of the loss.

4.2   Results

We compare the experimental results of MAM with some previous known research works, including Text-CNN, Bi-LSTM, the Bi-LSTM+Attention model, and the hierarchical RNN+Attention model. In Text-CNN, the feature maps obtained by the convolution kernel will be pooled and became the features used for classification [11]. Bi-LSTM is suitable for sequence problems, such as text and timing research, and can capture the context information. Bi-LSTM usually uses the output of the last position as a classification vector. The Bi-LSTM+Attention model adds an attention layer based on Bi-LSTM, and it calculates weights through the attention mechanism to integrate the information from every position of the text. The hierarchical RNN+Attention model firstly collects sentence representations from word features and then collects them into text features for classification. The experimental results are shown in Table 2, and all the metrics are based on the positive label.

It is clearly shown in Table 2 that Text-CNN achieves the best results in precision on depression and anorexia detection tasks, with 0.60 and 0.87. However, it has general experimental performance totally, typically in the recall of

Table 2. Experimental results on eRisk dataset.

		Depression	Anorexia	Suicide
Text-CNN	P	**0.60**	**0.87**	0.84
	R	0.29	0.48	0.33
	F1	0.39	0.61	0.48
Bi-LSTM	P	0.49	0.65	0.68
	R	0.38	0.61	0.44
	F1	0.43	0.63	0.53
Bi-LSTM+Att	P	0.59	0.71	0.68
	R	0.40	0.66	0.58
	F1	0.47	0.68	0.63
H_RNN+Att	P	0.59	0.74	**0.93**
	R	0.30	0.47	0.54
	F1	0.40	0.57	**0.68**
MAM	P	0.52	0.72	0.70
	R	**0.50**	**0.70**	**0.65**
	F1	**0.51**	**0.71**	0.67

positive samples. The hierarchical RNN+Attention model achieves the highest precision of 0.93 and the highest F1 score of 0.68 in suicide detection tasks, but its performance is not so good on the other two datasets. In general, our approach MAM performs the best as compared to the other mentioned models. Despite the fact, our approach MAM generally performs very well when considering precision alone, its highest recall scores on all the mental illness detection tasks make the overall performance the most competitive. The recall scores of MAM on depression, anorexia, and suicide datasets are respectively 0.50, 0.70, and 0.65. Furthermore, MAM has the best F1 scores on depression and anorexia detection tasks, with 0.51 and 0.71. Moreover, MAM has less time and space consumption than the standard RNN-based methods in processing long text data and achieves competent results.

Compared with standard attention mechanisms, the attention mechanism in MAM does not let the model learn attention weights by itself but assigns a reference benchmark for the model. This mechanism does not capture the text's specific expressions but mines the implicit information in the text. It is based on the idea that patients with mental illnesses use metaphors differently from normal people. The results show that this attention mechanism is sufficient.

4.3   Ablation Study

In this part, we study the influence of several important parts of our approach on performance. Table 3 shows the F1 scores in ablation study.

Metaphor Identification Method. We compare BERT with RNN_MHCA by replacing the original metaphor features with the metaphor features identified by BERT. To enlighten the study, as shown in Table 3 (Bert_metaphor), the F1 scores are 0.49 in depression detection, 0.68 in anorexia detection and 0.66 in suicide detection. These scores prove that RNN_MHCA can identify metaphors more effectively than BERT.

RNN Module. We remove the RNN module by directly combining the CNN output with metaphor features. The experimental results are shown in Table 3 (Remove RNN). The F1 scores are 0.45 in depression, 0.64 in anorexia, and 0.64 in suicide. Therefore, it proves that the RNN module is important for the better performance of our model.

Metaphor Features. We first remove the metaphor features in our model. As shown in Table 3 (Remove_metaphor), the experimental results (F1: 0.45, 0.64 and 0.55) show that removing the metaphor features will lead to a bad performance. Hence, we replace the metaphor features with the normal attention features in our model. As shown in Table 3 (Replace_metaphor), the experimental results (F1: 0.49, 0.69 and 0.66) are also not satisfactory. Thereby, it can achieve a creative performance to apply metaphor features into the text classification task for mental illness detection.

Table 3. The F1 scores in ablation study.

	Depression	Anorexia	Suicide
MAM	**0.51**	**0.70**	**0.67**
Bert_metaphor	0.49	0.68	0.66
Remove RNN	0.45	0.64	0.64
Remove metaphor	0.45	0.64	0.55
Replace metaphor	0.49	0.69	0.66

5   Conclusion

In this paper, we have proposed a novel metaphor-based approach (called MAM) to detect the mental illnesses of people who post their feelings on social media. We classify the social media texts based on their metaphors to detect the writers' mental health state. The CNN-RNN structure is proposed to achieve the representation of long texts. Metaphor features in the texts are extracted as the primary indicator of the text classification task. We apply an attention mechanism to calculate the attention weights based on the metaphorical features, and after that, it also integrates the text representation with the attention weights to get the text vectors for text classification. Experimental results show that MAM performs pretty well in depression, anorexia, and suicide detection.

In future work, we will deeply study the specific effect of metaphor features using bi-LSTM and various similar algorithms in mental illness detection, and demonstrate the specific inherent connection between metaphors and mental illness. We will compare other extraction frameworks to represent text features that are most suitable for metaphor information.

Acknowledgment. This work is partially supported by National Natural Science Foundation of China under Grants No. 62076051.

References

1. Aguilera, J., Farías, D.I.H., Ortega-Mendoza, R.M., Montes-y Gómez, M.: Depression and anorexia detection in social media as a one-class classification problem. Appl. Intell., 1–16 (2021)
2. Burdisso, S.G., Errecalde, M., Montes-y Gómez, M.: A text classification framework for simple and effective early depression detection over social media streams. Expert Syst. Appl. **133**, 182–197 (2019)
3. Chen, T., Xu, R., He, Y., Wang, X.: Improving sentiment analysis via sentence type classification using BILSTM-CRF and CNN. Expert Syst. Appl. **72**, 221–230 (2017)
4. Chen, X., Hai, Z., Wang, S., Li, D., Wang, C., Luan, H.: Metaphor identification: a contextual inconsistency based neural sequence labeling approach. Neurocomputing **428**, 268–279 (2021)
5. Gong, H., Gupta, K., Jain, A., Bhat, S.: Illinimet: illinois system for metaphor detection with contextual and linguistic information. In: Proceedings of the Second Workshop on Figurative Language Processing, pp. 146–153 (2020)
6. Guo, B., Zhang, C., Liu, J., Ma, X.: Improving text classification with weighted word embeddings via a multi-channel CNN model. Neurocomputing **363**, 366–374 (2019)
7. Gutierrez, E.D., Cecchi, G.A., Corcoran, C., Corlett, P.: Using automated metaphor identification to aid in detection and prediction of first-episode schizophrenia. In: Proceedings of the 2017 Conference on Empirical Methods in Natural Language Processing. pp. 2923–2930 (2017)
8. Hua, Q., Qundong, S., Dingchao, J., Lei, G., Yanpeng, Z., Pengkang, L.: A character-level method for text classification. In: 2018 2nd IEEE Advanced Information Management, Communicates, Electronic and Automation Control Conference (IMCEC) pp. 402–406. IEEE (2018)
9. Ive, J., Gkotsis, G., Dutta, R., Stewart, R., Velupillai, S.: Hierarchical neural model with attention mechanisms for the classification of social media text related to mental health. In: Proceedings of the Fifth Workshop on Computational Linguistics and Clinical Psychology: From Keyboard to Clinic, pp. 69–77 (2018)
10. Ji, S., Li, X., Huang, Z., Cambria, E.: Suicidal ideation and mental disorder detection with attentive relation networks. arXiv preprint arXiv:2004.07601 (2020)
11. Kim, Y.: Convolutional Neural Networks for Sentence Classification. arXiv e-prints arXiv:1408.5882, August 2014
12. Kumar, A., Sharma, K., Sharma, A.: Hierarchical deep neural network for mental stress state detection using iot based biomarkers. Pattern Recognition Letters (2021)

13. Llewellyn-Beardsley, J., et al.: Characteristics of mental health recovery narratives: systematic review and narrative synthesis. PloS one **14**(3), (2019)
14. Losada, D.E., Crestani, F., Parapar, J.: Overview of eRisk: early risk prediction on the internet. In: Bellot, P., et al. (eds.) CLEF 2018. LNCS, vol. 11018, pp. 343–361. Springer, Cham (2018). https://doi.org/10.1007/978-3-319-98932-7_30
15. Magaña, D.: Cultural competence and metaphor in mental healthcare interactions: a linguistic perspective. Patient Educ. Couns. **102**(12), 2192–2198 (2019)
16. Mao, R., Lin, C., Guerin, F.: End-to-end sequential metaphor identification inspired by linguistic theories. In: Proceedings of the 57th Annual Meeting of the Association for Computational Linguistics, pp. 3888–3898 (2019)
17. Preotiuc-Pietro, D., Sap, M., Schwartz, H.A., Ungar, L.H.: Mental illness detection at the world well-being project for the clpsych 2015 shared task. In: CLPsych@ HLT-NAACL, pp. 40–45 (2015)
18. Rivera, A.T., Oliver, A., Climent, S., Coll-Florit, M.: Neural metaphor detection with a residual BILSTM-CRF model. In: Proceedings of the Second Workshop on Figurative Language Processing, pp. 197–203 (2020)
19. Saravia, E., Chang, C.H., De Lorenzo, R.J., Chen, Y.S.: Midas: mental illness detection and analysis via social media. In: 2016 IEEE/ACM International Conference on Advances in Social Networks Analysis and Mining (ASONAM), pp. 1418–1421. IEEE (2016)
20. Sekulić, I., Strube, M.: Adapting deep learning methods for mental health prediction on social media. arXiv preprint arXiv:2003.07634 (2020)
21. Steen, G.: A method for linguistic metaphor identification: From MIP to MIPVU, vol. 14. John Benjamins Publishing (2010)
22. Wilks, Y.: A preferential, pattern-seeking, semantics for natural language inference. In: Words and Intelligence I, pp. 83–102. Springer (2007)
23. Yao, L., Mao, C., Luo, Y.: Clinical text classification with rule-based features and knowledge-guided convolutional neural networks. BMC Med. Inform. Decis. Mak. **19**(3), 71 (2019)
24. Yao, L., Mao, C., Luo, Y.: Graph convolutional networks for text classification. In: Proceedings of the AAAI Conference on Artificial Intelligence, vol. 33, pp. 7370–7377 (2019) https://doi.org/10.1609/aaai.v33i01.33017370
25. Zhang, D., Lin, H., Liu, X., Zhang, H., Zhang, S.: Combining the attention network and semantic representation for Chinese verb metaphor identification. IEEE Access **7**, 137103–137110 (2019)
26. Zhang, D., Zhang, M., Peng, C., Jung, J.J., Xia, F.: Metaphor research in the 21st century: a bibliographic analysis. Comput. Sci. Inf. Syst. **18**(1), 303–321 (2021)
27. Zhang, P., Huang, X., Wang, Y.; Jiang, C., He, S., Wang, H.: Semantic similarity computing model based on multi model fine-grained nonlinear fusion. IEEE Access **9**, 8433–8443 (2021)
28. Zheng, J., Zheng, L.: A hybrid bidirectional recurrent convolutional neural network attention-based model for text classification. IEEE Access **7**, 106673–106685 (2019)
29. Zhong, B., Huang, Y., Liu, Q.: Mental health toll from the coronavirus: social media usage reveals wuhan residents' depression and secondary trauma in the covid-19 outbreak. Computers in human behavior **114**, 106524 (2021)
30. Zirikly, A., Resnik, P., Uzuner, Ö., Hollingshead, K.: CLPsych 2019 shared task: Predicting the degree of suicide risk in Reddit posts. In: Proceedings of the Sixth Workshop on Computational Linguistics and Clinical Psychology, pp. 24–33. Association for Computational Linguistics, Minneapolis, Minnesot, June 2019. https://doi.org/10.18653/v1/W19-3003, https://www.aclweb.org/anthology/W19-3003

Feature Engineering with Process Mining Technique for Patient State Predictions

Liubov Elkhovskaya[1]([⊠]) [iD] and Sergey Kovalchuk[1,2] [iD]

[1] ITMO University, 197101 St. Petersburg, Russia
{lelkhovskaya,kovalchuk}@itmo.ru
[2] Almazov National Medical Research Centre, 197341 St. Petersburg, Russia

Abstract. Process mining is an emerging study area adopting a data-driven approach and classical model-based process analysis. Process mining techniques are applicable in different domains and may represent standalone tools or integrated solutions within other fields. In this paper, we propose an approach based on a meta-states concept to extract additional features from discovered process models for predictive modelling. We show how a simple assumption about cyclic process behaviours can not only help to structure and interpret the process model but to be used in machine learning tasks. We demonstrate the proposed approach for hypertension control status prognosis within a remote monitoring program. The results are potential for medical diagnosis and model interpretation.

Keywords: Process mining · Process discovery · Machine learning · Feature engineering · Health status prediction

1 Introduction

A wealth of data inevitably affects all aspects of life. This new oil is exploited in various domains of modelling. Today, there is no surprise in applying machine learning (ML) methods, e.g., in social networks or financial analytics. Awareness of what would happen in the short-term as well as long-term future by predictive modelling has become a valuable opportunity for most companies. Still, a better insight into what is currently happening remains in demand. A promising approach for the analysis of intraorganizational workflows has been found in process mining (PM). There are three types of PM techniques: process discovery, conformance checking, and process enhancement [1]. Process discovery techniques allow to automatically construct a (business) process model from routinely recorded data, an event log. Conformance checking aims to evaluate model compliance with data. After the analysis of the real process executions, its enhancement can be proposed.

PM shows the potential for utilizing and being applied in clustering, decision trees, deep learning, recommender systems, rule-based systems, etc. [2]. For example, process models with relevant information can be used for predictive modelling or simulation using ML and statistics, and vice versa. In this paper, we present an approach based on a meta-states concept and show how information derived from a discovered process

© Springer Nature Switzerland AG 2021
M. Paszynski et al. (Eds.): ICCS 2021, LNCS 12744, pp. 584–592, 2021.
https://doi.org/10.1007/978-3-030-77967-2_48

model can be used to enrich the feature space. The idea of the concept originated from the healthcare domain, where a patient is involved in the processes. Still, it is broadly considered as an extension of a process discovery technique. The proposed approach is demonstrated for the task of patient health status predictions. We have two interrelated datasets on in-home blood pressure (BP) measurements within a monitoring program for patients suffering from arterial hypertension (HT). A dataset with events triggered by measurements is exploited to construct an event log for monitoring process discovery, and patient-related data is for predictive modelling. Despite the concrete study case, we believe the approach is adaptable to other domains as well as PM is not limited to existing applications.

2 Related Works

Many approaches may benefit from the synergies of PM and ML. ML methods and PM techniques can be used supplementary to each other or collaboratively to solve a problem.

The most common trend is to apply ML in PM research. In [3], authors use a clustering algorithm to divide patient behaviours into dynamic obesity and HT groups as interactive process indicators. The K-Means algorithm with the Levenshtein distance is used in [4] also to cluster patient clinical pathways, and an alignment algorithm from bioinformatics is applied then to obtain general sequence templates for each cluster. A supervised learning technique based on Conditional Random Fields is proposed in [5] for a sequence labelling task, where event log data is considered as the feature space for classification. The supervised abstraction of events with high-level annotations has contributed to the discovery of more accurate and comprehendible process models.

Jans and Hosseinpour [6] propose a transactional verification framework where active learning and PM come together to reveal and classify process deviations in a continuous manner. In this framework, a conformance checking algorithm is used to compare real transaction traces against a normative process model, and a human expert helps justify uncertain deviations and enriches data for final classification. ML and PM can be both applied to supporting decision-making, e.g., in a product design process [7]. Here, the authors use ProM[1], a well-known and popular open-source mining tool, to perform process discovery and then predict resources/decisions for each activity by a supervised learning algorithm.

The next activity prediction task for runtime processes originates a new branch in PM, called predictive process monitoring [8]. The employment of ML techniques [9] or deep learning networks [10] in PM can also be useful for decision support in outcome-oriented predictive systems [11].

The study [12] is an example of utilizing PM techniques in ML tasks. The authors derive meta-features from the manufacturing process model discovered through PM: a failure rates feature composed of aggregated defect-rates of individual variables, a lag feature based on the id column, and a duplicate row feature as an indicator of data anomaly. Information retrieved from the process structure has increased the predictive

---

[1] www.promtools.org.

power of the model in a failure-detection system. In our study, we first search for potential features in a discovered process model and then deploy an event log.

3 Process Mining as Feature Engineering

Our approach has two steps, at which first a process model is discovered, and then meta-features are derived from its structure. Below we describe the implementation details of each step.

3.1 Process Discovery

We start by observing an event log to discover a general process flow. The event log contains information about process executions, where each record is an event described with several attributes (case id, activity, resource, timestamp, etc.). We use the ideas of Fuzzy Miner [13] to develop a tool[2] for log analysis as a Python package. The reasons for the algorithm choice are two-fold: (i) the algorithm is suitable for unstructured and complex processes, which exist in healthcare, due to constructing a model at different levels of details; (ii) a directly-follows graph (DFG) as an algorithm output permits cycles, which are crucial in our concept of meta-features, despite the DFG limitations [14].

The main idea of frequency-based miners is to find the most probable events and precedence relationships over them by evaluating their significance and filtering: more frequently observed events and transitions are deemed more significant and included in the model. We modify model construction by performing the depth-first search to check whether each node of the DFG is a descendant of the initial state and a predecessor of the terminal state. This way, we overcome the possibility of discovering an unsound model (without the option to complete the process).

3.2 Meta-states Search

We propose a new method of model abstraction based on the DFG structure rather than data properties. In healthcare, a cyclic behaviour of the process may represent a routine complex of procedures or repeated medical events, i.e., a patient is at some treatment stage, or a meta-state. We assume a cycle in the model to be a meta-state if the estimated probability of the repeating behaviour in the log exceeds the specified threshold. The mechanism is the same as DFG elements filtering. In this study, nodes included in meta-states are not allowed to present distinctly in a model, and all relationships of single events are redirected to corresponding meta-states. This may result in a different significance of elements and, therefore, different process representation.

The discovered meta-states are used further to enrich data for a prediction task. We derive meta-features as the relative lengths of a patient in a meta-state. The idea is that such latent states may correlate with general patient health status. The concept of meta-states can also help interpret both the process and predictive model. In the following section, we demonstrate our framework on the case of a monitoring program for patients with HT.

---

[2] https://github.com/Siella/ProFIT.

4   Case Study

Within a collaborative project [15], we have two interrelated datasets on in-home BP measurements and triggered events within remote monitoring provided by PMT Online[3], a company specialized in the development of medical information and telemedicine systems. We first present a summary of the datasets, and then apply our approach to the study case.

4.1   Description of Datasets

During the monitoring program, the HT patients regularly measure their BP in-home, and each record made by a toolkit is transferred to a server to be processed. There are several clinical and non-clinical events for medical staff that measurements may trigger. The main events are "Red zone" (RZ) and "Yellow zone" (YZ) that notify about exceeding critical and target levels of BP, respectively. These events have to be processed by operators and doctors, who may take some actions according to a scenario, e.g., contacting a patient instantly or by appointment. Usually, RZ events occur for patients who have no appropriate treatment plan yet. When a health state is normalized due to medications, YZ appears rather than RZ, and a patient can be transferred to a therapy control program to maintain its BP levels or to complete the monitoring when target levels are achieved.

The first dataset is a set of records reached the server. It contains information on patient measurements (systolic BP (SBP), diastolic BP (DBP), heart rate (HR)), sex, age, med program duration, living area, record and server response timestamps, and toolkit-related details. The second dataset is process-related data containing events with 18 types of activities. We combine activities and corresponding resource labels in the log because the same activities for different roles have different operational meanings. Unfortunately, both datasets contain only 53 common patients; this has led to a small data sample (Table 1).

Table 1. Datasets summary

	Measurement data	Event Log
Time period	2018/09/01 - 2018/12/30	
Num. of records	6,129	2,844
Min trace length	40	3
Avg trace length	120	53
Max trace length	251	242

---

[3] https://pmtonline.ru/ (in Russian).

4.2   Monitoring Process and Meta-states Discovery

First, we apply the extended discovery algorithm to the monitoring log. A manually adjusted process model is illustrated in Fig. 1 (left), where the green node indicates the beginning of the process and shows the total number of cases (patients), and the red node is related to the end of the process. The internal nodes and edges of the graph show the absolute frequencies of events and transitions, respectively: the more absolute value is, the darker or thicker element is.

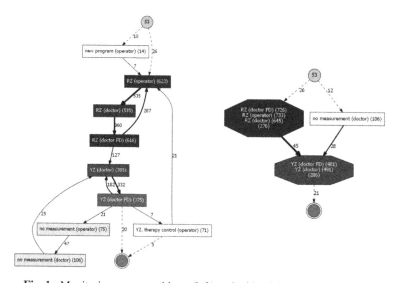

Fig. 1. Monitoring process without (left) and with (right) cycle aggregation

As seen, the process starts with patient registration in the program or with a triggered clinical event. A general process scheme reveals that most patients involved in monitoring are in bad health condition. Their regular BP measurements exceed the critical levels at the beginning of the treatment course. Additionally, one can see that events first occur for an operator, then for a physician, and at last for a physician of functional diagnostics (denoted as "doctor FD"). This behaviour complies with guidelines: an operator reacts to events that occur in the system, the notification is then transferred to a physician, who may contact a patient, and after that a functional diagnostics physician compiles a full report on the instance and actions taken.

The discovered meta-states (Fig. 1, right) correspond to RZ and YZ events, and it is possible to give some interpretation of the process. The cycles of such events are the main scenarios of the program realization, so they were identified as significant ones and therefore were folded into meta-states. In this case, the meta-states are implicitly related to a patient because the monitoring workers directly processed the events. However, patients initiate the process instances, and these meta-states can be seen as an impulse characteristic of the patient's health state. We use the relative time lengths of a patient being in some of the meta-states as meta-features in predictive modelling. We say *meta* because it is not explicit information, but it may correlate with the outcome and can be interpreted further.

4.3 Hypertension Control Status Prediction

Exploiting the measurements dataset, we address the task of binary classification, where a target is "in-control" or "out-of-control" HT status [16]. In-control state corresponds to remaining normal BP and HR levels during some period, and out-of-control state indicates instances of exceeding target or critical levels. According to the medical standards, HT is defined as a chronic condition of increased BP over 140/90 mm Hg (SBP/DBP). Since the measurements were performed at home, the threshold was lowered to 135/85 mm Hg. We also added an upper boundary for HR—80 bpm. Sequences of HT status assessments were combined into negative and positive episodes. It is required for a positive (in-control) episode to have at least two consecutive in-control status assessments. The dataset is finally prepared after applying aggregation functions (Table 2) to the selected features over a specified time window.

Table 2. Summary of aggregation functions applied to features in event sequences

Concept	Feature	Class	Aggregation
Vitals	SBP	Numeric (mm Hg)	Average
	DBP	Numeric (mm Hg)	Average
	HR	Numeric (bpm)	Average
Demographics	Age	Numeric (years)	Last
	Gender	Binary (male/female)	Last
Monitoring Program	Duration	Numeric (months)	Last
Meta-States	RZ Duration	Numeric (rel. time)	Ratio
	YZ Duration	Numeric (rel. time)	Ratio

Pre-processing steps also include removing outliers, data normalization/scaling and binarization, which are often essential before the modelling phase not to produce misleading results. We choose a logistic regression classifier for this task because: (i) it can handle both continuous and categorical variables; (ii) it is less susceptible to overfitting than, e.g., decision trees (especially in our case of small data); (iii) it can give a better insight into a new feature concept impact since logistic regression searches for a linear boundary in the feature space.

The number of out-of-control instances is about twice much as the number of in-control observations. To overcome class imbalance (and small data in addition), we utilize oversampling technique SMOTE. There are still too few observations that can produce untrusted results when evaluating classification performance. Thus, we measure uncertainties of evaluation metrics at various decision thresholds for a positive class to get results not affected by random train-test splitting. Mean precision-recall curves for the dataset without and with the meta-states feature concept are shown in Fig. 2.

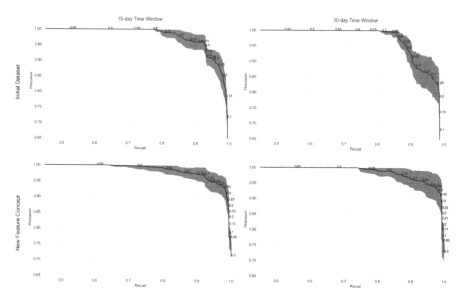

Fig. 2. Average precision-recall curve with standard deviation calculated from 20 random partitions

The results show that the new features extracted from the event log contribute to a better precision-recall trade-off, especially to a greater recall score. The average increase of the mean recall score is 0,045 for the 15-day time window and 0,059 for the 30-day time window. The mean precision score is not always enhanced by additional features. Its average increase is about 0,008 and 0,015 for the 15-day and 30-day time windows, respectively. Therefore, the inclusion of the meta-states features has improved the ability of the classifier to find positive class instances.

5   Conclusion

In this study, we have presented the extension of the process discovery algorithm and proposed the concept of meta-states based on the assumption of patient process behaviours. We have demonstrated how the PM technique can be utilized in ML tasks. Unfortunately, we were confined to small data, and a clinician's opinion is required. Despite the limitations, the case study results showed the potential of the proposed approach for diagnosis. Additionally, it helped interpret the modelled outcomes.

In further studies, we plan to continue the work on the project and extend its functionality. A promising direction is extending interpretability capabilities with different knowledge sources, including formal knowledge and data mining. For example, ML models or Hidden Markov models can be used to interpret meta-states found in the process models. Collaboration of PM and other communities can produce interesting and valuable results.

Acknowledgements. This research is financially supported by the Russian Science Foundation, Agreement #17–15-01177. Participation in the ICCS conference is supported by the NWO Science

Diplomacy Fund project #483.20.038 "Russian-Dutch Collaboration in Computational Science". The authors also wish to thank the colleagues from PMT Online for the data provided and valuable cooperation.

References

1. Van Der Aalst, W.M.P.: Process Mining: Data Science in Action. Springer, Heidelberg (2016)
2. Dos Santos Garcia, C., et al.: Process mining techniques and applications - A systematic mapping study. Expert Syst. Appl. **133**, 260–295 (2019)
3. Valero-Ramon, Z., Fernández-Llatas, C., Martinez-Millana, A., Traver, V.: Interactive process indicators for obesity modelling using process mining. In: Maglogiannis, I., Brahnam, S., Jain, L.C. (eds.) Advanced Computational Intelligence in Healthcare-7, vol. 891, pp. 45–64. Springer, Heidelberg (2020). https://doi.org/10.1007/978-3-662-61114-2_4
4. Kovalchuk, S.V., Funkner, A.A., Metsker, O.G., Yakovlev, A.N.: Simulation of patient flow in multiple healthcare units using process and data mining techniques for model identification. J. Biomed. Inform. **82**, 128–142 (2018)
5. Tax, N., Sidorova, N., Haakma, R., van der Aalst, W.M.P.: Event abstraction for process mining using supervised learning techniques. In: Bi, Y., Kapoor, S., Bhatia, R. (eds.) IntelliSys 2016, LNNS, vol. 15, pp. 251–269. Springer, Cham (2018). https://doi.org/10.1007/978-3-319-56994-9_18
6. Jans, M., Hosseinpour, M.: How active learning and process mining can act as continuous auditing catalyst. Int. J. Account. Inf. Syst. **32**, 44–58 (2019)
7. Es-Soufi, W., Yahia, E., Roucoules, L.: On the use of process mining and machine learning to support decision making in systems design. In: Harik, R., Rivest, L., Bernard, A., Eynard, B., Bouras, A. (eds.) PLM 2016, IFIPAICT, vol. 492, pp. 56–66. Springer, Cham (2016). https://doi.org/10.1007/978-3-319-54660-5_6
8. Kratsch, W., Manderscheid, J., Röglinger, M., Seyfried, J.: Machine learning in business process monitoring: a comparison of deep learning and classical approaches used for outcome prediction. Bus. Inf. Syst. Eng. 1–16 (2020).https://doi.org/10.1007/s12599-020-00645-0
9. Bozorgi, Z.D., Teinemaa, I., Dumas, M., La Rosa, M., Polyvyanyy, A.: Process mining meets causal machine learning: Discovering causal rules from event logs. In: van Dongen, B.F., Montali, M., Wynn, M.T. (eds.) ICPM 2020, pp. 129–136. IEEE (2020). https://doi.org/10.1109/ICPM49681.2020.00028
10. Pasquadibisceglie, V., Appice, A., Castellano, G., Malerba, D.: Predictive process mining meets computer vision. In: Fahland, D., Ghidini, C., Becker, J., Dumas, M. (eds.) BPM 2020, LNBIP, vol. 392, pp. 176-192. Springer, Cham (2020). https://doi.org/10.1007/978-3-030-58638-6_11
11. Teinemaa, I., Dumas, M., Rosa, M.L., Maggi, F.M.: Outcome-Oriented Predictive Process Monitoring: Review and Benchmark. ACM Trans. Knowl. Discov. Data. **13**(2), 17:1–17:57 (2019)
12. Flath, C.M., Stein, N.: Towards a data science toolbox for industrial analytics applications. Comput. Ind. **94**, 16–25 (2018)
13. Günther, C.W., van der Aalst, W.M.P.: Fuzzy mining - adaptive process simplification based on multi-perspective metrics. In: Alonso, G., Dadam, P., Rosemann, M. (eds.) BPM 2007, LNCS, vol. 4714, pp. 328–343. Springer, Berlin, Heidelberg (2007). https://doi.org/10.1007/978-3-540-75183-0_24
14. Van Der Aalst, W.M.P.: A practitioner's guide to process mining: limitations of the directly-follows graph. Procedia Comput. Sci. **164**, 321–328 (2019)

15. Elkhovskaya, L., Kabyshev, M., Funkner, A.A., Balakhontceva, M., Fonin, V., Kovalchuk, S.V.: Personalized assistance for patients with chronic diseases through multi-level distributed healthcare process assessment. In: Blobel, B., Giacomini, M. (eds.) pHealth 2019, vol. 261, pp. 309–312. IOS Press (2019)
16. Sun, J., McNaughton, C.D., Zhang, P., Perer, A., Gkoulalas-Divanis, A., Denny, J.C., Kirby, J., Lasko, T., Saip, A., Malin, B.A.: Predicting changes in hypertension control using electronic health records from a chronic disease management program. JAMIA **21**(2), 337–344 (2014)

Comparative Evaluation of Lung Cancer CT Image Synthesis with Generative Adversarial Networks

Alexander Semiletov[1], Aleksandra Vatian[1(✉)], Maksim Krychkov[1], Natalia Khanzhina[1], Anton Klochkov[1], Aleksey Zubanenko[2], Roman Soldatov[2], Anatoly Shalyto[1], and Natalia Gusarova[1]

[1] ITMO University, 49 Kronverksky Pr., St. Petersburg 197101, Russia
[2] Ltd International Diagnostic Center, 140 Leninsky Pr., St. Petersburg 198216, Russia

Abstract. Generative adversarial networks have already found widespread use for the formation of artificial, but realistic images of a wide variety of content, including medical imaging. Mostly they are considered to be used for expanding and augmenting datasets in order to improve accuracy of neural networks classification. In this paper we discuss the problem of evaluating the quality of computer tomography images of lung cancer, which is characterized by small size of nodules, synthesized using two different generative adversarial network, architectures – for 2D and 3D dimensions. We select the set of metrics for estimating the quality of the generated images, including Visual Turing Test, FID and MRR metrics; then we carry out a problem-oriented modification of the Turing test in order to adapt it both to the actually obtained images and to resource constraints. We compare the constructed GANs using the selected metrics; and we show that such a parameter as the size of the generated image is very important in the development of the GAN architecture. We consider that with this work we have for the first time shown that for small neo-plasms, direct scaling of the corresponding solutions used to generate large neo-plasms (for example, gliomas) is ineffective. Developed assessment methods have shown that additional techniques like MIP and special combinations of metrics are required to generate small neoplasms. In addition, an important conclusion can be considered that it is very important to use GAN networks not only, as is usually the case, for augmentation and expansion of the datasets, but for direct use in clinical practice by radiologists.

Keywords: Generative adversarial networks · 2D 3D GAN · CT image synthesis · Evaluation metrics · Lung cancer

1 Introduction

Generative adversarial networks (GANs), first proposed in [1], have already found widespread use for the formation of artificial, but realistic images of a wide variety of content, including in medicine [2, 3]. Initially, GANs in the field of medicine were used as an aid for augmentation of datasets for processing medical images based on

© Springer Nature Switzerland AG 2021
M. Paszynski et al. (Eds.): ICCS 2021, LNCS 12744, pp. 593–608, 2021.
https://doi.org/10.1007/978-3-030-77967-2_49

machine learning, primarily deep neural networks. But over the past two or three years, the range of scenarios for the use of GANs in medicine has dramatically expanded. With the help of GANs, it is possible to form medical images of various natures, which can be used as reference images when setting up automated classifiers of corresponding diseases, as well as for training less experienced pathologists and radiologists. For example, the paper [4] reports on Amyloid Brain PET Image Synthesis, which simulates changes in brain tissue in Alzheimer's disease. In the work [5], using GAN, images of plaques in coronary arteries, which are the main cause of atherosclerosis, are simulated. The authors [6], using GAN to simulate histological sections of liver tissue, experimentally confirmed the possibility of not storing natural tissue sections in glass and completely switching to digital histopathology, which is important for definitive diagnosis of non-fatty liver damage.

Accordingly, the requirements for assessing the effectiveness of the use of GAN in medicine have expanded. Of the wide variety of metrics proposed for the GAN Evaluation [7], only a few of them are used in medical applications. Initially, this role was played only by indicators of the effectiveness of training deep neural networks on datasets augmented with the help of GAN - such as ROC, FROC, AUC ROC etc. But now, great importance is attached to the visual qualities of the generated images and the degree of their similarity to the simulated prototype. Therefore, assessments of visual similarity entered the everyday life of the medical GAN developers. For this, both model-based methods, such as the visual Turing test and t-SNE, and model-agnoctic metrics, such as MMD, LOO, FID, DFD etc., are used. (A more detailed description of the listed methods and metrics is given in the next section of the article).

As practice shows and as literature reviews [2, 3, 8, 9] confirm, in recent years there has been an active and rapid development of GAN models, both in terms of improving architectural solutions and in terms of taking into account the specifics of the target area of medicine.

Adequate selection and problem-oriented adaptation of methods for assessing the quality of medical images generated with the help of GAN will allow not only assessing the prospects of a particular development at a fairly early stage, but also identifying key influence parameters that need to be highlighted during its further development. In oncological practice as a whole, the objects of interest are the neoplasms themselves and their structure, as well as the border zones between them and the surrounding tissues.

This article discusses the problem of evaluating the quality of CT images of lung cancer synthesized using GAN. The GAN methods are needed due to the fact that some specific cancer nodules are not presented well in the datasets like really small lung cancer nodules and etc. We describe our development of two problem-oriented GANs, which solve the problem of imitating malignant pulmonary nodes in 2D and 3D dimensions, respectively. We justify the selection of metrics for estimating the quality of the generated images, and of the parameters needed for their adaptation for lung tissue. For qualitative estimating we use Visual Turing test as well as d t-distributed Stochastic Neighbor Embedding (t-SNE). As quantitative metrics we use inception distance (FID). To evaluate the performance of the classifier trained on the basis of the augmented dataset, we use ROC-metrics. We present the results of a comparative evaluating of the constructed GAN using the selected metrics and show that from this point of view the development of the 2D approach seems to be more effective.

2 Background and Related Works

2.1 GAN's Specifics for Lung Tissue Imitation

Initially [1], the GAN scheme was a generator of random objects and a discriminator that distinguishes these objects from real ones, and a feedback loop was organized between both blocks through a back propagation mechanism. Modern GANs used in medical imaging, are far more sophisticated and more diverse in architecture [2], however, a problem-oriented analysis of the literature of recent years allows us to single out a number of main dominants.

Although the literature presents scenarios for generating medical images using GANs "from scratch", i.e. from random noise without any other conditional information [5, 10], however, in the last years the use of GANs for lung tissue imitation prevails in domain transformation scenarios, i.e. in image-to-image translation frameworks [11–16].

In general, in order to generate high-quality medical images, 3D GANs are applied, i.e. 3D fragments containing nodules and surrounding tissues are used for training [6, 13–15, 17]. This approach, of course, makes rather high demands on the complexity of the architecture and the level of computing resources. Meanwhile, the specifics of pulmonary nodes on CT scans is their rather small dimensions, i.e. a few (up to 2–3) slices of the CT image are enough to display an particular node. Therefore, in recent studies, there is also a 2D approach to generating images of pulmonary nodes [18]. The Maximum Intensity Projection (MIP) approach [19, 20] looks promising here. MIP is a method that projects 3-D voxels with maximum intensity to the plane of projection, thereby providing a transition to a 2D task.

As for up-to-date architecture for simulating pulmonary nodes, a plethora of variants are proposed here. Compared to the vanilla GAN [1], they apply various modifications of the loss function and approaches to normalization, as well as their combinations. For example, [11] employs WGAN, where the loss function is defined using the Wasserstein distance instead of the Jensen–Shannon divergence, [16] uses WGAN-GP, i.e. Wasserstein GAN with gradient penalty. The work of [15] is based on Conditional GAN (CGAN), where a discriminator is conditioned on an additional input. In [13] 3D-Multiconditional GAN is proposed containing two discriminators with different loss functions tailored for nodules and for context (surrounding tissues). In [17] a specialized CT-GAN based on a Conditional GAN is proposed.

[10] proposed to use Deep Convolutional Generative Adversarial Networks (DC-GANs), augmenting the standard GAN by using convolutional layers along with batch normalization. A wide spread solution for the generation of lung tissue is Progressive GAN, where the GAN is sequentially trained to create images of increasing dimension. For example, [12] implemented a Progressive Growing WGAN (PGGAN), with sliced Wasserstein distance loss, progressive growing, and pixel-wise normalization. [18] proposes a combined solution named Conditional Progressive Growing of GANs (CPGGANs), incorporating highly-rough bounding box conditions incrementally into PGGANs. Based on Progressive GAN and adding the Adaptive Image Normalization

(AdaIN), [21] developed StyleGAN architecture, which has been successfully used to generate medical images of different kind [6, 22], including pulmonary nodes [14].

2.2 Metrics for Evaluating the Quality of Synthesized Images

A short list of metrics fetched in the current literature for evaluating the quality of medical images, was presented earlier in the Introduction section. Now we present their consideration in more detail with an emphasis on their applicability to lung tissue imaging assessment.

Measures for Integral Effectiveness of CNN-Based Classifier. There are well-known indicators of the effectiveness of training deep neural networks on datasets augmented with the help of GAN - such as F-measure and its components, ROC, FROC, AUC ROC etc. Typically, they answer slightly different research questions. For example, the ROC method only involves stating the presence of an anomaly in the image, while the FROC method additionally requires the observer to detect anomalies [23]. However, as the analysis shows, in relation to the classification of lung cancer, these metrics are spread almost evenly in the literature. For example, as concerning to F-measure components, [10] uses False Recognition Rate (FRR) and True Recognition Rate (TRR), [11] applies accuracy, [24] and [15] estimate F-measure as a whole, [18] uses sensitivity in diagnosis with clinically acceptable additional False Positives (FPs), [16] adjusted False Positive per Scan vs Sensitivity. Such a diversity undoubtedly complicates a comparative assessment of the proposed solutions. At the same time, most researchers of the GUN-assisted lung cancer classification use ROC-curve [11, 12, 24] or FROC-curve as its variation [13].

Model-Based Methods. For an expert assessment of the synthetic images' realism, the Visual Turing Test is proposed in [25]. The full Visual Turing Test involves presenting real and virtual (generated) images to experienced radiologists, comparing them with the ground truth, and constructing a contingency table. However, the full procedure assumes a large amount of available statistical material for comparison [6], which is obviously lacking in pilot studies. Therefore, in most studies related to generative imaging of lung tissue, the Turing test is performed with different truncations without special justification [10, 12, 13, 15, 18] or not performed at all [11, 16].

One more model-based method is visualizing the data distribution via t-Distributed Stochastic Neighbor Embedding (t-SNE) [26]. The t-SNE method allows you to visually compare the distributions of real and generated images by translating high-dimensional data into a lower-dimensional space. Despite the rather low requirements for experimental and computational resources, as well as high information content [4, 27], the method is still used relatively rarely to assess the quality of lung imaging [12, 18].

Model-Agnoctic Metrics. According to [7, 28], in principle, the following model-independent metrics can be used to assess the quality of medical images generated by GANs.

The 1-NN classifier [29] assesses the similarity of the real image (labeled as 0) and the generated images (labeled as 1) by Leave-one-out cross-validation (LOOCV). However, this technique involves preliminary labeling of the images synthesized, for which, as the authors [29] themselves note, one must employ a "naturalness discriminator". For the time being, only a human expert can play this role for medical images, i.e. the technique becomes very resource-intensive and of little use for pilot projects.

The Maximum Mean Discrepancy (MMD) [30] is a distance-measure between two distributions which is defined as the squared distance between their embeddings in the a reproducing kernel Hilbert space F, i.e. is the distance between feature means μ_P and μ_Q of compared image data X and Y having probability measures P and Q respectively:

$$MMD^2(P, Q) = \|\mu_P - \mu_P\|_F^2. \tag{1}$$

The lower the result the more evidence that distributions are the same. Note that in CNN practice, when fulfilling the empirical estimation of MMD one usually limits to simple kernel functions [31].

The Fréchet inception distance (FID) [32] calculates the distance between the feature vectors of real images $N(\mu, C)$ and of images generated by the GAN $N(\mu_w, C_w)$. FID is based on the assumption of multidimensional Gaussian distributions of real and generated images, i.e. the mean and standard deviation is compared, so the formula is as follows:

$$d^2((m, C), (m_w, C_w)) = \|m - m_w\|_2^2 + Tr\left(C + C_w - 2(CC_w)^{1/2}\right), \tag{2}$$

where $Tr(.)$ is the trace of the covariance matrices of the feature vectors C and C_w.

Note that the comparison both in (1) and in (2) is not performed on the image itself, but on one of the deeper layers of CNN, which allows us to get away from the human perception of similarity in images. On the other hand, when using the Gaussian kernel function, expression (1) coincides with the first term of expression (2) up to notation, that is, FID and MMD metrics seem to some extent interchangeable. However, in the case of a strong discrepancy in the distributions, the use of the FID metric is criticized [5, 22], but for the assessment of lung imaging by the GANs, it is still used [12] as opposed to MMD.

Summarizing the review of the use of GAN for generating images of lung tissue, we see a motley picture of various architectures and approaches of GAN implementing, as well as of assessing the quality of the images obtained. This complicates the comparability of different projects and the possibility of assessing the prospects of the newly conducted research even at the pilot stage. In this regard, the authors of the article set themselves the following tasks:

– to carry out a pilot development of two GANs with different architectures and dimension approach, designed to generate images of lung tissue with cancerous nodes, under resource constraints;
– to select the set of metrics for estimating the quality of the generated images, and of the parameters needed for their adaptation for lung tissue;
– to compare the constructed GANs using the selected metrics and to identify key influence parameters that need to be highlighted during its further development.

3   Methods and Materials

3.1   Developing and Training GAN's Models

For comparison, we have developed two variants of GANs, which differ both in architecture and in approach of image forming (in 3D or 2D projection).

For 3D-GAN we chose CT-GAN from [17] as the baseline, but made some changes to it, which involved generator and discriminator weight updates frequency rationing and combining Wasserstein Loss (WL) with a baseline Mean Square Error (MSE) loss. Besides, we used AdaIN instead of Batch Normalization after each convolutional block, the normalization parameters being

$$AdaIN(x, y) = \sigma(y) \frac{x - \mu(x)}{\sigma(y)}, \tag{3}$$

where x is the previous layer output, y is the affine transformation. The modified blocks of 3D-GAN are depicted on Fig. 1a. As a detection model we used the solution of [33].

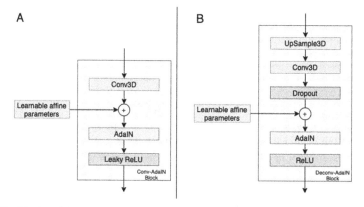

Fig. 1. The blocks of 3D-GAN modified in comparison of [Mirsky]: (A) Convolutional block, (B) Deconvolutional block.

For 2D-GAN we chose Syle-GAN [21] as the baseline fulfilling the MIP approach within. Since implementation of [21] is designed to work with 3-channel images, we converted the number of channels in the input and output layers to work with single-channel images. We also changed the Loss function type to BinaryCrossEntropy. We used the VGG11 model [34] being undemanding in terms of computing resources as a classifier.

To train both GAN models, we used the open Lung Image Database Consortium and Image Database Resource Initiative (LIDC-IDRI) dataset [35]. For 3D-GAN we employed its subset, namely LUNA-16 having the following properties compared to the base LIDC dataset: all the nodules are calibrated more precisely (the average size of nodules in LUNA16 is 8.3 mm with a standard deviation of 4.8 mm) compared to LIDC-IDRI (12.8 mm and 10.6 mm, respectively); each nodule is already labeled with a bounding box.

For 2D-GAN training we used LIDC-IDRI in general, but selecting the DICOM series only with tumor nodules. In order to extract the nodule we formed a circumscribing cube containing the nodule itself and the surrounding (context) tissues. The absolute dimensions of the cube were chosen in accordance with the size of the extracted nodule (from 1 mm to 40 mm), but were resampled to a single size of 128^3 pixels and to pixel values according to Hounsfield scale [36]:

$$x = \frac{2 \cdot (x_{in} - in_{\min})}{((in_{\max} - in_{\min}) - 1)}, \tag{4}$$

with boundary values of $in_{max} = 800$, $in_{min} = -1000$. In order to pass from 3D to 2D nodule image we performed a MIP lookup operation using the *numpy* package.

3.2 Evaluating the Quality of Synthesized Images of Lung Tissues

Measures for Integral Effectiveness. Although both GANs form images of nodules in the lungs, 3D-GAN is more focused on augmentation of datasets used in training DNN-based neoplasm detectors, while 2D-GAN is more focused on augmentation of datasets used in training DNN-based classifiers of neoplasm types. In this regard, in order to evaluate the classification model in both cases, we use ROC AUC (Area Under ROC Curve) metrics as measures for integral effectiveness of training DNN on datasets augmented with the help of GAN. In addition, for evaluation of 2D-GAN-based classifier we additionally use PR AUC (Precision-Recall Area Under Curve), and for 3D-GAN-based detection we use FROC in a modified form: instead of per-scan FROC calculations [33] we calculated sensitivity over average false positives per a scan crop of size 128^3. In this analysis sensitivity is defined as a percentage of crops on which the intersection over union of predicted and labeled bounding boxes is greater than 0.5.

Model-Based Methods. We used the Visual Turing Test, but made its problem-oriented modification. As with the traditional Visual Turing Test (see above), real and virtual (generated by GAN) images are demonstrated to N experienced radiologists. Each radiologist is presented with S sets containing 20 randomly selected images generated by a specific GAN, and we inform the radiologist that the exposed set can contain any mixture of real and generated images. Examples of the presented sets are shown in Fig. 2. The radiologist is asked to answer the following questions:

– If the presented set contains nodules, then which ones are real?
– If the presented set contains nodules, then which of them are solid (single), and which are subsolid (parietal)?

Based on the test results, the False Recognition Rate (FRR) is calculated as the proportion of nodules correctly identified by radiologists as generated among all generated nodes.

We calculated the t-SNE metric using the *scikit* package with default parameters[1].

---

[1] https://scikit-learn.org/stable/modules/generated/sklearn.manifold.TSNE.html.

We selected FID and MMD as **model-agnoctic metrics.** For implementation the FID metric, we used the code[2] with default parameters. The implementation of MMD is made according to the work[3]; as kernel functions we used radial basis function (instead of Gaussian kernel function, which is a standard practice for empirical estimates for CNN) and multiscale function.

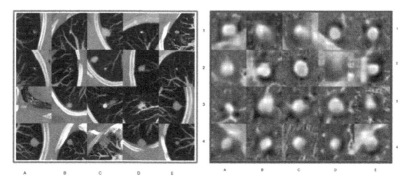

Fig. 2. a - An example of presentation for the evaluation of 2D-GAN. Images 1–10 contain nodules generated by 2D-GAN, images 11–20 are completely real. b - An example of presentation for the evaluation of 3D-GAN. All images contain generated nodules.

4   Experimental Results

Figure 3 shows the ROC-curves for the best learning epochs for 2D-GAN (a) and for 3D-GAN (b) respectively. For 3B-GAN we also give the FROC metric (Table 1).

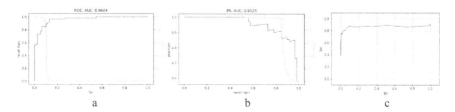

Fig. 3. ROC- curves for for 2D-GAN (a, b) and 3D-GAN (c)

Using the proposed approach of 2D-GAN for dataset augmentation we obtained the best values of ROCAUC = 0.9604 and PRAUC = 0.9625. This is better than the result of [11] on a comparable dataset and is only slightly inferior to the result of the same authors, obtained on much more powerful computing resources [37]. For 3D-GAN, the best value

---

[2] https://machinelearningmastery.com/how-to-implement-the-frechet-inception-distance-fid-from-scratch/.

[3] https://www.kaggle.com/onurtunali/maximum-mean-discrepancy.

Table 1. FROC metrics obtained for 3D-GAN from augmented and baseline model evaluation. Metrics mean and standard deviation are computed across 5-fold cross-validation experiments.

Average FP/crop	0.25	0.5	1	2	4	8	Average
Sensitivity (augmented)	0.330 ± 0.049	0.433 ± 0.056	0.555 ± 0.060	0.684 ± 0.058	0.794 ± 0.048	0.854 ± 0.037	0.608 ± 0.043
Sensitivity (baseline)	0.302 ± 0.042	0.414 ± 0.040	0.542 ± 0.041	0.647 ± 0.030	0.743 ± 0.020	0.822 ± 0.036	0.578 ± 0.028

Table 2. Examples of the results of the visual turing test.

2D-GAN	1	2	3	4	5	6	7	8	9	10	11	12	13	14	15	16	17	18	19	20
Radiologist 1	30	15	20	20	15	10	10	20	20	20	40	25	20	10	10	20	40	15	30	20
Radiologist 2	20	15	10	20	20	20	15	20	15	10	20	10	20	40	30	20	30	25	30	25
Real/generated nodules in the set	10/10	0/20	10/10	10/10	0/20	10/10	0/20	10/10	0/20	10/10	10/10	0/20	0/20	10/10	10/10	10/10	10/10	0/20	10/10	0/20
3D-GAN	1	2	3	4	5	6	7	8	9	10	11	12	13	14	15	16	17	18	19	20
Radiologist 1	25	20	40	30	25	20	20	30	50	40	15	20	25	10	20	15	20	30	40	30
Radiologist 2	30	25	70	80	70	80	20	30	50	40	55	30	35	20	40	25	50	60	40	60
Real/generated nodules in the set	0/20	0/20	10/10	10/10	0/20	10/10	0/20	10/10	10/10	10/10	0/20	10/10	0/20	0/20	10/10	0/20	10/10	10/10	10/10	0/20

of ROCAUC $= 0.95$. So (see also Table 1), the proposed 3D-GAN also surpasses the most modern of similar GAN implementations [13], chosen as a baseline.

When performing the Visual Turing Test, we recruited $N = 6$ radiologists, each of whom were presented with $S = 20$ test sets of 20 images each. Tables 2, 3 show examples of experimental data obtained from two radiologists. Table 4 contains the results of calculating the FRR metric, carried out by averaging over 6 radiologists, over 2 radiologists, and also for each of the two selected radiologists separately. Table 5 contains the analogous results concerning identifying of solid and subsolid nodules.

Table 3. FRR metrics calculated for different groups of radiologists.

	Radiologist 1	Radiologist 2	Average for 2 radiologists	Average for 6 radiologists
2D-GAN, $S = 20$	$20,5 \pm 8,6$	$20,7 \pm 7,4\%$	$20,6 \pm 8\%$	$19,4 \pm 5,4\%$
2D-GAN, $S = 5$	$20 \pm 5,4\%$	$17 \pm 4\%$	$18,5 \pm 4,7\%$	$17,9 \pm 7,2\%$
3D-GAN, $S = 20$	$45,5 \pm 18,9\%$	$26,25 \pm 9\%$	$35,8 \pm 14\%$	$41 \pm 11\%$
3D-GAN, $S = 5$	$28 \pm 6,7\%$	$55 \pm 22\%$	$41,5 \pm 14,3\%$	$47 \pm 15,3\%$

Table 4. FRR metrics calculated for identifying of solid and subsolid nodules

	Radiologist 1	Radiologist 2	Average for 2 radiologists	Average for 6 radiologists
2D-GAN	$92,5 \pm 7\%$	$94,6 \pm 5\%$	$93,6 \pm 6\%$	$94,5 \pm 3,5\%$
3D-GAN	$27,5 \pm 19,5\%$	$30,3 \pm 17.4\%$	$28,9 \pm 18.5\%$	$25.4 \pm 19.6\%$

Figure 4 shows the visualized t-SNE metrics for different combinations of pulmonary nodes (benign and malignant; real and virtual), as well as their summary, for 2D and 3D GANs. Table 5 summarizes the experimental data for the FID metrics, as well as for the MMD-metrics plotted for various kernel functions. The following designations are adopted in the Table 5: r-v is the value of the metric between real and synthesized data as a whole; rb-vb is the value of the metric between real and synthesized images of benign nodules; rm-vm is the value of the metric between real and synthesized images of malignant nodules. The advantages of using FID and MRR metrics is that we can make a comprehensive assessment of the generated nodules. By using Visual Turing Test we can carry out an integral assessment according to the context, taking into account the opinion of the doctor.

Table 5. FID and MMD metrics calculated for different groups of images (designations decoding - in the text)

	2D-GAN			3D-GAN		
	r-v	rb-vb	rm-vm	r-v	rb-vb	rm-vm
FID	51.61	74.98	37.35	171.10	130.93	670.66
MMD, radial basis kernel	0.0169	0.0214	0.0287	0.0193	0.0252	0.0375
MMD, multiscale kernel	0.0118	0.0208	0.0271	0.0123	0.0212	0.0280

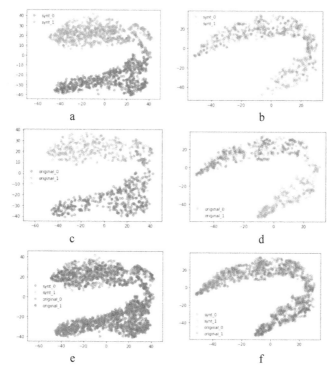

Fig. 4. Visualized t-SNE metrics for various combinations of embeddings of images with pulmonary nodes: a – benign vs malignant nodules, generated by 2D-GAN; b - benign vs malignant nodules, generated by 3D-GAN; c- benign vs malignant nodules, real images; d- benign vs malignant nodules, real images; e - all the nodules of dataset augmented by 2D-GAN; f - all the nodules of dataset augmented by 3D-GAN;

5 Discussion

The experimental data (see Fig. 4 and Table 1), as well as a comparison of these data with baselines show that from the point of view of augmentation of datasets used for CNN training, the developed GANs are at a completely conventional level, i.e. the chosen architectural and parametric solutions are quite successful. Note that 3DGAN was trained in obviously better conditions, since the nodule detection on LUNA16 is considered less challenging than on LIDC-IDRI due to the greater reliability of the data and smaller scatter in the size of nodules [17].

Let's move on to the analysis of the visual quality of the generated images (Fig. 2, 3 and Tables 2, 3, 4, 5). The number of radiologists performing a Visual Turing Test and their professional experience will obviously affect test results. Despite this, in most works related to the visual assessment of the quality of images generated by the GAN, the authors limit themselves, at best, to a statement of the number of experts involved, and, as a rule, there are only two [12, 13] or one [22] of them, and in [27], even an expert radiologist and a non-specialist are used. The results of our experiments, presented in Table 4, show that when passing from averaging the FRR values over 6 experts to averaging over 2 experts, the mathematical expectation, although it changed, remained within the standard deviation, and for some experts it was the significant change. This allows us to say that in pilot studies of the prospect of the GAN architecture, one can limit ourselves to two experts-radiologists, and the use of only one radiologist and, moreover, a non-specialist does not guarantee against the presence of emissions in the assessment and cannot be justified. At the same time, for large-scale studies, it is necessary to use stronger statistical measures of similarity, for example, the Fleiss kappa [38].

Analyzing the model-agnoctic metrics of image quality (Table 5 in comparison with Fig. 4), we can state that the MMD and PID metrics demonstrate different rankings of the quality of the resulting images. In this case, a change in the kernel function, although it affects the absolute values of the MMD metric, does not change its ratio for different groups of images. The gradation of the metrics is different - the FID metric, unlike MMD, takes into account not only average values, but also data variance. For example, on the t-SNE visualization (Fig. 4c), it can be seen that the real data has several outliers. Because of them, the average of real and generated data is different, but this practically does not affect the variance. And therefore, in this case, the FID metric can be considered fairer than MMD metric. In addition, it is necessary to emphasize the independent role of the t-SNE metric in assessing the quality of the generated images of pulmonary nodules, since it allows you to visually assess the degree of interpenetration of the generated objects of different classes (compare Fig. 4a and b).

Comparing the developed GANs using the selected metrics, we can draw the following conclusions. Although, as noted above, both GANs have a fairly high efficiency in terms of augmentation of datasets for CNN training, nevertheless, as can be seen from Table 3 and 4, as well as from Fig. 1 and 2, the visual quality of the images generated by 3D-GAN differ significantly. We can associate this with the fact that most of the existing GAN are focused on the generation of relatively large neoplasms, and the problem of their inserting into existing images is reduced to eliminating defects at the boundaries between the formed image and the substrate (native tissue) (see, for example, [6].

In the case of small nodules, as shown in Fig. 1 and 2, this effect is not observed, i.e. it is not possible to clearly identify the zones of the generated image that are most critical for visual assessment. Thus, direct scaling of solutions for large neoplasms to small sizes is hardly effective, and imitation of small neoplasms should be considered as a separate task when building a GAN. In our case, this was done using the MIR approach, but, of course, other options are also possible here. So, we consider the relative size of the simulated medical neoplasm to be one of the important influence parameters that need to be highlighted and estimated during the development of GUN architecture for medical purposes.

6   Conclusion

The specificity of lung cancer is that neoplasms are malignant nodules, which have small size (10–30 mm) and high morphological similarity with benign nodules normally present in the lungs. In this regard, according to the Lung Image Database Consortium, LIDC) [35], even experienced radiologists correctly classify only 75.1% of the nodes when compared with the results of biopsy. Therefore, there is a great need for an early assessment of the prospects for the development of one or another technology for the implementation of GAN for this area of medicine. This article is a contribution to solving this problem.

All the tasks set in the article have been solved, namely:

- we carried out a pilot development of two GANs of up-to-date level with different architectures and dimension approach, designed to generate images of lung tissue with cancerous nodes, under resource constraints;
- we selected the set of metrics for estimating the quality of the generated images, including Visual Turing Test, FID and MRR metrics; we carried out a problem-oriented modification of the Turing test in order to adapt it both to the actually obtained images and to resource constraints;
- we compared the constructed GANs using the selected metrics; we showed that such a parameter as the size of the generated image was very important in the development of the GAN architecture.

The proposed work may be useful for providing a baseline for future studies and implementing GANs for medical purposes, as well as for determining the direction of next future experiments in practical studies.

The work was supported by Russian Science Foundation, Grant 19-19-006–96.

References

1. Goodfellow I.J., et al.: Generative Adversarial Nets. arXiv:1406.2661v1 [stat.ML] 10 Jun 2014
2. Tschuchnig, M.E., Oostingh, G.J., Gadermayr, M.: Generative adversarial networks in digital pathology: a survey on trends and future potential. Patterns 1(6), 100089 (2020)
3. Yi, X., Walia, E., Babyn, P.: Generative adversarial network in medical imaging: a review. Med. Image Anal. 58, 101552 (2019)
4. Kang, H., et al.: Visual and quantitative evaluation of amyloid brain PET image synthesis with generative adversarial network. Appl. Sci. 10, 2628 (2020)
5. Bargsten, L., Schlaefer, A.: SpeckleGAN: a generative adversarial network with an adaptive speckle layer to augment limited training data for ultrasound image processing. Int. J. Comput. Assist. Radiol. Surg. 15(9), 1427–1436 (2020). https://doi.org/10.1007/s11548-020-02203-1

6. Levy, J.J., et al.: A large-scale internal validation study of unsupervised virtual trichrome staining technologies on nonalcoholic steatohepatitis liver biopsies. Mod. Pathol. **34**(4), 808–822 (2020). https://doi.org/10.1038/s41379-020-00718-1
7. Borji, A.: Pros and xons of GAN evaluation measures. arXiv:1802.03446v5 [cs.CV] 24 Oct 2018
8. Kazeminia, S., et al.: GANs for medical image analysis. arXiv.org > cs > arXiv:1809.062 22v3. 9 Oct 2019
9. Wang, T., et al.: A review on medical imaging synthesis using deep learning and its clinical applications. J. Appl. Clin. Vedical Phys. **22**(1), 11–36 (2021)
10. Chuquicusma, M.J.M., Hussein, S., Burt, J., Bagci, U.: How to fool radiologists with generative adversarial networks? A visual turing test for lung cancer diagnosis. In: IEEE International Symposium on Biomedical Imaging (ISBI) (2018)
11. Onishi, Y., et al.: Automated pulmonary nodule classification in computed tomography images using a deep convolutional neural network trained by generative adversarial networks. Hindawi BioMed. Res. Int. **2019**(6051939), 1–9 (2019)
12. Wang, Y., Zhou, L., Wang, M., et al.: Combination of generative adversarial network and convolutional neural network for automatic subcentimeter pulmonary adenocarcinoma classification. Quant. Imaging Med. Surg. **10**(6), 1249–1264 (2020)
13. Han, C., et al.: Synthesizing diverse lung nodules wherever massively: 3D multi-conditional GAN-based CT image augmentation for object detection. In: 2019 International Conference on 3D Vision (3DV), Quebec City, QC, Canada, pp. 729–737 (2019). https://doi.org/10.1109/3DV.2019.00085
14. Shi, H., Lu, J., Zhou, Q.: A novel data augmentation method using style-based GAN for robust pulmonary nodule segmentation. In: 2020 Chinese Control and Decision Conference (CCDC), Hefei, China, pp. 2486-2491 (2020)
15. Jin, D., Xu, Z., Tang, Y., Harrison, A.P., Mollura, D.J.: CT-realistic lung nodule simulation from 3D conditional generative adversarial networks for robust lung segmentation. In: Frangi, A.F., Schnabel, J.A., Davatzikos, C., Alberola-López, C., Fichtinger, G. (eds.) Medical Image Computing and Computer Assisted Intervention – MICCAI 2018. LNCS, vol. 11071, pp. 732–740. Springer, Cham (2018). https://doi.org/10.1007/978-3-030-00934-2_81
16. Gao, C., et al.: Augmenting LIDC dataset using 3D generative adversarial networks to improve lung nodule detection. Medical imaging 2019: computer-aided diagnosis. In: International Society for Optics and Photonics, vol. 10950 (2019)
17. Mirsky, Y. et al.: CT-GAN: malicious tampering of 3D medical imagery using deep learning. In: 28th {USENIX}. Security Symposium ({USENIX} Security 2019), pp. 461–478 (2019)
18. Han, C., et al.: Learning more with less: conditional PGGAN based data augmentation for brain metastases detection using highly-rough annotation on MR images. In: Proceedings of ACM International Conference on Information and Knowledge Management (CIKM) (2019)
19. Zhang, J., Xia, Y., Zeng, H., Zhang, Y.: NODULe: combining constrained multi-scale LoG filters with densely dilated 3D deep convolutional neural network for pulmonary nodule detection. Neurocomputing **317**, 159–167 (2018)
20. Zheng, S., Guo J., Cui,, X., Veldhuis, R.N.J., Matthijs, O., van Ooijen, P.M.A.: Automatic pulmonary nodule detection in CT scans using convolutional neural networks based on maximum intensity projection. arXiv:1904.05956 [cs.CV] 10 Jun 2019
21. Karras, T., Laine, S., Aila, T.: A style-based generator architecture for generative adversarial networks. arXiv:1812.04948v3 [cs.NE] 29 Mar 2019
22. Chang, A., Suriyakumar, V.M., Moturu, A., et al.: Using generative models for pediatric wbMRI. Medical Imaging with Deep Learning, pp. 1–7 (2020)
23. Hillis, S.L., Chakraborty, D.P., Orton, C.G.: ROC or FROC? It depends on the research question. Med. Phys. **44**(5), 1603–1606 (2017)

24. Ghosal, S.S., Sarkar, I., Hallaoui, I.E.: Lung nodule classification using convolutional autoencoder and clustering augmented learning method (CALM). http://ceur-ws.org/Vol-2551/paper-05.pdf. Accessed 05 Feb 2021

25. Salimans, I., Goodfellow, W., Zaremba, V., Cheung, A., Radford, X.: Chen. Improved techniques for training GANs. In: Advances in Neural Information Processing Systems (NIPS), pp. 2234–2242 (2016)

26. van der Maaten, L., Hinton, G.: Visualizing data using t-SNE. J. Mach. Learn. Res. **9**, 2579–2605 (2008)

27. Haarburger C., et al.: Multiparametric magnetic resonance image synthesis using generative adversarial networks. In: Eurographics Workshop on Visual Computing for Biology and Medicine (2019)

28. Xu, Q., et al.: An empirical study on evaluation metrics of generative adversarial networks. arXiv. arXiv:1806.07755 (2018)

29. Lopez-Paz, D., Oquab, M.: Revisiting classifier two-sample tests. arXiv. arXiv:1610.06545 (2016)

30. Gretton, A., Borgwardt, K.M., Rasch, M.J., Schölkopf, B., Smola, A.: A kernel two-sample test. J. Mach. Learn. Res. **13**, 723–773 (2012)

31. Dziugaite, G.K., Roy, D.M., Ghahramani, Z.: Training generative neural networks via maximum mean discrepancy optimization. arXiv preprint arXiv:1505.03906 (2015)

32. Dowson, D.C., Landau, B.V.: The Fréchet distance between multivariate normal distributions. J. Multivar. Anal. **12**(3), 450–455 (1982)

33. Li, Y., Fan, Y.: DeepSEED: 3D squeeze-and-excitation encoder-decoder convnets for pulmonary nodule detection. In: 2020 IEEE 17th International Symposium on Biomedical Imaging (ISBI). IEEE (2020)

34. Simonyan, K., Zisserman, A.: Very deep convolutional networks for large-scale image recognition. arXiv:1409.1556 [cs.CV] 10 Apr 2015

35. Armato, S.G.: The lung image database consortium (LIDC) and image database resource initiative (IDRI): a completed reference database of lung nodules on CT scan. Med. Phys. **38**(2), 915–931 (2011)

36. Feeman, T.G.: The Mathematics of Medical Imaging: A Beginner's Guide. Springer Undergraduate Texts in Mathematics and Technology. Springer, New York (2010).978-0387927114

37. Onishi, Y., et al.: Multiplanar analysis for pulmonary nodule classification in CT images using deep convolutional neural network and generative adversarial networks. Int. J. Comput. Assist. Radiol. Surg. **15**(1), 173–178 (2019). https://doi.org/10.1007/s11548-019-02092-z

38. Gwet, K.L.: Computing inter-rater reliability and its variance in the presence of high agreement. Br. J. Math. Stat. Psychol. **61**, 29–48 (2008)

Deep Convolutional Neural Networks in Application to Kidney Segmentation in the DCE-MR Images

Artur Klepaczko[1(✉)], Eli Eikefjord[2], and Arvid Lundervold[2,3,4]

[1] Lodz University of Technology, Łódź, Poland
`aklepaczko@p.lodz.pl`
[2] Department Health and Functioning, Western Norway University
of Applied Sciences, Bergen, Norway
[3] Department of Biomedicine, University of Bergen, Bergen, Norway
[4] Mohn Medical Imaging and Visualization Centre, Department of Radiology,
Haukeland University Hospital, Bergen, Norway

Abstract. This paper evaluates three convolutional neural network architectures – U-Net, SegNet, and Fully Convolutional (FC) DenseNets – in application to kidney segmentation in the dynamic contrast-enhanced magnetic resonance images (DCE-MRI). We found U-Net to outperform the alternative solutions with the Jaccard coefficient equal to 94% against 93% and 91% for SegNet and FCDenseNets, respectively. As a next step, we propose to classify renal mask voxels into cortex, medulla, and pelvis based on temporal characteristics of signal intensity time courses. We evaluate our computational framework on a set of 20 DCE-MRI series by calculating image-derived glomerular filtration rates (GFR) – an indicator of renal tissue state. Then we compare our calculated GFR with the available ground-truth values measured in the iohexol clearance tests. The mean bias between the two measurements amounts to -7.4 ml/min/$1.73\,m^2$ which proves the reliability of the designed segmentation pipeline.

Keywords: Semantic segmentation · Convolutional neural networks · Dynamic contrast-enhanced MRI · Kidney perfusion modeling

1 Introduction

1.1 Diagnostics of the Kidney

Contrast-enhanced magnetic resonance imaging (CE-MRI) is one of the methods routinely used in clinics for the diagnosis of renal impairments. It allows visualization of the kidney lesions such as tumor, cysts or focal segmental glomerulosclerosis. Moreover, if image acquisition is performed in a dynamic sequence, resulting in a temporal series of CE-MRI volumes, it is possible to determine the glomerular filtration rate (GFR) based on image data. GFR can be affected

© Springer Nature Switzerland AG 2021
M. Paszynski et al. (Eds.): ICCS 2021, LNCS 12744, pp. 609–622, 2021.
https://doi.org/10.1007/978-3-030-77967-2_50

by various renal diseases which lead to the loss of kidney filtration performance. Thus, quantification of renal perfusion provides an objective way for assessment of the potential of the kidney to restore its functional characteristics. In contrast to gold standard serum creatinine clearance test, image-derived GFR can be calculated for a single kidney, thus allowing lateral differentiation of kidney diseases, while providing spatially-resolved information on tissue damage.

In principle, the dynamic contrast-enhanced (DCE) MRI examination produces a series of T1-weighted volumes acquired at multiple time steps in the time interval covering the passage of a contrast agent (CA) through the abdominal arterial system. While the CA bolus enters the capillary bed and then tubular network of the kidneys, it effectively increases the T1 relaxation time of the penetrated tissues and modifies the contrast in the image. The temporal dynamics of this image signal intensity reflects physiological conditions of kidney function and constitutes the basis for pharmacokinetic (PK) modeling.

There have been numerous PK models proposed in the literature. Among those specifically dedicated to the kidney, one should mention the 2-compartment separable and filtration models proposed in [17] and [19], respectively. In this paper, the latter model will be used to estimate GFR for the experimental data available in our study. There were also more complex models proposed, such as e.g. the six-compartment formulation in [12]. However, these approaches became less popular due to unstable behavior of optimization procedure while fitting models parameters to image data.

In any case, application of the PK models require prior segmentation of the kidney parenchyma into cortex and medulla. Moreover, the pelvis needs to be separated as it does not contribute to the renal filtration process. Only voxels which contain renal nephrons have to be take into account. Thus, precise segmentation of the kidney regions is a crucial step in automated analysis of the dynamic contrast-enhanced MRI.

1.2   Related Work

The problem of kidney segmentation has been tackled by many authors. Frequently, voxels are classified based on their intensity time courses. For example in [23], voxels are clustered by the k-means algorithm. This approach was further developed in [13], where signal intensity time courses were preprocessed by the discrete wavelet transform. Image volumes were first manually cropped to cubic regions of interest (ROI) covering single kidneys. Then, the extracted wavelet coefficients of the ROI voxel time curves were submitted to the k-means clusterer. A problem which appears here is that there arises a need for the clustering algorithm to extract not only the three renal classes but also to separate them from the surrounding tissues. It is a frequent source of error since other tissues (liver, spleen and pancreas) exhibit signal dynamics similar to the kidney.

Therefore, a common strategy consists in firstly separating a whole kidney from other parts of an image. The delineated regions of interest should precisely fit kidney borders in order to get rid of all neighboring voxels. An example of such a solution are the area-under-the-curve maps (AUC) [3]. Those voxels in the DCE-MRI sequence which are penetrated by the tracer agent appear bright on AUC maps due to the largest area under their signal intensity time courses.

The coarse-to-fine segmentation strategy was also applied in [21], where the concept of Maximally Stable Temporal Volumes (MSTV) was introduced. The MSTV features allow to recognize kidneys by detecting spatially homogeneous and temporally stable structures. Spatial homogeneity is defined in terms of image binarization performed with a large span of thresholds. Independently from a threshold value renal voxels remain bright and possess bright neighbors in all 3 directions of the 3D space. Temporal stability, in turn, is reflected in the fact that spatial homogeneity of kidney voxels is observed in adjacent time frames of the imaging sequence. Fine-grained segmentation is obtained by reducing voxels time courses to vectors of principal components, which are next partitioned by k-means to multiple clusters.

Similarly, in [22], the first stage of the segmentation procedure employs Grub-Cut algorithm to create renal masks. Fine-tuning is achieved by classifying voxels with a pre-trained random forest classifier. Voxels are characterized by their respective image intensities in selected time frames of the dynamic sequence as well as their location within the ROIs constructed in the first stage. Although both MSTV- and GrubCut-based contributions seem to produce satisfactory results, they are rather conceptually complex algorithms, unavailable in open-source software. As such, they cannot be easily adopted by the clinical community.

On the other hand, with the advent of deep learning (DL) methods, semantic segmentation networks offer an attractive computational methodology for the problem of automatic kidney delineation in MR images. For example, in [10] various network architectures, i.e. fully convolutional network [16], SegNet [1], U-Net [15], and DeepLabV3+ [2], have been tested for segmentation of prostate cancer in T2-weighted MRI. Anatomical MR images of polycystic kidneys were segmented by a custom convolutional neural network (CNN) in [11]. Another approach has been presented in [20], where deep learning was employed for direct inference of brain perfusion maps from a DCE-MRI sequence without explicitly fitting a PK model to measured signals. However, there has been only a moderate number of DL-based approaches targeting DCE-MRI of the kidney. As one of few exceptions, the study described in [5] presents a cascade of two CNN networks. The first network roughly localizes the left and right kidney in a 4D DCE-MR image, whereas the second one performs fine delineation of renal borders.

1.3   Current Contribution

This paper presents a novel computational framework for automated segmentation of the kidney and its internal structures in the DCE-MR images using:

1. a convolutional neural network for delineation of the kidney parenchyma;
2. a classifier trained in supervised manner to partition parenchymal voxels into cortex, medulla and pelvis.

With regard to task 1, in order to find an optimal solution we have test three encoder-decoder CNN architectures: U-Net [15], SegNet [1] and 100-layers Fully Convolution DenseNets [9], referred to as *Tiramisu*. For the task 2, we have selected the support vector machine (SVM) classifier with the radial basis function kernel as it proved in our experiments to outperform other tested algorithms.

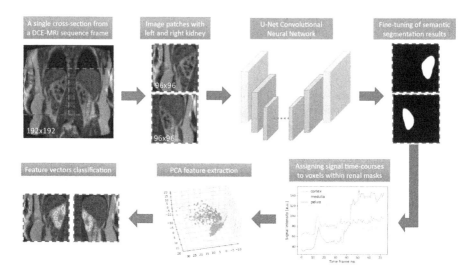

Fig. 1. Overview of the designed segmentation pipeline.

The proposed segmentation pipeline is visualized in Fig. 1. The initial coarse segmentation is accomplished by a neural network. This step is performed on subsequent two-dimensional cross-sections of a single volumetric image from the DCE-MRI sequence. This image corresponds to the frame of the highest signal enhancement in the cortex region, when the partitioning of the renal parenchyma into cortex and medulla is clearly visible.

Each cross-section is divided into left and right sides of 96-pixel width. On a given side, a centrally located image patch of 96-pixels height is selected. In this way, we ensure that left and right kidneys are processed separately. Prior to segmentation, we perform DCE image series matching in the time domain using B-splines deformable registration [8] to suppress motion artifacts. Thanks to image registration, kidney masks generated in one frame can be applied to all other frames of the dynamic series. Thus, renal voxels are prescribed feature vectors composed of MRI signal intensity values measured in subsequent time points. In order to obtain more general and compact characteristics of the signal dynamics, we extract feature aggregates using PCA transform. Eventually, a classifier trained to discriminate between temporal characteristics of cortex, medulla and pelvis regions assigns a voxel to an appropriate category.

In order to prove scalability of the designed framework, the proposed methodology was verified in the leave-one-subject-out manner based on the cohort of 10 healthy volunteers, each scanned twice with the DCE-MRI method. Hence, CNN network and classifier training was repeated 10 times, each time with one patient put apart. It was then possible, to objectively verify, how the system behaves in case of a subject not seen during the training phase. Between the scanning sessions, renal performance of each subject was evaluated using iohexol clearance procedure to establish the ground-truth value of the glomerular filtration rate and enable validation of the image-derived GFR measurements.

2 Semantic Segmentation of the Kidney

2.1 U-Net

The U-Net convolutional neural network was originally developed for segmentation of neuronal structures in electron microscopic stacks and proved effective in numerous other biomedical applications. As said, the input to our model is a 2D grey-level image – a 96×96-pixel patch of a DCE-MRI volume cross-section.

The U-Net network contains two symmetric paths – a contractive and an expansion one. The goal of the contractive path is to encode image pixel intensity patterns. It is accomplished by convolution with a series of 3×3 filters of trainable weights. Filters outputs activate the main processing components of the network – the neurons called rectified linear units (ReLU). They allow for modeling non-linear relationship between image features and the output segmentation map. It is followed by the 2×2 max-pooling operation which down-samples the feature maps.

Contraction is repeated four times to extract image descriptors on various scale levels. Each level is formed by two convolutional layers followed by a batch normalization layer. The convolution and normalization layer pairs are separated by the dropout layer, which randomly sets 20% of the input nodes to 0. This mechanism, active only during the training phase, prevents the network from overfitting [18].

The output of the last down-sampling block is passed on to the expansion path. It is built up from the same number of up-sampling levels as the contractive part and its main task is to recover original spatial resolution. In this study, up-sampling is realized by transposed convolution. Decoding blocks are also composed of two pairs of convolutional and batch normalization layers. However, no dropout mechanism is inserted in-between. Moreover, the high-resolution feature maps extracted in the down-sampling path not only feed the subsequent encoding layers but they are also concatenated to the inputs of the decoding layers at the respective levels of the up-sampling path. These additional connections help the decoding blocks to restore kidney segments localization more precisely.

The output of the last up-sampling block is connected to a convolutional layer with a 1×1-size filters. It performs pixel-wise convolution of the filter kernel with a 64-element feature vector and then submits the result to an output activation function. In our design, a sigmoid activation is used since the final decision is binary – a pixel belongs to renal parenchyma or background.

2.2 SegNet

SegNet is another encoder-decoder architecture [1], whose main characteristic is the application of the VGG16 [14] topology as the encoder backbone. Specifically it uses its first 13 convolulational layers to extract image features. Also, the method introduced in the decoder path to up-sample feature maps was different than the strategies used in fully convolutional networks. In this alternative approach, SegNet keeps record of the pixel indices selected by the max-pool operation and uses them in the corresponding decoder levels to perform non-linear up-sampling. Up-sampled maps are, in principle, zero-padded in positions not indicated by the memorized indices. Eventually, dense feature maps are created by convolution of up-sampled maps with trainable filter banks. Originally, SegNet, as majority of semantic segmentation neural networks, were developed for outdoor and indoor scene understanding, usually represented by the RGB image files. It conforms with the input of the VGG16 architecture pretrained for the color-coded images. Thus, its use for greyscale DCE-MR data requires duplication of the single image intensity channel to two other color channels, which is obviously a computational overhead.

2.3 Fully Convolutional DenseNets

The Tiramisu network builds on the concept of Densely Connected Convolutional Networks, which occurred effective in multiple classification tasks [7]. In this approach, both the encoding and decoding paths contain the so-called dense blocks. Each block is composed of batch normalization layer, ReLU activation, 3×3 convolution and dropout (with proability $= 0.2$). The input to each layer is concatenated with its output to feed the next layer. Also, each layer output is concatenated to the final output of the dense block. In between of the dense blocks there are transition down and transition up units which perform either max-pooling or up-sampling. The latter operation is conducted by transposed convolution, similarily to the U-Net architecture. The number of layers within each dense block can be adjusted to the needs of a given application. In our experiements we used the same configuration as it was proposed in the original paper. Therefore, we used 4, 5, 7, 10 and 12 layers in the contractive path, and the same number of layers in the reverse order in the expansion part of the network. Together with the transition down and up layers, there were 103 layers in total.

2.4 Network Training

In case of each network, trainable weights were initiated to random state by the method of He et al. [6]. The training process was conducted on image patches cropped from the DCE-MRI volumes, each containing a single, left or right kidney cross-section. As described above, 96×96-pixel image patches were extracted from volumes of the DCE sequence corresponding to the perfusion phase, i.e. time frames of the maximum signal contrast between cortex and medulla. In order to increase the number of training images, for each study we actually selected 3 such time frames – the one with maximum signal enhancement in the cortex region plus one preceding and one succeeding time frame. In each image volume, a single kidney is visible on 12 slices on average. It gives approximately 1440 training patches.

Additionally, we enlarged the training set through data augmentation. This was accomplished by picking 10 different vertical positions of the image patch and by randomly mirroring it in the horizontal direction. While selecting patch positions, we made sure that it embraced sufficiently large portion of the image center containing significant fragment of the renal parenchyma (see Fig. 2). Overall, the number of images available for training reached the value of 13964. One-third of the training images were separated for the validation purposes.

As said above, we have trained 10 different CNN models, one for every patient. While building a model dedicated to a given subject, its corresponding image patches (irrespectively of the examination session) were removed from the training and validation sets and used only for testing. Weights of the network were updated using the stochastic gradient descent algorithm with the constant learning rate = 0.01 and momentum = 0.99. The loss function chosen to optimize was the binary cross-entropy criterion, defined as

$$\mathcal{H} = -\frac{1}{N} \sum_{i=1}^{N} y_i \log{(p(y_i))} + (1 - y_i) \log{(1 - p(y_i))} \tag{1}$$

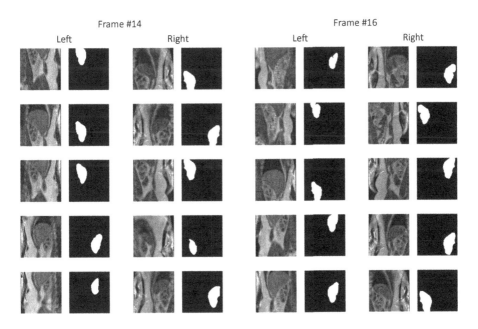

Fig. 2. Examples of training image patches extracted from left and right kidneys from two time frames of Subject 1. Data augmentation was realized by image flipping in horizontal direction and vertical shifting of patch location relative to image center.

where N is the number of voxels, y_i is the true voxel label, and $p(y_i)$ denotes the probability that an i-th voxel belongs to y_i category. Additionally, in order to monitor the quality of kidney segmentation over training epochs, we calculated the Jaccard coefficient, hereafter designated as IoU (intersection-over-union)

$$IoU = \frac{\sum_{i=1}^{K} y_i \wedge y_i^{pred}}{\sum_{i=1}^{N} y_i \vee y_i^{pred}}, \tag{2}$$

where K designates the number of pixels in a processed slice and y_i^{pred} is the predicted pixel category. Here, categories are Boolean-valued and a pixel is labeled *True* if it belongs to the kidney, *False* otherwise. In the case of each subject, the optimization algorithm was run for 50 epochs. The stored model corresponded to the epoch with the minimum score on the loss function obtained for the validation data set.

3   Renal Voxels Classification

3.1   Feature Extraction

Differentiation of voxels representing particular renal compartments could be based on raw signal intensity time courses. We propose, however, to transform signal waveforms into the space of reduced dimensionality using principal component analysis (PCA). The purpose of this transform is not only to decrease the complexity of the resulting classification model but also to extract a more general characteristics of the kidney tissue, representative for various subjects. We presumed that the extracted PCA components should explain at least 90% of the original data set variance. Therefore, in the case of our experimental data (see Sect. 5.1), where each dynamic sequence consisted of 74 time frames, vectors of 74 temporal features (i.e. image signal intensities in subsequent time steps) were transformed into the space of 20 PCA feature aggregates.

3.2   Feature Vectors Classification

Assignment of renal voxels to cortex, medulla or pelvis is performed by a classifier trained in the supervised manner. In our approach, historical data serve as patterns for building appropriate decision rules, later applied to new studies. The training vectors were acquired from regions of interest manually annotated in the respective parenchymal locations. The annotations were made only in voxels whose membership was unambiguous (see Fig. 3a–b), thus letting a trained classifier to decide about the dominating tissue category in case of voxels partially filled with various compartments. The number of training vectors collected from the 20 available examinations exceeded the value of 60,000. This data set was partitioned into 10 folds, each containing data vectors from all but one subject, left apart for testing purposes. In a given fold, the class distribution was approximately as follows: cortex – 58%, medulla – 31%, pelvis – 11%. In order to give

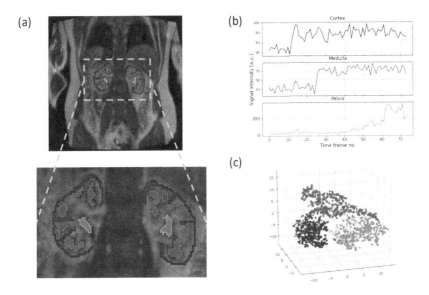

Fig. 3. Preparation of training data for supervised learning of classifiers: a) ROI placement in a DCE-MRI frame; b) signal time courses assigned to corresponding ROI voxels; c) 3-dimensional visualization of PCA feature vectors representing cortex (blue), medulla (red) and pelvis (magenta) ROIs. The visualization was obtained by transforming 20 PCA features using t-SNE method. (Color figure online)

classifiers a chance to learn to discriminate categories with equal accuracy, in each training fold the subsets representing cortex and medulla were resampled to match the size of the pelvis category. On average, the training set after resampling embraced over 16,000 vectors per fold. In a given training fold, data from both examination sessions were included. On the other hand, the testing folds contained from 600 to 4,800 vectors depending on the patient and examination session. Classifiers were evaluated using the balanced accuracy score calculated on the test sets. As previously noted, we used support vector machines with the radial basis function as the kernel to accomplish the classification task.

4   Pharmacokinetic Modeling

The 2-compartment filtration (2CFM) model assumes that signal measured in a given tissue voxel is a sum of contributions originating from intravascular (IV) and extracellular extravascular (EEV) spaces [19]. Furthermore, as in each PK model, the delivery of the gadolinium contrast agent through a feeding artery to the kidney, is encapsulated by the so-called arterial input function (AIF). Practically, AIF in case of the kidney studies, is the time-course of the contrast agent concentration in the abdominal aorta [4]. By convolving the AIF with a shifting and dispersion kernel one obtains tracer concentration in the IV compartment. Eventually, the time curve of the concentration in the EEV space is proportional to the integral of the concentration in the IV compartment. The proportionality coefficient, denoted as K^{trans}, controls the rate of CA transfer from IV to EEV compartment. K^{trans} multiplied by volume of the organ leads to calculation of GFR.

Formally, the CA concentration in the tissue is given by

$$C_{tissue}(t) = K^{trans} \int_0^t C_p^{kid}(\tau) + v_p C_p^{kid}(t), \tag{3}$$

with

$$C_p^{kid} = C_p^{art} \otimes g(t) = \int_0^t C_p^{art}(t - \tau) g(\tau) d\tau, \tag{4}$$

where C_p^{art} denotes the arterial input function, v_p – plasma volume fraction, and C_p^{kid} – CA concentration in the blood plasma. The first term in (3) represents the CA concentration in the EEV space, whereas the second term covers the concentration in the IV space obtained by convolving arterial input function with the vascular impulse response function (VIRF), defined as

$$g(t) = \begin{cases} 0 & t < \Delta \\ \frac{1}{T_g} e^{-\frac{t-\Delta}{T_g}} & t \geq \Delta \end{cases}. \tag{5}$$

Variables T_g – the dispersion time constant, and Δ – the delay interval, together with the volume fraction v_p and transfer constant K^{trans} form the complete set of the 2CFM model parameters, which we find using the Levenberg-Marquardt non-linear least squares curve-fitting procedure.

5   Experiments

5.1   MRI Data

Twenty DCE-MRI examinations were available for experiments. The datasets were collected for 10 healthy volunteers. Each subject was imaged twice, seven days apart (further, these examinations will be referred to as Session 1 and 2). The acquisition sequence used standard 3D FLASH spoiled gradient recalled echo technique with the following parameters: TR = 2.36 ms, TE = 0.8 ms, FA = 20°, in-plane resolution = 2.2 × 2.2 mm², slice thickness = 3 mm, acquisition matrix = 192 × 192, number of slices = 30. Prior to image acquisition, patients were administered 0.025 mmol/kg of GdDOTA at 3 mL/s flow rate. The contrast agent was injected intravenously. Then, 74 volumetric scans were gathered at 2.3 s time intervals. In order to validate the obtained estimates against ground truth values, volunteers underwent iohexol clearance tests. The measurement was carried out by administrating a dose of 5 mL of iohexol (300 mg I/mL; Omnipaque 300, GE Healthcare) and then by acquiring a venous blood sample after 4 h. All volunteers gave their written informed consent for participation in the examinations. The study protocol, including its ethical aspects, was approved by the Institutional Review Board at the Haukeland University Hospital Bergen, Norway.

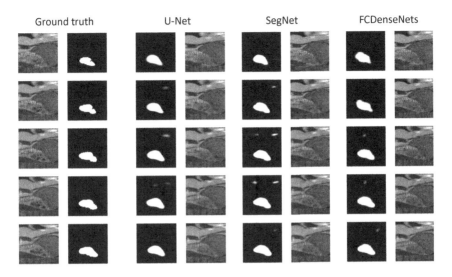

Fig. 4. Example outputs of the tested segmentation networks.

5.2 Results

Figure 4 shows example outputs of the tested semantic segmentation networks for one of the participating subjects along with the ground-truth annotation masks. In the selected image samples, it can be observed that SegNet in many cases produces false positive regions around actual renal tissue. Apparently, however, as also shown in Table 1, across all participating subjects, it was Tiramisu network which failed to precisely delineate parenchymal borders.

In the next stage, parenchymal voxels were classified into separate renal compartments. The results of classification for the test sets are presented in Table 2. All presented scores are mean values over 10 subjects. The SVM exhibits the balanced accuracy equal to 96% and also gains high rate of true positive detections, as well as it seems to be relatively robust against false predictions. Using our algorithm, we achieved the mean Jaccard coefficient for the cortex class in the left kidney equal to 93.2%. In case of the other regions the IoU equated approximately 91%, except for the pelvis class in the left kidney where it dropped to

Table 1. Mean Jaccard coefficients over all subjects and MR sessions.

| | IoU | |
CNN architecture	Left	Right
U-Net	0.941	0.940
SegNet	0.932	0.927
FCDenseNets	0.912	0.908

Table 2. Classification metrics averaged over the testing sets – subjects 1–10, both examination sessions.

Balanced accuracy	Cortex		Medulla		Pelvis	
	Recall	Precision	Recall	Precision	Recall	Precision
0.956	0.951	0.970	0.954	0.941	0.962	0.919

90.1%. The quality of fine segmentation can be visually confirmed by analyzing examples of the kidney decomposition into regions shown in Fig. 5.

Validity of the results was further verified by using the obtained segmentation masks in the process of GFR assessment. The mean signals in the cortex regions were fitted to the 2CFM pharmacokinetic model. Then, reproducibility of image-based GFR estimates were compared against iohexol-derived measurements using the Bland-Altman method. The mean difference μ_d for the MR examination Session 1 was equal to -7.4 ml/min/1.73 m². In the case of Session 2, the agreement with the reference method was found weaker (-12.9 versus -14.1 ml/min/1.73 m²).

Slices 13–16

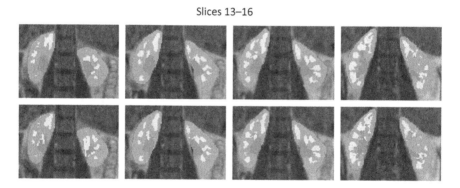

Fig. 5. Comparison of segmentation results obtained by the proposed method (bottom row) with ground truth annotations (top).

6   Conclusions

To conclude, in this paper we presented a computational framework for supporting quantitative assessment of kidney perfusion by providing an automated way of kidney parenchyma segmentation. We compared three CNN architectures for semantic segmentation. The obtained results demonstrated superior performance of the classic U-Net network over SegNet and FCDenseNets structures. Morever, we showed that classification of voxels belonging to the kidney masks automatically found by our designed U-Net network leads to reliable quantification of renal perfusion. These findings bring closer the clinical application of DCE-MR imaging as a routine method in kidney diagnostics. The designed segmentation method allows for increased objectivism of the image-derived perfusion parameters and also potentially faster diagnosis of renal impairments.

References

1. Badrinarayanan, V., Kendall, A., Cipolla, R.: SegNet: a deep convolutional encoder-decoder architecture for image segmentation. IEEE Trans. Pattern Anal. Mach. Intell. **39**(12), 2481–2495 (2017). https://doi.org/10.1109/TPAMI.2016. 2644615
2. Chen, L.-C., Zhu, Y., Papandreou, G., Schroff, F., Adam, H.: Encoder-decoder with atrous separable convolution for semantic image segmentation. In: Ferrari, V., Hebert, M., Sminchisescu, C., Weiss, Y. (eds.) ECCV 2018. LNCS, vol. 11211, pp. 833–851. Springer, Cham (2018). https://doi.org/10.1007/978-3-030-01234-2_49
3. Choi, Y., et al.: The initial area under the curve derived from dynamic contrast-enhanced MRI improves prognosis prediction in glioblastoma with unmethylated mgmt promoter. Am. J. Neuroradiol. **38**(8), 1528–1535 (2017)
4. Cutajar, M., Mendichovszky, I., Tofts, P., Gordon, I.: The importance of AIF ROI selection in DCE-MRI renography: reproducibility and variability of renal perfusion and filtration. Eur. J. Radiol. **74**(3), e154–e60 (2010)
5. Haghighi, M., Warfield, S.K., Kurugol, S.: Automatic renal segmentation in DCE-MRI using convolutional neural networks. In: 2018 IEEE 15th International Symposium on Biomedical Imaging (ISBI 2018), pp. 1534–1537 (2018). https://doi. org/10.1109/ISBI.2018.8363865
6. He, K., Zhang, X., Ren, S., Sun, J.: Delving deep into rectifiers: surpassing human-level performance on imagenet classification. In: Proceedings of the IEEE International Conference on Computer Vision (ICCV), pp. 1–11, December 2015
7. Huang, G., Liu, Z., Van Der Maaten, L., Weinberger, K.Q.: Densely connected convolutional networks. In: 2017 IEEE Conference on Computer Vision and Pattern Recognition (CVPR), pp. 2261–2269 (2017). https://doi.org/10.1109/CVPR. 2017.243
8. Johnson, H.J., McCormick, M.M., Ibanez, L.: The ITK Software Guide: Design and Functionality. Kitware (2020)
9. Jégou, S., Drozdzal, M., Vazquez, D., Romero, A., Bengio, Y.: The one hundred layers tiramisu: Fully convolutional densenets for semantic segmentation. In: 2017 IEEE Conference on Computer Vision and Pattern Recognition Workshops (CVPRW), pp. 1175–1183 (2017). https://doi.org/10.1109/CVPRW.2017.156
10. Khan, Z., Yahya, N., Alsaih, K., Ali, S.S.A., Meriaudeau, F.: Evaluation of deep neural networks for semantic segmentation of prostate in T2W MRI. Sensors **20**(11), 3183 (2020). https://doi.org/10.3390/s20113183
11. Kline, T.L., et al.: Performance of an artificial multi-observer deep neural network for fully automated segmentation of polycystic kidneys. J. Digit. Imaging **30**(4), 442–448 (2017). https://doi.org/10.1007/s10278-017-9978-1
12. Lee, V.S., et al.: Renal function measurements from MR renography and a simplified multicompartmental model. Am. J. Physiol.-Renal Physiol. **292**(5), F1548–F1559 (2007). https://doi.org/10.1152/ajprenal.00347.2006
13. Li, S., et al.: Wavelet-based segmentation of renal compartments in dce-mri of human kidney: Initial results in patients and healthy volunteers. Comput. Med. Imaging Graph. **36**, 108–18 (2012). https://doi.org/10.1016/j.compmedimag.2011. 06.005

14. Liu, S., Deng, W.: Very deep convolutional neural network based image classification using small training sample size. In: 2015 3rd IAPR Asian Conference on Pattern Recognition (ACPR), pp. 730–734 (2015). https://doi.org/10.1109/ACPR. 2015.7486599

15. Ronneberger, O., Fischer, P., Brox, T.: U-Net: convolutional networks for biomedical image segmentation. In: Navab, N., Hornegger, J., Wells, W.M., Frangi, A.F. (eds.) MICCAI 2015. LNCS, vol. 9351, pp. 234–241. Springer, Cham (2015). https://doi.org/10.1007/978-3-319-24574-4_28

16. Shelhamer, E., Long, J., Darrell, T.: Fully convolutional networks for semantic segmentation. IEEE Trans. Pattern Anal. Mach. Intell. **39**(4), 640–651 (2017). https://doi.org/10.1109/TPAMI.2016.2572683

17. Sourbron, S.P., Michaely, H.J., Reiser, M.F., Schoenberg, S.O.: MRI-measurement of perfusion and glomerular filtration in the human kidney with a separable compartment model. Investigative Radiology **43**(1), 40–48 (2008). https://doi.org/10. 1097/rli.0b013e31815597c5

18. Srivastava, N., Hinton, G., Krizhevsky, A., Sutskever, I., Salakhutdinov, R.: Dropout: a simple way to prevent neural networks from overfitting. J. Mach. Learn. Res. **15**(56), 1929–1958 (2014)

19. Tofts, P., Cutajar, M., Mendichovszky, I., Peters, A., Gordon, I.: Precise measurement of renal filtration and vascular parameters using a two-compartment model for dynamic contrast-enhanced MRI of the kidney gives realistic normal values. Eur. Radiol. **22**, 1320–30 (2012). https://doi.org/10.1007/s00330-012-2382-9

20. Ulas, C., et al.: Convolutional neural networks for direct inference of pharmacokinetic parameters: Application to stroke dynamic contrast-enhanced MRI. Frontiers Neurol. **9**, 1147 (2019). https://doi.org/10.3389/fneur.2018.01147

21. Yang, X., Le Minh, H., (Tim) Cheng, K.T., Sung, K.H., Liu, W.: Renal compartment segmentation in dce-mri images. Med. Image Anal. **32**(C), 269–280 (2016). https://doi.org/10.1016/j.media.2016.05.006

22. Yoruk, U., Hargreaves, B.A., Vasanawala, S.S.: Automatic renal segmentation for MR urography using 3D-GrabCut and random forests. Magn. Reson. Med. **79**(3), 1696–1707 (2018). https://doi.org/10.1002/mrm.26806

23. Zöllner, F., et al.: Assessment of 3d dce-mri of the kidneys using non-rigid image registration and segmentation of voxel time courses. Comput. Med. Imaging Graph. **33**, 171–81 (2009). https://doi.org/10.1016/j.compmedimag.2008.11.004

Comparison of Efficiency, Stability and Interpretability of Feature Selection Methods for Multiclassification Task on Medical Tabular Data

Ksenia Balabaeva[(✉)] and Sergey Kovalchuk

ITMO University, Saint-Petersburg, Russia

Abstract. Feature selection is an important step of machine learning pipeline. Certain models may select features intrinsically without human interactions or additional algorithms applied. Such algorithms usually belong to neural networks class. Others require help of a researcher or feature selection algorithms. However, it is hard to know beforehand which variables contain the most relevant information and which may cause difficulties for a model to learn the correct relations. In that respect, researchers have been developing feature selection algorithms. To understand what methods perform better on tabular medical data, we have conducted a set of experiments to measure accuracy, stability and compare interpretation capacities of different feature selection approaches. Moreover, we propose an application of Bayesian Inference to the task of feature selection that may provide more interpretable and robust solution. We believe that high stability and interpretability are as important as classification accuracy especially in predictive tasks in medicine.

Keywords: Feature selection · Bayesian inference · Explainable artificial intelligence · XAI · eXAI · Recursive feature elimination · Kbest

1 Introduction

Due to the recent advances in machine learning algorithms application to different domains, there is a huge demand of decision support systems based on learning methods. One of the most popular type of machine learning tasks is supervised learning which undermines the use of a dataset with corresponded labels to each instance. Such tasks are, for example, regression, binary and multiclassification, depending on the target variable type.

In supervised learning, the input data typically consist of a matrix of features and a target vector. Such matrix may take a form of an image in the tasks of image classification, object detection or semantic segmentation. Another example of feature matrix may be vectorized text representation. Such feature matrix is widely used in natural language processing tasks, such as, sentiment analysis, text classification, etc. Data in the form of images, video, plain text and audio is usually called unstructured data. However,

© Springer Nature Switzerland AG 2021
M. Paszynski et al. (Eds.): ICCS 2021, LNCS 12744, pp. 623–633, 2021.
https://doi.org/10.1007/978-3-030-77967-2_51

the most widely used type of the input data is a tabular data containing variables of different nature in each column. Estimates say that 20% of the data are structured and approximately 80% are unstructured [1].

Machine learning algorithms' performance strongly depend on the number and variability of samples provided in training datatset. However, the raise in the number of variables may lead to the curse of dimensionality [2]. This problem refers to a higher risk of overfitting especially if the number of features is higher than the number of samples. Another problem arising from a big dimensionality is that the observations in high dimensional space become equidistant which makes them harder to cluster or classify. To solve this problem, we have to reduce the feature space. There are several ways to deal with it: feature extraction (PCA [3], LDA [4], Transformer [5]) or feature selection, which will be discussed in more details in further sections. All in all, the aim of all techniques is in reducing the number of columns in a training dataset [6].

In present study we analyze only feature selection selection algorithms and there are two reasons why we eliminate feature extraction methods. First of all, we would like to compare the existing feature selection methods with the proposed application of Bayesian Inference to this task. Since the proposed algorithm select features it is clear that at first step, we have to compare its performance with analogous methods. Another reason is that feature extraction approaches compress the initial feature space to reduce the dimensionality. For instance, using PCA we get a number of principal components in which initial features are encoded. After such compression we can't explain what features are contained in a single component. Even though the number of such components may be low, this way of compression causes difficulties for the interpretability. Compared to feature extraction, feature selection techniques are more transparent and explainable, since the reduced feature space consist of the original variables in data. Moreover, such reduction may reduce the training time and contribute to the accuracy of the model.

In the present work we compare several feature selection techniques on the case of chronic heart failure stage prediction. We also present an approach of Bayesian Feature Selection application to the task of feature selection. As comparative standards, we evaluate the selection algorithms using f-score with macro averaging as performance indicator, stability of the model, using k-fold cross-validation and interpretability of the feature selection results.

The rest of this paper is structured in the following way. Section 2 provides background on feature selection studies in medical domain, describes feature selection concepts and the most wide spread methods. Section 3 presents an approach to feature selection using Bayesian Inference. Section 4 provides details on experimental pipeline. Section 5 describes the results and discussion. Section 6 concludes the work.

2   Related Works

The field of feature selection is a big part of machine learning domain. Therefore, there are plenty of algorithms appearing each year. However, there is lack of papers comparing the efficiency of feature selection techniques, their stability and interpretability. Therefore, it is quite hard for the practitioners to select the appropriate solution and understand the risk of overfitting using one or another method. Another issue is the lack of overviews on application feature selection methods to the medical domain.

As a rare example of such works, we may take the paper [6]. This work covers the field of medical imaging, DNA analysis, biomedical signals processing and testing the feature selection techniques on two tasks. As feature selection algorithms they compare CFS, Cons, INTERACT, InfoGain, ReliefF, SVM-RFE. There are also some works that address feature selection to one specific task. For instance, medical image retrieval [7] and In [8] the authors compare different information retrieval methods for medical image segmentation [8]. Such as thresholding based, clustering-based, watershed-based and graph-based, etc. In [9] authors apply Discrete Wavelet Transform to decompose an image into images with different scales in order to extract information. Another study concludes that biologically informed feature selection methods applied for Alzheimer diagnosis stages prediction are more efficient than uninformed [10]. Concerning medical signals processing, there are also a couple of works on EMG, EEG and ECG [11, 12]. For instance, in [11] authors extract features from EMG signals using time, frequency and time-frequency domain features. Another domain of feature selection application to medicine is microarray data classification [13–15]. In [13] authors provide a review on feature selection methods, and include the software overview for microarray data, which is primarily written for R and C programming languages.

2.1 Feature Selection Approaches

Feature selection is a procedure of processing initial dataset in the end of which a sample of features become eliminated due to their redundancy and lack of useful information. There are several types of classification of feature selection approaches, but we decided to use the classification proposed in this work [16].

2.1.1 Filter Methods

Filter methods are techniques in which only features characteristics are used without a learning model. [17]. Generally, such approaches consist of two stages: choosing a criteria to rank the feature, and selecting top-ranking features. Such algorithms are Correlation-based Feature Selection (CFS) [18], Variance threshold [19], F-test [20], etc.

Variance Threshold removes features with variation below a certain threshold. Variance here is treated as an indicator of information provided by the feature: those features that do not change much across observations have less information. The limitation of such method if that it doesn't undermine relations between features and labels.

Select K Best. This approach is associated with statistical F-test, which conducts a hypothesis testing. As a limitation of this method, we have to admit, that it only checks for linear relationships between features and target variable. Another specificity is that features with high correlation will be given a higher score and less correlated features will get lower score.

General limitation of filter techniques is the inability to get the information from model's performance while selecting the optimal feature set.

2.1.2 Wrapper Methods

The class of the wrapper methods define the best feature subset as a subset that leads to the higher performance of the learning model. Therefore, such methods use specific learning algorithm to select features [21].

The work of a basic wrapper model could be divided into three stages: feature set selection, feature set evaluation and induction algorithm (predefined learning model).

Forward Feature Selection. On the first iteration the model with no features. Then the features are added to the training dataset one by one to reach the highest score [21].

Backward Feature Elimination. On the first iteration the model is trained on the whole dataset. Iteratively features are eliminated one by one in the way to get highest score [21].

Recursive Feature Elimination (RFE). This is an optimization algorithm which aims to find the best performing feature subset. Unlike previous methods, this approach creates new model recursively [22].

2.1.3 Embedded Methods

There is a class of feature selection methods that took the advantages from both filterbased and wrapper methods. Embedded models incorporate the feature selection process inside the learning model [23]. Embedded models are more robust than wrapper methods, since the selection process is not evaluated by the learning model. For that reason, they are more stable and less susceptible to over-fitting. The most popular embedded methods are ridge and lasso regression.

Lasso Regression. This method incorporates L1 regularization by adding penalty equal to the absolute value of the magnitude of regression coefficients.

Ridge Regression. This method performs L2 regularization which by adding penalty equal to square of the magnitude of regression coefficients.

3 Methodology

As an alternative to existing feature selection algorithm, we propose to apply technique based on Bayesian Inference and probabilistic modeling. A similar procedure was applied to clustering results interpretation in our previous work [24]. However, the same technique can also contribute to feature selection in binary or multiclassification tasks.

The proposed approach is based on Bayesian inference [25]. The algorithm consists of three stages: posterior sampling, comparison matrix calculation and identification of features. The idea of this method is to select most typical features in each class by comparing their sampled distributions [24]. For instance, if the distribution of feature 1 significantly differs in class A compared to class B, we have to add feature 1 to the model. Based on this comparison we may select more relevant features that may help the classifier to build more accurate model.

Considering the methods classification provided in Chapter 2 the proposed approach is associated with filter methods, because it doesn't rely on score of the learning model while selecting the features.

3.1 BI Feature Selection Algorithm

Let $x_{n_i c_k}$ be a number of successes an observation belongs to class c_i with n_i being an overall number of observations for cluster c_i and p_{n_j, c_i} is the probability an observation belongs to class c_i.

Step 1. Posterior Sampling.

For each variable $f_j, j \in [0, M]$:

For each class $c_i, i \in [0, K]$:

Select the priors for $x_{n_i c_l}$ and $P(A)_{f_j c_l}$;

Calculate the posterior distribution $P(A/D)_{f_j c_l}$;

Sample W new observations from the posterior $P(A/D)_{f_j c_l}$;

Step 2. Feature comparison matrix.

Let I be a 2D matrix with the number of rows and columns equal to the number of classes. The value of each matrix element $i_{c_l c_{l+1}}$ is equal to the mean value of sampled probabilities comparison. The calculation depends on the hypothesis we want to check: whether the values of a feature in one class are higher or lower to the features in other class.

Step 3. Identification of features more typical for a class.

Output: dictionary with keys equal to class numbers and values equal to array of associated features with this class [24]. Finally, we can build a sample of more relevant features concatenating the output of the third step.

This approach has two main parameters: the significance level and the number of classes to which the distribution of a feature in current class must be significantly different. The significance level may vary from 0 to 1, where 0 means that there is no differences between distributions in classes and 1 means that the distributions are completely different. The number of comparison classes may vary from 0 to the maximal number of classes in target vector minus 0.

4 Experiments. Case of Chronic Heart Failure Stage Prediction

To test feature selection methods, we picked a multiclassification task, training ML models to predict the stage of congestive heart failure. Congestive heart failure (CHF) occurs when heart muscle struggles with pumping the blood as well as it should. This disease may be caused by narrowed arteries in heart or arterial hypertension which is a widely spread chronic disease. According to the clinical classification, there are 4 stages of CHF, where the first stage represents the weak disease and the fourth represents the severe progression of CHF.

The dataset consists of 1279 observations represented by patients. The target vector has 4 classes, representing the stages of CHF. Distribution of the target vector is depicted on Fig. 1. The most popular stage in the sample is the third one – there are more than 600 patients with this stage.

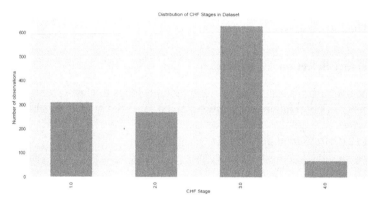

Fig. 1. Distribution of CHF stages

The feature dimension is represented by socio-demographic characteristics (age, gender), labs results (hemoglobin, neutrogene, etc.), blood pressure measures, main diagnosis, etc. In total, there are 178 features describing each patient, which is an extremely high dimensional space.

In order to reduce the number of features, we test different feature selection techniques (Table 1). Each of the selection algorithm has its own parameters that we optimized according to our quality metric F1_score.

Table 1. Feature selection algorithms and their parameters

Feature selection technique	Parameters
Variance threshold	Threshold
Bayesian feature selection	Significance level, num classes
KBest	Number of features selected
RFE	Minimal number of features to select
Lasso regression	–
Ridge regression	–

Since Ridge and Lasso Logistic Regressions are embedded methods, we used them only within Logistic regression, treating each modification as a single classifier.

After the selection is completed, we pass the new feature set to ML classifier. For this task three ML models were tested: Logistic Regression (Ridge and Lasso), Random Forest and Gradient Boosting. Further we check the quality of predictions using the test dataset and cross-validation with 5 folds calculating f1-score with macro-averaging. We chosen f1-score because it can be used for imbalanced classes (Fig. 1). The results of the experiments are presented in Sect. 5.

5 Results and Discussion

According to the experimental pipeline, we compared classifiers performance applying different feature selection techniques (Table 2). The scores presented in Table 2 were calculated on test data (33% of the initial dataset). The highest f1-score on test set for all classifiers was performed by recursive feature elimination (RFE). The second-best feature selectors were Bayesian Selection for Logistic Regressions and KBest (F-test) for Random forest and XGBoost.

Table 2. Comparison of the feature selection techniques efficiency according to the model's f1 score based on test data

Classifier/selection technique	No feature selection	Variance threshold	Bayesian selection	KBest	RFE
LogReg	0.36527	0.36527	**0.39669**	0.36527	**0.48699**
LogReg + L1	0.3793	0.3793	**0.39669**	0.3793	**0.4869**
LogReg + L2	0.36527	0.36527	**0.39669**	0.36527	**0.48699**
Random forest	0.45875	0.37113	0.4449	**0.46299**	**0.5839**
XGBoost	0.46013	0.39095	0.45019	**0.47764**	**0.63593**

We used K-fold validation to check stability and robustness of the classifiers trained on the selected feature sets. However, it is a useful tool to measure accuracy as well. The validation results are presented in Table 3. Here we see that almost all of the selection methods are losing the quality being checked on validation samples. However, the most significant drop is associated with RFE (0.15–0.20 f-1score decrease). According to the validation, the most accurate classification was performed using Bayesian Selection for logistic regression, K-Best selection for Random Forest and RFE for gradient boosting.

Concerning the stability, RFE is more exposed to the overfitting, since it is a wrapper method, and it exploits learning algorithms for feature evaluation. This issue finds confirmation in our experimental results due to the high score differences on training and validation sets. Other algorithms have relatively similar change in the score.

The third criteria of comparison id the optimal number of features found by each feature selection algorithm (Table 4). To a certain extent, the number of features represent the complexity, transparency and interpretability of the solution. The smaller the number of features – the easier it is to explain the results. For all classifiers Variance Threshold selected 27 features with threshold equal to 0.9–0.7. That is the smallest number of features selected by any algorithm. The size of the feature sample selected by Bayesian Selection varies from 47 to 73 for different models. KBest algorithm picked from 50 to 140 features. And RFE selected 27 features for Logistic Regression Model, 44 for

Table 3. Comparison of the feature selection techniques stability according to the model's mean validation f1-score (+− std).

Classifier/selection technique	No feature selection	Variance threshold	Bayesian selection	KBest	RFE
LogReg	0.34528 (+− 0.0442)	0.34528 (+−0.02214)	**0.3565**(+−0.0369)	0.34528 (+− 0.0442)	0.32973 (+− 0.0313)
LogReg + L1	0.34677(+−0.0221)	0.34677(+−0.04252)	**0.3565**(+−0.0369)	0.34677(+− 0.0442)	0.34677(+− 0.03132)
LogReg + L2	0.34528 (+− 0.02214)	0.34528 (+−0.02214)	**0.35652** (+−0.0369)	0.34528 (+− 0.0442)	0.32973 (+− 0.03132)
Random forest	0.38410 (+−0.04653)	0.39642 (+−0.1555)	0.404787 (+−0.08486)	**0.4105** (+−0.14035)	0.39829 (+−0.0329)
XGBoost	0.4220 (+−0.0270)	0.370087 (+−0.044835)	0.41719(+−0.0702)	0.3022 (+−0.06949)	**0.43086** (+−0.0632)

XGBoost and 90 for Random Forest. Even though for Logistic Regression both RFE and Variance Threshold picked the same number of features, the sample selected by Variance Threshold is more relevant, according to validation results (Table 3).

Table 4. Comparing the number of features selected by each FS technique

Classifier/selection technique	No feature selection	Variance threshold	Bayesian selection	KBest	RFE
LogReg	178	27	73	140	27
LogReg + L1	58	27	73	140	27
LogReg + L2	178	27	73	140	27
RandomForest	178	27	47	50	90
XGBoost	178	27	64	90	44

Concerning the interpretability of feature selection techniques, filter based methods and embedded methods are more transparent to users, since the logic of feature selection is simpler. For instance, variance threshold is just the elimination of uninformative features. Or Bayesian Selection is just the selection of features more specific for each particular class. Whereas RFE is a complex process of feature selection and evaluation requiring learning models. We do not include the nomenclature of selected features, since their initial number is high – almost 200 features and the limits of the paper won't allow us to discuss each selected set in detail.

6 Conclusion

As the main result, we have to say that the proposed objective of our work is reached. We suggested a filter-based approach to feature selection task based on Bayesian inference and probabilistic modeling. This method performs sufficient accuracy for multiclassification task and is quite robust to the problem of overfitting. In our experiments we compared this method with other feature selection algorithms and presented the results concerning their stability, accuracy and explainability.

All things considered, the obvious conclusion to be drawn is that feature selection may help to improve the ML models performance, reduce the learning time and foster the search of the optimal hyperparameters due to dimensionality reduction.

However, the choice of the feature selection algorithm strongly depends on data and machine learning model. In our case, Bayesian feature selection performed better on Logistic regression, K-best algorithm reached higher score working with Random forest and RFE booster the performance of XGBoost.

In future we would like to compare the proposed approach with other types of feature selection methods and test it on different datasets.

Acknowledgement. This research was supported by the by the Ministry of Science and Higher Education of Russian Federation, goszadanie no. 2019–1339. Participation in the ICCS conference was supported by the NWO Science Diplomacy Fund project # 483.20.038 "Russian-Dutch Collaboration in Computational Science".

References

1. https://www.cio.com/article/3406806/ai-unleashes-the-power-of-unstructured-data.html. Accessed 10 Feb 2021
2. Poler, R., Mula, J., Díaz-Madroñero, M.: Dynamic programming. In: Operations Research Problems, pp. 325–374. Springer, London (2014). https://doi.org/10.1007/978-1-4471-5577-5_9
3. Mishra, S., et al.: Principal component analysis. Int. J. Livestock Res. **1** (2017). https://doi.org/10.5455/ijlr.20170415115235.
4. Blei, D., Ng, A., Jordan, M.: Latent Dirichlet allocation. J. Mach. Learn. Res. **3**, 993–1022 (2003)
5. Polosukhin, I., Lukasz K., et al.: Attention Is All You Need (2017)
6. Remeseiro, B., Bolon-Canedo, V.: A review of feature selection methods in medical applications. Comput Biol Med. **112**, 103375 (2019). https://doi.org/10.1016/j.compbiomed.2019.103375. Epub 2019 Jul 31 PMID: 31382212
7. Kalpathy-Cramer, J.: Evaluating performance of biomedical image retrieval systemsan overview of the medical image retrieval task at imageclef 2004–2013. Comput. Med. Imag. Graph. **39**, 55–61 (2015)
8. Huang, Q., Luo, Y., Zhang, Q.: Breast ultrasound image segmentation: a survey. Int. J. Comput. Assist. Radiol. Surg. **12**(3), 493–507 (2017). https://doi.org/10.1007/s11548-016-1513-1
9. Sudarshan, V.K., et al.: Application of wavelet techniques for cancer diagnosis using ultrasound images: a review. Comput. Biol. Med. **69**, 97–111 (2016)
10. Rathore, S., Habes, M., Iftikhar, M.A., Shacklett, A., Davatzikos, C.: A review on neuroimaging-based classification studies and associated feature extraction methods for alzheimer's disease and its prodromal stages. Neuroimage **155**, 530–548 (2017)
11. Nazmi, N., Abdul Rahman, M., Yamamoto, S.-I., Ahmad, S., Zamzuri, H., Mazlan, S.: A review of classification techniques of emg signals during isotonic and isometric contractions. Sensors **16**(8), 1304 (2016)
12. Acharya, U.R., Fujita, H., Sudarshan, V.K., Bhat, S., Koh, J.E.: Application of entropies for automated diagnosis of epilepsy using EEG signals: a review. Knowl. Based Syst. **88**, 85–96 (2015)
13. Saeys, Y., Inza, I., Larrañaga, P.: A review of feature selection techniques in bioinformatics. Bioinformatics **23**(19), 2507–2517 (2007)
14. Remeseiro, B., Bolon-Canedo, V.: A review of feature selection methods in medical applications. Comput. Biol. Med. **112**, 103375 (2019)
15. Bolón-Canedo, V., Sánchez-Marono, N., Alonso-Betanzos, A., Benítez, J.M., Herrera, F.: A review of microarray datasets and applied feature selection methods. Inf. Sci. **28**(2), 111–135 (2014)
16. Colaco S.: Review on Feature Selection Algorithms (2019)
17. Liu, H., Motoda, H.: Computational Methods of Feature Selection. CRC Press, New York (2007)
18. Doshi, M., Chaturvedi, D.S.K.: Correlation based feature selection (cfs) technique to predict student performance. Int. J. Comput. Netw. Commun. (IJCNC). **6**(3), 197 (2014)

19. Guyon, I.: An introduction to variable and feature selection. J. Mach. Learn. Res. **3**, 1157–1182 (2003)
20. Elssied, N., Ibrahim, Assoc Prof. Dr. O., Osman, A.H.: A novel feature selection based on one-way ANOVA f-test for e-mail spam classification. Res. J. Appl. Sci. Eng. Technol. **7**(3), 625–638 (2014). https://doi.org/10.19026/rjaset.7.299
21. Kohavi, R., John, G.H.: Wrappers for feature subset selection. Artif. Intell. **97**(1), 273–324 (1997)
22. Sam, M.L., Camara, F., Ndiaye, S., Slimani, Y., Esseghir, M.A.: A Novel RFESVM-based Feature Selection Approach for Classification. Int. J. Adv. Sci. Technol. **43**(1), 27–36 (2012)
23. Kabir, M.M., Islam, M.M., Murase, K.: A new local search based hybrid genetic algorithm for feature selection. Neurocomputng **74**, 2194–2928 (2011)
24. Balabaeva, K., Kovalchuk, S.: Post-hoc interpretation of clinical pathways clustering using Bayesian inference. Procedia Computer Science **178**, 264–273 (2020)
25. Davidson-Pilon, C.: Bayesian Methods for Hackers: Probabilistic Programming and Bayesian Inference. Addison-Wesley (2019)

Side Effect Alerts Generation from EHR in Polish

Wojciech Jaworski[1] [iD], Małgorzata Marciniak[2(✉)] [iD],
and Agnieszka Mykowiecka[2] [iD]

[1] LekSeek Polska, Puławska 465, Warsaw, Poland
[2] Institute of Computer Science Polish Academy of Sciences, Jana Kazimierza 5,
Warsaw, Poland
{mm,agn}@ipipan.waw.pl

Abstract. The paper addresses the problem of extending an existing and widely used program for Polish public healthcare with a function for detecting possible occurrences of drug side effects. The task is performed in two steps. First, we extract information that binds names of drugs with side effects and their frequency. In the next step, we look for similar phrases in the list of side effect phrases. For all words in phrases, we use Polish Wordnet to find similar ones, and check if phrases with replaced words exist in the list. For long side effect phrases, which never occur in patient records, we look for simpler internal side effect phrases to generate alarms. Finally, we evaluate to what extent this action increases the efficiency of side effect alarms.

Keywords: Drug side effects · Electronic health records · Polish

1 Introduction

Electronic health record (EHR) systems are in common use all over the world in clinics, both for administrative services and to support the work of physicians. One of the basic required functionalities of such systems is to store information about patients' medical care: the reasons for patients' visits, the results of checkups, tests, diagnosis, and prescribed medications. Analysis of this data along with other resources can increase the efficiency of physicians and the accuracy of undertaken decisions. Clinical Decision Support Systems (CDSSs) are quite common for English data [26], and available for other languages e.g. Swedish [8], German [14], and Korean [3]. Polish EHR systems are focused on the organizational and administrative aspects of the clinic's functioning and on collecting information about patients, but do not analyze the data. A summary of the use of EHR systems in Poland in 2016 and their perspectives is given in [4].

The slow development of clinical decision support systems in Poland is due to the poor resources for the processing of medical data and there not yet being a national standard for the storage of medical information. Medical terminology resources in Polish are limited to International Classification for

© Springer Nature Switzerland AG 2021
M. Paszynski et al. (Eds.): ICCS 2021, LNCS 12744, pp. 634–647, 2021.
https://doi.org/10.1007/978-3-030-77967-2_52

Nursing Practice (INCP, https://www.icn.ch/); a small part of The Unified Medical Language System [15] (UMLS), i.e.: Medical Subject Headings (Mesh—a controlled biomedical vocabulary designed for medical literature indexing and searching) [27], and the International Statistical Classification of Diseases and Related Health Problems (ICD-10). Additionally, a list of drugs and supplements approved for use in Poland is publicly available. The Systematized Nomenclature of Medicine Clinical Terms [25] (SNOMED CT) is not translated into Polish. We are not aware of any general medical ontology of Polish medical data.

The paper addresses one of the EHR systems available on the Polish market, i.e. drWidget, which has been developed for 7 years and is implemented in over 16K outpatient clinics. The system collects data concerning patients' visits. Some of the information is given in a structured form such as: visit identification, doctor identification, and basic information about patient (sex, age, id). The records describing the patient's visit (interview and examination) are given in free texts. They usually contain a large number of spelling errors which makes them difficult to process. Nowadays, all prescriptions of medications are given in electronic form. Several records containing the structural description of a drug (and its dosage) can be added to a visit. This eliminates the problem of multiple variants of drug names being used in free texts. The system also provides doctors with Summaries of medical Product Characteristics (SmPCs) in text form.

In the paper, we describe a new functionality of the system, i.e., generating alerts when a symptom described in a free text about a patient visit might be a side effect of a drug being taken by the patient. Identification of situations when a drug causes undesirable secondary effects in addition to the desired therapeutic effect can help both a doctor and a patient by making it easier to make a proper diagnosis and sparing a patient unnecessary ailments. Systems which could facilitate a doctor's diagnoses can thus potentially be very advantageous for patients, but are frequently not very well received by the physicians themselves if they are not fully reliable. The general solution to this task is difficult, as it requires many aspects to be considered to recognize not only which drugs a patient takes now but what new symptoms he/she developed and for what reason. In the current version of the system, we limit generating alerts to the case when a medication prescribed in the previous patient visit has side effects which are similar to symptoms detected in the free text about the current visit.

Ultimately, the alerts will be generated in a system used by hundreds of doctors; therefore we want to minimize the number of false positives. Too many unjustified alarms would quickly lead to users ignore them [1].

2   Related Work

Medical texts, both scientific and clinical, constitute a vast amount of data which can be mined for different kinds of information. In [9], text mining was described as an emerging tool for leveraging underutilised data sources that can improve pharmacovigilance, including the objective of Adverse Drug Event (ADE) detection and assessment.

The impact of ADE detection on patient treatment is analysed in [23], while in [19], the authors make an analysis of alert types in order to improve their effectiveness in systems. In [24], the authors pay attention to the problem of ADE in older patients who take more drugs, as chronic diseases are more common and they experience more frequent ADEs. They review studies concerning usage of clinical decision support systems to reduce the prescribing of potentially inappropriate medications. The conclusions are that CDSSs are more effective in hospitals than in ambulatory care. But the authors expressed hope that more user friendly systems could improve their effectiveness.

A summary of 30 approaches published for ADE detection in the context of EHRs before 2017 is given in [6]. The problem of ADE was addressed in shared tasks. The Adverse Reaction Extraction from Drug Labels Track was organized during the Text Analysis Conference (TAC) in 2017. Based on annotated data, participants had to identify and normalize adverse reactions from drug labels. The best 10 teams taking part in the competition provided solutions based on machine learning methods: (Bi)LSTM, CRF, SVM, CNN; the best F1 score was 0.82 [22]. Another competition took place in 2018, when participants solved the problem of extraction of ADEs from clinical data. The organizers reported an F1 measure equal to 0.89 of the best systems "that process raw narrative text to discover concepts and find relations of those concepts to their medications" [10].

Most of the reports concern data analysis in English; however, the need for data processing in other languages is also noted. [21] describes the state of automatic patient data processing in Sweden in 2010, which is very close to the current situation in Poland. The authors claim that "the current structure, content and format of SmPCs make it difficult to incorporate them into CDSSs and link them to relevant patient information from the Electronic Health Records". Our paper addresses a method of incorporating data from SmPCs to support doctor's decisions based on EHRs in Polish.

3   Initial Drug Side Effects Identification Procedure

There is no official source of drug side effects in Polish and there are no corpora annotated with information on drug side effects such as the EU-ADR corpus [28] and that described in [7]. We are not able to construct the necessary resource with the help of UMLS as in [12]. The first step of the process was thus to construct a resource with possible drug side effects from SmPCs, which was the complete and up-to-date documentation of all prescription drugs authorized to be used in Poland. We extracted information about side effects associated with the frequency their occur for a given drug. It was carried out in two steps. First, we manually marked information on side effects in SmPCs. It allowed us to create a list of 13,347 various phrases. Most of them consisted of up to 4 tokens, but some were longer. We observed that in the actual visit description corpus, it was possible to match up to 5-tokens side effect phrases. Drug descriptions have free text form, but the fragments of possible side effects that are usually enumerated together with information about their frequency are of interest to us. It is expressed by very strict phrases, a list of which was prepared manually, e.g.: *często* 'often', *rzadko* 'rare', *bardzo rzadko* 'very rare'. For a given drug, we extracted pairs consisting of side effects and the frequency of their occurrence by simply matching the previously completed lists (of side effects and frequency phrases) with the drug description.

However, finding those side effects in a patient visit record is much harder. The main problems result from linguistic diversity:

- a side effect phrase may occur in inflected forms,
- a side effect may be expressed by various phrases,
- sometimes slightly more general or more precise information may be used in place of the term mentioned within the drug description,
- terminology used by a patient (who is usually not a doctor) differs from that used in drug descriptions,
- coordination is used quite often to describe side effects in SmPCs which is not common in patient records.

Side effect phrases which are listed within drug descriptions are in nominative form, so to be able to recognize all their forms in a text, we computed their inflected forms (noun phrases are declined by cases and numbers). This was done using a generator containing data from SGJP [31] and a guesser operating on the basis of the rules describing the inflection of the Polish language.

While creating an inflection model, we focused on the most productive Polish inflection rules. The model does not include irregular verbs and a small number of words that belong to other parts of irregular speech variety. They are not very numerous and their forms are just listed in the glossary attached to the model. The model also describes acronyms, frequently used words with a non-Polish spelling, and some dialect forms.

The list of symptoms is compared with the content of the interview and examination introduced by the doctor in the office program. If any symptom from the list occurs, the patient's history is checked for whether he or she was taking the drug causing the symptom. If this is the case, a warning is displayed about the possible occurrence of side effects. Comparison with the list of previously prescribed medications is necessary due to the fact that some undesirable symptoms, e.g. 'cough', are also common symptoms of diseases.

4   Looking for Semantically Similar Side Effect Phrases

In this section, we describe a method for identifying phrases that express a similar meaning and a more general one on the side effect list. For example, in the *Oritop* leaflet, the side effect phrase *ataki lęku* 'anxiety attacks' is mentioned, while for *Epitoram* the phrase *napad lęku* 'fit of anxiety' is used. The meaning of both phrases is the same, and both can be used interchangeably in patient records. Recognition of the latter phrase should generate information about a potential undesirable effect of the first drug. Long, complex phrases never occur in patient visits, so information about the occurrence of potential side effect should be generated if a shorter, more general phrase implied by the longer one occurs. For example, the phrase *reakcje nadwrażliwości na światło słoneczne i promieniowanie ultrafioletowe* 'hypersensitive reactions to sunlight and ultraviolet radiation' implies the following phrases: *nadwrażliwość na światło słoneczne* 'hypersensitivity to sunlight' and *nadwrażliwość na światło* 'hypersensitivity to light'.

Figure 1 is a diagram of the subsequent steps of the data processing and the flow of information during the search for semantically similar side effect phrases described in the section below.

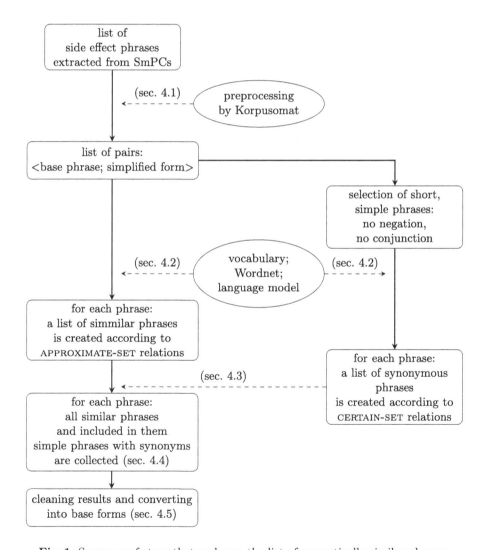

Fig. 1. Sequence of steps that make up the list of semantically similar phrases.

4.1 Preprocessing

The side effect phrases were analysed using the Korpusomat [13] service. The phrases were tagged by the Concraft tagger [29] which uses the Morfeusz analyser [30]. The tagger gives lemmas for all words which are present in the Morfeusz dictionary while for the out-of-vocabulary words, it guesses morphosyntactic descriptions. We obtained the list of pairs consisting of a phrase and the corresponding sequence of lemmas which we hereinafter refer to as a simplified base form. As Polish is a highly inflected language, the problem of morphological variants recognition is not a simple string comparison. For matching variants, we use the method described in [16] and operate on simplified base forms. For example, for the phrase *upośledzenie czynności nerek* 'impairment of renal function', the simplified base form is *upośledzenie czynność nerka* 'impairment function renal'. As we can see, only the first token in both phrase forms matches. Simplified base forms treat as equivalent phrases whose meanings are slightly different such as *świąd oka* 'ichy eye' and *świąd oczu* 'itchy eyes'. They have the same simplified base form *świąd oko*. As a result, both phrases are represented as one entry in the list of side effect phrases that now has 12,514 various entries (some of the phrases have been unified).

We have also created a token dictionary which consists of all lemmas used in the phrases. It consists of 4629 different tokens, which includes lemmas of words and fewer than 70 other tokens as punctuation marks, digits and a combination of them which were not segmented by the tagger (e.g.: *>3-krotnej* '>3-times').

4.2 Use of Wordnet Relations

In order to find similar phrases within the list of side effect phrases, we use methods similar to those described in [11]. We use Polish Wordnet [5,20] to find words and phrases similar to words from the token dictionary. For each word from the dictionary we found all words and phrases which are in the following relations: synonymy, inter-paradigmatic synonymy, and various types of derivatives e.g. pairs of nouns and adjectives *skóra – skórny* 'skin$_{noun}$ – skin$_{adj}$' (CERTAIN-SET). Moreover, we select the second set which includes pairs whose meaning is more distant. This set additionally contains: hipernyms, hiponyms and fuzzynyms (APPROXIMATE-SET).

As one word may refer to several synsets and there is no effective method to select which meaning may refer to a medical topic, we collect similar words and phrases for all synsets. We then select only those words and phrases that have all elements included in the token dictionary. For example, *noga* 'leg' refers to 7 synsets and only two of them are connected to human anatomy. One meaning is imprecise as it identifies 'leg' with 'foot' but that probably does not cause significant errors. The dictionary condition allows us to filter out meanings that are distant such as *piłka nożna* 'soccer', which is called colloquially *noga* 'leg'. The second test which we perform on the similar words/phrases collected from Wordnet is based on a distributional word2vec [17] language model. We use the model with vectors of 100 in length, calculated on the lemmatized texts of patient visits. We select for further processing, all similar words whose distance is at least 0.1. This criterion looks very mild but we want to remove only very distant pairs of meanings such as *senność* 'somnolence' and its colloquial synonym *śpiączka* 'coma'. If one or both lemmas are not in the dictionary, we accept such pairs assuming that they refer to a medical notion as they are present in the token dictionary.

After making the selection described above, we create all possible phrases where elements are substituted by all the selected synonyms for all side effect phrases. If such constructed phrase is present in the list of side effect phrases, we join them as variants. This procedure allows us to join phrases such as: *zanik skóry* 'disappearance of skin' and *atrofia skóry* 'skin atrophy'; *zamazane widzenie* 'blurred vision', *niejasne widzenie* 'unclear vision', and *niewyraźne widzenie* 'dim vision'.

Based on Wordnet, we select two sets of similar phrases that have almost certainly the same meaning and somehow similar ones. Such defined similarity might not be reciprocal and the first set consists of 1,079 phrases for which 1,303 phrases are similar. The second set consists of 2,329 entries for which 3,852 phrases are similar. The above numbers concern phrases in their simplified base forms. A large number of pairs are double counted (the relations are mainly reciprocal) and the effective number of encountered similar phrases is about half of the total.

4.3   Internal Phrases

As patient records contain side effect phrases up to 5 tokens, it is ineffective to look for the longer phrases which make up 20% of the side effect list. However, to make these phrases useful for generating alarms we recognize all simple phrases included in them. As a simple phrase we accept a phrase up to 5 tokens which does not include coordination (*i* 'and', *oraz* 'and', *lub* 'or', and characters: ',' , '/') and negation[1] (*nie* 'no', *bez* 'without', *wyjątek* 'exception'). The recognition of included phrases is limited to a very simple comparison of two bag-of-words. So phrase $\mathcal{A}$ is included by phrase $\mathcal{B}$ if all tokens of phrase $\mathcal{A}$ are in phrase $\mathcal{B}$. To perform this comparison, we use the simplified base form of phrases. If phrase $\mathcal{B}$ contains negation, it is shortened to the place where negation occurs. It allows us not to recognize 'aura' as a side effect entailed from the following phrase: *migrena bez aury* 'migraine without an aura'. In this case, from a logical point of view, the *migraine* itself does not follow either, but a patient may complain of migraine without stating that the aura does not occur.

The comparison of bag-of-words copes with coordinated information. If a patient uses a drug with the side effect *kwasica metaboliczna i ketonowa* 'metabolic and keto acidosis', an alarm is generated if her/his record also contains one of the following phrases included in the coordinated one: *kwasica ketonowa* 'keto acidosis' *kwasica metaboliczna* 'metabolic acidosis and just *kwasica* 'acidosis'.

---

[1] *brak* 'lack' is handled differently.

Polish is a free word order language, so for example, the side effect *choroba niedokrwienna serca* 'ischemic heart disease' is expressed by a phrase with a different word order: *niedokrwienna choroba serca*. As they consist of the same tokens, the comparison of the bag-of-words allows us to connect them as similar.

The comparison of bag-of-words allows us to recognize 16,240 pairs of side effect phrases (in simplified base forms) for 7,996 entries.

4.4   Unified List of Similar Phrases

For all phrases describing side effects we collect all variants in the following order. First, we generate all phrases according to the more distant Wordnet relations (APPROXIMATE-SET). Then, for each phrase and its variants, we find included phrases. In the next step, we add similar phrases for included phrases counted according to more restricted Wordnet relations (CERTAIN-SET). All repetitions are removed. The final list contains similar phrases for 9865 entries and consists of 38057 total variants.

4.5   Similar Phrases Update

The list of similar phrases elaborated using the algorithm described above has two potential shortcomings. Firstly, the lists of similar terms frequently include much broader concepts. In the example below, we see 'pain' as equivalent to 'muscle pain;' which potentially will cause many false alarms. Secondly, generalization can sometime be pursued further – the two lines can be combined together:

bóle mięśni   # bóle, bolesność, bóle
'muscles pain' #'pain', 'ache', 'pains'
ból mięśni   # bóle, bolesność, bóle
'muscles pain' # 'pain', 'ache', 'pains'

To overcome the above mentioned problems without manual effort, we proposed a strategy for cleaning and restructuring the list of similar terms. First, we eliminated equivalents that were shortened to just one word which is the main element of the phrase and occur independently on the symptoms list (as in the examples from the table above). Instead, we used these equivalents to exchange the first element of the phrase (removing repetitions which occur directly within one term description). The results for our example are given below:

bóe mięśni   # ból mięśni, bolesność mięśni
'muscles pain' #'muscles pain', 'muscles ache'
ból mięśni   # ból mięśni, bolesność mięśni
'muscles pain' # 'muscles pain', 'muscles ache'

The procedure introduces some new (potentially valid) terms (in our example: 'muscles ache'). It also introduces repetitions between terms. In this case, we use heuristics and we join terms which are lexicographically close, i.e. phrase elements are identical entirely, or at least they are identical for the first two letters, the Levenshtein distance is below 3 and words are longer than 4 letters, and a word is not written in capital letters. These conditions allow us to cover some plural forms (such as: 'ból mięśnia', 'bóle mięśnia') More liberal conditions could give us improper combinations such as '*podbrzusze* 'epigastrium' and *nadbrzusze* 'abdomen'. We also avoid identifying acronyms. The final result is:

bóle mięśni   # ból mięśni, bolesność mięśni
'muscles pain' #'muscles pain', 'muscles ache'

The problem which is not adequately solved at this moment is negation. Quite a lot of phrases in clinical notes are negated but there are no ready to use tools for Polish that are able to recognize them, similar to NegEX for English [2]. For Polish, the problem of negation in medical texts was addressed in [18]. We tested some simple methods in which we recognized several types of words introducing negation, such as *not*, *lack* and *without*, but in this particular case when texts consist mainly of noun phrases in the nominative, the simple method of recognizing only nominative forms of symptoms gave the best results. In Polish, negated phrases are in other cases, and their orthographic forms usually differ from the nominative ones. There are still some types of phrases which are incorrectly recognized, e.g. *duszność neguje* 'shortness of breath (he/she) negates', as they are abbreviated ways of expressing negation which are domain specific and were not identified in advance. In the future, we plan to addressed the problem in a more robust way.

5   Results and Evaluation

We applied the proposed algorithms to the set of data with 382,084 visits of 50,394 patients from different primary health care centers and specialist clinics in Poland which use the same software for data processing and storage. The documents are already segmented into various fields, but we were interested in two fields which have free text form and contain the exact text of examination and interview results written by a medical stuff member (usually by physicians themselves). 11,407 of patients had only one visit registered within this data, so there were 38,987 patients left for our evaluation. The average number of visits per patient was ten, but we limited ourselves to the simple case in which we have only a description of the current and the previous visits and we are looking for any potential side effects of the drugs prescribed (newly or as a continuation of a therapy) on the last by one visit as part of the symptoms reported by a patient during the current one. Both interview and examination fields from the previous and the last visits were analyzed and searched for symptoms which can be drug side effects. Newly occurring symptoms were identified and then, for all drugs administered during the last visit, all their possible side effects were compared with this new symptom list. The results of this procedure, using three versions of the possible side effect list, are given in Table 1.

We can observe in Table 1 that adding similar phrases from the Wordnet database did not introduce many new concepts to II list (less than 2%), but due to newly established similarity connections, many more symptoms were identified as possible drug side effects, hence much more (3.5 times more) such alarms were

Table 1. Side effects identified in the descriptions of the patient visits using different symptom lists. The first list (I) contains only symptoms extracted from textual drug descriptions. II list additionally contains terms obtained from Polish Wordnet as well as conjunct elements extracted from coordinated phrases (described in Sects. 4.1-4.4). The last list (III) is list II modified (as described in Sect. 4.5) to eliminate terms which are too general, and adds more phrase equivalents. The first row of the table contains the length of the symptom lists. The second presents the number of the symptoms which are identified in the drug descriptions as possible side effects of all drugs administered to patients during the first visit from the analyzed pair. In the third row, all types of symptoms identified during the last visits are shown. The last part of the table presents the final results, i.e. the number of symptoms identified as possible side effects of drugs used by a particular patient. To make the results more comparable, the numbers of different symptom types after merging their names based on the small Levenshtein distance (using the same criteria as described above) are shown in the last line.

	I.		II.		III.	
	Types	Occ.	Types	Occ.	Types	Occ.
Side effects of all drugs (lists lengths)	13,474	–	13,692	–	15,581	–
Possible side effects of drugs administered during the previous visit)	7,573	–	8,620	–	8,720	–
All symptoms registered during the last visits	1,372	–	1,374	–	1,373	–
Alarms: symptoms which could be drug side effects	original 143	1,309	284	4,677	215	2,027
	merged 126	–	225	–	186	–

raised. As was expected, in a small manually checked sample, quite a few of them were judged as evidently false (e.g. 'pain' identified as an occurrence of a 'back pain') which justified our next phase of list modification. As was also expected, III list contains significantly more new elements than both I and II lists (about 15%). At the same time, the number of symptoms recognized as potential side effects in the descriptions of the drugs used by the patients is only about 1% larger than in the case of list II. And finally, this time, the final list of the possible side effect alerts is only 50% larger than in the case of list I. These results look promising, as it is more probable that this amount of additional alarms is properly supported. The exact numbers for two specific connected side effects are given in Table 2. What is interesting is that in all three cases, the numbers of types of symptoms identified in the visit descriptions are almost identical. This supports the idea that visit descriptions are written using simpler language and use more typical phrases than drug description, hence it is much easier to cover the ways physicians express symptoms. What is challenging is how to match these symptoms to side effects listed in drug descriptions which are much more formal, complicated and detailed texts.

Table 2. Examples of symptoms that are potential drug side effects recognized using different side effect lists in a description of the last visit (the names of the lists are the same as in Table 1).

	I.		II.		III.	
	Org.	Merged	Org.	Merged	Org.	Merged
bóle głowy 'headache'	52	117	60	138	87	165
bóe głowy 'headaches'	65	–	78	–	78	–
bóle 'pain'	42	56	778	1951	78	138
bóle 'pains'	14	–	1173	–	60	–

Actual evaluation of our final solution will be possible only after deploying our method as part of the system used by physicians, which is planned. An introductory evaluation only covers evident false alarms which can be classified on the basis of general knowledge and the text itself, i.e. using a phrase in a negative scope or in the context of an improving status. Our aim was to determine the potential possibility of an undesirable symptom, while the final decision was left to the physician using the program. The evaluation was therefore performed by a person with experience in medical text annotation and not by a physician. The results on the first 20 examples are shown in the Table 3. Although the sample is small we can already observe that the coverage of symptoms is highly improved by using similar terms in list II. The further modifications make this list much more reliable. The increase in the symptom coverage after adding similar phrases

Table 3. Manual comparison of results for 20 patients, i.e. first 20 cases for which any of the methods recognized at least one side effect. The first part of the table presents the number and the percentage of cases in which the output of the method was correct. The number of times when the method correctly recognized the need of an alarm (TP) and the absence of such a situation (TN) is also added. Then, we give the number of alerts which were not raised at all, or were generated with incomplete lists of symptoms. The next two columns include alerts which are entirely wrong and such that have any additional (incorrect) symptom. Lastly, the F1 measure for the method is included.

	Correct outcome				Missing alerts (FN)				Erroneous recognition (FP)				F1
	nb	%	TP	TN	Entire alerts		Symptoms		Entire alerts		Symptoms		
I	8	.40	5	3	8	.40	3	.20	1	.05	0	.00	.45
II	9	.45	9	0	2	.10	2	.10	6	.30	1	.05	.62
III	17	.85	12	5	1	.05	0	.00	2	.10	0	.00	.89

is evident. It is also clear that updating the II list helps in reducing the number of false positives – symptoms wrongly recognized as possible side effects. In the case of list III, 2 false alarm are raised, while 1 is missing. One alarm is consequently wrongly generated by all the methods because of an error in the initial symptom list.

6   Conclusions

The proposed method for preparing a list of the potential side effect symptoms and their recognition in the actual visit description seem to work with an acceptable level of quality. Although the sample on which the initial evaluation was made is small, the F1=0.89 seems to be quite satisfactory, e.g. compare [10]. Even if in practice it will certainly turn out to be lower, we think that it would be possible to test out the method in a real environment to observe its practical value and shortcomings. In a situation when resources for medical text processing for Polish are very limited, using a general semantic resource such as Wordnet allowed us to improve the results of symptom identification. One of the problems that is not fully solved here, but will also be addressed, is spelling corrections as patients' records contain a large number of spelling errors which affects the recognition of side effect phrases. In the current version of the system, some of the errors are already taken into account by similarity measures, but the problem needs a more general solution.

There are currently 70,000 visits a day processed by the system. Each visit potentially generates an inquiry about side effects. The tests were performed which showed that the system was able to process such a high number of questions on-line. Looking for phrases which might indicate the side effects of drugs in patients' data is executed as separate thread in the EHR system. The dictionary is organized in the TRIE structure which means a quick search is possible.

For further investigation, it would be interesting to compare lists of side effect phrases with all phrases extracted from patient records in order to find other ways of expressing symptoms and side effects in patient records. We also plan to work on eliminating alerts in cases when the symptoms are most likely related with a new illness on the basis of other new symptoms identified simultaneously.

Acknowledgments. This work was financially supported by the National Centre for Research and Development in Poland, Grant POIR.01.01.01-00-0328/17.

References

1. Baker, D.: Medication alert fatigue: The potential for compromised patient safety. Hospital Pharmacy - HOSP PHARM 44, June 2009. https://doi.org/10.1310/hpj4406-460
2. Chapman, W., Bridewell, W., Hanbury, P., Cooper, G., Buchanan, B.: A simple algorithm for identifying negated findings and diseases in discharge summaries. J. Biomed. Inf. **34**(5), 301–310 (2001). https://doi.org/10.1006/jbin.2001.1029

3. Cho, I., Kim, J., Kim, J.H., Kim, H.Y., Kim, Y.: Design and implementation of a standards-based interoperable clinical decision support architecture in the context of the Korean EHR. Int. J. Med. Inf. **79**(9), 611–622 (2010)
4. Czerw, A., Fronczak, A., Witczak, K., Juszczyk, G.: Implementation of electronic health records in Polish outpatient health care clinics - starting point, progress, problems, and forecasts. Ann. Agric. Environ. Med. **23**(2), 329–334 (2016)
5. Dziob, A., Piasecki, M., Rudnicka, E.: plwordnet 4.1—a linguistically motivated, corpus-based bilingual resource. In: Fellbaum, C., Vossen, P., Rudnicka, E., Maziarz, M., Piasecki, M. (eds.) Proceedings of the 10th Global WordNet Conference: July 23–27, 2019, Wroclaw (Poland). pp. 353–362. Oficyna Wydawnicza Politechniki Wrocławskiej, Wrocław (2019)
6. Feng, C., Le, D., McCoy, A.: Using electronic health records to identify adverse drug events in ambulatory care: a systematic review. Appl. Clin. Inf. **10**, 123–128 (2019). https://doi.org/10.1055/s-0039-1677738
7. Gurulingappa, H., Rajput, A.M., Roberts, A., Fluck, J., Hofmann-Apitius, M., Toldo, L.: Development of a benchmark corpus to support the automatic extraction of drug-related adverse effects from medical case reports. J. Biomed. Inf. **45**(5), 885–892 (2012)
8. Hammar, T., Hellström, L., Ericson, L.: The use of a decision support system in swedish pharmacies to identify potential drug-related problems-effects of a national intervention focused on reviewing elderly patients' prescriptions. Pharmacy: J. Pharmacy Educ. Practice **8** (2020)
9. Harpaz, R., Callahan, A., Tamang, S., et al.: Text mining for adverse drug events: the promise, challenges, and state of the art. Drug Safety **37**, 777–790 (2014)
10. Henry, S., Buchan, K., Filannino, M., Stubbs, A., Uzuner, O.: 2018 n2c2 shared task on adverse drug events and medication extraction in electronic health records. J. Am. Med. Inf. Assoc. JAMIA **27**(1), 3–12 (2020)
11. Huang, K., Geller, J., Halper, M., Perl, Y., Xu, J.: Using WordNet synonym substitution to enhance UMLS source integration. Artif. Intell. Med. **46**(2), 97–109 (2009). https://doi.org/10.1016/j.artmed.2008.11.008
12. Kang, N., Singh, B., Bui, Q.C., Afzal, Z., van Mulligen, E.M., Kors, J.A.: Knowledge-based extraction of adverse drug events from biomedical text. BMC Bioinform. **15**, 1–8 (2014)
13. Kieraś, W., Kobyliński, Ł., Ogrodniczuk, M.: Korpusomat — a tool for creating searchable morphosyntactically tagged corpora. Comput. Methods Sci. Technol. **24**(1), 21–27 (2018)
14. Lemmen, C., Woopen, C., Stock, S.: Systems medicine 2030: a Delphi study on implementation in the German healthcare system. Health Policy **125**(1), 104–114 (2021)
15. Lindberg, D., Humphreys, B., McCray, A.: The unified medical language system. Yearbook Med. Inf **1**, 41–51 (1993)
16. Marciniak, M., Mykowiecka, A., Rychlik, P.: TermoPL — a flexible tool for terminology extraction. In: Proceedings of LREC pp. 2278–2284. ELRA, Portorož, Slovenia (2016)
17. Mikolov, T., Sutskever, I., Chen, K., Corrado, G., Dean, J.: Distributed representations of words and phrases and their compositionality. In: Advances in neural information processing systems. pp. 3111–3119 (2013)
18. Mykowiecka, A., Marciniak, M., Kupść, A.: Rule-based information extraction from patients' clinical data. J. Biomed. Inf. **42**(5), 923–936 (2009)

19. Page, N., Baysari, M., Westbrook, J.: A systematic review of the effectiveness of interruptive medication prescribing alerts in hospital CPOE systems to change prescriber behavior and improve patient safety. Int. J. Med. Inform. **105**, 22–30 (2017)
20. Piasecki, M., Szpakowicz, S., Broda, B.: A Wordnet from the Ground Up. Oficyna Wydawnicza Politechniki Wroclawskiej, Wroclaw (2009)
21. Rahmner, P., Eiermann, B., Korkmaz, S., Gustafsson, L., M, G., Maxwell, S., Eichle, H., Vég, A.: Physicians' reported needs of drug information at point of care in Sweden. Br. J. Clin. Pharmacol. **73**(1), 115–125 (2012)
22. Roberts, K., Demner-Fushman, D., Tonning, J.M.: Overview of the TAC 2017 adverse reaction extraction from drug labels track. In: Proceedings of the 2017 Text Analysis Conference, TAC 2017, Gaithersburg, Maryland, USA, November 13–14, 2017. NIST (2017)
23. Saxena, K., Lung, B.R., Becker, J.R.: Improving patient safety by modifying provider ordering behavior using alerts (CDSS) in CPOE system. Annual Symposium proceedings. AMIA Symposium **2011**, 1207–1216 (2011)
24. Scott, I.A., Pillans, P.I., Barras, M., Morris, C.: Using EMR-enabled computerized decision support systems to reduce prescribing of potentially inappropriate medications: a narrative review. Therapeutic Adv. Drug Safety **9**(9), 559–573 (2018)
25. Stearns, M.Q., Price, C., Spackman, K., Wang, A.Y.: Snomed clinical terms: overview of the development process and project status. In: Proceedings of the AMIA Symposium, pp. 662–6 (2001)
26. Sutton, R.T., Pincock, D., Baumgart, D.C., Sadowski, D.C., Fedorak, R.N., Kroeker, K.I.: An overview of clinical decision support systems: benefits, risks, and strategies for success. Digital Med. **3**(17), 1–10 (2020)
27. Ubysz, D., Fryzowska-Chrobot, I., Giermaziak, W.: Baza Tez-Mesh jako efektywne narzędzie do opracowania rzeczowego i wyszukiwania informacji z zakresu medycyny i nauk pokrewnych. Zarządzanie Biblioteką **11**(1), 59–73 (2019)
28. van Mulligen, E.M., Fourrier-Reglat, A., Gurwitz, D., Molokhia, M., Nieto, A., Trifiro, G., Kors, J.A., Furlong, L.I.: The EU-ADR corpus: Annotated drugs, diseases, targets, and their relationships. J. Biomed. Inf. **45**(5), 879–884 (2012)
29. Waszczuk, J.: Harnessing the CRF complexity with domain-specific constraints. The case of morphosyntactic tagging of a highly inflected language. In: Proceedings of COLING pp. 2789–2804 (2012)
30. Woliński, M.: Morfeusz reloaded. In: Calzolari, N., et al.(eds.) Proceedings of the Ninth International Conference on Language Resources and Evaluation, LREC 2014, pp. 1106–1111. ELRA, Reykjavík, Iceland (2014)
31. Woliński, M., Saloni, Z., Wołosz, R., Gruszczyński, W., Skowrońska, D., Bronk, Z.: Słownik gramatyczny języka polskiego, wyd. IV (2020). http://sgjp.pl

des-ist: A Simulation Framework to Streamline Event-Based *In Silico* Trials

Max van der Kolk[1]($\boxtimes$) , Claire Miller[1] , Raymond Padmos[1] , Victor Azizi[2], and Alfons Hoekstra[1]

[1] Computational Science Laboratory, Informatics Institute, University of Amsterdam, Science Park 904, 1098 XH Amsterdam, The Netherlands
m.vanderkolk@uva.nl
[2] Netherlands eScience Center, Science Park 140, 1098 XG Amsterdam, The Netherlands

Abstract. To popularise *in silico* trials for development of new medical devices, drugs, or treatment procedures, we present the modelling framework des-ist (Discrete Event Simulation framework for *In Silico* Trials). This framework supports discrete event-based simulations. Here, events are collected in an acyclic, directed graph, where each node corresponds to a component of the overall *in silico* trial. A simple API and data layout are proposed to easily couple numerous simulations by means of containerised environments, i.e. Docker and Singularity. An example *in silico* trial is highlighted studying treatment of acute ischemic stroke, as considered in the INSIST project.

The proposed framework enables straightforward coupling of the discrete models, reproducible outcomes by containerisation, and easy parallel execution by GNU Parallel. Furthermore, des-ist supports the user in creating, running, and analysing large numbers of virtual cohorts, automating repetitive user interactions. In future work, we aim to provide a tight integration with validation, verication and uncertainty quantication analyses, to enable sensitivity analysis of individual components of *in silico* trials and improve trust in the computational outcome to successfully augment classical medical trials and thereby enable faster development of treatment procedures.

Keywords: Event-based modelling · *In silico* trials · Ischemic stroke

1 Introduction

Research and development of new medical devices, drugs, or treatment procedures requires significant monetary and temporal resources [2]. Recent investigations estimate post-approval Research & Development (R&D) costs to average 985 million US dollars [21]. Regardless of careful planning, time-to-market [13] and R&D costs are increased by trial difficulties such as statistical uncertainties, trials lacking a clear understanding of their outcome, and other unforeseen side-effects caused by a newly proposed drug, device, or procedure [3].

© Springer Nature Switzerland AG 2021
M. Paszynski et al. (Eds.): ICCS 2021, LNCS 12744, pp. 648–654, 2021.
https://doi.org/10.1007/978-3-030-77967-2_53

To counteract these challenges, researchers are adopting computational bio-medicine to augment traditional *in vitro* and *in vivo* trials. These *in silico* experiments are enabled by recent developments in computational modelling in biomedicine [19]. The so-called *In Silico* Clinical Trials (ISCTs) [18] specifically consider cohorts of virtual patients representing highly specific population subsets and thereby enable (*in silico*) clinical trials considering rare diseases with reduced costs as development times decrease.

However, defining ISCTs is not without cost or complexity either [18]. Typically, accurate *in silico* simulations require detailed modelling across time and length-scales [6,16] and close collaboration of experts—i.e. clinicians, experimentalists, and software developers—spanning multiple fields of research [4]. Additionally, the numerical models require strict scrutinisation with advanced validation, verification and uncertainty quantification (VVUQ) analyses to gain sufficient trust in their outcome [20,22]. Finally, there are practical issues to address, e.g. interfacing multiple *in silico* models, reproducibility, and supporting various computational environments from personal workstations to large-scale cloud or High Performance Computing (HPC) environments [5].

In this work, we propose a simulation framework: des-ist, that addresses these difficulties in managing *In Silico* Clinical Trial. Specifically, it interfaces multiple *in silico* models using a predetermined data layout in combination with a simple, unified Application Programming Interface (API), which is enforced across each component of the *in silico* simulation pipeline. Furthermore, the simulation is split in a series of components, each captured in a containerised environment using Docker [11] or Singularity [9], to ensure reproducibility of the *in silico* trials. An additional benefit of these containerised environments is their ability to scale well towards cloud-based or HPC compute environments. Our application des-ist then orchestrates running these *in silico* trials, thereby enabling parallelism using GNU Parallel [17] and VVUQ analyses by interfacing EasyVVUQ [22].

To illustrate our proposed workflow, we present a numerical experiment from the *In Silico* clinical trials for acute Ischemic STroke (INSIST) project [8]. We show how des-ist supports the user in setting up and evaluating an ISCT pipeline. The example illustrates a detailed, event-based *in silico* pipeline studying acute ischemic stroke. The pipeline considers a containerised environment for each distinct event identified in the clinical trial, i.e. virtual patient generation, initial randomised clot placement, blood-flow and brain (re)perfusion analysis [7,12], chemical and mechanical treatments using thrombolysis [15] and thrombectomy [10], and finally the NIHSS scoring of each virtual patient [1].

The paper continues in Sect. 2 with a discussion on the chosen data layout and API definition as adopted in des-ist. Next, we run a single *in silico* pipeline in Sect. 3 and close with brief discussion and conclusions in Sect. 4.

2   Methods

In general, ISCTs can be separated into three distinct phases: *i)* generating statistical representative cohorts of virtual patients, *ii)* the analysis of a simulation driven pipeline *per* virtual patient, and *iii)* the (post-)processing of statistical data as generated by the *in silico* trials. The first and last steps are typically driven by statistical models. These are carefully formulated to ensure they generate sets of virtual patients that accurately represent a subset of interest of the studied population. The second step, i.e. evaluating the *in silico* experiment per patient, differs strongly depending on the type of the ISCT and chosen computational environment. Here, we will specifically consider *in silico* trials represented by a sequence of discrete events, where each event is assumed to instantaneously change the system's state. The events are stored in `des-ist` using a directed, acyclic graph, as conceptually illustrated in Fig. 1. The nodes in the graph represent to discrete events and each holds a reference to the matching container that needs to be invoked for the specific time instance. The graph is traversed by `des-ist` following the directed edges and thereby covering all individual events of the *in silico* trial.

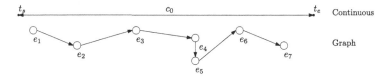

Fig. 1. A continuous time system c_0 is modelled in `des-ist` as a sequence of discrete events e_i, where each event is assumed to instantaneously change the system's state. The discrete events are stored in a directed, acyclic graph, implicitly storing the traversal order as well as a reference to the containerised simulation environment for each event e_i. Then, `des-ist` traverses the graph and invokes the simulation per event.

To combine various containers within a single ISCT, `des-ist` enforces an API and data layout requirements for each container. The API is written in Python—a programming language well suited for writing *glue* code [14]. The API contains three functions that need to be implemented for each container, i.e. `event`, `example`, and `test`. Most critically an implementation has to be written for the `event` call, which invokes the container's main simulation[1].

To ensure a container only has access to a given virtual patient's data, we enforce a specific data layout in combination with the presented API. Each trial has its own directory, e.g. `/trial`, containing a trial's configuration file and a subdirectory per virtual patient in the cohort, i.e. for patient i: `/trial/patient_i`. To ensure a container's access is limited to data of a specific virtual patient, we exploit *bind paths*—a feature present in both `Docker` [11] and `Singularity` [9]—where we map a constant path on the container to a specific path on the host system, e.g. `/patient/` $\Rightarrow$ `/host/trial/patient_i/`, simplifying the API. To control the working directories of the containers, `des-ist` only has to update these *bind paths* accordingly, leaving the containers unchanged.

---

[1] Non-Python code is included using Python's `subprocess` library https://docs.python.org/3.10/library/subprocess.html.

When creating a trial using `des-ist`'s command-line interface, the directory structures are automatically generated and provided with the corresponding trial and patient configuration files. For these files, we adopt the YAML format (https://yaml.org/), a format easy to modify by hand and automatically.

2.1 Containerised Environments

Adopting containerised environments provides numerous advantages in context of *In Silico* Clinical Trial. First of all, the separated containers allow for easy collaboration in large research consortia where various researchers—typically spread over many countries—can use their preferred environments without worrying about linking or interfacing of the multitude of simulation components, as used in INSIST [8]. As long as they adhere to the defined data layout and API, they can adopt any computational framework that is suited best for their specific *in silico* model.

Secondly, the strict container definitions imposed by `Docker` and `Singularity` ensure clear documentation on the container's package dependencies and requirements. These definitions directly enable reproducible research, as the containers can be shared and evaluated by others, thereby overcoming portability problems [9,11]. Concurrently, the containerised environments enable scaling towards cloud and HPC environments using `Docker` and `Singularity` technologies [9,11].

Another advantage by adhering to a unified API in event-based simulation, is the plug-and-play nature of the containers. This provides easy exchange of containers in a pipeline. For instance, a high fidelity model can be exchanged for a reduced order surrogate, simply by changing the trial's configuration file, and thereby the graph of events (Fig. 1). Alternatively, one can study the effect of treatment procedures by providing alternative containers for each variant of the treatment. Again, only the trial's configuration needs to be updated.

2.2 Parallelism

To decrease the wall time of *in silico* experiments, we can resort towards parallel evaluation of the *in silico* experiments per patient. This can be achieved in a variety of ways, considering small to large-scale cloud computing or HPC environments. In `des-ist` only the most basic parallelism has been considered so far, where each patient is evaluated on a single Central Processing Unit (CPU), where the distribution and balancing of tasks is assigned to `GNU Parallel` [17] to enable parallelism on single and multi-machine (HPC) architectures (Fig. 2).

3   Example

To illustrate `des-ist`'s use case, we consider the procedure of evaluating an event-based *in silico* trial. For this, we borrow a pipeline from the INSIST project [8], regarding the study of acute ischemic stroke, where the details of

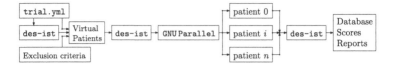

Fig. 2. Conceptual parallel execution approach in des-ist. The trial is initialised by des-ist considering a trial configuration, i.e. the trial.yml, to create a cohort of virtual patients using a specific statistical model. Then, des-ist traverses the event graph for each patient, where the simulations are executed (in parallel) by GNU Parallel [17]. Ultimately, the output of each simulation is collected and trial output is generated.

the *in silico* trial are omitted for brevity. A schematic illustration is shown in Fig. 3. These *in silico* trials are then generated automatically by des-ist, using a simple configuration file containing, for example, the cohort size, in/exclusion criteria, and random seeds. From there, des-ist automates all repetitive, and often error-prone, user interactions to generate the required data layout (Sect. 2). Furthermore, des-ist orchestrates simulation runs, automatically scheduled on the available resources through GNU Parallel. For each patient the sequence of discrete events (Fig. 3(e)) is evaluated, sequentially stepping through each simulation. Ultimately, all results are collected and statistical outputs are aggregated

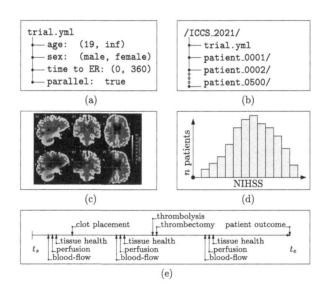

Fig. 3. Illustration of the steps in an *in silico* trial using des-ist. First, a trial configuration file is provided (a), containing properties of the trial. Then, des-ist creates a directory structure and data layout (b): a trial directory with a subdirectory per patient, after which each patient's simulation pipeline (e) is evaluated (Fig. 2). Finally, the results are aggregated providing per patient results (c) of brain perfusion and trial reports with statistical output measurements of the *in silico* trial (d).

over the considered cohort, providing insight into the trial's outcome, such as infarct volume, treatment outcome, and NIHSS distributions across the cohort of virtual patients.

4   Conclusion

To transition towards *In Silico* Clinial Trials for development of new drugs, medicine, and treatment procedures, we propose a numerical framework: `des-ist` to support users in running large numbers of *in silico* trials. The framework automates generating cohorts of virtual patients, ensures all *in silico* simulations are evaluated in the desired order by traversing an internal acyclic directed graph representing the *in silico* pipeline, enables parallel scheduling by `GNU Parallel`, and provides VVUQ analysis through `EasyVVUQ`. These features greatly reduce the repeated efforts of users in setting up, evaluating, and validating large *in silico* trials.

Additionally, we propose the adoption of containerised environments, using `Docker` or `Singularity`, to capture the individual, discrete simulation events. These containers are combined with a simple API and data layout to enable uniform data communication between the various simulation events. This has not only increased the reproduciblity of the *in silico* trials' simulations, it also reduces barriers for collaboration by reducing compatibility issues and enables scaling towards (large-scale) cloud or HPC compute environments.

Although only an illustrative *in silico* pipeline from the INSIST project was presented, we hope to convey the potential of orchestrating *in silico* trials using `des-ist`. In the near future, we aim to generalise the current implementation to support a variety of *in silico* pipelines and pursue validation of ISCTs. The `des-ist` application is to be released mid 2021 as open source software.

INSIST [8] has received funding from the European Union's Horizon 2020 research and innovation program under grant agreement No 777072.

References

1. Brott, T., et al.: Measurements of acute cerebral infarction: a clinical examination scale. Stroke **20**(7), 864–870 (1989). https://doi.org/10.1161/01.str.20.7.864
2. Dickson, M., Gagnon, J.P.: Key factors in the rising cost of new drug discovery and development. Nature Rev. Drug Discov. **3**(5), 417–429 (2004). https://doi.org/10.1038/nrd1382
3. Fogel, D.B.: Factors associated with clinical trials that fail and opportunities for improving the likelihood of success: a review. Contemp. Clin. Trials Commun. **11**, 156–164 (2018). https://doi.org/10.1016/j.conctc.2018.08.001
4. Groen, D., Zasada, S.J., Coveney, P.V.: Survey of multiscale and multiphysics applications and communities. Comput. Sci. Eng. **16**(2), 34–43 (2014). https://doi.org/10.1109/mcse.2013.47
5. Gupta, A., Milojicic, D.: Evaluation of HPC applications on cloud. In: 2011 Sixth Open Cirrus Summit, IEEE (October 2011). https://doi.org/10.1109/ocs.2011.10

6. Hoekstra, A., Chopard, B., Coveney, P.: Multiscale modelling and simulation: a position paper. Philos. Trans. Royal Soc. Math. Phys. Eng. Sci. **372**(2021), 20130377 (2014). https://doi.org/10.1098/rsta.2013.0377

7. Józsa, T.I., Padmos, R.M., Samuels, N., El-Bouri, W.K., Hoekstra, A.G., Payne, S.J.: A porous circulation model of the human brain for in silico clinical trials in ischaemic stroke. Interface Focus **11**(1), 20190127 (2020). https://doi.org/10.1098/rsfs.2019.0127

8. Konduri, P.R., Marquering, H.A., van Bavel, E.E., Hoekstra, A., Majoie, C.B.L.M.: In-silico trials for treatment of acute ischemic stroke. Front. Neurol. **11**, 1062 (2020). https://doi.org/10.3389/fneur.2020.558125

9. Kurtzer, G.M., Sochat, V., Bauer, M.W.: Singularity: scientific containers for mobility of compute. PLOS ONE **12**(5), 1–20 (2017). https://doi.org/10.1371/journal.pone.0177459

10. Luraghi, G., et al.: Applicability assessment of a stent-retriever thrombectomy finite-element model. Interface Focus **11**(1), 20190123 (2021). https://doi.org/10.1098/rsfs.2019.0123

11. Merkel, D.: Docker: Lightweight linux containers for consistent development and deployment. Linux J. **2014**(239), 2 (2014)

12. Padmos, R.M., Józsa, T.I., El-Bouri, W.K., Konduri, P.R., Payne, S.J., Hoekstra, A.G.: Coupling one-dimensional arterial blood flow to three-dimensional tissue perfusion models for in silico trials of acute ischaemic stroke. Interface Focus **11**(1), 20190125 (2020). https://doi.org/10.1098/rsfs.2019.0125

13. Prašnikar, J., Škerlj, T.: New product development process and time-to-market in the generic pharmaceutical industry. Ind. Market. Manage. **35**(6), 690–702 (2006). https://doi.org/10.1016/j.indmarman.2005.06.001

14. van Rossum, G.: Glue it all together with python (1998). https://www.python.org/doc/essays/omg-darpa-mcc-position/

15. Shibeko, A.M., Chopard, B., Hoekstra, A.G., Panteleev, M.A.: Redistribution of tpa fluxes in the presence of pai-1 regulates spatial thrombolysis. Biophys. J. **119**(3), 638–651 (2020). https://doi.org/10.1016/j.bpj.2020.06.020

16. Sloot, P.M., Hoekstra, A.G.: Multi-scale modelling in computational biomedicine. Briefings Bioinform. **11**(1), 142–152 (2009). https://doi.org/10.1093/bib/bbp038

17. Tange, O.: Gnu parallel - the command-line power tool.; login: The USENIX Magazine **36**(1), 42–47 (2011). https://doi.org/10.5281/zenodo.16303

18. Viceconti, M., Henney, A., Morley-Fletcher, E.: In silico clinical trials: how computer simulation will transform the biomedical industry. Int. J. Clin. Trials **3**(2), 37–46 (2016). https://www.ijclinicaltrials.com/index.php/ijct/article/view/105

19. Viceconti, M., Hunter, P.: The virtual physiological human: Ten years after. Ann. Rev. Biomed. Eng. **18**(1), 103–123 (2016). https://doi.org/10.1146/annurev-bioeng-110915-114742

20. Viceconti, M., Juarez, M.A., Curreli, C., Pennisi, M., Russo, G., Pappalardo, F.: Credibility of in silico trial technologies—a theoretical framing. IEEE J. Biomed. Health Inf. **24**(1), 4–13 (2020). https://doi.org/10.1109/jbhi.2019.2949888

21. Wouters, O.J., McKee, M., Luyten, J.: Estimated research and development investment needed to bring a new medicine to market, 2009–2018. JAMA **323**(9), 844 (2020). https://doi.org/10.1001/jama.2020.1166

22. Wright, D.W., et al.: Building confidence in simulation: Applications of easyvvuq. Adv. Theor. Simul. **3**(8), 1900246 (2020). https://doi.org/10.1002/adts.201900246

Identifying Synergistic Interventions to Address COVID-19 Using a Large Scale Agent-Based Model

Junjiang Li[ID] and Philippe J. Giabbanelli$^{(\boxtimes)}$[ID]

Department of Computer Science and Software Engineering, Miami University,
Oxford, OH 45056, USA
{lij111,giabbapj}@miamioh.edu
https://www.dachb.com

Abstract. There is a range of public health tools and interventions to address the global pandemic of COVID-19. Although it is essential for public health efforts to comprehensively identify *which* interventions have the largest impact on preventing new cases, most of the modeling studies that support such decision-making efforts have only considered a very small set of interventions. In addition, previous studies predominantly represented interventions as independent or examined a single scenario in which every possible intervention was applied. Reality has been more nuanced, as a *subset* of all possible interventions may be in effect for a given time period, in a given place. In this paper, we use cloud-based simulations and a previously published Agent-Based Model of COVID-19 (Covasim) to measure the individual and interacting contribution of interventions on reducing new infections in the US over 6 months. Simulated interventions include face masks, working remotely, stay-at-home orders, testing, contact tracing, and quarantining. Through a factorial design of experiments, we find that mask wearing together with transitioning to remote work/schooling has the largest impact. Having sufficient capacity to immediately and effectively perform contact tracing has a smaller contribution, primarily via interacting effects.

Keywords: Cloud-based simulations · Factorial analysis · Large-scale simulations · Synergistic interventions

1 Introduction

Mathematical models of COVID-19 commonly use 'compartmental models', which are systems of coupled differential equations that predict global quantities such as the number of infections at any given time [6]. Agent-based models (ABMs) were later used to capture heterogeneity in populations [18], representing how different individuals (e.g., in age, gender, or socio-economic factors)

The authors thank the Microsoft AI for Health program for supporting this research effort through a philanthropic grant.

© Springer Nature Switzerland AG 2021
M. Paszynski et al. (Eds.): ICCS 2021, LNCS 12744, pp. 655–662, 2021.
https://doi.org/10.1007/978-3-030-77967-2_54

have different risks, or willingness and abilities to comply with preventative measures [5]. In the absence of a widely used vaccine, our study and previous ABMs rely on *non-pharmaceutical interventions* including individual-level preventative measures (e.g., stay at home, social distance, using face masks) and subsequent interventions (e.g., contact tracing, quarantining). There are several limitations to these previous studies. First, they focused on the effects of a *small subset of interventions* commonly adopted by national governments, and not all interventions are applied in conjunction. In a sample of 10 studies [4], we noted that an average of only 2.3 interventions are used simultaneously, which is significantly less than the number of interventions that have been implemented or considered by governments [16]. Second, when several interventions are implemented, there is limited analysis to assess whether synergistic effects are obtained or whether most of the benefits can be attributed to only some of the policies. Identifying the right set of synergistic interventions is an important information for policy-making, particularly as compliance becomes an issue. Finally, we frequently note reporting issues such as an insufficiently motivated number of runs (since stochastic models need replicated runs to achieve a sufficient confidence interval) or a coarse resolution when the target population is large.

To support policymakers in identifying the right set of interventions and provide the necessary confidence in high-stake computations, our paper uses Design of Experiment (DoE) techniques and large-scale cloud-based simulations that measure the individual and interactive effects of interventions at a detailed level. Specifically, our contributions are twofold:

1) We measure the impact of *six interventions* on disease prevalence and mortality by accounting for interactive effects. Our interventions include face masks, social distancing, stay-at-home orders, testing, contact tracing, and quarantining. We simulates these interventions at *various levels of adherence*, thus accounting for possible variations in behavioral responses.
2) We provide all results at a very accurate population scale of 1 : 500 for the USA and within a confidence interval of at least 95%. This requires a massive number of computationally intensive experiments, which are performed through the cloud via the Microsoft Azure platform.

The remainder of the paper is structured as follows. In Sect. 2, we describe the procedures of our experiments, including the principles of factorial designs of experiments. Our results are discussed in Sect. 3. For an overview on modeling the biology and policies for COVID-19, or potential methodological limitations to this study, we refer the reader to our supplementary online materials [4].

2   Methods

Our simulations are performed **using the Covasim platform**. We ensure the accuracy of the simulations within our application context by using *(i)* a resolution[1] of 1 : 500 and *(ii)* a 95% Confidence Interval[2] [17, pp. 184–186]. Our simulation scenarios correspond to combinations of interventions at various levels. To quantify the effects of stipulated interventions, we report the total infections after 180 days. It is important to note that our simulated time period starts in September 2020 rather than at the beginning of the pandemic, when most of the population was susceptible and almost none had recovered. By choosing a starting time close to the writing to this article (December 2020), our simulated dynamics take into account the presence of recovered agents as well as individuals who are already at various stages of the infection. We made two modifications to Covasim accordingly: randomly selected agents are set to recovered based on the amount estimated by the CDC and scaled to our population resolution; and a separate random subset of agents are set to infected, with their date of infection chosen among the 14 days preceding the start of the simulation (based on CDC data) such that agents are realistically set to different stages of infection.

As Covasim already contains a large number of population parameters calibrated for the US, we used the same default values in our work. There were three exceptions, as the knowledge base upon which Covasim was built has since evolved and needed to reflect our current understanding of real-life disease dynamics. As detailed in our supplementary online materials [4], we updated the model regarding the distribution of incubation period, proportion of symptomatic cases, and testing delays.

We focused on **four _categories_ of interventions**: mask wearing (realized as direct reductions of the susceptibility of simulated agents), lockdowns, testing, and contact tracing. Several parameters are required for each category in order to precisely characterize how the intervention will unfold. For instance, testing is a matter of *how many* tests are available on a daily basis, *when* to test individuals who're entering quarantine as they were exposed to the virus, and the extent to which tests are *reliable* (i.e. test sensitivity). We list the parameters for each category of intervention in Table 1 together with the range of values that *could* be used and the specific subset that we *do* use. Our choice is motivated by the references provided in the table and is often limited to a binary due to the experimental set-up explained in the next paragraph. Every intervention is applied for the entire duration of the simulation. Each intervention takes place in certain networks, thus we account for the possibility that agents may wear masks in one network but not in another.

Our goal is to measure the impact of interventions on disease prevalence. We use a 2^k **factorial design** Design of Experiments (DoE) in which each parameter is set to two values, designated as 'low' and 'high' [13, pp. 233–303]. Our four categories of interventions result in 9 parameters (Fig. 1), which are listed as $X_1, \ldots, X_9$ in Table 1 together with their low and high values. A factorial design serves to investigate the synergistic effects of these parameters,

---

[1] Each simulated agent represents 500 real-world people. Given the US population of 328 million people, we have a simulated agent population of over 650,000.

[2] We identify and perform a sufficient number of replications in each scenario such that the average results are within a 95% Confidence Interval (95% CI).

Table 1. Covasim interventions used here, with the following shorthands: $w \to$ work network, $s \to$ school network, $h \to$ home network, and $c \to$ community network.

Intervention	Parameters	Possible values	Values chosen	Var.	Ref.
Mask wearing	Fractional reduction in transmissibility	$[0, 1]$	80%		[1]
	Applied networks	$\mathscr{P}(\{w, s, h, c\})$	$\{\{w, s, h, c\}, \{w, s\}, \{c\}\}$	X_1	
Lockdowns	Fractional reduction in contacts for $\{w, s\}$	$[0, 1]$	$\{5\%, 30\%\}$	X_2	[7]
	Fractional reduction in contacts for c	$[0, 1]$	$\{10\%, 30\%\}$	X_3	[7]
	Applied networks	$\mathscr{P}(\{w, s, h, c\})$	$X_1 \setminus h$		
Testing	Number of tests per day	Any integer	$\{6, 11.1\} \times 10^5$	X_4	[8]
	When to test quarantined individuals	{start, end, both, daily}	{start, end, both}	X_5	
	Test sensitivity	$[0, 1]$	$\{55\%, 100\%\}$	X_6	[10]
Contact tracing	% of contacts of a tested person that can be traced	$[0, 1]$	$\{0.2, 1\}$	X_7	[9]
	Delay in contact tracing	Any integer	$\{0, 7\}$	X_8	[9]
	Start contract tracing without waiting for results	{True, False}	{True, False}	X_9	

by measuring the response y (average number of infections at the end of the simulation) for each simulated combination. We obtain y over a number of replications necessary to fit the 95% confidence interval to within 5% of the average. To determine the contribution of each parameters (individually as well as in groups), we compute the variances in response contributed by each combination and perform an F-test. In other words, the variance is decomposed over individual parameters ($X_1, \ldots, X_9$) as well as interactions of parameters. We speak of 2nd order interaction when examining the joint effect of two parameters ($X_1 X_2, X_1 X_3, \ldots, X_8 X_9$), 3rd order interaction for three parameters, and so on. We include up to 3rd order interactions to scan for possible effects that would only happen when three interventions are applied jointly.

There are two special cases when applying the 2^k factorial design to COVID-19 interventions. First, we cannot assume which values of X_1 (networks for mask wearing) and X_5 (whether to test at the start, end, or both times in a quarantine) have the highest or lowest impact on disease incidence. Consequently, we performed simulations for each of the three possible values of X_1 and X_5. We identified the maximum and minimum number of infected individuals across these simulations, and thus set the high and low values accordingly (Table 2). For example, consider that we have 10 sick individuals when testing at the start of quarantine, 100 when testing at the end, and 1000 when testing at both times. In this case, 'start' is the intervention with highest impact ($X_5^{high} = start$) while 'both' does the least to control the spread of the virus ($X_5^{low} = both$). Second,

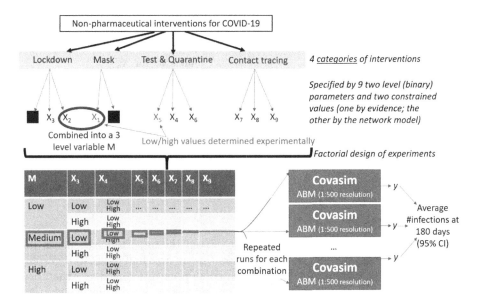

Fig. 1. Process flowchart from the identification of interventions (top) and their operationalization in the simulation to the factorial Design of Experiments (DoE) and repeated runs of a stochastic model targeting a 95% Confidence Interval.

not all $2^9 = 512$ combinations of the high and low levels are valid because of the interdependency between the networks for mask wearing and lockdown. If an intervention applies to work and school (e.g., working and learning remotely) then the fractional reduction of contact in the *community* should be 0 rather than $X_3^{low} = 0.1$, since the community network is not concerned by the intervention.

To resolve this issue, we first note that the highest impact intervention is the one that targets all networks, hence $X_1^{high} = \{w, s, h, c\}$ as seen in Table 2. The core question is thus to identify the *low* value, which is either $\{w, s\}$ or $\{c\}$. This choices sets either X_2 or X_3 (resp. the fractional reduction in contacts for $\{c\}$ and $\{w, s\}$) to 0%. Consequently, $2^7 = 128$ combinations are invalid and $384 = 3 \times 2^7$ remain. To avoid the complications of an incomplete parameter space, we combine two parameters into a new variable M with 3 levels (Table 3) and keep the rest unchanged. We can interpret this variable to be the level of "social distancing" in the broadest sense that is enforced. Moreover, the problem reduces to an ordinary factorial design analysis with one 3-level factor, which is analyzed using established methods [13, pp. 412–414].

3   Results and Discussion

Mask wearing together with transitioning to remote work/schooling has the largest impact (Table 4). It interacts with the constructs having the next largest impacts: ability to perform contact tracing and whether to start contact tracing

Table 2. Maximum responses caused by each choice of variables X_1 and X_5.

Variable	Value	Max response	Level
X_1	$\{w, s, c, h\}$	678.93	High
	$\{w, s\}$	278571.55	—
	c	359937.49	Low
X_5	Start	359862.78	—
	End	359606.72	High
	Both	359937.49	Low

Table 3. Correspondence between levels of M and combinations of levels of X_1 and X_2, given c as the low level for X_1.

M	X_1	X_2
High	$\{w, s, c, h\}$	30%
Med	$\{w, s, c, h\}$	5%
Low	c	0%

without waiting for test results. Our finding about the importance of masks and remote work/school differs from early works (in the first half of 2020) regarding COVID-19. Early systematic reviews released as preprints considered that "the evidence is not sufficiently strong to support widespread use of facemasks as a protective measure against COVID19" and that "masks alone have no significant effect in interrupting spread of [influenza-like illnesses ...] or influenza" (cited in [12]). More recently, a systematic review from December 2020 concluded that only four out of seventeen studies "supported the use of face masks [and] a meta-analysis of all 17 studies found no association between face mask intervention and respiratory infections" [14]. *However, once results are adjusted* for factors such as age and sex, the meta-analysis "suggests protective effect of the face mask intervention" [14]. Similarly, the most recent analyses and commentaries agree that reducing social network interactions in settings such as universities [3] is needed to avoid large outbreaks. Consequently, our result regarding the large effect of masks and remote work/schooling contributes to the more recent evidence base on interventions regarding COVID-19.

Although the preventative approach of using masks and shifting into remote work/school plays the largest role in reducing the likelihood of transmission (by lowering both the number of contacts and the virus transmissibility per contact), we do observe interacting effects with other intervention parameters. In particular, contact tracing is important to mitigate the pandemic, as demonstrated by the case of South Korea [15]. Our study contributes to understanding the specific parameters underlying contact tracing, as results stress the merits of having sufficient capacity to immediately and effectively perform contact tracing.

Limitations such as model validation are detailed in our supplementary online materials [4]. Our simulations used a factorial analysis in the artificial environment afforded by a model: these results could be contrasted to real data, as different US states and countries have implemented different combinations of interventions at various points in time. In addition, our results are reported using the confidence interval method to handle the stochastic nature of the simulations, but parameters may have different levels of uncertainty [2]. Finally, the scale of $1:500$ (resulting in over $650,000$ agents) was the maximum simulation size that we could perform given the number of repeats and hardware memory limitations [11]. Accuracy may thus be improved by simulating a full population.

Table 4. Factors responsible for $\geq 1\%$ of variance in the number of new infection cases

Variable(s)	Meaning	Contribution (%)
M	Masks + remote work/school	66.638
M and X_7	Masks + remote work/school, and ability to trace contacts of tested individuals	13.759
X_7	Ability to trace contacts of tested individuals	8.953
M and X_9	Masks + remote work/school, and starting contact tracing without waiting for test results	2.804
X_9	Contact tracing starts without waiting for test results	2.033
M and X_8	Masks + remote work/school + contact tracing delay	1.077

References

1. Chu, D.K., et al.: Physical distancing, face masks, and eye protection to prevent person-to-person transmission of SARS-CoV-2 and COVID-19: a systematic review and meta-analysis. Lancet **395**(10242), 1973–1987 (2020)
2. Edeling, W., et al.: Model uncertainty and decision making: predicting the impact of COVID-19 using the CovidSim epidemiological code (2020)
3. Ghaffarzadegan, N., et al.: Diverse computer simulation models provide unified lessons on university operation during a pandemic. BioScience **71**, 113–114 (2020)
4. Giabbanelli, P.J., Li, J.: Identifying synergistic interventions to address COVID-19 using a large scale agent-based model. medRxiv (2020)
5. Harper, C.A., Satchell, L.P., Fido, D., Latzman, R.D.: Functional Fear Predicts Public Health Compliance in the COVID-19 Pandemic. Int. J. Mental Health Addiction 1–14 (2020). https://doi.org/10.1007/s11469-020-00281-5
6. IHME COVID-19 Forecasting Team: Modeling COVID-19 scenarios for the United States. Nat. Med. **27**, 94–105 (2021)
7. Jacobsen, G.D., Jacobsen, K.H.: Statewide COVID-19 stay-at-home orders and population mobility in the united states. World Med. Health Policy **12**(4), 347–356 (2020)
8. John Hopkins University: Daily State-by-State Testing Trends (2020). https://coronavirus.jhu.edu/testing/individual-states
9. Kretzschmar, M.E., et al.: Impact of delays on effectiveness of contact tracing strategies for COVID-19: a modelling study. Lancet Public Health **5**(8), e452–e459 (2020)

10. Bastos, M.L., et al.: Diagnostic accuracy of serological tests for covid-19: Systematic review and meta-analysis. BMJ p. m2516 (2020)
11. Li, J., Giabbanelli, P.: Returning to a normal life via COVID-19 vaccines in the United States: a large-scale agent-based simulation study. JMIR Med. Inform. **9**(4), e27419 (2021)
12. Martin, G., Hanna, E., Dingwall, R.: Face masks for the public during COVID-19: an appeal for caution in policy (2020)
13. Montgomery, D.C.: Design and Analysis of Experiment. Wiley, Hoboken. 8 edn. (2012)
14. Ollila, H.M., et al.: Face masks prevent transmission of respiratory diseases: a meta-analysis of randomized controlled trials. medRxiv (2020)
15. Park, Y.J., et al.: Contact tracing during coronavirus disease outbreak, south korea, 2020. Emerg. Infect. Dis. **26**(10), 2465–2468 (2020)
16. Petherick, A., et al.: Variations in Government Responses to COVID-19 (2020)
17. Robinson, S.: Simulation: the practice of model development and use (2014)
18. Yang, J., et al.: Prevalence of comorbidities and its effects in patients infected with SARS-CoV-2: a systematic review and meta-analysis. Int. J. Infect. Dis. **94**, 91–95 (2020)

Modeling Co-circulation of Influenza Strains in Heterogeneous Urban Populations: The Role of Herd Immunity and Uncertainty Factors

Vasiliy N. Leonenko$^{(\boxtimes)}$ (iD)

ITMO University, 49 Kronverksky Pr, St. Petersburg 197101, Russia

Abstract. In this research, we aimed to assess the influence of herd immunity levels on the process of co-circulation of influenza strains in urban populations and to establish how the stochastic nature of epidemic processes might affect this influence. A spatially explicit individual-based model of multistrain epidemic dynamics was developed and simulations were performed using a 2010 synthetic population of Saint Petersburg, Russia. According to the simulation results, the largest influenza outbreaks are associated with low immunity levels to the virus strains which caused by these strains. At the same time, high immunity levels per se do not prevent outbreaks, although they might affect the resulting levels of disease prevalence. The results of the study will be used in the research of long-term immunity formation dynamics to influenza strains in Russian cities.

Keywords: Seasonal influenza · Herd immunity · Multiagent modeling · Stochastic processes · Strain co-circulation

1 Introduction

Outbreaks of influenza, one of the most widely spread human infectious diseases, result in 3 to 5 million cases of severe illness annually worldwide, and the mortality rate is from 250 to 640 thousand individuals per year [6]. One of the directions of influenza propagation studies, which help better understand the mechanics of disease dynamics and thus diminish its negative effects, is related to co-circulation of different influenza strains and its interplay with the levels of herd immunity to these strains. It is generally known that immunity level dynamics and disease incidence dynamics in the population are intertwined, but there are still open questions related to the quantification of their connection. In

This research was supported by The Russian Science Foundation, Agreement #20-71-00142. The participation in the ICCS conference was supported by the NWO Science Diplomacy Fund project #483.20.038 "Russian-Dutch Collaboration in Computational Science".

M. Paszynski et al. (Eds.): ICCS 2021, LNCS 12744, pp. 663–669, 2021.
https://doi.org/10.1007/978-3-030-77967-2_55

a conventional deterministic SEIR model, the population immunity level directly influences the outbreak size, and the onset of an outbreak is guaranteed when the number of susceptible and initially infected people are non-zero. At the same time, is it known that in real life the arrival of the infected individual in the population does not necessarily start an outbreak due to stochastic effects inherent to the initial stages of the epidemic onset [3,4]. These effects might significantly alter the properties of the outbreak, allowing it to gain momentum even if the level of population protection to the virus strain is high, or, alternatively, to make it die out in seemingly favorable conditions. As a result, it seems fair to ask to what extent the dominance of a particular influenza strain during the fixed epidemic season is defined by a mere effect of chance rather than by initial conditions, such as variation in immunity levels to different influenza strains. In the current study, we addressed this question by creating a spatially explicit individual-based model and examining the properties of artificial outbreaks caused by the introduction of several influenza strains into a synthetic population.

2 Experimental Setup

"Synthetic population" is an artificial spatially explicit human agent database representing the population of a city, a region or a country [2]. In this study, we have used a 2010 synthetic population of St Petersburg which was introduced in [9]. To simulate the circulation of multiple influenza strains, we employed a modified multiagent modeling framework first introduced in [8]. We do not take into account simultaneous co-infection, thus, if various strains are instantaneously transmitted to an individual at the place of contact, one of them is selected at random as the one causing the infection. Each agent in the population potentially interacts with other agents if they attend the same school (for schoolchildren), workplace (for working adults), or lives in the same household. The contacts in public transport are not considered. We take a simplifying assumption that the infection transmission coefficients are not dependent on the strain[1]. The infectivity of each individual is defined by a piecewise constant function g_τ which reflects the change of individual infectiousness over time from the moment of acquiring influenza [1]. Since, according to [7], a slight variation of g_τ values does not affect much the epidemic dynamics, we assumed the values of g_τ the same for all strains. Individuals recovered from the disease are considered immune to the particular influenza strain, that caused it, until the end of the simulation. Cross-immunity is not considered, i.e. the mentioned recovered individuals do not acquire immunity to other influenza strains.

---

[1] According to the published modeling results of other research groups, the virulence of A(H1N1) and A(H3N2) is almost similar, while the virulence of B strain might be slightly lower [5].

3   Simulations

In the course of simulations, we analyzed how different factors influence epidemic outbreaks. As a defining property of an outbreak, we used disease prevalence at day 15 (corresponds to two weeks after the virus introduction in the population). We assumed that zero prevalence at day 15 of the simulation run signifies the absence of an outbreak (a stable transmission chain was not formed and the disease died out).

Dependence on Initial Number of Infected and Herd Immunity Level. In this set of simulations, we considered that α_m, the fraction of individuals susceptible to the virus strain m, is equal for all the strains. Firstly, we tested the effect of changing the number of initially infected, taking $I_0^{(m)} = 1$, $I_0^{(m)} = 100$, and $I_0^{(m)} = 100$ for each strain $m \in \overline{1,3}$ (Fig. 1, left). Secondly, in the same way we tested the effect of changing $\alpha^{(m)}$ (Fig. 1, right). It can be seen that disease prevalence levels are largely defined by the values of $I_0^{(m)}$. The experiment demonstrated that the epidemics started from a single 'patient zero' have higher chances of dying out compared to the larger quantities of initially infected persons. The leftmost group of bars shows that the chance of an epidemic decline till day 15 is around 50% for the case of $I_0^{(m)} = 1$ and around 30% for the $I_0^{(m)} = 100$. As to the influence of herd immunity, its high level reduces the number of the occurred epidemic surges (although due to the stochastic nature of the outbreaks the dependency between the two variables might be not monotonic). The highest level of herd immunity, which corresponds to the smallest level of $\alpha^{(m)}$, prevented large outbreaks from happening (e.g., with $\alpha^{(m)} = 0.05$ no outbreaks with prevalence > 50000 were detected).

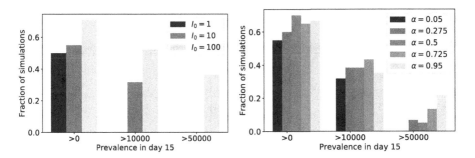

Fig. 1. Distribution of disease prevalence levels on day 15 depending on the fraction of initially susceptible individuals (left graph) and on the fraction of susceptible individuals (right graph)

Simulating Multistrain Outbreaks. The second set of experiments with the model
was dedicated to the simulation of co-circulation of three influenza strains,
A(H1N1), A(H3N2), and B, in the synthetic population with strain-specific herd
immunity levels. Unfortunately, the information on seroprevalence levels registered in the population of Saint Petersburg was not available to the author at
the time of the study. Two sets of corresponding values for the beginning of
2010–2011 epidemic season in Moscow and Voronezh were taken for the experiments, as these two cities are situated not far from Saint Petersburg and have the
fullest seroprevalence records. We assumed that the fraction of samples seropositive to the virus strain m from the provided dataset (Table 1) is equal to $1 - \alpha_m$,
which made it possible to calculate α_m. The resulting values of α_m for the
three strains were 0.57, 0.13, 0.1 for 'Voronezh' simulation and 0.78, 0.74, 0.6
for 'Moscow' simulation. According to the data, 'Voronezh' synthetic population possesses rather low susceptibility levels to A(H3N2) and B strains and has
slightly higher vulnerability to A(H1N1) strain. At the same time, 'Moscow' population is considerably less protected from the possible outbreak caused by any of
the three strains. The obtained distribution of the registered disease prevalence
is demonstrated in Fig. 2. As one can see, high levels of population protection
in 'Voronezh' experiment did not prevent the occurrence of epidemic outbreaks
of the strains A(H3N2) and B, and, moreover, their number is bigger than the
number of A(H1N1) outbreaks which have a larger reservoir of susceptible individuals. Nevertheless, the single case of considerably high disease prevalence
(>50000 cases at day 15) is attributed to A(H1N1). The results conform to the
previous experiment, where the value of $\alpha_m = 0.05$ (twice as low as the lowest
susceptibility level in this setting) still permitted full-fledged outbreaks in the
population. In the case of 'Moscow' experiment, the distribution of the number
of outbreaks of considerable sizes better conforms to the differences in levels of
susceptibility. The biggest registered outbreak is caused by the A(H3N2) strain,
which is also in fair agreement with the predefined values of α_m.

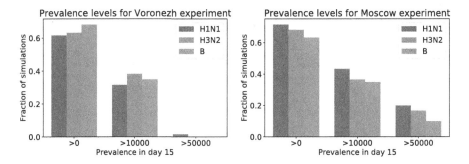

Fig. 2. Distribution of disease prevalence levels on day 15 for the immunity levels set
according to seroprevalence tests for Voronezh (left) and Moscow (right)

To assess the interrelation between the herd immunity to the particular virus strain and the possible chance that an epidemic caused by that strain will die out before getting the chance to gain momentum, we listed the corresponding data in Table 1. The numbers show that there might be a correlation between the immunity levels and percentage of 'failed' outbreaks in 'Moscow' experiment, whereas the data for Voronezh is contradictory. In the latter setting, high levels of immunity do not seem to negatively affect the probability of A(H3N3) and B strain–related outbreaks, as it was also shown in the above-mentioned Fig. 2.

Table 1. Percentage of samples seropositive to particular strains and number of epidemics with zero prevalence on day 15 in the simulations 'Voronezh' and 'Moscow'

	A(H1N1)	A(H3N3)	B
Voronezh, seropositives before the epidemic season	43%	87%	90%
'Voronezh' simulation, halted outbreaks	38.3%	36.7 %	31.7 %
Moscow, seropositives before the epidemic season	28.9%	37.4%	46%
'Moscow' simulation, halted outbreaks	28.3%	31.7%	36.7%

4   Results

In this research, we aimed at assessing the role of herd immunity in the epidemic progression through the population. The conclusion we can draw is the following: the actual influenza dynamics in the population, reproduced by stochastic models, is indeed influenced by herd immunity, but the factor of uncertainty inherent to the disease transmission process might somewhat reduce this influence. On one hand, we can see that the increased level of herd immunity apparently lessens the probability of a full-fledged outbreak, as well as lowers the prevalence of the outbreaks occurred. On the other hand, this impact is apparent only when we compare the experiments with dramatically different herd immunity levels. The change of the fraction of susceptible individuals from 0.95 to 0.05 resulted in the chance of an outbreak occurrence decreased only by 11% (from 66% to 55%). Also, as Fig. 1 demonstrates, the dependency between the mentioned two values is not monotonic.

A similar conclusion might be drawn from the experiment dedicated to the co–circulation of viruses in the population with the strain-specific levels of herd immunity. In both experimental settings, a single largest outbreak (prevalence level > 50000 for 'Voronezh' setting and > 100000 for 'Moscow' setting) was caused by the virus strain which was favored among the others due to decreased corresponding population protection. At the same time, the distribution of smaller outbreaks by strain types does not conform well to the levels of $\alpha^{(m)}$, nor is the proportion of 'successful' outbreaks in general (Fig. 2, Table 1).

What is more defining in the 'success' of the outbreak is the number of initially infected individuals introduced at the beginning of the simulation.

It is known that in SEIR-type deterministic models the introduction of infection is often modeled by one person, i.e. $I_0 = 1$, since in these types of models the initial surge of disease prevalence caused by a single individual is guaranteed, if the selected parameter values allow the infection propagation. At the same time, in the stochastic models starting the simulation from one 'patient zero' might lead to the infection dying out rapidly. In our case, for the population with 5% of susceptible individuals, a half of the started outbreaks dies out till day 15 (Fig. 1). In the author's opinion, this type of dynamics is more adequate, comparing with the output of mean-field approximations. A large number of the initially infected individuals introduced at once reduces the possibility of transmission chains being broken, but this way of initializing the simulations affects the final outbreak size and, besides that, seems unrealistic. The better option might be to allow an influx of infected individuals in the course of simulation [11]. In this case, the model will be able to demonstrate a time delay between the moment of the first introduction of the infected person in the population and the actual outbreak surge. This time delay takes place in real epidemic outbreaks—for instance, it was detected during the first wave of COVID-19 in Russia in spring of 2020. The surge of prevalence in deterministic SEIR models always starts at day 0, which might partially explain big biases in assessing the influenza outbreak peaks demonstrated by these models [10]. Additionally, as it is clear from the data, the surges of outbreaks of different strains are separated in time, so the reintroduction hypothesis is supported by this evidence as well.

5 Discussion

To conclude, the performed research showed that the percentage of individuals in the population who are immune to particular influenza strains might somewhat affect the properties of the forthcoming outbreaks—particularly, it might define the resulting dominant virus strain. At the same time, the data on immunity levels might be not enough to determine whether the outbreak itself will take place, because this largely depends on the external factors (possibly, on the dynamics of reintroduction of the infected people in the population with the migration influx). Also, a high level of registered immunity to a particular strain does not guarantee a total absence of disease cases caused by that strain. It might be still circulating in the population, although without causing a major outbreak. The obtained results might be sensitive to the structure of the regarded synthetic population,—particularly, to contact network structure defined by the population and the assumed rules of individual behavior,—and this matter should be considered thoroughly in separate research.

One of the drawbacks of the study is that we considered the seroprevalence levels obtained by laboratory testing of biological samples to be equal to the fraction of individuals in the population not vulnerable to influenza, which is, strictly speaking, not the case. The role of indirect protection of the individuals, which is also responsible for the herd immunity, was not considered and remains an aim for subsequent studies. Also, we plan to perform simulation runs with the increased modeling period ($T = 100$ instead of $T = 15$) to assess how the immunity levels affect the outbreak sizes, which at this point was not possible due to the limitations of the algorithm performance. The modeling framework developed in the course of the study will be used in the research of the long-term immunity formation dynamics to influenza strains in Russian cities. Besides that, it might be utilized to assess the effects of targeted vaccination campaigns and to simulate a co-circulation of different acute respiratory diseases in the populations, particularly, the co-circulation of influenza and COVID, which now draws wide attention of various research teams.

References

1. Baroyan, O., et al.: Computer modelling of influenza epidemics for large-scale systems of cities and territories. In: Proceedings of WHO Symposium on Quantitative Epidemiology, Moscow (1970)
2. Bates, S., et al.: Using synthetic populations to understand geospatial patterns in opioid related overdose and predicted opioid misuse. Comput. Math. Organ. Theory **25**(1), 36–47 (2019)
3. Brett, T., et al.: Detecting critical slowing down in high-dimensional epidemiological systems. PLOS Comput. Biol. **16**(3), 1–19 (2020)
4. Drake, J.M., et al.: The statistics of epidemic transitions. PLOS Comput. Biol. **15**(5), 1–14 (2019)
5. Hill, E.M., et al.: Seasonal influenza: Modelling approaches to capture immunity propagation. PLoS Comput. Biol. **15**(10), e1007096 (2019)
6. Iuliano, A.D., et al.: Estimates of global seasonal influenza-associated respiratory mortality: a modelling study. The Lancet **391**(10127), 1285–1300 (2018)
7. Ivannikov, Y., Ogarkov, P.: An experience of mathematical computing forecasting of the influenza epidemics for big territory. J. Infectol. **4**(3), 101–106 (2012). In Russian
8. Leonenko, V., Arzamastsev, S., Bobashev, G.: Contact patterns and influenza outbreaks in Russian cities: A proof-of-concept study via agent-based modeling. J. Comput. Sci. **44**, 101156 (2020)
9. Leonenko, V., Lobachev, A., Bobashev, G.: Spatial modeling of influenza outbreaks in saint petersburg using synthetic populations. In: Rodrigues, J.M.F., et al. (eds.) ICCS 2019. LNCS, vol. 11536, pp. 492–505. Springer, Cham (2019). https://doi.org/10.1007/978-3-030-22734-0_36
10. Leonenko, V.N., Novoselova, Y.K., Ong, K.M.: Influenza outbreaks forecasting in Russian cities: Is Baroyan-Rvachev approach still applicable? Procedia Comput. Sci. **101**, 282–291 (2016)
11. Pertsev, N., Leonenko, V.: Analysis of a stochastic model for the spread of tuberculosis with regard to reproduction and seasonal immigration of individuals. Russ. J. Numer. Anal. Math. Model. **29**(5), 285–295 (2014)

Two-Way Coupling Between 1D Blood Flow and 3D Tissue Perfusion Models

Raymond M. Padmos[1]([⊠])(ID), Tamás I. Józsa[2](ID), Wahbi K. El-Bouri[2,3](ID),
Gábor Závodszky[1](ID), Stephen J. Payne[2](ID), and Alfons G. Hoekstra[1](ID)

[1] Computational Science Laboratory, Informatics Institute, Faculty of Science,
University of Amsterdam, Science Park 904, Amsterdam 1098, XH, The Netherlands
r.m.padmos@uva.nl
[2] Institute of Biomedical Engineering, Department of Engineering Science,
University of Oxford, Parks Road, Oxford OX1 3PJ, UK
[3] Liverpool Centre for Cardiovascular Science, Department of Cardiovascular
and Metabolic Medicine, University of Liverpool, Liverpool, UK

Abstract. Accurately predicting brain tissue perfusion and infarct volume after an acute ischaemic stroke requires the two-way coupling of perfusion models on multiple scales. We present a method for such two-way coupling of a one-dimensional arterial blood flow model and a three-dimensional tissue perfusion model. The two-way coupling occurs through the pial surface, where the pressure drop between the models is captured using a coupling resistance. The two-way coupled model is used to simulate arterial blood flow and tissue perfusion during an acute ischaemic stroke. Infarct volume is estimated by setting a threshold on the perfusion change. By two-way coupling these two models, the effect of retrograde flow and its effect on tissue perfusion and infarct volume can be captured.

Keywords: Cerebral tissue perfusion · Acute ischaemic stroke · Multi-scale modelling · Infarct volume modelling · Blood flow simulations

1 Introduction

An acute ischaemic stroke (AIS) is caused by the sudden blockage of a major cerebral vessel by a thrombus. Every year, millions of people suffer an AIS, resulting in disability and possibly death [5]. The sudden loss of blood flow to tissue, i.e. perfusion, leads to the formation of a cerebral infarct. Understanding how an infarct forms during an AIS can help medical decision making and treatment development. To understand infarct formation, it is necessary to understand how an AIS affects brain tissue perfusion. Unfortunately, predicting brain tissue perfusion is not trivial.

One particular problem is the range of scale of the cerebral vasculature. The diameter of blood vessels ranges from micrometers in the capillaries to

© Springer Nature Switzerland AG 2021
M. Paszynski et al. (Eds.): ICCS 2021, LNCS 12744, pp. 670–683, 2021.
https://doi.org/10.1007/978-3-030-77967-2_56

millimeters in the large systemic arteries. There are billions of vessels in the microcirculation. Solving the Navier-Stokes (NS) equations for the entire vascular system is currently not feasible. As a result, approximations have to be made to solve parts of the vascular system using simplified equations. One-dimensional approximations of the NS equations accurately capture blood flow for the large vessels [1,20,21,25]. The microcirculation can be described as a three-dimensional porous medium [6,8,11,15].

There are multiple phenomena where effects on the small scale have an effect on the large scale, and vice versa. For instance, a growing infarct affects the flow of blood by the death of capillary pericytes, leading to vessel constriction [7]. In addition, retrograde blood flow beyond the thrombus is often observed [2,23]. Capturing these phenomena and their effect on arterial blood flow and tissue perfusion requires a fully coupled model that captures blood flow on multiple scales. Accurately predicting infarct volume after an AIS therefore requires two-way coupling of models of haemodynamic models describing blood flow both in the large arteries and in the microcirculation.

Here, we present a method for the two-way coupling between a one-dimensional blood flow model (1D BF) to a three-dimensional tissue perfusion model (3D perfusion). The coupling occurs through the pial surface where the pressure drop between the models is captured using a coupling resistance. First, a test model of the brain is used to illustrate the models, the coupling algorithm, and show the accuracy of the solutions. Then, the two-way coupled model is used to simulate arterial blood flow and cerebral tissue perfusion during an AIS. The change in tissue perfusion and infarct volume are quantified and compared to a one-way coupled case.

2   Methods

The main focus of this work is the coupling of a 1D blood flow and a 3D tissue perfusion model. The 1D BF and 3D perfusion models in this work are briefly described for completeness. We refer to our previous work for a more detailed description of the individual models [11,17].

2.1   One-Dimensional Blood Flow Model

Blood vessels are modelled as thin elastic tubes and blood as an in-compressible viscous fluid. Every vessel segment is modelled as a pressure-dependent resistance. The vessels are discretised to a minimum of three nodes, i.e. two terminal and one internal node, and to a maximum resolution of 2.5 mm along the length of the vessel.

The pressure in the network is calculated by solving the mass-balance equations given by

$$\sum_j G_{ij} \left(P_i - P_j \right) = S_i \tag{1}$$

with P_i the pressure at node i, G_{ij} the conductance, i.e. reciprocal of resistance, between nodes i and j, and S_i a source term for every node i. The conductance of a segment is given by $G = \frac{\pi R^4}{2(\zeta+2)\mu L}$, where R is the mean segment radius, L is the segment length, μ is the dynamic viscosity, and ζ is a constant related to the velocity profile, with 2 representing a parabolic profile, i.e. laminar flow, and 9 representing a flatter profile [22]. The larger constant is the result of a blunt velocity profile in the vessel [3]. We use the constant of a blunt profile, $\zeta = 9$.

The resulting system can be written as

$$\mathbf{G}\vec{P} = \begin{bmatrix} \sum G_{1j} & -G_{12} & \cdots & -G_{1N} \\ -G_{21} & \sum G_{2j} & \cdots & -G_{2N} \\ \vdots & \vdots & \ddots & \vdots \\ -G_{N1} & \cdots & \cdots & \sum G_{Nj} \end{bmatrix} \begin{bmatrix} P_i \\ \vdots \\ \vdots \\ P_N \end{bmatrix} = \begin{bmatrix} S_i \\ \vdots \\ \vdots \\ S_N \end{bmatrix} = \vec{S} \tag{2}$$

The pressure-area relationship of a vessel is given by [16]

$$P = P_0 + \frac{\sqrt{\pi}Eh}{A_0(1-\nu^2)}\left(\sqrt{A} - \sqrt{A_0}\right) \tag{3}$$

The volumetric flow rate in a segment, Q_{ij}, is calculated by

$$Q_{ij} = (P_i - P_j)G_{ij} \tag{4}$$

The resulting system is solved iteratively until

$$\varepsilon_P = \frac{\left|\vec{P_i} - \vec{P_{i-1}}\right|}{\left|\vec{P_i}\right|} < 10^{-12} \tag{5}$$

with $\vec{P_i}$ and $\vec{P_{i-1}}$ being the pressure vectors during the ith and $(i-1)^{th}$ iterations, respectively.

2.2   Three-Dimensional Tissue Perfusion Model

Tissue perfusion is simulated using a multi-compartmental porous medium approach [11]. Three compartments are used to simulate blood flow through the arterioles, capillaries and venules. The compartments are located at the same spatial location and are coupled locally.

The equations describing flow through the three compartments are given by

$$\nabla \cdot (\mathbf{K_a}\nabla p_a) - \beta_{ac}(p_a - p_c) = 0$$
$$\nabla \cdot (\mathbf{K_c}\nabla p_c) + \beta_{ac}(p_a - p_c) - \beta_{cv}(p_c - p_v) = 0$$
$$\nabla \cdot (\mathbf{K_v}\nabla p_v) + \beta_{cv}(p_c - p_v) = 0 \tag{6}$$

where p_a, p_c, p_v are the pressure corresponding to the arterial, capillary and venule compartment, $\mathbf{K_a}$, $\mathbf{K_c}$, and $\mathbf{K_v}$ are the permeability tensors of the respective compartment, and β_{ac}, and β_{cv} are the coupling coefficients between the arterial-capillary, capillary-venule compartments respectively. The brain is divided into white and grey matter regions with different coupling coefficients. The model parameters are optimised to achieve pre-set perfusion targets with a total cerebral perfusion of 600 mL/min. For the full derivation, solution method, and motivation, we refer to [11].

Tissue perfusion, in units of mL/min/100mL, is calculated as

$$F = 6000\beta_{ac}(p_a - p_c) \tag{7}$$

where 6000 is a unit conversion factor, the product of 60 s and 100 mL. Tissue perfusion drops during a stroke, the perfusion change is defined as

$$\Delta F = \frac{F^{\text{Stroke}} - F^{\text{Healthy}}}{F^{\text{Healthy}}} 100\% \tag{8}$$

An infarct can be determined by setting a threshold for the change in perfusion, a value of -70% is used in this paper [4].

2.3 Two-Way Coupling Between Blood Flow and Tissue Perfusion Models

We assume that between the outlets of the 1D BF model and the 3D perfusion model, vessels exist that are not included in either of the models and that result in a pressure drop. Furthermore, we assume that each outlet of the 1D model has its own perfusion territory on the pial surface. These regions are determined by a mapping algorithm, as described previously as part of our previous work [17]. During a baseline simulation, i.e. healthy scenario, the surface pressure of the 3D perfusion model is assumed to be $p_{surface}$. Setting the pressure at the surface to $p_{surface}$ closes the 3D perfusion model. The flow rate at the surface for each perfusion territory Q_i can then be calculated by integrating the velocity, (for example, $u_a = K_a \nabla p_a$), normal to the surface over the area. The coupling resistance between both models can be calculated as

$$R_i = \frac{P_i - P_{surface}}{Q_i} \tag{9}$$

where i corresponds to the outlet nodes of the 1D BF model. The coupling resistance is added to each outlet of the 1D BF model with the outlet pressure set to $p_{surface}$. The 1D BF model is solved and the outlet resistance updated until a relative tolerance of 10^{-9} is reached. To ensure convergence of the model, the coupling resistance is first calculated using the venous pressure. If the pressure at the outlets is less than $p_{surface}$, the inlet pressure is increased with the difference until the lowest outlet pressure is larger than $p_{surface}$. Finally, the coupling resistance is calculated using Eq. 9.

During a stroke, the assumption of a uniform surface pressure is lifted. The coupling problem becomes finding the surface pressures such that the two models agree on the volumetric flow rate at each outlet. This optimisation problem is solved using a Newton-Krylov solver minimising the difference in volumetric flow rate between the models to a relative tolerance less than 10^{-9}. A uniform surface pressure, i.e. $p_{surface}$ is used as an initial guess for the solver. The optimisation algorithm requires on the order of 3 iterations before convergence. The total run time is around 2.5 h on a AMD Ryzen 7 3700x 8-core processor with 80 GB of RAM. The main computational cost is running the 3D perfusion model to obtain an estimate of the Jacobian. The model parameters used in the simulations are listed in Table 1.

2.4   Test Model

A test model is created by combining a bifurcating tree with a cube. In the arterial compartment, four sides of the cube are coupled to the bifurcating tree, the top and bottom sides set as no flow boundaries. In the capillary compartment, all sides are set as no flow boundaries. In the venule compartment, the four sides coupled to the bifurcating tree are set to venous pressure, the top and bottom sides are set as no flow boundaries. The cube is divided into white and grey matter regions with different coupling coefficients. A symmetric bifurcating tree is generated using Murray's law, given by $R_0^3 = 2R_1^3$, starting from an initial vessel with a length of 200 mm, a radius of 10 mm and a Young's modules of 0.4 MPa. Daughter vessels have a Young's modules of 1.6 MPa, similar to a cerebral arteries. The model parameters used in the simulations are listed in Table 1. The parameters are optimised to obtain a pre-set perfusion in the healthy case of 600 mL/min according to [11]. A stroke is simulated by occluding one of the first generation vessels. The dimensions of the cube are 100 mm per edge, the inner white matter region has dimensions of 50 mm per edge with a cut-out cube of 25 mm per edge.

3   Results

3.1   Test Model

Figure 1 shows the test model designed to illustrate the models and the coupling algorithm. A section is removed to show the inner parts of the brain. The blood flow model is a symmetric bifurcating tree with four outlets and one inlet, Fig. 1a shows the pressure when a thrombus is occluding one of the vessels. Figure 1b shows the brain mesh used in this example. The brain consists of white matter and grey matter, which differ in their coupling coefficient, Fig. 1c shows the coupling coefficients for white and grey matter regions. Figure 1d, Fig. 1e, and Fig. 1f show the pressure in the arteriole, capillary, and venule compartments respectively during an occlusion. The volumetric flow rate through the top and the bottom surfaces is zero, the other four sides are coupled to the bifurcating tree shown in Fig. 1a. Tissue perfusion is calculated using Eq. 7. Figure 1g shows tissue perfusion during baseline while Fig. 1h shows tissue perfusion during an occlusion. The resulting perfusion change is shown in Fig. 1i.

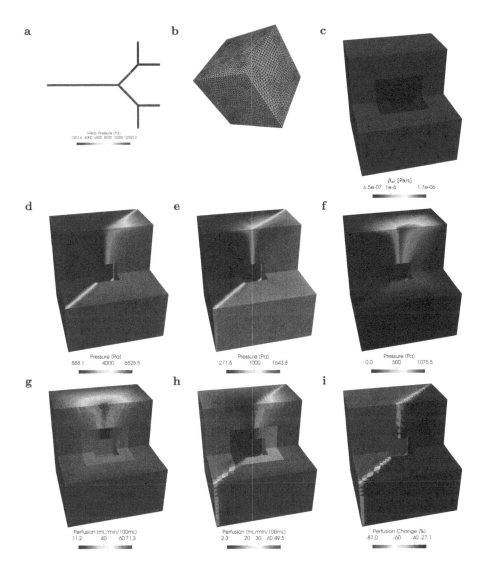

Fig. 1. (a) The arterial blood flow model consists of a symmetric bifurcating tree. Shown is the pressure during an occlusion of the first daughter vessel. (b) The brain mesh used in this example. (c) The permeability parameters of the perfusion model. White and grey matter regions have different values. (d, e, f) Pressure during an occlusion in the arteriole (d), capillary (e) and venule (f) compartments respectively. (g) Healthy tissue perfusion. (h) Tissue perfusion during an occlusion. (i) Perfusion change as a result of an occlusion.

Table 1. Model parameters, superscripts G and W indicate values corresponding to grey and white matter.

Parameter	Cube	Human brain	Unit
	Value	Value	
p_{in}	12500	12500	Pa
p_{venous}	0	0	Pa
$p_{surface}$	9000	8000	Pa
ν	0.5	0.5	–
μ	3.5	3.5	mPa
ζ	9	9	–
k_a	7.61×10^{-3}	1.987×10^{-3}	mm^{3}s/g
k_c [6]	4.28×10^{-7}	4.28×10^{-7}	mm^{3}s/g
k_v	1.52×10^{-2}	3.974×10^{-3}	mm^{3}s/g
β_{ac}^{G}	1.699×10^{-6}	1.624×10^{-6}	Pa/s
β_{cv}^{G}	5.947×10^{-6}	5.683×10^{-6}	Pa/s
β^{G}/β^{W}	2.61	2.58	–

3.2 Coupled Brain Model

The coupled model consists of a 1D BF model and a 3D perfusion model, as shown in Fig. 2. Figure 2a and Fig. 2b show the 1D BF model and the 3D perfusion model respectively. The two models are coupled through the pial surface, Fig. 2d, a coupling resistance captures the pressure drop caused by absent vessels. Figure 2e shows the pressure during a baseline simulation; Fig. 2f shows the resulting tissue perfusion during this baseline simulation.

3.3 Modelling Acute Ischaemic Stroke

An AIS can be simulated by occluding one of the major cerebral vessels. The occlusion in the simulations presented here is located in the right middle cerebral artery (MCA). Figure 3 shows the difference between the one-way and two-way coupled models during an AIS. Figures 3a and 3b show the pressure in the 1D BF model, while Figs. 3c and Fig. 3d show tissue perfusion during an occlusion of the right MCA for the one-way and two-way coupled models respectively. Figure 3e and Fig. 3f show the change in tissue perfusion during the same occlusion. If the models are not coupled, the predicted volumetric flow rate and pressure at the boundary regions downstream of the occlusion predicted by the 1D BF model are zero and venous pressure respectively. Simulating brain tissue perfusion with these values results in the worst-case scenario. Simulation brain tissue perfusion with two-way coupling results in smaller predicted infarct volumes. Table 2 lists tissue perfusion and infarct volumes values per simulation.

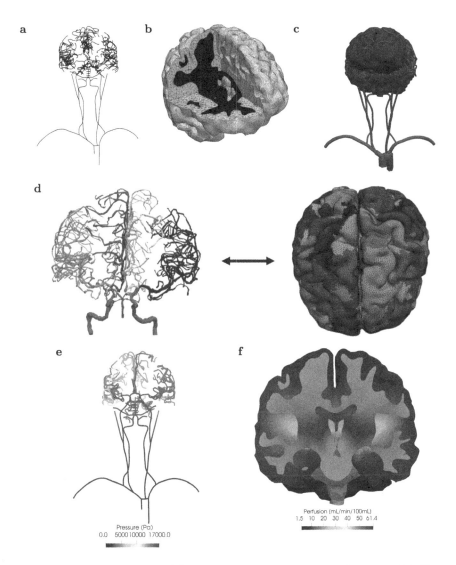

Fig. 2. (a) The 1D BF model. (b) The mesh used in the 3D perfusion model. Shown are the white and grey matter regions. A region is cut out to show the inner regions. (c) Merged 1D BF model and 3D perfusion model view. (d) The coupling method between the models. Each outlet of the 1D BF model is connected to a surface region on the boundary of the brain mesh. (e) Pressure in the 1D BF model during baseline. (f) Tissue perfusion in the 3D perfusion model during baseline.

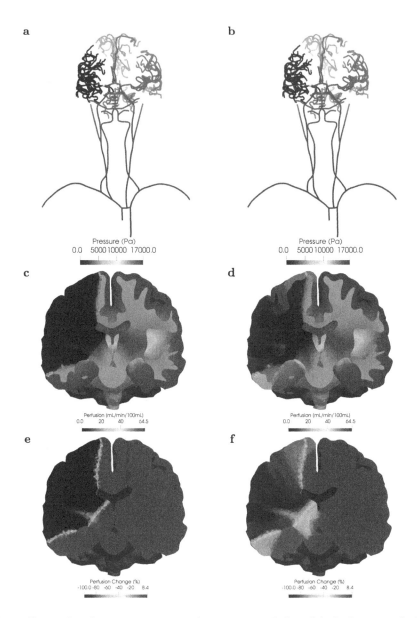

Fig. 3. Comparison between one-way and two-way coupled models during an occlusion. Front view of the brain, the brain is sliced to reveal the inner parts. (a, b) Pressure in the one-way (a) and two-way (b) coupled 1D BF model respectively. (c, d) Tissue perfusion in the one-way (c) and two-way (d) coupled 3D perfusion model respectively. (e, f) Tissue perfusion change in the one-way (e) and two-way (f) coupled 3D perfusion model respectively.

Table 2. Tissue perfusion and infarct volume values.

Simulation	Tissue Perfusion [mL/min/100mL]			Infarct Volume [mL]		
	Healthy	One-way	Two-way	Healthy	One-way	Two-way
Grey Matter	55.67	43.47	45.53	0.0	189.4	162.4
White Matter	20.62	15.47	16.36	0.0	120.0	99.2
Total	43.16	33.47	35.12	0.0	309.4	261.5

4 Discussion

Two-way coupling arterial blood flow and tissue perfusion models is not trivial. One difficulty occurs due to the difference of the mathematical formulation of the two models. The one-dimensional arterial blood flow model describes pressure and flow rates as discrete values in a single point, while the three-dimensional tissue model uses volume averages over the surface. This difficulty is resolved in our model by the use of a coupling surface at the cerebral cortex, or pial surface. An alternative is the use of volume source terms [9,15,19]. However, this requires resolving a large number of vessels to preserve anatomical connections between brain territories and large vessels. Instead, we utilise the 3D perfusion model to account for blood flow under the cortical surface. This coupling mirrors well the anatomical structure of the vasculature, namely that the human brain is perfused through the pial surface by the penetrating arterioles [10]. The preferred direction of flow due to the penetrating arteries is captured by using an anisotropic permeability tensor in the 3D perfusion model. The use of a coupling surface, i.e. boundary conditions, is therefore a valid alternative to volume source terms. In addition, this approach also simplifies the coupling. The surface pressure of the human brain is not known in detail and is assumed in this paper to be 8000 Pa (60 mmHg). The perfusion model parameters are optimised to achieve a pre-set total perfusion of 600 mL/min during the baseline simulation regardless of surface pressure.

Another difficulty arises due to the unavailability of data of the entire cerebral vasculature for a single patient. Between the two models, there is a lack of information by the absence of the pial surface vessels. Everything under 1 mm is missing, including pial vessels, because of the limited spatial resolution of medical imaging. We solve this problem by calculating a coupling resistance between the models. The resistance captures the effect of the missing vessels. Another approach would be to explicitly generate vessel networks with equivalent resistance [12,14,18]. However, the outlets of networks trees would also need to be explicitly coupled to the tissue model. The coupling resistances found depend on the value of assumed surface pressure, a higher pressure leads to a smaller value and therefore less resistance. The total resistance in the system determines the total amount of flow while the relative fraction at each outlet determines the distribution of flow at the outlets.

The velocity profile in the large vessels is not parabolic but rather blunt [22]. Models of the large vessels account for this by increasing the resistance. In the microcirculation, a laminar velocity profile is often assumed. In this paper, the 1D BF model uses the parameter of a blunt profile to calculate the pressure. This choice results in a larger pressure drop in the arteries. The blunt profile does not hold for the entire vasculature and decreases towards a parabolic profile, i.e. laminar flow in the microcirculation. However, it is not clear how this parameter changes and is therefore kept constant in this paper. The assumed pial surface pressure in the healthy scenario can be incompatible with the assumed velocity profile. To still achieve convergence, the inlet pressure is increased until the lowest pressure at an outlet matches the assumed surface pressure. For a laminar velocity profile, the inlet pressure does not need to be increased. It is worthwhile to note that the tissue perfusion and infarct volume values are not affected by the choice of velocity profile parameter within the tested range.

Figure 1 shows a simple example of a bifurcating tree representing the arterial vasculature and a cube representing the brain. The perfusion parameters are optimised to achieve a total perfusion of 600 mL/min during the baseline simulation. An occlusion is simulated by occluding one of the branches. The un-occluded branch then provides all blood flow to the downstream tissue. However, it is unable to perfuse the entire volume at the same level. As a result, the entire volume is perfused at a lower level with the perfusion deficit increasing with distance from the unblocked vessels. The infarct volume in this case would be half the volume. Note that tissue perfusion does not drop to zero anywhere in the volume. The surface pressure in the baseline simulation is chosen to be 9000 Pa as the pressure drop in the bifurcating tree is much smaller than for the human brain. Choosing a different value for the surface pressure will change the coupling resistances as a smaller pressure drop between the models means less resistance for the same volumetric flow rates.

Figures 2a and 2b show the 1D BF model and tissue perfusion model applied to a human cerebral vasculature and brain respectively. The outlets of the 1D BF model are coupled to the surface regions of the brain, as shown in Fig. 2d, the coupling resistances are not shown. By assuming a uniform surface pressure at the surface of the 3D perfusion model, the volumetric flow rates can be calculated and used to determine the coupling resistances. Figures 2e and 2f depict the resulting baseline simulation for the 1D BF and 3D perfusion models respectively. The baseline simulations are able to achieve realistic values for tissue perfusion, as shown in Table 2.

Figure 3 shows a comparison between the one-way and two-way coupled versions of the model. Coupling the two models provides a different estimate of the pressure and flow rates at the pial surface than the one-way model. The perfusion change, Figs. 3e and 3f, show that tissue perfusion drops less in the two-way coupled model. The surface pressure in the two-way coupled model is larger than in the one-way model. The surface pressure during an occlusion is not set to a certain value in the two-way coupled model, but is instead the solution of a root finding algorithm. This provides a more accurate solution of the surface pressure during an occlusion. In the two-way coupled model, the regions surrounding the infarcted region are perfused less as a result of retrograde flow. Table 2 lists the difference in mean tissue perfusion and infarct volumes for the healthy, one-way and two-way coupled simulations. The two-way coupled model results in a lower infarct volume at the chosen threshold.

The simulations presented in this paper represent a patient with a complete occlusion of the right MCA without any collateral vessels. These vessels provide flow through alternative pathways and maintain perfusion [13,24]. In addition, the effect of tissue death on blood flow is neglected. These effects are outside the scope of this paper. We are planning to investigate the neglected features in the future. The simulations presented here therefore represent a worst-case scenario. Two-way coupling is necessary to provide better estimates of infarct volume, capture the effect of collateral flow, and simulate the growth of the infarct core. In addition to these effects, model validation is also a direction of future research. Our aim is the creation of a model that can simulate the formation and growth of the infarct volume during an AIS. By two-way coupling these two types of models, we are one step closer to accurately predicting infarct volume and location.

5   Conclusion

Accurately predicting infarct volume after an AIS requires two-way coupling of models describing arterial blood flow and tissue perfusion. Here, we present a method for the two-way coupling between a one-dimensional arterial blood flow model and a three-dimensional tissue perfusion model. A test model is presented to showcase the models and algorithms. The two-way coupling allows for feedback between the models, thereby capturing retrograde flow, and leading to a smaller estimate of infarct volumes.

Funding. This project (INSIST; www.insist-h2020.eu) has received funding from the European Union's Horizon 2020 research and innovation programme under grant agreement No 777072.

Conflicts of Interest
The authors declare no conflict of interest.

References

1. Alastruey, J., Moore, S.M., Parker, K.H., David, T., Peiró, J., Sherwin, S.J.: Reduced modelling of blood flow in the cerebral circulation: Coupling 1-D, 0-D and cerebral auto-regulation models. Int. J. Numer. Methods Fluids **56**(8), 1061–1067 (2008). https://doi.org/10.1002/fld.1606
2. Arrarte Terreros, N., et al.: From perviousness to permeability, modelling and measuring intra-thrombus flow in acute ischemic stroke. J. Biomech. 110001 (2020). https://doi.org/10.1016/j.jbiomech.2020.110001

3. Boileau, E., et al.: A benchmark study of numerical schemes for one-dimensional arterial blood flow modelling. Int. J. Numer. Methods Biomed. Eng. **31**(10), e02732 (2015). https://doi.org/10.1002/cnm.2732

4. Chen, C., Bivard, A., Lin, L., Levi, C.R., Spratt, N.J., Parsons, M.W.: Thresholds for infarction vary between gray matter and white matter in acute ischemic stroke: a CT perfusion study. J. Cereb. Blood Flow Metab. **39**(3), 536–546 (2019). https://doi.org/10.1177/0271678X17744453

5. Donkor, E.S.: Stroke in the 21st century: a snapshot of the burden, epidemiology, and quality of life. Stroke Res. Treat. **2018**, (2018). https://doi.org/10.1155/2018/3238165

6. El-Bouri, W.K., Payne, S.J.: Multi-scale homogenization of blood flow in 3-dimensional human cerebral microvascular networks. J. Theor. Biol. **380**, 40–47 (2015). https://doi.org/10.1016/j.jtbi.2015.05.011

7. Hall, C.N., et al.: Capillary pericytes regulate cerebral blood flow in health and disease. Nature **508**(7494), 55–60 (2014). https://doi.org/10.1038/nature13165

8. Hodneland, E., et al.: A new framework for assessing subject-specific whole brain circulation and perfusion using MRI-based measurements and a multi-scale continuous flow model. PLOS Comput. Biol. **15**(6), e1007073 (2019). https://doi.org/10.1371/journal.pcbi.1007073

9. Hyde, E.R., et al.: Multi-scale parameterisation of a myocardial perfusion model using whole-organ arterial networks. Ann. Biomed. Eng. **42**(4), 797–811 (2014). https://doi.org/10.1007/s10439-013-0951-y

10. Iadecola, C.: The neurovascular unit coming of age: a journey through neurovascular coupling in health and disease. Neuron **96**(1), 17–42 (2017). https://doi.org/10.1016/j.neuron.2017.07.030

11. Józsa, T.I., Padmos, R.M., Samuels, N., El-Bouri, W.K., Hoekstra, A.G., Payne, S.J.: A porous circulation model of the human brain for in silico clinical trials in ischaemic stroke. Interface Focus **11**(1), 20190127 (2021). https://doi.org/10.1098/rsfs.2019.0127

12. Karch, R., Neumann, F., Neumann, M., Schreiner, W.: Staged growth of optimized arterial model trees. Ann. Biomed. Eng. **28**(5), 495–511 (2000). https://doi.org/10.1114/1.290, http://link.springer.com/10.1114/1.290

13. Kimmel, E.R., et al.: Absence of collaterals is associated with larger infarct volume and worse outcome in patients with large vessel occlusion and mild symptoms. J. Stroke Cerebrovasc. Dis. 1–6 (2019). https://doi.org/10.1016/j.jstrokecerebrovasdis.2019.03.032

14. Linninger, A., Hartung, G., Badr, S., Morley, R.: Mathematical synthesis of the cortical circulation for the whole mouse brain-part I. theory and image integration. Comput. Biol. Med. **110**, 265–275 (2019). https://doi.org/10.1016/j.compbiomed.2019.05.004

15. Michler, C., et al.: A computationally efficient framework for the simulation of cardiac perfusion using a multi-compartment Darcy porous-media flow model. Int. J. Numer. Methods Biomed. Eng. **29**(2), 217–232 (2013). https://doi.org/10.1002/cnm.2520

16. Olufsen, M.S.: Structured tree outflow condition for blood flow in larger systemic arteries. Am. J. Physiol. Heart Circulatory Physiol. **276**(1), H257–H268 (1999). https://doi.org/10.1152/ajpheart.1999.276.1.H257

17. Padmos, R.M., Józsa, T.I., El-Bouri, W.K., Konduri, P.R., Payne, S.J., Hoekstra, A.G.: Coupling one-dimensional arterial blood flow to three-dimensional tissue perfusion models for in silico trials of acute ischaemic stroke. Interface Focus **11**(1), 20190125 (2021). https://doi.org/10.1098/rsfs.2019.0125

18. Perdikaris, P., Grinberg, L., Karniadakis, G.E.: An effective fractal-tree closure model for simulating blood flow in large arterial networks. Ann. Biomed. Eng. **43**(6), 1432–1442 (2015). https://doi.org/10.1007/s10439-014-1221-3

19. Peyrounette, M., Davit, Y., Quintard, M., Lorthois, S.: Multiscale modelling of blood flow in cerebral microcirculation: details at capillary scale control accuracy at the level of the cortex. PLOS ONE **13**(1), e0189474 (2018). https://doi.org/10. 1371/journal.pone.0189474

20. Reymond, P., Merenda, F., Perren, F., Ru, D.: Validation of a one-dimensional model of the systemic arterial tree. Am. J. Physiol. Heart Circulatory Physiol. **297**, 208–222 (2009). https://doi.org/10.1152/ajpheart.00037.2009

21. Sherwin, S.J., Formaggia, L., Peirã O, J., Franke, V.: Computational modelling of 1D blood with variable mechanical properties and its application to the simulation of wave propagation in the human arterial system. Int. J. Numer. Meth. Fluids **43**, 673–700 (2003). https://doi.org/10.1002//d.543

22. Smith, N.P., Pullan, A.J., Hunter, P.J.: An anatomically based model of transient coronary blood flow in the heart. SIAM J. Appl. Math. **62**(3), 990–1018 (2002)

23. Sorimachi, T., Morita, K., Ito, Y., Fujii, Y.: Blood pressure measurement in the artery proximal and distal to an intra-arterial embolus during thrombolytic therapy. J. Neuro Interv. Surg. **3**(1), 43–46 (2011). https://doi.org/10.1136/jnis.2010. 003061

24. Tariq, N., Khatri, R.: Leptomeningeal collaterals in acute ischemic stroke. J. Vasc. Interv. Neurol. **1**(4), 91–5 (2008). http://www.ncbi.nlm.nih.gov/pubmed/ 22518231

25. Yu, H., Huang, G.P., Ludwig, B.R., Yang, Z.: An in-vitro flow study using an artificial circle of willis model for validation of an existing one-dimensional numerical model. Ann. Biomed. Eng. **47**(4), 1023–1037 (2019). https://doi.org/10.1007/ s10439-019-02211-6

Applying DCT Combined Cepstrum for the Assessment of the Arteriovenous Fistula Condition

Marcin Grochowina[1,2]([✉]) [iD] and Lucyna Leniowska[1] [iD]

[1] University of Rzeszów, al. Rejtana 16c, Rzeszów, Poland
gromar@ur.edu.pl
[2] Signum Sp z o.o., ul Podzwierzyniec 29, Łańcut, Poland
http://www.ur.edu.pl, http://signum.org.pl

Abstract. This paper focuses on a comparison of effectiveness of the artificial intelligence techniques in diagnosis of arteriovenous fistula condition. The use of discrete cosine transform (DTC) combined cepstrum in the feature extraction process made it possible to increase the value of classification quality indicators by about 10% (when compared to the previous approach based on averaged energy values in the third octave bands). This paper presents a methodology of extracting features from the acoustic signal emitted by the fistula. The supervised machine learning technique of k-NN, Multilayer Perceptron, RBF Network and Decision Tree C4.5 classifiers was applied to develop the classification model. For this, signals obtained from 38 patients on chronic hemodialysis were used. The results show that the cepstral analysis and obtained features provide an accuracy of above 90% in proper detection of vascular access stenosis.

Keywords: Cepstrum · Discrete Cosine Transformation (DCT) · Arteriovenous fistula · Supervised machine learning · Classification models

1 Introduction

The high quality of data sets used in the classification process has a significant impact on the outcome. This paper focuses on a comparison of effectiveness of the artificial intelligence techniques in diagnosis of arteriovenous fistula (AVF) condition by analyzing the acoustic signal emitted by the flowing blood. The AVF is an artificially formed connection between an artery and a vein, usually located in the wrist. It provides a bypass for the blood flow in one of the two arteries, which supply blood to the hand. A properly functioning AVF is extremely important for people who undergo a hemodialysis. Our diagnosis technique is based on the phono-angiography - a non-invasive method for evaluating the acoustic noise emitted from the vessel and produced by the local blood flow in the vascular

© Springer Nature Switzerland AG 2021
M. Paszynski et al. (Eds.): ICCS 2021, LNCS 12744, pp. 684–692, 2021.
https://doi.org/10.1007/978-3-030-77967-2_57

system. Based on the characteristic features of signals recorded, the condition of the fistula can be determined.

It has long been known that stenosis in the artery causes a swishing sound, which is heard as a bruit. According to the studies conducted, the character of the sound emitted by the blood flowing inside a fistula's vessel vary depending on the state of the fistula. The first mention of the numerical analysis of acoustic signals emitted by the AVF occurred in the mid-1980s [9]. And a detailed overview of various, more precise, methods for detecting vascular access stenosis was presented by Noor [7].

Classification is still one of the biggest challenges in the field of processing acoustic signal for the purpose of medical diagnosis. Researches published to date typically describe the fast Fourier transform (FFT) as a method to transform the acoustic signal into the frequency domain. This transform was also applied by the authors for finding features which are useful for the classification purposes. In the paper [4], an artificial neural network approach has been described for early detection of hemodialysis access problems. In [2], we compared the quality of classifications using SVM and k-NN classifiers and demonstrated the possibility and validity of the multi-class approach. We also compared several methods for feature selection with the developed joined-pairs method, dedicated to fistula problem classification [3]. In paper [6] we propose a similar system of the assessment of AVF condition, where parts of the sound signal corresponding to the cardiac cycle are subjected to the Wavelet transformation. This approach allowed for a bit higher quality of the results and a more reliable diagnosis.

Recently, an innovative idea has been proposed to develop and launch a low-cost diagnostic device that implements classification algorithms to identify patients with renal disease at risk of access stenosis, even before a full loss of access patency [5]. Applied classifiers have been implemented into a supervised learning process and the best ones have achieved 81% accuracy. On the other hand, development of useful data from a recorded acoustic signal for the purposes of classification and medical diagnosis is still difficult, labor-intensive and ambiguous. Therefore, alternative ways of extracting features are still being researched. In the last two decades, a lot of studies have been conducted on the use of digital orthogonal transforms for the analysis of acoustic signals. Therefore, a number of transformations have been introduced with declarations of having a better performance than others.

This paper focused on a comparison of effectiveness of the artificial intelligence techniques in diagnosis of the AVF condition. This diagnosis technique is based on a discrete cosine transform (DCT) [8] processing and data mining tools. To achieve the most realistic estimate of the AVF state we examined various options that are available for the estimation of the spectral envelope of the acoustic signal. A comparison of the various cepstral models showed that the cepstral based real envelope estimator has particularly favorable properties, except that its estimation is quite computationally demanding. In conclusion, we have compared the performance of the DCT with FFT transforms. Instead of FFT approach, DCT process is utilized to represent the time frequency domain approach for efficient signal registered from AVF. This new approach exceeds the performance of a formerly introduced method by almost 10%. This valuable result means that patients with kidney disease, who are at risk of losing patency, can be correctly identified with a high probability of 90%.

2 Materials

The sound emitted by the blood flowing through the AVF, which is the subject of the analysis, was recorded at the Clinical Dialysis Center of the St. Queen Jadwiga Provincial Hospital No. 2 in Rzeszów. Within a few months, recordings from 38 patients were obtained. Each patient was recorded several times, at intervals of 6–8 weeks. Each recording lasted for 30 s. The sampling rate was set to 8 kS/s, with a resolution of 16 bit.

The recordings were divided into fragments with a length of 8192 samples. Each being in line with the heart rhythm in such a way that the local maximum amplitude corresponding to the moment of heart contraction is located in the middle of the fragment. The fragments partially overlapped (Fig. 1). Each fragment was the basis for computation of the feature vector for the classification process.

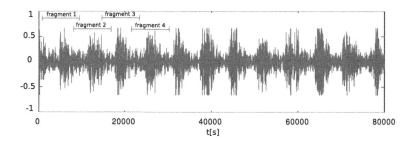

Fig. 1. Dividing the recording into fragments.

On the basis of Doppler ultrasound assessment, the medical staff of the dialysis center assigned each patient to one of the six classes with an increasing degree of AVF pathologisation assessed on the basis of Doppler ultrasound (A - proper AVF function; F - limited flow in the AVF). The numbers of patients, recordings and fragments obtained are presented in Table 1.

Table 1. Number of patients and vectors in classes

Class	A	B	C	D	E	F
Number of patients	3	5	7	10	9	4
Number of samples	11	23	29	42	37	14
Number of vectors	232	348	516	692	669	251

The study was approved by the local Bioethical Committee of the Regional Medical Chamber in Rzeszow (No. 17/B/2016).

3 Methods

The studies to date have shown that the condition of the fistula can be assessed by analyzing the frequency spectrum of the sound signal emitted by the blood [4] flowing through the fistula. Measurement of energy content in the third octave bands allowed to build a classification system with over 80% accuracy [5].

When analyzing the frequency spectrum characteristic of fistulas with progressive pathologisation, it can be seen that:

- for a functional fistula, almost all signal energy is concentrated in the 0–200 Hz band with a maximum at 150 Hz,
- with the progressive pathologisation, the frequency band in which the signal energy concentrates is extended towards higher frequencies
- with the extension of the frequency band, successive local maxima become visible at frequencies that are multiples of the first maximum.

To conclude, important information about the condition of the fistula can be obtained by analyzing the shape of the envelope of the frequency spectrum of the signal it emits. A tool that enables such analysis is cepstrum. In its original shape, the cepstrum is based on the Fourier transformation. Depending on the adopted strategy, it can operate on a complex frequency spectrum (complex cepstrum) or only on its module (real cepstrum). In case of the real cepstrum, the transform is affected by the loss of some information relating to the phase of the input signal.

For the analysis of the signal from the AVF, a modification of the cepstrum was applied using a discrete cosine transformation (DCT). It transforms a one-dimensional sequence of real numbers (signal) into another one-dimensional sequence of real numbers (spectrum), which is an extension of the input sequence into a series of cosine in the base of orthogonal Chebyshev polynomials.

This transformation can be described as a vector operation by the following formula:

$$\mathbf{C} = \mathbf{x} \cdot \mathbf{W}, \tag{1}$$

where

$$\mathbf{W} = [w_{k,n}]_{N \times N} = c(k) \cos \frac{k\pi(2n+1)}{2N}, \tag{2}$$

is the kernel transform and

$$c(0) = \sqrt{1/N}, \quad c(k) = \sqrt{2/N} \; for \; k > 0. \tag{3}$$

The signal handling operation to extract features for the classification process is shown in the code fragment (Octave [1]) below, and the results of each of the steps for exemplary signal fragments representing cases for classes A, C, and F are shown in Fig. 2.

```
f = dct(x(c-4096 : c+4095));
f = f ./ sum(f);
lf = log(abs(f));
clf = idct(lf);
r = clf(2:10);
```

First, local maximums of the signal amplitude are searched for, corresponding to the moments of contraction of the heart and dividing the recording into fragments (in the example above, the maximum index is stored in the c variable). Next, the fragment 8192 long is extracted and on its basis the DCT is calculated. The result is normalized to compensate for any differences in the volume of the recordings. Then, the logarithm of the absolute DCT value is calculated and the inverse transform is calculatedestimated. Significant information about the shape of the spectral envelope is contained in the first few fringes - in this case, the fringes with indices $k = 1 \div 9$ were used as features in the classification process. Using more features did not improve the quality of the results.

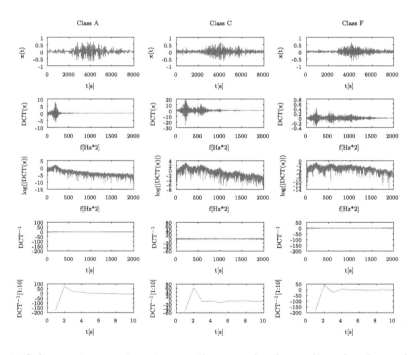

Fig. 2. Subsequent processing steps on the example of recordings for classes A, C, and F.

4   Results and Discussion

Based on the obtained data set, classification systems were built, tested and assessed for the quality. The division of the data set into training and testing subsets was conducted with Leave-one-out method, i.e. in each validation cycle, data derived from one patient was excluded from the set. Excluded data was a test set. This way the effect of 38-fold cross-validation was obtained.

The following classifiers were used in the research: Multilayer Perceptron, Radial Basis Function Network (RBF), k-NN and Decision Tree C4.5.

The settings of each of the algorithms were selected in an iterative tuning process based on the classification quality indicators. The results are presented in Table 2, 3, 4 and 5. Each of the table contains a confusion matrix in which the values are presented as a percentage and a list of classification quality indicators calculated on the basis of the confusion matrix. These indicators are presented separately for each class and jointly in the form of an average.

Compared to the results obtained up to date (Table 6), that is, based on the features calculated as averaged energy value in the third octave bands, a significant increase in the value of classification quality indicators has been observed.

The accuracy of vector assignment increased in each of the classes, allowing the global average F-score to increase from 0.805 to 0.951 for the Multilayer Perceptron. However, an undesirable phenomenon of erroneous classification of individual vectors into many, not only neighboring classes, was also observed. In the previous classification system, inter-class leaks were only observed within immediately adjacent classes. This means that the feature extraction algorithm

Table 2. Confusion matrix and quality indicators for Multilayer Perceptron

A	B	C	D	E	F	Classified as
97,9	0,3	1,0	0,3	0,2	0,2	A
4,9	87,3	6,9	0,0	1,0	0,0	B
2,1	1,6	89,9	2,1	3,2	1,1	C
0,0	0,0	1,3	96,1	2,6	0,0	D
0,0	0,0	3,1	0,5	95,3	1,0	E
0,0	0,0	0,0	0,0	3,8	96,2	F

TP rate	FP rate	Precision	Recall	F-score	ROC area	Class
0.979	0.015	0.984	0.979	0.982	0.994	A
0.873	0.005	0.947	0.873	0.908	0.945	B
0.899	0.02	0.894	0.899	0.897	0.964	C
0.961	0.006	0.914	0.961	0.937	0.998	D
0.953	0.012	0.938	0.953	0.946	0.997	E
0.962	0.004	0.911	0.962	0.936	0.999	F
0.940	0.011	0.934	0.940	0.937	0.983	Avg.

Table 3. Confusion matrix and quality indicators for RBF Network

A	B	C	D	E	F	Classified as
96,9	0,7	1,9	0,0	0,5	0,0	A
10,8	80,4	6,9	0,0	2,0	0,0	B
3,7	3,7	84,0	1,1	6,9	0,5	C
0,0	0,0	2,6	90,9	6,5	0,0	D
1,6	0,0	5,7	0,5	92,2	0,0	E
0,0	1,9	0,0	0,0	0,0	98,1	F

TP rate	FP rate	Precision	Recall	F-score	ROC area	Class
0.969	0.034	0.964	0.969	0.966	0.992	A
0.804	0.011	0.872	0.804	0.837	0.977	B
0.84	0.031	0.836	0.84	0.838	0.966	C
0.909	0.003	0.959	0.909	0.933	0.991	D
0.922	0.023	0.885	0.922	0.903	0.982	E
0.981	0.001	0.981	0.981	0.981	0.993	F
0.911	0.017	0.921	0.911	0.916	0.984	Avg.

Table 4. Confusion matrix and quality indicators for k-NN

A	B	C	D	E	F	Classified as
97,9	0,2	1,4	0,2	0,3	0,0	A
9,8	84,3	2,9	0,0	2,9	0,0	B
1,6	0,0	89,9	3,7	4,8	0,0	C
1,3	0,0	3,9	92,2	2,6	0,0	D
0,5	0,0	5,7	2,1	91,1	0,5	E
0,0	0,0	1,9	0,0	15,1	83,0	F

TP rate	FP rate	Precision	Recall	F-score	ROC area	Class
0.979	0.025	0.974	0.979	0.977	0.993	A
0.843	0.001	0.989	0.843	0.91	0.959	B
0.899	0.026	0.867	0.899	0.883	0.975	C
0.922	0.011	0.855	0.922	0.888	0.989	D
0.911	0.024	0.879	0.911	0.895	0.984	E
0.83	0.001	0.978	0.83	0.898	0.981	F
0.905	0.014	0.928	0.905	0.915	0.981	Avg.

based on the DCT cepstrum shows a better accuracy with increased sensitivity
to noise in the input signal.

Table 5. Confusion matrix and quality indicators for C4.5 Tree

A	B	C	D	E	F	Classified as
96,9	1,2	1,2	0,3	0,3	0,0	A
15,7	75,5	6,9	0,0	2,0	0,0	B
2,7	3,2	85,6	2,1	5,3	1,1	C
0,0	0,0	6,5	81,8	11,7	0,0	D
1,0	1,0	7,8	2,1	87,5	0,5	E
1,9	0,0	1,9	0,0	1,9	94,3	F

TP rate	FP rate	Precision	Recall	F-score	ROC area	Class
0.969	0.039	0.959	0.969	0.964	0.968	A
0.755	0.014	0.837	0.755	0.794	0.857	B
0.856	0.035	0.821	0.856	0.839	0.925	C
0.818	0.009	0.863	0.818	0.84	0.957	D
0.875	0.024	0.875	0.875	0.875	0.944	E
0.943	0.003	0.943	0.943	0.943	0.98	F
0.881	0.020	0.893	0.881	0.887	0.945	Avg.

Table 6. Classification quality indicators for averaging energy values in one third octave bands - k-NN [5]

Class	A	B	C	D	E	F	Avg.
Precision	0.91	0.68	0.72	0.81	0.78	0.92	0.803
Recall	0.91	0.69	0.67	0.81	0.83	0.92	0.805
F-score	0.91	0.69	0.69	0.81	0.81	0.92	0.805

5   Conclusion

An AVF is an artificially made connection between a vein and an artery. It must be well-functioning to ensure hemodialysis for patients with end-stage renal disease. In this paper, we reported on the process of finding the efficient diagnosis technique based on the artificial intelligence and data mining tools.

The feature extraction method we have described in this article is another in a series of attempts to improve the system for automatic diagnosis of AVF based on an acoustic signal. This activity was performed in a traditional way, as a sequence of following operations: feature extraction, selection and reduction of features, classification, and interpretation of the classification results. Reported solutions based on the FFT analysis, averaging energy in thirds, wavelet transformation and finally DCT combined cepstrum made it possible to achieve an efficiency of 90%.

We have also compared the quality of classifications using the supervised machine learning technique of k-NN, Multilayer Perceptron, RBF Network and Decision Tree C4.5 classifiers and demonstrated the possibility of the multi-class approach. These operations were performed on the acoustic signal recordings obtained from 38 patients on chronic hemodialysis. The obtained results of F-score indicator are above 0.9 and are higher than the results obtained so far by almost 10% for each of the classification constructed. This means that the modification of the feature extraction method and the use of DCT combined cepstrum allowed for a significant increase in the quality of the classification system. The only disadvantage of this method is that the increase in the level of effectiveness increases the sensitivity of the algorithm to noise, however, it can be solved by extending the recording time to obtain more fragments. More fragments mean more feature vectors in a single study that can be averaged, thus minimizing the effect of noise. The method based on the analysis of the spectral envelope shape has shown promise and further work is planned to optimize and improve it.

The next step will be to migrate the multiclass classification system to a regression model. Such an approach will enable the assessment of the fistula state in a continuous number of cases, and not only allocating to one of several classes, as before. This is important for tracking changes in the condition of fistula over time to detect potential hazards in advance.

References

1. Octave homepage. https://www.gnu.org/. Accessed 15 Jan 2021
2. Grochowina, M., Leniowska, L.: Comparison of SVM and K-NN classifiers in the estimation of the state of the arteriovenous fistula problem. In: 2015 Federated Conference on Computer Science and Information Systems (FedCSIS), pp. 249–254. IEEE (2015)
3. Grochowina, M., Leniowska, L.: The new method of the selection of features for the K-NN classifier in the arteriovenous fistula state estimation. In: 2016 Federated Conference on Computer Science and Information Systems (FedCSIS), pp. 281–285. IEEE (2016)
4. Grochowina, M., Leniowska, L., Dulkiewicz, P.: Application of artificial neural networks for the diagnosis of the condition of the arterio-venous fistula on the basis of acoustic signals. In: Ślęzak, D., Tan, A.-H., Peters, J.F., Schwabe, L. (eds.) BIH 2014. LNCS (LNAI), vol. 8609, pp. 400–411. Springer, Cham (2014). https://doi.org/10.1007/978-3-319-09891-3_37
5. Grochowina, M., Leniowska, L., Gala-Błądzińska, A.: The prototype device for non-invasive diagnosis of arteriovenous fistula condition using machine learning methods. Sci. Rep. **10**(1), 1–11 (2020)
6. Grochowina, M., Wojnar, K., Leniowska, L.: An application of wavelet transform for non-invasive evaluation of arteriovenous fistula state. In: 2019 Signal Processing: Algorithms, Architectures, Arrangements, and Applications (SPA), pp. 214–217. IEEE (2019)
7. Noor, A.M.: Non-imaging acoustical properties in monitoring arteriovenous hemodialysis access. A review. J. Eng. Technol. Sci. **47**(6), 658–673 (2015)
8. Randall, R., Antoni, J., Smith, W.: A survey of the application of the cepstrum to structural modal analysis. Mech. Syst. Sig. Process. **118**, 716–741 (2019)
9. Sekhar, L.N., Wasserman, J.F.: Noninvasive detection of intracranial vascular lesions using an electronic stethoscope. J. Neurosurg. **60**(3), 553–559 (1984)

Electrocardiogram Quality Assessment with Autoencoder

Jan Garus[1,2(✉)] ⓘ, Mateusz Pabian[1,2] ⓘ, Marcin Wisniewski[1] ⓘ, and Bartlomiej Sniezynski[2] ⓘ

[1] Healthcare Department, Comarch SA, al. Jana Pawła II 39a,
31-864 Krakow, Poland
jan.garus@comarch.pl
[2] AGH University of Science and Technology, al. Mickiewicza 30,
30-059 Krakow, Poland
https://www.comarch.pl/healthcare/
https://www.agh.edu.pl

Abstract. ECG recordings from wearable devices are affected with a relatively high amount of noise due to body motion and long time of the examination, which leads to many false alarms on ill-state detection and forces medical staff to spend more time on describing each recording. ECG quality assessment is hard due to impulse character of the signal and its high variability. In this paper we describe an anomaly detection algorithm based on the Autoencoder trained on good quality examples only. Once trained, this neural network reconstructs clean ECG signals more accurately than noisy examples, which allows to distinguish both classes. Presented method achieves a normalized F1 score of 93.34% on the test set extracted from public dataset of 2011 PhysioNet/Computing in Cardiology Challenge, outperforming the solution based on the best competition participants. In contrary to many state-of-the-art methods it can be applied even on short, single-channel ECG signals.

Keywords: ECG · Quality assessment · Autoencoder

1 Introduction

Electrocardiogram (ECG) signal quality (SQ) assessment has been a topic of 2011 PhysioNet/Computing in Cardiology (CinC) Challenge [1,2]. Despite a few years that have passed, it is still an issue, which emerges in the CinC Challenge 2017 [3], where a "noisy" class was among 4 output possibilities, causing the participants a lot of problems.

ECG quality classification is an important issue in a case of remote heart monitoring, especially when system is dedicated to simply record the data, without online analysis. In such a case, a patient does not have an immediate feedback about correctness of the lead placement etc. After a 24-h examination it may turn out that most of the signal is useless.

© Springer Nature Switzerland AG 2021
M. Paszynski et al. (Eds.): ICCS 2021, LNCS 12744, pp. 693–706, 2021.
https://doi.org/10.1007/978-3-030-77967-2_58

Automatic rhythm analysis and beat-to-beat classification of the ECG signal is also substantially affected with noisy recordings. Those signals often trigger false alarms, destroying even the best positive predictivity achieved on signals from standard databases (which are rather clean), like the MIT-BIH [4]. For example, a high-frequency noise may be incorrectly interpreted as heart beats, leading to false tachycardia alarm. Similarly, extremely low signal amplitude or sensor disconnection may be classified as an asystole.

Simple methods, like detection of lead disconnection are insufficient regarding a wide range of possible signal disruption.

The main contribution of this paper is to apply the anomaly detection approach to the ECG signal quality assessment. In this setting only typical ECG shapes are learned by the model, while noisy examples are treated literally as anomalies, in contrary to standard classifiers which requires a large, well annotated database containing all possible types of noise. Experimental results show that the proposed approach was more accurate than a classical one.

The paper is organized as follows. State-of-the-art methods are presented in Sect. 2. Section 3 contains brief introduction into ECG nomenclature and techniques used in our solution. Details of our one-class classifier approach are given in Sect. 4. The evaluation is presented in Sect. 5, including the comparison to challenge entries in Subsect. 5.5.

2   Related Work

As we have already mentioned, ECG signal quality assessment was a subject of the 2011 PhysioNet/CinC Challenge [2].

Among the challenge competitors, an excellent paper [5] from Gari Clifford *et al.*is worth an attention. A wide range of features (named *quality indices*) has been investigated, including the time and frequency domains, a statical distribution and strictly ECG-related aspects, *i.e.*differences between QRS-detection results gathered on separated channels and from two different algorithms. Proposed quality indices were used in many works, including an extension for arrhythmia [6] and more recent papers, like [7] which combines them with additional entropy-based and Lempel-Ziv compression-based features.

Two-step algorithm [8] utilizes covariance, mean variance, peak-to-peak value and maximum value in the first phase, then a covariance and time-delayed covariance supplies the SVM classifier. The same group of authors almost won the challenge with a procedure of 5 simple if-then rules [9], which exams an amplitude change over the time, the isoline drift and a standard deviation of the normalized signal. Speaking on the winners, most of the signal properties used in [10] also concern the amplitude: QRS vs. noise amplitude, peak-to-peak value, spikes, but also volume of flat signal fragments and number of crossing points in between channels.

Another rule-based algorithm was presented by [11], which utilizes 4 Signal Quality Indicators: straight line and huge impulse detection, the Gaussian noise detection and the detection of abnormalities *RR intervals*.

An approach of the DSP transformations followed by thresholding has been also proposed in many papers. In [12] low and high frequencies are separated and checked against arbitrary thresholds, as well as signal magnitude and a number of zero-passes. Follow-up model extends it with the modified Empirical Mode Decomposition [13,14]. Wavelet decomposition can be used in the same manner [15]. Thresholding of the signal saturation, energy and baseline change over the time was proposed in [16].

Simple signal autocorrelation within 5-s windows is utilized in [17].

Many methods utilize the cross-channel information to a much greater extend. A correlation between channels as well as the standard deviation of the signal autocorrelation is used in [18]. Similarly, the channel covariance matrix is used as random forest input in [19]. Authors of [20] compare each input channel to the reconstruction based on all others signal leads. In out method we also reconstruct the signal, but we use each separate channel do generate its own reconstruction. Therefore our approach can be applied for both multiple- and single-channel signals.

Some authors proposed yet another comparison-based algorithms, where the extracted QRS is compared against templates obtained with clustering [21] or statistical methods [22]. Another work [23] presents a method based on the cepstral analysis.

Neural networks are used in many papers, including Convolutional NN in [24] and Self Organizing Maps in [25]. Similarly to our work, an Autoencoder network was used in [26], however authors focused on ECG correction task (as a step of human identification procedure) rather than explicit assessment of the signal quality. In this paper we concentrate on the classification aspect. Furthermore, more complex AE architectures are evaluated in our work.

3   Materials and Methods

In this section we present a brief introduction into ECG signal, followed by key concepts and algorithms which form the foundations of our solution.

3.1   ECG Signal

Electrocardiogram (ECG) signal reflects how the heart is functioning. Data is acquired from a number of physical electrodes (varying from 2 to 10), then those physical readings are subtracted in pairs, forming signal channels used in further analysis, often referred as *leads*. In stationary examination, 12-lead record is a golden standard, but in case of mobile devices, fewer electrodes are usually available, resulting in a smaller number of data channels, sometimes even a single lead.

Healthy condition is called *the normal sinus rhythm*, which idealized waveform is illustrated in the Fig. 1. Consecutive fragments (waves) of the signal are named with letters P, Q, R, S and T.

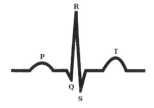

Fig. 1. ECG signal of the normal sinus rhythm.

Waves Q, R and S form the *QRS complex*, the most characteristic part of the heartbeat. Its presence, even disturbed by low signal quality or some medical condition is crucial in the ECG analysis. Waves Q and S may have too low amplitude to be visible, but the R should be easily recognized. In some heart conditions, the peak of the R wave is forked. The time distance between two successive R waves is commonly referred as the *RR interval*. It changes naturally in certain range as heart speeds up and slows down due to breathing and physical effort, but too big fluctuations may indicate both arrhythmia as well as problems with data acquisition. Waves P and T may be easily recognizable or barely visible or absent, depending on the signal quality and medical condition. Depending on the actual lead being analyzed, waves may be positive or negative in regard to the baseline. This is called the *polarization*.

Apart from medical problems, ideal ECG waveform may be disrupted by a number of causes, which includes: muscle movement, electrode shifts, skin contact changes and external electrical interference. Those disruptions or noises in the ECG signal are referred as *artifacts*.

Baseline of the raw ECG signal is not as flat as presented in the Fig. 1. In reality it is affected with a low frequency component, which arises from breathing, patient movement and electrical charge of electrodes. This so-called *baseline wander* not necessarily means bad signal quality and usually can be easily filtered out.

3.2   Anomaly Detection

Problem of the ECG signal quality assessment may be treated as a binary classification, with classes: *good quality* (GQ) and *bad quality* (BQ). An obvious solution is to build a dichotomous classifier and train it with a sufficient amount of examples of both classes.

However, it is hard to provide a balanced training set for this problem, which is required by a vast majority of classifiers to avoid classification bias [27]. There are too few BQ examples available. Moreover, due to irregularity and a variety of the artifacts' origin in the ECG signal, even smart duplication techniques, like *SMOTE* [28], are incapable of generating a sufficient training set which would cover the entire space of the BQ class.

Instead of working on some modifications of this dichotomous classifier, problem can be simplified by focusing on a GQ only. GQ is also a broad class, but can be described in a significantly more precise way than BQ, as a concatenation of all heart beat morphologies and rhythms being considered by physicians.

Such an approach is named an *anomaly detection* or a *one-class classification* [29]. As the first name emphasizes, this technique is primarily dedicated to detection of some extremely rare states. The latter explains what knowledge it achieves: it learns how to describe a single class and nothing more.

Using this method for SQ is a different way of thinking: rather than learning how to distinguish GQ and BQ, we are telling our system only how GQ examples look like. If it encounters some significantly different example, it would be BQ.

3.3   Autoencoders

An *autoencoder* (AE) or *Encoder-Decoder* [30–32] is a neural network, which can be used as an anomaly detection algorithm.

It consists of following layers: input, *encoder* layer(s), *latent* representation (aka *hidden* or the *code*), *decoder* layer(s), reconstruction (output). The encoder and decoder subnetworks are mirrors of each other, working in the opposite way. In order to speed up learning procedure their weights can be shared (also referred as tied). The entire network is trained to reproduce values from the input to the output, by penalizing differences between original and reconstructed signal in a cost function.

In order to prevent the AE from "cheating" by simply passing the values through entire network, additional restrictions are involved [33]. The simplest method is to form the network in a hourglass shape with a hidden representation much smaller than input and reconstruction layers. Another popular approach, the *Denoising AE* [34] corrupts the input data with a noise (in training only) while the cost function still compares the reconstruction against the original data.

Alternatively, the latent representation may be forced to be as simple as possible by minimizing the number of non-zero elements. It can be achieved by adding the sparsity measure [35] to the cost function which penalizes every non-zero element in the latent vector. Makhzani and Frey proposed the *k-Sparse Autoencoder* [36], which introduces an additional, non-linear layer, right after the latent vector. This layer passes only k greatest values from the hidden representation and replaces rest of them with zeros.

Strictly speaking, autoencoders belong to the *unsupervised learning* algorithms, as no class labels are required for training. However, if examples from one class only (GQ, in our case) would be used for training, we expect that AE will learn to reproduce other examples of this class in a significantly better extent than representatives of unseen class (BQ). This approach—applied to gamma ray readings—has been used by Sharma *et al.*in [29]. Some variants of this technique were introduced by An and Cho in [37].

Our idea is to detect BQ by comparing signal reconstruction error against threshold determined using a distinct validation set.

Algorithm 1: Algorithm overview

---

split signal into 10s-fragments
signal preprocessing
for *channel x* **in** *signal* **do**
   for *fragment z* **in** *x* **do**
      $\tilde{z} \leftarrow$ autoencoder(z)
      $d \leftarrow$ reconstruction_error(z, $\tilde{z}$)
      if d ¡ *threshold* **then** mark z as Good Quality
      else mark z as Bad Quality

---

4   Proposed Solution

General steps of the algorithm are presented in Algorithm 1, followed by subsections explaining details of this solution.

After initial split into 10 s-fragments (if database is not already in this form), signal preprocessing is applied. This step includes final extraction of 0.8 s, 1-channel fragments of the signal, positioned around supposed QRS complexes, but also few normalization operations, described in Subsect. 4.1.

Next, each 0.8 s fragment is processed with the autoencoder (Subsect. 4.2). Obtained output is then compared against the input signal (Subsect. 4.3). Finally, the quality of 0.8 s fragment is assigned using the BQ threshold calculated in the supervised phase of the training, as described in Subsect. 5.3.

4.1   Data Preprocessing

Our preliminary results showed, that it may be too hard for an autoencoder to reconstruct 10 s of ECG input. Moreover, according to our business requirements, higher time resolution of the results was needed. To address those circumstances, we decided to analyse a single-lead signal containing approximately a single heartbeat, which takes on average about 0.8 s .

As described in Algorithm 2, signal is resampled to 125 Hz and its amplitude is restricted to an interval of −6 mV to 6 mV, then the mean value is subtracted. The input 10 s-signal (one data channel at the time) is divided into 2.5 s-sections, from which 0.8 s-fragments (positioned around the largest peak) are finally extracted.

Few optional operations were also introduced: baseline cancellation and two types of normalization. For baseline cancellation we used a cascade of 600 ms and 200 ms median filters (similar to [38], but filters are in reversed order). Normalization of the QRS polarization ensures that the peak at the 20th sample is always positive (it may be different, depending on the actual ECG channel). Normalization to the interval of $[0, 1]$ assigns 0 to the minimum sample in the fragment, 1 to its maximum sample and accordingly modifies the rest of the samples to fit in this interval, keeping the signal shape.

Algorithm 2: Signal preprocessing of 10s-fragment

$\quad$ **for** *channel x* **in** *signal* **do**

$\quad\quad\mid\quad x \leftarrow$ resample x to $125Hz$

$\quad\quad\mid\quad$ **if** *option_cancel_baseline* **then** $\quad x \leftarrow$ cancel baseline in x

$\quad\quad\mid\quad x \leftarrow x - mean(x)$

$\quad\quad\mid\quad$ **for** *sample x_i* **in** *x* **do**

$\quad\quad\mid\quad\quad\mid\quad x_i \leftarrow max(x_i, -6mV)$

$\quad\quad\mid\quad\quad\mid\quad x_i \leftarrow min(x_i, 6mV)$

$\quad\quad\mid\quad$ **for** *i* **in** 0 .. 3 **do**

$\quad\quad\mid\quad\quad\mid\quad y \leftarrow x_{i*312 \,..\, (i+1)*312}$ // extract 2.5s fragment

$\quad\quad\mid\quad\quad\mid\quad j \leftarrow argmax(abs(y))$

$\quad\quad\mid\quad\quad\mid\quad z \leftarrow y_{j-20 \,..\, j+80}$ // extract 0.8s from each 2.5s fragment

$\quad\quad\mid\quad\quad\mid\quad z \leftarrow z - mean(z)$

$\quad\quad\mid\quad\quad\mid\quad$ **if** *option_normalize_polarization* **then**

$\quad\quad\mid\quad\quad\mid\quad\quad\mid\quad$ **if** $z_{20} < mean(z)$ **then** $\quad z \leftarrow -z + max(z)$

$\quad\quad\mid\quad\quad\mid\quad$ **if** *option_normalize_to_inverval_from_0_to_1* **then** $\quad z \leftarrow \frac{z-min(z)}{max(z)-min(z)}$

$\quad\quad\mid\quad\quad\mid\quad$ **save** z

4.2 Investigated AE Variants

We have implemented our anomaly detection model using following variants of the Autoencoder network: simple Autoencoder (no constraints) (AE), Denoising Autoencoder (DAE) [34], Sparsity Autoencoder (SpAE) [35] and k-Sparse Autoencder (kSpAE) [36].

Each model listed above has been manually optimized in terms of the size of the latent representation and its variant-specific parameters in order to maximize the accuracy of the signal quality assessment. The best configurations are listed in Table 1.

4.3 Reconstruction Error

We have defined and implemented following measures of dissimilarity between input signal x and it's reconstruction $\tilde{x}$:

$$E_{abs} = |x - \tilde{x}| \qquad E_{abs\_pf3} = E_{abs} + \sqrt{|E_{abs} - E_{med3}|}$$
$$E_{pow} = E_{abs}{}^2 \qquad\qquad E_{sqrt} = \sqrt{E_{abs}} \qquad\qquad (1)$$
$$E_{med3} = med_3(E_{abs}) \qquad E_{sqrt3} = \sqrt[3]{E_{abs}}$$

where med_k denotes a median filter of the length k.

Based on those element-wise errors, *signal reconstruction errors* are defined as the average value of corresponding measure:

$$d_* = \sum_{i=1}^{n} \frac{E_{*i}}{n} \qquad\qquad (2)$$

where n is the length of samples in signal x and E_{*i} is the value of measure E_* for i-th sample.

5   Evaluation

Details on the database preparation, training and numerical results are provided in this section. Comparison to the algorithm based on the state of the art solution is also given in the last Subsect. (5.5).

5.1   Databases

In the unsupervised part of the model (*i.e.* training the autoencoder) we were using the ECG examples obtained from the following databases: MIT-BIH [4], incartdb [1], ltafdb [39], svdb [40] and AHA DB [41]. Those databases are rather noise-free, so good signal quality was assumed for the entire set. We used channels 1 and 2 from those databases, except the *incartdb* where we have chosen leads 2 and 12. Sampling frequencies were normalized to $fs = 125$ Hz .

In the supervised stage, which was the threshold calculation, we used labelled signals extracted from the PhysioNet/CinC Challenge 2011 training set (1000 records) and *Event 2* open test set B (500 records). The concatenated set was randomly split into training (for threshold selection) and test (final algorithm effectiveness) sets in a ratio of 50:50. Original data was 12-lead, 10-s, $fs = 500$ Hz.

5.2   Reference Classification

Due to the signal manipulation (separating channels and sampling 0.8 s fragments), existing reference classification of the labelled database was outdated. We have manually classified those examples into 3 classes with following criterion: *GQ (good quality)*—visible QRS complex but also clear P and T waves (if present), *AQ (acceptable quality)*—visible heartbeats, but less clear shape of the QRS complexes, *BQ (bad quality)*—heartbeats unrecognizable or presence of high-amplitude distortions. Visualizations of the resulting sets are presented in Fig. 2.

During threshold calculation and final model evaluation AQ examples were incorporated into GQ, because there was no business need to distinguish them and their shapes were quite similar.

5.3   Training

Autoencoders were trainereconstructed as an g set (Subsect. 5.1), which was split into training, validation and testing subsets in a ratio of 70:15:15.

For each combination of the autoencoder variant (Sect. 4.2) and the measure of reconstruction error (Sect. 4.3) we calculated the BQ threshold using the labelled training set. It is set at the crossing point of the histograms made exclusively for GQ (including AQ) and BQ examples (using a half of the labelled set, leaving the other half for testing). This step is illustrated in Fig. 3. To avoid classification bias, number of signals from a minority class (BQ) was upscaled.

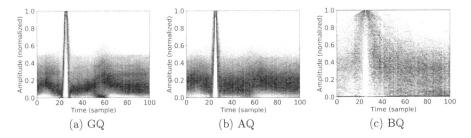

(a) GQ   (b) AQ   (c) BQ

Fig. 2. Visual summaries of the each class' signals in the quality-labelled set. Baseline cancellation, 0–1 normalization and the normalization of R-peak polarization were applied to all signals in each class (see Sect. 4.1). Then, for each time index (horizontal axis), histogram of values is calculated and drawn along the vertical axis. Based on those visualizations, signals from the GQ and AQ classes are quite similar within those subsets, in opposite to much more diverse BQ subset.

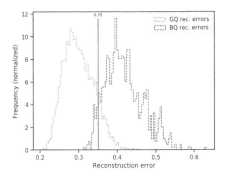

Fig. 3. Histogram of GQ vs BQ reconstruction error. Number of examples in BQ class is upscaled to fit the other class. Obtained for SpAE model with d_{sqrt3} as reconstruction error measure. BQ threshold is calculated at the crossing point of both histograms, *i.e.* at $d_{sqrt3} = 0.35$. This setting gives classification accuracy $F1_{norm} = 91.62\%$.

5.4   Results

Experimental results are summarized in this subsection. Examples of GQ and BQ signals and their reconstructions are illustrated in Fig. 4.

Table 1 presents selected results, namely the mean reconstruction errors (average of the *Mean Square Error* (MSE) along the set, *i.e.* $\overline{d_{pow}}$) on the autoencoder testing set (unlabelled) and classification results on the labelled test set. For the sake of readability results of only two measures are listed: d_{abs} and the one that gives the best classification results (for given architecture).

Due to imbalance in a number of GQ and BQ examples (the latter group is less than 15% of all available data), we used *normalized F1-score*, proposed in [42], to evaluate the classification. This accuracy measure gives a better insight in the anomaly detection problem, were anomalies are rare by definition, which

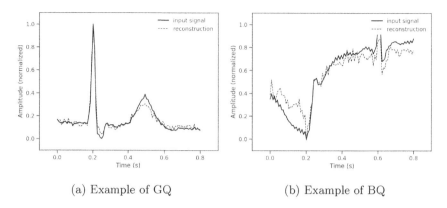

(a) Example of GQ   (b) Example of BQ

Fig. 4. Examples of the GQ and BQ signals and their reconstructions. GQ: The QRS complex, as well as the T wave are quite accurately reproduced which effect in a small reconstruction error. BQ: Signal drop before 0.2 s is reconstructed as an ill-formed QRS complex which is significantly different than input signal, leading to high reconstruction error. Interestingly, the AE correctly reproduces the signal around 0.6 s which may be a true QRS complex.

heavily affects standard F1-score. Following the suggestions in [42], we present both F1 and normalized F1 scores.

The best classification result $F1_{norm} = 93.34\%$ was obtained by the simplest model, which was the plain AE without any constraints, with the latent layer wider than the input (150 neurons vs. 100). Such a model was expected to copy values from the input to output without generalizing them, making no difference in reconstruction error between examples of both classes. Indeed, its mean reconstruction error is very low (0.0025), but surprisingly the error distribution allows to distinguish BQ examples from GQ with a relatively high effectiveness. The best results were obtained using the d_{abs} as reconstruction error measure.

Both Denoising AE (with Gausssian and Salt-and-Pepper noise) produced slightly higher reconstruction errors (0.0044 and 0.0104 respectively) and gave the classification results worse by almost 3%. The Sparsity AE's mean reconstruction error is also greater (0.0168), but model was better in classification task, achieving $F1_{norm} = 92.02\%$. Finally, the k-Sparse AE with only 5 non-zero values in 125-wide latent layer gave a significantly worse reconstruction (mean error is 0.4135) but almost the highest classification score of $F1_{norm} = 93.21\%$. In this model the most effective reconstruction error measure was the d_{abs_pf3}, which incorporates the penalty for local fluctuations of E_{abs}.

To summarize, presented results prove the correctness of proposed approach. The best models utilize the unconstrained AE and the k-Sparsity AE, both reaching over 93% accuracy expressed in $F1_{norm}$ and $F1 \approx 73\%$.

Table 1. Comparison of selected reconstruction errors and classification results on the *test set* for different variants of AE model. Following abbreviations were used: *AE* – autoencoder, *lat. w.* – latent width, *BC* – baseline cancellation, *NP* – normalize R-peak polarization, *rec. err* – reconstruction error ($\overline{d_{pow}}$) on unlabelled evaluation set, *F1-norm* – normalized F1-score [42], *simAE* – simple AE (without any constraints), *DAE-Ga* – Denoising AE with Gaussian noise of intensity α, *DAE-SP* – Denoising AE with Salt and Pepper noise of intensity α, *SpAE* – Sparsity AE with sparsity rate ρ and penalty weight w, *kSpAE* – k-Sparse AE with k non-zero elements in latent vector and regularization L2 of the latent layer parametrized by λ.

Model					AE rec.	Classification				
AE	lat. w.	BC	NP	params	err	d_{abs}		best d_*		
						$F1$	$F1_{norm}$	d_*	$F1$	$F1_{norm}$
simAE	150	Yes	No		0.0025	0.7347	0.9334	d_{abs}	0.7347	0.9334
DAE-Ga	100	Yes	No	$\alpha = 0.1$	0.0044	0.6662	0.9059	d_{abs}	0.6662	0.9059
DAE-SP	5	No	No	$\alpha = 0.1$	0.0104	0.6596	0.9076	d_{abs}	0.6596	0.9076
SpAE	150	No	Yes	$\rho = 0.005$ $w = 0.25$	0.0168	0.6809	0.9202	d_{abs}	0.6809	0.9202
kSpAE	125	Yes	Yes	$k = 5$ $\lambda = 0.0005$	0.4135	0.6300	0.9049	d_{abs_pf3}	0.7261	0.9321

5.5 Comparison to State of the Art

In contrary to 2011 PhysioNet/CinC Challenge, our solution operates on much shorter, and single-channel data (see Sect. 5.1). For this reason, a direct comparison of reported results is not possible, but we evaluated the challenge-winning algorithm [5] by building a similar Random Forest (RF) classifier using those features, which can be applied on our data, namely: frequency-domain (power spectral density in few bands and the relative power in QRS complex and in the baseline), time-domain (number of quantile crossings, percentage of flat signal, peak-to-peak amplitude) and statistical (skewness and kurtosis of the samples' distribution).

Labelled dataset was split identically as for the AE-based model (Subsect. 5.3) into validation (here: training) and test sets. After feature calculation, artificial BQ feature vectors were generated using the SMOTE algorithm [28], in order to balance the number of examples of both classes.

Finally, the RF classifier achieved $F1 = 0.3791$ and $F1_{norm} = 0.7231$, which is a significantly worse performance than results of proposed AE solution.

6 Conclusions and Further Work

We proposed a reliable ECG signal quality assessment algorithm, based on the anomaly detection approach, utilizing the autoencoder neural network.

Presented testing results confirm a high effectiveness of this approach. Its internal structure is rather simple, as no features nor rules has to be designed, so it can be easily applied to other kinds of signals. Thanks to unsupervised autoencoder training, only a small amount of data has to be annotated. This property is a significant advantage over classic classifier-based approach.

Our further efforts will concern on the relation between autoencoder reconstruction error during the training on the unlabelled data and the final classification accuracy. Other autoencoder architectures are also in our scope of interest.

Acknowledgements. The research presented in this paper was supported by the funds assigned to AGH University of Science and Technology by the Polish Ministry of Science and Higher Education. It has been carried out as part of the Industrial Doctorate Programme, in cooperation between Comarch and AGH University of Science and Technology. We would like to thank the management of Comarch e-Healthcare Department for a permission to publish this study.

References

1. Goldberger, A.L., et al.: Physiobank, physiotoolkit, and physionet. Circulation **101**(23), e215–e220 (2000)
2. Silva, I., Moody, G.B., Celi, L.: Improving the quality of ECGS collected using mobile phones: the PhysioNet/computing in cardiology challenge 2011. In: 2011 Computing in Cardiology, pp. 273–276, September 2011
3. Clifford, G.D., et al.: AF classification from a short single lead ECG recording: the PhysioNet/computing in cardiology challenge 2017. In: 2017 Computing in Cardiology (CinC), pp. 1–4 (2017)
4. Moody, G., Mark, R.: The impact of the MIT-BIH arrhythmia database. IEEE Eng. Med. Biol. Mag. **20**, 45–50 (2001)
5. Clifford, G.D., Behar, J., Li, Q., Rezek, I.: Signal quality indices and data fusion for determining clinical acceptability of electrocardiograms. Physiol. Meas. **33**(9), 1419 (2012)
6. Behar, J., Oster, J., Li, Q., Clifford, G.D.: ECG signal quality during arrhythmia and its application to false alarm reduction. IEEE Trans. Biomed. Eng. **60**(6), 1660–1666 (2013)
7. Liu, C., et al.: Signal quality assessment and lightweight QRS detection for wearable ECG SmartVest system. IEEE Internet Things J. **6**(2), 1363–1374 (2019)
8. Kuzilek, J., Huptych, M., Chudacek, V., Spilka, J., Lhotska, L.: Data driven approach to ECG signal quality assessment using multistep SVM classification. In: 2011 Computing in Cardiology, pp. 453–455, September 2011
9. Chudacek, V., Zach, L., Kuzilek, J., Spilka, J., Lhotska, L.: Simple scoring system for ECG quality assessment on android platform. In: 2011 Computing in Cardiology, pp. 449–451, September 2011
10. Hayn, D., Jammerbund, B., Schreier, G.: ECG quality assessment for patient empowerment in mHealth applications. In: 2011 Computing in Cardiology, pp. 353–356, September 2011
11. Liu, C., Li, P., Zhao, L., Liu, F., Wang, R.: Real-time signal quality assessment for ECGs collected using mobile phones. In: 2011 Computing in Cardiology, pp. 357–360, September 2011
12. Satija, U., Ramkumar, B., Manikandan, M.S.: A simple method for detection and classification of ECG noises for wearable ECG monitoring devices. In: 2015 2nd International Conference on Signal Processing and Integrated Networks (SPIN), pp. 164–169, February 2015

13. Satija, U., Ramkumar, B., Manikandan, M.S.: Automated ECG noise detection and classification system for unsupervised healthcare monitoring. IEEE J. Biomed. Health Inform. **22**(3), 722–732 (2018)
14. Satija, U., Ramkumar, B., Manikandan, M.S.: A new automated signal quality-aware ECG beat classification method for unsupervised ECG diagnosis environments. IEEE Sens. J. **19**(1), 277–286 (2019)
15. Hermawan, I.,et al.: Temporal feature and heuristics-based noise detection over classical machine learning for ECG signal quality assessment. In: 2019 International Workshop on Big Data and Information Security (IWBIS), pp. 1–8, October 2019
16. Redmond, S.J., Xie, Y., Chang, D., Basilakis, J., Lovell, N.H.: Electrocardiogram signal quality measures for unsupervised telehealth environments. Physiol. Meas. **33**(9), 1517 (2012)
17. Moeyersons, J., Testelmans, D., Buyse, B., Willems, R., Van Huffel, S., Varon, C.: Evaluation of a continuous ECG quality indicator based on the autocorrelation function. In: 2018 Computing in Cardiology Conference (CinC), vol. 45, pp. 1–4, September 2018
18. Martinez-Tabares, F.J., Espinosa-Oviedo, J., Castellanos-Dominguez, G.: Improvement of ecg signal quality measurement using correlation and diversity-based approaches. In: 2012 Annual International Conference of the IEEE Engineering in Medicine and Biology Society, pp. 4295–4298, August 2012
19. Morgado, E., et al.: Quality estimation of the electrocardiogram using cross-correlation among leads. BioMed. Eng. OnLine **14**, 59 (2015)
20. Naseri, H., Homaeinezhad, M.: Electrocardiogram signal quality assessment using an artificially reconstructed target lead. Comput. Meth. Biomech. Biomed. Eng. **18**(10), 1126–1141 (2015), pMID: 24460414
21. Shahriari, Y., Fidler, R., Pelter, M.M., Bai, Y., Villaroman, A., Hu, X.: Electrocardiogram signal quality assessment based on structural image similarity metric. IEEE Trans. Biomed. Eng. **65**(4), 745–753 (2018)
22. Shi, Y., et al.: Robust assessment of ECG signal quality for wearable devices. In: 2019 IEEE International Conference on Healthcare Informatics (ICHI), pp. 1–3, June 2019
23. Castiglioni, P., Meriggi, P., Faini, A., Rienzo, M.D.: Cepstral based approach for online quantification of ECG quality in freely moving subjects. In: 2011 Computing in Cardiology, pp. 625–628, September 2011
24. Zhou, X., Zhu, X., Nakamura, K., Mahito, N.: ECG quality assessment using 1D-convolutional neural network. In: 2018 14th IEEE International Conference on Signal Processing (ICSP), pp. 780–784, August 2018
25. Ghosal, P., Sarkar, D., Kundu, S., Roy, S., Sinha, A., Ganguli, S.: ECG beat quality assessment using self organizing map. In: 2017 4th International Conference on Opto-Electronics and Applied Optics (Optronix), pp. 1–5, November 2017
26. Karpinski, M., Khoma, V., Dudvkevych, V., Khoma, Y., Sabodashko, D.: Autoencoder neural networks for outlier correction in ECG- based biometric identification. In: 2018 IEEE 4th International Symposium on Wireless Systems within the International Conferences on Intelligent Data Acquisition and Advanced Computing Systems (IDAACS-SWS), pp. 210–215 (2018)
27. Bellinger, C., Drummond, C., Japkowicz, N.: Beyond the boundaries of SMOTE. In: Frasconi, P., Landwehr, N., Manco, G., Vreeken, J. (eds.) ECML PKDD 2016. LNCS (LNAI), vol. 9851, pp. 248–263. Springer, Cham (2016). https://doi.org/10.1007/978-3-319-46128-1_16
28. Chawla, N., Bowyer, K., Hall, L., Kegelmeyer, W.: SMOTE: synthetic minority over-sampling technique. J. Artif. Intell. Res. (JAIR) **16**, 321–357 (2002)

29. Sharma, S., Bellinger, C., Japkowicz, N., Berg, R., Ungar, K.: Anomaly detection in gamma ray spectra: a machine learning perspective. In: 2012 IEEE Symposium on Computational Intelligence for Security and Defence Applications, pp. 1–8, July 2012

30. Lecun, Y.: PhD thesis: Modeles connexionnistes de l'apprentissage (connectionist learning models). Universite P. et M. Curie (Paris 6) (6 1987)

31. Bourlard, H., Kamp, Y.: Auto-association by multilayer perceptrons and singular value decomposition. Biol. Cybern. **59**(4), 291–294 (1988)

32. Hinton, G.E., Zemel, R.S.: Autoencoders, minimum description length and Helmholtz free energy. In: Cowan, J.D., Tesauro, G., Alspector, J. (eds.) Advances in Neural Information Processing Systems, vol. 6, pp. 3–10. Morgan-Kaufmann (1994)

33. Goodfellow, I., Bengio, Y., Courville, A.: Deep Learning. MIT Press (2016). http://www.deeplearningbook.org

34. Vincent, P., Larochelle, H., Lajoie, I., Bengio, Y., Manzagol, P.A.: Stacked denoising autoencoders: learning useful representations in a deep network with a local denoising criterion. J. Mach. Learn. Res. **11**, 3371–3408 (2010)

35. Zhou, L., Yan, Y., Qin, X., Yuan, C., Que, D., Wang, L.: Deep learning-based classification of massive electrocardiography data. In: 2016 IEEE Advanced Information Management, Communicates, Electronic and Automation Control Conference (IMCEC), pp. 780–785 (2016)

36. Makhzani, A., Frey, B.J.: k-Sparse autoencoders. CoRR abs/1312.5663 (2013). http://arxiv.org/abs/1312.5663

37. An, J., Cho, S.: Variational autoencoder based anomaly detection using reconstruction probability. In: Special Lecture on IE (2015)

38. Chazal, P., Heneghan, C., Sheridan, E., Reilly, R., Nolan, P., O'Malley, M.: Automated processing of the single-lead electrocardiogram for the detection of obstructive sleep apnoea. IEEE Trans. Bio-med. Eng. **50**, 686–96 (2003)

39. Sahakian, A.V., Petrutiu, S., Swiryn, S.: Abrupt changes in fibrillatory wave characteristics at the termination of paroxysmal atrial fibrillation in humans. EP Europace **9**(7), 466–470 (2007)

40. Greenwald, S., Patil, R., Mark, R.: Improved detection and classification of arrhythmias in noise-corrupted electrocardiograms using contextual information (1990)

41. American Heart Association ECG Database. www.ecri.org/american-heart-association-ecg-database-usb/. Accessed 15 January 2019

42. Jeni, L., Cohn, J., De la Torre, F.: Facing imbalanced data - recommendations for the use of performance metrics. In: 2013 Humaine Association Conference on Affective Computing and Intelligent Interaction, vol. 2013, September 2013

Stenosis Assessment via Volumetric Flow Rate Calculation

Andrey I. Svitenkov[1] ⓘ, Pavel S. Zun[1,2] ⓘ, and Oleg A. Shramko[1(✉)] ⓘ

[1] ITMO University, Saint Petersburg 197101, Russia
{aisvitenkov,oashramko}@itmo.ru, p.zun@uva.nl
[2] University of Amsterdam, Amsterdam 1098 XH, The Netherlands

Abstract. Coronary artery stenosis is a condition that restricts blood flow to the myocardium, potentially leading to ischemia and acute coronary events. To decide whether an intervention is needed, different criteria can be used, e.g. calculation of fractional flow reserve (FFR). FFR can also be computed based on computer simulations of blood flow (virtual FFR, vFFR). Here we propose an alternative, more direct, metric for assessing the hemodynamic value of stenosis from computational models, the computed volumetric flow drop (VFD). VFD and vFFR are computed for several stenosis locations using a 1D model of the left coronary tree, and also an analytical model is presented to show why FFR value may differ from the true flow reduction. The results show that FFR = 0.8, which is often used as a criterion for stenting, may correspond to a reduction in volumetric flow from less than 10% to almost 30% depending on the stenosis location. The implications are that FFR-based assessment may overestimate the hemodynamic value of stenosis, and it's preferable to use a more direct metric for simulation-based estimation of stenosis value.

Keywords: Fractional flow reserve · Stenosis · Blood flow model · Lumped model · Coronary arteries

1 Introduction

Arterial stenosis is a pathological condition of an artery, characterized by a narrowing of its lumen. Stenting is often used to eliminate this defect. This invasive procedure consists in expanding the narrowed vessel and implanting a special mesh (a stent) into the wall of the vessel, which prevents the artery from re-narrowing.

In clinical practice it is important to assess the physiological importance of a particular arterial stenosis. This way, the treatment of coronary artery disease can be planned and improved. Currently, it is assessed by using a parameter called *fractional flow reserve* (FFR). According to the work [1], FFR is defined as the maximal blood flow in the presence of a stenosis in the artery, divided by the theoretical normal maximal flow in the same vessel. In practice FFR is calculated as the ratio of two pressure values: P_d, which is measured distally from the stenosis and P_p, measured proximally from it [2]. The question arises how to estimate the proximal and the distal pressures. In clinical

© Springer Nature Switzerland AG 2021
M. Paszynski et al. (Eds.): ICCS 2021, LNCS 12744, pp. 707–718, 2021.
https://doi.org/10.1007/978-3-030-77967-2_59

practice the mean pressure in the aorta (P_a) is usually substituted for P_p [3] and both pressures are measured during maximum hyperemia, which is an approximation [4–6].

In addition to invasive measurement, FFR can also be calculated from a numerical simulation of flow through the stenosis. By simulating the flow, the pressure across the stenosis (ΔP_{st}) can be obtained. If the proximal pressure is known, ΔP_{st} can be used to calculate the virtual FFR, also called vFFR.

The FFR is first mentioned in the article [4], where this concept is introduced to assess the severity of stenoses. In that work, a model of coronary circulation is considered, in which the stenosis is located in an artery. The artery is connected to the aorta on one side of the stenosis; on the other side it is connected to collateral vessels and the myocardial vascular bed. Starting with that work, aortic pressure has been used in the FFR calculations, and P_a is also used later in other works even with other topologies, with rare exceptions [2], although it would be more correct to use the proximal pressure.

There are many works in which vFFR is found through the pressure ratio. These works apply computational fluid dynamics to simulate hemodynamics and ultimately to obtain the pressures required to calculate vFFR. For this, first, computer anatomical models are created based on angiograms. Moreover, both non-invasive MRI data and invasive angiography can be used. The latter case seems less preferable, because one of the purposes of using vFFR is to avoid surgical interventions. In [7], a summary of vFFR models is given, comparing their accuracy against an invasively measured FFR value. Another study [8] presents four different computational methodologies for finding vFFR. In all cases, the required value is obtained as the ratio of P_d to P_a. Depending on the methodology P_a is either the mean aortic pressure, or a mean pressure at a point proximal to the stenosis. One of the methodologies states that P_a is found as the spatial average at the inlet region of approximately 2 mm length, and the region is defined manually.

When assessing the severity of stenosis, the main physiological quantity of interest (QOI) is the change in blood flow through the artery. However, a pressure-based assessment is used in clinical practice, because it is independent from the baseline flow, relatively simple and cost effective [9]. Following this method, vFFR is also calculated from pressures.

The idea of using volumetric flow rates ratios instead of pressure ratios in numerical calculations to assess the importance of stenosis is not new. In [10], a method for calculating the vFFR by constructing a lumped parameter model from angiograms was proposed. FFR$_{angio}$ is found as the ratio of the maximal flow rate in the stenosed artery and the maximal flow rate in the absence of the stenosis. The research results have shown that FFR$_{angio}$ has a high sensitivity, specificity, and accuracy when compared to clinical FFR (diagnostic accuracy is 92.2%).

Here we propose an alternative method, using the volumetric flow rate values in the artery with a stenosis (Q_s) and the flow rate in the same artery, but without a stenosis (Q_0). These values can be found from numerical modeling of blood circulation in the coronary vessels. In our study, a 1D model is built with boundary conditions specified by a 0D model. The proposed method is compared to the vFFR results.

Given the primary importance of the volumetric flow rate values and the fact that in any case, the calculation of vFFR requires modeling using the coronary artery system, the estimation of the change in flux seems preferable.

2   Illustration of the Idea

We have noted before that the decrease in the volumetric flow rate due to a stenosis has a direct physiological impact. Here we will present an analytical lumped model of coronary arteries to illustrate the relation between FFR and the volumetric flow rate in the artery with and without stenosis: Q_s and Q_0.

For this purpose, we consider a simplified steady state model of coronary arteries (CA). Represented by hydrodynamic resistance elements, the arteries form the hierarchical structure shown in Fig. 1A. The resistance R_σ is the resistance of the considered stenosis, R_i are for resistive arterial segments between bifurcations and R_{Ti} are for resistances of peripheral vessels.

We are looking for an expression of the pressure proximal and distal to the stenosis and of the flow through the stenosis as a function of all resistances in the model. Unfortunately, even for this simplified representation these expressions turn out to be very complex and do not have much illustrative use. For further simplification, we consider that the resistance of major arteries is much smaller than the resistance of peripheral arteries, arterioles and capillaries. Then the CA structure can be represented as a set of parallel branches (Fig. 1B) and, finally, just by two branches as in the Fig. 1C. The resistance R_S here represents all parallel branches of CA, it decreases with the number of branches N proportionally to $1/N$. The same is correct for R_{T0}: representing the resistance of all distal parts of CA, it is smaller for more proximal positions of the stenosis (because the number of capillaries downstream is larger).

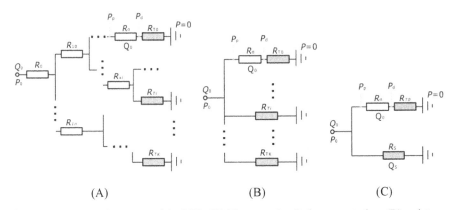

(A)   (B)   (C)

Fig. 1. Lumped steady-state model of CA. (A) The most detailed representation; (B) resistances of arterial segments are presumed negligible comparing with resistance of peripheral arteries; (C) the most simple representation.

For the model from Fig. 1C the values of P_d and P_p can be easy calculated, as well as the FFR value:

$$FFR = \frac{P_d}{P_p} = \frac{R_{T0}}{R_\sigma + R_{T0}}. \tag{1}$$

To estimate the flux through the stenosis we need to consider how the inlet boundary condition is set: that is, either by a prescribed pressure or by flux. In case of a prescribed pressure, the relation between the flux through the stenosis Q_σ and the flux for the same location without a stenosis $Q_{\sigma0}$ gives the same expression as (1). Thus, FFR can be considered an approximation of the relation $Q_\sigma/Q_{\sigma0}$ for fluxes, which has a direct physiological meaning. We will call this relation Volumetric Flux Drop (VFD) in the following. In practical terms it can be considered as a measure of *how much the flux through the stenosis decreases after stenting.*

Similar conclusions were stated by *Pijls et al.* with an introduction of FFR [4]. We, however, would also like to consider the case where the inlet flux is prescribed since we believe that it represents the real situation better. Cardiac compensation increases the arterial pressure in response to the increasing resistance of coronary arteries to provide the normal blood supply to the heart muscle [11]. We can emphasize this argument by considering the structure (C) as a part of structure (A) (see Fig. 1). Then the inlet point corresponds not to the aortic sinus but to some bifurcation of the CA. The inlet pressure then will change in response to a changing stenosis resistance as well as the inlet flux. Both choices are reasonable, but the choice of a constant flux is better since it takes the compensatory mechanisms into the account.

For the case of a prescribed inlet flux, the expression for the VFD value is different:

$$VFD = \frac{Q_\sigma}{Q_{\sigma0}} = \frac{R_{T0} + R_S}{R_\sigma + R_{T0} + R_S} = 1 - \frac{R_\sigma}{R_\sigma + R_{T0} + R_S}. \tag{2}$$

If the resistance R_S of all side branches is negligible compared to R_{T0}, then VFD tends to FFR. However, considering the last expression, we may see that the FFR value underestimates the VFD.

The presented illustration shows that even for a very simplified model of CA we clearly see the difference between VFD and FFR. We should expect an even larger discrepancy for the real situation due to a more complex structure of CA and a pulsatile blood flow.

3   1D Numerical Model

In this study, numerical modeling of hemodynamics in the system of coronary arteries was carried out. For this, the following 1D model was used.

The model formulation based on the equations relating the average velocity U and the area of the vessel lumen A was used:

$$\begin{cases} \frac{\partial A}{\partial t} + \frac{\partial AU}{\partial x} = 0 \\ \frac{\partial U}{\partial t} + U\frac{\partial U}{\partial x} + \frac{1}{\rho}\frac{\partial P}{\partial x} = \frac{8\mu\pi U}{\rho} \end{cases}, \tag{3}$$

where t is time, x is the longitudinal coordinate relative to the artery, ρ is the blood density, P is the pressure in the discretization point, μ is the dynamic blood viscosity.

The model takes into the account the elastic properties of the arteries, with Young's modulus $E = 225$ kPa. The thickness of arteries' walls h depends on the reference radius of a vessel [12] as $h = r_0(a \cdot \exp(br_0) + c \cdot \exp(dr_0))$, where $a = 0.2802$, $b = -5.053\,cm^{-1}$, $c = 0.1324$, $d = -0.1114\,cm^{-1}$, r_0 is an arterial radius at diastolic pressure.

All other parameters considered for the blood flow model are presented in the Table 1.

Table 1. General parameters of blood flow model for all simulations.

Property	Value
Blood density, ρ, kg·m^{-3}	1040
Blood viscosity, μ, mPa·s	3.5
Velocity profile order, ζ	9
Young's modulus, E, kPa	225.0
Space discretization step, mm	2.5
Timestep, ms	0.05

The outlet boundary conditions replacing the downstream vasculature were represented by RCR Windkessel elements [13]. A known flow rate with a ninth order velocity profile defines an inlet BC.

We consider a geometrical model of CA provided by Zygote (Zygote Media Group, Inc.) and constructed as a compilation of multiple CT models of coronary arteries for most typical anatomies (Fig. 2A).

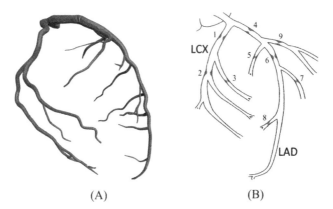

(A)   (B)

Fig. 2. Coronary arteries. (A) General appearance of the considered model of the LCA tree; (B) stenosis locations during the simulations (marked in red). (Color figure online)

Given the high variability of CA we should note insufficiency of such approach for reaching definitive conclusions. Population studies that consider a set of patient-specific geometries would be more preferable, but require the data we do not have. Nevertheless, considering a single and typical arterial structure provides an assessment of relation between vFFR and VFD for various positions of stenosis.

The 3D structure of CA has been translated to a 1D geometry by centerline detection followed by prescribing arterial radius in each discretization point along the centerline in accordance with the area of arterial lumen in the corresponding section.

The coronary arteries collectively form two networks, which originate from the right coronary artery (RCA) and the left coronary artery (LCA). LCA bifurcates into two arteries – the left circumflex coronary artery (LCX) and the left anterior descending artery (LAD). The calculations were performed for the LCA tree only. Figure 2 shows a general appearance of the considered LCA tree, and also indicates the locations of the stenosis during simulations. There are several stenoses on one sketch at the same time, but in fact the calculations were carried out several times for each stenosis one by one, i.e., only one stenotic site was present in the LCA branch during each simulation.

The stenosis was introduced into the model as a local narrowing of an artery with a constant radius (i.e., a cylinder), as shown in Fig. 3. The arterial radius was decreased at 3 discretization points, wherein none of them were the point of bifurcation.

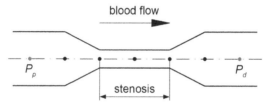

Fig. 3. Schematic model representing the stenosis. Dots show discretization points, red dots indicate points at which proximal pressure (P_p) and distal pressure (P_d) are calculated. (Color figure online)

Stenosis is always set in the middle of an arterial branch so there is at least one discretization point without a narrowing between the stenosis and bifurcations. It is important to avoid influence of boundary effects near bifurcations, which can lead to a misestimation of the hydrodynamic resistance caused by the narrowing. Furthermore, our simulations show almost no effect from shifting the stenosis along an arterial branch on the calculated vFFR and VFD.

The stenosis degree (*SD*) was determined by the cross-sectional area, i.e. the cross-sectional area of a stenotic site (A_{st}) was set using the formula:

$$A_{st} = A_n(1 - SD), \tag{4}$$

where A_n is the normal cross-sectional area (i.e. without the stenosis).

The simulations were performed with the stenosis degree values in range $0.20 \div 0.95$.

4   Results

The simulations provided VFD and vFFR values for 9 locations of stenosis with 9°, both for the prescribed inlet flow and for the prescribed pressure Fig. 4 presents the obtained data.

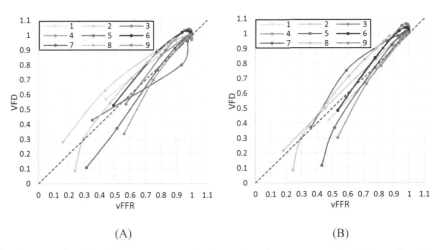

Fig. 4. Plots of relation between VFD and vFFR. The colors correspond to different locations of the stenosis. (A) Prescribed inlet flux scenario; (B) prescribed pressure scenario. (Color figure online)

It is clearly seen that, with rare exceptions, the graphs have a similar curved appearance. The points corresponding to lower degrees of the stenosis are located to the right. It can be seen that the greater the stenosis degree, the more points are located below the diagonal. Also, most of the points are to the left of the diagonal line, both for the prescribed inlet flow and for the prescribed pressure, especially in the first case, which indicates that for the specified range of stenosis degree, there are more situations when VFD exceeds vFFR.

The values of vFFR and VFD at approximately the same severe stenosis degree were also considered. Figure 5 shows the results for the same stenosis locations, but with a stenosis degree equal to 0.85 ± 0.005.

The points are located mainly to the left of the diagonal, especially in the high vFFR area $(0.8 \div 1.0)$. At low vFFR (< 0.4) the points are below the diagonal line. That is, Fig. 5 shows that for a specific degree of stenosis in the range of high vFFR, VFD more often exceeds vFFR, and vice versa for small vFFR. There is also a slight predominance of blue points to the left of the parity line, suggesting that more cases where VFD is greater than vFFR occur in the prescribed flow scenario.

Figures 4 and 5 show that in some cases the VFD exceeds 1, which contradicts the Eq. (2). The possible explanations for this will be given in the next section.

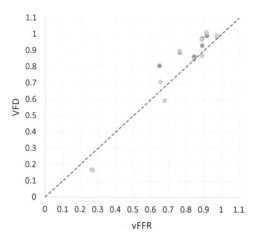

Fig. 5. VFD and vFFR for $SD = 0.85 \pm 0.005$. Blue color corresponds to the prescribed inlet flux scenario, orange color corresponds to the prescribed pressure scenario. (Color figure online)

5 Discussion

In this study we have compared virtual FFR-based stenosis assessment to a novel VFD-based assessment, which aims to provide a more accurate measure of stenosis' impact on hemodynamics by directly considering the flux reduction in the stenosed artery.

In the study, nine different stenosis locations were selected. To describe the stenosis location, we consider the number of bifurcations proximal to a stenosis (N_b), the peripherial arterial resistance distal to a stenosis (R_{T0}) and the peripherial arterial resistance of the entire LCA except the branch distal to a stenosis (R_S). The arterial peripheral resistances were calculated only from the terminal resistances, without taking compliance into account, but in the simulations it was also included. Table 2 shows the results obtained by both scenarios for approximately the same $SD = 0.85 \pm 0.005$. It can be seen that both scenarios show similar results, and in both cases there are strong differences between the vFFR and VFD for some stenosis locations.

For the scenario of prescribed inlet flux, for most cases VFD predicts a lower impact of the stenosis on hemodynamics than vFFR, which agrees with the analytical model presented in Sect. 2. In particular, for FFR = 0.7, which is associated with severe stenosis [14], there is quite a large spectrum of VFD values, depending on the specific vessel considered. With the exception of sites #5, #9 and #7, the VFD values are larger than vFFR. For FFR = 0.8, which is often recommended as a criterion for intervention [6, 8], the range of VFD values is smaller, and VFD is larger than vFFR for all sites except #5 and #9. This means that FFR-based assessment tends to overestimate the hemodynamic value of the stenosis. This difference between VFD and vFFR persists also for large arteries, for example #1, #4.

The different behavior of #5 in Fig. 4A compared to other plots is not predicted by our analytical estimation. However, the effect disappears in the prescribed pressure scenario that shows that this case is simpler for stenosis assessment. For the prescribed flux case we can only note large number of alternative drains and low total parallel resistance (R_S,

Table 2. Comparison table of the scenarios.

#	A_n, mm^2	N_b	R_{T0}, (N·s)/m^5	R_S, (N·s)/m^5	Prescr. inlet flux		Prescr. pressure	
					vFFR	VFD	vFFR	VFD
1	6.21	1	$1.01 \cdot 10^{10}$	$1.45 \cdot 10^{10}$	0.89	0.93	0.89	0.87
2	3.32	2	$1.17 \cdot 10^{10}$	$1.21 \cdot 10^{10}$	0.65	0.81	0.66	0.71
3	3.08	2	$7.58 \cdot 10^{10}$	$6.47 \cdot 10^{9}$	0.92	0.99	0.92	1.01
4	13.55	1	$1.45 \cdot 10^{10}$	$1.01 \cdot 10^{10}$	0.98	0.99	0.98	0.995
5	2.30	2	$8.77 \cdot 10^{10}$	$6.40 \cdot 10^{9}$	0.77	0.89	0.76	0.90
6	4.81	2	$2.57 \cdot 10^{10}$	$7.76 \cdot 10^{9}$	0.89	0.98	0.89	0.97
7	0.75	3	$2.05 \cdot 10^{11}$	$6.49 \cdot 10^{9}$	0.68	0.59	0.68	0.59
8	0.34	4	$1.18 \cdot 10^{11}$	$6.28 \cdot 10^{9}$	0.27	0.17	0.27	0.17
9	2.67	2	$5.31 \cdot 10^{10}$	$6.72 \cdot 10^{9}$	0.85	0.86	0.85	0.84

see the Table 2) of them, thus the blood flux can be easy distributed among them. Finally, the reason must be a combination of the location of this stenosis and the characteristics of the rest of the blood network in dynamics.

Interestingly, the difference between vFFR and VFD is the largest for intermediate values of vFFR between 0.4 and 0.8, while for very large and very small vFFR this difference is smaller. Also of note is the flow increase for very small stenosis degree values in several cases, so the VFD may be greater than 1. This effect persists for increased equilibration times and reduced timesteps, with the maximal volumetric flow increase of around 3% compared to the no-stenosis case. This may be caused by a numerical effect in the model, or it may be caused by dynamical properties of the flow, somewhat similar to wave impedance in electrodynamics.

For the prescribed pressure scenario, the difference between VFD and FFR is smaller. However, due to the dynamic effects and a more complex geometry than considered for the analytical model, these two measurements are not exactly the same, contrary to the analytical calculation results. VFD is higher than vFFR for the majority of points, except for some severely stenosed cases where it is lower. Similarly to the prescribed inlet flux case, in two cases there is a minor increase in flow through the stenosis site compared to the baseline. This increase also disappears for larger stenosis degrees.

It is also noticeable that the more distal the stenosis, the stronger the vFFR is correlated with the visual degree of narrowing. It is well-illustrated by Fig. 4.

Overall, this study shows that similar FFR values may correspond to significantly different decreases in pressure, depending on the stenosis location and the ability of the physiology to adapt to the increased resistance. For example, for FFR = 0.8 the model predicts the range of flow decrease from less than 10% to almost 30%. This means that using vFFR as a criterion for stenting may result in performing interventions when they are not really needed. The opposite (not stenting a problematic vessel) is less likely to happen, since we have found few cases where the actual flow was significantly less than what is expected from vFFR (two cases out of the nine considered stenosis positions).

Also, vFFR is still a big improvement over a simple visual assessment. A 50% diameter occlusion (a common visual guideline for stenting) can correspond to a very broad spectrum of both vFFR and VFD values.

VFD provides a more direct measure of stenosis impact than vFFR. However, despite that, it would be harder to get VFD-based assessment approved for use in clinical practice than it would be for virtual FFR-based assessment. This is because the quantity of interest provided by VFD is different from the (experimental) FFR used in clinical practice. Hence, to get approval from the regulatory bodies it will be necessary not only to convince them that the VFD values provided by the model are representative of the real artery, but also to demonstrate that VFD is a sensible measure of hemodynamic significance (which is non-trivial, despite VFD being a more direct measurement).

6 Conclusion

The study is devoted to assessment of the physiological importance of a particular arterial stenosis based on CA blood flow simulation. Currently, FFR values are widely used to address this problem. FFR can be either measured invasively or calculated from numerical simulations (vFFR). The latter way currently receives great attention of the modeling community.

However, the FFR does not provide a direct estimation of volumetric flow rate drop caused by a stenosis, which would be the most precise assessment of its physiological impact. Therefore, we have introduced a different parameter called Volumetric Flow Drop (VFD) which is calculated as the ratio of the flux through the stenosis to the flux through the same artery without a stenosis. It shows how the flux through the artery will increase after stenting of the stenosis.

These days many studies that use numerical simulations of blood flow for assessment of the stenosis impact in real clinical practices are being published. Thus, we would like to note that our study is of a different type and we are not presenting any novel methodology of blood flow modeling application. Our goal is to check the relation between FFR and VFD for various positions of stenosis and to detect the cases when the difference can be considerable. We are not considering a personalized blood flow model and thus all the related issues are out of the scope of this study.

We have shown that FFR and VFD parameters match for a simplified model of CA and only in the case when inlet pressure is prescribed. If the inlet flux is prescribed instead, the FFR value overestimates the VFD. The detailed investigation based on a 1D numerical blood flow model of CA shows that a real relation of FFR and VFD can be even more complex due to omitted complex structure of CA and the pulsatile nature of blood flow. In particular, for FFR = 0.7, which is associated with severe stenosis [13], there is a deviation of VFD values in range of $\pm$ 0.1 and more, that can have a direct influence on the treatment strategy.

We also would like to note that for the same model of stenosis and the same degree of stenosis a very wide variation of FFR and VFD has been found. This fact also emphasize that the physiological importance of a stenosis is rather related to the global flow pattern in the CA than to the local hydrodynamic resistance of the construction. Accurate calculation of vFFR requires modeling a considerable part of CA just as VFD does; that is confirmed by many related studies of vFFR calculation [8, 9].

For calculating the VFD, the 1D numerical model was used, verified using an analytical model. For the validation clinical data is needed, which is hard to obtain. Therefore, there is a field for further research on the importance of VFD.

Finally, we would like to conclude that the VFD may provide a more accurate estimation of the stenosis physiological importance and can be considerably different from the FFR value. The FFR was introduced as a method of invasive stenosis assessment and presents a compromise between accuracy and ease of *in vivo* measurement. VFD obviously cannot be measured *in vivo* directly, but there are no such limitations for simulation-based virtual measurements. Thus, the approach can be used if there is patient data sufficient to build a computational model. For the purpose of simulation results interpretation VFD provides a more accurate assessment of stenosis importance compared to vFFR.

Acknowledgements. This research was supported by The Russian Science Foundation, Agreement # 20–71-10108 (29.07.2020). Participation in the ICCS conference was supported by the NWO Science Diplomacy Fund project # 483.20.038 "Russian-Dutch Collaboration in Computational Science".

References

1. Pijls, N.H.J., et al.: Measurement of fractional flow reserve to assess the functional severity of coronary-artery stenoses. N Engl. J. Med. **334**, 1703–1708 (1996)
2. Saha, S., Purushotham, T., Prakash, K.A.: Numerical and experimental investigations of fractional flow reserve (FFR) in a stenosed coronary artery. In: E3S Web of Conferences, vol. 128, p. 02006 (2019)
3. Lotfi, A., et al.: Expert consensus statement on the use of fractional flow reserve, intravascular ultrasound, and optical coherence tomography: a consensus statement of the society of cardiovascular angiography and interventions. Catheter. Cardiovasc. Interv. **83**, 509–518 (2014)
4. Pijls, N.H.J., van Son, J.A.M., Kirkeeide, R.L., De Bruyne, B., Gould, K.L.: Experimental basis of determining maximum coronary, myocardial, and collateral blood flow by pressure measurements for assessing functional stenosis severit before and after percutaneous transluminal coronary angioplasty. Circulation **87**(4), 1354–1367 (1993)
5. Mehra, A., Mohan, B.: Value of FFR in clinical practice. Indian Heart J. **67**, 77–80 (2015)
6. Briceno, N., Lumley, M., Perera, D.: Fractional flow reserve: conundrums, controversies and challenges. Interv. Cardiol. **7**(6), 543–552 (2015)
7. Morris, P.D., van de Vosse, F.N., Lawford, P.V., Hose, D.R., Gunn, J.P.: "Virtual" (computed) fractional flow reserve: current challenges and limitations. JACC Cardiovasc Interv. **8**(8), 1009–1017 (2015)
8. Carson, J.M., et al.: Non-invasive coronary CT angiography-derived fractional flow reserve: a benchmark study comparing the diagnostic performance of four different computational methodologies. Int. J. Numer. Meth. Biomed. Eng. **35**, e3235 (2019)
9. Crystal, G.J., Klein, L.W.: Fractional flow reserve: physiological basis, advantages and limitations, and potential gender differences. Curr. Cardiol. Rev. **11**, 209–219 (2015)
10. Fearon, W.F., et al.: Accuracy of fractional flow reserve derived from coronary angiography. Circulation **139**(4), 477–484 (2019)

11. Levy, P.S., et al.: Limit to cardiac compensation during acute isovolemic hemodilution: influence of coronary stenosis. Am. J. Physiol. Heart Circulatory Physiol. **265**(1), H340–H349 (1993)
12. Boileau, E., et al.: A benchmark study of numerical schemes for one-dimensional arterial blood flow modeling. Int. J. Numer. Meth. Biomed. Eng. **31**(10), e02732 (2015)
13. Alastruey, J., et al.: Pulse wave propagation in a model human arterial network: assessment of 1-D visco-elastic simulations against in vitro measurements. J. Biomech. **44**(12), 2250–2258 (2011)
14. Fahmi, R., et al.: Dynamic myocardial perfusion in a porcine balloon-induced ischemia model using a prototype spectral detector CT. In: Proceedings of SPIE 9417, Medical Imaging 2015: Biomedical Applications in Molecular, Structural, and Functional Imaging, 94170Y (2015)

Fuzzy Ontology for Patient Emergency Department Triage

Khouloud Fakhfakh[1,2,3,4]([✉]), Sarah Ben Othman[1,4], Laetitia Jourdan[1,4],
Grégoire Smith[5], Jean Marie Renard[3,4,5], Slim Hammadi[1,2],
and Hayfa Zgaya-Biau[1,3,4]

[1] CRItAL Laboratory UML 9189, Villeneuve-d'Ascq, France
{Khouloud.fakhfakh,
sara.ben-othman,slim.hammadi}@centralelille.fr,
{laetitia.jourdan,hayfa.zgaya-biau}@univ-lille.fr
[2] Ecole-Central of Lille, Villeneuve-d'Ascq, France
[3] Cerim, Seyssinet-Pariset, France
jean-marie.renard@univ-lille.fr
[4] University of Lille, Lille, France
[5] CERIM EA2694 laboratory pôle recherche aile Est 2ème étage, LUHC Lille, Lille 1 Place de Verdun, 59045 cedex, France

Abstract. Triage in emergency department (ED) is adopted procedure in several countries using different emergency severity index systems. The objective is to subdivide patients into categories of increasing acuity to allow for prioritization and reduce emergency department congestion. However, while several studies have focused on improving the triage system and managing medical resources, the classification of patients depends strongly on nurse's subjective judgment and thus is prone to human errors. So, it is crucial to set up a system able to model, classify and reason about vague, incomplete and uncertain knowledge. Thus, we propose in this paper a novel fuzzy ontology based on a new Fuzzy Emergency Severity Index (F-ESI_2.0) to improve the accuracy of current triage systems. Therefore, we model some fuzzy relevant medical subdomains that influence the patient's condition. Our approach is based on continuous structured and unstructured textual data over more than two years collected during patient visits to the ED of the Lille University Hospital Center (LUHC) in France. The resulting fuzzy ontology is able to model uncertain knowledge and organize the patient's passage to the ED by treating the most serious patients first. Evaluation results shows that the resulting fuzzy ontology is a complete domain ontology which can improve current triage system failures.

Keywords: Fuzzy ontology · Uncertainty · Medical ontology · Patient triage

1 Introduction

An effective triage in emergency departments (ED) can limit overcrowding situations and enhance the care quality and the slightest mistake increases the risk of mortality [1, 2]. Recently, some studies have focused on improving the triage of patients by employing

© Springer Nature Switzerland AG 2021
M. Paszynski et al. (Eds.): ICCS 2021, LNCS 12744, pp. 719–734, 2021.
https://doi.org/10.1007/978-3-030-77967-2_60

new technologies [3, 4] refining collaboration and communication strategies [5, 6]. The major disadvantages of these systems are the processing of textual and unstructured data and the representation and manipulation of medical knowledge considering the imprecision of these data. Structured triage protocols are already being widely used in the ED [8]. These protocols are tools aimed at prioritizing patients according to some established criteria such us the emergency severity index (ESI). The ESI classification was created in order to facilitate the patient triage, and thus to improve the patient throughput and disposition decision [7]. Nevertheless, this triage system depends greatly on nurse's subjective judgment, then the risk of error is very high. In this context, Wunch et al. [13] have demonstrated that ontologies are a potential solution to the problem of patient triage. Ontologies describe knowledge in terms of concepts, objects and data properties and relationships between concepts. In the literature, to facilitate biomedical research and standardization of the medical vocabulary, several ontologies and knowledge bases have been defined, such as the Systematized Nomenclature of Medicine Clinical Terms (SNOMEDCT), International Classification of Diseases (ICD), Disease Ontology (DO), Symptoms Ontology (SYMP). In this context, some models for supporting healthcare professionals in patient triage process exist in the literature [10–12]. For example, the model of Farion et al. [9] deals with heterogeneous clinical decisions. The main objective of this kind of system is to improve medical decision making in triage and to facilitate data sharing between users. However, the problem of missing and imprecise data and knowledge is still one of the major limitations of medical systems. Therefore, the use of fuzzy subset theory and fuzzy logic is an intuitive solution to this problem, since the definition of a fuzzy ontology is based on the fuzzy subset theory to precise ontologies in order to represent uncertainties. Moreover, Zhaiet et al. try to define fuzzy ontologies based on the application of fuzzy logic [29], without distinction between precise and fuzzy components. However, Straccia limits itself to defining a fuzzy component from instances [30]. Ghorbel tried to define a fuzzy component based on the integration of uncertainty and imprecision in the definition of a precise component [31]. In the medical context, AlzFuzzyOnto [28] presents a fuzzy ontology specific to Alzheimer's disease (AD). Thus, with the help of experts, the points of uncertainty present in each concept and each relation of the ontology are analysed. Indeed, this ontology allows the generation of fuzzy concepts to represent fuzzy information and data in order to process imprecise knowledge and data and to refine the results obtained [28]. But, to the best of our knowledge, there is no fuzzy ontology in the literature that defines the triage system in ED. Besides, there are several techniques for developing classic ontology [13] but they are not sufficient to construct fuzzy ontologies [14]. Many methods are available to generate fuzzy ontologies such as map fuzzy model [14], FuzzyOntoMethodology [15], FONTO (Fuzzy for ONTOlogy) [17], FOGA (Fuzzy Ontology Generation FrAmework) [16]. The difference between these approaches is mainly in what aspects of the classical ontology are being fuzzified, and these aspects depend on the domain needs. However, the mentioned approaches do not guarantee the encoding of fuzzy in ontologies and the construction of fuzzy case-based domain ontologies at the same time. So, we propose in this paper a novel method which consists in fuzzifying an ontology by ensuring the application of fuzzy subset theory in an automatic way and in defining all the precise and fuzzy domain concepts with the case descriptions.

The fuzzification is a process of transforming a precise ontology into a fuzzy ontology in order to model forms of uncertainty [17]. So, in this work, we implement fuzzy ontologies that manage imprecise knowledge and data. We populate the implemented ontology with real data of patients, and we use reasoners for semantically querying the resulting fuzzy ontology. The paper has the following contributions:

- the first fuzzy ontology of the triage process to model and reason with medical fuzzy knowledge based on ED patient cases.
- a novel method for developing fuzzy ontology of domain.
- an extension of the ED severity index based on fuzzy concepts to better specify the severity of a patient's case and the possible waiting time.
- the validation of the method on real patient case scenarios.

We conduct this work in the context of a national project called Inter and Intra Hospital Logistics Optimization supported by the National Research Agency (2019–2022).

2 The Proposed Fuzzy Ontology for Triage System: FOTS

We propose a fuzzy ontology containing three basics modules: 1) conceptual model, 2) fuzzy-domain model and 3) case and reasoning model. In this section, we present the proposed methodology to develop the fuzzy ontology. Then, we define the novel fuzzy ESI based on fuzzy concepts, named F-ESI_2.0. Finally, we describe the current implemented instance of our proposed ontology based on these three models.

2.1 Proposed Methodology for Developing Fuzzy Ontologies: FOntoM

In this paper, we propose the novel FOntoM method in order to meet the needs of medical field which contains very complex vocabulary and knowledge. We drew on the Methontology method [18] adding the FONTO method of fuzzification [17] allowing the definition of the ontology fuzzy concepts. Three main models are created: the conceptual model, the fuzzy domain model and the case and reasoning (Fig. 1).

This method is composed of the following steps:

1. Specification to provide a clear description of the target glossary, defining the domain concepts.
2. Conceptualization to allow the organization and structuring of the knowledge acquired in the previous step by developing a conceptual model.
3. Formalization of the conceptual model developed in the previous step, using a formal ontology language.
4. Fuzzification to assign domain concepts that present a degree of uncertainty to fuzzy sets with a certain degree of membership (Truth).
5. Implementation to write the ontology in a machine-readable ontology language such as OWL 2[1]; integrate cases and implement semantic reasoners and reasoning rules.

---

[1] https://www.w3.org/TR/owl2-overview/

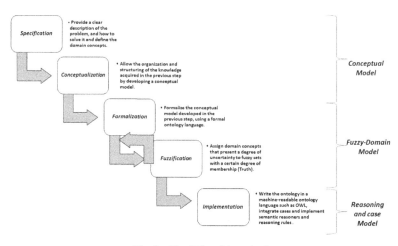

Fig. 1. The FOntoM method

The proposed ontology FOntoM contains both fuzzy and precise concepts and, as shown in Fig. 1, the specificity of this method is that an iterative loop is made between formalization and fuzzification steps, so the output of this loop is enabled if all the concepts of the fuzzy domain are defined. We choose the FONTO method [17] for fuzzification which is composed of three successive phases: the extraction of the fuzzy concepts, the determination of the fuzzy classes of the ontology and the calculation of the membership degrees. This method allows handling imprecise and vague knowledge through threshold values representing the different possible modalities, ensures quantitative knowledge modeling and codifies ill-defined knowledge. So, the fuzzification process consists in modeling a set of fuzzy concepts basing on the well-known fuzzy logic theory.

2.2 The FOTS Implementation Process

In this section, we present the FOTS ontology composed of three models: the conceptual model, the fuzzy domain model and the reasoning and case model. The implementation of FOTS is done using the FOntoM method (Sect. 2.1). We detail each model in the following subsections.

The FOTS Conceptual Model
(See Fig. 2). The proposed conceptual model is represented in Fig. 2 by an UML class diagram [19]. This conceptual model represents the triage process by defining the different triage actors and the relationships between them. To model the knowledges of our ontology (e.g. concept, property, classes), we use the publisher Modelio Open Source2. The classes of this diagram (Fig. 2) define the main elements on which depends the health condition of patients and that influence hospital management and subsequent clinical decision making. Generic classes of this model are presented in next section (Table 1).

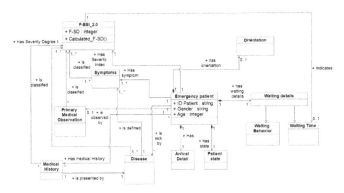

Fig. 2. The FOTS generic class diagram

The FOTS Fuzzy-Domain Model

This model represents fuzzy domain knowledges according the conceptual model defined in Fig.2. We describe our FOTS in OWL 2 Language and develop it in collaboration with the staff of LUHC. The resulting based on our proposed FOTS ontology contains 6 generic classes or concepts: Emergency patient, Disease, F-ESI_2.0 (Sect. 3.3), Primary Medical Observations, Symptoms and Medical History. The root class of these six classes is the Class Thing (Table 1).

Table 1. Generic concepts of FOTS fuzzy domain

Generic classes	Description
Emergency patient	All the specific data of a patient such as demographic data, mode of arrival, waiting time and way of waiting
Disease	Patient's illness
F-ESI_2.0	The health state of patient with a fuzzy severity index
Medical observation	The vital signs observed and revealed by the nurses
Symptoms	Patient's symptoms
Medical history	Patient's medical history

The OWL 2 classes are performed as sets of individuals (or sets of objects) and the Class Thing represents the set containing all individuals [19]. In this context, the FOTS fuzzy domain contains two levels of knowledge abstraction that we can qualify by knowledge of surface and deep knowledge:

– The generic level: contains the generic concepts of ED domain (Table 1).
– The fuzzy domain level: describes the triage field related to the patient severity defined by FOTS properties and concepts in the next subsection.

The FOTS Concepts/Properties

To define the domain concepts, we use standard medical ontologies such as SYMP and ICD ontologies to build a medical domain ontology according to the Medical Dictionary of Health[2]. Emergency physicians validate the SD of each element and the relations between classes, then we define all its elements with the OWL 2 language in ontology form. Table 2 presents an overview of the object properties defined in FOTS.

Table 2. The object properties

Object properties	Domain	Range
Has symptoms	Emergency patient	Symptoms
Is sick by	Emergency patient	Disease
Has severity index	Emergency patient	F-ESI_2.0
Has defined by	Disease	Symptoms, vital signs
Has state	Emergency patient	Patient state
Has wait time	Emergency patient	Waiting time

Each property is responsible for defining the relationships between domain concepts in order to create patient case scenarios in the triage process. Therefore, we define uncertain data and inaccurate knowledge in the field of hospital triage by fuzzy concepts and relationships such as medical observations, symptoms and diseases. Each fuzzy concept defines uncertain medical information, and through fuzzy description we can manage this uncertainty and formulate more accurate and correct results. In this context, the logical axioms define concepts by means of logical expressions. The Table 3 contains an overview of the axioms defined in the ontology with the mathematical expressions that allow to calculate the membership degree of instances to concepts and fuzzy relations. In order to define fuzzy concepts, three items are created for each of the numerical features: an abstract role (i.e., data property) for the numerical feature, a fuzzy data type for each linguistic term and a fuzzy concrete role (i.e., object property) for each linguistic term. For example, if we consider the variable Peripheral O2 Saturation, it's range of acceptable values would be [0, 99], the applicable linguistic terms would be: Very Severe [VSS (0, 86)], Severe [SS (85, 86, 91)], Little Tired [LTS (90, 91, 95)], and Normal [NS (94, 99)]. First, we create an abstract role named PSatO2. Second, a fuzzy data type is created for each of these fuzzy terms and finally, we have also defined fuzzy concrete roles: hasVSS_PSatO2, hasSS_PSatO2, hasLTS_PSatO2 and hasNS_PSatO2. The previously defined fuzzy datatypes are used as ranges for these roles. As shown in Fig. 3, a «Fuzzy Protégé» plugin was used to create the fuzzy datatype VSS_PSatO2.

Table 4 shows an example of a fuzzy property. This table provides an overview of the reported properties for each non-numeric fuzzy concept. These properties are created in order to define semantic relations and data. In order to exploit these fuzzy concepts in a decision support system and to have adequate results, we define fuzzy reasoners and

---

[2] https://www.health.harvard.edu/a-through-c.

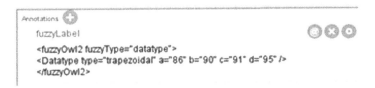

Fig. 3. Fuzzy Datatype Example

semantic rules ensuring a detailed description of the real case scenarios. Thus, we define the FOTS reasoning and case model in the following section.

The FOTS Reasoning and Case Model

After creating the fuzzy domain model (previous section), we define the reasoners to transform the model into a functional ontology and to integrate into a decision-support system. This model provides a detailed description for each emergency department patient case scenario containing all medical data. We describe the different relationships between fuzzy and precise medical concepts in our FOTS ontology and set up the case model by defining real patient case scenarios. In fact, Fuzzy DLs are extensions of classical Description Logics (DL) [30]. They have been proposed as languages that can represent and reason on vague or imprecise knowledge [26]. Thus, we use the Fuzzy Dl reasoner to define the severity index of each patient case based on the defined concepts. In this context, we apply the fuzzy disjunction rule to define and merge the SD for each symptom and consider the maximum degree in order to find the most accurate index. We use the SWRL (Semantic Web Rule Language) that is a rule language for the semantic web, combining the OWL-DL language and RuleML (Rule Markup Language) to create all possible scenarios of the patient state. These are integrated into the resulting ontology to reason semantically based on rules, for example: "Emergency_Patient (? x) ˆ Has_Age (?x, "very old") → Has_Waited_Behavior(?x, "Valid")". The exploitation of semantic rules can also be used to process missing data.

The FOTS Description

The resulting ontology contains 108 classes, 50 (fuzzy) object properties, 67 fuzzy datatype properties, 98 fuzzy datatypes, 917 axioms, 750 logical axioms, and 2489 concept instances for the 50 patient cases (Fig. 4). Each object property and each datatype property has an instance for every individual case. The implementation of our fuzzy ontology and specific concept cases is done by the Fuzzy OWL plugin. We have created an object property Has_Part, and its inverse Belongs_To to link all parts of a case to the case description concept Emergency_Patient. We have proposed some axioms to make sure that each case has one concept from each component. In the following section, we use performance measures to evaluate our ontology.

2.3   The Fuzzy Emergency Severity Index: F-ESI_2.0

The proposed F-ESI_2.0 presents an output of our FOTS ontology and is defined by the fuzzification of the Severity Degree (SD) total of four elements: age, symptoms,

Table 3. The overview of axioms

Concept Name	Description	Logical expression
Vital signs	Patient health indicators such as: Systolic blood pressure, heart rate, respiratory frequency,...	$\forall$ (X), Vital signs (X) $\Rightarrow$ Systolic Blood pressure (X) $\vee$ Heart Frequency (X) $\vee$ Respiratory Frequency (X) $\vee$ Body Temperature (X) $\vee$ Pain Level (X) $\vee$ Electrocardiogram (X) $\vee$ Glasgow_Score (X) $\vee$ Peripheral_O2_saturation (X)
Systolic_BP_Normal	Corresponds to the maximum pressure at the time of heart contraction. Its membership function is: SBPN (90, 100, 120, 130)	– SBPN(Value) = 1 if Value $\in$ [100–120[– SBPN(Value) = (Value - 90)/(100–90) if Value $\in$ [90–100[– SBPN(Value) = (130 - Value)/(130–120) if Value $\in$ [120–130[– SBPN(Value) = 0 else where
Hyper-Systolic_BP	When the maximum pressure at the time of heart contraction is between:]140–180[membership function is: SBPN (120, 130, 180)	– SBPN(Value) = 1 if Value $\in$]130–180[– SBPN (Value) = (Value - 120)/(130–120) if Value $\in$ [120–130[– SBPN (Value) = 0 elsewhere
Heart frequency_Normal	Represents tt of time (usually one minute). Its membership function is: FCN (60, 70, 90, 100)	– HFN(Value) = 1 if Value $\in$ [70–90[– HFN (Value) = (Value - 60)/(70–60) if Value $\in$ [60–70[– HFN (Value) = (100 - Value)/(100 - Value)/(100 - Value) 90) if Value $\in$ [90–100] – HFN (Value) = 0 elsewhere

medical history and medical observations, which are the primary references for defining a patient's health condition during triage according to health experts. Hence, based on the ESI used nowadays in hospital and in collaboration with the medical staff of the ED of Lille University Hospital Center (LUHC), we identified the SD of emergency patient as shown in Table 5.

The SD_x defines the SD linked the the x patient health indicator. For example, the SD_{age} defines the SD linked to the patient' age, i.e., if the age $\in$ [18, 40] then SDage = 0, if the age $\in$ [38, 55] then SDage = 1, if the age $\in$ [54, 76] then SDage = 2 and if the age > 75 then SDage = 3.

To find the F-ESI_2.0 for each patient, 3 steps are required:

Table 4. The fuzzy property Example

Fuzzy propriety	Domain	Range	Linguistic term	Syntax of DL	Type FM	FM
Has Systolic_BP_Normal	Emergency patient	Systolic_BP_Normal	Normal	Blood pressure ⊓ ∃ Normal_max.pressure	Trapezoidal	SBPN (90, 100, 120, 130)

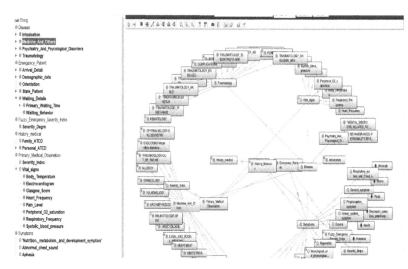

Fig. 4. The FOTS ontology

Table 5. The SD of patient

Severity Degree (SD)/patient state	Stable state	Moderate state	Urgent state	Very urgent state
SD age	0	1	2	3
SD history medical	0	1	2	3
SD symptoms	0	2	4	8
SD medical observations	0	2	4	8

Step 1: Calculate the SD for each item based on existing data.

Step 2: Calculate the sum of the Fuzzy Severity Degree (F-SD).

Step 3: Using the fuzzy concepts, we find the membership degree according to the class found and the corresponding F-ESI_2.0 (Fig. 5). In collaboration with health care experts, we select fuzzy classes that present the total F-SD for the elements that influence the patient's condition such as symptoms, age, reason for coming, history and nurses' medical observations. Fuzzy functions are selected according to the fuzzy data interval. Corresponding fuzzy classes are shown in Table 6.

For the calculation of severity regarding symptoms and medical observations, we rely on scientific medical documents that are subsequently validated by emergency physicians and we apply the fuzzy disjunction rules because we have a set of symptoms and vital signs for each patient. Thus, the F-ESI 2.0 can be a powerful tool to measure the severity status of patients, we will be able to evaluate its efficiency by testing it with a triage aid system based on the FOTS ontology but theoretically, this score is better than the existing scores because it is based on fuzzy logic which refines the results. In addition, it takes

Table 6. The F-ESI_2.0

F-SD	[0, 4[	]2, 7[	]5, 10[	]8, 13[	>11
F-ESI_2.0	5	4	3	2	1
ESI level	Nonurgent	Less urgent	Urgent	Emergent	Resuscitation
Fuzzification function	Left shoulder function	Trapezoidal function	Trapezoidal function	Trapezoidal function	Right shoulder function

into account all the medical elements such as medical history which are not considered in the past.

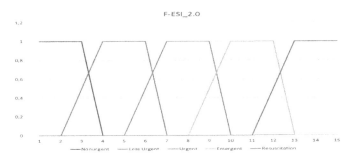

Fig. 5. The membership degree of F-ESI_2.0

3   The FOTS Evaluation

The resulting fuzzy ontology is evaluated regarding its syntax, semantics, and content coverage. The evaluation process assesses the conciseness, the correctness, the intelligibility and the adaptability of the ontology. There are no globally accepted evaluation mechanisms [20]. In fact, the ontology must be used, criticized and updated. According to Brewster et al. [21], precision and recall are not suitable for the evaluation because they depend on a comparison between evaluated ontology and a standard one [21]. We follow this method for evaluation.

3.1   Consistency Checking

This ontology is serialized in the OWL 2 format with the «Protégé³» 4.3 tool. It contains 108 fuzzy classes, 50 fuzzy object properties, 67 fuzzy datatype properties, 98 fuzzy datatypes, and 50 real cases. Consistency checking describes the syntactic-level evaluation. The SWRL rules and Fuzzy DL are developed by the «Protégé» editor to confirm that FOTS is consistent and free of errors. They do not reveal any discrepancies regarding this version of the ontology.

---

³ https://protege.stanford.edu/

3.2 Criteria of Evaluation

There is no benchmark ontology to measure its similarity with our ontology. Moreover, if a gold standard exists, then there will be no need to create other ontologies. But, a comparison with existing ontologies in the same domain is needed. However, there is no fuzzy ontology in the emergency triage to compare with. So, we consider the ontology of El Sappagh et al. [22] for diabetes diagnosis. Several criteria for ontology evaluation quality have been defined [23, 24]. We consider also the criteria of Djedidi and Aufaure [23] similar to the work of El Sappagh [22] with several metrics. These criteria concern complexity, cohesion, conceptualization, abstraction, completeness, and comprehension [23] (Table 7). The comparison between our ontology and the Diabetes ontology shows that FOTS is a complete, functional and semantically rich ontology.

3.3 Lexical, Vocabulary or Data Level Evaluation

Coverage is the completeness of terms or concepts to represent a domain [25]. So, our proposed ontology has to contain concepts and relations equal to those in the domain, and ontology instances identical to instances in the domain. Our ontology actually contains 50 cases, and it is open to other cases insertion. As our ontology has been defined in favor of the emergency triage process, all medical terms used in this medical system exist in the proposed ontology. Moreover, we use the ontologies of standardization such us ICD and SYMP ontology for defining the symptoms and diseases. Thus, the specific glossary of terms for triage are collected from the current system with the help of healthcare workers in LUHC. The coverage of FOTS is tested for all of these terms. FOTS has 100% concept coverage for all medical classes and relations required to describe ED patient cases. All needed concepts and relations to describe ED patient situations have been verified. Finally, the domain experts have evaluated the proposed ontology content regarding the clarity and conciseness. Since all concepts are extracted from the ED database and standardization ontologies, formal definitions are available for all terms. Therefore, FOTS complies with Gruber's three requirements such as clarity, including formal definition of classes, documentation of ontology, and use of classes as required [27].

3.4 Vagueness Evaluation

According to Alexopoulos et al. [25], the evaluation of the vagueness ontology quality has been defined by the set of metrics. These later include:

Vagueness Spread (VS): In a fuzzy ontology, the concepts, relations, attributes, and data types elements may be of a vague type. The VS measures the extent of vagueness representation in the ontology, and provides an indicator of the ontology's potential comprehensibility and shareability. An ontology with a high value of vagueness spread is less explicit and shareable than an ontology with a low value. As shown in Eq. (1), VS is the ratio of the number of vague ontology elements (classes, relations, and data types), noted by VOE and the total number of elements, noted by OE. We have C (Classes) = 108, OP (Object properties) = 50, FD (Fuzzy Datatypes) = 98, FDP = (Fuzzy Datatype Properties) = 67, and FOP (Fuzzy Object Properties) = 12.

Table 7. The comparative evaluation table between FOTS and diabetes ontology

Measure		Ontologies	
Criteria	Metrics	The proposed ontology	Diabetes ontology
Complexity	An average number of paths to reach a class from the root	3	3
	Average number of object properties per class	1.2	1.3
Abstraction	The average depth of the ontology	3	2
Cohesion	An average number of connected classes	54	27
Conceptualization	Semantic Richness: Ratio of the total number of semantic relations assigned to classes, divided by the total number of ontology relations (object properties and subsumption relations)	50/50 + 58 = 0.462	58/58 + 59 = 0.495
	Attribute Richness: Ratio of the total number of attributes (data properties describing ontology classes), divided by the total number of ontology classes	108/67 = 1.61	138/62 = 2.26
	Average number of subclasses per class	8	5
Completeness	There are no standard (fuzzy) case base ontologies to compare our ontology with it	Not Applicable	Not Applicable
Comprehension	Documentation of the properties	5%	2.04%
	Documentation of the classes	97%	88.71%

$$VS = \frac{|VOE|}{|OE|} = \frac{FD + FDP + FOP}{C + OP + FD + FDP} = 0.55 \tag{1}$$

Vagueness Explicitness (VE): It is the ratio of the number of vague ontological elements that are explicitly identified, noted by EVOE and VOE as in Eq. (2). The higher is the value of this metric, the better is the ontology. All fuzzy elements defined in the proposed ontology are explicitly defined, and fuzzy reasoner (Fuzzy DL) can infer other implicit elements at run time.

$$VE = \frac{|EVOE|}{|VOE|} = \frac{FD + FDP + FOP}{FOP + FD + FDP} = 1.0 \tag{2}$$

According to these indicators, we can prove that our ontology presents useful domain knowledge by considering the imprecision and uncertainty of medical information. The fuzzy elements of this domain are well defined explicitly with a very high comprehensibility. These characteristics are very important to put this ontology into a well-functioning decision support system.

4   Conclusion

In this paper, we proposed a novel fuzzy ontology (FOTS) for triage system in ED. The resulting ontology is enriched with multiple types of data, such as fuzzy, precise, text and semantic data. These different types of data facilitate the development of decision support system that contains fuzzy semantic-case retrieval algorithms and support queries expression by nurses. Thus, the FOntoM method allowed us to create a fuzzy ontology with a high coverage of triage domain in emergency services and to define all the useful fuzzy and precise knowledge. The FOTS is the unique fuzzy ontology at the ED and especially in the triage process domain. Moreover, in the evaluation section, we have proven that our functional fuzzy ontology presents triage domain knowledges and considers the imprecision and uncertainty of medical data in defining case scenarios. This representation of uncertainty helps to provide a decision support system with high performance and solve the problem of missing data. Therefore, it helps to improve triage and quality of care in the ED. In future work, we will study semantic retrieval algorithms that can be a potential solution for improving the integration of the ontology and solving the incompleteness of the annotations at querying time. Thus, we will focus to enhance the implementation of our ontology by using a programming language. The goal is to facilitate data retrieval and by setting up our ontology with the existing triage system. This will make it possible to add case-based reasoning and machine learning tools to improve the precision of our decision system.

References

1. Forero, R., McCarthy, S., Hillman, K.: Access block and emergency department overcrowding. Crit. Care **15**, 216 (2011). https://doi.org/10.1186/cc9998

2. Göransson, K.E., Ehrenberg, A., Marklund, B., Ehnfors, M.: Emergency department triage: is there a link between nurses' personal characteristics and accuracy in triage decisions? Accid. Emerg. Nurs. **14**(2), 83–88 (2006)
3. Sterling, R.E.P., Did, M., Schrager, J.D.: Prediction of emergency department patient disposition based on natural language processing of triage notes. Int. J. Med. Inf. **129**, 184–188 (2019)
4. Salman, O., Rasid, M., Saripan, M., Subramaniam, S.: Multisources data fusion framework for remote triage prioritization in telehealth. J. Med. Syst. **38**(9), 1–23 (2014)
5. Wang, S.-T.: Construct an optimal triage prediction model: a case study of the emergency department of a teaching hospital in Taiwan. J. Med. Syst. **37**(5), 1–11 (2013)
6. Dexheimer, J., et al.: An asthma management system in a pediatric emergency department. Int. J. Med. Inform. **82**(4), 230–238 (2013)
7. Jentsch, M., Ramirez, L., Wood, L., Elmasllari, E., The reconfiguration of triage by introduction of technology. In: Proceedings of the 15th International Conference on Human computer Interaction with Mobile Devices and Services, New York, NY, USA, pp. 55–64 (2013)
8. Christ, M., Grossmann, F., Winter, D., Bingisser, R., Platz, E.: Modern triage in the emergency department. Dtsch. Arztebl. Int. **107**(50), 892–898 (2010)
9. Farion, K., Michalowski, W., Wilk, S., O'Sullivan, D., Rubin, S., Weiss, D.: Clinical decision support system for point of care use: ontology driven design and software implementation. Meth. Inf. Med. **48**(4), 381–390 (2009)
10. Pedro, J., Burstein, F., Wassertheil, J., Arora, N., Churilov, L., Zaslavsky, A., On development and evaluation of prototype mobile decision support for hospital triage. In: Proceedings of the 38th Annual Hawaii International Conference on System Sciences, p. 157c (2005)
11. Jayaraman, P., Gunasekera, K., Burstein, F., Haghighi, P., Soetikno, H., Zaslavsky, A.: An ontology-based framework for real-time collection and visualization of mobile field triage data in mass gatherings. In: Proceedings of the 46th Annual Hawaii International Conference on System Sciences, Wailea, Maui, HI, pp. 146–155 (2013)
12. Wunsch, G., Costa, C.A., Righi, R.R.: A semantic-based model for Triage patients in emergency departments. J. Med. Syst. **41**(4), 1–12 (2017)
13. El-Sappagh, S., El-Masri, S., Elmogy, M., Riad, R., Saddik, B.: An ontological case base engineering methodology for diabetes management. J. Med. Syst. **38**(8), 1–14 (2014)
14. Zhang, F., Ma, Z., Yan, L., Cheng, J.: Construction of fuzzy OWL ontologies from fuzzy EER models: a semantics-preserving approach. Fuzzy Sets Syst. **229**, 1–32 (2013)
15. Maalej, S., Ghorbel, H., Bahri, A., Bouaziz, R.: Construction des composants ontologiques flousà partir de corpus de données sémantiques floues. In: Actes de la conférence Inforsid 2010, Marseille, France, pp. 361–376 (2010)
16. Quan, T, Hui, S, Cao, T.: FOGA: a fuzzy ontology generation framework for scholarly semantic web. In: Proceedings of the 2004 Knowledge Discovery and Ontologies Workshop, Pisa, Italy (2004)
17. Akremi, H., Zghal, S., Jouhet, V., Diallo, G.: FONTO: Une nouvelle méthode de la fuzzification d'ontologies. In: JFO2016 (2017)
18. Fernandez, M., Gómez-Pérez, A., Juristo, N.: METHONTOLOGY: from ontological art towards ontological engineering. In: Actes de AAAI 1997 (1997)
19. Cranefield, S., Purvis, M.: UML as an Ontology Modelling Language. Department of Information Science, University of Otago, New Zealand (1999)
20. Bright, T.J., Yoko Furuya, E., Kuperman, G.J., Cimino, J.J., Bakken, S.: Development and evaluation of an ontology for guiding appropriate antibiotic prescribing. J. Biomed. Inf. **45**(1), 120–128 (2012)
21. Alani, H., Dasmahapatra, S., Wilks, Y.: Data driven ontology evaluation. In: Proceedings of the International Conference on Language Resources and Evaluation, Lisbon, Portugal, pp. 164–168 (2004)

22. El-Sappagh, S., Elmogy, M.: A fuzzy ontology modeling for case base knowledge in diabetes mellitus domain. Eng. Sci. Technol. Int. J. **20**(3), 1025–1040 (2017)
23. Djedidi, R., Aufaure, M.-A.: ONTO-EVO A L an ontology evolution approach guided by pattern modeling and quality evaluation. In: Link, S., Prade, H. (eds.) 6th International Symposium on Foundations of Information and Knowledge Systems, FoIKS 2010, Sofia, Bulgaria, February 15-19, 2010. Proceedings, pp. 286–305. Springer, Heidelberg (2010). https://doi.org/10.1007/978-3-642-11829-6_19
24. Yu, J., Thom, J.A., Tam, A.: Evaluating ontology criteria for requirements in a geographic travel domain. In: Meersman, R., Tari, Z. (eds.) OTM 2005. LNCS, vol. 3761, pp. 1517–1534. Springer, Heidelberg (2005). https://doi.org/10.1007/11575801_36
25. Alexopoulos, P., Mylonas, P.: Towards vagueness-oriented quality assessment of ontologies. Artif. Intell. Meth. Appl. **8445**, 448–453 (2014)
26. Djellal, A.: Thèse pour l'obtention du diplôme de magister en informatique. Prise en compte de la notion de flou pour la représentation d'ontologies multi-points de vue en logique de descriptions. Université Mentouri Constantine, Algérie (2010)
27. Gruber, T.: Ontology. In: Liu, L., Özsu, M.T. (eds.) Encyclopedia of Database Systems. Springer, Boston (2009). https://doi.org/10.1007/978-0-387-39940-9_1318
28. Zekri, F., Turki, E., Bouaziz, R., AlzFuzzyOnto: Une ontologie floue pour l'aide à la décision dans le domaine de la maladie d'Alzheimer. In: Actes du XXXIIIème Congrès INFORSID, Biarritz, France, 26–29 May 2015, pp. 83–98 (2015)
29. Zhai, J., Liang, Y., Jiang, J., Yi, Y.: Fuzzy ontology models based on fuzzy linguistic variable for knowledge management and information retrieval. In: Shi, Z., Mercier-Laurent, E., Leake, D. (eds.) Intelligent Information Processing IV, pp. 58–67. Springer, Boston, MA (2008). https://doi.org/10.1007/978-0-387-87685-6_9
30. Straccia, U.: Reasoning with fuzzy description logics. J. Artif. Intell. **14**, 137–166 (2001)
31. Ghorbel, H., Bahri, A.B.R.: Fuzzy ontologies model for semantic web. In: The 2nd International Conference on Information, Process and Knowledge Management, eKNow 2010, St. Maarten, Netherlands Antilles (2010)

Ontology-Based Decision Support System for Dietary Recommendations for Type 2 *Diabetes Mellitus*

Maria Nisheva-Pavlova[✉], Iliyan Mihaylov, Stoian Hadzhiyski, and Dimitar Vassilev[✉]

FMI, Sofia University St. Kliment Ohridski, 5 James Bourchier Blvd, 1164 Sofia, Bulgaria
{marian,mihaylov,dimitar.vassilev}@fmi.uni-sofia.bg

Abstract. Decision support systems (DSS) play an increasingly important role in medical practice. By assisting physicians in making clinical decisions and subsequent recommendations, medical DSS are expected to improve the quality of healthcare. The role of DSS in diabetes treatment and in particular in post clinical treatment by organizing an improved regime of food balance and patient diets is the target area of the study. Based on the Diabetes Mellitus Treatment Ontology (DMTO), the developed DSS for dietary recommendations for patients with diabetes mellitus is aimed at improvement of patient care. Having into account the clinical history and the lab test profiles of the patients, these diet recommendations are automatically inferred using the DMTO subontologies for patient's lifestyle improvement and are based on reasoning on a set of newly developed production rules and the data from the patients records. The research presented in the paper is focused at intelligent integration of all data related to a particular patient and reasoning on them in order to generate personalized diet recommendations. A special-purpose knowledge base has been created, which enriches the DMTO with a set of original production rules and supports the elaboration of broader and more precise personalized dietary recommendations in the scope of the electronic health record services.

Keywords: Decision support system · Semantic interoperability · Knowledge base · Ontology · Type 2 diabetes mellitus · Diet recommendation

1 Introduction

Medical decision support systems and other intelligent applications in bio-medical practice and research depend on increasing amounts of digital information. Intelligent data integration in the biomedical domain is concerned as an instrument for combining data from different sources, creating a unified view and new knowledge as well as improving their interoperability and accessibility to a potential user [1]. Knowledge bases and in particular formal ontologies are being used to describe and organize shared biomedical knowledge [2]. Semantic interoperability is considered as important for a number of

M. Paszynski et al. (Eds.): ICCS 2021, LNCS 12744, pp. 735–741, 2021.
https://doi.org/10.1007/978-3-030-77967-2_61

healthcare activities including quality improvement programs, population health management and data management and is of great significance in electronic health records (EHR) information systems and their various services.

The major aim of this study is to develop a knowledge-based DSS for dietary recommendations for diabetes mellitus type 2. The principal tasks of the work are related to the intelligent integration of patient lab test, clinical data and food specifications aiming at development of automatically generated suggestions of a well-defined personal diet plan based on these data. The knowledge base is created using DMTO [4] and a set of related production rules defined specifically for the case. The main functionality of the inference engine is to generate an individual diet plan for each patient and to suggest some variants of particular menus covering this diet plan.

2 Problem Description

Patient Centered DSS. A properly designed DSS is an interactive software system intended to help decision makers to integrate useful information from a combination of raw data, documents, and personal knowledge, or business models to identify and solve problems and make decisions. Ontologies can add more power to clinical DSSs. An ontology can support knowledge sharing, easy maintenance, information integration and reuse in similar domains. The usage of production rules provides the differentiation of an extra layer of expert knowledge.

Diabetes Mellitus Application. Diabetes mellitus (DM) is a dangerous, complex, socially important chronic disease [4, 5]. Type 1 diabetes mellitus (T1DM) can only be treated with insulin, whereas patients with type 2 diabetes mellitus (T2DM, 90–95% of the cases) have a wide range of therapeutic options available, including lifestyle changes (mainly diet and food intake profile) and administration of multiple oral and/or injectable anti-diabetes drugs, including insulin.

DMTO creates a complete and consistent environment by enabling formal representation and integration of knowledge about treatment drugs, foods, education, lifestyle modifications, drug interactions, the patient profile, the patient's current conditions, and temporal aspects. DMTO introduces interesting features for T2DM treatment plans, and is expected to play a significant role in implementing intelligent, mobile, interoperable, and distributed DSS in diabetes therapies and post-clinical care [3].

The application of DSS in diabetes mellitus therapy has various examples including medicinal treatment, clinical survey of the patients, remote control and different levels of advising. The problems with post clinical activities, mainly the diet and lifestyle set of problems emerges as an important and challenging area. Our study is devoted mainly to this specific circle of problems related to the normal food intake considering specific diet requirements directly related to the patient personal ambulatory test profile.

3 Suggested Methodology

Conceptual Design. The applied methodology for developing the suggested ontology-based DSS is based on our conceptual model [6] of patient centered advising system for diet recommendations for T2DM. The system is with non-direct physician interaction for diet recommendation and utilizes a subject knowledge base which is implemented using DMTO and relevant production rules. DMTO provides the highest coverage and the most complete picture of coded knowledge about T2DM patients' current conditions, previous profiles, and T2DM-related aspects, including complications, symptoms, lab tests, interactions, treatment plan and diet frameworks, and glucose-related diseases and medications. The specific feature of the developed system could be related also to data integration between different clinical sources, food data and patient profiles.

Data Flow. As the designed ontology-based diet recommendation system is a patient-centered one, the main data comes from the patient medical records and in particular from the clinical laboratory tests (Fig. 1). Personal data protection and all related ethical issues are considered in the system. Data can also be integrated from non-clinical sources including the general practitioner and other medical diagnostic sources. The data normalization part of the system includes formatting and cleaning of the non-sense outliers in the data. Each import of patient data contains the patient's profile and all lab tests related to it. In addition to this information, a treatment plan, including lifestyle subplan and diet is created. As a result of appropriate reasoning, the amounts and proportions for macronutrients for each meal are set. Another part of the data flow in the system comes from the food specificity sources – here we use aside with the DMTO related libraries, also external sources such as US Food Data Central[1].

A structured model for patient's clinical lab test data is defined in the disease history record as part of the hospital record for each patient, holding information for example about glucose, glycated hemoglobin, cholesterol and uric acid.

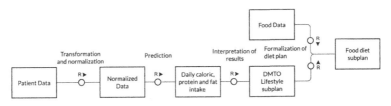

Fig. 1. System data flow

4 System Architecture

The suggested system consists of the following parts (Fig. 2): an input part (patient data, food data), a user integration point with RESTFul API service, a subsequent data integrator, a knowledge base and corresponding inference engine, and a storage part including: diet recommendations cache, food data storage and patient instance base. The user integration endpoint server is designed for the purposes of data integration, data normalization and interface development and application. The input part is based on

---

[1] https://fdc.nal.usda.gov/faq.html.

patient-centered data and some necessary food data for generating diet recommendations. The model for patient's data is determined on the medical laboratory check and the disease history as a part of the patient's health record. These patient data include all information from the laboratory tests, and its major components as glycated hemoglobin, glucose, cholesterol, uric acid. Patient data is used to create instances for the instance base. Food data comprises different geographic origin and energetic content for creating instances.

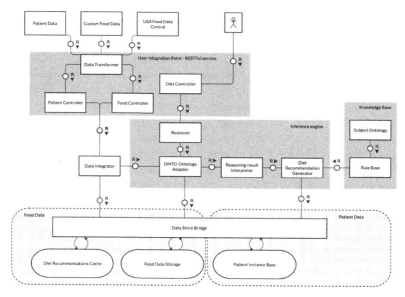

Fig. 2. Functional architecture of the system

The main component of our DSS for diet recommendations is the subject knowledge base, the core of which is DMTO. More precisely, the knowledge base of the DSS consists of two main parts an extendable copy of DMTO and a set of production rules (implemented in SWRL) describing specific knowledge for data analysis and decision making. The system has an application interface – server endpoints allowing the user to import patient data. Lab tests are linked with the patient profile via the property has_lab_test of the corresponding patient profile entity. When the ambulatory records are imported, the changes are saved and the generated IDs of all new patient profiles are returned.

The "heart" of the system is the inference engine, whose major function is the generation of diet recommendations based on the use of the knowledge of DMTO and decision-making rules. By processing a query for a patient diet suggestion, the server calls an appropriate reasoner which analyses the available data, performs suitable forms of inference (mainly forward checking) and generates a solution (Fig. 2).

A major contribution of this study is an appropriate extension of the DMTO by developing a model of DSS for dietary recommendations for T2DM. The first and most important patient specific modification is changing the data type from

RDF/XML to OWL/XML, in order to be able to create has_lifestyle_participant and has_(certain)_meal object properties. RDF/XML is a serialization syntax for RDF graphs. OWL/XML is a serialization syntax for the OWL 2 Structural Specification. RDF/XML ontologies could not be represented properly using standard XML tools.

The entity has_lifestyle_participant is an object property of the type diet. This property is owned by lifestyle subplan. has_(certain)_meal (breakfast for example) is an object property (breakfast for example) of type meal. It is owned by the diet class. These properties are essential for building the chain treatment plan – lifestyle subplan – diet – meal. A next significant modification is related to the extension of the "patient profile" to have more than one lab test.

In order to achieve a personalized diet, different proportions between fat, carbohydrates and proteins are set for each meal. For a healthy person with values for "total cholesterol" and "fasting plasma glucose" (FPG or Glucose) and "urine blood" (Uric Acid) in normal intervals, the proportions are fat 30%, carbohydrates 50% and proteins 20%. For a person with values out of norm – the proportions are fat 20%, carbohydrates 40% and proteins 40%, raising protein amount and lowering carbs and fats (Table 1). To identify if a patient has lab test results within the normal ranges or out of normal ranges, a set of rules are defined which check that and also set the required ratio between fat, proteins and carbohydrates according to the lab tests results. The ratio between fat, carbs and proteins is set for the particular meal of the diet for the patient. The improved lab test contains some elements representing a set of blood lab tests – total cholesterol, glucose and uric acid. In addition to its value, a lab test element has also the attributes min-threshold and max- threshold giving information about the range of its value.

Table 1. Ranges of blood lab tests.

Parameter	Measure unit	Min. threshold	Max. threshold
Glucose	mmol/l	3.3	6.2
Glycated hemoglobin	%	5.7	6.5
Total cholesterol	mmol/l	0	5.2
Uric acid	umol/l	208	428

5 Results and Discussion

To check whether a patient has lab test results within the normal ranges or out of normal ranges, a set of rules are defined. Rules are also set to meet the corresponding ratio between fat, proteins and carbohydrates according to the lab tests results.

The system initializes, integrates and gives values to the particular patient profile elements: patient plans (treatment, lifestyle, diet), patient total calories, total cholesterol lab test, FPG lab test, and diet referred to an example of a breakfast meal.

There is a check if these lab test values are in normal range or out of normal range. The normal proportions between carbs, fats and proteins are set up as follows 0.5, 0.3

and 0.2. Then the amounts of both calories and grams of a certain meal are calculated. The number of calories for a certain meal is calculated from the total calories multiplied by the proportion of macronutrients for the whole meal. The weight [grams] is calculated using the number of calories. The reasoning procedure uses all available patient data. The reasoner executes rules by setting proportions of the macronutrients. There is a rule for each lab test checking its values testing for belonging to the normal range. The total number of calories for a breakfast meal, for example, is calculated on the base of the total calories from the patient profile. The quantity per macronutrient is based on the calculated proportions.

Our first operation example of the system is based on importing patient data with lab tests in norm. An amount of 1700 total calories is set as a referent patient profile. The patient profile includes laboratory tests as: FPG – 4 mmol/l, total cholesterol – 5 mmol/l and uric acid – 379 umol/l. A treatment plan is created for the patient profile, including a lifestyle subplan, where a set of diets for breakfast is suggested.

The reasoner executes the rules setting the proportions for the macronutrients. There is a production (SWRL) rule for each type of lab test checking if its value is in the normal range. Rules are setting the following properties of a meal in the diet: calories – "carbohydrate per meal", "fat per meal", "protein per meal"; quantity – "carbohydrate grams", "fat grams" and "protein grams".

The total calories for the breakfast meal are calculated from the total calories from the patient profile as 1700 * 0.25 = 425. All values for the lab tests are in norm and consequently the proportions between the macronutrients are set to normal – 50% carbs, 30% fats, 20% proteins. The exact calories and amounts calculated after the reasoning for each macronutrient can be seen in Table 2 (first row).

Other tests of our DSS are based on patient data out of norm. The total calories are again 1700. We imported data of a few patients with different lab tests exceeding the normal range. One patient has total cholesterol – 6 mmol/l, another patient has FPG – 10.13 mmol/l and the last patient has urine acid – 500 umol/l. For each of these patients the proportions between the macronutrients are set to 40% carbs, 20% fats, 40% proteins and one can see the calories for each macronutrient is 22le 2 (second row).

Table 2. Suggested diet content.

Type	Carbohydrates (calories)	Carbohydrates (grams)	Fats (calories)	Fats (grams)	Proteins (calories)	Proteins (grams)
In norm	212.5	53.125	127.5	14.17	85.0	21.25
Out of norm	170.0	42.5	85.0	9.45	170.0	42.5

The main output of our system for diet recommendation is focused on the generation of alternative menu suggestion with a particularly fitted diet, in terms of solving a constraint satisfaction problem. The interface of the system is still in command line and the full completeness of a GUI is part of our future work of improving the functionalities of the system.

6 Conclusion

The paper discusses some results concerning an original ontology-based decision support system for dietary recommendations for T2DM. The created DSS is based on the development and use of an appropriate extension of DMTO with a set of production rules, for precise personalized dietary recommendations relevant to T2DM treatment. The workflow of our DSS is based on the successful integration of a number of modern semantic technologies and provides real semantic interoperability of the system with other healthcare information systems. An appropriate user interface that will provide personalized visualization of the generated results, oriented to the requests of end users with different profiles, is under development. We intend to develop an application interface module that will read the patient data from their EHRs in compliance with all requirements for personal data protection.

Acknowledgments. This research is supported by the National Scientific Program "eHealth". Logistical support was received from the National Scientific Program "Information and Communication Technologies for a Single Digital Market in Science, Education and Security (ICTinSES)".

References

1. Sutton, R.T., Pincock, D., Baumgart, D.C., Sadowski, D.C., Fedorak, R.N., Kroeker, K.I.: An overview of clinical decision support systems: benefits, risks, and strategies for success. NPJ. Digit. Med. **3**, 17 (2020). https://doi.org/10.1038/s41746-020-0221-y
2. Middleton, B., Sittig, D.F., Wright, A.: Clinical decision support: a 25 year retrospective and a 25 year vision. Yearbook Med. Inf. **2**(1), S103–16 (2016) https://doi.org/10.15265/IYS-2016-s034.
3. El-Sappagh, S., Kwak, D., Ali, F., Kwak, K.-S.: DMTO: a realistic ontology for standard diabetes mellitus treatment. J. Biomed. Semant. **9**, 8 (2018). https://doi.org/10.1186/s13326-018-0176-y
4. O'Connor, P.J., Sperl-Hillen, J.M.: Current status and future directions for electronic point-of-care clinical decision support to improve diabetes management in primary care. Diab. Technol. Ther. **21**(S2), S226–S234 (2019). https://doi.org/10.1089/dia.2019.0070
5. Jia, P., Zhao, P., Chen, J., Zhang, M.: Evaluation of clinical decision support systems for diabetes care: an overview of current evidence. J. Eval. Clin. Pract. **25**, 66–77 (2019). https://doi.org/10.1111/jep.12968
6. Nisheva-Pavlova, M., Hadzhiyski, S., Mihaylov, I., Avdjieva, I., Vassilev, D.: Linking Data for Ontology Based Advising in Healthcare. In: Proceedings of 2020 International Conference Automatics and Informatics (ICAI 2020 – Varna, Bulgaria, pp. 1 – 3. IEEE (2020). ISBN 978-172819308-3. https://doi.org/10.1109/ICAI50593.2020.9311382

Correction to: Predictability Classes for Forecasting Clients Behavior by Transactional Data

Elizaveta Stavinova, Klavdiya Bochenina, and Petr Chunaev

Correction to:
Chapter "Predictability Classes for Forecasting Clients Behavior by Transactional Data" in: M. Paszynski et al. (Eds.): *Computational Science – ICCS 2021*, **LNCS 12744, https://doi.org/10.1007/978-3-030-77967-2_16**

The chapter was inadvertently published with incomplete funding information in the acknowledgment. The missing funding information is now added and the chapter has been updated with the changes.

The updated version of this chapter can be found at
https://doi.org/10.1007/978-3-030-77967-2_16

© Springer Nature Switzerland AG 2021
M. Paszynski et al. (Eds.): ICCS 2021, LNCS 12744, p. C1, 2021.
https://doi.org/10.1007/978-3-030-77967-2_62

Author Index

Printed in the United States
by Baker & Taylor Publisher Services